ATOMS AND MOLECULES

ATOMS AND MOLECULES

Mitchel Weissbluth

Department of Applied Physics
Stanford University
Stanford, California

ACADEMIC PRESS New York San Francisco London 1978

A Subsidiary of Harcourt Brace Jovanovich, Publishers

ACADEMIC PRESS, INC.
111 Fifth Avenue, New York, New York 10003

United Kingdom Edition published by
ACADEMIC PRESS, INC. (LONDON) LTD.
24/28 Oval Road, London NW1 7DX

Library of Congress Cataloging in Publication Data

Weissbluth, Mitchel.
Atoms and molecules.

Bibliography: p.
1. Atoms. 2. Molecules. I. Title.
QC173.W433 539 76-55979
ISBN 0-12-744450-5

PRINTED IN THE UNITED STATES OF AMERICA

To

Margaret

Steve, Marc, and Tom

CONTENTS

PREFACE

Group theoretical methods in atomic and molecular physics were employed very early in the history of quantum mechanics, notably by H. Weyl, E. P. Wigner, and H. Bethe, although on the whole widespread acceptance was not achieved for some thirty years. An important impetus toward a renewed interest in group theory on the part of physicists was the work of G. Racah who introduced the formalism of irreducible tensor operators and demonstrated their utility in the evaulation of atomic matrix elements. Extensions to molecular systems followed within a relatively short time. It is the purpose of this book to discuss the basic properties of atoms and molecules, taking full advantage of these powerful methods.

Part I contains the essential mathematics pertaining to angular mometum properties, finite and continuous rotation groups, tensor operators, the Wigner–Eckart theorem, vector fields, and vector spherical harmonics. Part II provides the quantum mechanical background on specialized topics, it being assumed that the student has had at least an undergraduate course in quantum mechanics. Included are symmetry considerations, second quantization, density matrices, and several types of time-dependent and time-independent approximation methods. Discussion of atomic structure begins in Part III. Starting with the Dirac equation, its nonrelativistic approximation provides the basis for the derivation of the Hamiltonians for all important interactions, e.g., spin–orbit, external fields, hyperfine, etc. Multielectron atoms are discussed in Part IV, which treats multiplet theory and the Hartree–Fock formulation. Electromagnetic radiation fields and their interactions with atoms in first and higher orders are treated in Part V, which also includes topics of relevance to spectroscopy. Finally, Part VI is devoted to molecules and complexes, including such topics as the Born–Oppenheimer approximation, molecular orbitals, the self-consistent field method, electronic states, vibrational and rotational states, molecular spectra, and ligand field theory.

The quantum mechanics of atoms and molecules, once the exclusive domain of physicists, has in recent years proliferated into other fields, primarily chemistry and several branches of engineering. In recognition of this wider interest, a full year graduate course in atomic and molecular physics has been taught in the Department of Applied Physics at Stanford University. Attendees consisted of students working in diverse fields such as spectroscopy, magnetic resonance, Mössbauer resonance, quantum electronics, solid state electronics, astrophysics, and biological physics. The present volume is an outgrowth of this course.

PART I

Mathematical Background

CHAPTER 1

ANGULAR MOMENTUM

1.1 Orbital Angular Momentum

The orbital angular momentum operator **L** is defined by

$$\mathbf{L} = \frac{1}{\hbar}(\mathbf{r} \times \mathbf{p}) \tag{1.1-1}$$

where **r** is a vector whose components r_i are x, y, z (or x_1, x_2, x_3) and

$$\mathbf{p} = -i\hbar\boldsymbol{\nabla} \tag{1.1-2}$$

is the linear momentum operator; the rectangular components of the gradient operator $\boldsymbol{\nabla}$ are $\partial/\partial x, \partial/\partial y, \partial/\partial z$. Expanding (1.1-1),

$$L_x = \frac{1}{\hbar}(yp_z - zp_y) = -i\left(y\frac{\partial}{\partial z} - z\frac{\partial}{\partial y}\right)$$

$$= i\left(\sin\varphi\frac{\partial}{\partial\theta} + \cot\theta\cos\varphi\frac{\partial}{\partial\varphi}\right), \tag{1.1-3a}$$

$$L_y = \frac{1}{\hbar}(zp_x - xp_z) = -i\left(z\frac{\partial}{\partial x} - x\frac{\partial}{\partial z}\right)$$

$$= i\left(-\cos\varphi\frac{\partial}{\partial\theta} + \cot\theta\sin\varphi\frac{\partial}{\partial\varphi}\right), \tag{1.1-3b}$$

$$L_z = \frac{1}{\hbar}(xp_y - yp_x) = -i\left(x\frac{\partial}{\partial y} - y\frac{\partial}{\partial x}\right)$$

$$= -i\frac{\partial}{\partial\varphi}. \tag{1.1-3c}$$

In (1.1-3) the angles θ and φ are the polar and azimuth angles, respectively.

The operators L_x, L_y, and L_z are Hermitian, i.e.,

$$L_i^\dagger = L_i \qquad (i = x, y, z), \tag{1.1-4}$$

and, as functions of the coordinates, L_x, L_y, and L_z are pure imaginary operators.

It will often be convenient to use *spherical components* of **L**; these are defined as

$$\begin{aligned}
L_{+1} &= -\frac{1}{\sqrt{2}}(L_x + iL_y) = -\frac{1}{\sqrt{2}}\,e^{i\varphi}\left(\frac{\partial}{\partial\theta} + i\cot\theta\,\frac{\partial}{\partial\varphi}\right),\\
L_{-1} &= \frac{1}{\sqrt{2}}(L_x - iL_y) = -\frac{1}{\sqrt{2}}\,e^{-i\varphi}\left(\frac{\partial}{\partial\theta} - i\cot\theta\,\frac{\partial}{\partial p}\right),\\
L_0 &= L_z.
\end{aligned} \tag{1.1-5}$$

The inverse relations are

$$L_x = -\frac{1}{\sqrt{2}}(L_{+1} - L_{-1}), \qquad L_y = \frac{i}{\sqrt{2}}(L_{+1} + L_{-1}), \qquad L_z = L_0. \tag{1.1-6}$$

In contrast to the rectangular components of **L**, L_{+1} and L_{-1} are *not* Hermitian since

$$L_{+1}^\dagger = -L_{-1}, \qquad L_{-1}^\dagger = -L_{+1}. \tag{1.1-7}$$

The components of **r** and **p** satisfy certain commutation relations:

$$[r_i, p_j] = i\hbar\,\delta_{ij}, \tag{1.1-8a}$$

$$[r_i, r_j] = [p_i, p_j] = 0, \tag{1.1-8b}$$

$$[r_i, p^2] = 2i\hbar p_i, \tag{1.1-8c}$$

$$[p_i, p^2] = 0 \tag{1.1-8d}$$

in which $r_i, r_j = x, y, z$; $p_i, p_j = p_x, p_y, p_z$, and $p^2 = {p_x}^2 + {p_y}^2 + {p_z}^2$. The definition of **L** (1.1-1) together with (1.1-8) imply that

$$[L_x, L_y] = iL_z, \qquad [L_y, L_z] = iL_x, \qquad [L_z, L_x] = iL_y. \tag{1.1-9}$$

These may be written in any of the compact forms:

$$[L_i, L_j] = iL_k \qquad (i, j, k \text{ cyclic}), \tag{1.1-10a}$$

$$\mathbf{L} \times \mathbf{L} = i\mathbf{L}, \tag{1.1-10b}$$

$$[L_i, L_j] = i\varepsilon_{ijk}L_k, \tag{1.1-10c}$$

in which ε_{ijk} is the antisymmetric unit tensor of rank 3 defined by

$$\varepsilon_{ijk} = \begin{cases} +1, & i, j, k \text{ in cyclic order,} \\ -1, & i, j, k \text{ not in cyclic order,} \\ 0, & \text{two indices alike.} \end{cases} \tag{1.1-11}$$

The three statements (1.1-10a)–(1.1-10c) are equivalent in all respects. Additional commutator relations among the components of **L**, **r**, and **p** are

$$[L_i, r_j] = i\varepsilon_{ijk} r_k, \tag{1.1-12a}$$

$$[L_i, p_j] = i\varepsilon_{ijk} p_k, \tag{1.1-12b}$$

$$[L_0, L_{\pm 1}] = \pm L_{\pm 1}, \qquad [L_{+1}, L_{-1}] = -L_0. \tag{1.1-13}$$

Another important operator is L^2, also known as the *total orbital angular momentum operator*. It may be expressed in various equivalent forms:

$$\begin{aligned} L^2 &= L_x{}^2 + L_y{}^2 + L_z{}^2 \\ &= -\left[\frac{\partial^2}{\partial\theta^2} + \cot\theta \frac{\partial}{\partial\theta} + (1 + \cot^2\theta)\frac{\partial^2}{\partial\varphi^2}\right] \\ &= -\left[\frac{1}{\sin\theta}\frac{\partial}{\partial\theta}\left(\sin\theta\frac{\partial}{\partial\theta}\right) + \frac{1}{\sin^2\theta}\frac{\partial^2}{\partial\varphi^2}\right] \\ &= -L_{+1}L_{-1} + L_0{}^2 - L_{-1}L_{+1} \\ &= \sum_q (-1)^q L_q L_{-q} \qquad (q = 1, 0, -1). \end{aligned} \tag{1.1-14}$$

Employing relations (1.1-13) we also have

$$L^2 = -2L_{+1}L_{-1} + L_0(L_0 - 1) = -2L_{-1}L_{+1} + L_0(L_0 + 1). \tag{1.1-15}$$

L^2 commutes with all components of **L**, i.e.,

$$[L^2, L_\mu] = 0 \tag{1.1-16}$$

where L_μ refers to either rectangular components (L_x, L_y, L_z) or spherical components (L_{+1}, L_0, L_{-1}) of **L**.

1.2 Spherical Harmonics and Related Functions

The spherical harmonics $Y_{lm}(\theta, \varphi)$ are defined by

$$Y_{lm}(\theta, \varphi) = \sqrt{(-1)^{m+|m|}}\sqrt{\frac{2l+1}{4\pi}}\sqrt{\frac{(l-|m|)!}{(l+|m|)!}}\, P_l^{|m|}(\cos\theta) e^{im\varphi} \tag{1.2-1}$$

TABLE 1.1

Spherical Harmonics[a]

l	m	$r^l Y_{lm}(x, y, z)$	$Y_{lm}(\theta, \varphi)$
0	0	$\sqrt{\frac{1}{4\pi}}$	$\sqrt{\frac{1}{4\pi}}$
1	0	$\sqrt{\frac{3}{4\pi}}\, z$	$\sqrt{\frac{3}{4\pi}} \cos\theta$
1	± 1	$\mp\sqrt{\frac{3}{8\pi}}\,(x \pm iy)$	$\mp\sqrt{\frac{3}{8\pi}} \sin\theta\, e^{\pm i\varphi}$
2	0	$\sqrt{\frac{5}{4\pi}}\sqrt{\frac{1}{4}}\,(3z^2 - r^2)$	$\sqrt{\frac{5}{4\pi}}\sqrt{\frac{1}{4}}\,(3\cos^2\theta - 1)$
2	± 1	$\mp\sqrt{\frac{5}{4\pi}}\sqrt{\frac{3}{2}}\, z(x \pm iy)$	$\mp\sqrt{\frac{5}{4\pi}}\sqrt{\frac{3}{2}} \cos\theta \sin\theta\, e^{\pm i\varphi}$
2	± 2	$\sqrt{\frac{5}{4\pi}}\sqrt{\frac{3}{8}}\,(x \pm iy)^2$	$\sqrt{\frac{5}{4\pi}}\sqrt{\frac{3}{8}} \sin^2\theta\, e^{\pm 2i\varphi}$
3	0	$\sqrt{\frac{7}{4\pi}}\sqrt{\frac{1}{4}}\, z(5z^2 - 3r^2)$	$\sqrt{\frac{7}{4\pi}}\sqrt{\frac{1}{4}}\,(2\cos^3\theta - 3\cos\theta\sin^2\theta)$
3	± 1	$\mp\sqrt{\frac{7}{4\pi}}\sqrt{\frac{3}{16}}\,(x \pm iy)(5z^2 - r^2)$	$\mp\sqrt{\frac{7}{4\pi}}\sqrt{\frac{3}{16}}\,(4\cos^2\theta\sin\theta - \sin^3\theta)e^{\pm i\varphi}$
3	± 2	$\sqrt{\frac{7}{4\pi}}\sqrt{\frac{15}{8}}\, z(x \pm iy)^2$	$\sqrt{\frac{7}{4\pi}}\sqrt{\frac{15}{8}} \cos\theta\sin^2\theta\, e^{\pm 2i\varphi}$
3	± 3	$\mp\sqrt{\frac{7}{4\pi}}\sqrt{\frac{5}{16}}\,(x \pm iy)^3$	$\mp\sqrt{\frac{7}{4\pi}}\sqrt{\frac{5}{16}} \sin^3\theta\, e^{\pm 3i\varphi}$
4	0	$\sqrt{\frac{9}{4\pi}}\sqrt{\frac{1}{64}}\,(35z^4 - 30z^2r^2 + 3r^4)$	$\sqrt{\frac{9}{4\pi}}\sqrt{\frac{1}{64}}\,(35\cos^4\theta - 30\cos^2\theta + 3)$
4	± 1	$\mp\sqrt{\frac{9}{4\pi}}\sqrt{\frac{5}{16}}\,(x \pm iy)(7z^3 - 3zr^2)$	$\mp\sqrt{\frac{9}{4\pi}}\sqrt{\frac{5}{16}} \sin\theta(7\cos^3\theta - 3\cos\theta)e^{\pm i\varphi}$
4	± 2	$\sqrt{\frac{9}{4\pi}}\sqrt{\frac{5}{32}}\,(x \pm iy)^2(7z^2 - r^2)$	$\sqrt{\frac{9}{4\pi}}\sqrt{\frac{5}{32}} \sin^2\theta(7\cos^2\theta - 1)e^{\pm 2i\varphi}$
4	± 3	$\mp\sqrt{\frac{9}{4\pi}}\sqrt{\frac{35}{16}}\, z(x \pm iy)^3$	$\mp\sqrt{\frac{9}{4\pi}}\sqrt{\frac{35}{16}} \sin^3\theta\cos\theta\, e^{\pm 3i\varphi}$
4	± 4	$\sqrt{\frac{9}{4\pi}}\sqrt{\frac{35}{128}}\,(x \pm iy)^4$	$\sqrt{\frac{9}{4\pi}}\sqrt{\frac{35}{128}} \sin^4\theta\, e^{\pm 4i\varphi}$

[a] In spectroscopic notation, functions that are proportional to Y_{lm} with $l = 0, 1, 2, 3, \ldots$ are called s, p, d, f, ... functions.

with

$$l = 0, 1, 2, \ldots, \tag{1.2-2a}$$

$$m = l, l-1, \ldots, -l, \tag{1.2-2b}$$

and $P_l^{|m|}(\cos\theta)$ an associated Legendre polynomial. The phase convection in (1.2-1) is not universal; the one adopted here is known as the *Condon–Shortley convention.* Some of the commonly used spherical harmonics are listed in Table 1.1; among their properties are:

$$Y_{l\,-m}(\theta, \varphi) = (-1)^m Y_{lm}^*(\theta, \varphi), \tag{1.2-3a}$$

$$Y_{lm}(\pi - \theta, \pi + \varphi) = (-1)^l Y_{lm}(\theta, \varphi). \tag{1.2-3b}$$

The change from (θ, φ) to $(\pi - \theta, \pi + \varphi)$ corresponds to an inversion, that is, a change from (x, y, z) to $(-x, -y, -z)$. From (1.2-3b) it is seen that $Y_{lm}(\theta, \varphi)$ changes sign under inversion when l is an odd integer; when l is even, there is no change in sign. In the former case, $Y_{lm}(\theta, \varphi)$ is said to have *odd parity* and in the latter, *even parity*. The quantity $(-1)^l$, which is equal to $+1$ for l even and -1 for l odd is called the *parity factor.*

When $\theta = 0$,

$$Y_{lm}(0, \varphi) = \begin{cases} 0 & \text{for} \quad m \neq 0, \\ \sqrt{\dfrac{2l+1}{4\pi}} & \text{for} \quad m = 0. \end{cases} \tag{1.2-4}$$

The spherical harmonics satisfy an orthogonality relation

$$\int Y_{lm}^*(\theta, \varphi) Y_{l'm'}(\theta, \varphi) \sin\theta \, d\theta \, d\varphi \equiv \int Y_{lm}^*(\theta, \varphi) Y_{l'm'}(\theta, \varphi) \, d\Omega = \delta_{ll'} \, \delta_{mm'} \tag{1.2-5}$$

in which $d\Omega = \sin\theta \, d\theta \, d\varphi$ is an element of solid angle. An arbitrary function $f(\theta, \varphi)$, satisfying the usual criteria for expansion in terms of an orthonormal set, may be expanded in terms of spherical harmonics as

$$f(\theta, \varphi) = \sum_{l=0}^{\infty} \sum_{m=-l}^{l} a_{lm} Y_{lm}(\theta, \varphi), \tag{1.2-6a}$$

$$a_{lm} = \int Y_{lm}^*(\theta, \varphi) f(\theta, \varphi) \, d\Omega. \tag{1.2-6b}$$

It is often desirable to work with real functions constructed as linear combinations of the (complex) spherical harmonics. Several examples are listed in Table 1.2 and are shown in the form of polar diagrams in Fig. 1.1.

Orbital angular momentum operators and spherical harmonics are intimately related. This may be seen from the standpoint of a central force

TABLE 1.2

Real Combinations of Spherical Harmonics

Notation	Cartesian coordinates	Polar coordinates	Spherical harmonics
s	1	1	$\sqrt{4\pi}\,Y_{0\,0}$
p_x	x	$r\sin\theta\cos\varphi$	$\sqrt{\frac{4\pi}{3}}\sqrt{\frac{1}{2}}(-Y_{1\,1}+Y_{1\,-1})r$
p_y	y	$r\sin\theta\sin\varphi$	$\sqrt{\frac{4\pi}{3}}\sqrt{\frac{1}{2}}\,i(Y_{1\,1}+Y_{1\,-1})r$
p_z	z	$r\cos\theta$	$\sqrt{\frac{4\pi}{3}}Y_{1\,0}r$
d_{z^2}	$\frac{1}{2}(3z^2-r^2)$	$\frac{1}{2}r^2(3\cos^2\theta-1)$	$\sqrt{\frac{4\pi}{5}}Y_{2\,0}r^2$
$d_{x^2-y^2}$	$\frac{1}{2}\sqrt{3}(x^2-y^2)$	$\frac{1}{2}\sqrt{3}r^2\sin^2\theta\cos2\varphi$	$\sqrt{\frac{4\pi}{5}}\sqrt{\frac{1}{2}}(Y_{2\,2}+Y_{2\,-2})r^2$
d_{xy}	$\sqrt{3}xy$	$\sqrt{3}r^2\sin^2\theta\cos\varphi\sin\varphi$	$\sqrt{\frac{4\pi}{5}}\sqrt{\frac{1}{2}}\,i(-Y_{2\,2}+Y_{2\,-2})r^2$
d_{yz}	$\sqrt{3}yz$	$\sqrt{3}r^2\sin\theta\cos\theta\sin\varphi$	$\sqrt{\frac{4\pi}{5}}\sqrt{\frac{1}{2}}\,i(Y_{2\,1}+Y_{2\,-1})r^2$
d_{zx}	$\sqrt{3}zx$	$\sqrt{3}r^2\sin\theta\cos\theta\cos\varphi$	$\sqrt{\frac{4\pi}{5}}\sqrt{\frac{1}{2}}(-Y_{2\,1}+Y_{2\,-1})r^2$

problem. Let the Hamiltonian of a particle of mass m and momentum $\mathbf{p}$ be

$$\mathscr{H}=\frac{p^2}{2m}+V \tag{1.2-7}$$

where V is the potential energy. The Schrödinger equation

$$\nabla^2\psi+\frac{2m}{\hbar^2}(E-V)\psi=0 \tag{1.2-8}$$

may be transformed into spherical coordinates as

$$r^2\left(\frac{\partial^2}{\partial r^2}+\frac{2}{r}\frac{\partial}{\partial r}\right)\psi+\frac{2mr^2}{\hbar^2}(E-V)\psi=-\left[\frac{1}{\sin\theta}\frac{\partial}{\partial\theta}\left(\sin\theta\frac{\partial}{\partial\theta}\right)+\frac{1}{\sin^2\theta}\frac{\partial^2}{\partial\varphi^2}\right]\psi. \tag{1.2-9}$$

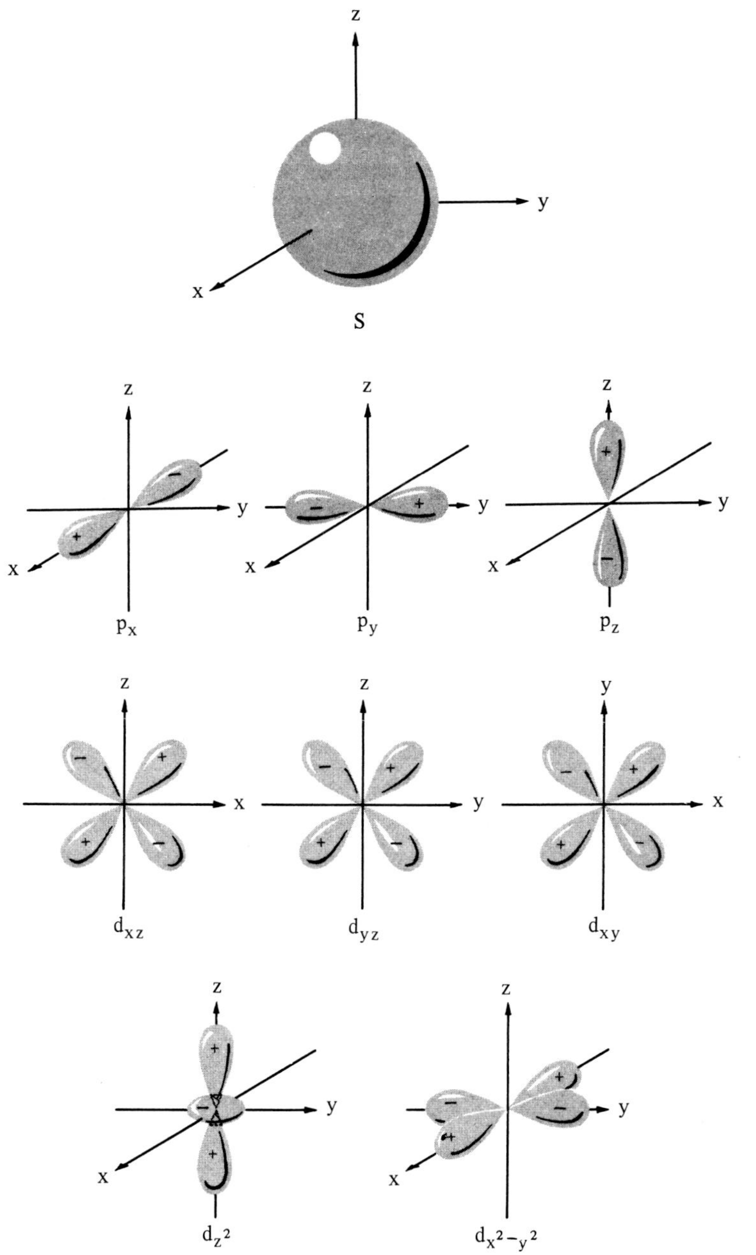

FIG. 1.1 Polar diagrams of s, p, and d functions. (From C. J. Ballhausen and H. B. Gray, "Molecular Orbital Theory," copyright © 1964 by W. A. Benjamin, Inc., Menlo Park, California.)

In a central field for which $V = V(r)$, (1.2-9) is separable into two equations one of which depends on r only and the other on θ and φ. Thus let

$$\psi(r, \theta, \varphi) = R(r)\Theta(\theta, \varphi), \tag{1.2-10}$$

$$R(r) = \frac{1}{r} P(r). \tag{1.2-11}$$

The Schrödinger equation (1.2-9) now separates into

$$\frac{d^2P(r)}{dr^2} + \frac{2m}{\hbar^2}[E - V(r)]P(r) = \frac{\lambda}{r^2} P(r), \tag{1.2-12}$$

$$-\left[\frac{1}{\sin\theta}\frac{\partial}{\partial\theta}\left(\sin\theta\frac{\partial}{\partial\theta}\right) + \frac{1}{\sin^2\theta}\frac{\partial^2}{\partial\varphi^2}\right]\Theta(\theta, \varphi) = \lambda\Theta(\theta, \varphi), \tag{1.2-13}$$

in which λ is a separation constant. From (1.1-14) it is seen that the operator on the left-hand side of (1.2-13) is just L^2; thus

$$L^2\Theta(\theta, \varphi) = \lambda\Theta(\theta, \varphi). \tag{1.2-14}$$

Quantum-mechanical wave functions and their first derivatives must be everywhere continuous, single-valued, and finite. When these conditions are imposed on $\psi(r, \theta, \varphi)$, it is found that

$$\Theta(\theta, \varphi) = Y_{lm}(\theta, \varphi), \tag{1.2-15}$$

$$\lambda = l(l + 1). \tag{1.2-16}$$

In other words, the solutions to the quantum-mechanical central force problem are products of radial functions and angular functions and the latter are the spherical harmonics $Y_{lm}(\theta, \varphi)$ which satisfy

$$L^2 Y_{lm}(\theta, \varphi) = l(l + 1)Y_{lm}(\theta, \varphi). \tag{1.2-17}$$

The relation expressed by (1.2-17) lends itself to the interpretation that $Y_{lm}(\theta, \varphi)$ is an eigenfunction of the operator L^2 and the corresponding eigenvalue is $l(l + 1)$. Alternatively, (1.2-17) is derivable from the basic definition for L^2 (1.1-14) and $Y_{lm}(\theta,\varphi)$ (1.2-1). The restrictions on l and m are contained in (1.2-2a) and (1.2-2b); in particular, it should be noted that l and m, known as *quantum numbers* in physical terminology, are integers. In older quantum mechanical formulations $\sqrt{l(l + 1)}$ was regarded as the magnitude of the vector $\mathbf{L}$ and m was the projection of $\mathbf{L}$ on the z axis. Although this description is somewhat lacking in rigor, it does provide a useful pictorial representation (Fig. 1.2) which serves as the origin for the designation of m as a *projection* quantum number. Also, because m degeneracies are removed by a magnetic field (see Section 17.1), m is also known as a *magnetic* quantum number.

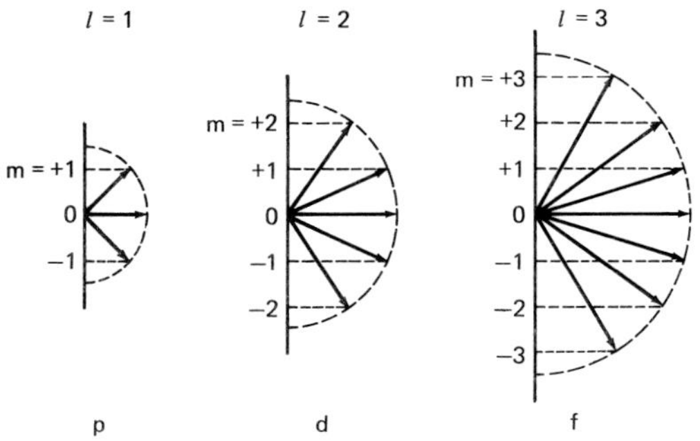

FIG. 1.2 Geometrical relation between the quantum numbers l and m. In these diagrams $\sqrt{l(l+1)}$ is regarded as the magnitude of **L**.

According to (1.1-16), L^2 commutes with all components of **L** and in particular

$$[L^2, L_z] = 0. \tag{1.2-18}$$

We therefore expect an eigenfunction of L^2 to be simultaneously an eigenfunction of L_z. Since the φ-dependence of $Y_{lm}(\theta, \varphi)$ is entirely confined to $e^{im\varphi}$ and L_z is given by (1.1-3c), we have

$$L_z Y_{lm}(\theta, \varphi) = m Y_{lm}(\theta, \varphi). \tag{1.2-19}$$

Equations (1.2-17) and (1.2-19) exhibit the basic connections between orbital angular momentum operators and spherical harmonics. It is important to note that because of the noncommutativity of the components of **L**, simultaneous eigenfunctions of L^2 and L_z, in general, will not be eigenfunctions of any other component of **L**.

We now list several useful formulas involving spherical harmonics. The reciprocal distance between two points whose position vectors are $\mathbf{r}_1$ and $\mathbf{r}_2$ (Fig. 1.3) is given by

$$\frac{1}{|\mathbf{r}_1 - \mathbf{r}_2|} = \sum_{l=0}^{\infty} \frac{r_<^{\,l}}{r_>^{\,l+1}} P_l(\cos\omega), \tag{1.2-20}$$

in which $r_<$ stands for the smaller of the two distances $|\mathbf{r}_1|$ and $|\mathbf{r}_2|$, $r_>$ is the greater of the two distances, and $P_l(\cos\omega)$ is a Legendre polynomial. If $\mathbf{r}_1$ is in the direction (θ_1, φ_1) and $\mathbf{r}_2$ is in the direction (θ_2, φ_2), then the angle ω is the angle between the two directions. The *addition theorem*

$$P_l(\cos\omega) = \frac{4\pi}{2l+1} \sum_{m=-l}^{l} Y_{lm}^*(\theta_1, \varphi_1) Y_{lm}(\theta_2, \varphi_2) \tag{1.2-21}$$

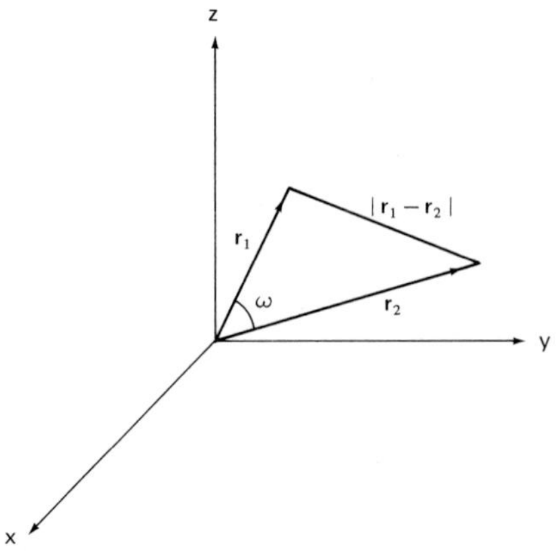

FIG. 1.3 Notation and coordinate system for Eq. (1.2-20).

permits us to replace (1.2-20) by

$$\frac{1}{|\mathbf{r}_1 - \mathbf{r}_2|} = \sum_{l=0}^{\infty} \sum_{m=-l}^{l} \frac{4\pi}{2l+1} \frac{r_<^{\,l}}{r_>^{l+1}} Y_{lm}^*(\theta_1, \varphi_1) Y_{lm}(\theta_2, \varphi_2). \quad (1.2\text{-}22)$$

Another variation of (1.2-22) is obtained by writing

$$\begin{aligned} \mathbf{Y}_1^{(l)} \cdot \mathbf{Y}_2^{(l)} &\equiv \mathbf{Y}^{(l)}(\theta_1, \varphi_1) \cdot \mathbf{Y}^{(l)}(\theta_2, \varphi_2) \\ &= \sum_{m=-l}^{l} (-1)^m Y_{l\,-m}(\theta_1, \varphi_1) Y_{lm}(\theta_2, \varphi_2) \\ &= \sum_{m=-l}^{l} Y_{lm}^*(\theta_1, \varphi_1) Y_{lm}(\theta_2, \varphi_2). \end{aligned} \quad (1.2\text{-}23)$$

Substitution in (1.2-22) yields

$$\frac{1}{|\mathbf{r}_1 - \mathbf{r}_2|} = \sum_{l=0}^{\infty} \frac{4\pi}{2l+1} \frac{r_<^{\,l}}{r_>^{l+1}} \mathbf{Y}_1^{(l)} \cdot \mathbf{Y}_2^{(l)}. \quad (1.2\text{-}24)$$

When $l = 1$ in (1.2-21),

$$P_1(\cos\omega) = \cos\omega = \frac{4\pi}{3} \sum_{m=-l}^{l} Y_{1m}^*(\theta_1, \varphi_1) Y_{1m}(\theta_2, \varphi_2), \quad (1.2\text{-}25)$$

which then provides an expression for the cosine of the angle between $\mathbf{r}_1$ and $\mathbf{r}_2$ (Fig. 1.3). Alternatively, if ω is set equal to zero in (1.2-21),

$$P_l(1) = \frac{4\pi}{2l+1} \sum_{m=-l}^{l} |Y_{lm}(\theta, \varphi)|^2.$$

Since $P_l(1) = 1$,

$$\sum_{m=-l}^{l} |Y_{lm}(\theta, \varphi)|^2 = \frac{2l+1}{4\pi}. \tag{1.2-26}$$

Also, setting $Y_{l'm'}(\theta, \varphi) = Y_{00} = 1/\sqrt{4\pi}$ in (1.2-5) yields

$$\int Y_{lm}(\theta, \varphi)\, d\Omega = \sqrt{4\pi}\, \delta(l, 0)\, \delta(m, 0). \tag{1.2-27}$$

A plane wave may be expanded in terms of spherical harmonics as

$$e^{i\mathbf{k}\cdot\mathbf{r}} = 4\pi \sum_{l=0}^{\infty} \sum_{m=-l}^{l} i^l j_l(kr) Y_{lm}(\theta_r, \varphi_r) Y^*_{lm}(\theta_k, \varphi_k), \tag{1.2-28}$$

in which $j_l(kr)$ is a spherical Bessel function (Appendix 5), (r, θ_r, φ_r) the coordinates of the point of observation, and (θ_k, φ_k) the direction of the wave vector (Fig. 1.4).

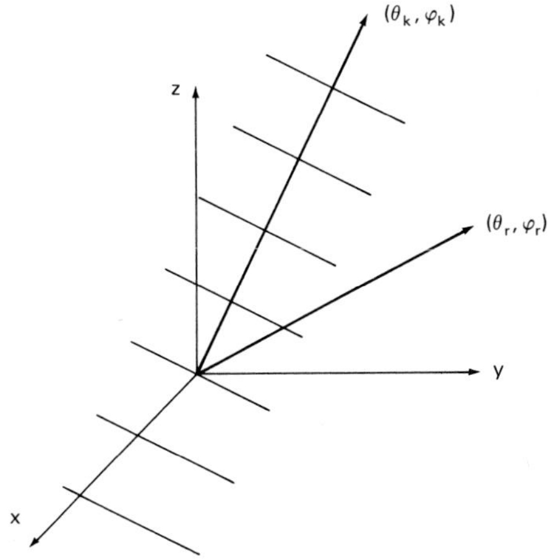

FIG. 1.4 Notation and coordinate system for Eq. (1.2-28).

The integral of the product of three spherical harmonics is given by

$$\begin{aligned} &\int_0^\pi \sin\theta\, d\theta \int_0^{2\pi} d\varphi\; Y^*_{l'm'}(\theta, \varphi) Y_{LM}(\theta, \varphi) Y_{lm}(\theta, \varphi) \\ &= \int Y^*_{l'm'}(\theta, \varphi) Y_{LM}(\theta, \varphi) Y_{lm}(\theta, \varphi)\, d\Omega \\ &\equiv \langle l'm' | Y_{LM} | lm \rangle \\ &= (-1)^{m'} \sqrt{\frac{(2l'+1)(2L+1)(2l+1)}{4\pi}} \begin{pmatrix} l' & L & l \\ -m' & M & m \end{pmatrix} \begin{pmatrix} l' & L & l \\ 0 & 0 & 0 \end{pmatrix}. \end{aligned} \tag{1.2-29}$$

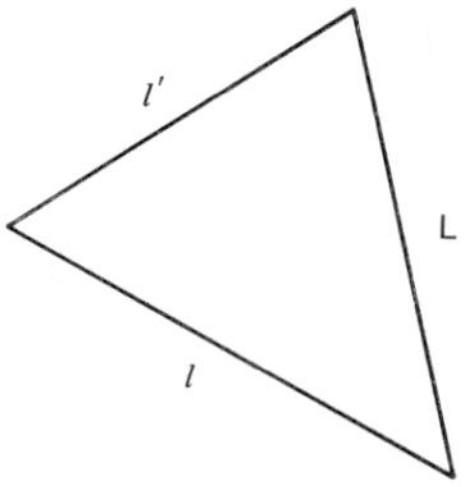

FIG. 1.5 The triangle relation for angular momenta.

This is known as the *Gaunt formula*; the quantities $\left(\begin{smallmatrix} a & b & c \\ d & e & f \end{smallmatrix}\right)$ are numerical coefficients called *3j symbols* whose properties are described in Section 1.5. The integral (1.2-29) vanishes unless the conditions

$$-m' + M + m = 0, \tag{1.2-30}$$

$$l' + L + l \text{ is an even integer}, \tag{1.2-31}$$

$$\left.\begin{array}{r} l' + L - l \\ l' - L + l \\ -l' + L + l \end{array}\right\} \geqslant 0 \tag{1.2-32}$$

are satisfied. The symbol $\triangle(l'Ll)$ is often used as shorthand for (1.2-32) together with the condition that $l' + L + l$ is an integer (not necessarily even). These are also known as the triangle conditions (Fig. 1.5). Selected numerical values of (1.2-29) are given in Table 11.1. When the triangle conditions are satisfied,

$$Y_{LM}(\theta, \varphi) = (-1)^{l'-l-M}\sqrt{2L+1}\sum_{mm'}\begin{pmatrix} l & l' & L \\ m & m' & -M \end{pmatrix} Y_{lm}(\theta, \varphi) Y_{l'm'}(\theta, \varphi), \tag{1.2-33}$$

$$Y_{lm}(\theta, \varphi) Y_{l'm'}(\theta, \varphi) = \sum_{LM}\sqrt{\frac{(2l+1)(2l'+1)(2L+1)}{4\pi}} \times \begin{pmatrix} l & l' & L \\ m & m' & M \end{pmatrix}\begin{pmatrix} l & l' & L \\ 0 & 0 & 0 \end{pmatrix} Y^*_{LM}(\theta, \varphi), \tag{1.2-34}$$

$$Y^*_{lm}(\theta, \varphi) Y_{l'm'}(\theta, \varphi) = \sum_{LM}(-1)^{m'}\sqrt{\frac{(2l+1)(2l'+1)(2L+1)}{4\pi}} \times \begin{pmatrix} l & l' & L \\ m & -m' & M \end{pmatrix}\begin{pmatrix} l & l' & L \\ 0 & 0 & 0 \end{pmatrix} Y_{LM}(\theta, \varphi). \tag{1.2-35}$$

Equations (1.2-34) and (1.2-35) are equivalent.

1.3 Generalized Angular Momentum

The commutation rules (1.1-10) for the components of orbital angular momentum operators followed from definition (1.1-1) and the commutation rules (1.1-8a) and (1.1-8b). This development led to the conclusion that the orbital angular momentum quantum numbers, l and m, were integers. However, other kinds of angular momenta are encountered in physical problems and the quantum numbers associated with such angular momenta are not necessarily integers. It is therefore necessary to extend the formalism in such a way as to permit the appearance of nonintegral quantum numbers but without invalidating any of the previous results pertaining to orbital angular momentum.

For this purpose we take the commutation rules (1.1-10) as the starting point of the development. The generalized angular momentum operator $\mathbf{J}$ is then defined as a vector operator with Hermitian components J_x, J_y, and J_z which satisfy

$$\mathbf{J} \times \mathbf{J} = i\mathbf{J}. \tag{1.3-1}$$

By analogy with (1.1-5), the spherical components are defined as

$$J_{+1} = -\frac{1}{\sqrt{2}}(J_x + iJ_y), \qquad J_0 = J_z, \qquad J_{-1} = \frac{1}{\sqrt{2}}(J_x - iJ_y), \tag{1.3-2}$$

with the inverse relations

$$J_x = -\frac{1}{\sqrt{2}}(J_{+1} - J_{-1}), \qquad J_y = \frac{i}{\sqrt{2}}(J_{+1} + J_{-1}), \qquad J_z = J_0. \tag{1.3-3}$$

It should be remarked that the structural resemblance between (1.3-2) and the spherical harmonics $Y_{1m}(\theta, \varphi)$ (Table 1.1) is not accidental. This aspect will be further explored in the discussion on irreducible tensors in Section 6.1. We also have

$$J_{\pm 1}^{\dagger} = -J_{\mp 1}, \tag{1.3-4}$$

and, as a direct consequence of (1.3-1),

$$[J_0, J_{\pm 1}] = \pm J_{\pm 1}, \qquad [J_{+1}, J_{-1}] = -J_0. \tag{1.3-5}$$

Since J_x, J_y, and J_z are Hermitian, the total angular momentum operator

$$J^2 = J_x^2 + J_y^2 + J_z^2 \tag{1.3-6a}$$

must also be Hermitian. It may be written in various forms as

$$J^2 = -J_{+1}J_{-1} + J_0^2 - J_{-1}J_{+1} \tag{1.3-6b}$$

$$= \sum_q (-1)^q J_q J_{-q} \qquad (q = 1, 0, -1) \tag{1.3-6c}$$

$$= -2J_{+1}J_{-1} + J_0(J_0 - 1) \tag{1.3-6d}$$

$$= -2J_{-1}J_{+1} + J_0(J_0 + 1) \tag{1.3-6e}$$

in which the last two expressions are based on (1.3-5). J^2 commutes with all rectangular and spherical components of **J**, i.e.,

$$[J^2, J_\mu] = 0. \tag{1.3-7}$$

So far, all the relations that have been written in terms of **J** are duplicates of corresponding relations in terms of **L** given in the previous section. However, at this stage the development proceeds in a new direction. Since J^2 commutes with all components of **J** and, in particular, with J_0, there exist simultaneous eigenfunctions of the two operators. Using the Dirac notation (Appendix 1), let such eigenfunctions, represented symbolically be $|\lambda m\rangle$, satisfy

$$J^2|\lambda m\rangle = \lambda|\lambda m\rangle, \qquad J_0|\lambda m\rangle = m|\lambda m\rangle. \tag{1.3-8}$$

It is seen that in this notation, the eigenfunctions (eigenkets) are labeled by the eigenvalues. Since J^2 and J_0 are both Hermitian, λ and m must be real and

$$\langle\lambda' m'|\lambda m\rangle = \delta_{\lambda'\lambda}\delta_{m'm}. \tag{1.3-9}$$

To proceed further, we invoke a basic postulate of quantum mechanics, namely, that the scalar product of any state vector f with itself is positive definite, i.e.,

$$\langle f|f\rangle \geqslant 0, \qquad \langle f|f\rangle = 0 \quad \text{only if} \quad f = 0. \tag{1.3-10}$$

Using the "turn-over" rule ((A2-4) Appendix 2), the Hermitian property of J_x, and (1.3-10), it is seen that

$$\langle\lambda m|J_x^2|\lambda m\rangle = \langle J_x^\dagger\lambda m|J_x\lambda m\rangle = \langle J_x\lambda m|J_x\lambda m\rangle \geqslant 0. \tag{1.3-11}$$

Similarly,

$$\langle\lambda m|J_y^2|\lambda m\rangle \geqslant 0 \qquad \text{and} \qquad \langle\lambda m|J_z^2|\lambda m\rangle \geqslant 0. \tag{1.3-12}$$

It follows that

$$\langle\lambda m|J^2|\lambda m\rangle \geqslant 0. \tag{1.3-13}$$

But, from (1.3-8) and (1.3-9),

$$\langle\lambda m|J^2|\lambda m\rangle = \lambda;$$

therefore, in view of (1.3-13),

$$\lambda \geqslant 0. \tag{1.3-14}$$

Thus λ is not only real but is positive or zero.

It will now be shown that m has an upper and a lower bound. From (1.3-6e),

$$J_{-1}J_{+1} = \tfrac{1}{2}(J_0{}^2 + J_0 - J^2).$$

Therefore,

$$\langle\lambda m|J_{-1}J_{+1}|\lambda m\rangle = \tfrac{1}{2}(m^2 + m - \lambda);$$

but

$$\langle\lambda m|J_{-1}J_{+1}|\lambda m\rangle = -\langle J_{+1}\lambda m|J_{+1}\lambda m\rangle, \tag{1.3-15}$$

where the right-hand side of (1.3-15) has again been obtained by the "turn-over" rule. As before,

$$\langle J_{+1}\lambda m|J_{+1}\lambda m\rangle \geqslant 0, \tag{1.3-16}$$

so that

$$\tfrac{1}{2}(m^2 + m - \lambda) \leqslant 0 \tag{1.3-17}$$

or

$$\lambda \geqslant m^2 + m \equiv f(m). \tag{1.3-18}$$

A plot of $f(m)$ as a function of m is shown in Fig. 1.6. Since λ is positive, it is

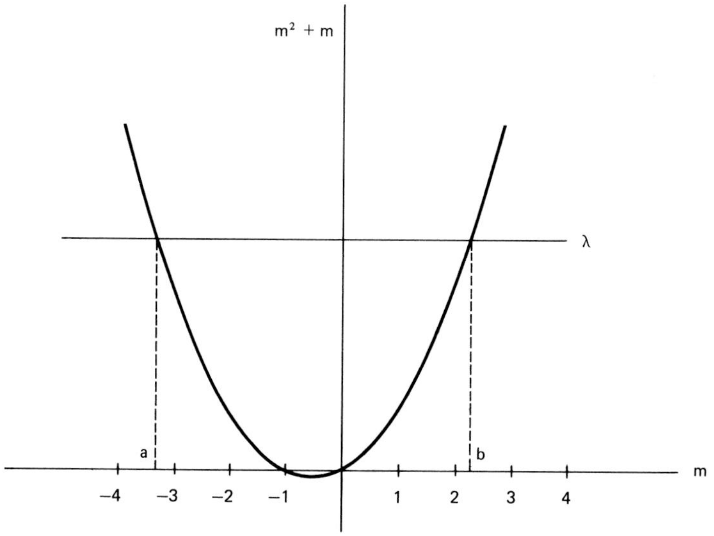

FIG. 1.6 Plot of Eq. (1.3-18).

evident that m possesses an upper bound, say, $b \geqslant 0$ and a lower bound $a \leqslant 0$. Both bounds depend on λ and there are no eigenvalues of J_0 outside of the interval (a, b).

Using the commutator relations (1.3-5),

$$J_0 J_{+1}|\lambda m\rangle = J_{+1} J_0|\lambda m\rangle + J_{+1}|\lambda m\rangle = (m+1)J_{+1}|\lambda m\rangle,$$

from which it is concluded that $J_{+1}|\lambda m\rangle$ is an eigenfunction of J_0 with eigenvalue $(m+1)$. Thus J_{+1}, acting on $|\lambda m\rangle$, has the effect of displacing m upward by one unit. This may be expressed by writing

$$J_{+1}|\lambda m\rangle = c|\lambda\, m+1\rangle, \tag{1.3-19}$$

where c is a constant. Repeating this process, it is found that $J_{+1}^n|\lambda m\rangle$ is an eigenfunction of J_0 with eigenvalue $(m+n)$. Similarly,

$$J_0 J_{-1}|\lambda m\rangle = J_{-1} J_0|\lambda m\rangle - J_{-1}|\lambda m\rangle = (m-1)J_{-1}|\lambda m\rangle,$$

which indicates that

$$J_{-1}|\lambda m\rangle = c'|\lambda\, m-1\rangle, \tag{1.3-20}$$

where c' is another constant. In this case $J_{-1}^n|\lambda m\rangle$ is an eigenfunction of J_0 with eigenvalue $(m-n)$. Because of (1.3-19) and (1.3-20), J_{+1} and J_{-1} are also known as *ladder operators*. We now have the two sequences (or ladders)

$$\begin{aligned} &\vdots \\ J_0 J_{+1}^2|\lambda m\rangle &= (m+2)J_{+1}^2|\lambda m\rangle \\ J_0 J_{+1}|\lambda m\rangle &= (m+1)J_{+1}|\lambda m\rangle \\ J_0|\lambda m\rangle &= m|\lambda m\rangle \\ J_0 J_{-1}|\lambda m\rangle &= (m-1)J_{-1}|\lambda m\rangle \\ J_0 J_{-1}^2|\lambda m\rangle &= (m-2)J_{-1}^2|\lambda m\rangle \\ &\vdots \end{aligned} \tag{1.3-21}$$

However, the sequences do not continue indefinitely in both directions; there is, in fact, an upper and lower bound. To see this, we note that both J_{+1} and J_{-1} commute with J^2 so that

$$J^2 J_{\pm 1}^n|\lambda m\rangle = J_{\pm 1}^n J^2|\lambda m\rangle = \lambda J_{\pm 1}^n|\lambda m\rangle. \tag{1.3-22}$$

This means that $J_{\pm 1}^n|\lambda m\rangle$ is an eigenfunction of J^2 and the corresponding eigenvalue is λ. Thus we have a fixed value of λ and all the eigenvalues of J_0 are confined to an interval such as (a, b) in Fig. 1.6. Let these eigenvalues be

$$m_l, \quad m_l+1, \quad m_l+2, \ldots, m_u,$$

where m_l is the lowest eigenvalue in the sequence and does not necessarily coincide with the endpoint a. Similarly m_u is the highest eigenvalue and does not necessarily coincide with the endpoint b. But to ensure that the eigenvalues m_l, $m_l + 1, \ldots, m_u$ remain within the interval (a, b), it is necessary to impose the conditions

$$J_{+1}|\lambda m_u\rangle = 0, \tag{1.3-23a}$$

$$J_{-1}|\lambda m_l\rangle = 0. \tag{1.3-23b}$$

However, from (1.3-6e) and (1.3-8),

$$J_{-1}J_{+1}|\lambda m_u\rangle = \tfrac{1}{2}(J_0{}^2 + J_0 - J^2)|\lambda m_u\rangle = \tfrac{1}{2}(m_u{}^2 + m_u - \lambda)|\lambda m_u\rangle. \tag{1.3-24}$$

Condition (1.3-23a) therefore implies that

$$m_u{}^2 + m_u - \lambda = 0. \tag{1.3-25}$$

In the same fashion,

$$J_{+1}J_{-1}|\lambda m_l\rangle = \tfrac{1}{2}(J_0{}^2 - J_0 - J^2)|\lambda m_l\rangle = \tfrac{1}{2}(m_l{}^2 - m_l - \lambda)|\lambda m_l\rangle = 0 \tag{1.3-26}$$

or

$$m_l{}^2 - m_l - \lambda = 0. \tag{1.3-27}$$

In order to satisfy both (1.3-25) and (1.3-27) we must have

$$m_u = -m_l. \tag{1.3-28}$$

It then follows that it is impossible to have a sequence $m_l, m_l + 1, \ldots, m_u$ that satisfies (1.3-28) unless all members of the sequence are either integral or half-integral.

It is customary to replace m_u by j; we then have, from (1.3-25),

$$\lambda = j(j + 1).$$

Equations (1.3-8) and (1.3-9) may now be written as

$$\begin{aligned} J^2|jm\rangle &= j(j + 1)|jm\rangle, \\ J_z|jm\rangle &= J_0|jm\rangle = m|jm\rangle, \\ \langle j'm'|jm\rangle &= \delta_{j'j}\delta_{m'm}. \end{aligned} \tag{1.3-29}$$

Since m_u $(=j)$ must be integral or half-integral, the possible values of j are

$$j = 0, \quad \tfrac{1}{2}, \quad 1, \quad \tfrac{3}{2}, \ldots. \tag{1.3-30}$$

Only positive values of j appear because of (1.3-14). Also, since the possible values of m lie between m_l and m_u, we have, in view of (1.3-28),

$$m = j, \quad j - 1, \ldots, -j. \tag{1.3-31}$$

As before, m is called the *projection* or *magnetic* quantum number.

The result embodied in (1.3-30) contains the basic distinction between generalized angular momentum and orbital angular momentum. If, in (1.3-30), we were to allow j to assume only integral values, it would merely be necessary to replace **J** by **L** and j by l to reproduce all the results pertaining to orbital angular momentum operators. However, (1.3-30) also permits j to have half-integral values. This is a new result and suggests the possible existence of angular momentum operators whose properties differ in certain respects from those associated with orbital angular momentum. Indeed, this turns out to be the case and leads to far-reaching physical consequences.

Matrix elements of the various angular momentum operators may now be calculated. From (1.3-29),

$$\langle j'm'|J^2|jm\rangle = j(j+1)\,\delta_{j'j}\,\delta_{m'm}, \tag{1.3-32}$$

$$\langle j'm'|J_0|jm\rangle = m\,\delta_{j'j}\,\delta_{m'm}. \tag{1.3-33}$$

To obtain matrix elements of J_{+1}, we refer to (1.3-24); in the present notation

$$J_{-1}J_{+1}|jm\rangle = \tfrac{1}{2}(J_0{}^2 + J_0 - J^2)|jm\rangle = \tfrac{1}{2}[m(m+1) - j(j+1)]|jm\rangle, \tag{1.3-34}$$

so that

$$\langle jm|J_{-1}J_{+1}|jm\rangle = \tfrac{1}{2}[m(m+1) - j(j+1)]. \tag{1.3-35}$$

From the definition of matrix multiplication (or the closure property) we also have

$$\langle jm|J_{-1}J_{+1}|jm\rangle = \sum_{j'm'} \langle jm|J_{-1}|j'm'\rangle\langle j'm'|J_{+1}|jm\rangle. \tag{1.3-36}$$

The sum in (1.3-36) may be simplified by application of the following theorem: If $[A, B] = 0$, A Hermitian, $A\psi_1 = a_1\psi_1$, $A\psi_2 = a_2\psi_2$, and $a_1 \neq a_2$, then $\langle\psi_1|B|\psi_2\rangle = 0$. In the present case J_{-1} commutes with J^2 and both $|jm\rangle$ and $|j'm'\rangle$ are eigenfunctions of J^2 with eigenvalues $j(j+1)$ and $j'(j'+1)$. Therefore $\langle jm|J_{-1}|j'm'\rangle = 0$ when $j \neq j'$. The same argument applies to J_{+1}; hence

$$\sum_{j'm'} \langle jm|J_{-1}|j'm'\rangle\langle j'm'|J_{+1}|jm\rangle = \sum_{m'} \langle jm|J_{-1}|jm'\rangle\langle jm'|J_{+1}|jm\rangle. \tag{1.3-37}$$

The sum over m' cannot be simplified in the same way because J_{-1} and J_{+1} do not commute with J_0. Nevertheless the sum over m' reduces to one term because J_{-1} acting on $|jm'\rangle$ displaces m' downward by one unit as in (1.3-20), while J_{+1} acting on $|jm'\rangle$ displaces m' upward by one unit as in (1.3-19). The orthogonality condition in (1.3-29) then eliminates all terms from the sum except $\langle jm|J_{-1}|j\,m+1\rangle\langle j\,m+1|J_{+1}|jm\rangle$. By the "turn-over" rule and (1.3-4)

$$\begin{aligned}&\langle jm|J_{-1}|j\,m+1\rangle\langle j\,m+1|J_{+1}|jm\rangle\\ &\quad = -\langle J_{+1}jm|j\,m+1\rangle\langle j\,m+1|J_{+1}jm\rangle = -|\langle j\,m+1|J_{+1}|jm\rangle|^2. \end{aligned} \tag{1.3-38}$$

Combining (1.3-35), (1.3-36), and (1.3-38), we have

$$\langle j\,m+1|J_{+1}|jm\rangle = -\sqrt{\tfrac{1}{2}[j(j+1)-m(m+1)]} \tag{1.3-39}$$

in which the arbitrary phase factor has been chosen in conformity with the Condon–Shortley convention. All other matrix elements are zero. By a similar development it is found that the nonvanishing matrix elements of J_{-1} are

$$\langle j\,m-1|J_{-1}|jm\rangle = \sqrt{\tfrac{1}{2}[j(j+1)-m(m-1)]}. \tag{1.3-40}$$

Matrix elements of J_x and J_y follow immediately from (1.3-3) in combination with (1.3-39) and (1.3-40). Some numerical values are listed in Table 1.3.

BLE 1.3

atrix Elements of $\langle j\,m\pm1|J_{\pm1}|jm\rangle = \mp\sqrt{\tfrac{1}{2}[j(j+1)-m(m\pm1)]}$

m	$\langle j\,m+1\|J_{+1}\|jm\rangle$	$\langle j\,m-1\|J_{-1}\|jm\rangle$	j	m	$\langle j\,m+1\|J_{+1}\|jm\rangle$	$\langle j\,m-1\|J_{-1}\|jm\rangle$
$\frac{1}{2}$	0	$\sqrt{\frac{1}{2}}$	2	1	$-\sqrt{2}$	$\sqrt{3}$
$-\frac{1}{2}$	$-\sqrt{\frac{1}{2}}$	0	2	0	$-\sqrt{3}$	$\sqrt{3}$
1	0	1	2	-1	$-\sqrt{3}$	$\sqrt{2}$
0	-1	1	2	-2	$-\sqrt{2}$	0
-1	-1	0	$\frac{5}{2}$	$\frac{5}{2}$	0	$\sqrt{\frac{5}{2}}$
$\frac{3}{2}$	0	$\sqrt{\frac{3}{2}}$	$\frac{5}{2}$	$\frac{3}{2}$	$-\sqrt{\frac{5}{2}}$	2
$\frac{1}{2}$	$-\sqrt{\frac{3}{2}}$	$\sqrt{2}$	$\frac{5}{2}$	$\frac{1}{2}$	-2	$\frac{3}{\sqrt{2}}$
$-\frac{1}{2}$	$-\sqrt{2}$	$\sqrt{\frac{3}{2}}$	$\frac{5}{2}$	$-\frac{1}{2}$	$-\frac{3}{\sqrt{2}}$	2
$-\frac{3}{2}$	$-\sqrt{\frac{3}{2}}$	0	$\frac{5}{2}$	$-\frac{3}{2}$	-2	$\sqrt{\frac{5}{2}}$
2	0	$\sqrt{2}$	$\frac{5}{2}$	$-\frac{5}{2}$	$-\sqrt{\frac{5}{2}}$	0

In the interest of avoiding excessively cumbersome language and where misunderstanding is unlikely, it is not uncommon to refer to **J** or its components simply as "angular momenta" rather than "angular momentum operators." Similarly, one may speak of a state as having an angular momentum **J** as a substitute for saying that the wave function of the state is an eigenfunction of J^2 and J_z with eigenvalues $j(j+1)$ and m, respectively. At times the state may simply be labeled by j and m.

1.4 Spin

The eigenfunctions of J^2 and J_0 in (1.3-29) have been symbolized by $|jm\rangle$ with j amd m given by (1.3-30) and (1.3-31). If j is integral, it is known from the properties of orbital angular momentum that the spherical harmonics are eigenfunctions of J^2 $(=L^2)$ and J_0 $(=L_0)$, that is,

$$|jm\rangle = |lm\rangle = Y_{lm}(\theta, \varphi). \tag{1.4-1}$$

When j is half-integral, the eigenfunctions $|jm\rangle$ are not functions of the coordinates—they must be specified in other ways. A case in point is $j = \frac{1}{2}$ which is associated with the spin angular momentum properties of an electron (as well as of other particles, e.g., proton, neutron, etc.)

In discussing the spin properties of a particle it is customary to adopt a notation in which $J = S$ and $j = s$. For a fixed value of s,

$$\begin{aligned} S^2|sm\rangle &= s(s+1)|sm\rangle, \\ S_0|sm\rangle &= m|sm\rangle, \\ \langle sm'|sm\rangle &= \delta_{m'm}, \end{aligned} \tag{1.4-2}$$

from (1.3-29). Also, in conformity with the definition of angular momentum operators

$$\mathbf{S} \times \mathbf{S} = i\mathbf{S}. \tag{1.4-3}$$

We now give the matrices for the various operators when $s = \frac{1}{2}$. Columns will be labeled by the value of m starting with the highest value and progressing to the lowest value; rows will be labeled by m', also in the same sequence. Thus from (1.3-32), the matrix elements of S^2 are

$\langle sm'|S^2|sm\rangle =$ (1.4-4)

m' \ m	$\frac{1}{2}$	$-\frac{1}{2}$
$\frac{1}{2}$	$\frac{3}{4}$	0
$-\frac{1}{2}$	0	$\frac{3}{4}$

or, more compactly,

$$S^2 = \begin{pmatrix} \frac{3}{4} & 0 \\ 0 & \frac{3}{4} \end{pmatrix}; \tag{1.4-5}$$

that is, the matrix in (1.4-5) is the matrix representation of the operator S^2 in the basis set

$$|sm\rangle = |\tfrac{1}{2}\tfrac{1}{2}\rangle, |\tfrac{1}{2} - \tfrac{1}{2}\rangle.$$

Similarly, from (1.3-33), (1.3-39), and (1.3-40),

$$S_0 = S_z = \begin{pmatrix} \frac{1}{2} & 0 \\ 0 & -\frac{1}{2} \end{pmatrix}, \tag{1.4-6}$$

$$S_{+1} = \begin{pmatrix} 0 & -\sqrt{\frac{1}{2}} \\ 0 & 0 \end{pmatrix}, \tag{1.4-7}$$

$$S_{-1} = \begin{pmatrix} 0 & 0 \\ \sqrt{\frac{1}{2}} & 0 \end{pmatrix}. \tag{1.4-8}$$

Also, on combining (1.4-7) and (1.4-8) as in (1.3-3),

$$S_x = \begin{pmatrix} 0 & \frac{1}{2} \\ \frac{1}{2} & 0 \end{pmatrix}, \tag{1.4-9}$$

$$S_y = i\begin{pmatrix} 0 & -\frac{1}{2} \\ \frac{1}{2} & 0 \end{pmatrix}. \tag{1.4-10}$$

The eigenfunctions $|sm\rangle$ may also be written as

$$|sm\rangle = \begin{cases} |\tfrac{1}{2}\tfrac{1}{2}\rangle = \begin{pmatrix} 1 \\ 0 \end{pmatrix} \equiv \alpha, \\ |\tfrac{1}{2} -\tfrac{1}{2}\rangle = \begin{pmatrix} 0 \\ 1 \end{pmatrix} \equiv \beta, \end{cases} \tag{1.4-11}$$

in which $\binom{1}{0}$ and $\binom{0}{1}$ are to be understood as column matrices. These expressions have been designed explicitly to satisfy (1.4-2); thus

$$S^2|\tfrac{1}{2}\tfrac{1}{2}\rangle = \begin{pmatrix} \frac{3}{4} & 0 \\ 0 & \frac{3}{4} \end{pmatrix}\begin{pmatrix} 1 \\ 0 \end{pmatrix} = \tfrac{3}{4}\begin{pmatrix} 1 \\ 0 \end{pmatrix} = \tfrac{1}{2}(\tfrac{1}{2} + 1)|\tfrac{1}{2}\tfrac{1}{2}\rangle$$

or

$$S^2\alpha = \tfrac{1}{2}(\tfrac{1}{2} + 1)\alpha.$$

Similarly

$$S_0|\tfrac{1}{2}\tfrac{1}{2}\rangle = \begin{pmatrix} \tfrac{1}{2} & 0 \\ 0 & -\tfrac{1}{2} \end{pmatrix}\begin{pmatrix} 1 \\ 0 \end{pmatrix} = \tfrac{1}{2}\begin{pmatrix} 1 \\ 0 \end{pmatrix} = \tfrac{1}{2}|\tfrac{1}{2}\tfrac{1}{2}\rangle$$

or

$$S_0\alpha = \tfrac{1}{2}\alpha.$$

The orthogonality relations in (1.4-2) are also satisfied as, for example, by

$$\begin{aligned} \langle\tfrac{1}{2}\tfrac{1}{2}|\tfrac{1}{2}\tfrac{1}{2}\rangle &= \langle\alpha|\alpha\rangle = (1 \quad 0)\begin{pmatrix} 1 \\ 0 \end{pmatrix} = 1, \\ \langle\tfrac{1}{2}\tfrac{1}{2}|\tfrac{1}{2} \; -\tfrac{1}{2}\rangle &= \langle\alpha|\beta\rangle = (1 \quad 0)\begin{pmatrix} 0 \\ 1 \end{pmatrix} = 0. \end{aligned} \tag{1.4-12}$$

In still another notation, the eigenfunction $|sm\rangle$ is written in the form $\xi(m)$ or ξ_m to suggest that ξ is a function of a "spin coordinate" or "spin variable" m, the latter being the projection quantum number. With $m = \pm\frac{1}{2}$

$$\begin{aligned} \xi(\tfrac{1}{2}) &= \xi_{1/2} = |\tfrac{1}{2}\tfrac{1}{2}\rangle = \alpha, \\ \xi(-\tfrac{1}{2}) &= \xi_{-1/2} = |\tfrac{1}{2} \; -\tfrac{1}{2}\rangle = \beta. \end{aligned} \tag{1.4-13}$$

Evidently, $|sm\rangle$ is not a function of coordinates; mathematically, it is known as a *spinor*.

The *Pauli spin matrices* are defined by

$$\sigma_z = \begin{pmatrix} 1 & 0 \\ 0 & -1 \end{pmatrix}, \qquad \sigma_x = \begin{pmatrix} 0 & 1 \\ 1 & 0 \end{pmatrix}, \qquad \sigma_y = \begin{pmatrix} 0 & -i \\ i & 0 \end{pmatrix}. \tag{1.4-14}$$

Apart from a numerical factor these matrices are the same as those in (1.4-6), (1.4-9), and (1.4-10); in fact

$$\boldsymbol{\sigma} = 2\mathbf{S}. \tag{1.4-15}$$

The difference between $\boldsymbol{\sigma}$ and $\mathbf{S}$ appears to be trivial; nevertheless, it is important to recognize that because $\boldsymbol{\sigma}$ satisfies

$$\boldsymbol{\sigma} \times \boldsymbol{\sigma} = 2i\boldsymbol{\sigma}, \tag{1.4-16}$$

which is *not* of the same form as $\mathbf{S} \times \mathbf{S} = i\mathbf{S}$, the operator $\boldsymbol{\sigma}$ does *not* qualify as an angular momentum operator in contrast to S, which *is* an angular momentum operator.

1.5 Coupling of Two Angular Momenta

In this section we shall explain the sense in which two angular momentum operators, $\mathbf{J}_1$ and $\mathbf{J}_2$, are coupled to form a new angular momentum operator $\mathbf{J}$, and how the respective eigenfunctions and eigenvalues are related.

It will be assumed that $\mathbf{J}_1$ and $\mathbf{J}_2$ operate in different spaces, by which it is meant that any component of $\mathbf{J}_1$ commutes with any component of $\mathbf{J}_2$, or that

$$[J_{1i}, J_{2j}] = 0, \qquad \text{all} \quad i, j. \tag{1.5-1}$$

Thus $\mathbf{J}_1$ may be a spin angular momentum operator while $\mathbf{J}_2$ is associated with orbital angular momentum, or $\mathbf{J}_1$ and $\mathbf{J}_2$ may be angular momentum operators belonging to two different particles.

According to the general definition of angular momentum operators (1.3-1), $\mathbf{J}_1$ and $\mathbf{J}_2$ satisfy

$$\mathbf{J}_1 \times \mathbf{J}_1 = i\mathbf{J}_1, \qquad \mathbf{J}_2 \times \mathbf{J}_2 = i\mathbf{J}_2, \tag{1.5-2}$$

and there exist sets of orthornormal eigenfunctions such that

$$J_1^2|j_1m_1\rangle = j_1(j_1+1)|j_1m_1\rangle, \qquad J_{1z}|j_1m_1\rangle = m_1|j_1m_1\rangle, \tag{1.5-3}$$

$$J_2^2|j_2m_2\rangle = j_2(j_2+1)|j_2m_2\rangle, \qquad J_{2z}|j_2m_2\rangle = m_2|j_2m_2\rangle, \tag{1.5-4}$$

with

$$\begin{gathered} j_1, j_2 = 0, \ \tfrac{1}{2}, \ 1, \ \tfrac{3}{2}, \ldots, \\ m_1 = j_1, \ j_1 - 1, \ldots, -j_1, \qquad m_2 = j_2, \ j_2 - 1, \ldots, -j_2. \end{gathered} \tag{1.5-5}$$

We now define a new operator $\mathbf{J}$ by

$$\mathbf{J} = \mathbf{J}_1 + \mathbf{J}_2 \tag{1.5-6}$$

with the understanding that each component of $\mathbf{J}$ is the sum of the corresponding components of $\mathbf{J}_1$ and $\mathbf{J}_2$, i.e.,

$$J_x = J_{1x} + J_{2x}, \qquad J_{+1} = J_{1\,+1} + J_{2\,+1},$$

with similar relations for other components. Is $\mathbf{J}$ an angular momentum operator? The commutation properties provide the answer; thus

$$\begin{aligned} [J_x, J_y] &= [J_{1x} + J_{2x}, J_{1y} + J_{2y}] \\ &= [J_{1x}, J_{1y}] + [J_{1x}, J_{2y}] + [J_{2x}, J_{1y}] + [J_{2x}, J_{2y}]. \end{aligned} \tag{1.5-7}$$

The second and third commutators vanish because of (1.5-1); Eq. (1.5-7) then becomes

$$[J_x, J_y] = [J_{1x}, J_{1y}] + [J_{2x}, J_{2y}] = i(J_{1z} + J_{2z}) = iJ_z. \tag{1.5-8}$$

Other commutators of the components of J are evaluated in similar fashion; it may therefore be concluded that

$$\mathbf{J} \times \mathbf{J} = i\mathbf{J}, \tag{1.5-9}$$

which is sufficient to identify $\mathbf{J}$ as an angular momentum operator. We regard (1.5-6) as the defining relation for the coupling of two angular momentum operators, $\mathbf{J}_1$ and $\mathbf{J}_2$, to form a new angular momentum operator $\mathbf{J}$. Parenthetically, it may be remarked that arbitrary linear combinations of $\mathbf{J}_1$ and $\mathbf{J}_2$ do not necessarily produce angular momentum operators.

In view of (1.5-9) there exist orthonormal eigenfunctions $|j_1 j_2 jm\rangle$, also abbreviated to $|jm\rangle$, which satisfy

$$\begin{aligned} J^2|jm\rangle &\equiv J^2|j_1 j_2 jm\rangle = j(j+1)|jm\rangle \equiv j(j+1)|j_1 j_2 jm\rangle, \\ J_z|jm\rangle &\equiv J_z|j_1 j_2 jm\rangle = m|jm\rangle \equiv m|j_1 j_2 jm\rangle, \end{aligned} \tag{1.5-10a}$$

$$j = 0,\ \tfrac{1}{2},\ 1,\ \tfrac{3}{2}, \ldots; \qquad m = j,\ j-1, \ldots, -j. \tag{1.5-10b}$$

The mathematical problem is to establish the relationships between the eigenfunctions and eigenvalues of $J_1{}^2, J_{1z}, J_2{}^2, J_{2z}$ on one hand and the eigenfunctions and eigenvalues of J^2, J_z on the other. For this purpose we construct products of $|j_1 m_1\rangle$ and $|j_2 m_2\rangle$ which are written as

$$|j_1 j_2 m_1 m_2\rangle \equiv |j_1 m_1\rangle |j_2 m_2\rangle. \tag{1.5-11}$$

Clearly, such products as well as their linear combinations are still eigenfunctions of $J_1{}^2, J_{1z}, J_2{}^2, J_{2z}$ with the same eigenvalues as in (1.5-3) and (1.5-4) since $\mathbf{J}_1$ and $\mathbf{J}_2$ operate exclusively in their individual spaces. Thus

$$\begin{aligned} J_1{}^2|j_1 j_2 m_1 m_2\rangle &= j_1(j_1+1)|j_1 j_2 m_1 m_2\rangle, \\ J_{1z}|j_1 j_2 m_1 m_2\rangle &= m_1|j_1 j_2 m_1 m_2\rangle, \\ J_2{}^2|j_1 j_2 m_1 m_2\rangle &= j_2(j_2+1)|j_1 j_2 m_1 m_2\rangle, \\ J_{2z}|j_1 j_2 m_1 m_2\rangle &= m_2|j_1 j_2 m_1 m_2\rangle, \end{aligned} \tag{1.5-12}$$

from which it follows that

$$\begin{aligned} J_z|j_1 j_2 m_1 m_2\rangle &= (J_{1z} + J_{2z})|j_1 j_2 m_1 m_2\rangle \\ &= (m_1 + m_2)|j_1 j_2 m_1 m_2\rangle \\ &= m|j_1 j_2 m_1 m_2\rangle \end{aligned} \tag{1.5-13}$$

where

$$m = m_1 + m_2. \tag{1.5-14}$$

Equation (1.5-13) shows that $|j_1 j_2 m_1 m_2\rangle$ is an eigenfunction of J_z. Since J_z commutes with

$$J^2 = (J_{1x} + J_{2x})^2 + (J_{1y} + J_{2y})^2 + (J_{1z} + J_{2z})^2, \tag{1.5-15}$$

we should like an eigenfunction of J_z to be simultaneously an eigenfunction of J^2. Unfortunately, $|j_1 j_2 m_1 m_2\rangle$ is not an eigenfunction of J^2; however, it is possible to construct linear combinations of the form

$$|j_1 j_2 jm\rangle = \sum_{m_1 m_2} |j_1 j_2 m_1 m_2\rangle\langle j_1 j_2 m_1 m_2 | j_1 j_2 jm\rangle \tag{1.5-16}$$

such that $|j_1 j_2 jm\rangle$ is simultaneously an eigenfunction of J_z and J^2. The quantities $\langle j_1 j_2 m_1 m_2 | j_1 j_2 jm\rangle$ are numerical coefficients which are known as Clebsch–Gordan (CG) coefficients or vector addition coefficients. In a commonly employed terminology one refers to $|j_1 j_2 jm\rangle$ as an eigenfunction in the coupled representation and to $|j_1 j_2 m_1 m_2\rangle$ as an eigenfunction in the uncoupled representation.

The quantum numbers in the coupled representation must be related in some fashion to those in the uncoupled representation. To establish these relations we write

$$\begin{aligned} J_z|j_1 j_2 jm\rangle &= (J_{1z} + J_{2z}) \sum_{m_1 m_2} |j_1 j_2 m_1 m_2\rangle\langle j_1 j_2 m_1 m_2 | j_1 j_2 jm\rangle \\ &= \sum_{m_1 m_2} (m_1 + m_2)|j_1 j_2 m_1 m_2\rangle\langle j_1 j_2 m_1 m_2 | j_1 j_2 jm\rangle. \end{aligned} \tag{1.5-17}$$

If it is stipulated that

$$J_z|j_1 j_2 jm\rangle = m|j_1 j_2 jm\rangle = m \sum_{m_1 m_2} |j_1 j_2 m_1 m_2\rangle\langle j_1 j_2 m_1 m_2 | j_1 j_2 jm\rangle, \tag{1.5-18}$$

then

$$\langle j_1 j_2 m_1 m_2 | j_1 j_2 jm\rangle = 0 \qquad \text{when} \quad m \neq m_1 + m_2, \tag{1.5-19}$$

and the expression for $|j_1 j_2 jm\rangle$ in (1.5-16) may be rewritten as

$$|j_1 j_2 jm\rangle = \sum_{m_1} |j_1 j_2 j\, m - m_1\rangle\langle j_1 j_2 j\, m - m_1 | j_1 j_2 jm\rangle, \tag{1.5-20a}$$

or

$$|j_1 j_2 jm\rangle = \sum_{m_2} |j_1 j_2 j\, m - m_2\rangle\langle j_1 j_2 j\, m - m_2 | j_1 j_2 jm\rangle. \tag{1.5-20b}$$

To find the possible values of j it is observed, from (1.5-5), that

$$-j_1 \leqslant m_1 \leqslant j_1, \qquad -j_2 \leqslant m_2 \leqslant j_2 \tag{1.5-21}$$

or, since $m_1 + m_2 = m$,

$$-j_1 \leqslant m - m_2 \leqslant j_1, \qquad -j_2 \leqslant m - m_1 \leqslant j_2. \tag{1.5-22}$$

Now let m assume its maximum value, namely j, and let m_1 and m_2 assume their respective maximum values, j_1 and j_2. For this case, (1.5-22) becomes

$$-j_1 \leqslant j - j_2 \leqslant j_1, \qquad -j_2 \leqslant j - j_1 \leqslant j_2 \tag{1.5-23}$$

or

$$j_2 - j_1 \leqslant j \leqslant j_1 + j_2, \qquad j_1 - j_2 \leqslant j \leqslant j_1 + j_2, \tag{1.5-24}$$

which may be combined into the single expression

$$|j_1 - j_2| \leqslant j \leqslant j_1 + j_2. \tag{1.5-25}$$

Now consider an example: Suppose $j_1 = \frac{1}{2}$ and $j_2 = 1$. From (1.5-10b), the values of j are restricted to integral and half-integral positive values; therefore (1.5-25) will be satisfied by $j = \frac{1}{2}, 1, \frac{3}{2}$. When $j = 1$, the values of m are $1, 0, -1$; but it is also necessary to satisfy $m = m_1 + m_2$ when $m_1 = \pm\frac{1}{2}$ and $m_2 = 1, 0, -1$. This is obviously impossible and the value $j = 1$ must be eliminated. To avoid such inconsistencies it is necessary to supplement (1.5-25) with the condition

$$j_1 + j_2 + j = n \tag{1.5-26}$$

where n is an integer. Since (1.5-25) is equivalent to

$$\left.\begin{array}{r} j_1 + j_2 - j \\ j_1 - j_2 + j \\ -j_1 + j_2 + j \end{array}\right\} \geqslant 0, \tag{1.5-27}$$

the two conditions (1.5-26) and (1.5-27) taken together are the triangle conditions $\triangle(j_1 j_2 j)$ which we have already encountered in Section 1.2. An equivalent statement for the allowed values of j is

$$j = j_1 + j_2, \quad j_1 + j_2 - 1, \ldots, |j_1 - j_2|. \tag{1.5-28a}$$

The possible values of m must satisfy (1.5-10b) as well as (1.5-14); hence

$$m = m_1 + m_2 = j, \quad j - 1, \ldots, -j. \tag{1.5-28b}$$

We shall now illustrate the derivation of the CG coefficients in a simple case. Let a system with $j_1 = \frac{1}{2}$ be coupled to another system with $j_2 = \frac{1}{2}$. Then $m_1 = \pm\frac{1}{2}$, $m_2 = \pm\frac{1}{2}$. On the basis of (1.5-27) and (1.5-28),

$$j = \begin{cases} 0, & m = 0, \\ 1, & m = 1, 0, -1. \end{cases} \tag{1.5-29}$$

Starting with the maximum value of j $(=1)$ and the maximum value of m $(=1)$,

$$|jm\rangle = |1\ 1\rangle = \overset{j_1 j_2 m_1 m_2}{|\tfrac{1}{2}\tfrac{1}{2}\tfrac{1}{2}\tfrac{1}{2}\rangle}. \tag{1.5-30}$$

The right side of (1.5-30) is the only product of $|j_1 m_1\rangle$ and $|j_2 m_2\rangle$ which satisfies $m = m_1 + m_2$ when $m = 1$. Operating on $|1\ 1\rangle$ with J_{-1} we obtain (Table 1.3)

$$J_{-1}|1\ 1\rangle = |1\ 0\rangle. \tag{1.5-31}$$

But

$$\begin{aligned} J_{-1}|1\ 1\rangle &= J_{-1}|\tfrac{1}{2}\tfrac{1}{2}\tfrac{1}{2}\tfrac{1}{2}\rangle = (J_{1\ -1} + J_{2\ -1})|\tfrac{1}{2}\tfrac{1}{2}\tfrac{1}{2}\tfrac{1}{2}\rangle \\ &= \frac{1}{\sqrt{2}}|\tfrac{1}{2}\tfrac{1}{2}\ -\tfrac{1}{2}\tfrac{1}{2}\rangle + \frac{1}{\sqrt{2}}|\tfrac{1}{2}\tfrac{1}{2}\tfrac{1}{2}\ -\tfrac{1}{2}\rangle; \end{aligned} \tag{1.5-32}$$

therefore

$$|1\ 0\rangle = \frac{1}{\sqrt{2}}|\tfrac{1}{2}\tfrac{1}{2}\ -\tfrac{1}{2}\tfrac{1}{2}\rangle + \frac{1}{\sqrt{2}}|\tfrac{1}{2}\tfrac{1}{2}\tfrac{1}{2}\ -\tfrac{1}{2}\rangle. \tag{1.5-33}$$

The process is repeated by operating on $|1\ 0\rangle$ with J_{-1}; the result is

$$|1\ -1\rangle = |\tfrac{1}{2}\tfrac{1}{2}\ -\tfrac{1}{2}\ -\tfrac{1}{2}\rangle. \tag{1.5-34}$$

This takes care of the case $j = 1$. For $j = 0$, $m = 0$ there is only one eigenstate $|00\rangle$ which must be orthogonal to $|1\ 1\rangle$, $|1\ 0\rangle$, and $|1\ -1\rangle$; this is satisfied by

$$|00\rangle = \frac{1}{\sqrt{2}}|\tfrac{1}{2}\tfrac{1}{2}\tfrac{1}{2}\ -\tfrac{1}{2}\rangle - \frac{1}{\sqrt{2}}|\tfrac{1}{2}\tfrac{1}{2}\ -\tfrac{1}{2}\tfrac{1}{2}\rangle. \tag{1.5-35}$$

The CG coefficients may now be organized in tabular form as in the first part of Table 1.4.

All the functions $|jm\rangle$ given by (1.5-30), (1.5-33)–(1.5-35) are eigenfunctions of J_z and J^2. To illustrate, take $|1\ 1\rangle = |\tfrac{1}{2}\tfrac{1}{2}\tfrac{1}{2}\tfrac{1}{2}\rangle$ as in (1.5-30). Quite clearly $|\tfrac{1}{2}\tfrac{1}{2}\tfrac{1}{2}\tfrac{1}{2}\rangle$ is an eigenfunction of J_z with an eigenvalue equal to one. To verify that $|\tfrac{1}{2}\tfrac{1}{2}\tfrac{1}{2}\tfrac{1}{2}\rangle$ is also an eigenfunction of J^2, write

$$J^2 = (\mathbf{J}_1 + \mathbf{J}_2)\cdot(\mathbf{J}_1 + \mathbf{J}_2) = J_1{}^2 + J_2{}^2 + 2\mathbf{J}_1\cdot\mathbf{J}_2. \tag{1.5-36a}$$

The scalar product $\mathbf{J}_1\cdot\mathbf{J}_2$ can be expressed in terms of the spherical components (1.3-2) (or on the basis of the general form (6.1-18)):

$$\begin{aligned} \mathbf{J}_1\cdot\mathbf{J}_2 &= J_{1x}J_{2x} + J_{1y}J_{2y} + J_{1z}J_{2z} \\ &= -J_{1\ +1}J_{2\ -1} + J_{10}J_{20} - J_{1\ -1}J_{2\ +1}. \end{aligned} \tag{1.5-36b}$$

The evaluation of $J^2|\tfrac{1}{2}\tfrac{1}{2}\tfrac{1}{2}\tfrac{1}{2}\rangle$ then gives $2|\tfrac{1}{2}\tfrac{1}{2}\tfrac{1}{2}\tfrac{1}{2}\rangle = 1(1+1)|\tfrac{1}{2}\tfrac{1}{2}\tfrac{1}{2}\tfrac{1}{2}\rangle$ as required by (1.5-10).

This procedure, in which the CG coefficients are generated by the ladder operators (J_{-1} and J_{+1}), becomes quite tedious in more complicated situations. A general formula for these coefficients is (Wigner, 1959):

$$\begin{aligned} \langle j_1j_2m_1m_2|j_1j_2jm\rangle &= \delta(m, m_1+m_2)\sqrt{\frac{(j_1+j_2-j)!(j+j_1-j_2)!(j+j_2-j_1)!(2j+1)}{(j+j_1+j_2+1)!}} \\ &\times\sum_k \frac{(-1)^k\sqrt{(j_1+m_1)!(j_1-m_1)!(j_2+m_2)!(j_2-m_2)!(j+m)!(j-m)!}}{k!(j_1+j_2-j-k)!(j_1-m_1-k)!(j_2+m_2-k)!(j-j_2+m_1+k)!(j-j_1-m_2+k)!} \end{aligned} \tag{1.5-37}$$

TABLE 1.4

Clebsch–Gordan Coefficients $\langle j_1 j_2 m_1 m_2 | j_1 j_2 jm \rangle$.
$\langle j_1 j_2 m_1 m_2 | j_1 j_2 jm \rangle = (-1)^{j_1+j_2-j} \langle j_2 j_1 m_2 m_1 | j_2 j_1 jm \rangle$.

$j_1 = \frac{1}{2}$	$j_2 = \frac{1}{2}$	$j = 1$			$j = 0$
m_1	m_2	$m = 1$	$m = 0$	$m = -1$	$m = 0$
1/2	1/2	1			
1/2	−1/2		$\sqrt{1/2}$		$\sqrt{1/2}$
−1/2	1/2		$\sqrt{1/2}$		$-\sqrt{1/2}$
−1/2	−1/2			1	

$j_1 = 1$	$j_2 = \frac{1}{2}$	$j = \frac{3}{2}$				$j = \frac{1}{2}$	
m_1	m_2	$m = \frac{3}{2}$	$m = \frac{1}{2}$	$m = -\frac{1}{2}$	$m = -\frac{3}{2}$	$m = \frac{1}{2}$	$m = -\frac{1}{2}$
1	1/2	1					
1	−1/2		$\sqrt{1/3}$			$\sqrt{2/3}$	
0	1/2		$\sqrt{2/3}$			$-\sqrt{1/3}$	
0	−1/2			$\sqrt{2/3}$			$\sqrt{1/3}$
−1	1/2			$\sqrt{1/3}$			$-\sqrt{2/3}$
−1	−1/2				1		

$j_1 = \frac{3}{2}$	$j_2 = \frac{1}{2}$	$j = 2$					$j = 1$		
m_1	m_2	$m = 2$	$m = 1$	$m = 0$	$m = -1$	$m = -2$	$m = 1$	$m = 0$	$m = -1$
3/2	1/2	1							
3/2	−1/2		$\sqrt{1/4}$				$\sqrt{3/4}$		
1/2	1/2		$\sqrt{3/4}$				$-\sqrt{1/4}$		
1/2	−1/2			$\sqrt{1/2}$				$\sqrt{1/2}$	
−1/2	1/2			$\sqrt{1/2}$				$-\sqrt{1/2}$	
−1/2	−1/2				$\sqrt{3/4}$				$\sqrt{1/4}$
−3/2	1/2				$\sqrt{1/4}$				$-\sqrt{3/4}$
−3/2	−1/2					1			

TABLE 1.4 *(continued)*

$j_1 = 1$	$j_2 = 1$	$j = 2$					$j = 1$			$j = 0$
m_1	m_2	$m = 2$	$m = 1$	$m = 0$	$m = -1$	$m = -2$	$m = 1$	$m = 0$	$m = -1$	$m = 0$
1	1	1								
1	0		$\sqrt{1/2}$				$\sqrt{1/2}$			
1	−1			$\sqrt{1/6}$				$\sqrt{1/2}$		$\sqrt{1/3}$
0	1		$\sqrt{1/2}$				$-\sqrt{1/2}$			
0	0			$\sqrt{2/3}$				0		$-\sqrt{1/3}$
0	−1				$\sqrt{1/2}$				$\sqrt{1/2}$	
−1	1			$\sqrt{1/6}$				$-\sqrt{1/2}$		$\sqrt{1/3}$
−1	0				$\sqrt{1/2}$				$-\sqrt{1/2}$	
−1	−1					1				

$j_1 = 2$	$j_2 = \frac{1}{2}$	$j = \frac{5}{2}$						$j = \frac{3}{2}$			
m_1	m_2	$m = \frac{5}{2}$	$m = \frac{3}{2}$	$m = \frac{1}{2}$	$m = -\frac{1}{2}$	$m = -\frac{3}{2}$	$m = -\frac{5}{2}$	$m = \frac{3}{2}$	$m = \frac{1}{2}$	$m = -\frac{1}{2}$	$m = -\frac{3}{2}$
2	1/2	1									
2	−1/2		$\sqrt{1/5}$					$\sqrt{4/5}$			
1	1/2		$\sqrt{4/5}$					$-\sqrt{1/5}$			
1	−1/2			$\sqrt{2/5}$					$\sqrt{3/5}$		
0	1/2			$\sqrt{3/5}$					$-\sqrt{2/5}$		
0	−1/2				$\sqrt{3/5}$					$\sqrt{2/5}$	
−1	1/2				$\sqrt{2/5}$					$-\sqrt{3/5}$	
−1	−1/2					$\sqrt{4/5}$					$\sqrt{1/5}$
−2	1/2					$\sqrt{1/5}$					$-\sqrt{4/5}$
−2	−1/2						1				

TABLE 1.4 *(continued)*

$j_1 = \frac{3}{2}$	$j_2 = 1$	$j = \frac{5}{2}$						$j = \frac{3}{2}$				$j = \frac{1}{2}$	
m_1	m_2	$m = \frac{5}{2}$	$m = \frac{3}{2}$	$m = \frac{1}{2}$	$m = -\frac{1}{2}$	$m = -\frac{3}{2}$	$m = -\frac{5}{2}$	$m = \frac{3}{2}$	$m = \frac{1}{2}$	$m = -\frac{1}{2}$	$m = -\frac{3}{2}$	$m = \frac{1}{2}$	$m = -\frac{1}{2}$
3/2	1	1											
3/2	0		$\sqrt{2/5}$					$\sqrt{3/5}$					
3/2	−1			$\sqrt{1/10}$					$\sqrt{2/5}$			$\sqrt{1/2}$	
1/2	1		$\sqrt{3/5}$					$-\sqrt{2/5}$					
1/2	0			$\sqrt{3/5}$					$\sqrt{1/15}$			$-\sqrt{1/3}$	
1/2	−1				$\sqrt{3/10}$					$\sqrt{8/15}$			$\sqrt{1/6}$
−1/2	1			$\sqrt{3/10}$					$-\sqrt{8/15}$			$\sqrt{1/6}$	
−1/2	0				$\sqrt{3/5}$					$-\sqrt{1/15}$			$-\sqrt{1/3}$
−1/2	−1					$\sqrt{3/5}$					$\sqrt{2/5}$		
−3/2	1				$\sqrt{1/10}$					$-\sqrt{2/5}$			$\sqrt{1/2}$
−3/2	0					$\sqrt{2/5}$					$-\sqrt{3/5}$		
−3/2	−1						1						

$j_1 = 2$	$j_2 = 1$	$j = 3$							$j = 2$					$j = 1$		
m_1	m_2	$m = 3$	$m = 2$	$m = 1$	$m = 0$	$m = -1$	$m = -2$	$m = -3$	$m = 2$	$m = 1$	$m = 0$	$m = -1$	$m = -2$	$m = 1$	$m = 0$	$m = -1$
2	1	1														
2	0		$\sqrt{1/3}$						$\sqrt{2/3}$							
2	−1			$\sqrt{1/15}$						$\sqrt{1/3}$				$\sqrt{3/5}$		
1	1		$\sqrt{2/3}$						$-\sqrt{1/3}$							
1	0			$\sqrt{8/15}$						$\sqrt{1/6}$				$-\sqrt{3/10}$		
1	−1				$\sqrt{1/5}$						$\sqrt{1/2}$				$\sqrt{3/10}$	
0	1			$\sqrt{6/15}$						$-\sqrt{1/2}$				$\sqrt{1/10}$		
0	0				$\sqrt{3/5}$						0				$-\sqrt{2/5}$	
0	−1					$\sqrt{6/15}$						$\sqrt{1/2}$				$\sqrt{1/10}$
−1	1				$\sqrt{1/5}$						$-\sqrt{1/2}$				$\sqrt{3/10}$	
−1	0					$\sqrt{8/15}$						$-\sqrt{1/6}$				$-\sqrt{3/10}$
−1	−1						$\sqrt{2/3}$						$\sqrt{1/3}$			
−2	1					$\sqrt{1/15}$						$-\sqrt{1/3}$				$\sqrt{3/5}$
−2	0						$\sqrt{1/3}$						$-\sqrt{2/3}$			
−2	−1							1								

with

$$j_1, j_2, j = 0, \tfrac{1}{2}, 1, \tfrac{3}{2}, \ldots, \tag{1.5-38}$$

$$\triangle\,(j_1 j_2 j): \begin{cases} j_1 + j_2 + j \ = n \quad \text{(an integer)}, & (1.5\text{-}39) \\ \left.\begin{matrix} j_1 + j_2 - j \\ j_1 - j_2 + j \\ -j_1 + j_2 + j \end{matrix}\right\} \geqslant 0, & (1.5\text{-}40) \end{cases}$$

$$m_1 = j_1, j_1 - 1, \ldots, -j_1, \quad m_2 = j_2, j_2 - 1, \ldots, -j_2, \quad m = j, j - 1, \ldots, -j. \tag{1.5-41}$$

Numerical values of some of the coupling coefficients are given in Table 1.4 (Heine, 1960); we list a few of their properties:

$$\langle j_1 j_2 m_1 m_2 | j_1 j_2 jm \rangle = 0 \qquad \text{unless} \quad m = m_1 + m_2, \tag{1.5-42}$$

$$\langle j_1 j_2 m_1 m_2 | j_1 j_2 jm \rangle \text{ is real,} \tag{1.5-43}$$

$$\sum_{m_1 m_2} \langle j_1 j_2 m_1 m_2 | j_1 j_2 jm \rangle \langle j_1 j_2 m_1 m_2 | j_1 j_2 j'm' \rangle = \delta_{jj'}\,\delta_{mm'}, \tag{1.5-44a}$$

$$\sum_{jm} \langle j_1 j_2 m_1 m_2 | j_1 j_2 jm \rangle \langle j_1 j_2 m_1' m_2' | j_1 j_2 jm \rangle = \delta_{m_1 m_1'}\,\delta_{m_2 m_2'}, \tag{1.5-44b}$$

$$\begin{aligned} &\sqrt{j(j+1) - m(m+1)}\,\langle j_1 j_2 m_1 m_2 | j_1 j_2 j\; m+1 \rangle \\ &\quad = \sqrt{j_1(j_1+1) - m_1(m_1-1)}\,\langle j_1 j_2\; m_1 - 1\; m_2 | j_1 j_2 jm \rangle \\ &\qquad + \sqrt{j_2(j_2+1) - m_2(m_2-1)}\,\langle j_1 j_2 m_1\; m_2 - 1 | j_1 j_2 jm \rangle, \end{aligned} \tag{1.5-45a}$$

$$\begin{aligned} &\sqrt{j(j+1) - m(m-1)}\,\langle j_1 j_2 m_1 m_2 | j_1 j_2 j\; m-1 \rangle \\ &\quad = \sqrt{j_1(j_1+1) - m_1(m_1+1)}\,\langle j_1 j_2\; m_1 + 1\; m_2 | j_1 j_2 jm \rangle \\ &\qquad + \sqrt{j_2(j_2+1) - m_2(m_2+1)}\,\langle j_1 j_2 m_1\; m_2 + 1 | j_1 j_2 jm \rangle, \end{aligned} \tag{1.5-45b}$$

$$\langle j_1 j_2 m_1 m_2 | j_1 j_2 jm \rangle = (-1)^{j_1 + j_2 - j} \langle j_2 j_1 m_2 m_1 | j_2 j_1 jm \rangle. \tag{1.5-46}$$

The $3j$ symbols encountered in the Gaunt formula (1.2-29) are closely related to the CG coefficients; they are defined by

$$\begin{pmatrix} j_1 & j_2 & j \\ m_1 & m_2 & m \end{pmatrix} = \frac{(-1)^{j_1 - j_2 - m}}{\sqrt{2j+1}} \langle j_1 j_2 m_1 m_2 | j_1 j_2 j\; -m \rangle \tag{1.5-47}$$

in which the left side is a $3j$ symbol whose general form is written as

$$\begin{pmatrix} j_1 & j_2 & j_3 \\ m_1 & m_2 & m_3 \end{pmatrix}.$$

Among its properties are (Rotenberg *et al.*, 1959)

$$\begin{pmatrix} j_1 & j_2 & j_3 \\ m_1 & m_2 & m_3 \end{pmatrix} = \begin{pmatrix} j_2 & j_3 & j_1 \\ m_2 & m_3 & m_1 \end{pmatrix} = \begin{pmatrix} j_3 & j_1 & j_2 \\ m_3 & m_1 & m_2 \end{pmatrix}$$

$$= \begin{pmatrix} j_2 & j_3 & j_1 \\ m_3 & m_1 & m_2 \end{pmatrix} = \begin{pmatrix} j_3 & j_1 & j_2 \\ m_2 & m_3 & m_1 \end{pmatrix}, \quad (1.5\text{-}48)$$

$$(-1)^{j_1+j_2+j_3}\begin{pmatrix} j_1 & j_2 & j_3 \\ m_1 & m_2 & m_3 \end{pmatrix} = \begin{pmatrix} j_2 & j_1 & j_3 \\ m_2 & m_1 & m_3 \end{pmatrix} = \begin{pmatrix} j_1 & j_3 & j_2 \\ m_1 & m_3 & m_2 \end{pmatrix}$$

$$= \begin{pmatrix} j_3 & j_2 & j_1 \\ m_3 & m_2 & m_1 \end{pmatrix} = \begin{pmatrix} j_1 & j_2 & j_3 \\ -m_1 & -m_2 & -m_3 \end{pmatrix}, \quad (1.5\text{-}49)$$

$$\sum_{m_1 m_2} \begin{pmatrix} j_1 & j_2 & j_3 \\ m_1 & m_2 & m_3 \end{pmatrix}\begin{pmatrix} j_1 & j_2 & j_3' \\ m_1 & m_2 & m_3' \end{pmatrix} = \frac{\delta(j_3, j_3')\,\delta(m_3, m_3')}{(2j_3+1)}, \quad (1.5\text{-}50)$$

$$\sum_{j_3 m_3} (2j_3+1)\begin{pmatrix} j_1 & j_2 & j_3 \\ m_1 & m_2 & m_3 \end{pmatrix}\begin{pmatrix} j_1 & j_2 & j_3 \\ m_1' & m_2' & m_3 \end{pmatrix} = \delta(m_1, m_1')\,\delta(m_2, m_2'), \quad (1.5\text{-}51)$$

$$\begin{pmatrix} j_1 & j_2 & j_3 \\ m_1 & m_2 & m_3 \end{pmatrix} = 0 \qquad \text{unless} \begin{cases} m_1 + m_2 + m_3 = 0, \\ \triangle(j_1 j_2 j_3). \end{cases} \quad (1.5\text{-}52)$$

Equation (1.5-16) for the eigenfunction in the coupled representation, written in terms of $3j$ symbols, is

$$|j_1 j_2 jm\rangle = (-1)^{j_2 - j_1 - m} \sum_{m_1 m_2} \sqrt{2j+1}\begin{pmatrix} j_1 & j_2 & j \\ m_1 & m_2 & -m \end{pmatrix} |j_1 j_2 m_1 m_2\rangle. \quad (1.5\text{-}53)$$

It is also possible to express $|j_1 j_2 m_1 m_2\rangle$ in terms of $|j_1 j_2 jm\rangle$:

$$|j_1 j_2 m_1 m_2\rangle = \sum_{jm} (-1)^{j_2 - j_1 - m} \sqrt{2j+1}\begin{pmatrix} j_1 & j_2 & j \\ m_1 & m_2 & -m \end{pmatrix} |j_1 j_2 jm\rangle. \quad (1.5\text{-}54)$$

A number of special formulas for $3j$ symbols (Edmonds, 1960) are listed in Table 1.5. Extensive numerical tables are given by Rotenberg *et al.* (1959); a short list of numerical values is given in Table 1.6.

To recapitulate, an eigenfunction $|jm\rangle$ (or $|j_1 j_2 jm\rangle$) in the coupled representation is related to the eigenfunctions $|j_1 j_2 m_1 m_2\rangle$ in the uncoupled representation by

$$|jm\rangle \equiv |j_1 j_2 jm\rangle = \sum_{m_1 m_2} |j_1 j_2 m_1 m_2\rangle\langle j_1 j_2 m_1 m_2 | j_1 j_2 jm\rangle$$

$$= (-1)^{j_2 - j_1 - m}\sqrt{2j+1}\sum_{m_1 m_2}\begin{pmatrix} j_1 & j_2 & j \\ m_1 & m_2 & -m \end{pmatrix}|j_1 j_2 m_1 m_2\rangle \quad (1.5\text{-}55)$$

TABLE 1.5

Special Formulas for 3j Symbols

$$\begin{pmatrix} j_1 & j_2 & j_3 \\ 0 & 0 & 0 \end{pmatrix} = 0 \quad \text{if } j_1 + j_2 + j_3 \text{ is odd}$$

$$\begin{pmatrix} j+\frac{1}{2} & j & \frac{1}{2} \\ m & -m-\frac{1}{2} & \frac{1}{2} \end{pmatrix} = (-1)^{j-m-\frac{1}{2}} \sqrt{\frac{j-m+\frac{1}{2}}{(2j+2)(2j+1)}}$$

$$\begin{pmatrix} j+1 & j & 1 \\ m & -m-1 & 1 \end{pmatrix} = (-1)^{j-m-1} \sqrt{\frac{(j-m)(j-m+1)}{(2j+3)(2j+2)(2j+1)}}$$

$$\begin{pmatrix} j+1 & j & 1 \\ m & -m & 0 \end{pmatrix} = (-1)^{j-m-1} \sqrt{\frac{(j+m+1)(j-m+1)}{(2j+3)(j+1)(2j+1)}}$$

$$\begin{pmatrix} j & j & 1 \\ m & -m-1 & 1 \end{pmatrix} = (-1)^{j-m} \sqrt{\frac{(j-m)(j+m+1)}{(j+1)(2j+1)(2j)}}$$

$$\begin{pmatrix} j & j & 1 \\ m & -m & 0 \end{pmatrix} = (-1)^{j-m} \frac{m}{\sqrt{(2j+1)(j+1)j}}$$

$$\begin{pmatrix} j & j & 0 \\ m & -m & 0 \end{pmatrix} = (-1)^{j-m} \frac{1}{\sqrt{2j+1}}$$

$$\begin{pmatrix} j & j & 2 \\ m & -m & 0 \end{pmatrix} = (-1)^{j-m} \frac{3m^2 - j(j+1)}{\sqrt{(2j+3)(j+1)(2j+1)2j(2j-1)}}$$

with

$$m = m_1 + m_2, \qquad j = j_1 + j_2,\ j_1 + j_2 - 1, \ldots, |j_1 - j_2|.$$

The quantum number j is confined to integral or half-integral values and $m = j, j-1, \ldots, -j$; $|j_1 j_2 m_1 m_2\rangle$ is an eigenfunction of $J_1{}^2$, $J_2{}^2$, J_{1z}, J_{2z}, and $J_z\,(=J_{1z} + J_{2z})$ with eigenvalues $j_1(j_1+1)$, $j_2(j_2+1)$, m_1, m_2, and $m\,(=m_1 + m_2)$, respectively. However, $|j_1 j_2 m_1 m_2\rangle$ is not an eigenfunction of J^2. Note that j_1, j_2, m_1, and m_2 serve to label the eigenfunction as well as to specify the eigenvalues; for this reason, j_1, j_2, m_1, and m_2 are often said to be "good" quantum numbers (in the uncoupled representation). Also, $|j_1 j_2 jm\rangle$ is an eigenfunction of $J_1{}^2$, $J_2{}^2$, J^2, and J_z with eigenvalues $j_1(j_1+1)$, $j_2(j_2+1)$, $j(j+1)$, and m, respectively, but $|j_1 j_2 jm\rangle$ is not an eigenfunction of J_{1z} and J_{2z}. The "good" quantum numbers in the coupled representation are therefore j_1, j_2, j, and m.

Henceforth, *coupling coefficients* will be understood to be either $3j$ symbols or Clebsch–Gordan (CG) coefficients.

TABLE 1.6

Numerical Values of 3j Symbols.[a,b]

j_1	j_2	j_3	m_1	m_2	m_3		j_1	j_2	j_3	m_1	m_2	m_3	
1	1	0	0	0	0	*0$\underline{1}$	3/2	3/2	1	1/2	-3/2	1	*$\underline{1}$0$\underline{1}$
2	1	1	0	0	0	1$\underline{11}$	3/2	3/2	1	1/2	-1/2	0	*$\underline{211}$
2	2	0	0	0	0	00$\underline{1}$	3/2	3/2	1	3/2	-3/2	0	$\underline{2}$1$\underline{1}$
2	2	2	0	0	0	*10$\underline{11}$,	3/2	3/2	1	3/2	-1/2	-1	*$\underline{1}$0$\underline{1}$
3	2	1	0	0	0	*01$\underline{11}$,	2	1	1	-1	0	1	*$\underline{1}$0$\underline{1}$
3	3	0	0	0	0	*000$\underline{1}$,	2	1	1	0	-1	1	$\underline{111}$
3	3	2	0	0	0	2$\underline{111}$,	2	1	1	0	0	0	1$\underline{11}$
4	2	2	0	0	0	10$\underline{11}$,	2	1	1	1	-1	0	*$\underline{1}$0$\underline{1}$
4	3	1	0	0	0	2$\underline{2}$0$\underline{1}$,	2	1	1	1	0	-1	*$\underline{1}$0$\underline{1}$
4	3	3	0	0	0	*100$\underline{1}$,$\underline{1}$	2	1	1	2	-1	-1	00$\underline{1}$
4	4	0	0	0	0	0$\underline{2}$	2	3/2	1/2	0	-1/2	1/2	*$\underline{1}$0$\underline{1}$
4	4	2	0	0	0	*2$\underline{2}$1$\underline{1}$,$\underline{1}$	2	3/2	1/2	1	-3/2	1/2	$\underline{2}$0$\underline{1}$
4	4	4	0	0	0	120$\underline{1}$,$\underline{11}$	2	3/2	1/2	1	-1/2	-1/2	$\underline{2}$1$\underline{1}$
1/2	1/2	0	1/2	-1/2	0	$\underline{1}$	2	3/2	1/2	2	-3/2	-1/2	*00$\underline{1}$
1	1/2	1/2	0	-1/2	1/2	$\underline{11}$	2	3/2	3/2	-1	-1/2	3/2	$\underline{1}$0$\underline{1}$
1	1/2	1/2	1	-1/2	-1/2	*0$\underline{1}$	2	3/2	3/2	0	-3/2	3/2	*$\underline{2}$0$\underline{1}$
1	1	0	0	0	0	*0$\underline{1}$	2	3/2	3/2	0	-1/2	1/2	*$\underline{2}$0$\underline{1}$
1	1	0	1	-1	0	0$\underline{1}$	2	3/2	3/2	1	-3/2	1/2	$\underline{1}$0$\underline{1}$
1	1	1	-1	0	1	$\underline{11}$	2	3/2	3/2	1	-1/2	-1/2	0
1	1	1	0	-1	1	*$\underline{11}$	2	3/2	3/2	2	-3/2	-1/2	*$\underline{1}$0$\underline{1}$
1	1	1	0	0	0	0	2	3/2	3/2	2	-1/2	-3/2	$\underline{1}$0$\underline{1}$
1	1	1	1	-1	0	$\underline{11}$	2	2	0	0	0	0	00$\underline{1}$
1	1	1	1	0	-1	*$\underline{11}$	2	2	0	1	-1	0	*00$\underline{1}$
3/2	1	1/2	-1/2	0	1/2	*$\underline{11}$	2	2	0	2	-2	0	00$\underline{1}$
3/2	1	1/2	1/2	-1	1/2	$\underline{21}$	2	2	1	-1	0	1	*$\underline{1}$0$\underline{1}$
3/2	1	1/2	1/2	0	-1/2	$\underline{11}$	2	2	1	0	-1	1	$\underline{1}$0$\underline{1}$
3/2	1	1/2	3/2	-1	-1/2	*$\underline{2}$	2	2	1	0	0	0	0
3/2	3/2	0	1/2	-1/2	0	*$\underline{2}$	2	2	1	1	-2	1	*0$\underline{11}$
3/2	3/2	0	3/2	-3/2	0	$\underline{2}$	2	2	1	1	-1	0	*$\underline{111}$
3/2	3/2	1	-1/2	-1/2	1	1$\underline{11}$	2	2	1	1	0	-1	$\underline{1}$0$\underline{1}$

[a] The table gives values of $\begin{pmatrix} j_1 & j_2 & j_3 \\ m_1 & m_2 & m_3 \end{pmatrix}^2$ in prime notation which lists only the exponents of the prime numbers in the order 2, 3, 5, 7, 11, ... Negative exponents are underscored. For example,

$$210\underline{2} = \frac{2^2 \times 3^1 \times 5^0}{7^2} = \frac{12}{49}.$$

An asterisk indicates a negative radical. To look up the value of a 3j symbol use the symmetry properties

TABLE 1.6 (*continued*)

j_1	j_2	j_3	m_1	m_2	m_3		j_1	j_2	j_3	m_1	m_2	m_3	
2	2	1	2	-2	0	111	5/2	2	3/2	1/2	-1	1/2	*2111,
2	2	1	2	-1	-1	*011	5/2	2	3/2	1/2	0	-1/2	1011,
2	2	2	-2	0	2	1011,	5/2	2	3/2	3/2	-2	1/2	3111,
2	2	2	-1	-1	2	*0111,	5/2	2	3/2	3/2	-1	-1/2	1111,
2	2	2	-1	0	1	1011,	5/2	2	3/2	3/2	0	-3/2	*0111,
2	2	2	0	-2	2	1011,	5/2	2	3/2	5/2	-2	-1/2	*1101,
2	2	2	0	-1	1	1011,	5/2	2	3/2	5/2	-1	-3/2	1001,
2	2	2	0	0	0	*1011,	5/2	5/2	0	1/2	-1/2	0	11
2	2	2	1	-2	1	*0111,	5/2	5/2	0	3/2	-3/2	0	*11
2	2	2	1	-1	0	1011,	5/2	5/2	0	5/2	-5/2	0	11
2	2	2	1	0	-1	1011,	5/2	5/2	1	-1/2	-1/2	1	*0111,
2	2	2	2	-2	0	1011	5/2	5/2	1	1/2	-3/2	1	3111,
2	2	2	2	-1	-1	*0111	5/2	5/2	1	1/2	-1/2	0	1111,
2	2	2	2	0	-2	1011	5/2	5/2	1	3/2	-5/2	1	*0101,
5/2	3/2	1	-1/2	-1/2	1	*201	5/2	5/2	1	3/2	-3/2	0	*1111,
5/2	3/2	1	1/2	-3/2	1	211	5/2	5/2	1	3/2	-1/2	-1	3111,
5/2	3/2	1	1/2	-1/2	0	101	5/2	5/2	1	5/2	-5/2	0	1111,
5/2	3/2	1	3/2	-3/2	0	*011	5/2	5/2	1	5/2	-3/2	-1	*0101,
5/2	3/2	1	3/2	-1/2	-1	*101	5/2	5/2	2	-3/2	-1/2	2	2211,
5/2	3/2	1	5/2	-3/2	-1	11	5/2	5/2	2	-1/2	-3/2	2	*2211,
5/2	2	1/2	-1/2	0	1/2	101	5/2	5/2	2	-1/2	-1/2	1	0
5/2	2	1/2	1/2	-1	1/2	*011	5/2	5/2	2	1/2	-5/2	2	2001,
5/2	2	1/2	1/2	0	-1/2	*101	5/2	5/2	2	1/2	-3/2	1	0011,
5/2	2	1/2	3/2	-2	1/2	111	5/2	5/2	2	1/2	-1/2	0	*2111,
5/2	2	1/2	3/2	-1	-1/2	111	5/2	5/2	2	3/2	-5/2	1	*1001,
5/2	2	1/2	5/2	-2	-1/2	*11	5/2	5/2	2	3/2	-3/2	0	2111,
5/2	2	3/2	-3/2	0	3/2	*0111,	5/2	5/2	2	3/2	-1/2	-1	0011,
5/2	2	3/2	-1/2	-1	3/2	2211,	5/2	5/2	2	5/2	-5/2	0	2111,
5/2	2	3/2	-1/2	0	1/2	1011,	5/2	5/2	2	5/2	-3/2	-1	*1001,
5/2	2	3/2	1/2	-2	3/2	*0011,	5/2	5/2	2	5/2	-1/2	-2	2001,

(1.5-48) and (1.5-49) to (a) interchange the columns so that $j_1 \geqslant j_2 \geqslant j_3$ and (b) change the signs (if necessary) of m_1, m_2, and m_3 so that $m_2 \leqslant 0$.

[b] Reprinted from "The 3-*j* and 6-*j* Symbols," by Rotenberg, Bivins, Metropolis, and Wooten by permission of the M.I.T. Press, Cambridge, Massachusetts. Copyright ©, 1959 by The Massachusetts Institute of Technology.

1.6 Coupling of Three Angular Momenta

The methods developed for the coupling of two angular momentum operators may, in principle, be extended to any number of operators—but not without additional complications. The coupling of three angular momenta

$$\mathbf{J} = \mathbf{J}_1 + \mathbf{J}_2 + \mathbf{J}_3 \tag{1.6-1}$$

serves as a useful illustration.

Clearly, (1.6-1) may be written as

$$\mathbf{J}_1 + \mathbf{J}_2 = \mathbf{J}_{12}, \qquad \mathbf{J}_{12} + \mathbf{J}_3 = \mathbf{J}. \tag{1.6-2}$$

Each step then involves two angular momenta and these may be handled by the methods of the last section. Thus suppose

$$j_1 = 1, \qquad j_2 = \tfrac{1}{2}, \qquad j_3 = \tfrac{1}{2}.$$

When $\mathbf{J}_1$ and $\mathbf{J}_2$ are coupled to form $\mathbf{J}_{12}$, the possible values of $\mathbf{J}_{12}$ are $\frac{1}{2}$ and $\frac{3}{2}$. We indicate these by writing

$$j_1 j_2(j_{12}) = \begin{cases} 1\tfrac{1}{2}(\tfrac{1}{2}), \\ 1\tfrac{1}{2}(\tfrac{3}{2}). \end{cases} \tag{1.6-3}$$

On further coupling of $\mathbf{J}_{12}$ with $\mathbf{J}_3$ to form $\mathbf{J}$, the possible values of j may be written in a notation which keeps track of the coupling sequence:

$$j_1 j_2(j_{12}) j_3; j = \begin{cases} 1\tfrac{1}{2}(\tfrac{1}{2})\tfrac{1}{2};\ 0, \\ 1\tfrac{1}{2}(\tfrac{1}{2})\tfrac{1}{2};\ 1, \\ 1\tfrac{1}{2}(\tfrac{3}{2})\tfrac{1}{2};\ 1, \\ 1\tfrac{1}{2}(\tfrac{3}{2})\tfrac{1}{2};\ 2. \end{cases} \tag{1.6-4}$$

We may also construct the eigenfunctions $|jm\rangle$ from $|j_1 m_1\rangle$, $|j_2 m_2\rangle$, and $|j_3 m_3\rangle$. As a first step $|j_{12} m_{12}\rangle$ is written as a linear combinations of $|j_1 m_1\rangle$ and $|j_2 m_2\rangle$:

$$\begin{aligned} |j_{12} m_{12}\rangle = |\tfrac{1}{2}\tfrac{1}{2}\rangle_{12} &= \sqrt{\tfrac{2}{3}}|1\ 1\rangle|\tfrac{1}{2}\ -\tfrac{1}{2}\rangle - \sqrt{\tfrac{1}{3}}|1\ 0\rangle|\tfrac{1}{2}\tfrac{1}{2}\rangle, \\ |\tfrac{1}{2}\ -\tfrac{1}{2}\rangle_{12} &= \sqrt{\tfrac{1}{3}}|1\ 0\rangle|\tfrac{1}{2}\ -\tfrac{1}{2}\rangle - \sqrt{\tfrac{2}{3}}|1\ -1\rangle|\tfrac{1}{2}\tfrac{1}{2}\rangle, \\ |\tfrac{3}{2}\tfrac{3}{2}\rangle_{12} &= |1\ 1\rangle|\tfrac{1}{2}\tfrac{1}{2}\rangle, \\ |\tfrac{3}{2}\tfrac{1}{2}\rangle_{12} &= \sqrt{\tfrac{1}{3}}|1\ 1\rangle|\tfrac{1}{2}\ -\tfrac{1}{2}\rangle + \sqrt{\tfrac{2}{3}}|1\ 0\rangle|\tfrac{1}{2}\tfrac{1}{2}\rangle, \\ |\tfrac{3}{2}\ -\tfrac{1}{2}\rangle_{12} &= \sqrt{\tfrac{2}{3}}|1\ 0\rangle|\tfrac{1}{2}\ -\tfrac{1}{2}\rangle + \sqrt{\tfrac{1}{3}}|1\ -1\rangle|\tfrac{1}{2}\tfrac{1}{2}\rangle, \\ |\tfrac{3}{2}\tfrac{3}{2}\rangle_{12} &= |1\ -1\rangle|\tfrac{1}{2}\ -\tfrac{1}{2}\rangle. \end{aligned} \tag{1.6-5}$$

Now suppose we wish to calculate $|jm\rangle = |1\ 1\rangle$. Again it is possible to find the required linear combinations of $|j_{12} m_{12}\rangle$ and $|j_3 m_3\rangle$. Indeed, on the basis of (1.6-4) we expect to find two such expressions; they are

$$\begin{aligned}|j_1j_2(j_{12})j_3; jm\rangle &= |1\tfrac{1}{2}(\tfrac{1}{2})\tfrac{1}{2}; 1\ 1\rangle \\ &= |\tfrac{1}{2}\tfrac{1}{2}\rangle_{12}|\tfrac{1}{2}\tfrac{1}{2}\rangle \\ &= \sqrt{\tfrac{2}{3}}|1\ 1\rangle|\tfrac{1}{2}\ -\tfrac{1}{2}\rangle|\tfrac{1}{2}\tfrac{1}{2}\rangle - \sqrt{\tfrac{1}{3}}|1\ 0\rangle|\tfrac{1}{2}\tfrac{1}{2}\rangle|\tfrac{1}{2}\tfrac{1}{2}\rangle, \qquad (1.6\text{-}6a)\end{aligned}$$

$$\begin{aligned}|j_1j_2(j_{12})j_3; jm\rangle &= |1\tfrac{1}{2}(\tfrac{3}{2})\tfrac{1}{2}; 1\ 1\rangle \\ &= \sqrt{\tfrac{3}{4}}|\tfrac{3}{2}\tfrac{3}{2}\rangle_{12}|\tfrac{1}{2}\ -\tfrac{1}{2}\rangle - \sqrt{\tfrac{1}{4}}|\tfrac{3}{2}\tfrac{1}{2}\rangle_{12}|\tfrac{1}{2}\tfrac{1}{2}\rangle \\ &= \sqrt{\tfrac{3}{4}}|1\ 1\rangle|\tfrac{1}{2}\tfrac{1}{2}\rangle|\tfrac{1}{2}\ -\tfrac{1}{2}\rangle - \tfrac{1}{2}\sqrt{\tfrac{3}{2}}|1\ 1\rangle|\tfrac{1}{2}\ -\tfrac{1}{2}\rangle|\tfrac{1}{2}\tfrac{1}{2}\rangle \\ &\quad -\sqrt{\tfrac{1}{6}}|1\ 0\rangle|\tfrac{1}{2}\tfrac{1}{2}\rangle|\tfrac{1}{2}\tfrac{1}{2}\rangle. \qquad (1.6\text{-}6b)\end{aligned}$$

The sequential scheme (1.6-2) has led us in a stepwise fashion to (1.6-6a) and (1.6-6b) which provide two expressions for $|jm\rangle = |1\ 1\rangle$ in terms of the products $|j_1m_1\rangle|j_2m_2\rangle|j_3m_3\rangle$. However, the replacement of (1.6-1) by (1.6-2) is certainly not unique. Another possible coupling scheme is

$$\mathbf{J}_2 + \mathbf{J}_3 = \mathbf{J}_{23}, \qquad \mathbf{J}_1 + \mathbf{J}_{23} = \mathbf{J}. \qquad (1.6\text{-}7)$$

In place of (1.6-3) and (1.6-4) we now have

$$j_2j_3(j_{23}) = \begin{cases}\tfrac{1}{2}\tfrac{1}{2}(0), \\ \tfrac{1}{2}\tfrac{1}{2}(1),\end{cases}$$

and

$$j_1, j_2j_3(j_{23}); j = \begin{cases}1, \tfrac{1}{2}\tfrac{1}{2}(0);\ 1 \\ 1, \tfrac{1}{2}\tfrac{1}{2}(1);\ 0 \\ 1, \tfrac{1}{2}\tfrac{1}{2}(1);\ 1 \\ 1, \tfrac{1}{2}\tfrac{1}{2}(1);\ 2.\end{cases} \qquad (1.6\text{-}8)$$

The possible values of j are the same as those in (1.6-4) even though the coupling has been carried out in a different sequence. The eigenfunctions $|j_{23}m_{23}\rangle$ are

$$\begin{aligned}|j_{23}m_{23}\rangle = |0\ 0\rangle_{23} &= \sqrt{\tfrac{1}{2}}|\tfrac{1}{2}\tfrac{1}{2}\rangle|\tfrac{1}{2}\ -\tfrac{1}{2}\rangle - \sqrt{\tfrac{1}{2}}|\tfrac{1}{2}\ -\tfrac{1}{2}\rangle|\tfrac{1}{2}\tfrac{1}{2}\rangle \\ |1\ 1\rangle_{23} &= |\tfrac{1}{2}\tfrac{1}{2}\rangle|\tfrac{1}{2}\tfrac{1}{2}\rangle \\ |1\ 0\rangle_{23} &= \sqrt{\tfrac{1}{2}}|\tfrac{1}{2}\tfrac{1}{2}\rangle|\tfrac{1}{2}\ -\tfrac{1}{2}\rangle + \sqrt{\tfrac{1}{2}}|\tfrac{1}{2}\ -\tfrac{1}{2}\rangle|\tfrac{1}{2}\tfrac{1}{2}\rangle \\ |1\ -1\rangle_{23} &= |\tfrac{1}{2}\ -\tfrac{1}{2}\rangle|\tfrac{1}{2}\ -\tfrac{1}{2}\rangle,\end{aligned} \qquad (1.6\text{-}9)$$

and $|jm\rangle = |1\ 1\rangle$ now assumes the two forms

$$\begin{aligned}|j_1, j_2j_3(j_{23}); jm\rangle &= |1, \tfrac{1}{2}\tfrac{1}{2}(0); 1\ 1\rangle \\ &= |1\ 1\rangle|00\rangle_{23} \\ &= \sqrt{\tfrac{1}{2}}|1\ 1\rangle|\tfrac{1}{2}\tfrac{1}{2}\rangle|\tfrac{1}{2}\ -\tfrac{1}{2}\rangle - \sqrt{\tfrac{1}{2}}|1\ 1\rangle|\tfrac{1}{2}\ -\tfrac{1}{2}\rangle|\tfrac{1}{2}\tfrac{1}{2}\rangle, \qquad (1.6\text{-}10a)\end{aligned}$$

$$
\begin{aligned}
|j_1,j_2j_3(j_{23});jm\rangle &= |1,\tfrac{1}{2}\tfrac{1}{2}(1);1\,1\rangle \\
&= \sqrt{\tfrac{1}{2}}|1\,1\rangle|1\,0\rangle_{23} - \sqrt{\tfrac{1}{2}}|1\,0\rangle|1\,1\rangle_{23} \\
&= \tfrac{1}{2}|1\,1\rangle|\tfrac{1}{2}\tfrac{1}{2}\rangle|\tfrac{1}{2}\,-\tfrac{1}{2}\rangle + \tfrac{1}{2}|1\,1\rangle|\tfrac{1}{2}\,-\tfrac{1}{2}\rangle|\tfrac{1}{2}\tfrac{1}{2}\rangle \\
&\quad -\sqrt{\tfrac{1}{2}}|1\,0\rangle|\tfrac{1}{2}\tfrac{1}{2}\rangle|\tfrac{1}{2}\tfrac{1}{2}\rangle.
\end{aligned} \tag{1.6-10b}
$$

We now have four separate expressions for $|jm\rangle = |1\,1\rangle$, namely (1.6-6a) and (1.6-6b) and (1.6-10a) and (1.6-10b).

This example illustrates the following properties:

1. When three angular momentum operators J_1, J_2, and J_3 are coupled to form $\mathbf{J}$, the possible values of j are the same regardless of the coupling scheme.
2. The eigenfunctions $|jm\rangle$ are not uniquely determined by specifying $|j_1m_1\rangle$, $|j_2m_2\rangle$, and $|j_3m_3\rangle$ but depend on the details of the coupling scheme.

Thus, in general,

$$|j_1j_2(j_{12})j_3;j\rangle \neq |j_1,j_2j_3(j_{23});j\rangle.$$

However, the two eigenfunctions are related by a unitary transformation (Sobelman, 1972)

$$|j_1,j_2j_3(j_{23});j\rangle = \sum_{j_{12}} |j_1j_2(j_{12})j_3;j\rangle\langle j_1j_2(j_{12})j_3;j|j_1,j_2j_3(j_{23});j\rangle \tag{1.6-11}$$

in which

$$
\begin{aligned}
&\langle j_1j_2(j_{12})j_3;j|j_1,j_2j_3(j_{23});j\rangle \\
&\quad = (-1)^{j_1+j_2+j_3+j}\sqrt{(2j_{12}+1)(2j_{23}+1)}\begin{Bmatrix} j_1 & j_2 & j_{12} \\ j_3 & j & j_{23} \end{Bmatrix}.
\end{aligned} \tag{1.6-12}
$$

The quantity in the braces is a $6j$ symbol whose general form is

$$\begin{Bmatrix} j_1 & j_2 & j_3 \\ l_1 & l_2 & l_3 \end{Bmatrix}$$

and whose definition is given by (Rotenberg *et al.*, 1959)

$$
\begin{aligned}
\begin{Bmatrix} j_1 & j_2 & j_3 \\ l_1 & l_2 & l_3 \end{Bmatrix} &= (-1)^{j_1+j_2+l_1+l_2}\,\Delta(j_1j_2j_3)\,\Delta(l_1l_2j_3)\,\Delta(l_1j_2l_3)\,\Delta(j_1l_2l_3) \\
&\quad \times \sum_k \frac{(-1)^k(j_1+j_2+l_1+l_2+1-k)!}{k!(j_1+j_2-j_3-k)!(l_1+l_2-j_3-k)!(j_1+l_2-l_3-k)!} \\
&\qquad (l_1+j_2-l_3-k)!(-j_1-l_1+j_3+l_3+k)!(-j_2-l_2+j_3+l_3+k)!
\end{aligned} \tag{1.6-13}
$$

where

$$\triangle(abc) = \sqrt{\frac{(a+b-c)!(a-b+c)!(b+c-a)!}{(a+b+c+1)!}}.$$

The four triangle relations may be represented symbolically as

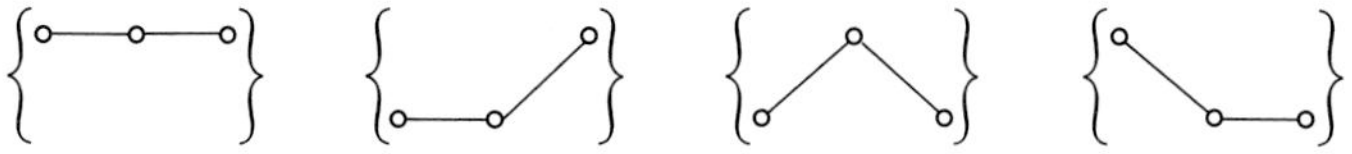

$\triangle(j_1 j_2 j_3)$ $\quad$ $\triangle(l_1 l_2 j_3)$ $\quad$ $\triangle(l_1 j_2 l_3)$ $\quad$ $\triangle(j_1 l_2 l_3)$

The $6j$ symbols are invariant under

(a) an interchange of columns,
(b) an interchange of any two numbers in the bottom row with the corresponding two numbers in the top row.

Thus

$$\begin{Bmatrix} j_1 & j_2 & j_3 \\ l_1 & l_2 & l_3 \end{Bmatrix} = \begin{Bmatrix} j_2 & j_1 & j_3 \\ l_2 & l_1 & l_3 \end{Bmatrix} = \begin{Bmatrix} j_1 & j_3 & j_2 \\ l_1 & l_3 & l_2 \end{Bmatrix} = \cdots$$
$$= \begin{Bmatrix} j_1 & j_2 & j_3 \\ l_1 & l_2 & l_3 \end{Bmatrix} = \begin{Bmatrix} l_1 & l_2 & j_3 \\ j_1 & j_2 & l_3 \end{Bmatrix} = \cdots. \tag{1.6-14}$$

Among their properties are

$$\begin{Bmatrix} j_1 & j_2 & j_3 \\ l_1 & l_2 & l_3 \end{Bmatrix} = (-1)^{j_1+j_2+l_1+l_2} W(j_1 j_2 l_2 l_1; j_3 l_3) \tag{1.6-15}$$

in which $W(j_1 j_2 l_2 l_1; j_3 l_3)$ is the Racah W coefficient,

$$\sum_{j_3} (2j_3+1) \begin{Bmatrix} j_1 & j_2 & j_3 \\ l_1 & l_2 & l_3 \end{Bmatrix} \begin{Bmatrix} j_1 & j_2 & j_3 \\ l_1 & l_2 & l_3' \end{Bmatrix} = \frac{\delta(l_3, l_3')}{2l_3+1}, \tag{1.6-16}$$

$$\sum_{l_3} (-1)^{j+j_3+l_3}\, (2l_3+1) \begin{Bmatrix} j_1 & j_2 & j_3 \\ l_1 & l_2 & l_3 \end{Bmatrix} \begin{Bmatrix} j_1 & l_1 & j \\ j_2 & l_2 & l_3 \end{Bmatrix} = \begin{Bmatrix} j_1 & j_2 & j_3 \\ l_2 & l_1 & j \end{Bmatrix}. \tag{1.6-17}$$

A number of special formulas are listed in Table 1.7; numerical values are given in Table 1.8 which is also taken from the tables of Rotenberg *et al.* (1959).

To illustrate the use of (1.6-11) and (1.6-12) it is possible to write

$$|1, \tfrac{1}{2}\tfrac{1}{2}(0); 1\ 1\rangle = -\sqrt{\tfrac{1}{3}}\,|1\tfrac{1}{2}(\tfrac{1}{2})\tfrac{1}{2}; 1\ 1\rangle + \sqrt{\tfrac{2}{3}}\,|1\tfrac{1}{2}(\tfrac{3}{2})\tfrac{1}{2}; 1\ 1\rangle \tag{1.6-18}$$

TABLE 1.7

Special Formulas for 6j Symbols[a]

$$\begin{Bmatrix} a & b & c \\ 0 & c & b \end{Bmatrix} = (-1)^{a+b+c}\,[(2b+1)(2c+1)]^{-1/2},$$

$$\begin{Bmatrix} a & b & c \\ \frac{1}{2} & c-\frac{1}{2} & b+\frac{1}{2} \end{Bmatrix} = (-1)^{a+b+c}\left[\frac{(a+c-b)(a+b-c+1)}{(2b+1)(2b+2)\,2c(2c+1)}\right]^{1/2},$$

$$\begin{Bmatrix} a & b & c \\ \frac{1}{2} & c-\frac{1}{2} & b-\frac{1}{2} \end{Bmatrix} = (-1)^{a+b+c}\left[\frac{(a+b+c+1)(b+c-a)}{2b(2b+1)\,2c(2c+1)}\right]^{1/2},$$

$$s = a + b + c.$$

$$X = a(a+1) - b(b+1) - c(c+1).$$

$$\begin{Bmatrix} a & b & c \\ 1 & c-1 & b-1 \end{Bmatrix} = (-1)^{s}\left[\frac{s(s+1)(s-2a-1)(s-2a)}{(2b-1)\,2b\,(2b+1)(2c-1)\,2c(2c+1)}\right]^{1/2},$$

$$\begin{Bmatrix} a & b & c \\ 1 & c-1 & b \end{Bmatrix} = (-1)^{s}\left[\frac{2(s+1)(s-2a)(s-2b)(s-2c+1)}{2b(2b+1)(2b+2)(2c-1)\,2c(2c+1)}\right]^{1/2},$$

$$\begin{Bmatrix} a & b & c \\ 1 & c-1 & b+1 \end{Bmatrix} = (-1)^{s}\left[\frac{(s-2b)(s-2b-1)(s-2c+1)(s-2c+2)}{(2b+1)(2b+2)(2b+3)(2c-1)\,2c(2c+1)}\right]^{1/2},$$

$$\begin{Bmatrix} a & b & c \\ 1 & c & b \end{Bmatrix} = (-1)^{s}\frac{2X}{[2b(2b+1)(2b+2)\,2c(2c+1)(2c+2)]^{1/2}},$$

$$\begin{Bmatrix} a & b & c \\ 2 & c-2 & b \end{Bmatrix} = (-1)^{s}(s-2b-1)^{1/2}$$

$$\times\left[\frac{6s(s+1)(s-2a-1)(s-2a)(s-2b)(s-2c+1)(s-2c+2)}{(2b-1)\,2b(2b+1)(2b+2)(2b+3)(2c-3)(2c-2)(2c-1)\,2c(2c+1)}\right]^{1/2},$$

$$\begin{Bmatrix} a & b & c \\ 2 & c-2 & b+1 \end{Bmatrix} = (-1)^{s}\,2$$

$$\times\left[\frac{(s+1)(s-2a)(s-2b-2)(s-2b-1)(s-2b)(s-2c+1)(s-2c+2)(s-2c+3)}{2b(2b+1)(2b+2)(2b+3)(2b+4)(2c-3)(2c-2)(2c-1)\,2c(2c+1)}\right]^{1/2}$$

[a] (Sobelman, 1972)

TABLE 1.7 (*continued*)

$$\begin{Bmatrix} a & b & c \\ 2 & c-2 & b+2 \end{Bmatrix} = (-1)^s [s-2b-3]^{1/2}$$

$$\times \left[\frac{(s-2b-2)(s-2b-1)(s-2b)(s-2c+1)(s-2c+2)(s-2c+3)(s-2c+4)}{(2b+1)(2b+2)(2b+3)(2b+4)(2b+5)(2c-3)(2c-2)(2c-1)2c(2c+1)}\right]^{1/2},$$

$$\begin{Bmatrix} a & b & c \\ 2 & c-1 & b-1 \end{Bmatrix} = (-1)^s$$

$$\times \frac{4[(a+b)(a-b+1)-(c-1)(c-b+1)][s(s+1)(s-2a-1)(s-2a)]^{1/2}}{[(2b-2)(2b-1)2b(2b+1)(2b+2)(2c-2)(2c-1)2c(2c+1)(2c+2)]^{1/2}},$$

$$\begin{Bmatrix} a & b & c \\ \frac{3}{2} & c-\frac{1}{2} & b-\frac{1}{2} \end{Bmatrix}$$

$$= (-1)^s \frac{[2(s-2b)(s-2c)-(s+2)(s-2a-1)][(s+1)(s-2a)]^{1/2}}{[(2b-1)2b(2b+1)(2b+2)(2c-1)2c(2c+1)(2c+2)]^{1/2}},$$

$$\begin{Bmatrix} a & b & c \\ \frac{3}{2} & c-\frac{1}{2} & b+\frac{1}{2} \end{Bmatrix}$$

$$= (-1)^s \frac{[(s-2b-1)(s-2c)-2(s+2)(s-2a)][(s-2b)(s-2c+1)]^{1/2}}{[2b(2b+1)(2b+2)(2b+3)2c(2c+1)(2c+2)(2c+3)]^{1/2}},$$

$$\begin{Bmatrix} a & b & c \\ 2 & c-2 & b-2 \end{Bmatrix}$$

$$= (-1)^s \left[\frac{(s-2)(s-1)s(s+1)(s-2a-3)(s-2a-2)(s-2a-1)(s-2a)}{(2b-3)(2b-2)(2b-1)2b(2b+1)(2c-3)(2c-2)(2c-1)2c(2c+1)}\right]^{1/2},$$

$$\begin{Bmatrix} a & b & c \\ 2 & c-2 & b-1 \end{Bmatrix} = (-1)^s 2$$

$$\times \left[\frac{(s-1)s(s+1)(s+2a-2)(s-2a-1)(s-2a)(s-2b)(s-2c+1)}{(2b-2)(2b-1)2b(2b+1)(2b+2)(2c-3)(2c-2)(2c-1)2c(2c+1)}\right]^{1/2},$$

TABLE 1.7 (*continued*)

$$\begin{Bmatrix} a & b & c \\ 2 & c-1 & b \end{Bmatrix} = (-1)^s 2$$

$$\times \frac{[(a+b+1)(a-b) - c^2 + 1][6(s+1)(s-2a)(s-2b)(s-2c+1)]^{1/2}}{[(2b-1)2b(2b+1)(2b+2)(2b+3)(2c-2)(2c-1)2c(2c+1)(2c+2)]^{1/2}},$$

$$\begin{Bmatrix} a & b & c \\ 2 & c-1 & b+1 \end{Bmatrix} = (-1)^s$$

$$\times \frac{4[(a+b+2)(a-b-1)-(c-1)(b+c+2)][(s-2b-1)(s-2b)(s-2c+1)(s-2c+2)]^{1/2}}{[2b(2b+1)(2b+2)(2b+3)(2b+4)(2c-2)(2c-1)2c(2c+1)(2c+2)]^{1/2}}$$

$$\begin{Bmatrix} a & b & c \\ 2 & c & b \end{Bmatrix}$$

$$= (-1)^s \frac{2[3X(X+1) - 4b(b+1)c(c+1)]}{[(2b-1)2b(2b+1)(2b+2)(2b+3)(2c-1)2c(2c+1)(2c+2)(2c+3)]^{1/2}}$$

in which numerical values of the $6j$ symbols have been obtained from Table 1.8. Similarly

$$|1, \tfrac{1}{2}\tfrac{1}{2}(1); 1\ 1\rangle = \sqrt{\tfrac{2}{3}}|1\tfrac{1}{2}(\tfrac{1}{2})\tfrac{1}{2}; 1\ 1\rangle + \sqrt{\tfrac{1}{3}}|1\tfrac{1}{2}(\tfrac{3}{2})\tfrac{1}{2}; 1\ 1\rangle. \tag{1.6-19}$$

On substituting (1.6-6a) and (1.6-6b) into (1.6-18) and (1.6-19) it is readily verified that the results are those given by (1.6-10a) and (1.6-10b).

1.7 Summary and Examples

Definition J_x, J_y, and J_z are components of an angular momentum operator $\mathbf{J}$ if

(a) J_x, J_y, and J_z are Hermitian,
(b) J_x, J_y, and J_z satisfy

$$\begin{aligned} \mathbf{J} \times \mathbf{J} &= i\mathbf{J}, \\ [J_i, J_j] &= i\varepsilon_{ijk} J_k, \\ [J_i, J_j] &= iJ_k \quad (i, j, k \text{ cyclic}), \end{aligned} \tag{1.7-1}$$

where ε_{ijk} is the antisymmetric unit tensor of rank 3. The three forms of (1.7-1) are equivalent.

TABLE 1.8

Numerical Values of 6j Symbols[a,b]

j_1	j_2	j_3	l_1	l_2	l_3		j_1	j_2	j_3	l_1	l_2	l_3	
1/2	1/2	0	0	0	1/2	*$\underline{1}$	2	1	1	1	1	1	$\underline{22}$
1/2	1/2	0	1/2	1/2	0	*$\underline{2}$	2	1	1	2	1	1	$\underline{222}$
1	1/2	1/2	0	1/2	1/2	$\underline{2}$	2	3/2	1/2	0	1/2	3/2	$\underline{3}$
1	1/2	1/2	1	1/2	1/2	$\underline{22}$	2	3/2	1/2	1/2	1	1	$\underline{21}$
1	1	0	0	0	1	$0\underline{1}$	2	3/2	1/2	1	1/2	3/2	$\underline{3}0\underline{1}$
1	1	0	1/2	1/2	1/2	$\underline{11}$	2	3/2	1/2	1	3/2	1/2	$\underline{4}$
1	1	0	1	1	0	$0\underline{2}$	2	3/2	1/2	1	3/2	3/2	*$\underline{2}0\underline{1}$
1	1	0	1	1	1	*$0\underline{2}$	2	3/2	1/2	3/2	1	1	$\underline{311}$
1	1	1	1/2	1/2	1/2	*$0\underline{2}$	2	3/2	1/2	2	3/2	1/2	$\underline{4}0\underline{2}$
1	1	1	1	0	1	*$0\underline{2}$	2	3/2	1/2	2	3/2	3/2	*$\underline{2}0\underline{2}$
1	1	1	1	1	0	*$0\underline{2}$	2	3/2	3/2	0	3/2	3/2	*$\underline{4}$
1	1	1	1	1	1	$\underline{22}$	2	3/2	3/2	1/2	1	1	*$\underline{31}$
3/2	1	1/2	0	1/2	1	*$\underline{11}$	2	3/2	3/2	1	1/2	3/2	*$\underline{2}0\underline{1}$
3/2	1	1/2	1/2	1	1/2	*$0\underline{2}$	2	3/2	3/2	1	3/2	1/2	*$\underline{2}0\underline{1}$
3/2	1	1/2	1	1/2	1	*$\underline{22}$	2	3/2	3/2	1	3/2	3/2	$\underline{4}0\underline{2}$
3/2	1	1/2	3/2	1	1/2	*$\underline{42}$	2	3/2	3/2	3/2	1	1	*$1\underline{12}$
3/2	3/2	0	0	0	3/2	*$\underline{2}$	2	3/2	3/2	2	1/2	3/2	*$\underline{2}0\underline{2}$
3/2	3/2	0	1/2	1/2	1	*$\underline{3}$	2	3/2	3/2	2	3/2	1/2	*$\underline{2}0\underline{2}$
3/2	3/2	0	1	1	1/2	*$\underline{21}$	2	3/2	3/2	2	3/2	3/2	$\underline{4}2\underline{2}$
3/2	3/2	0	1	1	3/2	$\underline{21}$	2	2	0	0	0	2	$00\underline{1}$
3/2	3/2	0	3/2	3/2	0	*$\underline{4}$	2	2	0	1/2	1/2	3/2	$\underline{1}0\underline{1}$
3/2	3/2	0	3/2	3/2	1	$\underline{4}$	2	2	0	1	1	1	$0\underline{11}$
3/2	3/2	1	1/2	1/2	1	$\underline{32}1$	2	2	0	1	1	2	*$0\underline{11}$
3/2	3/2	1	1	0	3/2	$\underline{21}$	2	2	0	3/2	3/2	1/2	$\underline{2}0\underline{1}$
3/2	3/2	1	1	1	1/2	$\underline{32}1$	2	2	0	3/2	3/2	3/2	*$\underline{2}0\underline{1}$
3/2	3/2	1	1	1	3/2	*$\underline{121}$	2	2	0	2	2	0	$00\underline{2}$
3/2	3/2	1	3/2	1/2	1	$\underline{22}$	2	2	0	2	2	1	*$00\underline{2}$
3/2	3/2	1	3/2	3/2	0	$\underline{4}$	2	2	0	2	2	2	$00\underline{2}$
3/2	3/2	1	3/2	3/2	1	*$\underline{4220}$,2	2	2	1	1/2	1/2	3/2	*$\underline{2}0\underline{1}$
2	1	1	0	1	1	$0\underline{2}$	2	2	1	1	0	2	*$0\underline{11}$

TABLE 1.8 (*continued*)

j_1	j_2	j_3	l_1	l_2	l_3		j_1	j_2	j_3	l_1	l_2	l_3	
2	2	1	1	1	1	*201	2	2	2	1	1	1	2121,
2	2	1	1	1	2	221	2	2	2	1	1	2	2121,
2	2	1	3/2	1/2	3/2	*301	2	2	2	3/2	1/2	3/2	3021,
2	2	1	3/2	3/2	1/2	*322	2	2	2	3/2	3/2	1/2	3021,
2	2	1	3/2	3/2	3/2	102	2	2	2	3/2	3/2	3/2	0
2	2	1	2	1	1	*202	2	2	2	2	0	2	002
2	2	1	2	1	2	2121,	2	2	2	2	1	1	2121,
2	2	1	2	2	0	*002	2	2	2	2	1	2	*202
2	2	1	2	2	1	22	2	2	2	2	2	0	002
2	2	1	2	2	2	*202	2	2	2	2	2	1	*202
							2	2	2	2	2	2	*2222,

^{a} The table lists $\begin{Bmatrix} j_1 & j_2 & j_3 \\ l_1 & l_2 & l_3 \end{Bmatrix}^2$ in prime notation (see Table 1.6). To find the value of a 6j symbol use the symmetry properties (1.6-14) to (a) place the largest of the six parameters in the upper left-hand corner (j_1 position), (b) place the largest of the remaining four parameters in the middle of the top row (j_2 position), and (c) make $l_1 > l_2$ if $j_1 = j_2$.

^{b} Reprinted from "The 3-j and 6-j Symbols," by Rotenberg, Bivins, Metropolis, and Wooten by permission of the M.I.T. Press, Cambridge, Massachusetts. Copyright ©, 1959 by the Massachusetts Institute of Technology.

Spherical Components

$$J_{+1} = -\frac{1}{\sqrt{2}}[J_x + iJ_y] = -J^\dagger_{-1}, \qquad J_{-1} = \frac{1}{\sqrt{2}}[J_x - iJ_y] = -J^\dagger_{+1},$$

$$\text{(1.7-2)}$$

$$J_x = -\frac{1}{\sqrt{2}}[J_{+1} - J_{-1}], \qquad J_y = \frac{i}{\sqrt{2}}[J_{+1} + J_{-1}], \qquad J_z = J_0$$

J^2

$$J^2 = J_x^2 + J_y^2 + J_z^2 = -J_{+1}J_{-1} + J_0 - J_{-1}J_{+1}$$
$$= \sum_q (-1)^q J_q J_{-q} \qquad (q = 1, 0, -1)$$
$$= -2J_{+1}J_{-1} + J_0(J_0 - 1) = -2J_{-1}J_{+1} + J_0(J_0 + 1). \quad \text{(1.7-3)}$$

Commutators

$$[J_0, J_{\pm 1}] = \pm J_{\pm 1}, \qquad [J_{+1}, J_{-1}] = -J_0. \qquad \text{(1.7-4)}$$

In terms of the CG coefficients, (1.7-4) has the form

$$[J_p, J_q] = -\sqrt{2}\langle 1\,1\; pq|1\;1\;1\; p+q\rangle J_{p+q} \qquad (p, q = 1, 0, -1), \tag{1.7-5}$$

$$[J_\mu, J^2] = 0 \qquad (J_\mu = J_x, J_y, J_z, J_{+1}, J_{-1}). \tag{1.7-6}$$

Operators and Eigenstates

$$\begin{aligned} J^2|jm\rangle &= j(j+1)|jm\rangle, \\ J_z|jm\rangle &= J_0|jm\rangle = m|jm\rangle, \\ J_{\pm 1}|jm\rangle &= \mp\sqrt{\tfrac{1}{2}[j(j+1)-(m\pm 1)m]}\,|j\; m\pm 1\rangle \end{aligned} \tag{1.7-7}$$

$$j = 0, \tfrac{1}{2}, 1, \tfrac{3}{2}, 2, \tfrac{5}{2}, \ldots, \qquad m = j, j-1, \ldots, -j. \tag{1.7-8}$$

Matrix Elements

$$\langle j'm'|J^2|jm\rangle = j(j+1)\,\delta_{j'j}\,\delta_{m'm}, \tag{1.7-9}$$

$$\langle j'm'|J_0|jm\rangle = m\,\delta_{j'j}\,\delta_{m'm} = \langle j'm'|J_z|jm\rangle, \tag{1.7-10}$$

$$\langle j'm'|J_{+1}|jm\rangle = -\sqrt{\tfrac{1}{2}[j(j+1)-(m+1)m]}\,\delta_{j'j}\,\delta_{m'\,m+1}, \tag{1.7-11}$$

$$\langle j'm'|J_{-1}|jm\rangle = \sqrt{\tfrac{1}{2}[j(j+1)-(m-1)m]}\,\delta_{j'j}\,\delta_{m'\,m-1}. \tag{1.7-12}$$

An alternative expression for the matrix element of J_{+1}, J_0, or J_{-1} is derived in Section 6.3:

$$\begin{aligned} &\langle j'm'|J_q|jm\rangle \\ &= (-1)^{j-m'}\begin{pmatrix} j' & 1 & j \\ -m' & q & m \end{pmatrix}\sqrt{(2j'+1)(j'+1)}\,\delta_{j'j} \qquad (q = 1, 0, -1). \end{aligned} \tag{1.7-13}$$

For the rectangular components

$$\begin{aligned} \langle j\; m+1|J_x|jm\rangle &= \tfrac{1}{2}\sqrt{j(j+1)-m(m+1)}, \\ \langle j\; m-1|J_x|jm\rangle &= \tfrac{1}{2}\sqrt{j(j+1)-m(m-1)}, \end{aligned} \tag{1.7-14}$$

$$\begin{aligned} \langle j\; m+1|J_y|jm\rangle &= -\tfrac{i}{2}\sqrt{j(j+1)-m(m+1)}, \\ \langle j\; m-1|J_y|jm\rangle &= \tfrac{i}{2}\sqrt{j(j+1)-m(m-1)}. \end{aligned} \tag{1.7-15}$$

Examples (see (1.4-4) for format)

$j \equiv s = \frac{1}{2}$; $\mathbf{J} \equiv \mathbf{S}$

$$\begin{aligned} &S_x = \frac{1}{2}\begin{pmatrix} 0 & 1 \\ 1 & 0 \end{pmatrix}, \qquad S_y = \frac{1}{2}\begin{pmatrix} 0 & -i \\ i & 0 \end{pmatrix}, \quad S_z = \frac{1}{2}\begin{pmatrix} 1 & 0 \\ 0 & -1 \end{pmatrix} = S_0, \\ &S_{+1} = \frac{1}{\sqrt{2}}\begin{pmatrix} 0 & -1 \\ 0 & 0 \end{pmatrix}, \quad S_{-1} = \frac{1}{\sqrt{2}}\begin{pmatrix} 0 & 0 \\ 1 & 0 \end{pmatrix}, \quad S^2 = \frac{3}{4}\begin{pmatrix} 1 & 0 \\ 0 & 1 \end{pmatrix}. \end{aligned} \tag{1.7-16}$$

$$|jm\rangle \equiv |sm\rangle = \begin{cases} |\tfrac{1}{2}\,\tfrac{1}{2}\rangle \equiv \alpha \equiv \xi_{1/2} \equiv \xi(\tfrac{1}{2}) = \begin{pmatrix}1\\0\end{pmatrix}, \\ |\tfrac{1}{2}\,-\tfrac{1}{2}\rangle \equiv \beta \equiv \xi_{-1/2} \equiv \xi(-\tfrac{1}{2}) = \begin{pmatrix}0\\1\end{pmatrix}, \end{cases} \tag{1.7-17}$$

$$\begin{aligned} S_x\alpha &= \beta/2, & S_x\beta &= \alpha/2, \\ S_y\alpha &= i\beta/2, & S_y\beta &= -i\alpha/2, \\ S_z\alpha &= S_0\alpha = \alpha/2, & S_z\beta &= S_0\beta = -\beta/2, \\ S_{+1}\alpha &= 0, & S_{+1}\beta &= -\alpha/\sqrt{2}, \\ S_{-1}\alpha &= \beta/\sqrt{2}, & S_{-1}\beta &= 0, \\ S^2\alpha &= \tfrac{3}{4}\alpha, & S^2\beta &= \tfrac{3}{4}\beta. \end{aligned} \tag{1.7-18}$$

$j = 1$

$$\begin{aligned} J_x &= \frac{1}{\sqrt{2}}\begin{pmatrix}0&1&0\\1&0&1\\0&1&0\end{pmatrix}, & J_y &= \frac{i}{\sqrt{2}}\begin{pmatrix}0&-1&0\\1&0&-1\\0&1&0\end{pmatrix}, \\ J_0 = J_z &= \begin{pmatrix}1&0&0\\0&0&0\\0&0&-1\end{pmatrix}, & J^2 &= \begin{pmatrix}2&0&0\\0&2&0\\0&0&2\end{pmatrix}, \\ J_{+1} &= \begin{pmatrix}0&-1&0\\0&0&-1\\0&0&0\end{pmatrix}, & J_{-1} &= \begin{pmatrix}0&0&0\\1&0&0\\0&1&0\end{pmatrix}. \end{aligned} \tag{1.7-19}$$

Coupling of Angular Momentum Operators The coupling of two angular momentum operators $\mathbf{J}_1$ and $\mathbf{J}_2$ is expressed by the relation

$$\mathbf{J} = \mathbf{J}_1 + \mathbf{J}_2 \tag{1.7-20}$$

where $\mathbf{J}$ is also an angular momentum operator, and

$$\begin{aligned} J_1^2|j_1m_1\rangle &= j_1(j_1+1)|j_1m_1\rangle, & J_{1z}|j_2m_2\rangle &= j_2(j_2+1)|j_1m_1\rangle, \\ J_2^2|j_2m_2\rangle &= j_2(j_2+1)|j_1m_1\rangle, & J_{2z}|j_2m_2\rangle &= m_2|j_2m_2\rangle, \end{aligned} \tag{1.7-21}$$

with

$$j_1, j_2 = 0, \tfrac{1}{2}, 1, \tfrac{3}{2}, \ldots, \quad m_1 = j_1, j_1 - 1, \ldots, -j_1, \quad m_2 = j_2, j_2 - 1, \ldots, -j_2. \tag{1.7-22}$$

The coupled and uncoupled representations are related by

$$|jm\rangle \equiv |j_1j_2jm\rangle = \sum_{m_1m_2} |j_1j_2m_1m_2\rangle\langle j_1j_2m_1m_2|j_1j_2jm\rangle$$

$$= (-1)^{j_2-j_1-m}\sqrt{2j+1}\sum_{m_1m_2}\begin{pmatrix} j_1 & j_2 & j \\ m_1 & m_2 & -m \end{pmatrix}|j_1j_2m_1m_2\rangle \tag{1.7-23}$$

where

$$\begin{aligned} j &= j_1+j_2,\quad j_1+j_2-1,\ldots,|j_1-j_2| \\ &= 0,\tfrac{1}{2},1,\tfrac{3}{2},\ldots, \end{aligned} \tag{1.7-24}$$

$$m = m_1+m_2 = j,\quad j-1,\ldots,-j,$$

$$j_1+j_2+j = n \qquad \text{(an integer).}$$

Example Coupling of two electronic spins

$$j_1 \equiv s_1 = \tfrac{1}{2},\qquad m_1 = \pm\tfrac{1}{2},\qquad |s_1m_1\rangle = \begin{cases} |\tfrac{1}{2}\,\tfrac{1}{2}\rangle_1 = \alpha(1), \\ |\tfrac{1}{2}\,-\tfrac{1}{2}\rangle_1 = \beta(1), \end{cases}$$

$$j_2 = s_2 = \tfrac{1}{2},\qquad m_2 = \pm\tfrac{1}{2},\qquad |s_2m_2\rangle = \begin{cases} |\tfrac{1}{2}\,\tfrac{1}{2}\rangle_2 = \alpha(2), \\ |\tfrac{1}{2}\,-\tfrac{1}{2}\rangle_2 = \beta(2), \end{cases}$$

$$j \equiv S = \begin{cases} 1, & m \equiv M = 1, 0, -1, \\ 0, & m \equiv M = 0. \end{cases}$$

$$\begin{aligned} |SM\rangle = |1\;1\rangle &= \alpha(1)\alpha(2), \\ |1\;0\rangle &= \frac{1}{\sqrt{2}}[\alpha(1)\beta(2) + \beta(1)\alpha(2)], \\ |1\;-1\rangle &= \beta(1)\beta(2), \\ |0\;0\rangle &= \frac{1}{\sqrt{2}}[\alpha(1)\beta(2) - \beta(1)\alpha(2)]. \end{aligned} \tag{1.7-25}$$

Three angular momentum operators $\mathbf{J}_1, \mathbf{J}_2$, and $\mathbf{J}_3$ may be coupled to form a resultant $\mathbf{J}$

$$\mathbf{J} = \mathbf{J}_1 + \mathbf{J}_2 + \mathbf{J}_3. \tag{1.7-26}$$

However, the eigenfunctions of $\mathbf{J}$ depend on the coupling sequence and are related through unitary transformations of the form

$$|j_1,j_2j_3(j_{23});j\rangle = \sum_{j_{12}} (-1)^{j_1+j_2+j_3+j}\sqrt{(2j_{12}+1)(2j_{23}+1)}$$

$$\times \begin{Bmatrix} j_1 & j_2 & j_{12} \\ j_3 & j & j_{23} \end{Bmatrix} |j_1j_2(j_{12})j_3;j\rangle. \tag{1.7-27}$$

CHAPTER 2

ROTATIONS

2.1 Coordinate Rotations and Scalar Functions

A rotation is defined as a linear transformation which leaves the scalar product of two vectors invariant and whose determinant is +1. Consider two coordinate systems (x, y, z) and (x', y', z') inclined with respect to one another at an angle α about the z axis as shown in Fig. 2.1.

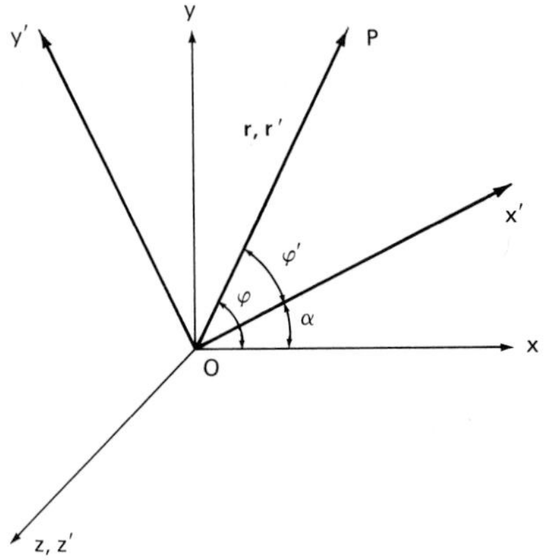

FIG. 2.1 Rotation of coordinates.

The coordinates of a point P can be specified either with reference to (x, y, z) or to (x', y', z'). Suppose, for example, that the temperature T varies from point to point according to some law described by

$$T = f(x, y, z) \equiv f(\mathbf{r}). \tag{2.1-1}$$

There is, of course, nothing special about the coordinate system (x, y, z); the law expressing the variation of temperature from point to point could equally well be referred to the (x', y', z') coordinate system. However, the form of the law, i.e., the dependence on the coordinates, will generally be different in the two systems. Let

$$T = g(x', y', z') \equiv g(\mathbf{r}') \tag{2.1-2}$$

in the (x', y', z') system. Clearly, the temperature at any point such as P, when calculated from (2.1-1) or (2.1-2), must be the same. We, therefore, require that

$$f(\mathbf{r}) = g(\mathbf{r}'). \tag{2.1-3}$$

To illustrate these points, let

$$f(\mathbf{r}) = a(x^2 - y^2) \tag{2.1-4}$$

where a is a constant. The transformation connecting (x, y, z) to (x', y', z') is given by

$$x' = x\cos\alpha + y\sin\alpha, \qquad y' = -x\sin\alpha + y\cos\alpha, \qquad z' = z \tag{2.1-5}$$

or

$$\mathbf{r}' = R(\alpha, z)\mathbf{r} \tag{2.1-6}$$

in which

$$R(\alpha, z) = \begin{pmatrix} \cos\alpha & \sin\alpha & 0 \\ -\sin\alpha & \cos\alpha & 0 \\ 0 & 0 & 1 \end{pmatrix} \tag{2.1-7}$$

and

$$r' = \begin{pmatrix} x' \\ y' \\ z' \end{pmatrix}, \qquad r = \begin{pmatrix} x \\ y \\ z \end{pmatrix}. \tag{2.1-8}$$

If, say, $\alpha = 45°$, $f(\mathbf{r})$ is transformed by (2.1-5) to

$$g(\mathbf{r}') = -2ax'y' \tag{2.1-9}$$

which, clearly, differs from (2.1-4) in its dependence on the coordinates. At the point $(x = 3, y = 2)$, $f(\mathbf{r}) = 5a$. In the primed system the same point has coordinates $x' = 5/\sqrt{2}$, $y' = -1\sqrt{2}$ and $g(\mathbf{r}') = 5a$, as is to be expected.

It may happen that the functional forms of $f(\mathbf{r})$ and $g(\mathbf{r}')$ are exactly the same. Thus, if $f(\mathbf{r}) = x^2 + y^2$, transformation (2.1-5) yields $g(\mathbf{r}') = x'^2 + y'^2$. A function such as $x^2 + y^2$ is said to be an *invariant function* under the coordinate transformation $R(\alpha, z)$.

We now ask the following question: Given a scalar function $f(\mathbf{r}) = f(x, y, z)$ and a coordinate transformation $\mathbf{r}' = R\mathbf{r}$, what is the formal method for finding $g(\mathbf{r}')$ such that $f(\mathbf{r}) = g(\mathbf{r}')$? For this purpose we define an operation P_R by the relation

$$P_R f(\mathbf{r}) = f(R^{-1}\mathbf{r}), \tag{2.1-10}$$

assuming, of course, that the inverse transformation R^{-1} exists. The operator P_R accomplishes the desired result. This may be seen as follows: The transformation $\mathbf{r}' = R\mathbf{r}$ can also be expressed as $\mathbf{r} = R^{-1}\mathbf{r}'$; therefore,

$$f(\mathbf{r}) = f(R^{-1}\mathbf{r}'). \tag{2.1-11}$$

But, by (2.1-10),

$$f(R^{-1}\mathbf{r}') = P_R f(\mathbf{r}'). \tag{2.1-12}$$

If, then, $g(\mathbf{r}')$ is identified with $P_R f(\mathbf{r}')$ we have

$$g(\mathbf{r}') \equiv P_R f(\mathbf{r}') = f(R^{-1}\mathbf{r}') = f(\mathbf{r}). \tag{2.1-13}$$

Thus P_R acting on a function $f(\mathbf{r})$ produces a new function $g(\mathbf{r})$ such that $g(\mathbf{r}')$ is the transformed function under the coordinate transformation $\mathbf{r}' = R\mathbf{r}$. The main point to be kept in mind is that R transforms coordinates while P_R transforms functions of coordinates. The distinction is an important one.

A situation that occurs quite frequently is one in which

$$f(\mathbf{r}) = \mathscr{H}(\mathbf{r})\varphi(\mathbf{r}) \tag{2.1-14}$$

where $\mathscr{H}(\mathbf{r})$ is some operator which changes $\varphi(\mathbf{r})$ into $f(\mathbf{r})$. Thus $\mathscr{H}(\mathbf{r})$ could be another scalar function that operates on $\varphi(\mathbf{r})$ simply by multiplication or $\mathscr{H}(\mathbf{r})$ could be a differential operator like ∇^2, etc. Now, according to (2.1-10)

$$P_R f(\mathbf{r}) = \mathscr{H}(R^{-1}\mathbf{r})\varphi(R^{-1}\mathbf{r}). \tag{2.1-15}$$

But

$$\begin{aligned} P_R f(\mathbf{r}) &= P_R \mathscr{H}(\mathbf{r})\varphi(\mathbf{r}) = P_R \mathscr{H}(\mathbf{r}) P_R^{-1} P_R \varphi(\mathbf{r}) \\ &= P_R \mathscr{H}(\mathbf{r}) P_R^{-1} \varphi(R^{-1}\mathbf{r}). \end{aligned} \tag{2.1-16}$$

Comparing (2.1-15) with (2.1-16), it is seen that

$$\mathscr{H}(R^{-1}\mathbf{r}) = P_R \mathscr{H}(\mathbf{r}) P_R^{-1}. \tag{2.1-17}$$

We regard (2.1-17) as the defining relation for the transformation of operators.

The special case of an invariant function has already been mentioned. A function $f(\mathbf{r})$ is said to be an invariant function under a coordinate transformation $\mathbf{r}' = R\mathbf{r}$ if

$$P_R f(\mathbf{r}) = f(R^{-1}\mathbf{r}) = f(\mathbf{r}). \tag{2.1-18}$$

It should be carefully noted that the invariance of a function refers to its dependence on the coordinates. All functions, invariant or not, yield the same value at any particular physical point provided the transformation from one coordinate system to another is carried out properly.

In similar fashion, an operator $\mathscr{H}(\mathbf{r})$ is said to be an *invariant operator* under a coordinate transformation $\mathbf{r}' = R\mathbf{r}$ if

$$P_R \mathscr{H}(\mathbf{r}) P_R^{-1} = \mathscr{H}(R^{-1}\mathbf{r}) = \mathscr{H}(\mathbf{r}). \tag{2.1-19}$$

From (2.1-19) it is seen that an invariant operator $\mathscr{H}(\mathbf{r})$ commutes with P_R, i.e.,

$$[P_R, \mathscr{H}(\mathbf{r})] = 0. \tag{2.1-20}$$

Assuming the transformation $\mathbf{r}' = R\mathbf{r}$ to be linear, $\det R$ is just the Jacobian and since

$$|\det R| = 1, \tag{2.1-21}$$

the volume element $d\mathbf{r}' \equiv dx'\, dy'\, dz'$ is equal to the volume element $d\mathbf{r} \equiv dx\, dy\, dz$. This simply means that lengths and angles are preserved during the transformation. In that case, since

$$f(\mathbf{r}) = g(\mathbf{r}')$$

we have

$$\int f(\mathbf{r})\, d\mathbf{r} = \int g(\mathbf{r}')\, d\mathbf{r}' = \int g(\mathbf{r})\, d\mathbf{r} = \int P_R f(\mathbf{r})\, d\mathbf{r}. \tag{2.1-22}$$

2.2 Rotations and Angular Momenta

Angular momentum operators have been defined in Section 1.3 on the basis of the commutation rules (1.3-1). Another important aspect of angular momentum operators is their close relationship to rotations.

Consider once more the coordinate rotation (2.1-5), but instead of a finite rotation angle α about the z axis, let the angle be an infinitesimal one, $\delta\alpha$. In that case, to first order,

$$x' = x + y\,\delta\alpha, \qquad y' = -x\,\delta\alpha + y, \qquad z' = z, \tag{2.2-1}$$

or

$$\mathbf{r}' = R(\delta\alpha, z)\mathbf{r}, \tag{2.2-2}$$

where

$$R(\delta\alpha, z) = \begin{pmatrix} 1 & \delta\alpha & 0 \\ -\delta\alpha & 1 & 0 \\ 0 & 0 & 1 \end{pmatrix}. \tag{2.2-3}$$

The operator $P_R(\delta\alpha, z)$ associated with the coordinate transformation $R(\delta\alpha, z)$ gives

$$P_R(\delta\alpha, z)f(\mathbf{r}) = f[R^{-1}(\delta\alpha, z)\mathbf{r}]. \tag{2.2-4}$$

Since $R(\delta\alpha, z)$ is an orthogonal matrix, $R^{-1}(\delta\alpha, z)$ is simply the transpose of $R(\delta\alpha, z)$, so that

$$P_R(\delta\alpha, z)f(\mathbf{r}) = f(x - y\,\delta\alpha, y + x\,\delta\alpha, z). \tag{2.2-5}$$

To first order in $\delta\alpha$,

$$\begin{aligned} P_R(\delta\alpha, z)f(\mathbf{r}) &= f(\mathbf{r}) - y\,\delta\alpha\,\frac{\partial f(\mathbf{r})}{\partial x} + x\,\delta\alpha\,\frac{\partial f(\mathbf{r})}{\partial y} \\ &= f(\mathbf{r}) + \delta\alpha\left(x\frac{\partial}{\partial y} - y\frac{\partial}{\partial x}\right)f(\mathbf{r}). \end{aligned} \tag{2.2-6}$$

But according to (1.1-3c),

$$x\frac{\partial}{\partial y} - y\frac{\partial}{\partial x} = iL_z, \tag{2.2-7}$$

therefore,

$$P_R(\delta\alpha, z)f(\mathbf{r}) = (1 + i\,\delta\alpha\,L_z)f(\mathbf{r}). \tag{2.2-8}$$

For a finite rotation through an angle α, it is merely necessary to subdivide α into $n\,\delta\alpha$ in such a way that α remains constant as $n \to \infty$ and $\delta\alpha \to 0$. Then

$$\begin{aligned} P_R(\alpha, z) &= \lim_{n\to\infty,\,\delta\alpha\to 0} P_R{}^n(\delta\alpha, z) \\ &= \lim_{n\to\infty,\,\delta\alpha\to 0} (1 + i\,\delta\alpha\,L_z)^n \\ &= e^{i\alpha L_z} \end{aligned} \tag{2.2-9}$$

in which the exponential operator is to be understood as a power series

$$e^{i\alpha L_z} = 1 + i\alpha L_z + \frac{1}{2!}(i\alpha L_z)^2 + \cdots. \tag{2.2-10}$$

The development leading to (2.2-9) may be repeated for any rotation axis whose direction is specified by a unit vector $\hat{\mathbf{n}}$, in which case

$$P_R(\delta\omega, \hat{\mathbf{n}}) = 1 + i\,\delta\omega\,\hat{\mathbf{n}} \cdot \mathbf{L}, \tag{2.2-11}$$

$$P_R(\omega, \hat{\mathbf{n}}) = e^{i\omega\hat{n}\cdot\mathbf{L}}, \tag{2.2-12}$$

where $\delta\omega$ and ω are the infinitesimal and finite angles of rotation, respectively. The connection between rotation and orbital angular momentum is contained in (2.2-12). As an extension it is possible to define a general angular momentum operator **J** (which includes **L** as a special case) as a vector operator that satisfies

$$P_R(\omega,\hat{\mathbf{n}}) = e^{i\omega\hat{n}\cdot\mathbf{J}}, \qquad (2.2\text{-}13)$$

and whose components J_x, J_y, and J_z are Hermitian. It follows at once from the latter property that P_R is a unitary operator. Also, by considering two rotations in succession and then reversing the order it may be shown that the commutation rules $\mathbf{J} \times \mathbf{J} = i\mathbf{J}$ follow directly from (2.2-13).

If $\mathscr{H}$ commutes with J_z, it also commutes with $e^{i\alpha J_z}$ since the latter is defined as a power series in J_z as in (2.2-10). Therefore

$$[\mathscr{H}, P_R(\alpha, z)] = 0 \qquad (2.2\text{-}14)$$

or

$$P_R(\alpha, z)\mathscr{H}P_R^{-1}(\alpha, z) = \mathscr{H}, \qquad (2.2\text{-}15)$$

which shows that if an operator commutes with J_z, it is invariant under a rotation about the z axis. Generalizing this result to a rotation about an arbitrary axis, it may be said that an operator which commutes with all the components of **J** is invariant under rotations. Such an operator is a *scalar operator*. Conversely, a scalar operator commutes with all components of **J**.

It is possible to express $R(\omega, \hat{\mathbf{n}})$ in exponential form. The reader should, however, be warned that many pitfalls await him unless extreme care is exercised in maintaining the distinction between $R(\omega, \hat{\mathbf{n}})$ and $P_R(\omega, \hat{\mathbf{n}})$. As emphasized in the previous section, $R(\omega, \hat{\mathbf{n}})$ rotates coordinates through an angle ω about an axis in the direction of the unit vector $\hat{\mathbf{n}}$; on the other hand, $P_R(\omega, \hat{\mathbf{n}})$ operates on functions of coordinates and gives the new form of the function when the coordinates have been rotated by $R(\omega, \hat{\mathbf{n}})$.

2.3 Transformation Properties of Angular Momentum Eigenfunctions

The spherical harmonics $Y_{lm}(\theta, \varphi)$ are eigenfunctions of L^2 and L_z as shown in (1.2-17) and (1.2-19). The intimate connection between angular momentum operators and rotations, discussed in the previous section, provides the motivation for investigating the form of these eigenfunctions when the coordinates are rotated. It will be sufficient to confine the discussion to proper rotations, since an improper rotation is equivalent to a combination of a proper rotation and an inversion, and the properties of spherical harmonics under inversion are already known from (1.2-3b).

The general rotation in three dimensions requires three parameters for its specification, although the parameters may be chosen in various ways. For example, one may specify the direction of the rotation axis and the angle of rotation about that axis, as was done in the last section. Another common mode of description is in terms of Euler angles. We use the latter to discuss the rotational properties of the spherical harmonics. Unfortunately, there is no standard convention as to the definitions and nomenclature of the Euler angles. Among the possible numerous choices we follow those of Hamermesh (1962): Assuming a right-handed coordinate system, a positive rotation about a given axis is defined as one which would carry a right-handed screw in the positive direction along the axis. The successive rotations are:

1. $R(\gamma, z)$, a rotation γ $(0 \leqslant \gamma < 2\pi)$ about the z axis; the new axes are labeled x_1, y_1, z_1; $z_1 = z$.
2. $R(\beta, y_1)$, a rotation β $(0 \leqslant \beta \leqslant \pi)$ about the y_1 axis; the new axes are labeled x_2, y_2, z_2; $y_2 = y_1$.
3. $R(\alpha, z_2)$, a rotation α $(0 \leqslant \alpha < 2\pi)$ about the z_2 axis; the new axes are labeled x_3, y_3, z_3; $z_3 = z_2$.

The three operations taken in sequence correspond to the coordinate transformation

$$\mathbf{r}' = R(\alpha, \beta, \gamma)\mathbf{r} = R(\alpha, z_2)R(\beta, y_1)R(\gamma, z)\mathbf{r}. \tag{2.3-1}$$

When $\beta = 0$, it is necessary to specify $\alpha + \gamma$, while for $\beta = \pi$, $\alpha - \gamma$ is to be specified.

The Euler angles may also be expressed relative to fixed axes. For this purpose we note that

$$R(\beta, y_1) = R(\gamma, z)R(\beta, y)R(-\gamma, z) = R(\gamma, z)R(\beta, y)R^{-1}(\gamma, z), \tag{2.3-2}$$

which may be regarded as a geometrical relation or as an expression for the transform of $R(\beta, y)$. Similarly,

$$R(\alpha, z_2) = R(\beta, y_1)R(\alpha, z)R(-\beta, y_1) = R(\beta, y_1)R(\alpha, z)R^{-1}(\beta, y_1). \tag{2.3-3}$$

When (2.3-2) and (2.3-3) are inserted into (2.3-1), the result is

$$\begin{aligned} R(\alpha, \beta, \gamma) &= R(\beta, y_1)R(\alpha, z)R^{-1}(\beta, y_1)R(\beta, y_1)R(\gamma, z) \\ &= R(\beta, y_1)R(\alpha, z)R(\gamma, z) \\ &= R(\gamma, z)R(\beta, y)R^{-1}(\gamma, z)R(\alpha, z)R(\gamma, z) \\ &= R(\gamma, z)R(\beta, y)R(\alpha, z). \end{aligned} \tag{2.3-4}$$

Therefore, an alternative expression to (2.3-1) is

$$\mathbf{r}' = R(\alpha, \beta, \gamma)\mathbf{r} = R(\gamma, z)R(\beta, y)R(\alpha, z)\mathbf{r}. \tag{2.3-5}$$

The rotation operator $R(\alpha, \beta, \gamma)$ may also be expressed as a matrix. Thus

$$R(\gamma, z) = \begin{pmatrix} \cos\gamma & \sin\gamma & 0 \\ -\sin\gamma & \cos\gamma & 0 \\ 0 & 0 & 1 \end{pmatrix},$$
$$R(\alpha, z_2) = \begin{pmatrix} \cos\alpha & \sin\alpha & 0 \\ -\sin\alpha & \cos\alpha & 0 \\ 0 & 0 & 1 \end{pmatrix}. \tag{2.3-6}$$

By permuting the variables we obtain

$$R(\beta, y_1) = \begin{pmatrix} \cos\beta & 0 & -\sin\beta \\ 0 & 1 & 0 \\ \sin\beta & 0 & \cos\beta \end{pmatrix}. \tag{2.3-7}$$

On multiplying the three matrices in the order in which they are written in (2.3-1) we have

$$R(\alpha, \beta, \gamma) = \begin{pmatrix} \cos\alpha & \sin\alpha & 0 \\ -\sin\alpha & \cos\alpha & 0 \\ 0 & 0 & 1 \end{pmatrix} \begin{pmatrix} \cos\beta & 0 & -\sin\beta \\ 0 & 1 & 0 \\ \sin\beta & 0 & \cos\beta \end{pmatrix} \begin{pmatrix} \cos\gamma & \sin\gamma & 0 \\ -\sin\gamma & \cos\gamma & 0 \\ 0 & 0 & 1 \end{pmatrix}$$
$$= \begin{pmatrix} \cos\alpha\cos\beta\cos\gamma - \sin\alpha\sin\gamma & \cos\alpha\cos\beta\sin\gamma + \sin\alpha\cos\gamma & -\cos\alpha\sin\beta \\ -\sin\alpha\cos\beta\cos\gamma - \cos\alpha\sin\gamma & -\sin\alpha\cos\beta\sin\gamma + \cos\alpha\cos\gamma & \sin\alpha\sin\beta \\ \sin\beta\cos\gamma & \sin\beta\sin\gamma & \cos\beta \end{pmatrix}. \tag{2.3-8}$$

The same result is obtained on the basis of (2.3-4) in which $R(\alpha, \beta, \gamma)$ is expressed as a product of rotations referred to fixed axes but the individual matrices entering into the product are more complicated.

We now turn to the effect on the spherical harmonics when the coordinate transformation is $R(\alpha, \beta, \gamma)$. As in (2.1-10),

$$P_R(\alpha, \beta, \gamma) f(\mathbf{r}) = f[R^{-1}(\alpha, \beta, \gamma)\mathbf{r}]. \tag{2.3-9}$$

Since the matrix for $R(\alpha, \beta, \gamma)$ is given explicitly in (2.3-8) it is possible, in principle, to apply (2.3-9) to the spherical harmonics. Thus suppose $\alpha = 0$, $\beta = \pi/2, \gamma = 0$. From (2.3-8),

$$R(0, \pi/2, 0) = \begin{pmatrix} 0 & 0 & -1 \\ 0 & 1 & 0 \\ 1 & 0 & 0 \end{pmatrix}, \tag{2.3-10a}$$

and

$$R^{-1}(0, \pi/2, 0) = \begin{pmatrix} 0 & 0 & 1 \\ 0 & 1 & 0 \\ -1 & 0 & 0 \end{pmatrix}. \tag{2.3-10b}$$

Applying (2.3-9),

$$P_R(0, \pi/2, 0)f(x, y, z) = f(z, y, -x). \tag{2.3-11}$$

The effect of the coordinate transformation $R(0, \pi/2, 0)$ on $Y_{1\,m}(\theta, \varphi)$, for example, may now be readily calculated:

$$P_R\left(0, \frac{\pi}{2}, 0\right) Y_{11} = P_R\left[-\sqrt{\frac{3}{4\pi}}\frac{1}{\sqrt{2}}\frac{x+iy}{r}\right] = -\sqrt{\frac{3}{4\pi}}\frac{1}{\sqrt{2}}\frac{z+iy}{r}$$

$$= \frac{1}{2}Y_{11} - \frac{1}{\sqrt{2}}Y_{10} + \frac{1}{2}Y_{1\,-1}, \tag{2.3-12a}$$

$$P_R\left(0, \frac{\pi}{2}, 0\right) Y_{10} = P_R\left[\sqrt{\frac{3}{4\pi}}\frac{z}{r}\right] = -\sqrt{\frac{3}{4\pi}}\frac{x}{r}$$

$$= \frac{1}{\sqrt{2}}Y_{11} - \frac{1}{\sqrt{2}}Y_{1\,-1}, \tag{2.3-12b}$$

$$P_R\left(0, \frac{\pi}{2}, 0\right) Y_{1\,-1} = P_R\left[\sqrt{\frac{3}{4\pi}}\frac{1}{\sqrt{2}}\frac{x-iy}{r}\right] = \sqrt{\frac{3}{4\pi}}\frac{1}{\sqrt{2}}\frac{z-iy}{r}$$

$$= \frac{1}{2}Y_{11} + \frac{1}{\sqrt{2}}Y_{10} + \frac{1}{2}Y_{1\,-1}. \tag{2.3-12c}$$

The relationships in (2.3-12a)–(2.3-12c) can be written more compactly in the form

$$P_R\left(0, \frac{\pi}{2}, 0\right) Y_{1m}(\theta, \varphi) = \sum_{m'} Y_{1m'}(\theta, \varphi) D^{(1)}_{m'm}\left(0, \frac{\pi}{2}, 0\right) \tag{2.3-13}$$

with

$$D^{(1)}_{m'm}\left(0, \frac{\pi}{2}, 0\right) = \begin{array}{c|ccc} m' \backslash m & 1 & 0 & -1 \\ \hline 1 & \frac{1}{2} & \frac{1}{\sqrt{2}} & \frac{1}{2} \\ 0 & -\frac{1}{\sqrt{2}} & 0 & \frac{1}{\sqrt{2}} \\ -1 & \frac{1}{2} & -\frac{1}{\sqrt{2}} & \frac{1}{2} \end{array} \equiv \begin{pmatrix} \frac{1}{2} & \frac{1}{\sqrt{2}} & \frac{1}{2} \\ -\frac{1}{\sqrt{2}} & 0 & \frac{1}{\sqrt{2}} \\ \frac{1}{2} & \frac{1}{\sqrt{2}} & \frac{1}{2} \end{pmatrix}. \tag{2.3-14}$$

To take another example, let $\beta = \gamma = 0$. In this case the coordinate transformation $R(\alpha, 0, 0)$ is more conveniently represented by (Fig 2.1)

$$R(\alpha, 0, 0) = \begin{cases} \theta' = \theta, \\ \varphi' = \varphi - \alpha, \end{cases} \tag{2.3-15}$$

so that

$$P_R(\alpha, 0, 0) f(\theta, \varphi) = f[R^{-1}(\alpha, 0, 0)(\theta, \varphi)] = f(\theta, \varphi + \alpha). \tag{2.3-16}$$

Therefore,

$$P_R(\alpha, 0, 0) Y_{lm}(\theta, \varphi) = e^{im\alpha} Y_{lm}(\theta, \varphi). \tag{2.3-17}$$

If $P_R(\alpha, 0, 0)$ is written $e^{i\alpha L_z}$ as in (2.2-9), the result (2.3-17) is obtained directly since $Y_{lm}(\theta, \varphi)$ is an eigenfunction of L_z with the eigenvalue m.

When $l = 1$, for example,

$$P_R(\alpha, 0, 0)(Y_{11}, Y_{10}, Y_{1\,-1}) = (Y_{11}, Y_{10}, Y_{1\,-1}) \begin{pmatrix} e^{i\alpha} & 0 & 0 \\ 0 & 1 & 0 \\ 0 & 0 & e^{-i\alpha} \end{pmatrix} \tag{2.3-18}$$

in which $(Y_{11}, Y_{10}, Y_{1\,-1})$ is regarded as a row matrix. Using the notation in (2.3-13) and (2.3-14),

$$D^{(1)}_{m'm}(\alpha, 0, 0) = \begin{pmatrix} e^{i\alpha} & 0 & 0 \\ 0 & 1 & 0 \\ 0 & 0 & e^{-i\alpha} \end{pmatrix}, \tag{2.3-19}$$

and for arbitrary l,

$$D^{(l)}_{m'm}(\alpha, 0, 0) = \begin{pmatrix} e^{il\alpha} & 0 & 0 & \cdots & 0 \\ 0 & e^{i(l-1)\alpha} & 0 & \cdots & 0 \\ \vdots & \vdots & \vdots & \vdots & \vdots \\ 0 & 0 & 0 & \cdots & e^{-il\alpha} \end{pmatrix}. \tag{2.3-20}$$

Equations (2.3-13) and (2.3-17) are examples of the more general relationship

$$P_R(\alpha, \beta, \gamma) Y_{lm}(\theta, \varphi) = \sum_{m'=-l}^{l} Y_{lm'}(\theta, \varphi) D^{(l)}_{m'm}(\alpha, \beta, \gamma) \tag{2.3-21}$$

in which the $D^{(l)}_{m'm}(\alpha, \beta, \gamma)$ are numerical coefficients. A very important property of spherical harmonics is expressed by (2.3-21), namely, that under rotations of coordinates, spherical harmonics of order l transform among themselves. Thus if $l = 1$, the linear combination on the right side of (2.3-21) contains, at most, Y_{11}, Y_{10}, and $Y_{1\,-1}$ as was illustrated in (2.3-12). No other spherical harmonic can appear in the transformation of any member of the set Y_{1m}.

To prove (2.3-21) we note that the Laplacian operator ∇^2 is an invariant operator under an orthogonal transformation of coordinates; therefore,

$$P_R \nabla^2 P_R^{-1} = \nabla^2. \tag{2.3-22}$$

In spherical coordinates, using (1.1-14)

$$\nabla^2 = \frac{\partial^2}{\partial r^2} + \frac{2}{r}\frac{\partial}{\partial r} - \frac{L^2}{r^2}. \tag{2.3-23}$$

Since r remains fixed under an orthogonal transformation, (2.3-22) implies that

$$P_R L^2 P_R^{-1} = L^2 \tag{2.3-24}$$

or

$$[P_R, L^2] = 0. \tag{2.3-25}$$

But, from (1.2-17),

$$L^2 Y_{lm}(\theta, \varphi) = l(l+1) Y_{lm}(\theta, \varphi). \tag{2.3-26}$$

Therefore

$$P_R L^2 Y_{lm}(\theta, \varphi) = L^2 P_R Y_{lm}(\theta, \varphi) \tag{2.3-27a}$$

$$= l(l+1) P_R Y_{lm}(\theta, \varphi). \tag{2.3-27b}$$

Comparing (2.3-26) with (2.3-27), it is seen that both $Y_{lm}(\theta, \varphi)$ and $P_R Y_{lm}(\theta, \varphi)$ are eigenfunctions of L^2 with the same eigenvalue $l(l+1)$. For a given value of l, however, there are just $2l+1$ independent eigenfunctions of L^2; these are the spherical harmonics $Y_{lm}(\theta, \varphi)$. Therefore $P_R Y_{lm}(\theta, \varphi)$ can only be a linear combination of the $Y_{lm}(\theta, \varphi)$ that belong to the same value of l. The mathematical form of this statement is given by (2.3-21).

The spherical harmonics may act as *operators* as for example, in the product $Y_{lm}\varphi(\mathbf{r}) = f(\mathbf{r})$ where Y_{lm} acts on $\varphi(\mathbf{r})$ (by multiplication) to produce another function $f(\mathbf{r})$. In that case, the transformation law (2.3-21) must be modified to read

$$P_R(\alpha, \beta, \gamma) Y_{lm}(\theta, \varphi) P_R^{-1}(\alpha, \beta\, \gamma) = \sum_{m'=-l}^{l} Y_{lm'}(\theta, \varphi) D^{(l)}_{m'm}(\alpha, \beta, \gamma) \tag{2.3-28}$$

so as to be consistent with the general transformation law for operators (2.1-17).

The general form of $D^{(l)}_{m'm}(\alpha, \beta, \gamma)$ will be derived in Section 4.5 by group theoretical methods. By analogy with (2.3-21) it is postulated that the transformation of generalized angular momentum eigenfunctions $|jm\rangle$ under a rotation of coordinates is given by

$$P_R(\alpha, \beta, \gamma)|jm\rangle = \sum_{m'} |jm'\rangle D^{(j)}_{m'm}(\alpha, \beta, \gamma). \tag{2.3-29}$$

This aspect, too, will receive further attention from the standpoint of group theory.

CHAPTER 3

ELEMENTS OF GROUP THEORY

In this chapter we present an introduction to certain portions of group theory that enter into the description of atoms and molecules. More complete treatments of group theory are given in various books; for example, those by Wigner (1959), Heine (1960), Hamermesh (1962), McWeeny (1963), Tinkham 1964), and Hochstrasser (1966).

3.1 Definitions and Basic Properties

Suppose we have a set of elements $g_1, g_2, \ldots, g_h$ and a specified operation that tells us how to combine two elements. The result of combining two elements g_i and g_j is called a *product* and is written $g_i g_j$. Both the elements and the operations may be quite general, e.g., numbers combined by arithmetical operations, matrices combined by arithmetical operations, matrices combined by matrix multiplication, coordinate transformations, permutations performed sequentially, etc. A group of *order h* is defined as a set of *h* elements $g_1, g_2, \ldots, g_h$ which, together with the specified operation, satisfies the conditions (or postulates):

1. If g_i and g_j are members of the set, the product $g_i g_j$ is also a member of the set. The order may be important so that $g_i g_j$ is not necessarily the same as $g_j g_i$.
2. The associative law applies to the product of three elements, i.e., $g_i(g_j g_k) = (g_i g_j)g_k$.
3. The set contains one, and only one, element called the *identity element e* such that $eg_i = g_i e = g_i$, where g_i is any element of the group.
4. Each element of the set has an inverse (or reciprocal) which is also a member of the set, i.e., for every element g_i there is a unique element g_j such that $g_i g_j = g_j g_i = e$ or $g_j = g_i^{-1}$ where g_i^{-1} denotes the inverse of g_i.

As an example of a group we consider the specific set of coordinate transformations shown in Fig. 3.1; they are all of the form

$$\mathbf{r}' = R\mathbf{r} \tag{3.1-1}$$

in which $\mathbf{r}'$ and $\mathbf{r}$ are given by (2.1-8) and R is a 3×3 orthogonal matrix. The transformation and the matrix will be denoted by the same symbol.

E Identity transformation

$$x' = x, \qquad y' = y, \qquad z' = z, \tag{3.1-2a}$$

$$E = \begin{pmatrix} 1 & 0 & 0 \\ 0 & 1 & 0 \\ 0 & 0 & 1 \end{pmatrix}. \tag{3.1-2b}$$

A A rotation of coordinates about the x axis through 180°

$$x' = x, \qquad y' = -y, \qquad z' = -z, \tag{3.1-3a}$$

$$A = \begin{pmatrix} 1 & 0 & 0 \\ 0 & -1 & 0 \\ 0 & 0 & -1 \end{pmatrix}. \tag{3.1-3b}$$

B A rotation of coordinates about O2 through 180°

$$x' = -\tfrac{1}{2}x + \tfrac{1}{2}\sqrt{3}y, \qquad y' = \tfrac{1}{2}\sqrt{3}x + \tfrac{1}{2}y, \qquad z' = -z, \tag{3.1-4a}$$

$$B = \begin{pmatrix} -\frac{1}{2} & \frac{1}{2}\sqrt{3} & 0 \\ \frac{1}{2}\sqrt{3} & \frac{1}{2} & 0 \\ 0 & 0 & -1 \end{pmatrix}. \tag{3.1-4b}$$

C A rotation of coordinates about O3 through 180°

$$x' = -\tfrac{1}{2}x - \tfrac{1}{2}\sqrt{3}y, \qquad y' = -\tfrac{1}{2}\sqrt{3}x + \tfrac{1}{2}y, \qquad z' = -z, \tag{3.1-5a}$$

$$C = \begin{pmatrix} -\frac{1}{2} & -\frac{1}{2}\sqrt{3} & 0 \\ -\frac{1}{2}\sqrt{3} & \frac{1}{2} & 0 \\ 0 & 0 & -1 \end{pmatrix}. \tag{3.1-5b}$$

D A positive (counterclockwise) rotation of coordinates about the z axis through 120°

$$x' = -\tfrac{1}{2}x + \tfrac{1}{2}\sqrt{3}y, \qquad y' = -\tfrac{1}{2}\sqrt{3}x - \tfrac{1}{2}y, \qquad z' = z, \tag{3.1-6a}$$

$$D = \begin{pmatrix} -\frac{1}{2} & \frac{1}{2}\sqrt{3} & 0 \\ -\frac{1}{2}\sqrt{3} & -\frac{1}{2} & 0 \\ 0 & 0 & 1 \end{pmatrix}. \tag{3.1-6b}$$

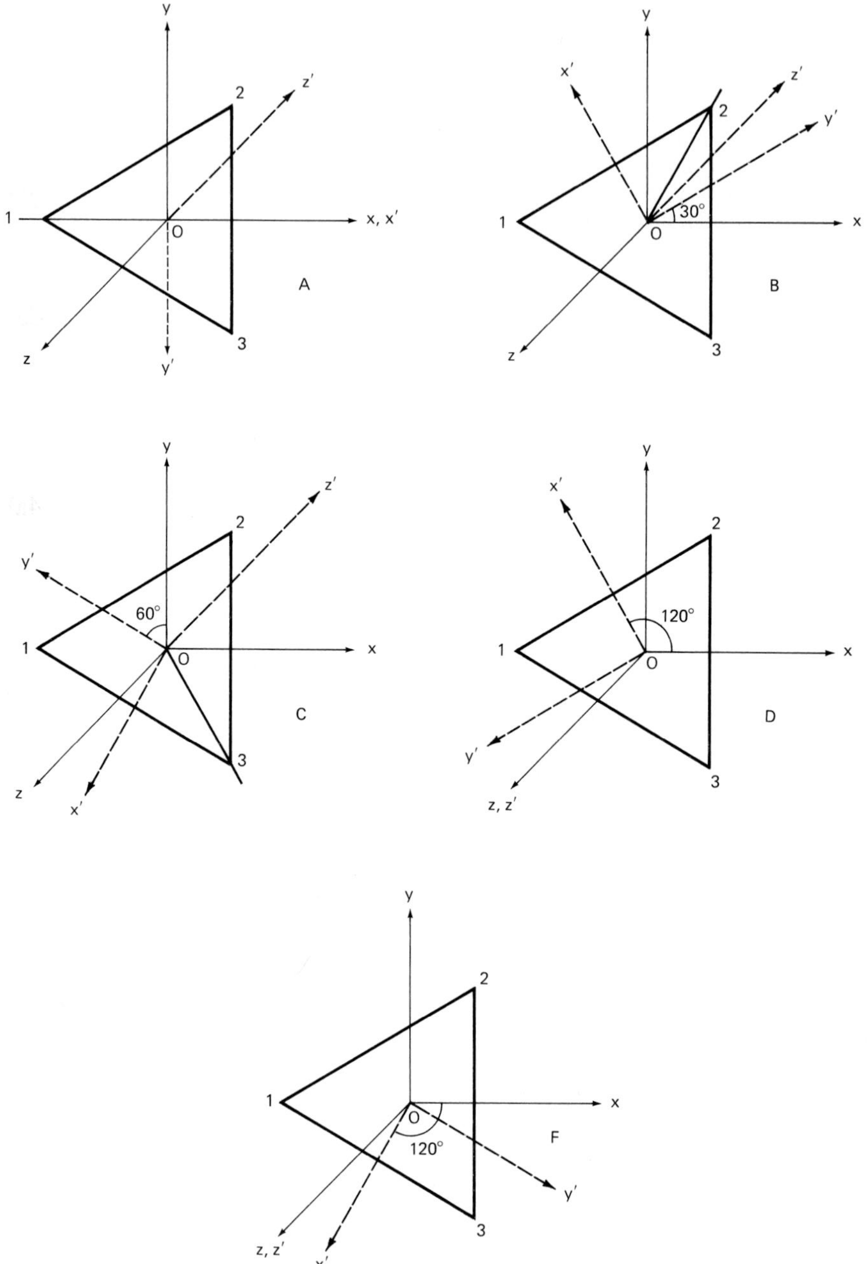

FIG. 3.1 Coordinate transformations A–F in D_3.

F A negative (clockwise) rotation of coordinates about the *z* axis through 120°

$$x' = -\tfrac{1}{2}x - \tfrac{1}{2}\sqrt{3}y, \qquad y' = \tfrac{1}{2}\sqrt{3}x - \tfrac{1}{2}y, \qquad z' = z, \tag{3.1-7a}$$

$$F = \begin{pmatrix} -\frac{1}{2} & -\frac{1}{2}\sqrt{3} & 0 \\ \frac{1}{2}\sqrt{3} & -\frac{1}{2} & 0 \\ 0 & 0 & 1 \end{pmatrix}. \tag{3.1-7b}$$

Suppose, now, that the coordinate transformations are performed in succession as, for example,

$$\mathbf{r}' = C\mathbf{r}, \tag{3.1-8a}$$

followed by

$$\mathbf{r}'' = A\mathbf{r}'. \tag{3.1-8b}$$

The two rotations are equivalent to a single rotation

$$\mathbf{r}'' = A\mathbf{r}' = AC\mathbf{r} = F\mathbf{r} \tag{3.1-9}$$

in which the equality $AC = F$ is readily verified from (3.1-3), (3.1-5), and (3.1-7). Thus the matrix product AC is completely equivalent to two coordinate transformations performed in the order C followed by A. One may then adopt either of two equivalent viewpoints:

1. $E, A, \ldots, F$ are coordinate transformations; the combination or product of two elements, e.g., AC, consists of two successive transformations in the order C followed by A.
2. $E, A, \ldots, F$ are matrices; the combination or product of two elements, e.g., AC, is the matrix product of matrix A and matrix C in the order AC.

It will now be shown that the set $E, A, \ldots, F$ together with the law of combination (successive transformations or matrix multiplication) constitutes a group. To facilitate this demonstration, it is convenient to construct a table of all possible products of pairs of elements. This is shown in Table 3.1. We note that some pairs commute while others do not, e.g., $AC = F$, but $CA = D$.

Quite clearly, the set $E, A, \ldots, F$ satisfies the group postulates since, from Table 3.1, it is seen that the product of any two elements is a member of the set; the associative law holds as, for example,

$$A(BC) = AD = B, \qquad (AB)C = DC = B; \tag{3.1-10}$$

also, there is an identity element E and each element has an inverse:

$$E^{-1} = E, \quad A^{-1} = A, \quad B^{-1} = B, \quad C^{-1} = C, \quad D^{-1} = F, \quad F^{-1} = D. \tag{3.1-11}$$

TABLE 3.1

Multiplication Table for the Elements of D_3

	E	A	B	C	D	F
E	E	A	B	C	D	F
A	A	E	D	F	B	C
B	B	F	E	D	C	A
C	C	D	F	E	A	B
D	D	C	A	B	F	E
F	F	B	C	A	E	D

Table 3.1, often called a multiplication table, completely defines the group which will be called D_3 in anticipation of the nomenclature for point groups (Section 5.1). Note that each element appears once and only once in each column and each row.

It is well to keep in mind that the transformations (rotations) previously described are transformations of *coordinates*. In some discussions the physical object—in our case, the triangle—is rotated, in which case $E, A, \ldots, F$ represent covering operations. There is no advantage to one point of view over the other but it is of utmost importance to maintain consistency. In Section 2.1 we introduced two operators, R and P_R; the former represented a transformation of coordinates and the latter described the effect on a function when the coordinates were transformed. When R corresponds to a rotation through an angle α about a particular axis, P_R may be regarded as an operator which rotates an *object* through an angle $-\alpha$ about the same axis.

It may also be shown that if a set of coordinate transformations R consisting of $R_1, R_2, \ldots, R_h$ constitutes a group, then the set P_R consisting of elements $P_{R_1}, P_{R_2}, \ldots, P_{R_h}$ is also a group. To prove this let

$$P_{R_j} f(\mathbf{r}) = g(\mathbf{r}). \tag{3.1-12}$$

Then

$$P_{R_i} P_{R_j} f(\mathbf{r}) = P_{R_i} g(\mathbf{r}) = g(R_i^{-1}\mathbf{r}). \tag{3.1-13}$$

But, from (3.1-12),

$$g(R_i^{-1}\mathbf{r}) = P_{R_j} f(R_i^{-1}\mathbf{r}) \tag{3.1-14}$$

and

$$P_{R_j} f(R_i^{-1}\mathbf{r}) = f(R_j^{-1}R_i^{-1}\mathbf{r}) = f[(R_iR_j)^{-1}\mathbf{r}] = P_{R_iR_j} f(\mathbf{r}). \tag{3.1-15}$$

From (3.1-13)–(3.1-15) we conclude that

$$P_{R_j} P_{R_i} = P_{R_j R_i}. \tag{3.1-16}$$

In other words the operators in the set $\{P_R\}$ satisfy the same multiplication table as the coordinate transformations in the set $\{R\}$ so that if $\{R\}$ is a group, so is $\{P_R\}$. We have here an example of *isomorphism* which has the following meaning: Two groups G and G' are isomorphic when: (a) to each element g_i of G there corresponds one, and only one, element g_i' of G' and, conversely, to each element g_i' of G' there corresponds one, and only one, element g_i of G and (b) if in group G there exists the relation $g_i g_j = g_k$, then, in group G', $g_i' g_j' = g_k'$. It may happen that two groups G and G' are related in such a way that to each element g_i of G there corresponds one, and only one, element g_i' of G' but to each element g_i' of G' there corresponds at least one and possibly more than one element of G. When this occurs and the law of combination is preserved, that is, when $g_i g_j = g_k$ implies $g_i' g_j' = g_k'$, then G is said to be *homomorphic* to G'. Homomorphism is not a reciprocal property. If G is homomorphic to G', then G' is not necessarily homomorphic to G. The number of elements of G must be equal to or greater than the number of elements of G'; if the number is equal, the homomorphism becomes an isomorphism and the latter *is* a reciprocal relation.

A group may contain elements which, taken by themselves, satisfy the group postulates. Such a set of elements is called a *subgroup*. Within D_3 we can identify the subgroups E; E, A; E, B; E, C; E, D, F.

An element g_j is said to be *conjugate* to an element g_i if

$$g_j = x g_i x^{-1} \tag{3.1-17}$$

where x is some member of the group. If g_j is conjugate to g_i, then g_i is conjugate to g_j. If two elements g_i and g_j are conjugate to a third element g_k, they are also conjugate to one another.

The set of conjugates of a given element g_i, including g_i, is called a *class*. Each element belongs to one class and one class only and the identity element always forms a class by itself. A group can therefore be subdivided into classes. For example, D_3 consists of three classes

$$C_1 = E, \qquad C_2 = A, B, C, \qquad C_3 = D, F. \tag{3.1-18}$$

Suppose the elements of a group commute with each other so that $g_i g_j = g_j g_i$ for all elements of the group. Such a group is called *Abelian*. Then $x g_i x^{-1} = g_i x x^{-1} = g_i$. This means that g_i is its own conjugate element. Since this holds for all elements, the number of classes is equal to the order (number of elements) h of the group.

3.2 Representations and Characters

A *matrix representation* of a group is defined as a set of square, nonsingular matrices (matrices with nonvanishing determinants) that satisfy the multiplication table of the group when the matrices are multiplied by the ordinary rules of matrix multiplication. There are other kinds of representations (see Section 6.1) but unless there is a specific statement to that effect, it will be understood that a representation means a *matrix* representation.

We give several examples of representations of the group D_3:

$$\Gamma^{(1)}(E) = \Gamma^{(1)}(A) = \Gamma^{(1)}(B) = \Gamma^{(1)}(C) = \Gamma^{(1)}(D) = \Gamma^{(1)}(F) = 1. \quad (3.2\text{-}1)$$

This is a one-dimensional representation (1×1 matrices) which obviously satisfies the group multiplication table. It appears to be trivial but nevertheless plays an important role in later developments. This representation is called the *totally symmetric representation* or the *unit representation*. Another one-dimensional representation is

$$\Gamma^{(2)}(E)=1, \quad \Gamma^{(2)}(A)=\Gamma^{(2)}(B)=\Gamma^{(2)}(C)=-1, \quad \Gamma^{(2)}(D)=\Gamma^{(2)}(F)=1. \quad (3.2\text{-}2)$$

A two-dimensional representation consisting of 2×2 matrices is:

$$\Gamma^{(3)}(E) = \begin{pmatrix} 1 & 0 \\ 0 & 1 \end{pmatrix}, \qquad \Gamma^{(3)}(A) = \begin{pmatrix} 1 & 0 \\ 0 & -1 \end{pmatrix},$$

$$\Gamma^{(3)}(B) = \begin{pmatrix} -\frac{1}{2} & \frac{1}{2}\sqrt{3} \\ \frac{1}{2}\sqrt{3} & \frac{1}{2} \end{pmatrix}, \qquad \Gamma^{(3)}(C) = \begin{pmatrix} -\frac{1}{2} & -\frac{1}{2}\sqrt{3} \\ -\frac{1}{2}\sqrt{3} & \frac{1}{2} \end{pmatrix}, \quad (3.2\text{-}3)$$

$$\Gamma^{(3)}(D) = \begin{pmatrix} -\frac{1}{2} & \frac{1}{2}\sqrt{3} \\ -\frac{1}{2}\sqrt{3} & -\frac{1}{2} \end{pmatrix}, \qquad \Gamma^{(3)}(F) = \begin{pmatrix} -\frac{1}{2} & -\frac{1}{2}\sqrt{3} \\ \frac{1}{2}\sqrt{3} & -\frac{1}{2} \end{pmatrix}.$$

The set of matrices $E, A, \ldots, F$ [(3.1-2b)–(3.1-7b)] have been shown to satisfy the group multiplication table. We may therefore regard these matrices as a three-dimensional representation of D_3:

$$\Gamma^{(4)}(E)=E=\begin{pmatrix} 1 & 0 & 0 \\ 0 & 1 & 0 \\ 0 & 0 & 1 \end{pmatrix}, \qquad \Gamma^{(4)}(A)=A=\begin{pmatrix} 1 & 0 & 0 \\ 0 & -1 & 0 \\ 0 & 0 & -1 \end{pmatrix},$$

$$\Gamma^{(4)}(B)=B=\begin{pmatrix} -\frac{1}{2} & \frac{1}{2}\sqrt{3} & 0 \\ \frac{1}{2}\sqrt{3} & \frac{1}{2} & 0 \\ 0 & 0 & -1 \end{pmatrix}, \quad \Gamma^{(4)}(C)=C=\begin{pmatrix} -\frac{1}{2} & -\frac{1}{2}\sqrt{3} & 0 \\ -\frac{1}{2}\sqrt{3} & \frac{1}{2} & 0 \\ 0 & 0 & -1 \end{pmatrix}, \quad (3.2\text{-}4)$$

$$\Gamma^{(4)}(D)=D=\begin{pmatrix} -\frac{1}{2} & \frac{1}{2}\sqrt{3} & 0 \\ -\frac{1}{2}\sqrt{3} & -\frac{1}{2} & 0 \\ 0 & 0 & 1 \end{pmatrix}, \quad \Gamma^{(4)}(F)=F=\begin{pmatrix} -\frac{1}{2} & -\frac{1}{2}\sqrt{3} & 0 \\ \frac{1}{2}\sqrt{3} & -\frac{1}{2} & 0 \\ 0 & 0 & 1 \end{pmatrix}.$$

Another three-dimensional representation is the set

$$\Gamma^{(5)}(E) = \begin{pmatrix} 1 & 0 & 0 \\ 0 & 1 & 0 \\ 0 & 0 & 1 \end{pmatrix}, \qquad \Gamma^{(5)}(A) = \begin{pmatrix} 1 & 0 & 0 \\ 0 & 0 & 1 \\ 0 & 1 & 0 \end{pmatrix},$$

$$\Gamma^{(5)}(B) = \begin{pmatrix} 0 & 0 & 1 \\ 0 & 1 & 0 \\ 1 & 0 & 0 \end{pmatrix}, \qquad \Gamma^{(5)}(C) = \begin{pmatrix} 0 & 1 & 0 \\ 1 & 0 & 0 \\ 0 & 0 & 1 \end{pmatrix}, \tag{3.2-5}$$

$$\Gamma^{(5)}(D) = \begin{pmatrix} 0 & 0 & 1 \\ 1 & 0 & 0 \\ 0 & 1 & 0 \end{pmatrix}, \qquad \Gamma^{(5)}(F) = \begin{pmatrix} 0 & 1 & 0 \\ 0 & 0 & 1 \\ 1 & 0 & 0 \end{pmatrix}.$$

The number of representations that may be constructed is without limit; as a final example we show a six-dimensional representation:

$$\Gamma^{(6)}(E) = \begin{pmatrix} 1 & 0 & 0 & 0 & 0 & 0 \\ 0 & 1 & 0 & 0 & 0 & 0 \\ 0 & 0 & 1 & 0 & 0 & 0 \\ 0 & 0 & 0 & 1 & 0 & 0 \\ 0 & 0 & 0 & 0 & 1 & 0 \\ 0 & 0 & 0 & 0 & 0 & 1 \end{pmatrix}, \qquad \Gamma^{(6)}(A) = \begin{pmatrix} 1 & 0 & 0 & 0 & 0 & 0 \\ 0 & -1 & 0 & 0 & 0 & 0 \\ 0 & 0 & 1 & 0 & 0 & 0 \\ 0 & 0 & 0 & -1 & 0 & 0 \\ 0 & 0 & 0 & 0 & 1 & 0 \\ 0 & 0 & 0 & 0 & 0 & 1 \end{pmatrix},$$

$$\Gamma^{(6)}(B) = \begin{pmatrix} -\tfrac{1}{2} & \tfrac{1}{2}\sqrt{3} & 0 & 0 & 0 & 0 \\ \tfrac{1}{2}\sqrt{3} & \tfrac{1}{2} & 0 & 0 & 0 & 0 \\ 0 & 0 & -\tfrac{1}{2} & \tfrac{1}{2}\sqrt{3} & 0 & 0 \\ 0 & 0 & \tfrac{1}{2}\sqrt{3} & \tfrac{1}{2} & 0 & 0 \\ 0 & 0 & 0 & 0 & 1 & 0 \\ 0 & 0 & 0 & 0 & 0 & 1 \end{pmatrix}, \qquad \Gamma^{(6)}(C) = \begin{pmatrix} -\tfrac{1}{2} & -\tfrac{1}{2}\sqrt{3} & 0 & 0 & 0 & 0 \\ -\tfrac{1}{2}\sqrt{3} & \tfrac{1}{2} & 0 & 0 & 0 & 0 \\ 0 & 0 & -\tfrac{1}{2} & -\tfrac{1}{2}\sqrt{3} & 0 & 0 \\ 0 & 0 & -\tfrac{1}{2}\sqrt{3} & \tfrac{1}{2} & 0 & 0 \\ 0 & 0 & 0 & 0 & 1 & 0 \\ 0 & 0 & 0 & 0 & 0 & 1 \end{pmatrix},$$

$$\Gamma^{(6)}(D) = \begin{pmatrix} -\tfrac{1}{2} & \tfrac{1}{2}\sqrt{3} & 0 & 0 & 0 & 0 \\ -\tfrac{1}{2}\sqrt{3} & -\tfrac{1}{2} & 0 & 0 & 0 & 0 \\ 0 & 0 & -\tfrac{1}{2} & \tfrac{1}{2}\sqrt{3} & 0 & 0 \\ 0 & 0 & -\tfrac{1}{2}\sqrt{3} & -\tfrac{1}{2} & 0 & 0 \\ 0 & 0 & 0 & 0 & 1 & 0 \\ 0 & 0 & 0 & 0 & 0 & 1 \end{pmatrix}, \qquad \Gamma^{(6)}(F) = \begin{pmatrix} -\tfrac{1}{2} & -\tfrac{1}{2}\sqrt{3} & 0 & 0 & 0 & 0 \\ \tfrac{1}{2}\sqrt{3} & -\tfrac{1}{2} & 0 & 0 & 0 & 0 \\ 0 & 0 & -\tfrac{1}{2} & -\tfrac{1}{2}\sqrt{3} & 0 & 0 \\ 0 & 0 & \tfrac{1}{2}\sqrt{3} & -\tfrac{1}{2} & 0 & 0 \\ 0 & 0 & 0 & 0 & 1 & 0 \\ 0 & 0 & 0 & 0 & 0 & 1 \end{pmatrix}. \tag{3.2-6}$$

The symbol Γ will be used to denote a general representation i.e., the entire set of matrices that satisfy the group multiplication table; a particular matrix belonging to the jth representation will be indicated by $\Gamma^{(j)}(R)$.

If the matrices belonging to a representation Γ are subjected to a similarity transformation, the result is a new representation Γ'. The two representations Γ and Γ' are said to be *equivalent*. If Γ and Γ' cannot be transformed into one another by a similarity transformation, Γ and Γ' are then said to be *inequivalent*. It may be shown (Tinkham, 1964) that for a finite group every representation is equivalent to a unitary representation (i.e., consisting entirely of unitary matrices). We shall assume henceforth that all representations are unitary unless the contrary is explicitly stated (see also Section 3.4).

The *character* of a representation matrix $\Gamma^{(j)}(R)$ is the sum of the diagonal elements (trace). The character is denoted by $\chi^{(j)}(R)$; thus

$$\chi^{(j)}(R) = \sum_{\alpha} \Gamma_{\alpha\alpha}^{(j)}(R). \tag{3.2-7}$$

In view of the fact that the trace of a matrix is invariant under a similarity transformation and the representation matrices associated with two conjugate elements are, by definition, related by a similarity transformation, it follows that elements of a group belonging to the same class have the same character in any given representation. To illustrate the foregoing, it has been shown in (3.1-18) that D_3 consists of three classes: $C_1(E)$, $C_2(A, B, C)$, and $C_3(D, F)$. Examination of any of the representations $\Gamma^{(1)}, \ldots, \Gamma^{(6)}$ listed above, say $\Gamma^{(5)}$ [as in (3.2-5)], reveals that $\chi^{(5)}(E) = 3$; $\chi^{(5)}(A) = \chi^{(5)}(B) = \chi^{(5)}(C) = 1$; $\chi^{(5)}(D) = \chi^{(5)}(F) = 0$. For $\Gamma^{(6)}$ [Eq. (3.2-6)], we have $\chi^{(6)}(E) = 6$; $\chi^{(6)}(A) = \chi^{(6)}(B) = \chi^{(6)}(C) = 2$; $\chi^{(6)}(D) = \chi^{(6)}(F) = 0$.

Since equivalent representations are related by a similarity transformation, they must have the same characters. This condition is both necessary and sufficient; in other words, if two representations have the same set of characters, they are equivalent or equal (see also Section 3.3).

The character of the unit element (identity) $\chi(E)$ gives the dimension of the representation, and when the representation is one-dimensional, the representation matrices and the characters obviously coincide.

A representation of a group G is also a representation of any subgroup of G since any set of matrices which satisfy the multiplication table of G must also satisfy the multiplication table of the subgroup the latter table being part of the former.

3.3 Reducible and Irreducible Representations

An important notion concerning representations is that of reducibility. Consider, for example $\Gamma^{(6)}(B)$ in (3.2-6):

$$\Gamma^{(6)}(B) = \left(\begin{array}{cc|cc|c|c} -\frac{1}{2} & \frac{1}{2}\sqrt{3} & 0 & 0 & 0 & 0 \\ \frac{1}{2}\sqrt{3} & \frac{1}{2} & 0 & 0 & 0 & 0 \\ \hline 0 & 0 & -\frac{1}{2} & \frac{1}{2}\sqrt{3} & 0 & 0 \\ 0 & 0 & \frac{1}{2}\sqrt{3} & \frac{1}{2} & 0 & 0 \\ \hline 0 & 0 & 0 & 0 & 1 & 0 \\ \hline 0 & 0 & 0 & 0 & 0 & 1 \end{array}\right), \tag{3.3-1}$$

This matrix consists of four separate blocks along the main diagonal—two of the blocks are two-dimensional and the other two are one-dimensional. Moreover, examination of the separate blocks reveals that the two-dimensional blocks are identical with the representation matrix $\Gamma^{(3)}(B)$ [Eq. (3.2-3)] and the one-dimensional blocks coincide with $\Gamma^{(1)}(B)$ [Eq. (3.2-1)]. Therefore $\Gamma^{(6)}(B)$ is said to be *reducible* into $2\Gamma^{(3)}(B)$ and $2\Gamma^{(1)}(B)$. Symbolically this is expressed by

$$\Gamma^{(6)}(B) = 2\Gamma^{(3)}(B) + 2\Gamma^{(1)}(B) \tag{3.3-2}$$

in which the right side of the equation signifies that $\Gamma^{(3)}(B)$ appears twice and $\Gamma^{(1)}(B)$ appears twice along the main diagonal in (3.3-1). In this context the symbol + clearly has nothing to do with addition.‡ All the other representation matrices in $\Gamma^{(6)}$ can be subdivided into blocks that are identifiable as belonging to $\Gamma^{(3)}$ and $\Gamma^{(1)}$. Hence (3.3-2) may be generalized to

$$\Gamma^{(6)} = 2\Gamma^{(3)} + 2\Gamma^{(1)}. \tag{3.3-3}$$

The order in which the terms on the right-hand side of (3.3-3) are written is immaterial.

The reduction (or decomposition) of the representation $\Gamma^{(6)}$ into $2\Gamma^{(3)}$ and $2\Gamma^{(1)}$ was achieved simply by inspecting each of the representation matrices belonging to $\Gamma^{(6)}$. We shall now extend the notion of reducibility as follows: If by a similarity transformation a representation can be put into block form—as exemplified by (3.3-1)—then the representation is reducible. Consider, for example, $\Gamma^{(5)}$ [Eq. (3.2-5)] which, in contrast to $\Gamma^{(6)}$, consists of representation matrices that cannot all be subdivided by inspection into smaller matrices along the main diagonals. It will now be shown that under the extended definition of reducibility $\Gamma^{(5)}$ is nevertheless a reducible representation. Let

$$M = \begin{pmatrix} \sqrt{\frac{2}{3}} & -\sqrt{\frac{1}{6}} & -\sqrt{\frac{1}{6}} \\ 0 & \sqrt{\frac{1}{2}} & -\sqrt{\frac{1}{2}} \\ -\sqrt{\frac{1}{3}} & -\sqrt{\frac{1}{3}} & -\sqrt{\frac{1}{3}} \end{pmatrix}. \tag{3.3-4}$$

If we carry out the operation $M\Gamma^{(5)}(C)M^{-1}$, for example, the result is

$$\begin{pmatrix} \sqrt{\frac{2}{3}} & -\sqrt{\frac{1}{6}} & -\sqrt{\frac{1}{6}} \\ 0 & \sqrt{\frac{1}{2}} & -\sqrt{\frac{1}{2}} \\ -\sqrt{\frac{1}{3}} & -\sqrt{\frac{1}{3}} & -\sqrt{\frac{1}{3}} \end{pmatrix} \begin{pmatrix} 0 & 1 & 0 \\ 1 & 0 & 0 \\ 0 & 0 & 1 \end{pmatrix} \begin{pmatrix} \sqrt{\frac{2}{3}} & 0 & -\sqrt{\frac{1}{3}} \\ -\sqrt{\frac{1}{6}} & \sqrt{\frac{1}{2}} & -\sqrt{\frac{1}{3}} \\ -\sqrt{\frac{1}{6}} & -\sqrt{\frac{1}{2}} & -\sqrt{\frac{1}{3}} \end{pmatrix}$$

$$= \begin{pmatrix} -\frac{1}{2} & -\frac{1}{2}\sqrt{3} & 0 \\ -\frac{1}{2}\sqrt{3} & \frac{1}{2} & 0 \\ 0 & 0 & 1 \end{pmatrix}. \tag{3.3-5}$$

‡ The symbol $\oplus$ is also used.

The right-hand side of (3.3-5) is a matrix which can be subdivided into a two-dimensional and a one-dimensional matrix along the main diagonal. The two-dimensional matrix is $\Gamma^{(3)}(C)$ [Eq. (3.2-3)] and the one-dimensional matrix is $\Gamma^{(1)}(C)$ [Eq. (3.2-1)]. In similar fashion $M\Gamma^{(5)}(R)M^{-1}$, where $\Gamma^{(5)}(R)$ is any representation matrix belonging to $\Gamma^{(5)}$, results in a matrix that contains $\Gamma^{(3)}(R)$ and $\Gamma^{(1)}(R)$ along the main diagonal. $\Gamma^{(5)}$ is then said to be a reducible representation which reduces into $\Gamma^{(3)}$ and $\Gamma^{(1)}$. Symbolically,

$$\Gamma^{(5)} = \Gamma^{(1)} + \Gamma^{(3)}, \tag{3.3-6}$$

and since the character is invariant under a similarity transformation, $\chi^{(5)}(R) = \chi^{(1)}(R) + \chi^{(3)}(R)$. [Here, of course, the symbol + has the usual meaning of addition].

A general representation Γ is reducible if there exists a matrix M such that for each representation matrix $\Gamma(R)$, $M\Gamma(R)M^{-1}$ is a matrix consisting of blocks along the main diagonal and the dimensions of the blocks are smaller than the dimension of Γ. The reduction of Γ is expressed by writing

$$\Gamma = \sum_i a_i \Gamma^{(i)} \tag{3.3-7}$$

where each a_i is a positive integer that indicates how many times the representation matrix $\Gamma^{(i)}(R)$ appears along the main diagonal of the transformed matrix $M\Gamma(R)M^{-1}$ as is shown schematically:

$$M\Gamma(R)M^{-1} = \begin{pmatrix} \cdot\,\boxed{} & & \\ & \boxed{\Gamma^{(i)}(R)} & \\ & & \boxed{}\,\ddots \end{pmatrix}. \tag{3.3-8}$$

An irreducible representation is one whose matrices cannot be reduced in the sense described above by a similarity transformation. Such representations are of central importance in the application of group theory to physical problems by virtue of a number of theorems which we shall now state without proof.

1. Although a group may have an infinite number of representations, the number of inequivalent, irreducible representations of a group is severely restricted; it is equal to the number of classes. Furthermore, there is a restriction on the dimensions of the irreducible representations. If l_i is the dimension

of the ith irreducible representation and h the order of the group, then

$$\sum_i l_i^2 = h. \tag{3.3-9}$$

The group D_3 has three classes; hence there are three irreducible representations. Also, for D_3, $h = 6$. Condition (3.3-9) can only be satisfied by

$$\sum_i l_i^2 = 1^2 + 1^2 + 2^2 = 6. \tag{3.3-10}$$

Hence, of the three irreducible representations in D_3, two are one-dimensional and one is two-dimensional.

2. For inequivalent, irreducible representations of a group there is an orthogonality theorem (also known as the *great orthogonality theorem*) which states that

$$\sum_R \Gamma_{\mu\nu}^{(i)}(R)^* \Gamma_{\alpha\beta}^{(j)}(R) = \frac{h}{l_i} \delta_{ij} \delta_{\mu\alpha} \delta_{\nu\beta}. \tag{3.3-11}$$

In (3.3-11) the sum is taken over all elements R of the group; i, j are labels on the irreducible representations.

3. If a matrix commutes with all matrices of an irreducible representation, the matrix must be a multiple of the unit matrix (that is, a constant matrix). Also, if the only matrix that commutes with all matrices of a representation is a constant matrix, then the representation is irreducible. This is known as Schur's lemma.

Henceforth, an irreducible representation will be understood to mean an inequivalent, irreducible representation.

The orthogonality theorem (3.3-11) provides a method whereby a representation Γ may be reduced without actually constructing the matrix M and performing the operation $M\Gamma M^{-1}$ as was done in (3.3-4) and (3.3-5). If $\mu = \nu$ and $\alpha = \beta$, (3.3-11) becomes

$$\sum_R \Gamma_{\mu\mu}^{(i)}(R)^* \Gamma_{\alpha\alpha}^{(j)}(R) = \frac{h}{l_i} \delta_{ij} \delta_{\mu\alpha}. \tag{3.3-12}$$

Summing over μ and α,

$$\sum_R \left(\sum_\mu \Gamma_{\mu\mu}^{(i)}(R)^* \right) \left(\sum_\alpha \Gamma_{\alpha\alpha}^{(j)}(R) \right) = \sum_R \chi^{(i)}(R)^* \chi^{(j)}(R)$$

$$= \sum_{\mu=1}^{l_i} \sum_{\alpha=1}^{l_j} \frac{h}{l_i} \delta_{ij} \delta_{\mu\alpha} = \frac{h}{l_i} \delta_{ij} \sum_{\mu=1}^{l_i} \sum_{\alpha=1}^{l_j} \delta_{\mu\alpha}.$$

But

$$\sum_{\mu=1}^{l_i} \sum_{\alpha=1}^{l_j} \delta_{\mu\alpha} = l_i \text{ or } l_j \qquad \text{(whichever is smaller).}$$

Therefore,

$$\sum_R \chi^{(i)}(R)^* \chi^{(j)}(R) = h\,\delta_{ij}, \tag{3.3-13}$$

and we see that the characters of irreducible representations also satisfy an orthogonality theorem. Now let Γ be a reducible representation and let $\chi(R)$ be the character of $\Gamma(R)$. Since the reduction of Γ is, by definition, accomplished by a similarity transformation and the trace of a matrix is invariant under a similarity transformation, then

$$\chi(R) = \sum_j a_j \chi^{(j)}(R). \tag{3.3-14}$$

The right-hand side of (3.3-14) is simply the trace of the transformed matrix. It is assumed that the reduction has been carried out as far as possible so that the reducible representation Γ has been decomposed into irreducible representations. Multiplying (3.3-14) on the left by $\chi^{(i)}(R)^*$ and summing over R, we obtain

$$\sum_R \chi^{(i)}(R)^* \chi(R) = \sum_R \sum_j a_j \chi^{(i)}(R)^* \chi^{(j)}(R). \tag{3.3-15}$$

But, by the orthogonality theorem (3.3-13), the right-hand side is just ha_i. We then have the important result that

$$a_i = \frac{1}{h} \sum_R \chi^{(i)}(R)^* \chi(R). \tag{3.3-16}$$

Assuming that the characters are known, the value of the integer a_i, which is the number of times the irreducible representation $\Gamma^{(i)}$ is contained in Γ, may now be calculated. The utility of (3.3-16) is considerably enhanced by the existence of character tables (Table 5.2) which list the characters of the irreducible representations belonging to various groups. Regardless of the method employed to reduce a representation into its irreducible components, the result will be the same, except perhaps for the order in which the irreducible representation matrices appear along the main diagonal.

The result expressed by (3.3-16) may also be employed to show that if two representations have the same set of characters, they are equivalent as was mentioned in Section 3.2. This follows from the fact that the coefficients a_i are the same for the two equivalent representations.

It is instructive to show how the characters of the irreducible representations are obtained in a simple case. For the group D_3, it has already been shown (3.3-10) that there are three irreducible representations with dimensions 1, 1, and 2. An obvious one-dimensional representation is $\Gamma^{(1)}(R) = \chi^{(1)}(R) = 1$ for every element R. This is the unit representation (3.2-1). For the second one-dimensional representation, we have

$$\chi^{(2)}(E) = 1, \qquad \chi^{(2)}(A) = \chi^{(2)}(B) = \chi^{(2)}(C), \qquad \chi^{(2)}(D) = \chi^{(2)}(F). \quad (3.3\text{-}17)$$

These relations are based on (3.1-18) and the theorem that elements belonging to the same class have the same character. To satisfy the orthogonality theorem (3.3-13) and the requirement that the absolute value of the character of a one-dimensional representation matrix be equal to one (unitarity requirement), we have

$$\chi^{(2)}(E) = 1, \qquad \chi^{(2)}(A) = \chi^{(2)}(B) = \chi^{(2)}(C) = -1, \qquad \chi^{(2)}(D) = \chi^{(2)}(F) = 1 \quad (3.3\text{-}18)$$

which is identical to (3.2-2). Finally, for the two-dimensional representation $\chi(E) = 2$; imposing the orthogonality condition (3.3-13) leads to

$$\chi^{(3)}(E) = 2, \qquad \chi^{(3)}(A) = \chi^{(3)}(B) = \chi^{(3)}(C) = 0, \qquad \chi^{(3)}(D) = \chi^{(3)}(F) = -1. \quad (3.3\text{-}19)$$

All the characters may be collected and displayed as in Table 3.2.

TABLE 3.2

Character Table for D_3

	E	A	B	C	D	F
$\Gamma^{(1)}$	1	1	1	1	1	1
$\Gamma^{(2)}$	1	−1	−1	−1	1	1
$\Gamma^{(3)}$	2	0	0	0	−1	−1

Previously it was shown that $\Gamma^{(5)}$ [Eq. (3.2-5)] is reducible. To accomplish the reduction by means of (3.3-16) we first enumerate the characters of the $\Gamma^{(5)}$ matrices; these are

$$\chi^{(5)}(E) = 3, \qquad \chi^{(5)}(A) = \chi^{(5)}(B) = \chi^{(5)}(C) = 1, \qquad \chi^{(5)}(D) = \chi^{(5)}(F) = 0.$$

Using (3.3-16) and Table 3.2 it is found that

$$a_1 = 1, \qquad a_2 = 0, \qquad a_3 = 1. \quad (3.3\text{-}20)$$

Hence

$$\Gamma^{(5)} = \Gamma^{(1)} + \Gamma^{(3)}, \quad (3.3\text{-}21)$$

as in (3.3-6).

In the case of Abelian groups the number of classes is equal to the order of the group. Hence the number of irreducible representations is equal to the order of the group. To satisfy (3.3-9) all the irreducible representations of Abelian groups must be one-dimensional.

It has been stated that a representation Γ of a group G is also a representation of any subgroup G_{sub} of G. The subgroup has its own irreducible representations; let these be denoted by Γ_{sub}. The representation Γ is generally reducible in terms of Γ_{sub} even if Γ is an irreducible representation of G.

Suppose each element of a group is multiplied on the left by some particular element of the group. In D_3, for example, let $E, A, \ldots, F$ be multiplied by A. From the multiplication table of the group (Table 3.1) one finds

$$A(E, A, B, C, D, F) = (A, E, D, F, B, C). \tag{3.3-22}$$

This relation can be expressed in matrix form if we regard $(E, A, \ldots, F)$ and $(A, E, \ldots, C)$ as row vectors. Thus

$$(E, A, B, C, D, F)\begin{pmatrix} 0 & 1 & 0 & 0 & 0 & 0 \\ 1 & 0 & 0 & 0 & 0 & 0 \\ 0 & 0 & 0 & 0 & 1 & 0 \\ 0 & 0 & 0 & 0 & 0 & 1 \\ 0 & 0 & 1 & 0 & 0 & 0 \\ 0 & 0 & 0 & 1 & 0 & 0 \end{pmatrix} = (A, E, D, F, B, C). \tag{3.3-23}$$

By using all the elements of the group as multipliers we generate a set of matrices with dimensions equal to the order of the group. Such a set of matrices form a representation known as the *regular representation* Γ_{reg}.

There is an easy way to obtain the regular representation. This is accomplished by first constructing the multiplication table of the elements with the inverse elements. For D_3, with the help of Table 3.1, we obtain the following tabulation:

	E^{-1}	A^{-1}	B^{-1}	C^{-1}	D^{-1}	F^{-1}
E	E	A	B	C	F	D
A	A	E	D	F	C	B
B	B	F	E	D	A	C
C	C	D	F	E	B	A
D	D	C	A	B	E	F
F	F	B	C	A	D	E

To find $\Gamma_{\text{reg}}(A)$ it is merely necessary to replace A by 1 and all other elements by zero. This immediately gives the 6×6 matrix in (3.3-23). Similarly, to find $\Gamma_{\text{reg}}(B)$ all the B's are replaced by 1's and the other elements by zero, etc.

It is evident from this construction that the characters of the regular representation are given by

$$\chi_{\text{reg}}(E) = h, \qquad \chi_{\text{reg}}(R) = 0 \qquad (R \neq E). \tag{3.3-24}$$

The only representation matrix which has a nonvanishing character is $\chi_{\text{reg}}(E)$ and the character is equal to the order of the group. When the regular representation is decomposed into the irreducible representations of the group [by means of (3.3-16)] it is found that each irreducible representation appears a number of times equal to the dimensionality of the irreducible representation. For D_3 we have

$$\Gamma_{\text{reg}} = \Gamma^{(1)} + \Gamma^{(2)} + 2\Gamma^{(3)}. \tag{3.3-25}$$

3.4 Basis Functions

The representations of a group are intimately connected with sets of functions called *basis functions*. A few examples will help to establish the central idea. Let

$$\varphi_1(\mathbf{r}) = x, \qquad \varphi_2(\mathbf{r}) = y. \tag{3.4-1}$$

We now inquire as to how these functions are altered under the coordinate transformations $E, A, \ldots, F$ which are elements of the group D_3. Under any coordinate transformation described by $\mathbf{r}' = R\mathbf{r}$, a function $f(\mathbf{r})$ transforms in accordance with

$$P_R f(\mathbf{r}) = f(R^{-1}\mathbf{r}) \tag{3.4-2}$$

as in (2.1-10). For $R = E$, we obtain, trivially,

$$P_E\varphi_1(\mathbf{r}) = \varphi_1(\mathbf{r}) = x, \qquad P_E\varphi_2(\mathbf{r}) = \varphi_2(\mathbf{r}) = y$$

or, in matrix form,

$$P_E(\varphi_1, \varphi_2) = (\varphi_1, \varphi_2)\begin{pmatrix} 1 & 0 \\ 0 & 1 \end{pmatrix} \tag{3.4-3}$$

in which (φ_1, φ_2) is a row matrix. For $R = A$, from (3.1-3),

$$A = \begin{pmatrix} 1 & 0 & 0 \\ 0 & -1 & 0 \\ 0 & 0 & -1 \end{pmatrix}, \qquad A^{-1} = \begin{pmatrix} 1 & 0 & 0 \\ 0 & -1 & 0 \\ 0 & 0 & -1 \end{pmatrix}.$$

Therefore

$$P_A f(\mathbf{r}) = P_A f(x, y, z) = f(x, -y, -z),$$

$$P_A \varphi_1(\mathbf{r}) = P_A x = x = \varphi_1(\mathbf{r}), \qquad P_A \varphi_2(\mathbf{r}) = P_A y = -y = -\varphi_2(\mathbf{r})$$

or

$$P_A(\varphi_1, \varphi_2) = (\varphi_1, \varphi_2)\begin{pmatrix} 1 & 0 \\ 0 & -1 \end{pmatrix}. \tag{3.4-4}$$

Proceeding in the same fashion with the rest of the coordinate transformations in D_3, it is found that

$$P_B(\varphi_1, \varphi_2) = (\varphi_1, \varphi_2)\begin{pmatrix} -\frac{1}{2} & \frac{1}{2}\sqrt{3} \\ \frac{1}{2}\sqrt{3} & \frac{1}{2} \end{pmatrix}, \tag{3.4-5}$$

$$P_C(\varphi_1, \varphi_2) = (\varphi_1, \varphi_2)\begin{pmatrix} -\frac{1}{2} & -\frac{1}{2}\sqrt{3} \\ -\frac{1}{2}\sqrt{3} & \frac{1}{2} \end{pmatrix}, \tag{3.4-6}$$

$$P_D(\varphi_1, \varphi_2) = (\varphi_1, \varphi_2)\begin{pmatrix} -\frac{1}{2} & \frac{1}{2}\sqrt{3} \\ -\frac{1}{2}\sqrt{3} & -\frac{1}{2} \end{pmatrix}, \tag{3.4-7}$$

$$P_F(\varphi_1, \varphi_2) = (\varphi_1, \varphi_2)\begin{pmatrix} -\frac{1}{2} & -\frac{1}{2}\sqrt{3} \\ \frac{1}{2}\sqrt{3} & -\frac{1}{2} \end{pmatrix}. \tag{3.4-8}$$

Examination of (3.4-3)–(3.4-8) reveals that the 2×2 matrices are precisely those belonging to $\Gamma^{(3)}$ in (3.2-3). In other words, the two functions $\varphi_1(\mathbf{r}) = x$ and $\varphi_2(\mathbf{r}) = y$ are capable of generating the $\Gamma^{(3)}$ representation of D_3 by successive application of the operators $P_E, P_A, \ldots, P_F$. The pair of functions $\varphi_1(\mathbf{r})$ and $\varphi_2(\mathbf{r})$ are therefore said to be basis functions for the $\Gamma^{(3)}$ representation of D_3. In other forms of expression, $\varphi_1(\mathbf{r})$ and $\varphi_2(\mathbf{r})$ are said to transform according to $\Gamma^{(3)}$ or to belong to $\Gamma^{(3)}$. The general statement is that a set of linearly independent functions‡ $\varphi_1^{(j)}(\mathbf{r}), \varphi_2^{(j)}(\mathbf{r}), \ldots, \varphi_n^{(j)}(\mathbf{r})$ are basis functions for the $\Gamma^{(j)}$ representation of a group if

$$P_R \varphi_k^{(j)}(\mathbf{r}) = \sum_{\lambda=1}^{n} \varphi_\lambda^{(j)}(\mathbf{r}) \Gamma_{\lambda k}^{(j)}(R) \tag{3.4-9}$$

for all the elements R of the group. One may also speak of members of the same basis set as *partners*.

The important point here is that members (partners) of a basis set transform among themselves. Specifically, when P_R is applied to any member of a basis set, the general result is a linear combination of the members of the set. The

‡ A set of functions φ_k is linearly independent if $\sum_k a_k \varphi_k = 0$ only when all the coefficients a_k are zero.

coefficients in the linear combinations are matrix elements of representation matrices of the group whose elements are the operators P_R (or the group whose elements are the coordinate transformations R, since the two groups are isomorphic). From (3.4-9) it is seen that each member in the basis set is associated with a particular column in the representation matrices so that the number of functions in the basis set is equal to the dimension of the representation.

There is no single unique basis set for a given representation, but neither can the set be chosen arbitrarily. As an example, another basis set for $\Gamma^{(3)}$ is the set $\varphi_1(\mathbf{r}) = yz$ and $\varphi_2(\mathbf{r}) = -xz$. Other examples of basis sets are

$$\Gamma^{(2)}: \varphi(\mathbf{r}) = z, \tag{3.4-10}$$

$$\begin{aligned} \Gamma^{(6)}: \varphi_1(\mathbf{r}) &= x^2 - y^2, & \varphi_2(\mathbf{r}) &= -2xy; \\ \varphi_3(\mathbf{r}) &= yz, & \varphi_4(\mathbf{r}) &= -xz \\ \varphi_5(\mathbf{r}) &= x^2 + y^2 + z^2, & \varphi_6(\mathbf{r}) &= 2z^2 - x^2 - y^2. \end{aligned} \tag{3.4-11}$$

It may also be verified that z^2 and $x^2 + y^2$ are each basis functions for $\Gamma^{(1)}$. This is nothing more than a statement that z^2 and $x^2 + y^2$ are each invariant under the group of coordinate transformations belonging to D_3. Indeed, any function which belongs to $\Gamma^{(1)}$ (the unit representation) is an invariant function since, for $\Gamma^{(1)}$,

$$P_R\varphi(\mathbf{r}) = \varphi(\mathbf{r})$$

for all elements R belonging to the group.

The basis set for a representation consists of linearly independent functions. These can always be chosen so as to be orthonormal, i.e.,

$$\langle \varphi_l^{(j)}(\mathbf{r}) | \varphi_k^{(j)}(\mathbf{r}) \rangle = \delta_{lk}. \tag{3.4-12}$$

From (3.4-9),

$$\langle \varphi_l^{(j)}(\mathbf{r}) | P_R | \varphi_k^{(j)}(\mathbf{r}) \rangle = \sum_\lambda \langle \varphi_l^{(j)}(\mathbf{r}) | \varphi_\lambda^{(j)}(\mathbf{r}) \rangle \Gamma_{\lambda k}^{(j)}(R) = \Gamma_{lk}^{(j)}(R). \tag{3.4-13}$$

Since P_R has been shown to be a unitary operator, $\Gamma_{lk}^{(j)}(R)$ is a matrix element of a unitary representation matrix $\Gamma^{(j)}(R)$. Thus representations generated by orthonormal basis functions are unitary. This is consistent with the previous statement in Section 3.2 that any representation can be transformed into a unitary representation. The assumption of unitary representations then implies that the corresponding basis sets are orthonormal.

According to (3.4-12), two basis functions, $\varphi_l^{(j)}(\mathbf{r})$ and $\varphi_k^{(j)}(\mathbf{r})$, that belong to different rows of the same (unitary) representation are orthogonal. This statement holds for both reducible and irreducible representations; however, in the latter case the theorem may be extended. Let $\varphi_k^{(j)}$ and $\varphi_{k'}^{(j')}$ belong to two different irreducible representations. We have

$$\langle P_R\varphi_k^{(j)}|P_R\varphi_{k'}^{(j')}\rangle = \langle\varphi_k^{(j)}|P_R{}^\dagger P_R|\varphi_{k'}^{(j')}\rangle = \langle\varphi_k^{(j)}|\varphi_{k'}^{(j')}\rangle, \tag{3.4-14}$$

since P_R is unitary. But, from (3.4-9),

$$\langle P_R\varphi_k^{(j)}|P_R\varphi_{k'}^{(j')}\rangle = \sum_{\lambda\lambda'} \Gamma_{\lambda k}^{(j)}(R)^*\Gamma_{\lambda' k'}^{(j')}(R)\langle\varphi_\lambda^{(j)}|\varphi_{\lambda'}^{(j')}\rangle. \tag{3.4-15}$$

Comparing (3.4-14) and (3.4-15),

$$\langle\varphi_k^{(j)}|\varphi_{k'}^{(j')}\rangle = \sum_{\lambda\lambda'} \Gamma_{\lambda k}^{(j)}(R)^*\Gamma_{\lambda' k'}^{(j')}(R)\langle\varphi_\lambda^{(j)}|\varphi_{\lambda'}^{(j')}\rangle. \tag{3.4-16}$$

On summing over R, the left side becomes

$$\sum_R \langle\varphi_k^{(j)}|\varphi_{k'}^{(j')}\rangle = h\langle\varphi_k^{(j)}|\varphi_{k'}^{(j')}\rangle, \tag{3.4-17}$$

and the right side, by virtue of the orthogonality theorem (3.3-11), is

$$\begin{aligned}\sum_R\sum_{\lambda\lambda'} \Gamma_{\lambda k}^{(j)}(R)^*\Gamma_{\lambda' k'}^{(j')}(R)\langle\varphi_\lambda^{(j)}|\varphi_{\lambda'}^{(j')}\rangle &= \sum_{\lambda\lambda'} \frac{h}{l_j}\,\delta_{jj'}\,\delta_{\lambda\lambda'}\,\delta_{kk'}\langle\varphi_\lambda^{(j)}|\varphi_{\lambda'}^{(j')}\rangle \\ &= \frac{h}{l_j}\,\delta_{jj'}\,\delta_{kk'}\sum_{\lambda=1}^{l_j}\langle\varphi_\lambda^{(j)}|\varphi_\lambda^{(j')}\rangle \\ &= h\,\delta_{jj'}\,\delta_{kk'}.\end{aligned} \tag{3.4-18}$$

We then have, from (3.4-17) and (3.4-18), that

$$\langle\varphi_k^{(j)}|\varphi_{k'}^{(j')}\rangle = \delta_{jj'}\,\delta_{kk'} \tag{3.4-19}$$

which says that two functions belonging to two different irreducible representations are orthogonal. It is now possible to give a more complete statement concerning the orthogonality of basis functions: Two functions that belong to different irreducible representations or to different columns of the same unitary, but not necessarily irreducible, representation are orthogonal.

3.5 Projection Operators

Let us return to (3.3-21) where it was found that

$$\Gamma^{(5)} = \Gamma^{(1)} + \Gamma^{(3)}. \tag{3.5-1}$$

It is pertinent to inquire what relations, if any, exist between the basis functions of $\Gamma^{(5)}$ and the basis functions of the component irreducible representations $\Gamma^{(1)}$ and $\Gamma^{(3)}$. Suppose (f_1, f_2, f_3) is the set of (normalized) basis functions of $\Gamma^{(5)}$ [Eq. (3.2-5)]. It follows, from the definition of basis functions (3.4-9), that

$$P_E(f_1, f_2, f_3) = (f_1, f_2, f_3)\begin{pmatrix}1 & 0 & 0\\ 0 & 1 & 0\\ 0 & 0 & 1\end{pmatrix} = (f_1, f_2, f_3),$$

$$P_A(f_1, f_2, f_3) = (f_1, f_2, f_3)\begin{pmatrix}1 & 0 & 0\\ 0 & 0 & 1\\ 0 & 1 & 0\end{pmatrix} = (f_1, f_3, f_2),$$

$$P_B(f_1, f_2, f_3) = (f_1, f_2, f_3)\begin{pmatrix}0 & 0 & 1\\ 0 & 1 & 0\\ 1 & 0 & 0\end{pmatrix} = (f_3, f_2, f_1),$$

$$P_C(f_1, f_2, f_3) = (f_1, f_2, f_3)\begin{pmatrix}0 & 1 & 0\\ 1 & 0 & 0\\ 0 & 0 & 1\end{pmatrix} = (f_2, f_1, f_3), \tag{3.5-2}$$

$$P_D(f_1, f_2, f_3) = (f_1, f_2, f_3)\begin{pmatrix}0 & 0 & 1\\ 1 & 0 & 0\\ 0 & 1 & 0\end{pmatrix} = (f_2, f_3, f_1),$$

$$P_F(f_1, f_2, f_3) = (f_1, f_2, f_3)\begin{pmatrix}0 & 1 & 0\\ 0 & 0 & 1\\ 1 & 0 & 0\end{pmatrix} = (f_3, f_1, f_2).$$

We now construct a set of three new orthonormal functions which are linear combinations of f_1, f_2, and f_3, as:

$$g_1 = \frac{1}{\sqrt{3}}(f_1 + f_2 + f_3), \qquad g_2 = \frac{1}{\sqrt{6}}(2f_1 - f_2 - f_3), \qquad g_3 = \frac{1}{\sqrt{2}}(-f_2 + f_3). \tag{3.5-3}$$

Using (3.5-2) it is seen that

$$P_E g_1 = P_A g_1 = \cdots = P_F g_1 = g_1. \tag{3.5-4}$$

Therefore, g_1 is a basis function for the unit representation $\Gamma^{(1)}$. Similarly,

$$P_A(g_2, g_3) = (g_2, -g_3) = (g_2, g_3)\begin{pmatrix}1 & 0\\ 0 & -1\end{pmatrix}, \tag{3.5-5}$$

$$\begin{aligned} P_B(g_2, g_3) &= (-\tfrac{1}{2}g_2 + \tfrac{1}{2}\sqrt{3}g_3, \tfrac{1}{2}\sqrt{3}g_2 + \tfrac{1}{2}g_3) \\ &= (g_2, g_3)\begin{pmatrix}-\frac{1}{2} & \frac{1}{2}\sqrt{3}\\ \frac{1}{2}\sqrt{3} & \frac{1}{2}\end{pmatrix}. \end{aligned} \tag{3.5-6}$$

Proceeding in the same fashion with the operators $P_C, \ldots, P_F$, it is established that (g_2, g_3) is a basis set for $\Gamma^{(3)}$ [Eq. (3.2-3)].

The general result is that it is possible to construct basis functions for the irreducible representations contained in a reducible representation by forming linear combinations of the basis functions of the reducible representation. The formal procedure is based on the use of projection operators which will now be defined.

Let $\varphi_k^{(j)}$ be a basis function belonging to the jth irreducible representation $\Gamma^{(j)}$. Then, by definition,

$$P_R \varphi_k^{(j)} = \sum_{\lambda=1}^{l_j} \varphi_\lambda^{(j)} \Gamma_{\lambda k}^{(j)}(R). \tag{3.5-7}$$

Multiplying through by $\Gamma_{\lambda' k'}^{(j')}(R)^*$ and summing over R, we obtain

$$\begin{aligned} \sum_R \Gamma_{\lambda' k'}^{(j')}(R)^* P_R \varphi_k^{(j)} &= \sum_R \sum_{\lambda=1}^{l_j} \varphi_\lambda^{(j)} \Gamma_{\lambda' k'}^{(j')}(R)^* \Gamma_{\lambda k}^{(j)}(R) \\ &= \frac{h}{l_j} \sum_\lambda \varphi_\lambda^{(j)} \delta_{jj'} \delta_{\lambda\lambda'} \delta_{kk'} \\ &= \frac{h}{l_j} \varphi_{\lambda'}^{(j)} \delta_{jj'} \delta_{kk'}, \end{aligned} \tag{3.5-8}$$

in which the second and third equalities are a consequence of the orthogonality theorem (3.3-11). We may now define an operator

$$\rho_{\lambda k}^{(j)} = \frac{l_j}{h} \sum_R \Gamma_{\lambda k}^{(j)}(R)^* P_R \tag{3.5-9}$$

having the property

$$\rho_{\lambda k}^{(j)} \varphi_l^{(i)} = \varphi_\lambda^{(i)} \delta_{ij} \delta_{lk} \tag{3.5-10}$$

or, when $i = j$ and $l = k$,

$$\rho_{\lambda k}^{(j)} \varphi_k^{(j)} = \varphi_\lambda^{(j)}. \tag{3.5-11}$$

Thus, given one member $\varphi_k^{(j)}$ of a basis set for an irreducible representation, it is possible to generate other members of the set by means of the operator $\rho_{\lambda k}^{(j)}$.

When $\lambda = k$, (3.5-10) becomes

$$\rho_{kk}^{(j)} \varphi_l^{(i)} = \varphi_k^{(i)} \delta_{ij} \delta_{lk} \tag{3.5-12a}$$

or

$$\rho_{kk}^{(j)} \varphi_k^{(j)} = \varphi_k^{(j)}, \tag{3.5-12b}$$

$$\rho_{kk}^{(j)} \varphi_l^{(i)} = 0, \qquad i \neq j, \quad l \neq k. \tag{3.5-12c}$$

From (3.5-12b) it is seen that $\varphi_k^{(j)}$ is an eigenfunction of $\rho_{kk}^{(j)}$ with the eigenvalue equal to one. Also

$$(\rho_{kk}^{(j)})^2\varphi_k^{(j)} = \rho_{kk}^{(j)}\varphi_k^{(j)} = \varphi_k^{(j)}, \tag{3.5-13}$$

so that

$$(\rho_{kk}^{(j)})^2 = \rho_{kk}^{(j)} = \frac{l_j}{h}\sum_R \Gamma_{kk}^{(j)}(R)^* P_R. \tag{3.5-14}$$

$\rho_{kk}^{(j)}$ is known as a projection operator; operators that obey a relation of the type $O^2 = O$ are said to be *idempotent*. From (3.5-14) we see that projection operators are idempotent.

If one now sums $\rho_{kk}^{(j)}$ over k, then, from (3.5-14),

$$\sum_k \rho_{kk}^{(j)} \equiv \rho^{(j)} = \frac{l_j}{h}\sum_R\sum_k \Gamma_{kk}^{(j)}(R)^* P_R = \frac{l_j}{h}\sum_R \chi^{(j)}(R)^* P_R \tag{3.5-15}$$

where the second equality arises from definition (3.2-7); $\rho^{(j)}$ is also a projection operator.

To illustrate the use of (3.5-15) we will generate the basis functions for $\Gamma^{(3)}$ in the decomposition of $\Gamma^{(5)}$ as given by (3.5-1). From (3.2-3),

$$\chi^{(3)}(E) = 2, \qquad \chi^{(3)}(A) = \chi^{(3)}(B) = \chi^{(3)}(C) = 0, \qquad \chi^{(3)}(D) = \chi^{(3)}(F) = -1. \tag{3.5-16}$$

Therefore,

$$\rho^{(3)} f_1 = \tfrac{2}{6}[2P_E f_1 - P_D f_1 - P_F f_1] = \tfrac{1}{3}[2f_1 - f_2 - f_3] \equiv h_1 \tag{3.5-17a}$$

from (3.5-2). Similarly,

$$\rho^{(3)} f_2 = \tfrac{1}{3}[2f_2 - f_3 - f_1] \equiv h_2, \tag{3.5-17b}$$

$$\rho^{(3)} f_3 = \tfrac{1}{3}[2f_3 - f_1 - f_2] \equiv h_3. \tag{3.5-17c}$$

The three functions h_1, h_2, and h_3 are not independent since $h_1 + h_2 + h_3 = 0$. Hence there are only two independent functions and these may be constructed in an infinite number of ways. The choice corresponding to (3.5-3), after normalization, is $g_2 = h_1$ and $g_3 = h_3 - h_2$.

Another useful property of projection operators is associated with the "the resolution of the identity." If $\rho^{(j)}$ is summed over j,

$$\sum_j \rho^{(j)}\varphi_l^{(i)} = \sum_{jk} \rho_{kk}^{(j)}\varphi_l^{(i)} = \sum_{jk} \varphi_k^{(i)}\,\delta_{ij}\,\delta_{lk}$$

from (3.5-12a). But,

$$\sum_{jk} \varphi_k^{(i)}\,\delta_{ij}\delta_{ik} = \varphi_l^{(i)}.$$

Therefore,

$$\sum_j \rho^{(j)} = P_E. \tag{3.5-18}$$

For example, in D_3, with the help of the characters given in Table 3.2, it is found that

$$\begin{aligned}\sum_j \rho^{(j)} &= \rho^{(1)} + \rho^{(2)} + \rho^{(3)} \\ &= \frac{l_1}{h}\sum_R \chi^{(1)}(R)P_R + \frac{l_2}{h}\sum_R \chi^{(2)}(R)P_R + \frac{l_3}{h}\sum_R \chi^{(3)}(R)P_R \\ &= \tfrac{1}{6}[P_E + P_A + P_B + P_C + P_D + P_F] \\ &\quad + \tfrac{1}{6}[P_E - P_A - P_B - P_C + P_D + P_F] + \tfrac{2}{6}[2P_E - P_D - P_F] = P_E.\end{aligned} \tag{3.5-19}$$

Let us now suppose that, in (3.5-12a), $\varphi_l^{(i)}$ is replaced by an arbitrary function φ. In view of (3.5-12b) and (3.5-12c), we shall assume that

$$\rho_{kk}^{(j)}\varphi = \varphi_k^{(j)}, \tag{3.5-20}$$

that is, the operator $\rho_{kk}^{(j)}$ can only project $\varphi_k^{(j)}$ (which belongs to the kth row of the irreducible representation $\Gamma^{(j)}$) and nothing else. The relationship between φ and $\varphi_k^{(j)}$ is obtained by summing (3.5-20) over j and k. Thus

$$\sum_{jk} \rho_{kk}^{(j)}\varphi = \sum_j \rho^{(j)}\varphi = P_E\varphi = \varphi. \tag{3.5-21}$$

Comparing (3.5-20) and (3.5-21), it is seen that

$$\varphi = \sum_{jk} \varphi_k^{(j)}, \tag{3.5-22}$$

which shows that an arbitrary function can be expressed as a sum of the basis functions for the irreducible representations of the group.‡ It is instructive to regard (3.5-22) in the geometric (or Hilbert space) sense. From this point of view the $\varphi_k^{(j)}$ are basis vectors that span an orthogonal (more generally, a unitary) space of dimensionality equal to $\sum_j l_j$. An arbitrary function φ is a vector in this space with "components" $\varphi_k^{(j)}$.

As a simple illustration of (3.5-22) consider the coordinate transformations

$$\mathbf{r}' = E\mathbf{r} = \mathbf{r}, \qquad \mathbf{r}' = i\mathbf{r} = -\mathbf{r}, \tag{3.5-23}$$

‡ A general proof of (3.5-22) may be constructed as follows (Wigner, 1959): Consider an arbitrary function φ and a group G consisting of elements $R_1, R_2, \ldots, R_h$. Construct the functions $f_1 = P_{R_1}\varphi$, $f_2 = P_{R_2}\varphi, \ldots, f_h = P_{R_h}\varphi$ and discard enough functions until the set $\{f_1, f_2, \ldots, f_g\}$ is linearly independent. But $\{f_1, f_2, \ldots, f_g\}$ is a basis set for a (generally reducible) representation Γ of G; hence let M be the matrix that reduces Γ into $\sum_j a_j\Gamma^{(j)}$. The basis functions $\varphi_k^{(j)}$ belonging to the irreducible representations $\Gamma^{(j)}$ are linear combinations of $f_1, f_2, \ldots, f_i, \ldots, f_g$. Since M has an inverse, each f_i is a linear combination of the $\varphi_k^{(j)}$ which then implies that $\varphi(=P_{R_i}^{-1} f_i)$ is also expressible as a linear combination of the $\varphi_k^{(j)}$.

where E is the identity transformation (three-dimensional unit matrix) and i an inversion:

$$i = \begin{pmatrix} -1 & 0 & 0 \\ 0 & -1 & 0 \\ 0 & 0 & -1 \end{pmatrix}. \tag{3.5-24}$$

The two operators E and i are elements of the Abelian group C_i whose character table is

C_i	E	i
$\Gamma^{(1)}$	1	1
$\Gamma^{(2)}$	1	-1

(3.5-25)

For an arbitrary function $f(\mathbf{r})$,

$$P_E f(\mathbf{r}) = f(\mathbf{r}), \qquad P_i f(\mathbf{r}) = f(i^{-1}\mathbf{r}) = f(i\mathbf{r}) = f(-\mathbf{r}).$$

Let us now project basis functions for $\Gamma^{(1)}$ and $\Gamma^{(2)}$ using $f(\mathbf{r})$ and (3.5-15):

$$\begin{aligned} f^{(1)}(\mathbf{r}) &\equiv \rho^{(1)} f(\mathbf{r}) = \tfrac{1}{2}[P_E + P_i] f(\mathbf{r}) = \tfrac{1}{2}[f(\mathbf{r}) + f(-\mathbf{r})], \\ f^{(2)}(\mathbf{r}) &= \rho^{(2)} f(\mathbf{r}) = \tfrac{1}{2}[P_E - P_i] f(\mathbf{r}) = \tfrac{1}{2}[f(\mathbf{r}) - f(-\mathbf{r})], \end{aligned} \tag{3.5-26}$$

and

$$f^{(1)}(\mathbf{r}) + f^{(2)}(\mathbf{r}) = f(\mathbf{r}). \tag{3.5-27}$$

It should be remarked that a function generated by a projection operator is usually not normalized; when necessary, the normalization is carried out separately.

3.6 Product Representations

Consider the two matrices

$$A = \begin{pmatrix} a_{11} & a_{12} \\ a_{21} & a_{22} \end{pmatrix} \quad \text{and} \quad B = \begin{pmatrix} b_{11} & b_{12} \\ b_{21} & b_{22} \end{pmatrix}.$$

The direct (or Kronecker) product of A and B, written $A \times B$, is given by

$$A \times B = \begin{pmatrix} a_{11}\begin{pmatrix} b_{11} & b_{12} \\ b_{21} & b_{22} \end{pmatrix} & a_{12}\begin{pmatrix} b_{11} & b_{12} \\ b_{21} & b_{22} \end{pmatrix} \\ a_{21}\begin{pmatrix} b_{11} & b_{12} \\ b_{21} & b_{22} \end{pmatrix} & a_{22}\begin{pmatrix} b_{11} & b_{12} \\ b_{21} & b_{22} \end{pmatrix} \end{pmatrix}$$

$$= \begin{pmatrix} a_{11}b_{11} & a_{11}b_{12} & a_{12}b_{11} & a_{12}b_{12} \\ a_{11}b_{21} & a_{11}b_{22} & a_{12}b_{21} & a_{12}b_{22} \\ a_{21}b_{11} & a_{21}b_{12} & a_{22}b_{11} & a_{22}b_{12} \\ a_{21}b_{21} & a_{21}b_{22} & a_{22}b_{21} & a_{22}b_{22} \end{pmatrix}. \tag{3.6-1}$$

In general, the direct product of two square matrices A and B (which need not be of the same dimension) is a matrix $A \times B$ whose elements are all the possible products of an element of A with an element of B. Each element of $A \times B$ bears a double set of indices. The characters of A, B, and $A \times B$ are evidently related by

$$\chi(A \times B) = \chi(A)\chi(B). \tag{3.6-2}$$

Now let (ψ_1, ψ_2) be a set of basis functions for a two-dimensional irreducible representation $\Gamma^{(\mu)}$ of a group G and let (φ_1, φ_2) similarly belong to $\Gamma^{(\nu)}$ of G. This means, according to (3.4-9), that

$$P_R\psi_1 = \psi_1\Gamma_{11}^{(\mu)}(R) + \psi_2\Gamma_{21}^{(\mu)}(R), \qquad P_R\psi_2 = \psi_1\Gamma_{12}^{(\mu)}(R) + \psi_2\Gamma_{22}^{(\mu)}(R),$$

with analogous relations for (φ_1, φ_2). Therefore,

$$\begin{aligned} P_R(\psi_1\varphi_1) &= [\psi_1\Gamma_{11}^{(\mu)}(R) + \psi_2\Gamma_{21}^{(\mu)}(R)][\varphi_1\Gamma_{11}^{(\nu)}(R) + \varphi_2\Gamma_{21}^{(\nu)}(R)] \\ &= \psi_1\varphi_1\Gamma_{11}^{(\mu)}(R)\Gamma_{11}^{(\nu)}(R) + \psi_1\varphi_2\Gamma_{11}^{(\mu)}(R)\Gamma_{21}^{(\nu)}(R) \\ &\quad + \psi_2\varphi_1\Gamma_{21}^{(\mu)}(R)\Gamma_{11}^{(\nu)}(R) + \psi_2\varphi_2\Gamma_{21}^{(\mu)}(R)\Gamma_{21}^{(\nu)}(R), \end{aligned}$$

and continuing in the same fashion it is found that

$$\begin{aligned} &P_R(\psi_1\varphi_1, \psi_1\varphi_2, \psi_2\varphi_1, \psi_2\varphi_2) \\ &= (\psi_1\varphi_1, \psi_1\varphi_2, \psi_2\varphi_1, \psi_2\varphi_2)\begin{pmatrix} \Gamma_{11}^{(\mu)}\Gamma_{11}^{(\nu)} & \Gamma_{11}^{(\mu)}\Gamma_{12}^{(\nu)} & \Gamma_{12}^{(\mu)}\Gamma_{11}^{(\nu)} & \Gamma_{12}^{(\mu)}\Gamma_{12}^{(\nu)} \\ \Gamma_{11}^{(\mu)}\Gamma_{21}^{(\nu)} & \Gamma_{11}^{(\mu)}\Gamma_{22}^{(\nu)} & \Gamma_{12}^{(\mu)}\Gamma_{21}^{(\nu)} & \Gamma_{12}^{(\mu)}\Gamma_{22}^{(\nu)} \\ \Gamma_{21}^{(\mu)}\Gamma_{11}^{(\nu)} & \Gamma_{21}^{(\mu)}\Gamma_{12}^{(\nu)} & \Gamma_{22}^{(\mu)}\Gamma_{11}^{(\nu)} & \Gamma_{22}^{(\mu)}\Gamma_{12}^{(\nu)} \\ \Gamma_{21}^{(\mu)}\Gamma_{21}^{(\nu)} & \Gamma_{21}^{(\mu)}\Gamma_{22}^{(\nu)} & \Gamma_{22}^{(\mu)}\Gamma_{21}^{(\nu)} & \Gamma_{22}^{(\mu)}\Gamma_{22}^{(\nu)} \end{pmatrix}. \end{aligned} \tag{3.6-3}$$

Hence it may be concluded that the four product functions $\psi_1\varphi_1$, $\psi_1\varphi_2$, $\psi_2\varphi_1$, $\psi_2\varphi_2$ are basis functions for a representation—called a *product representation*—of the group G. The product representation $\Gamma^{(\mu\times\nu)} \equiv \Gamma^{(\mu)} \times \Gamma^{(\nu)}$ is generally reducible even if $\Gamma^{(\mu)}$ and $\Gamma^{(\nu)}$ are irreducible. As in (3.6-2),

$$\chi^{(\mu\times\nu)}(R) = \chi^{(\mu)}(R)\chi^{(\nu)}(R). \tag{3.6-4}$$

These results can be formalized by writing

$$\begin{aligned} P_R(\psi_j^{(\mu)}\varphi_l^{(\nu)}) &= \sum_{ik} \psi_i^{(\mu)}\varphi_k^{(\nu)}\Gamma_{ij}^{(\mu)}(R)\Gamma_{kl}^{(\nu)}(R) \\ &= \sum_{ik} \psi_i^{(\mu)}\varphi_k^{(\nu)}[\Gamma^{(\mu)}(R) \times \Gamma^{(\nu)}(R)]_{ik,jl} \end{aligned} \tag{3.6-5}$$

and

$$\chi^{(\mu\times\nu)}(R) = \sum_{ij} \Gamma_{ii}^{(\mu)}(R)\Gamma_{jj}^{(\nu)}(R). \tag{3.6-6}$$

It is sometimes useful to regard (ψ_1, ψ_2) and (φ_1, φ_2)—the basis sets for $\Gamma^{(\mu)}$ and $\Gamma^{(\nu)}$, respectively—as the components of two-dimensional vectors.

The four product functions $\psi_1\varphi_1$, $\psi_1\varphi_2$, $\psi_2\varphi_1$, and $\psi_2\varphi_2$ would then be, from this point of view, a tensor of rank 2 with respect to the group G. As previously indicated such a tensor is generally reducible. We shall further develop this approach in Section 6.5.

An important case arises when $\mu = \nu$; then the two representations forming the product representation are the same. As before, let both (ψ_1, ψ_2) and (φ_1, φ_2) belong to $\Gamma^{(\mu)}$ $(= \Gamma^{(\nu)})$. It should be kept in mind that ψ_1 and ψ_2 are partners, as are φ_1 and φ_2, but there is no partnership relation between any of the ψ's with any of the φ's. From (3.6-3) we find (with the superscript μ suppressed)

$$P_R(\psi_1\varphi_1) = \psi_1\varphi_1[\Gamma_{11}(R)]^2 + (\psi_1\varphi_2 + \psi_2\varphi_1)\Gamma_{11}(R)\Gamma_{21}(R) + \psi_2\varphi_2[\Gamma_{21}(R)]^2, \tag{3.6-7}$$

$$P_R(\psi_1\varphi_2 + \psi_2\varphi_1) = 2\psi_1\varphi_1\Gamma_{11}(R)\Gamma_{12}(R) + (\psi_1\varphi_2 + \psi_2\varphi_1)[\Gamma_{11}(R)\Gamma_{22}(R) + \Gamma_{12}(R)\Gamma_{21}(R)] + 2\psi_2\varphi_2\Gamma_{21}(R)\Gamma_{22}(R), \tag{3.6-8}$$

$$P_R(\psi_2\varphi_2) = \psi_1\varphi_1[\Gamma_{12}(R)]^2 + (\psi_1\varphi_2 + \psi_2\varphi_1)\Gamma_{12}(R)\Gamma_{22}(R) + \psi_2\varphi_2[\Gamma_{22}(R)]^2, \tag{3.6-9}$$

$$P_R(\psi_1\varphi_2 - \psi_2\varphi_1) = (\psi_1\varphi_2 - \psi_2\varphi_1)[\Gamma_{11}(R)\Gamma_{22}(R) - \Gamma_{12}(R)\Gamma_{21}(R)]. \tag{3.6-10}$$

The products $\psi_1\varphi_1$, $\psi_1\varphi_2 + \psi_2\varphi_1$, $\psi_2\varphi_2$ are symmetric products in the sense that they remain unchanged under an interchange of indices. The product $\psi_1\varphi_2 - \psi_2\varphi_1$ is antisymmetric. From (3.6-7)–(3.6-9) it is seen that the symmetric products transform among themselves and do not mix with the antisymmetric product and vice versa. This feature may be expressed by writing

$$\Gamma^{(\mu)} \times \Gamma^{(\mu)} = (\Gamma^{(\mu)} \times \Gamma^{(\mu)})^+ + (\Gamma^{(\mu)} \times \Gamma^{(\mu)})^- \tag{3.6-11}$$

where $(\Gamma^{(\mu)} \times \Gamma^{(\mu)})^+$ stands for the product representation whose basis functions are symmetric products; $(\Gamma^{(\mu)} \times \Gamma^{(\mu)})^+$ is called the *symmetric product representation*; $(\Gamma^{(\mu)} \times \Gamma^{(\mu)})^-$, the product representation whose basis functions are antisymmetric, is called the *antisymmetric product representation.*

In this example $(\Gamma^{(\mu)} \times \Gamma^{(\mu)})^+$ is three-dimensional and $(\Gamma^{(\mu)} \times \Gamma^{(\mu)})^-$ is one-dimensional. If $\varphi_i = \psi_i$, $(\Gamma^{(\mu)} \times \Gamma^{(\mu)})^-$ is identically zero, as may be seen from (3.6-10). The general statement is that if l_μ is the dimension of $\Gamma^{(\mu)}$, the symmetric and antisymmetric product representations are characterized in the manner shown in Table 3.3.

Another important theorem pertaining to product representations is the following: Let $\Gamma^{(\mu)}$ and $\Gamma^{(\nu)}$ be two irreducible representations. The characters of the product representation matrices $\Gamma^{(\mu)}(R)^* \times \Gamma^{(\nu)}(R)$ are $\chi^{(\mu)}(R)^*\chi^{(\nu)}(R)$.

TABLE 3.3

Characters of Product Representations

	Dimension	Character
$\Gamma^{(\mu)} \times \Gamma^{(\mu)}$	l_μ^2	$[\chi^{(\mu)}(R)]^2$
$(\Gamma^{(\mu)} \times \Gamma^{(\mu)})^+$	$\frac{1}{2}l_\mu(l_\mu + 1)$	$\frac{1}{2}\{[\chi^{(\mu)}(R)]^2 + \chi^{(\mu)}(R^2)\}$
$(\Gamma^{(\mu)} \times \Gamma^{(\mu)})^-$	$\frac{1}{2}l_\mu(l_\mu - 1)$	$\frac{1}{2}\{[\chi^{(\mu)}(R)]^2 - \chi^{(\mu)}(R^2)\}$

Using (3.3-16), the number of times $\Gamma^{(1)}(R)$ will appear in the decomposition of $\Gamma^{(\mu)}(R)^* \times \Gamma^{(\nu)}(R)$ will be given by

$$a_1 = \frac{1}{h}\sum_R \chi^{(1)}(R)^*[\chi^{(\mu)}(R)^*\chi^{(\nu)}(R)] = \frac{1}{h}\sum_R \chi^{(\mu)}(R)^*\chi^{(\nu)}(R) = \delta_{\mu\nu} \qquad (3.6\text{-}12)$$

in which the Kronecker δ arises as a consequence of the orthogonality theorem (3.3-13). This means that $\Gamma^{(\mu)*} \times \Gamma^{(\nu)}$ contains $\Gamma^{(1)}$ only if $\mu = \nu$ and then only once.

For illustrative purposes consider the product representation $\Gamma^{(3)} \times \Gamma^{(3)}$ in D_3. Using Tables 3.2 and 3.3 the characters are given in the accompanying tabulation.

	E	A	B	C	D	F
$\Gamma^{(3)} \times \Gamma^{(3)}$	4	0	0	0	1	1
$(\Gamma^{(3)} \times \Gamma^{(3)})^+$	3	1	1	1	0	0
$(\Gamma^{(3)} \times \Gamma^{(3)})^-$	1	-1	-1	-1	1	1

Upon reducing these representation according to (3.3-16) we obtain

$$\begin{aligned} \Gamma^{(3)} \times \Gamma^{(3)} &= \Gamma^{(1)} + \Gamma^{(2)} + \Gamma^{(3)}, \\ (\Gamma^{(3)} \times \Gamma^{(3)})^+ &= \Gamma^{(1)} + \Gamma^{(3)}, \\ (\Gamma^{(3)} \times \Gamma^{(3)})^- &= \Gamma^{(2)}. \end{aligned} \qquad (3.6\text{-}13)$$

Further insight may be gained by considering the behavior of the basis functions. Let (ψ_1, ψ_2) and (φ_1, φ_2) be two sets of basis functions for $\Gamma^{(3)}$. For the element B of D_3 we have, from (3.2-3),

$$\begin{aligned} P_B(\psi_1, \psi_2) &= (\psi_1, \psi_2)\begin{pmatrix} -\frac{1}{2} & \frac{1}{2}\sqrt{3} \\ \frac{1}{2}\sqrt{3} & \frac{1}{2} \end{pmatrix}, \\ P_B(\varphi_1, \varphi_2) &= (\varphi_1, \varphi_2)\begin{pmatrix} -\frac{1}{2} & \frac{1}{2}\sqrt{3} \\ \frac{1}{2}\sqrt{3} & \frac{1}{2} \end{pmatrix}, \end{aligned} \qquad (3.6\text{-}14)$$

and

$$P_B(\psi_1\varphi_1, \psi_1\varphi_2, \psi_2\varphi_1, \psi_2\varphi_2)$$
$$= (\psi_1\varphi_1, \psi_1\varphi_2, \psi_2\varphi_1, \psi_2\varphi_2)\begin{pmatrix} \frac{1}{4} & -\frac{1}{4}\sqrt{3} & -\frac{1}{4}\sqrt{3} & \frac{3}{4} \\ -\frac{1}{4}\sqrt{3} & -\frac{1}{4} & \frac{3}{4} & \frac{1}{4}\sqrt{3} \\ -\frac{1}{4}\sqrt{3} & \frac{3}{4} & -\frac{1}{4} & \frac{1}{4}\sqrt{3} \\ \frac{3}{4} & \frac{1}{4}\sqrt{3} & \frac{1}{4}\sqrt{3} & \frac{1}{4} \end{pmatrix} \quad (3.6\text{-}15)$$

in which the 4×4 matrix is just the direct product $\Gamma^{(3)}(B) \times \Gamma^{(3)}(B)$. From (3.6-15) one may also obtain

$$P_B(\psi_1\varphi_1, \psi_1\varphi_2 + \psi_2\varphi_1, \psi_2\varphi_2, \psi_1\varphi_2 - \psi_2\varphi_1)$$
$$= (\psi_1\varphi_1, \psi_1\varphi_2 + \psi_2\varphi_1, \psi_2\varphi_2, \psi_1\varphi_2 - \psi_2\varphi_1)\begin{pmatrix} \frac{1}{4} & -\frac{1}{2}\sqrt{3} & \frac{3}{4} & 0 \\ -\frac{1}{4}\sqrt{3} & \frac{1}{2} & \frac{1}{4}\sqrt{3} & 0 \\ \frac{3}{4} & \frac{1}{2}\sqrt{3} & \frac{1}{4} & 0 \\ 0 & 0 & 0 & -1 \end{pmatrix}. \quad (3.6\text{-}16)$$

The matrix in (3.6-16) consists of two matrices along the main diagonal. One is a three-dimensional matrix which is identifiable as belonging to the symmetric representation $(\Gamma^{(3)} \times \Gamma^{(3)})^+$ and the other is a one-dimensional matrix belonging to the antisymmetric representation $(\Gamma^{(3)} \times \Gamma^{(3)})^-$. In this manner, (3.6-11) may be verified for all the representation matrices.

Situations arise in which product functions are not all independent. For example, in Section 3.4 we saw that

$$(\psi_1, \psi_2) = (x, y), \qquad (\varphi_1, \varphi_2) = (yz, -xz)$$

were each basis sets for $\Gamma^{(3)}$. However, $\psi_1\varphi_1 + \psi_2\varphi_2 = 0$ so that there are only three, instead of four, independent product functions. In that event it is not possible to construct bases for all the irreducible representations into which the product representation decomposes. Thus, in the previous example, if one attempts to construct linear combinations of $\psi_i\varphi_j$ to serve as bases for the representations $\Gamma^{(1)}$, $\Gamma^{(2)}$, and $\Gamma^{(3)}$ into which $\Gamma^{(3)} \times \Gamma^{(3)}$ decomposes, as shown in (3.6-13), it is found that the basis for $\Gamma^{(1)}$ is identically zero.

Finally, let

$$\psi_1 = \varphi_1 = x, \qquad \psi_2 = \varphi_2 = y;$$

then $\psi_1\varphi_2 - \psi_2\varphi_1 = 0$ and (3.6-16) assumes the form

$$P_B(\psi_1{}^2, 2\psi_1\psi_2, \psi_2{}^2) = (\psi_1{}^2, 2\psi_1\psi_2, \psi_2{}^2)\begin{pmatrix} \frac{1}{4} & -\frac{1}{2}\sqrt{3} & \frac{3}{4} \\ -\frac{1}{4}\sqrt{3} & \frac{1}{2} & \frac{1}{4}\sqrt{3} \\ \frac{3}{4} & \frac{1}{2}\sqrt{3} & \frac{1}{4} \end{pmatrix}. \quad (3.6\text{-}17)$$

In this case only the symmetric representation exists.

Suppose $\mu_i^{(\alpha)}$, with $i = 1, 2, \ldots, n_\alpha$, is a basis set for an irreducible representation $\Gamma^{(\alpha)}$ of dimension n_α and $v_j^{(\beta)}$, with $j = 1, 2, \ldots, n_\beta$, is a basis set for an irreducible representation $\Gamma^{(\beta)}$ of dimension n_β. The direct product $\Gamma^{(\alpha)} \times \Gamma^{(\beta)}$, we noted, is generally reducible as

$$\Gamma^{(\alpha)} \times \Gamma^{(\beta)} = \sum_\gamma a_\gamma \Gamma^{(\gamma)}. \tag{3.6-18}$$

If is often necessary to derive basis functions $\omega_k^{(\gamma)}$ for the irreducible representations $\Gamma^{(\gamma)}$ in terms of the $u_i^{(\alpha)}$ and $v_j^{(\beta)}$. The general result, as discussed in Section 3.4, is that $\omega_k^{(\gamma)}$ is a linear combination of the $u_i^{(\alpha)}$ and $v_j^{(\beta)}$. This may be expressed formally by writing

$$\omega_k^{(\gamma)} = \sum_{ij} u_i^{(\alpha)} v_j^{(\beta)} \langle \alpha\beta ij | \gamma k \rangle \tag{3.6-19}$$

in which the $\langle \alpha\beta ij | \gamma k \rangle$ are numerical coefficients known as coupling coefficients.‡ In principle the coupling coefficients may be derived by means of projection operators as described in the previous section. In practice this becomes quite tedious; tables of coupling coefficients for common groups are given by Koster *et al.* (1963) (see also Tables 5.9 and 5.11).

3.7 Matrix Elements

Group theoretical considerations are often helpful in establishing selection rules, that is, conditions under which a particular matrix element must vanish. For this purpose we return to the orthogonality theorem for irreducible representations (3.3-11):

$$\sum_R \Gamma_{\mu\nu}^{(i)}(R)^* \Gamma_{\alpha\beta}^{(j)}(R) = \frac{h}{l_i} \delta_{ij} \delta_{\mu\alpha} \delta_{\nu\beta} \tag{3.7-1}$$

where h is the order of the group and l_i the dimensionality of $\Gamma^{(i)}(R)$. In particular, if i and j refer to two nonequivalent, irreducible representations and $i \neq j$, the orthogonality theorem takes the form

$$\sum_R \Gamma_{\mu\nu}^{(i)}(R)^* \Gamma_{\alpha\beta}^{(j)}(R) = 0. \tag{3.7-2}$$

Now let $\Gamma^{(i)}(R)$ be the totally symmetric (unit) representation. All the elements of this representation are simply equal to $+1$. Therefore, (3.7-2) becomes

$$\sum_R \Gamma_{\alpha\beta}^{(j)}(R) = 0 \tag{3.7-3}$$

‡ In some of the literature general coefficients of the type $\langle \alpha\beta ij | \gamma k \rangle$ are called Clebsch–Gordan coefficients. We do not use this terminology in order to avoid confusion with the Clebsch–Gordan coefficients defined in Section 1.5.

where $\Gamma^{(j)}$ is any irreducible representation of the group other than the unit representation.

If $\psi_\mu^{(j)}(\mathbf{r})$ is a member of a basis set for $\Gamma^{(j)}$, we have, as a consequence of (2.1-22),

$$\int \psi_\mu^{(j)}(\mathbf{r})\, d\mathbf{r} = \int P_R \psi_\mu^{(j)}(\mathbf{r})\, d\mathbf{r} \tag{3.7-4}$$

$$= \sum_\alpha \Gamma_{\alpha\mu}^{(j)}(R) \int \psi_\alpha^{(j)}(\mathbf{r})\, d\mathbf{r} \tag{3.7-5}$$

in which the second expression on the right is based on the definition of a basis function. Summing over all elements of the group

$$\sum_R \int \psi_\mu^{(j)}(\mathbf{r})\, d\mathbf{r} = \sum_\alpha \sum_R \Gamma_{\alpha\mu}^{(j)}(R) \int \psi_\alpha^{(j)}(\mathbf{r})\, d\mathbf{r}. \tag{3.7-6}$$

In view of (3.7-3), the right-hand side of (3.7-6) is zero; hence

$$\sum_R \int \psi_\mu^{(j)}(\mathbf{r})\, d\mathbf{r} = h \int \psi_\mu^{(j)}(\mathbf{r})\, d\mathbf{r} = 0$$

or

$$\int \psi_\mu^{(j)}(\mathbf{r})\, d\mathbf{r} = 0 \tag{3.7-7}$$

provided $\Gamma^{(j)}$ is not the unit representation.

Now consider the matrix element

$$M = \langle \psi_\alpha^{(i)} | Q_\beta^{(j)} | \varphi_\gamma^{(k)} \rangle = \int \psi_\alpha^{(i)}(\mathbf{r})^* Q_\beta^{(j)}(\mathbf{r}) \varphi_\gamma^{(k)}(\mathbf{r})\, d\mathbf{r} \tag{3.7-8}$$

where $\psi_\alpha^{(i)}$, $Q_\beta^{(j)}$ and $\varphi_\gamma^{(k)}$ transform according to the irreducible representations $\Gamma^{(i)}$, $\Gamma^{(j)}$ and $\Gamma^{(k)}$, respectively. The product $\psi_\alpha^{(i)*} Q_\beta^{(j)} \varphi_\gamma^{(k)}$ (the integrand) belongs to the product representation $\Gamma^{(i)*} \times \Gamma^{(j)} \times \Gamma^{(k)}$. Suppose, initially, that the product representation is irreducible. In that case, the integrand is a basis function for an irreducible representation of the group and the matrix element will vanish, according to (3.7-7), unless the irreducible representation is the unit representation. More generally, the product representation will be reducible. Assuming this to be the case, let

$$\Gamma^{(i)*} \times \Gamma^{(j)} \times \Gamma^{(k)} = \sum_\mu a_\mu \Gamma^{(\mu)}. \tag{3.7-9}$$

The integrand, according to (3.5-22) can then be written as a sum of basis functions for the irreducible representations $\Gamma^{(\mu)}$ contained in the (reducible) product representation. This mean that the matrix element M can be expressed as a sum of matrix elements $M^{(\mu)}$ in which the individual integrands belong to the irreducible representations $\Gamma^{(\mu)}$. Once again the theorem expressed by (3.7-7) may be invoked: all the matrix elements $M^{(\mu)}$ will vanish except those whose integrands belong to the unit representation. The net result is that *the matrix element* (3.7-8) *will be different from zero only if* $\Gamma^{(i)*} \times \Gamma^{(j)} \times \Gamma^{(k)}$ *contains the unit representation.* We also saw in the previous section that the unit representation was contained in $\Gamma^{(\mu)*} \times \Gamma^{(\nu)}$ only if $\mu = \nu$. Therefore an equivalent statement of the matrix element theorem is that $M \neq 0$ only if $\Gamma^{(j)} \times \Gamma^{(k)}$ contains $\Gamma^{(i)}$. For the common situation in which all three representations $\Gamma^{(i)}$, $\Gamma^{(j)}$, and $\Gamma^{(k)}$ have real characters, it may be shown (Section 5.2) that $M = 0$ unless the product of any two of the three representations contains the third.

The consequences of the matrix element theorem are usually known as selection rules. For example, if an operator Q is invariant under the transformations of a group or, in other words, if Q transforms according to the unit representations, $\Gamma^{(i)}$ and $\Gamma^{(k)}$ are complex conjugates, or, for representations with real characters the theorem requires that $\Gamma^{(i)} = \Gamma^{(k)}$ which means that matrix elements of Q between basis functions belonging to different irreducible representations will vanish.

In writing (3.7-8) it is presumed that the functions are independent, that is, either $i \neq k$ so that $\psi_\alpha^{(i)}$ and $\varphi_\gamma^{(k)}$ belong to two different irreducible representations or, if $i = k$, then $\psi_\alpha^{(i)}$ and $\varphi_\gamma^{(i)}$ are two independent sets of partners. Now suppose we are interested in diagonal matrix elements with $\psi_\alpha^{(i)} = \varphi_\gamma^{(k)}$. For representations with real characters the theorem stated above would say that the matrix element $\langle \psi_\alpha^{(i)} | Q_\beta^{(j)} | \psi_\alpha^{(i)} \rangle$ vanishes unless $\Gamma^{(i)} \times \Gamma^{(i)}$ contains $\Gamma^{(j)}$. Now, from (3.6-11),

$$\Gamma^{(i)} \times \Gamma^{(i)} = (\Gamma^{(i)} \times \Gamma^{(i)})^+ + (\Gamma^{(i)} \times \Gamma^{(i)})^- \tag{3.7-10}$$

and $(\Gamma^{(i)} \times \Gamma^{(i)})^-$ is identically zero when the two sets of basis functions are the same, as was shown in the previous section. Therefore, for diagonal matrix elements, the stronger form of the selection rule theorem is that

$$\langle \psi_\alpha^{(i)} | Q_\beta^{(j)} | \psi_\alpha^{(i)} \rangle = 0 \tag{3.7-11}$$

unless $(\Gamma^{(i)} \times \Gamma^{(i)})^+$—the symmetric product representation—contains $\Gamma^{(j)}$. An equivalent statement is that the matrix will vanish unless the unit representation is contained in the symmetric direct product $([\Gamma^{(i)} \times \Gamma^{(j)}] \times \Gamma^{(i)})^+$. Finally, if the integrand contains partner functions, only the symmetric product exists, as was shown in Section 3.6, and the selection rules are then based entirely on the symmetric product.

CHAPTER 4

CONTINUOUS ROTATION GROUPS

Thus far we have been concerned with groups consisting of a finite number of elements. It is now necessary to extend the discussion to continuous groups containing an infinite number of elements. Of particular interest in atomic and molecular physics are the rotation groups in two and three dimensions.

4.1 Rotation Group in Two Dimensions, C_∞

A pure rotation of coordinates through an angle α in a plane perpendicular to a fixed axis is described by the transformation

$$\mathbf{r}' = R(\alpha)\mathbf{r} \tag{4.1-1}$$

with

$$R(\alpha) = \begin{pmatrix} \cos\alpha & \sin\alpha \\ -\sin\alpha & \cos\alpha \end{pmatrix}. \tag{4.1-2}$$

For each value of α there is a matrix $R(\alpha)$, although all the distinct matrices $R(\alpha)$ are obtained with α in the range $-\pi \leqslant \alpha \leqslant \pi$. We now show that transformations (4.1-1) satisfy the group postulates (Section 3.1).

Two rotations $R(\alpha_1)$ and $R(\alpha_2)$ performed in sequence are equivalent to a single rotation $R(\alpha_1 + \alpha_2)$ which is also a member of the set. Since

$$[R(\alpha_1)R(\alpha_2)]R(\alpha_3) = R(\alpha_1)[R(\alpha_2)R(\alpha_3)] = R(\alpha_1 + \alpha_2 + \alpha_3), \tag{4.1-3}$$

the associative law holds. When $\alpha = 0$,

$$R(0) = \begin{pmatrix} 1 & 0 \\ 0 & 1 \end{pmatrix}, \tag{4.1-4}$$

and $R(\alpha)R(0) = R(0)R(\alpha) = R(\alpha)$; therefore $R(0)$ is the identity element. Finally,

$$R(\alpha)R(-\alpha) = R(-\alpha)R(\alpha) = R(0), \tag{4.1-5}$$

so that $R(-\alpha)$ is the inverse to $R(\alpha)$, i.e.,

$$R^{-1}(\alpha) = R(-\alpha). \tag{4.1-6}$$

The group postulates having been satisfied, the infinite set of elements $R(\alpha)$ constitutes a group which is called C_∞. Each element of the group is a function of a single continuous variable α; hence C_∞ is said to be a one-parameter continuous group.

C_∞ is an Abelian group because the sequence in which two rotations are executed is inconsequential, that is,

$$R(\alpha_1)R(\alpha_2) = R(\alpha_2)R(\alpha_1) = R(\alpha_1 + \alpha_2). \tag{4.1-7}$$

According to the discussion of Section 3.3, the irreducible representations of an Abelian group are one-dimensional and the number of such representations is equal to the number of elements in the group. Therefore C_∞ has an infinite number of one-dimensional (irreducible) representations. To find these representations, assume that the rotations take place in the xy plane. The vector $\mathbf{r}$ in (4.1-1) then has components (x, y) or $(r\cos\varphi, r\sin\varphi)$ as shown in Fig. 2.1. Using the latter components, the transformation $\mathbf{r}' = R(\alpha)\mathbf{r}$ is equivalent to $r' = r$, $\varphi' = \varphi - \alpha$ as in (2.3-15) or

$$(r', \varphi') = R(\alpha)(r, \varphi) = (r, \varphi - \alpha). \tag{4.1-8}$$

Similarly,

$$R^{-1}(\alpha)(r, \varphi) = (r, \varphi + \alpha). \tag{4.1-9}$$

We then have

$$P_R(\alpha)f(r, \varphi) = f[R^{-1}(\alpha)(r, \varphi)] = f(r, \varphi + \alpha), \tag{4.1-10}$$

which, for

$$f(r, \varphi) = e^{im\varphi},$$

gives

$$P_R(\alpha)e^{im\varphi} = e^{im\alpha}e^{im\varphi}. \tag{4.1-11}$$

This result is a special case of (3.4-9); hence $e^{im\varphi}$ may be regarded as a basis function for the one-dimensional (irreducible) representation $e^{im\alpha}$. Writing

$$\Gamma^{(m)}(\alpha) = e^{im\alpha}, \tag{4.1-12}$$

it is seen that

$$\Gamma^{(m)}(\alpha_1)\Gamma^{(m)}(\alpha_2) = \Gamma^{(m)}(\alpha_1 + \alpha_2), \tag{4.1-13}$$

which, by comparison with (4.1-7), establishes the isomorphism between the representations $\Gamma^{(m)}(\alpha)$ and the coordinate rotations $R(\alpha)$. But $R(-\pi) = R(\pi)$; therefore

$$\Gamma^{(m)}(-\pi) = \Gamma^{(m)}(\pi). \tag{4.1-14}$$

It then follows, in view of (4.1-12), that

$$m = 0, \pm 1, \pm 2, \ldots . \tag{4.1-15}$$

Representations $\Gamma^{(m)}$ and $\Gamma^{(-m)}$ (or characters $\chi^{(m)}$ and $\chi^{(-m)}$) are complex conjugates. In certain applications it is convenient to combine the two complex conjugate representations to form a two-dimensional representation (Section 5.2). However, we shall continue to regard the irreducible representations of C_∞ as one-dimensional unless it is noted otherwise.

The properties of C_∞ have been developed on the basis of rotations in the xy plane. Clearly, the coordinate transformation expressed by (4.1-8) does not depend on the orientation of the rotation axis. The character of a rotation therefore depends only on the angle of rotation and is independent of the orientation of the rotation axis. In other words, rotations through the same angle, regardless of the axis about which they occur, belong to the same class.

4.2 Rotation Group in Three Dimensions, $O^+(3)$

Proper rotations in three dimensions have been discussed in Sections 2.2 and 2.3. To recapitulate, a general three-dimensional proper rotation of coordinates may be described by

$$\mathbf{r}' = R(\alpha, \beta, \gamma)\mathbf{r} \tag{4.2-1}$$

where α, β, γ are the Euler angles (Section 2.3) and $R(\alpha, \beta, \gamma)$ is a real orthogonal matrix whose determinant is equal to $+1$ and whose specific form is given by (2.3-8). The set of all possible (proper) three-dimensional rotations which may be regarded as the set of all possible matrices $R(\alpha, \beta, \gamma)$ constitutes an infinite continuous group called $O^+(3)$. The proof of this statement is merely an extension of that given for C_∞ in the previous section. There are, however, important differences between the two groups. In the first place, C_∞ is a one-parameter group while $O^+(3)$ is a group whose elements depend on the three parameters α, β, γ; that is, $O^+(3)$ is a three-parameter group. Secondly; the *order* in which rotations are executed in a plane is of no consequence so that C_∞ is Abelian and the irreducible representations are all one-dimensional.

But in three-dimensional space, two rotations performed in sequence are generally not equivalent to the same two rotations performed in the reverse order. Therefore $O^+(3)$ is non-Abelian and the irreducible representations are not all one-dimensional.

The transformation properties of the spherical harmonics provide some information concerning the representations of $O^+(3)$. According to (2.3-21):

$$P_R(\alpha,\beta,\gamma)Y_{lm}(\theta,\varphi) = \sum_{m'} Y_{lm'}(\theta,\varphi)D^{(l)}_{m'm}(\alpha,\beta,\gamma). \tag{4.2-2}$$

In Section 2.3 this relation was regarded as a statement describing the manner in which the spherical harmonics transform under a coordinate transformation $R(\alpha,\beta,\gamma)$. In the present group theoretical context, however, it is seen that (4.2-2) is a special case of (3.4-9) which then implies that the $Y_{lm}(\theta,\varphi)$ may be regarded as basis functions for the representations $D^{(l)}(\alpha,\beta,\gamma)$ of $O^+(3)$. At this stage it is not known whether the $D^{(l)}(\alpha,\beta,\gamma)$ are reducible or irreducible. In fact they are the latter, as will be shown in Section 4.5. Since there are $(2l+1)$ spherical harmonics associated with each value of l and $l = 0,1,2,3,\ldots$, we have the result that for a given integral value of l, $D^{(l)}(\alpha,\beta,\gamma)$ is an irreducible representation of dimension $(2l+1)$, that is, with dimensions $1,3,5,7,\ldots$. But this is not the whole story; as will be shown in Section 4.5, there are also irreducible representations with dimensions $2,4,6,\ldots$. In this case, the spherical harmonics can no longer serve as basis functions.

Thus far the general form of $D^{(l)}_{m'm}(\alpha,\beta,\gamma)$ has not been developed (although a few specific examples have been given in Section 2.3); it is nevertheless possible to give the general form of the character of $D^{(l)}(\alpha,\beta,\gamma)$. Consider the coordinate transformation $R(\alpha,0,0)$ which is a rotation of coordinates through an angle α about the z axis. From (2.3-20) we have an expression for $D^{(l)}_{m'm}(\alpha,0,0)$; hence the character is immediately given by

$$\begin{aligned}\chi^{(l)}(\alpha,0,0) &= e^{il\alpha} + e^{i(l-1)\alpha} + \cdots + e^{-il\alpha}\\ &= \frac{(e^{il\alpha} + \cdots + e^{-il\alpha})(e^{i\alpha}-1)e^{-i\alpha/2}}{(e^{i\alpha}-1)e^{-i\alpha/2}}\\ &= \frac{\sin(l+\frac{1}{2})\alpha}{\sin\frac{1}{2}\alpha}.\end{aligned} \tag{4.2-3}$$

This expression is actually more general than may appear at first since the z axis may always be chosen to coincide with the axis of rotation. We may therefore generalize (4.2-3) to

$$\chi^{(l)}(\omega) = \frac{\sin(l+\frac{1}{2})\omega}{\sin\frac{1}{2}\omega}, \tag{4.2-4}$$

where ω is the angle of rotation about an axis whose orientation is entirely arbitrary. Moreover, since elements having the same character belong to the same class, it is seen that, for the group $O^+(3)$, rotations through a given angle about any axis belong to the same class, as has previously been noted.

4.3 Special Unitary Group in Two Dimensions, $SU(2)$

We must now digress from the discussion of the group $O^+(3)$ in order to discuss the unitary group $SU(2)$. It will be shown that the two groups are related and that it is possible to obtain information on the representations of $O^+(3)$ from those of $SU(2)$.

Consider the set of matrices

$$U = \begin{pmatrix} a & b \\ -b^* & a^* \end{pmatrix}, \tag{4.3-1}$$

in which a and b are complex numbers and

$$\det U \equiv |U| = |a|^2 + |b|^2 = 1. \tag{4.3-2}$$

The inverse of U is

$$U^{-1} = \begin{pmatrix} a^* & -b \\ b^* & a \end{pmatrix} = U^\dagger ; \tag{4.3-3}$$

therefore U is unitary. It is readily verified that the set of matrices U satisfy the group postulates with respect to matrix multiplication. The group is known as $SU(2)$, the special unitary group in two dimensions.

Suppose we write

$$u' = au + bv, \qquad v' = -b^*u + a^*v, \tag{4.3-4}$$

or in matrix form with

$$\boldsymbol{\rho} = \begin{pmatrix} u \\ v \end{pmatrix}, \tag{4.3-5}$$

$$\boldsymbol{\rho}' = U\boldsymbol{\rho} = \begin{pmatrix} au & + bv \\ -b^*u + a^*v \end{pmatrix}. \tag{4.3-6}$$

From (4.3-3),

$$U^{-1}\boldsymbol{\rho} = \begin{pmatrix} a^*u - bv \\ b^*u + av \end{pmatrix}. \tag{4.3-7}$$

Also, let

$$f_m^{(j)}(\boldsymbol{\rho}) = \frac{u^{j+m}v^{j-m}}{\sqrt{(j+m)!(j-m)!}} \tag{4.3-8}$$

with $j = 0, \frac{1}{2}, 1, \frac{3}{2}, \ldots$ and $m = j, j-1, \ldots, -j$. If u and v transform as in (4.3-4) or (4.3-6), then $f_m^{(j)}(\boldsymbol{\rho})$ transforms according to

$$P_u f_m^{(j)}(\boldsymbol{\rho}) = f_m^{(j)}(U^{-1}\boldsymbol{\rho}) = \frac{(a^*u - bv)^{j+m}(b^*u + av)^{j-m}}{\sqrt{(j+m)!(j-m)!}}. \tag{4.3-9}$$

Examination of the right side of (4.3-9) reveals that it is a linear combination of the $f_m^{(j)}(\boldsymbol{\rho})$. For example, let $j = \frac{1}{2}$; then

$$f_{1/2}^{(1/2)} = u, \qquad f_{-1/2}^{(1/2)} = v, \tag{4.3-10}$$

and

$$\begin{aligned} P_u f_{1/2}^{(1/2)} &= a^*u - bv = a^* f_{1/2}^{(1/2)} - b f_{-1/2}^{(1/2)}, \\ P_u f_{-1/2}^{(1/2)} &= b^*u + av = b^* f_{1/2}^{(1/2)} + a f_{-1/2}^{(1/2)}. \end{aligned} \tag{4.3-11}$$

In matrix form

$$P_u(f_{1/2}^{(1/2)}, f_{-1/2}^{(1/2)}) = (f_{1/2}^{(1/2)}, f_{-1/2}^{(1/2)}) \begin{pmatrix} a^* & b^* \\ -b & a \end{pmatrix}. \tag{4.3-12}$$

For $j = 1$,

$$f_1^{(1)} = \frac{u^2}{\sqrt{2}}, \qquad f_0^{(1)} = uv, \qquad f_{-1}^{(1)} = \frac{v^2}{\sqrt{2}}; \tag{4.3-13}$$

$$P_u f_1^{(1)} = \frac{1}{\sqrt{2}}(a^*u - bv)^2 = a^{*2} f_1^{(1)} - \sqrt{2}a^*b f_0^{(1)} + b^2 f_{-1}^{(1)},$$

$$P_u f_0^{(1)} = (a^*u - bv)(b^*u + av) = \sqrt{2}a^*b^* f_1^{(1)} + (aa^* - bb^*) f_0^{(1)} - \sqrt{2}ab f_{-1}^{(1)},$$

$$P_u f_{-1}^{(1)} = \frac{1}{\sqrt{2}}(b^*u + av)^2 = b^{*2} f_1^{(1)} + \sqrt{2}ab^* f_0^{(1)} + a^2 f_{-1}^{(1)},$$

or in matrix form

$$P_u(f_1^{(1)}, f_0^{(1)}, f_{-1}^{(1)}) = (f_1^{(1)}, f_0^{(1)}, f_{-1}^{(1)}) \times \begin{pmatrix} a^{*2} & \sqrt{2}a^*b^* & b^{*2} \\ -\sqrt{2}a^*b & aa^* - bb^* & \sqrt{2}ab^* \\ b^2 & -\sqrt{2}ab & a^2 \end{pmatrix}. \tag{4.3-14}$$

These are but two examples which demonstrate that the polynomials $f_m^{(j)}(\boldsymbol{\rho})$ transform among themselves under the transformation U. By expanding the right side of (4.3-9) by the binomial theorem it may be shown (Wigner, 1959) that

$$P_u f_m^{(j)}(\boldsymbol{\rho}) = \sum_{m'} f_{m'}^{(j)}(\boldsymbol{\rho}) U_{m'm}^{(j)}(a, b), \tag{4.3-15}$$

where

$$U^{(j)}_{m'm}(a,b) = \sum_{\lambda} \frac{(-1)^{\lambda}\sqrt{(j+m)!(j-m)!(j+m')!(j-m')!}}{\lambda!(j+m-\lambda)!(j-m'-\lambda)!(m'-m+\lambda)!} \times a^{j-m'-\lambda} a^{*j+m-\lambda} b^{\lambda} b^{*m'-m+\lambda} \tag{4.3-16a}$$

$$= \frac{1}{(m'-m)!}\sqrt{\frac{(j-m)!(j+m')!}{(j+m)!(j-m')!}}\, a^{j-m'} a^{*j+m} b^{*m'-m} + \sum_{\lambda \neq 0} \frac{(-1)^{\lambda}\sqrt{(j+m)!(j-m)!(j+m')!(j-m')!}}{\lambda!(j+m-\lambda)!(j-m'-\lambda)!(m'-m+\lambda)!} \times a^{j-m'-\lambda} a^{*j+m-\lambda} b^{\lambda} b^{*m'-m+\lambda}. \tag{4.3-16b}$$

The summation runs over all (integral) values of λ for which the denominator in (4.3-16a) is finite. Negative values of λ are therefore excluded. The structure of (4.3-15) permits the interpretation that the polynomials $f^{(j)}_m(\boldsymbol{\rho})$ are basis functions for the (unitary) representation $U^{(j)}(a, b)$ of the group $SU(2)$.

Two special cases of (4.3-16) are of interest. First, suppose that $b = 0$. All the terms in $U^{(j)}_{m'm}$ with $\lambda \neq 0$ are then zero and, when $\lambda = 0$, all terms with $m' - m \neq 0$ are zero; the only nonvanishing term is the one with $\lambda = 0$ and $m = m'$. Therefore, from (4.3-16b),

$$U^{(j)}_{m'm}(a, 0) = \delta_{m'm} a^{j-m} a^{*j+m}. \tag{4.3-17}$$

When

$$a = e^{-i\alpha/2}$$

we have

$$U^{(j)}_{m'm}(e^{-i\alpha/2}, 0) = \delta_{m'm} e^{im\alpha}. \tag{4.3-18}$$

The second special case is the one in which $j = m'$. In the denominator of (4.3-16a) there is the term $(j - m' - \lambda)!$. When $j = m'$, this term remains finite only for $\lambda = 0$; therefore

$$U^{(j)}_{jm}(a,b) = \sqrt{\frac{(2j)!}{(j+m)!(j-m)!}}\, a^{*j+m} b^{*j-m}. \tag{4.3-19}$$

It is now possible to prove that $U^{(j)}(a, b)$ is an irreducible representation of $SU(2)$. The proof is based on Schur's lemma (Section 3.3). Assume that there exists a matrix M which commutes with $U^{(j)}(a, b)$ so that

$$MU^{(j)}(a,b) = U^{(j)}(a,b)M, \tag{4.3-20}$$

or, in terms of the matrix elements,

$$\sum_k M_{m'k} U^{(j)}_{km}(a,b) = \sum_k U^{(j)}_{m'k}(a,b) M_{km}. \tag{4.3-21}$$

If (4.3-21) holds generally, it must also hold in the special case when $U^{(j)}(a, b)$ is the diagonal matrix $U^{(j)}(e^{-i\alpha/2}, 0)$ given by (4.3-18). For this case, (4.3-21) becomes

$$\sum_k M_{m'k}\,\delta_{km}e^{im\alpha} = \sum_k \delta_{m'k}e^{ik\alpha}M_{km}, \tag{4.3-22}$$

or

$$M_{m'm}e^{im\alpha} = e^{im'\alpha}M_{m'm}. \tag{4.3-23}$$

Therefore $M_{m'm}$ is identically zero unless $m = m'$, that is, M must be a diagonal matrix, in which case (4.3-21) may be written as

$$M_{m'm'}U^{(j)}_{m'm}(a, b) = U^{(j)}_{m'm}(a, b)M_{mm}. \tag{4.3-24}$$

If, further, $j = m'$,

$$M_{jj}U^{(j)}_{jm}(a, b) = U^{(j)}_{jm}(a, b)M_{mm}. \tag{4.3-25}$$

But $U^{(j)}_{jm}(a, b)$ is given by (4.3-19) which, in general, is not equal to zero. Therefore $M_{jj} = M_{mm}$ for all m. This means that M is not only a diagonal matrix but all the diagonal elements are the same. In other words, M is a multiple of the unit matrix. Thus it has been shown that the only matrix that commutes with the representation matrices $U^{(j)}(a, b)$ is a constant matrix. According to Schur's lemma, $U^{(j)}(a, b)$ must then be an irreducible representation.

4.4 Connection between $O^+(3)$ and $SU(2)$

Consider the 2×2 matrices

$$A = \begin{pmatrix} -z & x + iy \\ x - iy & z \end{pmatrix}, \qquad A' = \begin{pmatrix} -z' & x' + iy' \\ x' - iy' & z' \end{pmatrix} \tag{4.4-1}$$

where A' is related to A by the unitary transformation

$$A' = UAU^{-1}, \tag{4.4-2}$$

and U is given by (4.3-1). Upon performing the matrix multiplication in (4.4-2) it is found that

$$\begin{aligned} x' &= \tfrac{1}{2}(a^2 + a^{*2} - b^2 - b^{*2})x + \tfrac{1}{2}i(a^2 - a^{*2} + b^2 - b^{*2})y \\ &\quad + (ab + a^*b^*)z, \\ y' &= \tfrac{1}{2}i(a^{*2} - a^2 + b^2 - b^{*2})x + \tfrac{1}{2}(a^2 + a^{*2} + b^2 + b^{*2})y \\ &\quad + i(a^*b^* - ab)z, \\ z' &= -(a^*b + ab^*)x + i(a^*b - ab^*)y + (aa^* - bb^*)z. \end{aligned} \tag{4.4-3}$$

The transformation represented by (4.4-3) is of the form $\mathbf{r}' = R\mathbf{r}$. All the elements in the matrix R are real; the determinant of R is $+1$ and $\tilde{R} = R^{-1}$. Therefore R is a real orthogonal matrix and (4.4-3) represents a pure rotation. Evidently, there is some correspondence between the matrices U which are elements of $SU(2)$ and proper rotations which are elements of $O^+(3)$. A few examples will illustrate the nature of this correspondence. Let

$$a = e^{-i\alpha/2}, \qquad b = 0. \tag{4.4-4}$$

From (4.3-1),

$$U(\alpha) = \begin{pmatrix} e^{-i\alpha/2} & 0 \\ 0 & e^{i\alpha/2} \end{pmatrix}, \tag{4.4-5}$$

and, from (4.4-3),

$$x' = x\cos\alpha + y\sin\alpha, \qquad y' = -x\sin\alpha + y\cos\alpha, \qquad z' = z, \tag{4.4-6}$$

or

$$\mathbf{r}' = R(\alpha)\mathbf{r} \tag{4.4-7}$$

with

$$R(\alpha) = \begin{pmatrix} \cos\alpha & \sin\alpha & 0 \\ -\sin\alpha & \cos\alpha & 0 \\ 0 & 0 & 1 \end{pmatrix}. \tag{4.4-8}$$

Now, suppose that

$$a = -e^{-i\alpha/2}, \qquad b = 0; \tag{4.4-9}$$

then

$$\begin{pmatrix} -e^{-i\alpha/2} & 0 \\ 0 & -e^{i\alpha/2} \end{pmatrix} = -U(\alpha). \tag{4.4-10}$$

Substituting (4.4-9) into (4.4-3) we obtain precisely the same transformation (4.4-6) or the same matrix $R(\alpha)$ as given in (4.4-8). Thus there are two elements $U(\alpha)$ and $-U(\alpha)$ in $SU(2)$ which correspond to the single element $R(\alpha)$ in $O^+(3)$.

We consider another example; let

$$a = a^* = \cos\tfrac{1}{2}\beta, \qquad b = b^* = -\sin\tfrac{1}{2}\beta. \tag{4.4-11}$$

From (4.3-1),

$$U(\beta) = \begin{pmatrix} \cos\frac{1}{2}\beta & -\sin\frac{1}{2}\beta \\ \sin\frac{1}{2}\beta & \cos\frac{1}{2}\beta \end{pmatrix}, \tag{4.4-12}$$

and, from (4.4-3),

$$R(\beta) = \begin{pmatrix} \cos\beta & 0 & -\sin\beta \\ 0 & 1 & 0 \\ \sin\beta & 0 & \cos\beta \end{pmatrix}. \tag{4.4-13}$$

A rotation through an angle β about the y axis may therefore be described by $\pm U(\beta)$ or $R(\beta)$.

Finally, let

$$a = e^{-i(\alpha+\gamma)/2} \cos \tfrac{1}{2}\beta, \qquad b = -e^{-i(\alpha-\gamma)/2} \sin \tfrac{1}{2}\beta. \tag{4.4-14}$$

Then

$$U(\alpha, \beta, \gamma) = \begin{pmatrix} e^{-i(\alpha+\gamma)/2} \cos \tfrac{1}{2}\beta & -e^{-i(\alpha-\gamma)/2} \sin \tfrac{1}{2}\beta \\ e^{i(\alpha-\gamma)/2} \sin \tfrac{1}{2}\beta & e^{i(\alpha+\gamma)/2} \cos \tfrac{1}{2}\beta \end{pmatrix}, \tag{4.4-15}$$

and the corresponding rotation matrix $R(\alpha, \beta, \gamma)$, on the basis of (4.4-3), is

$$R(\alpha, \beta, \gamma) = \begin{pmatrix} \cos\alpha\cos\beta\cos\gamma - \sin\alpha\sin\gamma & \cos\alpha\cos\beta\sin\gamma + \sin\alpha\cos\gamma & -\cos\alpha\sin\beta \\ -\sin\alpha\cos\beta\cos\gamma - \cos\alpha\sin\gamma & -\sin\alpha\cos\beta\sin\gamma + \cos\alpha\cos\gamma & \sin\alpha\sin\beta \\ \sin\beta\cos\gamma & \sin\beta\sin\gamma & \cos\beta \end{pmatrix}. \tag{4.4-16}$$

This is identical with (2.3-8) and indicates that a rotation of coordinates through an arbitrary set of Euler angles may be expressed either by $R(\alpha, \beta, \gamma)$ or by $\pm U(\alpha, \beta, \gamma)$ as given by (4.4-15).

We see, then, that the correspondence between $SU(2)$ and $O^+(3)$ is not simply an isomorphism. Since there are two elements in $SU(2)$ that correspond to a single element in $O^+(3)$, the group $SU(2)$ is *homomorphic* to $O^+(3)$. This relationship may be represented schematically:

$SU(2)$ $\qquad$ $O^+(3)$

$U(\alpha, \beta, \gamma)$ ⟶ $R(\alpha, \beta, \gamma)$

$-U(\alpha, \beta, \gamma)$ ⟶ $R(\alpha, \beta, \gamma)$

4.5 Irreducible Representations of $O^+(3)$

The homomorphism between $SU(2)$ and $O^+(3)$ provides an elegant method for deriving a general expression for the irreducible representation of $O^+(3)$. In the previous section it was shown that a general three-dimensional rotation can be described by $\pm U(a, b)$ with a and b given by (4.4-14) and $U(a, b)$ an element of $SU(2)$ whose irreducible representations are calculated by the formula (4.3-16). Because $SU(2)$ is homomorphic to $O^+(3)$, it is merely necessary to substitute (4.4-14) into (4.3-16) to obtain the irreducible representations of $O^+(3)$. We then have

$$D^{(j)}_{m'm}(\alpha, \beta, \gamma) = U^{(j)}_{m'm}(e^{-i(\alpha+\gamma)/2}\cos\tfrac{1}{2}\beta, \quad -e^{-i(\alpha-\gamma)/2}\sin\tfrac{1}{2}\beta). \qquad (4.5\text{-}1)$$

There is a slight complication with phase factors associated with our adoption of the Condon–Shortley phase convention for the spherical harmonics. Direct evaluation of (4.5-1) does not lead to quite the same results as those obtained by transforming the spherical harmonics as in Section 2.3, but exact consistency is easily restored by transforming (4.5-1) with the unitary matrix $\delta_{m'm}(-1)^m$. The final result is

$$D^{(j)}_{m'm}(\alpha, \beta, \gamma) = \sum_{\lambda} \frac{(-1)^{\lambda}\sqrt{(j+m)!(j-m)!(j+m')!(j-m')!}}{\lambda!(j+m-\lambda)!(j-m'-\lambda)!(m'-m+\lambda)!}$$
$$\times\, e^{im'\alpha}(\cos\tfrac{1}{2}\beta)^{2j+m-m'-2\lambda}(\sin\tfrac{1}{2}\beta)^{m'-m+2\lambda}e^{im\gamma}. \qquad (4.5\text{-}2)$$

It is important to note the values of j. Since (4.5-2) is a direct consequence of (4.3-16) and the latter is based on the polynomial $f^{(j)}_{m}(\mathbf{p})$ in (4.3-8) with $j = 0, \frac{1}{2}, 1, \ldots$, the permissible values of j in (4.5-2) are also integral and half-integral values of j. Had we attempted to derive a general expression for the irreducible representations of $O^+(3)$ based on the transformation properties of the spherical harmonics, we would have been entirely unaware of the existence of representations with half-integral values of j.

The basis functions for representations with half-integral values of j cannot be the spherical harmonics. The situation here is reminiscent of that encountered in the discussion of generalized angular momentum where it was found that $|jm\rangle$, the eigenfunctions of J_z, and J^2 could not, in general (with j integral or half-integral), be written as a function of the space coordinates only. Rather, $|jm\rangle$ is generally a function of both space and spin coordinates. In the same way that $|lm\rangle = Y_{lm}(\theta, \varphi)$, the eigenfunctions of L_z and L^2, turned out to be basis functions for the irreducible representations $D^{(l)}(\alpha, \beta, \gamma)$ with $l = 0, 1, 2, \ldots$ (Section 4.2), we now have $|jm\rangle$, the eigenfunctions of J_z and J^2, as basis functions for the irreducible representations $D^{(j)}(\alpha, \beta, \gamma)$, i.e.,

$$P_R(\alpha, \beta, \gamma)|jm\rangle = \sum_{m'} |jm'\rangle D^{(j)}_{m'm}(\alpha, \beta, \gamma) \qquad (4.5\text{-}3a)$$

with $j = 0, \frac{1}{2}, 1, \frac{3}{2}, \ldots$. Since $m = j, j-1, \ldots, -j$, the dimension of $D^{(j)}(\alpha, \beta, \gamma)$ is $(2j+1)$. In (4.5-3a) the coordinate rotations are expressed in terms of the Euler angles. This is convenient but not mandatory since, as shown in Section 2.3, there exist other sets of three parameters that serve the same purpose. If R represents a general rotation in three dimensions, one may write in place of (4.5-3a)

$$P_R|jm\rangle = \sum_{m'} |jm'\rangle D^{(j)}_{m'm}(R). \qquad (4.5\text{-}3b)$$

The character of a representation $D^{(j)}$ is given by (4.2-4) with l replaced by j:

$$\chi^{(j)}(\omega) = \sin(j + \tfrac{1}{2})\omega / \sin \tfrac{1}{2}\omega \tag{4.5-4}$$

Suppose we form the direct product of two irreducible representations $D^{(j_1)}$ and $D^{(j_2)}$. In general, the direct product $D^{(j_1)} \times D^{(j_2)}$ is reducible and we should like to know how to decompose it into irreducible representations. The method described in Section 3.3 and, in particular, expression (3.3-16) is not applicable because $O^+(3)$ is a continuous group and therefore contains an infinite number of elements. However, it is possible to employ another method based on the fact that two representations with the same set of characters are either equivalent or equal (Section 3.2).

From (4.2-3),

$$\chi^{(j_1)}(\omega) = \sum_{m_1=-j_1}^{j_1} e^{im_1\omega}, \tag{4.5-5}$$

where ω is the angle of rotation and $m_1 = j_1, j_1 - 1, \ldots, -j_1$. A similar expression holds for $\chi^{(j_2)}(\omega)$. The character of $D^{(j_1)} \times D^{(j_2)}$ is $\chi^{(j_1 \times j_2)}(\omega)$ which, according to (3.6-4), is the product of $\chi^{(j_1)}(\omega)$ and $\chi^{(j_2)}(\omega)$, or

$$\chi^{(j_1 \times j_2)}(\omega) = \chi^{(j_1)}(\omega)\chi^{(j_2)}(\omega) = \sum_{m_1=-j_1}^{j_1} e^{im_1\omega} \sum_{m_2=-j_2}^{j_2} e^{im_2\omega}. \tag{4.5-6}$$

An example will illustrate the structure of (4.5-6). Let $j_1 = 1$ and $j_2 = 2$. Then

$$\sum_{m_1=-j_1}^{j_1} e^{im_1\omega} = e^{i\omega} + 1 + e^{-i\omega}, \tag{4.5-7}$$

$$\sum_{m_2=-j_2}^{j_2} e^{im_2\omega} = e^{2i\omega} + e^{i\omega} + 1 + e^{-i\omega} + e^{-2i\omega}, \tag{4.5-8}$$

$$\begin{aligned}\sum_{m_1=-j_1}^{j_1} e^{im_1\omega} \sum_{m_2=-j_2}^{j_2} e^{im_2\omega} &= (e^{i\omega} + 1 + e^{-i\omega})(e^{2i\omega} + e^{i\omega} + 1 + e^{-i\omega} + e^{-2i\omega}) \\ &= (e^{3i\omega} + e^{2i\omega} + e^{i\omega} + 1 + e^{-i\omega} + e^{-2i\omega} + e^{-3i\omega}) \\ &\quad + (e^{2i\omega} + e^{i\omega} + 1 + e^{-i\omega} + e^{-2i\omega}) + (e^{i\omega} + 1 + e^{-i\omega}) \\ &= \chi^{(j_1+j_2)} + \chi^{(j_1+j_2-1)} + \chi^{(|j_1-j_2|)}. \end{aligned} \tag{4.5-9}$$

The computation may be readily generalized to

$$\chi^{(j_1 \times j_2)}(\omega) = \sum_{j=|j_1-j_2|}^{j_1+j_2} \chi^{(j)}(\omega), \tag{4.5-10}$$

which can only be satisfied by

$$D^{(j_1)} \times D^{(j_2)} = \sum_{j=|j_1-j_2|}^{j_1+j_2} D^{(j)}. \tag{4.5-11}$$

Thus the decomposition of the product representation $D^{(j_1)} \times D^{(j_2)}$ consists of the irreducible representations $D^{(j_1+j_2)}, D^{(j_1+j_2-1)}, \ldots, D^{(|j_1-j_2|)}$ with no irreducible representation appearing more than once.

We have seen that the functions $|j_1 m_1\rangle$ form a basis for a $(2j_1 + 1)$-dimensional irreducible representation $D^{(j_1)}$ of $O^+(3)$; similarly, the functions $|j_2 m_2\rangle$ form a basis for a $(2j_2 + 1)$-dimensional irreducible representation $D^{(j_2)}$. The product functions $|j_1 m_1\rangle|j_2 m_2\rangle \equiv |j_1 j_2 m_1 m_2\rangle$ form a basis for a $(2j_1 + 1)(2j_2 + 1)$-dimensional representation $D^{(j_1)} \times D^{(j_2)}$. But the latter is generally reducible as in (4.5-11); therefore the basis functions $|jm\rangle$ for the irreducible representations $D^{(j)}$, the components in the reduction of $D^{(j_1)} \times D^{(j_2)}$, must be expressible as linear combinations of $|j_1 m_1\rangle$ and $|j_2 m_2\rangle$. It may, in fact, be shown (Wigner, 1959; Hamermesh, 1962) that

$$|jm\rangle = \sum_{m_1 m_2} |j_1 j_2 m_1 m_2\rangle\langle j_1 j_2 m_1 m_2 | j_1 j_2 jm\rangle, \tag{4.5-12}$$

in which $\langle j_1 j_2 m_1 m_2 | j_1 j_2 j_m\rangle$ are the same coefficients that appear in the coupling of two angular momentum operators $\mathbf{J}_1$ and $\mathbf{J}_2$ (1.5-16).

In Section 3.7 we obtained a general selection rule on a matrix element

$$M = \langle \psi_\alpha^{(i)} | Q_\beta^{(j)} | \varphi_\gamma^{(k)} \rangle \tag{4.5-13}$$

where $\psi_\alpha^{(i)}$, $Q_\beta^{(j)}$, and $\varphi_\gamma^{(j)}$ transform according to the irreducible representations $\Gamma^{(i)}$, $\Gamma^{(j)}$, and $\Gamma^{(k)}$ of a group G. The rule was that M vanishes unless $\Gamma^{(i)*} \times \Gamma^{(j)} \times \Gamma^{(k)}$ contains the unit representation. Now suppose G is the group $O^+(3)$. In that case M will be different from zero only if $D^{(i)*} \times D^{(j)} \times D^{(k)}$ contains the unit representation. Since the characters of D are real, $D^{(i)} \times D^{(j)}$ must contain $D^{(k)}$ (or the direct product of any two of the three irreducible representations $D^{(i)}$, $D^{(j)}$, $D^{(k)}$ must contain the third). In view of (4.5-11) an equivalent statement is that $M = 0$ unless the triangle condition $\triangle\,(ijk)$ is satisfied.

Our discussion of the three-dimensional rotation group has so far been confined to the group of proper rotations $O^+(3)$ which consists of all three-dimensional orthogonal matrices with determinants equal to $+1$. The general or *full* three-dimensional rotation group $O(3)$ also includes all real three-dimensional orthogonal matrices with determinants equal to -1. The latter correspond to rotation–reflections, that is, elements that are equivalent to a proper rotation combined with an inversion. Such combinations are also known as *improper* rotations and we note that proper rotations and inversions commute. Therefore the elements of $O(3)$ consist of all elements $R(\alpha, \beta, \gamma)$

shown in (4.4-16) together with the elements $iR(\alpha, \beta, \gamma) = R(\alpha, \beta, \gamma)i$ where i is the inversion matrix (3.5-24).

In Section 3.5 we encountered the group C_i which contains just two elements: the identity element E—in this case a three-dimensional unit matrix—and the inversion element i. Clearly the elements of $O(3)$ consist of all possible combinations of elements in $O^+(3)$ with elements in C_i. Symbolically we may then write

$$O(3) = O^+(3) \times C_i \tag{4.5-14}$$

to indicate that the group $O(3)$ is the direct product of the groups $O^+(3)$ and C_i.

Since the group representations, including the irreducible ones, must also satisfy (4.5-14), each irreducible representation of $O(3)$ may be regarded as the direct product of an irreducible representation of $O^+(3)$ with an irreducible representation of C_i. We already know the irreducible representations of $O^+(3)$ from (4.5-2). As for C_i—an Abelian group with two elements—there are just two one-dimensional representations $\Gamma^{(1)}$ and $\Gamma^{(2)}$ shown in (3.5-25). Hence

$$D^{(j)} \times \Gamma^{(1)} = D^{(j)}, \tag{4.5-15}$$

$$D^{(j)} \times \Gamma^{(2)} = \pm D^{(j)}, \tag{4.5-16}$$

in which the positive sign in (4.5-16) is for elements $ER = R$ and the negative sign for elements iR, with R an element of $O^+(3)$.

In summary, for each irreducible representation $D^{(j)}$ belonging to $O^+(3)$ there are two irreducible representations $D_{\pm}^{(j)}$ belonging to $O(3)$; they are also called *even* $(+)$ and *odd* $(-)$ representations. The dimensions are $2j + 1$, and j can assume integral or half-integral values. For pure rotations R through an angle ω about some axis, the characters of the irreducible representations in $O(3)$ are the same as those in $O^+(3)$, i.e.,

$$\chi_{\pm}^{(j)}(R) = \chi^{(j)}(R), \tag{4.5-17}$$

but for elements iR

$$\chi_{+}^{(j)}(iR) = \chi^{(j)}(R), \qquad \chi_{-}^{(j)}(iR) = -\chi^{(j)}(R). \tag{4.5-18}$$

Thus formula (4.5-4) may also be used to evaluate the characters of the irreducible representations of $O(3)$.

4.6 Summary and Examples

A proper rotation of coordinates in three dimensions can be described by

$$\mathbf{r}' = R(\alpha, \beta, \gamma)\mathbf{r}, \tag{4.6-1}$$

where α, β, γ are Euler angles (Section 2.3) and $R(\alpha, \beta, \gamma)$ is the orthogonal matrix (2.3-8) whose determinant is equal to $+1$. By varying α, β, γ in a continuous fashion, an infinite set of matrices $R(\alpha, \beta, \gamma)$ is obtained. These matrices are the elements of the (continuous) rotation group in three dimensions $O^+(3)$.

The irreducible representations of $O^+(3)$ are $D^{(j)}(\alpha, \beta, \gamma)$ with $j = 0, \frac{1}{2}, 1, \frac{3}{2}, \ldots$; their matrix elements $D^{(j)}_{m'm}(\alpha, \beta, \gamma)$ are given by (4.5-2) with $m', m = j, j-1, \ldots, -j$. The dimension of $D^{(j)}(\alpha, \beta, \gamma)$ is $(2j+1)$ and the character is

$$\chi^{(j)}(\omega) = \sin(j + \tfrac{1}{2})\omega / \sin \tfrac{1}{2}\omega, \tag{4.6-2}$$

in which ω is the angle of rotation about any axis. A few properties of $D^{(j)}(\alpha, \beta, \gamma)$ are:

$$[D^{(j)}(\alpha, \beta, \gamma)]^\dagger = [D^{(j)}(\alpha, \beta, \gamma)]^{-1} = D^{(j)}(-\gamma, -\beta, -\alpha), \tag{4.6-3}$$

$$\det D^{(j)}(\alpha, \beta, \gamma) = 1, \tag{4.6-4}$$

$$D^{(l)}_{0m}(\alpha, \beta, \gamma) = \sqrt{\frac{4\pi}{2l+1}}\, Y_{lm}(\beta, \gamma). \tag{4.6-5}$$

Several examples of $D^{(j)}(\alpha, \beta, \gamma)$ are shown in Tables 4.1–4.3.

TABLE 4.1

$D^{(1/2)}_{m'm}(\alpha, \beta, \gamma)$

m' \ m	$\frac{1}{2}$	$-\frac{1}{2}$
$\frac{1}{2}$	$e^{i\alpha/2} \cos \frac{1}{2}\beta\, e^{i\gamma/2}$	$e^{i\alpha/2} \sin \frac{1}{2}\beta\, e^{-i\gamma/2}$
$-\frac{1}{2}$	$-e^{-i\alpha/2} \sin \frac{1}{2}\beta\, e^{i\gamma/2}$	$e^{-i\alpha/2} \cos \frac{1}{2}\beta\, e^{-i\gamma/2}$

TABLE 4.2

$D^{(1)}_{m'm}(\alpha, \beta, \gamma)$

m' \ m	1	0	-1
1	$e^{i\alpha}\left(\frac{1+\cos\beta}{2}\right)e^{i\gamma}$	$e^{i\alpha}\frac{\sin\beta}{\sqrt{2}}$	$e^{i\alpha}\left(\frac{1-\cos\beta}{2}\right)e^{-i\gamma}$
0	$-\frac{\sin\beta}{\sqrt{2}}e^{i\gamma}$	$\cos\beta$	$\frac{\sin\beta}{\sqrt{2}}e^{-i\gamma}$
-1	$e^{-i\alpha}\left(\frac{1-\cos\beta}{2}\right)e^{i\gamma}$	$-e^{-i\alpha}\frac{\sin\beta}{\sqrt{2}}$	$e^{-i\alpha}\left(\frac{1+\cos\beta}{2}\right)e^{-i\gamma}$

TABLE 4.3

$D^{(2)}_{m'm}(\alpha, \beta, \gamma)$

m' \ m	2	1	0	−1	−2
2	$\frac{1}{4}(1+\cos\beta)^2 \times e^{2i\alpha+2i\gamma}$	$\frac{1}{2}\sin\beta(1+\cos\beta) \times e^{2i\alpha+i\gamma}$	$\sqrt{\frac{3}{8}}\sin^2\beta\, e^{2i\alpha}$	$\frac{1}{2}\sin\beta(1-\cos\beta) \times e^{2i\alpha-i\gamma}$	$\frac{1}{4}(1-\cos\beta)^2 \times e^{2i\alpha-2i\gamma}$
1	$-\frac{1}{2}\sin\beta(1+\cos\beta) \times e^{i\alpha+2i\gamma}$	$\frac{1}{2}(2\cos^2\beta+\cos\beta-1) \times e^{i\alpha+i\gamma}$	$\sqrt{\frac{3}{2}}\sin\beta\cos\beta\, e^{i\alpha}$	$-\frac{1}{2}(2\cos^2\beta-\cos\beta-1) \times e^{i\alpha-i\gamma}$	$\frac{1}{2}\sin\beta(1-\cos\beta) \times e^{i\alpha-2i\gamma}$
0	$\sqrt{\frac{3}{8}}\sin^2\beta\, e^{2i\gamma}$	$-\sqrt{\frac{3}{2}}\sin\beta\cos\beta\, e^{i\gamma}$	$\frac{3}{2}\cos^2\beta-\frac{1}{2}$	$\sqrt{\frac{3}{2}}\sin\beta\cos\beta\, e^{-i\gamma}$	$\sqrt{\frac{3}{8}}\sin^2\beta\, e^{-2i\gamma}$
−1	$-\frac{1}{2}\sin\beta(1-\cos\beta) \times e^{-i\alpha+2i\gamma}$	$-\frac{1}{2}(2\cos^2\beta-\cos\beta-1) \times e^{-i\alpha+i\gamma}$	$-\sqrt{\frac{3}{2}}\sin\beta\cos\beta\, e^{-i\alpha}$	$\frac{1}{2}(2\cos^2\beta+\cos\beta-1) \times e^{-i\alpha-i\gamma}$	$\frac{1}{2}\sin\beta(1+\cos\beta) \times e^{-i\alpha-2i\gamma}$
−2	$\frac{1}{4}(1-\cos\beta)^2 \times e^{-2i\alpha+2i\gamma}$	$-\frac{1}{2}\sin\beta(1-\cos\beta) \times e^{-2i\alpha+i\gamma}$	$\sqrt{\frac{3}{8}}\sin^2\beta\, e^{-2i\alpha}$	$-\frac{1}{2}\sin\beta(1+\cos\beta) \times e^{-2i\alpha-i\gamma}$	$\frac{1}{4}(1+\cos\beta)^2 \times e^{-2i\alpha-2i\gamma}$

The basis functions for $D^{(j)}(\alpha, \beta, \gamma)$ are symbolized by $|jm\rangle$ which, in general, are functions of space and spin coordinates. This means that

$$P_R(\alpha, \beta, \gamma)|jm\rangle = \sum_{m'} |jm'\rangle D^{(j)}_{m'm}(\alpha, \beta, \gamma). \tag{4.6-6}$$

The spherical harmonics $Y_{lm}(\theta, \varphi) = |lm\rangle$ are basis functions for $D^{(j)}(\alpha, \beta, \gamma)$ when $j = l = 0, 1, 2, \ldots$, or

$$P_R(\alpha, \beta, \gamma) Y_{lm}(\theta, \varphi) = \sum_{m'} Y_{lm'}(\theta, \varphi) D^{(l)}_{m'm}(\alpha, \beta, \gamma). \tag{4.6-7}$$

Equation (4.6-7) is the basic relationship for the transformation of the spherical harmonics under rotations of coordinates.

The transformation properties of the spin functions are obtained directly from $D^{(1/2)}(\alpha, \beta, \gamma)$ (Table 4.1). To avoid the notational confusion between the Euler angles α, β, γ and the spin functions α, β we shall write the latter, as in Section 1.4,

$$\alpha = |\tfrac{1}{2}\tfrac{1}{2}\rangle, \qquad \beta = |\tfrac{1}{2}\ -\tfrac{1}{2}\rangle.$$

Then

$$\begin{aligned} P_R(\alpha, \beta, \gamma)|\tfrac{1}{2}\tfrac{1}{2}\rangle &= e^{i\alpha/2} \cos \tfrac{1}{2}\beta\, e^{i\gamma/2} |\tfrac{1}{2}\tfrac{1}{2}\rangle \\ &\quad - e^{i\alpha/2} \sin \tfrac{1}{2}\beta\, e^{i\gamma/2} |\tfrac{1}{2}\ -\tfrac{1}{2}\rangle, \end{aligned} \tag{4.6-8a}$$

$$\begin{aligned} P_R(\alpha, \beta, \gamma)|\tfrac{1}{2}\ -\tfrac{1}{2}\rangle &= e^{i\alpha/2} \sin \tfrac{1}{2}\beta\, e^{-i\gamma/2} |\tfrac{1}{2}\tfrac{1}{2}\rangle \\ &\quad + e^{-i\alpha/2} \cos \tfrac{1}{2}\beta\, e^{-i\gamma/2} |\tfrac{1}{2}\ -\tfrac{1}{2}\rangle. \end{aligned} \tag{4.6-8b}$$

For the special case of a coordinate transformation $R(\alpha, \beta, \gamma) = R(\alpha, 0, 0)$,

$$\begin{aligned} P_R(\alpha, 0, 0)|\tfrac{1}{2}\tfrac{1}{2}\rangle &= e^{i\alpha/2}|\tfrac{1}{2}\tfrac{1}{2}\rangle, \\ P_R(\alpha, 0, 0)|\tfrac{1}{2}\ -\tfrac{1}{2}\rangle &= e^{-i\alpha/2}|\tfrac{1}{2}\ -\tfrac{1}{2}\rangle, \end{aligned} \tag{4.6-9}$$

and

$$\chi^{(1/2)}(\alpha, 0, 0) = e^{i\alpha/2} + e^{-i\alpha/2} = \frac{\sin(\tfrac{1}{2} + \tfrac{1}{2})\alpha}{\sin \tfrac{1}{2}\alpha} = 2 \cos \tfrac{1}{2}\alpha. \tag{4.6-10}$$

If α is increased by 2π, the relations in (4.6-9) become

$$\begin{aligned} P_R(\alpha + 2\pi, 0, 0)|\tfrac{1}{2}\tfrac{1}{2}\rangle &= -e^{i\alpha/2}|\tfrac{1}{2}\tfrac{1}{2}\rangle, \\ P_R(\alpha + 2\pi, 0, 0)|\tfrac{1}{2}\ -\tfrac{1}{2}\rangle &= -e^{-i\alpha/2}|\tfrac{1}{2}\ -\tfrac{1}{2}\rangle, \end{aligned} \tag{4.6-11}$$

and in place of (4.6-10) we obtain

$$\chi^{(1/2)}(\alpha + 2\pi, 0, 0) = -2 \cos \tfrac{1}{2}\alpha. \tag{4.6-12}$$

The double-valued feature of the spin functions which emerges from a comparison of (4.6-9) and (4.6-10) with (4.6-11) and (4.6-12) should be carefully noted.

TABLE 4.4

Summary of Properties of Angular Momentum Operators and the Three-Dimensional Rotation Group

1. The irreducible representations $D^{(j)}(\alpha, \beta, \gamma)$ of $O^+(3)$ are of dimension $2j + 1$ with $j = 0, \frac{1}{2}, 1, \frac{3}{2}, \ldots$.
2. For a rotation through an angle ω about an axis whose direction is defined by a unit vector $\hat{\mathbf{n}}$,

$$P_R(\omega, \hat{n}) = e^{i\omega\hat{\mathbf{n}}\cdot\mathbf{J}} \qquad \chi^{(j)}(\omega) = \frac{\sin(j + \frac{1}{2})\omega}{\sin\frac{1}{2}\omega}.$$

$\chi^{(j)}(\omega)$ is single-valued for integral values of j and double-valued for half-integral values.

3. The angular momentum eigenfunctions $|jm\rangle$ are basis functions for $D^{(j)}(\alpha, \beta, \gamma)$:

$$P_R(\alpha, \beta, \gamma)|jm\rangle = \sum_{m'} |jm'\rangle\, D^{(j)}_{m'm}(\alpha, \beta, \gamma).$$

The spherical harmonics $Y_{lm}(\theta, \varphi)$ are basis functions for $D^{(l)}(\alpha, \beta, \gamma)$ with $l = 0, 1, 2, \ldots$.

4. The product representation $D^{(j_1)} \times D^{(j_2)}$ is reducible:

$$D^{(j_1)} \times D^{(j_2)} = \sum_{j=|j_1-j_2|}^{j_1+j_2} D^{(j)}$$

and the basis functions for $D^{(j)}$ are given by

$$\begin{aligned} |jm\rangle &= \sum_{m_1 m_2} |j_1 j_2 m_1 m_2\rangle\langle j_1 j_2 m_1 m_2 | j_1 j_2 jm\rangle \\ &= (-1)^{j_2 - j_1 - m}\sqrt{2j+1} \sum_{m_1 m_2} \begin{pmatrix} j_1 & j_2 & j \\ m_1 & m_2 & -m \end{pmatrix} |j_1 j_2 m_1 m_2\rangle \end{aligned}$$

with

$$\begin{aligned} j &= j_1 + j_2, j_1 + j_2 - 1, \ldots, |j_1 - j_2| \qquad & m &= m_1 + m_2 \\ &= 0, \tfrac{1}{2}, 1, \tfrac{3}{2}, \ldots, & &= j, j - 1, \ldots, -j. \end{aligned}$$

Rotation Group $O^+(3)$

Angular Momentum

1. $\mathbf{J}$ is an angular momentum operator if J_x, J_y, and J_z are Hermitian and $\mathbf{J} \times \mathbf{J} = i\mathbf{J}$
2. $|jm\rangle$ is a simultaneous eigenfunction of J_z and J^2 satisfying

$$J_z|jm\rangle = m|jm\rangle \qquad J^2|jm\rangle = j(j+1)|jm\rangle$$

with $j = 0, \frac{1}{2}, 1, \frac{3}{2}, \ldots$ and $m = j, j-1, \ldots, -j$. For a fixed value of j there are $2j+1$ eigenfunctions $|jm\rangle$.

3. Under a rotation of coordinates $\mathbf{r}' = R(\alpha, \beta, \gamma)\mathbf{r}$, the eigenfunctions $|jm\rangle$ transform according to

$$P_R(\alpha, \beta, \gamma)|jm\rangle = \sum_{m'} |jm'\rangle\, D^{(j)}_{m'm}(\alpha, \beta, \gamma).$$

4. The spherical harmonics $Y_{lm}(\theta, \varphi) \equiv |lm\rangle$ satisfy

$$L_z|lm\rangle = m|lm\rangle, \qquad L^2|lm\rangle = l(l+1)|lm\rangle$$

with $l = 0, 1, 2, \ldots$ and $m = l, l-1, \ldots, -l$ where L_x, L_y, and L_z are Hermitian and $\mathbf{L} \times \mathbf{L} = i\mathbf{L}$. $\mathbf{L}$ is the orbital angular momentum operator. Under a rotation of coordinates $\mathbf{r}' = R(\alpha, \beta, \gamma)\mathbf{r}$, the eigenfunctions $|lm\rangle$ transform according to

$$P_R(\alpha, \beta, \gamma)|lm\rangle = \sum_{m'} |lm'\rangle\, D^{(l)}_{m'm}(\alpha, \beta, \gamma).$$

5. Given that $\mathbf{J}_1$ and $\mathbf{J}_2$ are two angular momentum operators, $|j_1 m_1\rangle$ is a simultaneous eigenfunction of J_{1z} and $J_1{}^2$, $|j_2 m_2\rangle$ is a simultaneous eigenfunction of J_{2z} and $J_2{}^2$, and

$$[J_{1i}, J_{2j}] = 0, \qquad |j_1 m_1\rangle|j_2 m_2\rangle \equiv |j_1 j_2 m_1 m_2\rangle.$$

The coupling of $\mathbf{J}_1$ and $\mathbf{J}_2$ is defined by $\mathbf{J} = \mathbf{J}_1 + \mathbf{J}_2$; the eigenfunctions $|jm\rangle$ which satisfy $J_z|jm\rangle = m|jm\rangle$ and $J^2|jm\rangle = j(j+1)|jm\rangle$ are given by

$$\begin{aligned} |jm\rangle &= \sum_{m_1 m_2} |j_1 j_2 m_1 m_2\rangle\langle j_1 j_2 m_1 m_2|j_1 j_2 jm\rangle \\ &= (-1)^{j_2 - j_1 - m}\sqrt{2j+1} \sum_{m_1 m_2} \begin{pmatrix} j_1 & j_2 & j \\ m_1 & m_2 & -m \end{pmatrix} |j_1 j_2 m_1 m_2\rangle \end{aligned}$$

with

$$\begin{aligned} j &= j_1 + j_2, j_1 + j_2 - 1, \ldots, |j_1 - j_2| \qquad & m &= m_1 + m_2 \\ &= 0, \tfrac{1}{2}, 1, \tfrac{3}{2}, \ldots, & &= j, j-1, \ldots, -j. \end{aligned}$$

The full rotation group $O(3)$ consists of all real orthogonal matrices whose determinants are ± 1. For every irreducible representation $D^{(j)}$ of $O^+(3)$ there are two irreducible representations $D_{\pm}^{(j)}$ of $O(3)$ with $D_{+}^{(j)}$ and $D_{-}^{(j)}$ even and odd, respectively, under inversion.

Table 4.4 summarizes the properties of angular momentum operators, the rotation group $O^+(3)$, and their interrelationships. The latter may be summarized by the statement that the angular momentum operators, by virtue of their commutation properties, are the *generators* of the rotation group or, alternatively, that they form the *algebra* of the rotation group.

CHAPTER 5

FINITE GROUPS

5.1 Point Groups—Symmetry Operations and Nomenclature

Points groups are defined as groups consisting of elements whose axes and planes of symmetry have at least one common point of intersection. All possible symmetry operations for point groups can be represented as a combination of

(a) a rotation through a definite angle about some axis and
(b) a reflection in some plane.

Definitions of Symmetry Operations:

E identity transformation.

C_n a rotation through an angle $2\pi/n$ about a given axis; $n = 1, 2, 3, \ldots$. When $n = 1$, the angle of rotation is 2π (or 0) which is simply an identity transformation E. $C_n^2, C_n^3, \ldots$ are rotations through $2(2\pi/n)$, $3(2\pi/n), \ldots$. Also, $C_n^n = E$.

σ reflection in a plane; $\sigma^2 = E$.

σ_h reflection in a horizontal plane which is defined as a plane perpendicular to the principal axis of rotation (axis with highest n).

σ_v reflection in a vertical plane which is a plane containing the principal axis.

σ_d a special case of σ_v in which the vertical plane bisects the angle between the twofold axes perpendicular to the symmetry axis.

i inversion or reflection in the origin.

S_n rotary reflection. This is a rotation through an angle of $2\pi/n$ together with a reflection in a plane perpendicular to the axis. S_n is also called an improper rotation, $S_2 = i$, $S_n = C_n\sigma_h = \sigma_h C_n$; also $i\sigma_h = C_2$, $iC_2 = \sigma_h$.

We note that σ_h and C_n commute; other operations that commute are two rotations about the same axis, two reflections in perpendicular planes, and two rotations by π about perpendicular axes.

A common nomenclature for point groups is the Schonflies notation: C (cyclic), D (dihedral), O (octahedral), and T (tetrahedral). These are further subdivided as follows:

C_n n rotations about an n-fold symmetry axis; the elements are C_n, $C_n^2, \ldots, C_n^{n-1}, C_n^n (=E)$; the group C_n is Abelian.

C_{nv} n rotations of the group C_n and n reflections σ_v in vertical planes intersecting at angles of π/n. C_{3v} contains the elements E, C_3, C_3^2, $3\sigma_v$ and the angle between reflection planes is $\pi/3$.

C_{nh} n rotations of the group C_n and n rotary–reflection operations of the type $C_n^k\sigma_h$, $k = 1, 2, \ldots, n$. C_{3h} contains the elements E, C_3, C_3^2, σ_h, $C_3\sigma_h (= S_3)$, $C_3^2\sigma_h (= S_3^5)$. C_{nh} is Abelian. If n is even, C_{nh} contains the inversion operator i. The group C_s ($\equiv C_{1h}$) contains the elements E and σ_h.

S_{2n} $2n$ rotary reflections. S_6 contains the elements E, $C_3(S_6^2)$, $C_3^2(S_6^4)$, $i(S_6^3)$, S_6, S_6^5. The group S_n with n odd is the same as C_{nh}. The group S_2 ($\equiv C_i$) contains the elements E and i.

D_n n rotations of the group C_n and n rotations through an angle π about horizontal axes (perpendicular to the n-fold axis). D_3 contains E, $2C_3$, $3C_2$.

D_{nd} contains $4n$ elements: $2n$ elements of the group D_n and $2n$ elements obtained by multiplying each element of D_n by σ_d. D_{3d} contains E, $2S_6$, $2C_3$, i, $3C_2$, $3\sigma_d$. The reflection planes σ_d bisect the angles between the three twofold axes in D_3.

D_{nh} contains $4n$ elements: $2n$ elements of the group D_n and $2n$ elements obtained by multiplying each element of D_n by σ_h. D_{3h} contains E, σ_h, $2C_3$, $2S_3$, $3C_2$, $3\sigma_v$.

T the group of proper rotations of a regular tetrahedron: E, $4C_3$, $4C_3^2$, $3C_2$.

T_d contains the elements of T plus those obtained by multiplication with σ_d. The 24 elements are E, $8C_3$, $6\sigma_d$, $6S_4$, $3C_2$.

T_h contains the elements of T plus those obtained by multiplication with the inversion operator i. The 24 elements are E, $3C_2$, $4C_3$, $4C_3^2$, i, $3\sigma_h$, $4iC_3$, and $4S_6$.

O The symmetry group of proper rotations of a cube. There are 24 elements: E, $8C_3$, $3C_2$, $6C_4$, $6C_2'$. A detailed description of the group is given in Section 5.3.

O_h The elements of O plus those obtained by multiplication with i. The 48 elements are E, $8C_3$, $3C_2$, $6C_4$, $6C_2'$, i, $8S_6$, $3\sigma_h$, $6S_4$, $6\sigma_d$. O_h is the full symmetry group of the cube or octahedron including reflections and improper rotations.

Previously it was shown (Section 4.5) that the general rotation group $O(3)$ can be regarded as the direct product of the groups $O^+(3)$ and C_i. The same is true for other point groups. Also, irreducible representations of the product group are given by the direct product of the irreducible representations of the component groups. For example, D_3 contains the elements E, $2C_3$ and $3C_2$; C_i contains E and i. By taking all possible products of the elements of both groups we get E, $2C_3$, $3C_2$, i, $2C_3i$ $(= 2S_6)$ and $3C_2i$ $(= 3\sigma_d)$. These are then the elements of the direct product group $D_3 \times C_i = D_{3d}$ and the irreducible representations of D_{3d} are the direct products of $\Gamma^{(1)}$, $\Gamma^{(2)}$ and $\Gamma^{(3)}$ of D_3 with $\Gamma^{(1)}$ and $\Gamma^{(2)}$ of C_i, i.e.,

$$\Gamma^{(i)}(D_3) \times \Gamma^{(1)}(C_i) = \Gamma_+^{(i)}, \qquad \Gamma^{(i)}(D_3) \times \Gamma^{(2)}(C_i) = \Gamma_-^{(i)}, \qquad (i = 1, 2, 3)$$

with the characters

$$\chi_+^{(i)} = \chi^{(i)}(D_3)\chi^{(1)}(C_i), \qquad \chi_-^{(i)} = \chi^{(i)}(D_3)\chi^{(2)}(C_i)$$

Other direct product groups are

$$\begin{array}{lll} C_{3h} = C_3 \times C_s, & D_{2h} = D_2 \times C_i, & T_h = T \times C_i, \\ C_{4h} = C_4 \times C_i, & D_{4h} = D_4 \times C_i, & O_h = O \times C_i. \\ C_{6h} = C_6 \times C_i, & D_{6h} = D_6 \times C_i, & \\ S_6 = C_3 \times C_i, & D_{3d} = D_3 \times C_i, & \end{array} \qquad (5.1\text{-}1)$$

Two other groups are of importance in molecules. These are the groups $C_{\infty v}$ and $D_{\infty h}$; both are infinite groups.

$C_{\infty v}$ the group of a general linear molecule. The elements are E, $2C_\varphi$, and σ_v; $2C_\varphi$ are rotations through $\pm\varphi$ about the molecular axis and σ_v is a reflection in a plane containing the axis.

$D_{\infty h}$ The symmetry group of a homonuclear diatomic molecule or a symmetric linear molecule (like O=C=O). The group contains E, $2C_\varphi$, i, $2iC_\varphi$ $(=2S_\varphi)$, σ_v, and $i\sigma_v$ $(=C_2)$.

Crystallographic point groups are restricted to those containing axes of symmetry of order $n = 1, 2, 3, 4$ and 6. There are 32 groups of this type (Table 5.1); their detailed properties are described by Koster *et al.* (1963).

TABLE 5.1

Classification of the 32 Crystallographic Point Groups

Cubic	O, O_h, T_d, T, T_h
Tetragonal	$C_4, S_4, D_{2d}, C_{4v}, C_{4h}, D_4, D_{4h}$
Hexagonal	$D_{3h}, D_6, D_{6h}, C_{3h}, C_6, C_{6h}, C_{6v}$
Trigonal	$C_{3v}, D_{3d}, D_3, C_3, S_6$
Rhombic	$C_{2v}, D_2, D_{2h},$
Triclinic	$C_1, C_i\,(S_2)$
Monoclinic	$C_{1h}\,(C_s), C_2, C_{2h}$

Table 5.2 is a character table for the more common point groups. Included in the table are the characters of the irreducible representations as well as the linear and bilinear combinations of x, y, z and R_x, R_y, R_z that constitute basis functions for the irreducible representations. R_x, R_y and R_z are components of an *axial* (pseudo-) vector. Under proper rotations (R_x, R_y, R_z) transform like (x, y, z) which are components of a *polar* vector, but under inversion the components of an axial vector remain unchanged while the components of a polar vector change sign. For example, if **A** and **B** are polar vectors, $\mathbf{A} \times \mathbf{B}$, $\nabla \times \mathbf{A}$, $\nabla \times \mathbf{B}$ are all axial vectors. We note (Table 5.2) that the characters of the representation matrices generated by (R_x, R_y, R_z) are the same as those generated by (x, y, z) when the symmetry element is a proper rotation, but, for improper rotations, there is a difference in sign.

All the point groups are subgroups of $O(3)$, the full three-dimensional rotation group (including proper and improper rotations). In turn, most point groups have (nontrivial) subgroups which are also point groups but with fewer symmetry elements. Thus O_h is a subgroup of $O(3)$; D_{4h} is a subgroup of O_h; D_4 is a subgroup of D_{4h}; etc. It has previously been noted (Sections 3.2 and 3.3) that a representation Γ of a group G is also a representation of every subgroup of G. If, however, Γ is an irreducible representation of G, it is not necessarily an irreducible representation of the subgroup.

As an illustration, consider the irreducible representation $D^{(2)}$ of $O^+(3)$. $D^{(2)}$ is also a representation of the subgroup D_3 but in the latter group $D^{(2)}$ is no longer an irreducible representation because, as will be shown, $D^{(2)}$

TABLE 5.2

Character Tables and Basis Functions[a,b,c]

C_1	E
A	1

C_s	E	σ_h		
A'	1	1	x, y, R_z	x^2, y^2, z^2, xy
A''	1	-1	z, R_x, R_y	yz, xz

C_i	E	i		
A_g	1	1	R_x, R_y, R_z	$x^2, y^2, z^2, xy, xz, yz$
A_u	1	-1	x, y, z	

C_2	E	C_2		
A	1	1	z, R_z	x^2, y^2, z^2, xy
B	1	-1	x, y, R_x, R_y	yz, xz

C_3	E	C_3	C_3^2		$\epsilon = \exp(2\pi i/3)$
A	1	1	1	z, R_z	$x^2 + y^2, z^2$
E	$\begin{Bmatrix} 1 \\ 1 \end{Bmatrix}$	$\begin{matrix} \epsilon \\ \epsilon^* \end{matrix}$	$\begin{matrix} \epsilon^* \\ \epsilon \end{matrix}$	$(x, y)(R_x, R_y)$	$(x^2 - y^2, xy)(yz, xz)$

C_4	E	C_4	C_2	C_4^3		
A	1	1	1	1	z, R_z	$x^2 + y^2, z^2$
B	1	-1	1	-1		$x^2 - y^2, xy$
E	$\begin{Bmatrix} 1 \\ 1 \end{Bmatrix}$	$\begin{matrix} i \\ -i \end{matrix}$	$\begin{matrix} -1 \\ -1 \end{matrix}$	$\begin{matrix} -i \\ i \end{matrix}$	$(x, y)(R_x, R_y)$	(yz, xz)

C_6	E	C_6	C_3	C_2	C_3^2	C_6^5		$\epsilon = \exp(2\pi i/6)$
A	1	1	1	1	1	1	z, R_z	$x^2 + y^2, z^2$
B	1	-1	1	-1	1	-1		
E_1	$\begin{Bmatrix} 1 \\ 1 \end{Bmatrix}$	$\begin{matrix} \epsilon \\ \epsilon^* \end{matrix}$	$\begin{matrix} -\epsilon^* \\ -\epsilon \end{matrix}$	$\begin{matrix} -1 \\ -1 \end{matrix}$	$\begin{matrix} -\epsilon \\ -\epsilon^* \end{matrix}$	$\begin{matrix} \epsilon^* \\ \epsilon \end{matrix}$	(x, y) (R_x, R_y)	(xz, yz)
E_2	$\begin{Bmatrix} 1 \\ 1 \end{Bmatrix}$	$\begin{matrix} -\epsilon^* \\ -\epsilon \end{matrix}$	$\begin{matrix} -\epsilon \\ -\epsilon^* \end{matrix}$	$\begin{matrix} 1 \\ 1 \end{matrix}$	$\begin{matrix} -\epsilon^* \\ -\epsilon \end{matrix}$	$\begin{matrix} -\epsilon \\ -\epsilon^* \end{matrix}$		$(x^2 - y^2, xy)$

[a] A and B refer to one-dimensional representations; E and T are labels for two- and three-dimensional representations, respectively; g and u indicate even and odd representations under inversion.

[b] From Cotton, F. A., "Chemical Applications of Group Theory," 2nd ed., Copyright © 1971, Wiley (Interscience) Publishers. Reprinted by permission of John Wiley & Sons, Inc.

[c] Tables T, T_d, T_h, O, O_h, (Atkins, 1970).

TABLE 5.2 *(continued)*

D_2	E	$C_2(z)$	$C_2(y)$	$C_2(x)$		
A	1	1	1	1		x^2, y^2, z^2
B_1	1	1	-1	-1	z, R_z	xy
B_2	1	-1	1	-1	y, R_y	xz
B_3	1	-1	-1	1	x, R_x	yz

D_3	E	$2C_3$	$3C_2$		
A_1	1	1	1		$x^2 + y^2, z^2$
A_2	1	1	-1	z, R_z	
E	2	-1	0	$(x, y)(R_x, R_y)$	$(x^2 - y^2, xy)(xz, yz)$

D_4	E	$2C_4$	$C_2(= C_4^2)$	$2C_2'$	$2C_2''$		
A_1	1	1	1	1	1		$x^2 + y^2, z^2$
A_2	1	1	1	-1	-1	z, R_z	
B_1	1	-1	1	1	-1		$x^2 - y^2$
B_2	1	-1	1	-1	1		xy
E	2	0	-2	0	0	$(x, y)(R_x, R_y)$	(xz, yz)

D_6	E	$2C_6$	$2C_3$	C_2	$3C_2'$	$3C_2''$		
A_1	1	1	1	1	1	1		$x^2 + y^2, z^2$
A_2	1	1	1	1	-1	-1	z, R_z	
B_1	1	-1	1	-1	1	-1		
B_2	1	-1	1	-1	-1	1		
E_1	2	1	-1	-2	0	0	$(x, y)(R_x, R_y)$	(xz, yz)
E_2	2	-1	-1	2	0	0		$(x^2 - y^2, xy)$

C_{2v}	E	C_2	$\sigma_v(xz)$	$\sigma_v'(yz)$		
A_1	1	1	1	1	z	x^2, y^2, z^2
A_2	1	1	-1	-1	R_z	xy
B_1	1	-1	1	-1	x, R_y	xz
B_2	1	-1	-1	1	y, R_x	yz

C_{3v}	E	$2C_3$	$3\sigma_v$		
A_1	1	1	1	z	$x^2 + y^2, z^2$
A_2	1	1	-1	R_z	
E	2	-1	0	$(x, y)(R_x, R_y)$	$(x^2 - y^2, xy)(xz, yz)$

TABLE 5.2 (continued)

C_{4v}	E	$2C_4$	C_2	$2\sigma_v$	$2\sigma_d$		
A_1	1	1	1	1	1	z	$x^2 + y^2, z^2$
A_2	1	1	1	−1	−1	R_z	
B_1	1	−1	1	1	−1		$x^2 - y^2$
B_2	1	−1	1	−1	1		xy
E	2	0	−2	0	0	$(x, y)(R_x, R_y)$	(xz, yz)

C_{6v}	E	$2C_6$	$2C_3$	C_2	$3\sigma_v$	$3\sigma_d$		
A_1	1	1	1	1	1	1	z	$x^2 + y^2, z^2$
A_2	1	1	1	1	−1	−1	R_z	
B_1	1	−1	1	−1	1	−1		
B_2	1	−1	1	−1	−1	1		
E_1	2	1	−1	−2	0	0	$(x, y)(R_x, R_y)$	(xz, yz)
E_2	2	−1	−1	2	0	0		$(x^2 - y^2, xy)$

C_{2h}	E	C_2	i	σ_h		
A_g	1	1	1	1	R_z	x^2, y^2, z^2, xy
B_g	1	−1	1	−1	R_x, R_y	xz, yz
A_u	1	1	−1	−1	z	
B_u	1	−1	−1	1	x, y	

C_{3h}	E	C_3	C_3^2	σ_h	S_3	S_3^5		$\epsilon = \exp(2\pi i/3)$
A'	1	1	1	1	1	1	R_z	$x^2 + y^2, z^2$
E'	1	ϵ	ϵ^*	1	ϵ	ϵ^*	(x, y)	$(x^2 - y^2, xy)$
	1	ϵ^*	ϵ	1	ϵ^*	ϵ		
A''	1	1	1	−1	−1	−1	z	
E''	1	ϵ	ϵ^*	−1	$-\epsilon$	$-\epsilon^*$	(R_x, R_y)	(xz, yz)
	1	ϵ^*	ϵ	−1	$-\epsilon^*$	$-\epsilon$		

C_{4h}	E	C_4	C_2	C_4^3	i	S_4^3	σ_h	S_4		
A_g	1	1	1	1	1	1	1	1	R_z	$x^2 + y^2, z^2$
B_g	1	−1	1	−1	1	−1	1	−1		$x^2 - y^2, xy$
E_g	1	i	−1	$-i$	1	i	−1	$-i$	(R_x, R_y)	(xz, yz)
	1	$-i$	−1	i	1	$-i$	−1	i		
A_u	1	1	1	1	−1	−1	−1	−1	z	
B_u	1	−1	1	−1	−1	1	−1	1		
E_u	1	i	−1	$-i$	−1	$-i$	1	i	(x, y)	
	1	$-i$	−1	i	−1	i	1	$-i$		

C_{6h}	E	C_6	C_3	C_2	C_3^2	C_6^5	i	S_3^5	S_6^5	σ_h	S_6	S_3		$\epsilon = \exp(2\pi i/6)$
A_g	1	1	1	1	1	1	1	1	1	1	1	1	R_z	$x^2 + y^2, z^2$
B_g	1	−1	1	−1	1	−1	1	−1	1	−1	1	−1		
E_{1g}	1	ϵ	$-\epsilon^*$	−1	$-\epsilon$	ϵ^*	1	ϵ	$-\epsilon^*$	−1	$-\epsilon$	ϵ^*	(R_x, R_y)	(xz, yz)
	1	ϵ^*	$-\epsilon$	−1	$-\epsilon^*$	ϵ	1	ϵ^*	$-\epsilon$	−1	$-\epsilon^*$	ϵ		
E_{2g}	1	$-\epsilon^*$	$-\epsilon$	1	$-\epsilon^*$	$-\epsilon$	1	$-\epsilon^*$	$-\epsilon$	1	$-\epsilon^*$	$-\epsilon$		$(x^2 - y^2, xy)$
	1	$-\epsilon$	$-\epsilon^*$	1	$-\epsilon$	$-\epsilon^*$	1	$-\epsilon$	$-\epsilon^*$	1	$-\epsilon$	$-\epsilon^*$		
A_u	1	1	1	1	1	1	−1	−1	−1	−1	−1	−1	z	
B_u	1	−1	1	−1	1	−1	−1	1	−1	1	−1	1		
E_{1u}	1	ϵ	$-\epsilon^*$	−1	$-\epsilon$	ϵ^*	−1	$-\epsilon$	ϵ^*	1	ϵ	$-\epsilon^*$	(x, y)	
	1	ϵ^*	$-\epsilon$	−1	$-\epsilon^*$	ϵ	−1	$-\epsilon^*$	ϵ	1	ϵ^*	$-\epsilon$		
E_{2u}	1	$-\epsilon^*$	$-\epsilon$	1	$-\epsilon^*$	$-\epsilon$	−1	ϵ^*	ϵ	−1	ϵ^*	ϵ		
	1	$-\epsilon$	$-\epsilon^*$	1	$-\epsilon$	$-\epsilon^*$	−1	ϵ	ϵ^*	−1	ϵ	ϵ^*		

TABLE 5.2 (continued)

D_{2h}	E	$C_2(z)$	$C_2(y)$	$C_2(x)$	i	$\sigma(xy)$	$\sigma(xz)$	$\sigma(yz)$		
A_g	1	1	1	1	1	1	1	1		x^2, y^2, z^2
B_{1g}	1	1	−1	−1	1	1	−1	−1	R_z	xy
B_{2g}	1	−1	1	−1	1	−1	1	−1	R_y	xz
B_{3g}	1	−1	−1	1	1	−1	−1	1	R_x	yz
A_u	1	1	1	1	−1	−1	−1	−1		
B_{1u}	1	1	−1	−1	−1	−1	1	1	z	
B_{2u}	1	−1	1	−1	−1	1	−1	1	y	
B_{3u}	1	−1	−1	1	−1	1	1	−1	x	

D_{3h}	E	$2C_3$	$3C_2$	σ_h	$2S_3$	$3\sigma_v$		
A_1'	1	1	1	1	1	1		x^2+y^2, z^2
A_2'	1	1	−1	1	1	−1	R_z	
E'	2	−1	0	2	−1	0	(x, y)	(x^2-y^2, xy)
A_1''	1	1	1	−1	−1	−1		
A_2''	1	1	−1	−1	−1	1	z	
E''	2	−1	0	−2	1	0	(R_x, R_y)	(xz, yz)

D_{4h}	E	$2C_4$	C_2	$2C_2'$	$2C_2''$	i	$2S_4$	σ_h	$2\sigma_v$	$2\sigma_d$		
A_{1g}	1	1	1	1	1	1	1	1	1	1		x^2+y^2, z^2
A_{2g}	1	1	1	−1	−1	1	1	1	−1	−1	R_z	
B_{1g}	1	−1	1	1	−1	1	−1	1	1	−1		x^2-y^2
B_{2g}	1	−1	1	−1	1	1	−1	1	−1	1		xy
E_g	2	0	−2	0	0	2	0	−2	0	0	(R_x, R_y)	(xz, yz)
A_{1u}	1	1	1	1	1	−1	−1	−1	−1	−1		
A_{2u}	1	1	1	−1	−1	−1	−1	−1	1	1	z	
B_{1u}	1	−1	1	1	−1	−1	1	−1	−1	1		
B_{2u}	1	−1	1	−1	1	−1	1	−1	1	−1		
E_u	2	0	−2	0	0	−2	0	2	0	0	(x, y)	

D_{6h}	E	$2C_6$	$2C_3$	C_2	$3C_2'$	$3C_2''$	i	$2S_3$	$2S_6$	σ_h	$3\sigma_d$	$3\sigma_v$		
A_{1g}	1	1	1	1	1	1	1	1	1	1	1	1		x^2+y^2, z^2
A_{2g}	1	1	1	1	−1	−1	1	1	1	1	−1	−1	R_z	
B_{1g}	1	−1	1	−1	1	−1	1	−1	1	−1	1	−1		
B_{2g}	1	−1	1	−1	−1	1	1	−1	1	−1	−1	1		
E_{1g}	2	1	−1	−2	0	0	2	1	−1	−2	0	0	(R_x, R_y)	(xz, yz)
E_{2g}	2	−1	−1	2	0	0	2	−1	−1	2	0	0		(x^2-y^2, xy)
A_{1u}	1	1	1	1	1	1	−1	−1	−1	−1	−1	−1		
A_{2u}	1	1	1	1	−1	−1	−1	−1	−1	−1	1	1	z	
B_{1u}	1	−1	1	−1	1	−1	−1	1	−1	1	−1	1		
B_{2u}	1	−1	1	−1	−1	1	−1	1	−1	1	1	−1		
E_{1u}	2	1	−1	−2	0	0	−2	−1	1	2	0	0	(x, y)	
E_{2u}	2	−1	−1	2	0	0	−2	1	1	−2	0	0		

D_{2d}	E	$2S_4$	C_2	$2C_2'$	$2\sigma_d$		
A_1	1	1	1	1	1		x^2+y^2, z^2
A_2	1	1	1	−1	−1	R_z	
B_1	1	−1	1	1	−1		x^2-y^2
B_2	1	−1	1	−1	1	z	xy
E	2	0	−2	0	0	(x, y); (R_x, R_y)	(xz, yz)

D_{3d}	E	$2C_3$	$3C_2$	i	$2S_6$	$3\sigma_d$		
A_{1g}	1	1	1	1	1	1		x^2+y^2, z^2
A_{2g}	1	1	−1	1	1	−1	R_z	
E_g	2	−1	0	2	−1	0	(R_x, R_y)	(x^2-y^2, xy), (xz, yz)
A_{1u}	1	1	1	−1	−1	−1		
A_{2u}	1	1	−1	−1	−1	1	z	
E_u	2	−1	0	−2	1	0	(x, y)	

TABLE 5.2 (*continued*)

S_4	E	S_4	C_2	S_4^3		
A	1	1	1	1	R_z	x^2+y^2, z^2
B	1	-1	1	-1	z	x^2-y^2, xy
E	$\begin{cases}1 \\ 1\end{cases}$	$\begin{matrix}i \\ -i\end{matrix}$	$\begin{matrix}-1 \\ -1\end{matrix}$	$\begin{rcases}-i \\ i\end{rcases}$	$(x, y); (R_x, R_y)$	(xz, yz)

S_6	E	C_3	C_3^2	i	S_6^5	S_6		$\epsilon = \exp(2\pi i/3)$
A_g	1	1	1	1	1	1	R_z	x^2+y^2, z^2
E_g	$\begin{cases}1 \\ 1\end{cases}$	$\begin{matrix}\epsilon \\ \epsilon^*\end{matrix}$	$\begin{matrix}\epsilon^* \\ \epsilon\end{matrix}$	$\begin{matrix}1 \\ 1\end{matrix}$	$\begin{matrix}\epsilon \\ \epsilon^*\end{matrix}$	$\begin{rcases}\epsilon^* \\ \epsilon\end{rcases}$	(R_x, R_y)	(x^2-y^2, xy); (xz, yz)
A_u	1	1	1	-1	-1	-1	z	
E_u	$\begin{cases}1 \\ 1\end{cases}$	$\begin{matrix}\epsilon \\ \epsilon^*\end{matrix}$	$\begin{matrix}\epsilon^* \\ \epsilon\end{matrix}$	$\begin{matrix}-1 \\ -1\end{matrix}$	$\begin{matrix}-\epsilon \\ -\epsilon^*\end{matrix}$	$\begin{rcases}-\epsilon^* \\ -\epsilon\end{rcases}$	(x, y)	

T	E	$4C_3$	$4C_3^2$	$3C_2$		$\epsilon = \exp(2\pi i/3)$
A	1	1	1	1		$x^2+y^2+z^2$
E	$\begin{cases}1 \\ 1\end{cases}$	$\begin{matrix}\epsilon \\ \epsilon^*\end{matrix}$	$\begin{matrix}\epsilon^* \\ \epsilon\end{matrix}$	$\begin{rcases}1 \\ 1\end{rcases}$		$(x^2-y^2, 2z^2-x^2-y^2)$
T	3	0	0	-1	(x, y, z) (R_x, R_y, R_z)	(xy, xz, yz)

T_d	E	$8C_3$	$3C_2$	$6S_4$	$6\sigma_d$		
A_1	1	1	1	1	1		$x^2+y^2+z^2$
A_2	1	1	1	-1	-1		
E	2	-1	2	0	0		$(2z^2-x^2-y^2,$ $x^2-y^2)$
T_1	3	0	-1	1	-1	(R_x, R_y, R_z)	
T_2	3	0	-1	-1	1	(x, y, z)	(xy, xz, yz)

T_h	E	$4C_3$	$4C_3^2$	$3C_2$	i	$4S_6$	$4S_6^2$	$3\sigma_d$		$\epsilon = \exp(2\pi i/3)$
A_g	1	1	1	1	1	1	1	1		$x^2+y^2+z^2$
E_g	$\begin{cases}1 \\ 1\end{cases}$	$\begin{matrix}\epsilon \\ \epsilon^*\end{matrix}$	$\begin{matrix}\epsilon^* \\ \epsilon\end{matrix}$	$\begin{matrix}1 \\ 1\end{matrix}$	$\begin{matrix}1 \\ 1\end{matrix}$	$\begin{matrix}\epsilon \\ \epsilon^*\end{matrix}$	$\begin{matrix}\epsilon^* \\ \epsilon\end{matrix}$	$\begin{rcases}1 \\ 1\end{rcases}$		$(2z^2-x^2-y^2,$ $x^2-y^2)$
T_g	3	0	0	-1	3	0	0	-1	(R_x, R_y, R_z)	(xy, yz, xz)
A_u	1	1	1	1	-1	-1	-1	-1		
E_u	$\begin{cases}1 \\ 1\end{cases}$	$\begin{matrix}\epsilon \\ \epsilon^*\end{matrix}$	$\begin{matrix}\epsilon^* \\ \epsilon\end{matrix}$	$\begin{matrix}1 \\ 1\end{matrix}$	$\begin{matrix}-1 \\ -1\end{matrix}$	$\begin{matrix}-\epsilon \\ -\epsilon^*\end{matrix}$	$\begin{matrix}-\epsilon^* \\ -\epsilon\end{matrix}$	$\begin{rcases}-1 \\ -1\end{rcases}$		
T_u	3	0	0	-1	-3	0	0	1	(x, y, z)	

TABLE 5.2 (*continued*)

O	E	$8C_3$	$3C_2$	$6C_4$	$6C_2'$		
A_1	1	1	1	1	1		$x^2+y^2+z^2$
A_2	1	1	1	−1	−1		
E	2	−1	2	0	0		$(2z^2-x^2-y^2, x^2-y^2)$
T_1	3	0	−1	1	−1	(x, y, z) (R_x, R_y, R_z)	
T_2	3	0	−1	−1	1		(xy, xz, yz)

O_h	E	$8C_3$	$6C_2'$	$6C_4$	$3C_2$ $(=C_4^2)$	i	$6S_4$	$8S_6$	$3\sigma_h$	$6\sigma_d$		
A_{1g}	1	1	1	1	1	1	1	1	1	1		$x^2+y^2+z^2$
A_{2g}	1	1	−1	−1	1	1	−1	1	1	−1		
E_g	2	−1	0	0	2	2	0	−1	2	0		$(2z^2-x^2-y^2, x^2-y^2)$
T_{1g}	3	0	−1	1	−1	3	1	0	−1	−1	(R_x, R_y, R_z)	
T_{2g}	3	0	1	−1	−1	3	−1	0	−1	1		(xz, yz, xy)
A_{1u}	1	1	1	1	1	−1	−1	−1	−1	−1		
A_{2u}	1	1	−1	−1	1	−1	1	−1	−1	1		
E_u	2	−1	0	0	2	−2	0	1	−2	0		
T_{1u}	3	0	−1	1	−1	−3	−1	0	1	1	(x, y, z)	
T_{2u}	3	0	1	−1	−1	−3	1	0	1	−1		

$C_{\infty v}$	E	$2C_\varphi$	$\cdots$	σ_v		
$A_1 \equiv \Sigma^+$	1	1	$\cdots$	1	z	x^2+y^2, z^2
$A_2 \equiv \Sigma^-$	1	1	$\cdots$	−1	R_z	
$E_1 \equiv \Pi$	2	$2\cos\varphi$	$\cdots$	0	$(x, y); (R_x, R_y)$	(xz, yz)
$E_2 \equiv \Delta$	2	$2\cos 2\varphi$	$\cdots$	0		(x^2-y^2, xy)
$E_3 \equiv \Phi$	2	$2\cos 3\varphi$	$\cdots$	0		
$\cdots$	$\cdots$	$\cdots$	$\cdots$	$\cdots$		

$D_{\infty h}$	E	$2C_\varphi$	$\cdots$	σ_v	i	$2S_\varphi$	$\cdots$	C_2		
Σ_g^+	1	1	$\cdots$	1	1	1	$\cdots$	1		x^2+y^2, z^2
Σ_g^-	1	1	$\cdots$	−1	1	1	$\cdots$	−1	R_z	
Π_g	2	$2\cos\varphi$	$\cdots$	0	2	$-2\cos\varphi$	$\cdots$	0	(R_x, R_y)	(xz, yz)
Δ_g	2	$2\cos 2\varphi$	$\cdots$	0	2	$2\cos 2\varphi$	$\cdots$	0		(x^2-y^2, xy)
$\cdots$	$\cdots$	$\cdots$	$\cdots$	$\cdots$	$\cdots$	$\cdots$	$\cdots$	$\cdots$		
Σ_u^+	1	1	$\cdots$	1	−1	−1	$\cdots$	−1	z	
Σ_u^-	1	1	$\cdots$	−1	−1	−1	$\cdots$	1		
Π_u	2	$2\cos\varphi$	$\cdots$	0	−2	$2\cos\varphi$	$\cdots$	0	(x, y)	
Δ_u	2	$2\cos 2\varphi$	$\cdots$	0	−2	$-2\cos 2\varphi$	$\cdots$	0		
$\cdots$	$\cdots$	$\cdots$	$\cdots$	$\cdots$	$\cdots$	$\cdots$	$\cdots$	..		

can be decomposed in terms of the irreducible representations of D_3. The characters of $D^{(2)}$ for the elements $E, A, \ldots F$ belonging to D_3 may be evaluated by means of (4.5-4):

$$\begin{aligned} \chi^{(2)}(E) &= 5, \\ \chi^{(2)}(A) &= \chi^{(2)}(B) = \chi^{(2)}(C) = 1, \\ \chi^{(2)}(D) &= \chi^{(2)}(F) = -1. \end{aligned} \tag{5.1-2}$$

With the aid of (3.3-16) and the character table for D_3 (Table 5.2), it is found that

$$D^{(2)} = \Gamma^{(1)} + 2\Gamma^{(3)}. \tag{5.1-3}$$

Suppose now that we are interested in decomposing an irreducible representation like $D^{(3/2)}$. A complication arises because of the double-valued property of $\chi^{(3/2)}$ (and all other characters of $D^{(j)}$ when j is half-integral). Thus

$$\chi^{(3/2)}(\omega) = \sin 2\omega / \sin \tfrac{1}{2}\omega, \tag{5.1-4}$$

whereas

$$\chi^{(3/2)}(\omega + 2\pi) = -\sin 2\omega / \sin \tfrac{1}{2}\omega. \tag{5.1-5}$$

A systematic procedure for handling such cases is based on the formalism of *double groups* whose properties are described in the next section.

5.2 Double Groups

An ordinary finite point group G may be converted to a double group $\bar{G}$ by treating rotations of $2n\pi$ with n even (including zero) as separate symmetry operations from rotations of $2n\pi$ with n odd (Bethe, 1929). Actually, such groups are nothing more than subgroups of $SU(2)$. Adopting the notation

$$\begin{aligned} E &= \text{rotation through } 2n\pi, \qquad n = 0, 2, 4, \ldots, \\ \bar{E} &= \text{rotation through } 2n\pi, \qquad n = 1, 3, 5, \ldots, \end{aligned} \tag{5.2-1}$$

the effect of introducing the additional element $\bar{E}$ into the group is to enlarge the group not only by $\bar{E}$ but by all new elements obtained as products of $\bar{E}$ with the original elements of the group. In D_3, for example, we have the elements E, A, B, C, D, F. With $\bar{E}$ included, the group is enlarged to E, $\bar{E}$, A, $\bar{A}$, B, $\bar{B}$, C, $\bar{C}$, D, $\bar{D}$, F, $\bar{F}$ where the bar over a particular element denotes the product with $\bar{E}$, e.g., $\bar{A} = \bar{E}A = A\bar{E}$, etc. The new enlarged group is known as a *double* (*or extended*) *group*.

The number of elements in a double group is often less than twice the number of elements in the original group. This happens when the product of $\bar{E}$ with some element of the original group coincides with an element already contained in the original group. The number of classes in the double group is also increased which means, then, that the number of irreducible representations must also be increased.

Using the notation of Table 5.2, the character table of $\bar{D}_3$ (the double group) is now enlarged to that shown in Table 5.3 in which $2C_3$ stands for D, F; $2\bar{C}_3$ for $\bar{D}, \bar{F}$; $3C_2$ for A, B, C; $3\bar{C}_2$ for $\bar{A}, \bar{B}, \bar{C}$. The irreducible representations $\Gamma^{(1)}$, $\Gamma^{(2)}$, and $\Gamma^{(3)}$ are the same as those in Table 5.2; the irreducible representations $\Gamma^{(4)}$, $\Gamma^{(5)}$, and $\Gamma^{(6)}$ are characteristic of the double group.

TABLE 5.3

Character Table for the Double Groups $\bar{C}_{3v}$ and $\bar{D}_3$

$\bar{C}_{3v}$		E	$\bar{E}$	$2C_3$	$2\bar{C}_3$	$3\sigma_v$	$3\bar{\sigma}_v$
$\bar{D}_3$		E	$\bar{E}$	$2C_3$	$2\bar{C}_3$	$3C_2$	$3\bar{C}_2$
$\Gamma^{(1)}$	A_1	1	1	1	1	1	1
$\Gamma^{(2)}$	A_2	1	1	1	1	-1	-1
$\Gamma^{(3)}$	E	2	2	-1	-1	0	0
$\Gamma^{(4)}$	$E_{1/2}$	2	-2	1	-1	0	0
$\Gamma^{(5)}$		1	-1	-1	1	i	$-i$
$\Gamma^{(6)}$		1	-1	-1	1	$-i$	i
	$E_{3/2}$	2	-2	-2	2	0	0

Now that we have the complete table of characters for $\bar{D}_3$ (Table 5.3) it is possible to decompose $D^{(3/2)}$ in terms of the irreducible representations of $\bar{D}_3$. Again using (4.5-4) it is found that

$$\begin{aligned} &\chi^{(3/2)}(E) = 4, && \chi^{(3/2)}(\bar{E}) = -4, \\ &\chi^{(3/2)}(C_3) = -1, && \chi^{(3/2)}(\bar{C}_3) = 1, \\ &\chi^{(3/2)}(C_2) = 0, && \chi^{(3/2)}(\bar{C}_2) = 0, \end{aligned} \tag{5.2-2}$$

and, from (3.3-16),

$$D^{(3/2)} = \Gamma^{(4)} + \Gamma^{(5)} + \Gamma^{(6)}. \tag{5.2-3}$$

Examination of the character tables reveals that some characters are real while others are complex (including pure imaginaries). If a character $\chi(R)$ is complex, the representation matrix $\Gamma(R)$ must be complex since a matrix with real elements cannot have a complex trace. Moreover, if Γ is a complex representation, Γ^* must also be a representation of the group since if $\Gamma(R)\Gamma(S) = \Gamma(RS)$, then $\Gamma^*(R)\Gamma^*(S) = \Gamma^*(RS)$, i.e., the representation matrices Γ^* must also satisfy the group multiplication table. Also, if Γ is irreducible, so is Γ^* and if the character of $\Gamma(R)$ is $\chi(R)$, then the character of $\Gamma^*(R)$ is $\chi^*(R)$. Such representations are said to be of type b (Koster *et al.*, 1963).

Type b representations occur in pairs as do their basis functions. Thus if φ is a basis function for Γ, φ must be complex and φ^* is a basis function for Γ^*. In the group C_3, for example, $\varphi = x + iy$ is a basis function for $\Gamma^{(2)}$; $\varphi^* = x - iy$ is a basis function for $\Gamma^{(3)}$, and $\Gamma^{(2)} = \Gamma^{(3)*}$. If we construct the pair of real functions

$$x = \tfrac{1}{2}(\varphi + \varphi^*), \qquad y = -\tfrac{1}{2}i(\varphi - \varphi^*), \tag{5.2-4}$$

then the set (x, y) is a basis set for a real two-dimensional representation; the latter is reducible.

If a character $\chi(R)$ is real, the representation matrix $\Gamma(R)$ may be real and if it is, it is classified as type a. But $\Gamma(R)$ may also be complex provided the imaginary parts of the diagonal elements cancel when the trace is taken. In that event $\Gamma(R)$ and $\Gamma^*(R)$ have the same (real) character $\chi(R)$. Since two representations with the same characters are either equal or equivalent, $\Gamma(R)$ and $\Gamma^*(R)$ are equivalent, i.e., one representation can be transformed into the other by a similarity transformation. Such representations are type c. To summarize the classification, we have

(a) Γ is real or can be transformed to real form,
(b) Γ and Γ^* are not equivalent,
(c) Γ and Γ^* are equivalent but cannot be transformed to real form.

There exists a test which can be applied to the characters of an irreducible representation in order to establish the class to which the representation belongs (Wigner, 1959):

$$\frac{1}{h}\sum_R \chi^{(j)}(R^2) = \begin{cases} +1, & \Gamma^{(j)} \text{ belongs to type a}, \\ 0, & \Gamma^{(j)} \text{ belongs to type b}, \\ -1, & \Gamma^{(j)} \text{ belongs to type c}. \end{cases} \tag{5.2-5}$$

Taking the group $\bar{D}_3$ as an example we have

$$E^2 = \bar{E}^2 = E,$$
$$A^2 = \bar{A}^2 = B^2 = \bar{B}^2 = C^2 = \bar{C}^2 = \bar{E},$$
$$D^2 = \bar{D}^2 = \bar{F},$$
$$F^2 = \bar{F}^2 = \bar{D}.$$

With the help of (5.2-5) it is found that $\Gamma^{(1)}, \Gamma^{(2)}, \Gamma^{(3)}$ are type a, $\Gamma^{(4)}$ is type c, and $\Gamma^{(5)}, \Gamma^{(6)}$ are type b.

In Section 3.6 it was shown that $\Gamma^{(\mu)*} \times \Gamma^{(\nu)}$ contains the unit representation $\Gamma^{(1)}$ only when $\mu = \nu$. For representations of type a and c—because of their real characters—it is permissible to replace $\Gamma^{(\mu)*} \times \Gamma^{(\nu)}$ by $\Gamma^{(\mu)} \times \Gamma^{(\nu)}$ but for type b representations such a replacement is not valid.

5.3 The Groups O, D_4 and D_{6h}

The group O, known as the *octahedral* group, is part of the cubic system of groups (Table 5.1). The symmetry elements of O may be visualized as the proper rotations of a cube. Referring to Fig. 5.1, the operations are defined as:

E identity,
$8C_3$ rotations of $\pm 120°$ about the four body diagonals, such as AB,
$3C_2$ rotations of $180°$ about the coordinate axes,
$6C_2'$ rotations of $180°$ about the six axes bisecting opposite sides, such as CD,
$6C_4$ rotations of $\pm 90°$ about the coordinate axes.

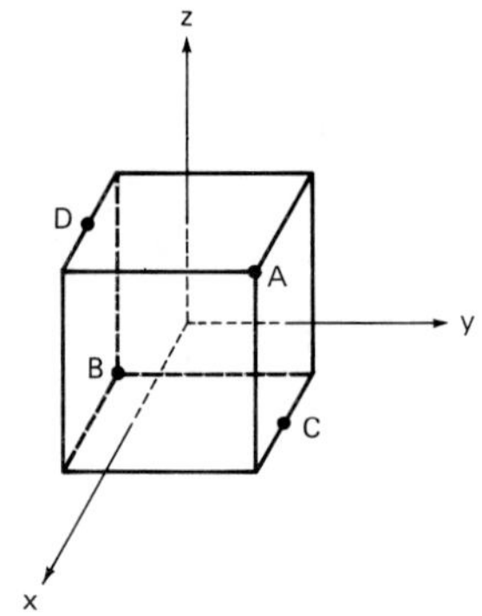

FIG. 5.1 Coordinate system and symmetry axes for the group O

For the double group we also have the operation $\bar{E}$ and the products of $\bar{E}$ with the operations just listed. The characters of the irreducible representations of the double group are given in Table 5.4 which also includes $\bar{T}_d$ and several sets of notations employed by various authors. The reduction of direct products of irreducible representations is shown in Table 5.5; the reduction of symmetric products is also shown.

TABLE 5.4

Character Table for the Double Groups $\bar{T}_d$ and $\bar{O}$

$\bar{T}_d$				E	$\bar{E}$	$8C_3$	$8\bar{C}_3$	$3C_2$ $3\bar{C}_2$	$6S_4$	$6\bar{S}_4$	$6\sigma_d$ $6\bar{\sigma}_d$
$\bar{O}$				E	$\bar{E}$	$8C_3$	$8\bar{C}_3$	$3C_2$ $3\bar{C}_2$	$6C_4$	$6\bar{C}_4$	$6C_2'$ $6\bar{C}_2'$
$\Gamma^{(1)}$			A_1	1	1	1	1	1	1	1	1
$\Gamma^{(2)}$			A_2	1	1	1	1	1	-1	-1	-1
$\Gamma^{(3)}$			E	2	2	-1	-1	2	0	0	0
$\Gamma^{(4)}$		F_1	T_1	3	3	0	0	-1	1	1	-1
$\Gamma^{(5)}$		F_2	T_2	3	3	0	0	-1	-1	-1	1
$\Gamma^{(6)}$	E_1	$E_{1/2}$	E'	2	-2	1	-1	0	$\sqrt{2}$	$-\sqrt{2}$	0
$\Gamma^{(7)}$	E_2	$E_{3/2}$	E''	2	-2	1	-1	0	$-\sqrt{2}$	$\sqrt{2}$	0
$\Gamma^{(8)}$	G	$G_{3/2}$	U'	4	-4	-1	1	0	0	0	0

An irreducible representation of $\bar{O}$ is also a representation of each subgroup of $\bar{O}$ but with respect to a subgroup, the representation may or may not be irreducible. Table 5.6 lists the reduction of the irreducible representations of $\bar{O}$ in terms of the irreducible representations of the subgroups $\bar{D}_4$ and $\bar{D}_3$. Since $\bar{O}$ is a subgroup of $O^+(3)$, the irreducible representations of the latter $D^{(j)}$ are generally reducible in terms of the irreducible representations of $\bar{O}$. The reduction is shown in Table 5.7.

Basis functions for the irreducible representations of O, constructed as linear combinations of angular momentum eigenfunctions, are shown in Table 5.8 up to $j = 4$ (Watanabe, 1966). Finally, coupling coefficients (Section 3.6) for O are listed in Table 5.9 in the notation and phase conventions of Koster *et al.* (1963).

TABLE 5.5

Decomposition of Product Representations for the Groups $\bar{T}_d$ and $\bar{O}$

	$\Gamma^{(8)}$	$\Gamma^{(7)}$	$\Gamma^{(6)}$	$\Gamma^{(5)}$	$\Gamma^{(4)}$	$\Gamma^{(3)}$	$\Gamma^{(2)}$	$\Gamma^{(1)}$
$\Gamma^{(1)}$	$\Gamma^{(8)}$	$\Gamma^{(7)}$	$\Gamma^{(6)}$	$\Gamma^{(5)}$	$\Gamma^{(4)}$	$\Gamma^{(3)}$	$\Gamma^{(2)}$	$\Gamma^{(1)}$
$\Gamma^{(2)}$	$\Gamma^{(8)}$	$\Gamma^{(6)}$	$\Gamma^{(7)}$	$\Gamma^{(4)}$	$\Gamma^{(5)}$	$\Gamma^{(3)}$	$\Gamma^{(1)}$	
$\Gamma^{(3)}$	$\Gamma^{(6)} + \Gamma^{(7)} + \Gamma^{(8)}$	$\Gamma^{(8)}$	$\Gamma^{(8)}$	$\Gamma^{(4)} + \Gamma^{(5)}$	$\Gamma^{(4)} + \Gamma^{(5)}$	$\Gamma^{(1)} + \Gamma^{(2)} + \Gamma^{(3)}$		
$\Gamma^{(4)}$	$\Gamma^{(6)} + \Gamma^{(7)} + 2\Gamma^{(8)}$	$\Gamma^{(7)} + \Gamma^{(8)}$	$\Gamma^{(6)} + \Gamma^{(8)}$	$\Gamma^{(2)} + \Gamma^{(3)} + \Gamma^{(4)} + \Gamma^{(5)}$	$\Gamma^{(1)} + \Gamma^{(3)} + \Gamma^{(4)} + \Gamma^{(5)}$			
$\Gamma^{(5)}$	$\Gamma^{(6)} + \Gamma^{(7)} + 2\Gamma^{(8)}$	$\Gamma^{(6)} + \Gamma^{(8)}$	$\Gamma^{(7)} + \Gamma^{(8)}$	$\Gamma^{(1)} + \Gamma^{(3)} + \Gamma^{(4)} + \Gamma^{(5)}$				
$\Gamma^{(6)}$	$\Gamma^{(3)} + \Gamma^{(4)} + \Gamma^{(5)}$	$\Gamma^{(2)} + \Gamma^{(5)}$	$\Gamma^{(1)} + \Gamma^{(4)}$					
$\Gamma^{(7)}$	$\Gamma^{(3)} + \Gamma^{(4)} + \Gamma^{(5)}$	$\Gamma^{(1)} + \Gamma^{(4)}$						
$\Gamma^{(8)}$	$\Gamma^{(1)} + \Gamma^{(2)} + \Gamma^{(3)} + 2\Gamma^{(4)} + 2\Gamma^{(5)}$							

Symmetric products

$$(\Gamma^{(3)} \times \Gamma^{(3)})^+ = \Gamma^{(1)} + \Gamma^{(3)}$$
$$(\Gamma^{(4)} \times \Gamma^{(4)})^+ = (\Gamma^{(5)} \times \Gamma^{(5)})^+ = \Gamma^{(1)} + \Gamma^{(3)} + \Gamma^{(5)}$$
$$(\Gamma^{(6)} \times \Gamma^{(6)})^+ = (\Gamma^{(7)} \times \Gamma^{(7)})^+ = \Gamma^{(4)}$$
$$(\Gamma^{(8)} \times \Gamma^{(8)})^+ = \Gamma^{(2)} + 2\Gamma^{(4)} + \Gamma^{(5)}$$

TABLE 5.6

Decomposition of the Irreducible Representations of $\bar{O}$ into Irreducible Representations of the Subgroups $\bar{D}_4$ and $\bar{D}_3$

$\bar{O}$	$\bar{D}_4$ [a]	$\bar{D}_3$ [b]
$\Gamma^{(1)}$	$\Gamma^{(1)}$	$\Gamma^{(1)}$
$\Gamma^{(2)}$	$\Gamma^{(3)}$	$\Gamma^{(2)}$
$\Gamma^{(3)}$	$\Gamma^{(1)} + \Gamma^{(3)}$	$\Gamma^{(3)}$
$\Gamma^{(4)}$	$\Gamma^{(2)} + \Gamma^{(5)}$	$\Gamma^{(2)} + \Gamma^{(3)}$
$\Gamma^{(5)}$	$\Gamma^{(4)} + \Gamma^{(5)}$	$\Gamma^{(1)} + \Gamma^{(3)}$
$\Gamma^{(6)}$	$\Gamma^{(6)}$	$\Gamma^{(4)}$
$\Gamma^{(7)}$	$\Gamma^{(7)}$	$\Gamma^{(4)}$
$\Gamma^{(8)}$	$\Gamma^{(6)} + \Gamma^{(7)}$	$\Gamma^{(4)} + \Gamma^{(5)} + \Gamma^{(6)}$

[a] See Table 5.10
[b] See Table 5.3

TABLE 5.7

Decomposition of Irreducible Representations of $O^+(3)$ into Irreducible Representations of $\bar{O}$

$O^+(3)$	$\bar{O}$	$O^+(3)$	$\bar{O}$
$D^{(0)}$	$\Gamma^{(1)}$	$D^{(3)}$	$\Gamma^{(2)} + \Gamma^{(4)} + \Gamma^{(5)}$
$D^{(1/2)}$	$\Gamma^{(6)}$	$D^{(7/2)}$	$\Gamma^{(6)} + \Gamma^{(7)} + \Gamma^{(8)}$
$D^{(1)}$	$\Gamma^{(4)}$	$D^{(4)}$	$\Gamma^{(1)} + \Gamma^{(3)} + \Gamma^{(4)} + \Gamma^{(5)}$
$D^{(3/2)}$	$\Gamma^{(8)}$	$D^{(9/2)}$	$\Gamma^{(6)} + 2\Gamma^{(8)}$
$D^{(2)}$	$\Gamma^{(3)} + \Gamma^{(5)}$	$D^{(5)}$	$\Gamma^{(3)} + 2\Gamma^{(4)} + \Gamma^{(5)}$
$D^{(5/2)}$	$\Gamma^{(7)} + \Gamma^{(8)}$		

The group D_4 is a tetragonal group. Assuming that the z axis is oriented along the four-fold symmetry axis, the operations are defined as:

- E identity.
- $2C_4$ rotations of $\pm 90°$ about the z axis.
- C_2 rotation of $180°$ about the z axis.
- $2C_2'$ rotations of $180°$ about the x and y axes.
- $2C_2''$ rotations of $180°$ about two diagonals (in the xy plane).

The properties of the group are summarized in Tables 5.10 and 5.11.

TABLE 5.8

Basis Functions for the Irreducible Representations of $\bar{O}$

Irreducible representation	Name	Basis functions	
$\Gamma^{(1)}$	a_1	$Y_{0\,0}$	$\sqrt{\frac{1}{4\pi}}$
	a_1	$\frac{1}{2\sqrt{6}}\{\sqrt{14}Y_{4\,0} + \sqrt{5}(Y_{4\,4} + Y_{4\,-4})\}$	$\frac{1}{r^4}\sqrt{\frac{9}{4\pi}}\sqrt{\frac{7}{3}\frac{5}{4}}\left(x^4 + y^4 + z^4 - \frac{3r^4}{5}\right)$
$\Gamma^{(2)}$	a_2	$\frac{i}{\sqrt{2}}(-Y_{3\,2} + Y_{3\,-2}$	$\frac{1}{r^3}\sqrt{\frac{7}{4\pi}}\sqrt{15}xyz$
$\Gamma^{(3)}$	$d\gamma$ $e\theta, eu$	$Y_{2\,0}$	$\frac{1}{r^2}\sqrt{\frac{5}{4\pi}}\frac{1}{2}(3z^2 - r^2)$
	$d\gamma$ $e\varepsilon, ev$	$\frac{1}{\sqrt{2}}(Y_{2\,2} + Y_{2\,-2})$	$\frac{1}{r^2}\sqrt{\frac{5}{4\pi}}\frac{\sqrt{3}}{2}(x^2 - y^2)$
		$d_{z^2} = \frac{1}{2}(3z^2 - r^2) = \frac{1}{2}r^2(3\cos^2\theta - 1)$	
		$d_{x^2-y^2} = \frac{\sqrt{3}}{2}(x^2 - y^2) = \frac{\sqrt{3}}{2}r^2\sin^2\theta\cos 2\varphi$	
	$e\theta, eu$	$\frac{1}{2\sqrt{6}}\{\sqrt{10}Y_{4\,0} - \sqrt{7}(Y_{4\,4} + Y_{4\,-4})\}$	$\frac{1}{r^4}\sqrt{\frac{9}{4\pi}}7\sqrt{5}\left[z^4 - \frac{x^4 + y^4}{2} - \frac{6}{7}\left\{z^2 - \frac{1}{2}(x^2 + y^2)\right\}r^2\right]$
	$e\varepsilon, ev$	$\frac{1}{\sqrt{2}}(-Y_{4\,2} - Y_{4\,-2})$	$\frac{1}{r^4}\sqrt{\frac{9}{4\pi}\frac{7}{4}}\sqrt{5}\left\{(x^4 - y^4) - \frac{6}{7}z^2(x^2 - y^2)\right\}$

$\Gamma^{(4)}$	t_1x	$\frac{1}{\sqrt{2}}(-Y_{1\,1}+Y_{1\,-1})$	$\frac{1}{r}\sqrt{\frac{3}{4\pi}}\,x=\sqrt{\frac{3}{4\pi}}\sin\theta\cos\varphi$
	t_1y	$\frac{i}{\sqrt{2}}(Y_{1\,1}+Y_{1\,-1})$	$\frac{1}{r}\sqrt{\frac{3}{4\pi}}\,y=\sqrt{\frac{3}{4\pi}}\sin\theta\sin\varphi$
	t_1z	$Y_{1\,0}$	$\frac{1}{r}\sqrt{\frac{3}{4\pi}}\,z=\sqrt{\frac{3}{4\pi}}\cos\theta$
		$p_x=x,\quad p_y=y,\quad p_z=z$	
	t_1x	$\frac{1}{4}\{\sqrt{5}(-Y_{3\,3}+Y_{3\,-3})+\sqrt{3}(Y_{3\,1}-Y_{3\,-1})\}$	$\frac{1}{r^3}\sqrt{\frac{7}{4\pi}\frac{5}{2}}\left(x^3-\frac{3xr^2}{5}\right)$
	t_1y	$-\frac{i}{4}\{\sqrt{5}(Y_{3\,3}+Y_{3\,-3})+\sqrt{3}(Y_{3\,1}+Y_{3\,-1})\}$	$\frac{1}{r^3}\sqrt{\frac{7}{4\pi}\frac{5}{2}}\left(y^3-\frac{3yr^2}{5}\right)$
	t_1z	$Y_{3\,0}$	$\frac{1}{r^3}\sqrt{\frac{7}{4\pi}\frac{5}{2}}\left(z^3-\frac{3zr^2}{5}\right)$
	t_1x	$-\frac{i}{4}\{\sqrt{7}(Y_{4\,1}+Y_{4\,-1})+(Y_{4\,3}+Y_{4\,-3})\}$	$\frac{1}{r^4}\sqrt{\frac{9}{4\pi}}\frac{\sqrt{35}}{2}\,yz(y^2-z^2)$
	t_1y	$\frac{1}{4}\{\sqrt{7}(-Y_{4\,1}+Y_{4\,-1})+(Y_{4\,3}-Y_{4\,-3})\}$	$\frac{1}{r^4}\sqrt{\frac{9}{4\pi}}\frac{\sqrt{35}}{2}\,zx(z^2-x^2)$
	t_1z	$-\frac{i}{\sqrt{2}}(Y_{4\,4}-Y_{4\,-4})$	$\frac{1}{r^4}\sqrt{\frac{9}{4\pi}}\frac{\sqrt{35}}{2}\,xy(x^2-y^2)$

$$|t_1 1\rangle=-\frac{1}{\sqrt{2}}(|t_1x\rangle+i|t_1y\rangle);\ |t_1 0\rangle=|t_1z\rangle;\ |t_1-1\rangle=\frac{1}{\sqrt{2}}(|t_1x\rangle-i|t_1y\rangle)$$

$$|t_1x\rangle=-\frac{1}{\sqrt{2}}(|t_1 1\rangle-|t_1-1\rangle);\ |t_1y\rangle=\frac{i}{\sqrt{2}}(|t_1 1\rangle+|t_1-1\rangle);\ |t_1z\rangle=|t_1 0\rangle$$

TABLE 5.8 *(continued)*

Irreducible representation	Name	Basis functions	
$\Gamma^{(5)}$	$d\varepsilon$ $t_2\xi$	$\frac{i}{\sqrt{2}}(Y_{2\,1} + Y_{2\,-1})$	$\frac{1}{r^2}\sqrt{\frac{5}{4\pi}}\sqrt{3}yz$
	$d\varepsilon$ $t_2\eta$	$\frac{1}{\sqrt{2}}(-Y_{2\,1} + Y_{2\,-1})$	$\frac{1}{r^2}\sqrt{\frac{5}{4\pi}}\sqrt{3}zx$
	$d\varepsilon$ $t_2\zeta$	$\frac{i}{\sqrt{2}}(-Y_{2\,2} + Y_{2\,-2})$	$\frac{1}{r^2}\sqrt{\frac{5}{4\pi}}\sqrt{3}xy$
		$d_{xy} = \sqrt{3}xy = \sqrt{3}r^2\sin^2\theta\cos\varphi\sin\varphi$	
		$d_{yz} = \sqrt{3}yz = \sqrt{3}r^2\sin\theta\cos\theta\sin\varphi$	
		$d_{zx} = \sqrt{3}zx = \sqrt{3}r^2\sin\theta\cos\theta\cos\varphi$	
	$t_2\xi$	$-\frac{1}{4}\{\sqrt{5}(-Y_{3\,1} + Y_{3\,-1}) + \sqrt{3}(-Y_{3\,3} + Y_{3\,-3})\}$	$\frac{1}{r^3}\sqrt{\frac{7}{4\pi}\frac{1}{2}}\sqrt{15}x(y^2 - z^2)$
	$t_2\eta$	$\frac{i}{2}\{\sqrt{5}(Y_{3\,1} + Y_{3\,-1}) - \sqrt{3}(Y_{3\,3} + Y_{3\,-3})\}$	$\frac{1}{r^3}\sqrt{\frac{7}{4\pi}\frac{1}{2}}\sqrt{15}y(z^2 - x^2)$
	$t_2\zeta$	$\frac{1}{\sqrt{2}}(Y_{3\,2} + Y_{3\,-2})$	$\frac{1}{r^3}\sqrt{\frac{7}{4\pi}\frac{1}{2}}\sqrt{15}z(x^2 - y^2)$
	$t_2\xi$	$-\frac{i}{4}\{\sqrt{7}(Y_{4\,3} + Y_{4\,-3}) - (Y_{4\,1} + Y_{4\,-1})\}$	$\frac{1}{r^4}\sqrt{\frac{9}{4\pi}\frac{7}{2}}\sqrt{5}yz\left(x^2 - \frac{r^2}{7}\right)$
	$t_2\eta$	$\frac{1}{4}\{\sqrt{7}(Y_{4\,3} - Y_{4\,-3}) + (Y_{4\,1} - Y_{4\,-1})\}$	$\frac{1}{r^4}\sqrt{\frac{9}{4\pi}\frac{7}{2}}\sqrt{5}zx\left(y^2 - \frac{r^2}{7}\right)$

	$t_2\zeta$	$-\frac{i}{\sqrt{2}}(Y_{4\,2} - Y_{4\,-2})$		$\frac{1}{r^4}\sqrt{\frac{9}{4\pi}}\frac{7}{2}\sqrt{5}xy\left(z^2 - \frac{r^2}{7}\right)$
		$\lvert t_2 1\rangle = -\frac{1}{\sqrt{2}}(\lvert t_2\xi\rangle + i\lvert t_2\eta\rangle)$; $\lvert t_2 0\rangle = \lvert t_2\zeta\rangle$; $\lvert t_2 -1\rangle = \frac{1}{\sqrt{2}}(\lvert t_2\xi\rangle - i\lvert t_2\eta\rangle)$		
		$\lvert t_2\xi\rangle = -\frac{1}{\sqrt{2}}(\lvert t_2 1\rangle - \lvert t_2 -1\rangle)$; $\lvert t_2\eta\rangle = \frac{i}{\sqrt{2}}(\lvert t_2 1\rangle + \lvert t_2 -1\rangle)$; $\lvert t_2\zeta\rangle = \lvert t_2 0\rangle$		
$\Gamma^{(6)}$	$\psi^6_{1/2}$	$\left\lvert\frac{1}{2}\frac{1}{2}\right\rangle$		$\sqrt{\frac{5}{12}}\left\lvert\frac{7}{2}\,-\frac{7}{2}\right\rangle + \sqrt{\frac{7}{12}}\left\lvert\frac{7}{2}\frac{1}{2}\right\rangle$
	$\psi^6_{-1/2}$	$\left\lvert\frac{1}{2}\,-\frac{1}{2}\right\rangle$		$-\sqrt{\frac{5}{12}}\left\lvert\frac{7}{2}\frac{7}{2}\right\rangle - \sqrt{\frac{7}{12}}\left\lvert\frac{7}{2}\,-\frac{1}{2}\right\rangle$
$\Gamma^{(7)}$	$\psi^7_{1/2}$		$\frac{1}{\sqrt{6}}\left\lvert\frac{5}{2}\frac{5}{2}\right\rangle - \sqrt{\frac{5}{6}}\left\lvert\frac{5}{2}\,-\frac{3}{2}\right\rangle$	$\frac{\sqrt{3}}{2}\left\lvert\frac{7}{2}\frac{5}{2}\right\rangle - \frac{1}{2}\left\lvert\frac{7}{2}\,-\frac{3}{2}\right\rangle$
	$\psi^7_{-1/2}$		$\frac{1}{\sqrt{6}}\left\lvert\frac{5}{2}\,-\frac{5}{2}\right\rangle - \sqrt{\frac{5}{6}}\left\lvert\frac{5}{2}\frac{3}{2}\right\rangle$	$-\frac{\sqrt{3}}{2}\left\lvert\frac{7}{2}\,-\frac{5}{2}\right\rangle + \frac{1}{2}\left\lvert\frac{7}{2}\frac{3}{2}\right\rangle$
$\Gamma^{(8)}$	$\psi^8_{3/2}$	$\left\lvert\frac{3}{2}\frac{3}{2}\right\rangle$	$-\frac{1}{\sqrt{6}}\left\lvert\frac{5}{2}\frac{3}{2}\right\rangle - \sqrt{\frac{5}{6}}\left\lvert\frac{5}{2}\,-\frac{5}{2}\right\rangle$	$\frac{1}{2}\left\lvert\frac{7}{2}\,-\frac{5}{2}\right\rangle + \frac{\sqrt{3}}{2}\left\lvert\frac{7}{2}\frac{3}{2}\right\rangle$
	$\psi^8_{1/2}$	$\left\lvert\frac{3}{2}\frac{1}{2}\right\rangle$	$\left\lvert\frac{5}{2}\frac{1}{2}\right\rangle$	$\sqrt{\frac{7}{12}}\left\lvert\frac{7}{2}\,-\frac{7}{2}\right\rangle - \sqrt{\frac{5}{12}}\left\lvert\frac{7}{2}\frac{1}{2}\right\rangle$
	$\psi^8_{-1/2}$	$\left\lvert\frac{3}{2}\,-\frac{1}{2}\right\rangle$	$-\left\lvert\frac{5}{2}\,-\frac{1}{2}\right\rangle$	$\sqrt{\frac{7}{12}}\left\lvert\frac{7}{2}\frac{7}{2}\right\rangle - \sqrt{\frac{5}{12}}\left\lvert\frac{7}{2}\,-\frac{1}{2}\right\rangle$
	$\psi^8_{-3/2}$	$\left\lvert\frac{3}{2}\,-\frac{3}{2}\right\rangle$	$\sqrt{\frac{5}{6}}\left\lvert\frac{5}{2}\frac{5}{2}\right\rangle + \frac{1}{\sqrt{6}}\left\lvert\frac{5}{2}\,-\frac{3}{2}\right\rangle$	$\frac{1}{2}\left\lvert\frac{7}{2}\frac{5}{2}\right\rangle + \frac{\sqrt{3}}{2}\left\lvert\frac{7}{2}\,-\frac{3}{2}\right\rangle$

TABLE 5.9

Coupling Coefficients for the Irreducible Representations of O and T_d[a,b]

	$u_2v_1^3$	$u_2v_2^3$
ψ_1^3	0	1
ψ_2^3	−1	0

	$u_2v_{-1/2}^6$	$u_2v_{1/2}^6$
$\psi_{-1/2}^7$	1	0
$\psi_{1/2}^7$	0	1

	$u_2v_{-1/2}^7$	$u_2v_{1/2}^7$
$\psi_{-1/2}^6$	1	0
$\psi_{1/2}^6$	0	1

	$u_2v_x^4$	$u_2v_y^4$	$u_2v_z^4$
ψ_{yx}^5	1	0	0
ψ_{xz}^5	0	1	0
ψ_{yz}^5	0	0	1

	$u_2v_{yz}^5$	$u_2v_{xz}^5$	$u_2v_{xy}^5$
ψ_x^4	1	0	0
ψ_y^4	0	1	0
ψ_z^4	0	0	1

	$u_2v_{-3/2}^8$	$u_2v_{-1/2}^8$	$u_2v_{1/2}^8$	$u_2v_{3/2}^8$
$\psi_{-3/2}^8$	0	0	1	0
$\psi_{-1/2}^8$	0	0	0	−1
$\psi_{1/2}^8$	−1	0	0	0
$\psi_{3/2}^8$	0	1	0	0

	$u_1^3v_1^3$	$u_1^3v_2^3$	$u_2^3v_1^3$	$u_2^3v_2^3$
ψ_1	$1/\sqrt{2}$	0	0	$1/\sqrt{2}$
ψ_2	0	$1/\sqrt{2}$	$-1/\sqrt{2}$	0
ψ_1^3	$-1/\sqrt{2}$	0	0	$1/\sqrt{2}$
ψ_2^3	0	$1/\sqrt{2}$	$1/\sqrt{2}$	0

	$u_1^3v_x^4$	$u_1^3v_y^4$	$u_1^3v_z^4$	$u_2^3v_x^4$	$u_2^3v_y^4$	$u_2^3v_z^4$
ψ_x^4	$-1/2$	0	0	$\sqrt{3}/2$	0	0
ψ_y^4	0	$-1/2$	0	0	$-\sqrt{3}/2$	0
ψ_z^4	0	0	1	0	0	0
ψ_{yz}^5	$-\sqrt{3}/2$	0	0	$-1/2$	0	0
ψ_{xz}^5	0	$\sqrt{3}/2$	0	0	$-1/2$	0
ψ_{xy}^5	0	0	0	0	0	1

[a] u_1^3 transforms as $3z^2 - r^2$; u_2^3 transforms as $\sqrt{3}(x^2 - y^2)$.

[b] Reprinted from "Properties of the Thirty-two Point Groups," by Koster, Dimmock, Wheeler, and Schatz by permission of The M.I.T. Press, Cambridge, Massachusetts.

TABLE 5.9 *(continued)*

	$u^3_1v^5_{yz}$	$u^3_1v^5_{xz}$	$u^3_1v^5_{xy}$	$u^3_2v^5_{yz}$	$u^3_2v^5_{xz}$	$u^3_2v^5_{xy}$
ψ^4_x	$-\sqrt{3}/2$	0	0	$-1/2$	0	0
ψ^4_y	0	$\sqrt{3}/2$	0	0	$-1/2$	0
ψ^4_z	0	0	0	0	0	1
ψ^5_{yz}	$-1/2$	0	0	$\sqrt{3}/2$	0	0
ψ^5_{xz}	0	$-1/2$	0	0	$-\sqrt{3}/2$	0
ψ^5_{xy}	0	0	1	0	0	0

	$u^3_1v^6_{-1/2}$	$u^3_1v^6_{1/2}$	$u^3_2v^6_{-1/2}$	$u^3_2v^6_{1/2}$
$\psi^8_{-3/2}$	0	0	0	1
$\psi^8_{-1/2}$	1	0	0	0
$\psi^8_{1/2}$	0	-1	0	0
$\psi^8_{3/2}$	0	0	-1	0

	$u^3_1v^7_{-1/2}$	$u^3_1v^7_{1/2}$	$u^3_2v^7_{-1/2}$	$u^3_2v^7_{1/2}$
$\psi^8_{-3/2}$	0	-1	0	0
$\psi^8_{-1/2}$	0	0	1	0
$\psi^8_{1/2}$	0	0	0	-1
$\psi^8_{3/2}$	1	0	0	0

	$u^3_1v^8_{-3/2}$	$u^3_1v^8_{-1/2}$	$u^3_1v^8_{1/2}$	$u^3_1v^8_{3/2}$	$u^3_2v^8_{-3/2}$	$u^3_2v^8_{-1/2}$	$u^3_2v^8_{1/2}$	$u^3_2v^8_{3/2}$
$\psi^6_{-1/2}$	0	$1/\sqrt{2}$	0	0	0	0	0	$1/\sqrt{2}$
$\psi^6_{1/2}$	0	0	$-1\sqrt{2}$	0	$-1/\sqrt{2}$	0	0	0
$\psi^7_{-1/2}$	0	0	0	$1/\sqrt{2}$	0	$-1/\sqrt{2}$	0	0
$\psi^7_{-3/2}$	$-1/\sqrt{2}$	0	0	0	0	0	$1/\sqrt{2}$	0
$\psi^8_{-3/2}$	$1/\sqrt{2}$	0	0	0	0	0	$1/\sqrt{2}$	0
$\psi^8_{-1/2}$	0	$-1/\sqrt{2}$	0	0	0	0	0	$1/\sqrt{2}$
$\psi^8_{1/2}$	0	0	$-1/\sqrt{2}$	0	$1/\sqrt{2}$	0	0	0
$\psi^8_{3/2}$	0	0	0	$1/\sqrt{2}$	0	$1/\sqrt{2}$	0	0

TABLE 5.9 (continued)

	$u_x^4v_x^4$	$u_x^4v_y^4$	$u_x^4v_z^4$	$u_y^4v_x^4$	$u_y^4v_y^4$	$u_y^4v_z^4$	$u_z^4v_x^4$	$u_z^4v_y^4$	$u_z^4v_z^4$
ψ_1	$1/\sqrt{3}$	0	0	0	$1/\sqrt{3}$	0	0	0	$1/\sqrt{3}$
ψ_1^3	$-1/\sqrt{6}$	0	0	0	$-1/\sqrt{6}$	0	0	0	$\sqrt{2}/\sqrt{3}$
ψ_2^3	$1/\sqrt{2}$	0	0	0	$-1/\sqrt{2}$	0	0	0	0
ψ_x^4	0	0	0	0	0	$1/\sqrt{2}$	0	$-1/\sqrt{2}$	0
ψ_y^4	0	0	$-1/\sqrt{2}$	0	0	0	$1/\sqrt{2}$	0	0
ψ_z^4	0	$1/\sqrt{2}$	0	$-1/\sqrt{2}$	0	0	0	0	0
ψ_{yz}^5	0	0	0	0	0	$1/\sqrt{2}$	0	$1/\sqrt{2}$	0
ψ_{xz}^5	0	0	$1/\sqrt{2}$	0	0	0	$1/\sqrt{2}$	0	0
ψ_{xy}^5	0	$1/\sqrt{2}$	0	$1/\sqrt{2}$	0	0	0	0	0

	$u_x^4v_{yz}^5$	$u_x^4v_{xz}^5$	$u_x^4v_{xy}^5$	$u_y^4v_{yz}^5$	$u_y^4v_{xz}^5$	$u_y^4v_{xy}^5$	$u_z^4v_{yz}^5$	$u_z^4v_{xz}^5$	$u_z^4v_{xy}^5$
ψ_2	$1/\sqrt{3}$	0	0	0	$1/\sqrt{3}$	0	0	0	$1/\sqrt{3}$
ψ_1^3	$1/\sqrt{2}$	0	0	0	$-1/\sqrt{2}$	0	0	0	0
ψ_2^3	$1/\sqrt{6}$	0	0	0	$1/\sqrt{6}$	0	0	0	$-\sqrt{2}/\sqrt{3}$
ψ_x^4	0	0	0	0	0	$1/\sqrt{2}$	0	$1/\sqrt{2}$	0
ψ_y^4	0	0	$1/\sqrt{2}$	0	0	0	$1/\sqrt{2}$	0	0
ψ_z^4	0	$1/\sqrt{2}$	0	$1/\sqrt{2}$	0	0	0	0	0
ψ_{yz}^5	0	0	0	0	0	$1/\sqrt{2}$	0	$-1/\sqrt{2}$	0
ψ_{xz}^5	0	0	$-1/\sqrt{2}$	0	0	0	$1/\sqrt{2}$	0	0
ψ_{xy}^5	0	$1/\sqrt{2}$	0	$-1/\sqrt{2}$	0	0	0	0	0

	$u_x^4v_{-1/2}^6$	$u_x^4v_{1/2}^6$	$u_y^4v_{-1/2}^6$	$u_y^4v_{1/2}^6$	$u_z^4v_{-1/2}^6$	$u_z^4v_{1/2}^6$
$\psi_{-1/2}^6$	0	$-i/\sqrt{3}$	0	$-1/\sqrt{3}$	$i/\sqrt{3}$	0
$\psi_{1/2}^6$	$-i/\sqrt{3}$	0	$1/\sqrt{3}$	0	0	$-i/\sqrt{3}$
$\psi_{-3/2}^8$	$i/\sqrt{2}$	0	$1/\sqrt{2}$	0	0	0
$\psi_{-1/2}^8$	0	$i/\sqrt{6}$	0	$1/\sqrt{6}$	$i\sqrt{2}/\sqrt{3}$	0
$\psi_{1/2}^8$	$-i/\sqrt{6}$	0	$1/\sqrt{6}$	0	0	$i\sqrt{2}/\sqrt{3}$
$\psi_{3/2}^8$	0	$-i/\sqrt{2}$	0	$1/\sqrt{2}$	0	0

	$u_x^4v_{-1/2}^7$	$u_x^4v_{1/2}^7$	$u_y^4v_{-1/2}^7$	$u_y^4v_{1/2}^7$	$u_x^4v_{-1/2}^7$	$u_z^4v_{1/2}^7$
$\psi_{-1/2}^7$	0	$-i/\sqrt{3}$	0	$-1/\sqrt{3}$	$i/\sqrt{3}$	0
$\psi_{1/2}^7$	$-i/\sqrt{3}$	0	$1/\sqrt{3}$	0	0	$-i/\sqrt{3}$
$\psi_{-3/2}^8$	$-i/\sqrt{6}$	0	$1/\sqrt{6}$	0	0	$i\sqrt{2}/\sqrt{3}$
$\psi_{-1/2}^8$	0	$i/\sqrt{2}$	0	$-1/\sqrt{2}$	0	0
$\psi_{1/2}^8$	$-i/\sqrt{2}$	0	$-1/\sqrt{2}$	0	0	0
$\psi_{3/2}^8$	0	$i/\sqrt{6}$	0	$1/\sqrt{6}$	$i\sqrt{2}/\sqrt{3}$	0

TABLE 5.9 (*continued*)

	$u_x^4 v_{-3/2}^8$	$u_x^4 v_{-1/2}^8$	$u_x^4 v_{1/2}^8$	$u_x^4 v_{3/2}^8$	$u_y^4 v_{-3/2}^8$	$u_y^4 v_{-1/2}^8$	$u_y^4 v_{1/2}^8$	$u_y^4 v_{3/2}^8$	$u_z^4 v_{-3/2}^8$	$u_z^4 v_{-1/2}^8$	$u_z^4 v_{1/2}^8$	$u_z^4 v_{3/2}^8$
$\psi_{-3/2}^8$	0	$i/\sqrt{5}$	0	0	0	$1/\sqrt{5}$	0	0	$-i\sqrt{3}/\sqrt{5}$	0	0	0
$\psi_{-1/2}^8$	$i/\sqrt{5}$	0	$i\sqrt{4}/\sqrt{15}$	0	$-1/\sqrt{5}$	0	$\sqrt{4}/\sqrt{15}$	0	0	$-i/\sqrt{15}$	0	0
$\psi_{1/2}^8$	0	$i\sqrt{4}/\sqrt{15}$	0	$i/\sqrt{5}$	0	$-\sqrt{4}/\sqrt{15}$	0	$1/\sqrt{5}$	0	0	$i/\sqrt{15}$	0
$\psi_{3/2}^8$	0	0	$i/\sqrt{5}$	0	0	0	$-1/\sqrt{5}$	0	0	0	0	$i\sqrt{3}/\sqrt{5}$
$\phi_{-3/2}^8$	0	$i/\sqrt{20}$	0	$-i\sqrt{5}/\sqrt{12}$	0	$1/\sqrt{20}$	0	$\sqrt{5}/\sqrt{12}$	$i/\sqrt{15}$	0	0	0
$\phi_{-1/2}^8$	$i/\sqrt{20}$	0	$-i\sqrt{3}/\sqrt{20}$	0	$-1/\sqrt{20}$	0	$-\sqrt{3}/\sqrt{20}$	0	0	$-i\sqrt{3}/\sqrt{5}$	0	0
$\phi_{1/2}^8$	0	$-i\sqrt{3}/\sqrt{20}$	0	$i/\sqrt{20}$	0	$\sqrt{3}/\sqrt{20}$	0	$1/\sqrt{20}$	0	0	$i\sqrt{3}/\sqrt{5}$	0
$\phi_{3/2}^8$	$-i\sqrt{5}/\sqrt{12}$	0	$i/\sqrt{20}$	0	$-\sqrt{5}/\sqrt{12}$	0	$-1/\sqrt{20}$	0	0	0	0	$-i/\sqrt{15}$
$\psi_{-1/2}^6$	$-i/2$	0	$i/\sqrt{12}$	0	$1/2$	0	$1/\sqrt{12}$	0	0	$-i/\sqrt{3}$	0	0
$\psi_{1/2}^6$	0	$-i/\sqrt{12}$	0	$i/2$	0	$1/\sqrt{12}$	0	$1/2$	0	0	$-i/\sqrt{3}$	0
$\psi_{-1/2}^7$	$i/\sqrt{12}$	0	$i/2$	0	$1/\sqrt{12}$	0	$-1/2$	0	0	0	0	$-i/\sqrt{3}$
$\psi_{1/2}^7$	0	$-i/2$	0	$-i/\sqrt{12}$	0	$-1/2$	0	$1/\sqrt{12}$	$-i/\sqrt{3}$	0	0	0

TABLE 5.9 (*continued*)

	$u^5_{yz}v^5_{yz}$	$u^5_{yz}v^5_{xz}$	$u^5_{yz}v^5_{xy}$	$u^5_{xz}v^5_{yz}$	$u^5_{xz}v^5_{xz}$	$u^5_{xz}v^5_{xy}$	$u^5_{xy}v^5_{yz}$	$u^5_{xy}v^5_{xz}$	$u^5_{xy}v^5_{xy}$
ψ_1	$1/\sqrt{3}$	0	0	0	$1/\sqrt{3}$	0	0	0	$1/\sqrt{3}$
ψ^3_1	$-1/\sqrt{6}$	0	0	0	$-1/\sqrt{6}$	0	0	0	$\sqrt{2}/\sqrt{3}$
ψ^3_2	$1/\sqrt{2}$	0	0	0	$-1/\sqrt{2}$	0	0	0	0
ψ^4_x	0	0	0	0	0	$1/\sqrt{2}$	0	$-1/\sqrt{2}$	0
ψ^4_y	0	0	$-1/\sqrt{2}$	0	0	0	$1/\sqrt{2}$	0	0
ψ^4_z	0	$1/\sqrt{2}$	0	$-1/\sqrt{2}$	0	0	0	0	0
ψ^5_{yz}	0	0	0	0	0	$1/\sqrt{2}$	0	$1/\sqrt{2}$	0
ψ^5_{xz}	0	0	$1/\sqrt{2}$	0	0	0	$1/\sqrt{2}$	0	0
ψ^5_{xy}	0	$1/\sqrt{2}$	0	$1/\sqrt{2}$	0	0	0	0	0

	$u^5_{yz}v^6_{-1/2}$	$u^5_{yz}v^6_{1/2}$	$u^5_{xz}v^6_{-1/2}$	$u^5_{xz}v^6_{1/2}$	$u^5_{xy}v^6_{-1/2}$	$u^5_{xy}v^6_{1/2}$
$\psi^7_{-1/2}$	0	$-i/\sqrt{3}$	0	$-1/\sqrt{3}$	$i/\sqrt{3}$	0
$\psi^7_{1/2}$	$-i/\sqrt{3}$	0	$1/\sqrt{3}$	0	0	$-i/\sqrt{3}$
$\psi^8_{-3/2}$	$-i/\sqrt{6}$	0	$1/\sqrt{6}$	0	0	$i\sqrt{2}/\sqrt{3}$
$\psi^8_{-1/2}$	0	$i/\sqrt{2}$	0	$-1/\sqrt{2}$	0	0
$\psi^8_{1/2}$	$-i/\sqrt{2}$	0	$-1/\sqrt{2}$	0	0	0
$\psi^8_{3/2}$	0	$i/\sqrt{6}$	0	$1/\sqrt{6}$	$i\sqrt{2}/\sqrt{3}$	0

	$u^5_{yz}v^7_{-1/2}$	$u^5_{yz}v^7_{1/2}$	$u^5_{xz}v^7_{-1/2}$	$u^5_{xz}v^7_{1/2}$	$u^5_{xy}v^7_{-1/2}$	$u^5_{xy}v^7_{1/2}$
$\psi^6_{-1/2}$	0	$-i/\sqrt{3}$	0	$-1/\sqrt{3}$	$i/\sqrt{3}$	0
$\psi^6_{1/2}$	$-i/\sqrt{3}$	0	$1/\sqrt{3}$	0	0	$-i/\sqrt{3}$
$\psi^8_{-3/2}$	$i/\sqrt{2}$	0	$1/\sqrt{2}$	0	0	0
$\psi^8_{-1/2}$	0	$i/\sqrt{6}$	0	$1/\sqrt{6}$	$i\sqrt{2}/\sqrt{3}$	0
$\psi^8_{1/2}$	$-i/\sqrt{6}$	0	$1/\sqrt{6}$	0	0	$i\sqrt{2}/\sqrt{3}$
$\psi^8_{3/2}$	0	$-i/\sqrt{2}$	0	$1/\sqrt{2}$	0	0

TABLE 5.9 (*continued*)

	$u^5_{yz}v^8_{-3/2}$	$u^5_{yz}v^8_{-1/2}$	$u^5_{yz}v^8_{1/2}$	$u^5_{yz}v^8_{3/2}$	$u^5_{xz}v^8_{-3/2}$	$u^5_{xz}v^8_{-1/2}$	$u^5_{xz}v^8_{1/2}$	$u^5_{xz}v^8_{3/2}$	$u^5_{xy}v^8_{-3/2}$	$u^5_{xy}v^8_{-1/2}$	$u^5_{xy}v^8_{1/2}$	$u^5_{xy}v^8_{3/2}$
$\psi^8_{-3/2}$	0	$i\sqrt{4}/\sqrt{15}$	0	$i/\sqrt{5}$	0	$-\sqrt{4}/\sqrt{15}$	0	$1/\sqrt{5}$	0	0	$i/\sqrt{15}$	0
$\psi^8_{-1/2}$	0	0	$-i/\sqrt{5}$	0	0	0	$1/\sqrt{5}$	0	0	0	0	$-i\sqrt{3}/\sqrt{5}$
$\psi^8_{1/2}$	0	$-i/\sqrt{5}$	0	0	0	$-1/\sqrt{5}$	0	0	$i\sqrt{3}/\sqrt{5}$	0	0	0
$\psi^8_{3/2}$	$i/\sqrt{5}$	0	$i\sqrt{4}/\sqrt{15}$	0	$-1/\sqrt{5}$	0	$\sqrt{4}/\sqrt{15}$	0	0	$-i/\sqrt{15}$	0	0
$\phi^8_{-3/2}$	0	$-i\sqrt{3}/\sqrt{20}$	0	$i/\sqrt{20}$	0	$\sqrt{3}/\sqrt{20}$	0	$1/\sqrt{20}$	0	0	$i\sqrt{3}/\sqrt{5}$	0
$\phi^8_{-1/2}$	$i\sqrt{5}/\sqrt{12}$	0	$-i/\sqrt{20}$	0	$\sqrt{5}/\sqrt{12}$	0	$1/\sqrt{20}$	0	0	0	0	$i/\sqrt{15}$
$\phi^8_{1/2}$	0	$-i/\sqrt{20}$	0	$i\sqrt{5}/\sqrt{12}$	0	$-1/\sqrt{20}$	0	$-\sqrt{5}/\sqrt{12}$	$-i/\sqrt{15}$	0	0	0
$\phi^8_{3/2}$	$i/\sqrt{20}$	0	$-i\sqrt{3}/\sqrt{20}$	0	$-1/\sqrt{20}$	0	$-\sqrt{3}/\sqrt{20}$	0	0	$-i\sqrt{3}/\sqrt{5}$	0	0
$\psi^7_{-1/2}$	$-i/2$	0	$i/\sqrt{12}$	0	$1/2$	0	$1/\sqrt{12}$	0	0	$-i/\sqrt{3}$	0	0
$\psi^7_{1/2}$	0	$-i/\sqrt{12}$	0	$i/2$	0	$1/\sqrt{12}$	0	$1/2$	0	0	$-i/\sqrt{3}$	0
$\psi^6_{-1/2}$	$i/\sqrt{12}$	0	$i/2$	0	$1/\sqrt{12}$	0	$-1/2$	0	0	0	0	$-i/\sqrt{3}$
$\psi^6_{1/2}$	0	$-i/2$	0	$-i/\sqrt{12}$	0	$-1/2$	0	$1/\sqrt{12}$	$-i/\sqrt{3}$	0	0	0

TABLE 5.9 *(continued)*

	$u^6_{-1/2}v^6_{-1/2}$	$u^6_{-1/2}v^6_{1/2}$	$u^6_{1/2}v^6_{-1/2}$	$u^6_{1/2}v^6_{1/2}$
ψ_1	0	$-1/\sqrt{2}$	$1/\sqrt{2}$	0
ψ^4_x	$-i/\sqrt{2}$	0	0	$i/\sqrt{2}$
ψ^4_y	$1/\sqrt{2}$	0	0	$1/\sqrt{2}$
ψ^4_z	0	$-i/\sqrt{2}$	$-i/\sqrt{2}$	0

	$u^6_{-1/2}v^7_{-1/2}$	$u^6_{-1/2}v^7_{1/2}$	$u^6_{1/2}v^7_{-1/2}$	$u^6_{1/2}v^7_{1/2}$
ψ_2	0	$-1/\sqrt{2}$	$1/\sqrt{2}$	0
ψ^5_{yz}	$-i/\sqrt{2}$	0	0	$i/\sqrt{2}$
ψ^5_{xz}	$1/\sqrt{2}$	0	0	$1/\sqrt{2}$
ψ^5_{xy}	0	$-i/\sqrt{2}$	$-i/\sqrt{2}$	0

	$u^7_{-1/2}v^7_{-1/2}$	$u^7_{-1/2}v^7_{1/2}$	$u^7_{1/2}v^7_{-1/2}$	$u^7_{1/2}v^7_{1/2}$
ψ_1	0	$-1/\sqrt{2}$	$1/\sqrt{2}$	0
ψ^4_x	$-i/\sqrt{2}$	0	0	$i/\sqrt{2}$
ψ^4_y	$1/\sqrt{2}$	0	0	$1/\sqrt{2}$
ψ^4_z	0	$-i/\sqrt{2}$	$-i/\sqrt{2}$	0

	$u^6_{-1/2}v^8_{-3/2}$	$u^6_{-1/2}v^8_{-1/2}$	$u^6_{-1/2}v^8_{1/2}$	$u^6_{-1/2}v^8_{3/2}$	$u^6_{1/2}v^8_{-3/2}$	$u^6_{1/2}v^8_{-1/2}$	$u^6_{1/2}v^8_{1/2}$	$u^6_{1/2}v^8_{3/2}$
ψ^4_x	0	$-i/\sqrt{8}$	0	$i\sqrt{3}/\sqrt{8}$	$i\sqrt{3}/\sqrt{8}$	0	$-i/\sqrt{8}$	0
ψ^4_y	0	$1/\sqrt{8}$	0	$\sqrt{3}/\sqrt{8}$	$-\sqrt{3}/\sqrt{8}$	0	$-1/\sqrt{8}$	0
ψ^4_z	0	0	$-i/\sqrt{2}$	0	0	$i/\sqrt{2}$	0	0
ψ^5_{yz}	0	$-i\sqrt{3}/\sqrt{8}$	0	$-i/\sqrt{8}$	$-i/\sqrt{8}$	0	$-i\sqrt{3}/\sqrt{8}$	0
ψ^5_{xz}	0	$-\sqrt{3}/\sqrt{8}$	0	$1/\sqrt{8}$	$-1/\sqrt{8}$	0	$\sqrt{3}/\sqrt{8}$	0
ψ^5_{xy}	$-i/\sqrt{2}$	0	0	0	0	0	0	$i/\sqrt{2}$
ψ^3_1	0	0	$-1/\sqrt{2}$	0	0	$-1/\sqrt{2}$	0	0
ψ^3_2	$-1/\sqrt{2}$	0	0	0	0	0	0	$-1/\sqrt{2}$

TABLE 5.9 *(continued)*

	$u^7_{-1/2}v^8_{-3/2}$	$u^7_{-1/2}v^8_{-1/2}$	$u^7_{-1/2}v^8_{1/2}$	$u^7_{-1/2}v^8_{3/2}$	$u^7_{1/2}v^8_{-3/2}$	$u^7_{1/2}v^8_{-1/2}$	$u^7_{1/2}v^8_{1/2}$	$u^7_{1/2}v^8_{3/2}$
ψ^4_x	0	$-i\sqrt{3}/\sqrt{8}$	0	$-i/\sqrt{8}$	$-i/\sqrt{8}$	0	$-i\sqrt{3}/\sqrt{8}$	0
ψ^4_y	0	$-\sqrt{3}/\sqrt{8}$	0	$1/\sqrt{8}$	$-1/\sqrt{8}$	0	$\sqrt{3}/\sqrt{8}$	0
ψ^4_z	$-i/\sqrt{2}$	0	0	0	0	0	0	$i/\sqrt{2}$
ψ^5_{yz}	0	$-i/\sqrt{8}$	0	$i\sqrt{3}/\sqrt{8}$	$i\sqrt{3}/\sqrt{8}$	0	$-i/\sqrt{8}$	0
ψ^5_{xz}	0	$1/\sqrt{8}$	0	$\sqrt{3}/\sqrt{8}$	$-\sqrt{3}/\sqrt{8}$	0	$-1/\sqrt{8}$	0
ψ^5_{xy}	0	0	$-i/\sqrt{2}$	0	0	$i/\sqrt{2}$	0	0
ψ^3_1	$-1/\sqrt{2}$	0	0	0	0	0	0	$-1/\sqrt{2}$
ψ^3_2	0	0	$1/\sqrt{2}$	0	0	$1/\sqrt{2}$	0	0

	$u^8_{-3/2}v^8_{-3/2}$	$u^8_{-3/2}v^8_{-1/2}$	$u^8_{-3/2}v^8_{1/2}$	$u^8_{-3/2}v^8_{3/2}$	$u^8_{-1/2}v^8_{-3/2}$	$u^8_{-1/2}v^8_{-1/2}$	$u^8_{-1/2}v^8_{1/2}$	$u^8_{-1/2}v^8_{3/2}$
ψ_1	0	0	0	$-1/2$	0	0	$1/2$	0
ψ_2	0	$-1/2$	0	0	$-1/2$	0	0	0
ψ^3_1	0	0	0	$1/2$	0	0	$1/2$	0
ψ^3_2	0	$1/2$	0	0	$-1/2$	0	0	0
ψ^4_x	0	0	$-i\sqrt{3}/\sqrt{20}$	0	0	$i/\sqrt{5}$	0	$i\sqrt{3}/\sqrt{20}$
ψ^4_y	0	0	$\sqrt{3}/\sqrt{20}$	0	0	$-1/\sqrt{5}$	0	$\sqrt{3}/\sqrt{20}$
ψ^4_z	0	0	0	$-i3/\sqrt{20}$	0	0	$i/\sqrt{20}$	0
ϕ^4_x	$-i\sqrt{5}/4$	0	$i\sqrt{3}/\sqrt{80}$	0	0	$i3/\sqrt{80}$	0	$-i\sqrt{3}/\sqrt{80}$
ϕ^4_y	$-\sqrt{5}/4$	0	$-\sqrt{3}/\sqrt{80}$	0	0	$-3/\sqrt{80}$	0	$-\sqrt{3}/\sqrt{80}$
ϕ^4_z	0	0	0	$-i/\sqrt{20}$	0	0	$-i3/\sqrt{20}$	0
ψ^5_{yz}	0	0	$i/2$	0	0	0	0	$i/2$
ψ^5_{xz}	0	0	$1/2$	0	0	0	0	$-1/2$
ψ^5_{xy}	0	$i/2$	0	0	$-i/2$	0	0	0
ϕ^5_{yz}	$-i\sqrt{3}/4$	0	$-i/4$	0	0	$-i\sqrt{3}/4$	0	$i/4$
ϕ^5_{xz}	$\sqrt{3}/4$	0	$-1/4$	0	0	$-\sqrt{3}/4$	0	$-1/4$
ϕ^5_{xy}	0	$i/2$	0	0	$i/2$	0	0	0

TABLE 5.9 (*continued*)

	$u^8_{1/2}v^8_{-3/2}$	$u^8_{1/2}v^8_{-1/2}$	$u^8_{1/2}v^8_{1/2}$	$u^8_{1/2}v^8_{3/2}$	$u^8_{3/2}v^8_{-3/2}$	$u^8_{3/2}v^8_{-1/2}$	$u^8_{3/2}v^8_{1/2}$	$u^8_{3/2}v^8_{3/2}$
ψ_1	0	$-1/2$	0	0	$1/2$	0	0	0
ψ_2	0	0	0	$1/2$	0	0	$1/2$	0
ψ_1^3	0	$-1/2$	0	0	$-1/2$	0	0	0
ψ_2^3	0	0	0	$1/2$	0	0	$-1/2$	0
ψ_x^4	$-i\sqrt{3}/\sqrt{20}$	0	$-i/\sqrt{5}$	0	0	$i\sqrt{3}/\sqrt{20}$	0	0
ψ_y^4	$\sqrt{3}/\sqrt{20}$	0	$-1/\sqrt{5}$	0	0	$\sqrt{3}/\sqrt{20}$	0	0
ψ_z^4	0	$i/\sqrt{20}$	0	0	$-i3/\sqrt{20}$	0	0	0
ϕ_x^4	$i\sqrt{3}/\sqrt{80}$	0	$-i3/\sqrt{80}$	0	0	$-i\sqrt{3}/\sqrt{80}$	0	$i\sqrt{5}/4$
ϕ_y^4	$-\sqrt{3}/\sqrt{80}$	0	$-3/\sqrt{80}$	0	0	$-\sqrt{3}/\sqrt{80}$	0	$-\sqrt{5}/4$
ϕ_z^4	0	$-i3/\sqrt{20}$	0	0	$-i/\sqrt{20}$	0	0	0
ψ_{yz}^5	$-i/2$	0	0	0	0	$-i/2$	0	0
ψ_{xz}^5	$-1/2$	0	0	0	0	$1/2$	0	0
ψ_{xy}^5	0	0	0	$-i/2$	0	0	$i/2$	0
ϕ_{yz}^5	$-i/4$	0	$i\sqrt{3}/4$	0	0	$i/4$	0	$i\sqrt{3}/4$
ϕ_{xz}^5	$-1/4$	0	$-\sqrt{3}/4$	0	0	$-1/4$	0	$\sqrt{3}/4$
ϕ_{xy}^5	0	0	0	$i/2$	0	0	$i/2$	0

TABLE 5.10

Character Table for the Double Groups $\bar{D}_4$, $\bar{C}_{4v}$, and $\bar{D}_{2d}$

	$\bar{D}_4$	E	$\bar{E}$	$2C_4$	$2\bar{C}_4$	C_2 $\bar{C}_2$	$2C_2'$ $2\bar{C}_2'$	$2C_2''$ $2\bar{C}_2''$
	$\bar{C}_{4v}$	E	$\bar{E}$	$2C_4$	$2\bar{C}_4$	C_2 $\bar{C}_2$	$2\sigma_v$ $2\bar{\sigma}_v$	$2\sigma_d$ $2\bar{\sigma}_d$
	$\bar{D}_{2d}$	E	$\bar{E}$	$2S_4$	$2\bar{S}_4$	C_2 $\bar{C}_2$	$2C_2'$ $2\bar{C}_2'$	$2\sigma_d$ $2\bar{\sigma}_d$
$\Gamma^{(1)}$	A_1	1	1	1	1	1	1	1
$\Gamma^{(2)}$	A_2	1	1	1	1	1	-1	-1
$\Gamma^{(3)}$	B_1	1	1	-1	-1	1	1	-1
$\Gamma^{(4)}$	B_2	1	1	-1	-1	1	-1	1
$\Gamma^{(5)}$	E	2	2	0	0	-2	0	0
$\Gamma^{(6)}$	E'	2	-2	$\sqrt{2}$	$-\sqrt{2}$	0	0	0
$\Gamma^{(7)}$	E''	2	-2	$-\sqrt{2}$	$\sqrt{2}$	0	0	0

TABLE 5.11

Coupling Coefficients for the Double Groups $\bar{D}_4$, $\bar{C}_{4v}$, and $\bar{D}_{2d}$[a]

	$u_2v_x^5$	$u_2v_y^5$
ψ_x^5	0	1
ψ_y^5	−1	0

	$u_2v_{-1/2}^6$	$u_2v_{1/2}^6$
$\psi_{-1/2}^6$	i	0
$\psi_{1/2}^6$	0	−i

	$u_2v_{-1/2}^7$	$u_2v_{1/2}^7$
$\psi_{-1/2}^7$	i	0
$\psi_{1/2}^7$	0	−i

	$u_3v_x^5$	$u_3v_y^5$
ψ_x^5	0	1
ψ_y^5	1	0

	$u_3v_{-1/2}^6$	$u_3v_{1/2}^6$
$\psi_{-1/2}^7$	1	0
$\psi_{1/2}^7$	0	1

	$u_3v_{-1/2}^7$	$u_3v_{1/2}^7$
$\psi_{-1/2}^6$	1	0
$\psi_{1/2}^6$	0	1

	$u_4v_x^5$	$u_4v_y^5$
ψ_x^5	0	1
ψ_y^5	1	0

	$u_4v_{-1/2}^6$	$u_4v_{1/2}^6$
$\psi_{-1/2}^7$	i	0
$\psi_{1/2}^7$	0	−i

	$u_4v_{-1/2}^7$	$u_4v_{1/2}^7$
$\psi_{-1/2}^6$	i	0
$\psi_{1/2}^6$	0	−i

	$u_x^5v_x^5$	$u_x^5v_y^5$	$u_y^5v_x^5$	$u_y^5v_y^5$
ψ_1	$1/\sqrt{2}$	0	0	$1/\sqrt{2}$
ψ_2	0	$1/\sqrt{2}$	$-1/\sqrt{2}$	0
ψ_3	$1/\sqrt{2}$	0	0	$-1/\sqrt{2}$
ψ_4	0	$1/\sqrt{2}$	$1/\sqrt{2}$	0

	$u_x^5v_{-1/2}^6$	$u_x^5v_{1/2}^6$	$u_y^5v_{-1/2}^6$	$u_y^5v_{1/2}^6$
$\psi_{-1/2}^6$	0	$i/\sqrt{2}$	0	$-1/\sqrt{2}$
$\psi_{1/2}^6$	$i/\sqrt{2}$	0	$1/\sqrt{2}$	0
$\psi_{-1/2}^7$	0	$i/\sqrt{2}$	0	$1/\sqrt{2}$
$\psi_{1/2}^7$	$i/\sqrt{2}$	0	$-1/\sqrt{2}$	0

	$u_x^5v_{-1/2}^7$	$u_x^5v_{1/2}^7$	$u_y^5v_{-1/2}^7$	$u_y^5v_{1/2}^7$
$\psi_{-1/2}^6$	0	$i/\sqrt{2}$	0	$1/\sqrt{2}$
$\psi_{1/2}^6$	$i/\sqrt{2}$	0	$-1/\sqrt{2}$	0
$\psi_{-1/2}^7$	0	$i/\sqrt{2}$	0	$-1/\sqrt{2}$
$\psi_{1/2}^7$	$i/\sqrt{2}$	0	$1/\sqrt{2}$	0

[a] Reprinted from "Properties of the Thirty-two Point Groups," by Koster, Dimmock, Wheeler, and Schatz by permission of The M.I.T. Press, Cambridge, Massachusetts. Copyright © 1963 by The Massachusetts Institute of Technology.

TABLE 5.11 (*continued*)

	$u^6_{-1/2}v^6_{-1/2}$	$u^6_{-1/2}v^6_{1/2}$	$u^6_{1/2}v^6_{-1/2}$	$u^6_{1/2}v^6_{1/2}$
ψ_1	0	$-1/\sqrt{2}$	$1/\sqrt{2}$	0
ψ_2	0	$-i/\sqrt{2}$	$-i/\sqrt{2}$	0
ψ^5_x	$-i/\sqrt{2}$	0	0	$i/\sqrt{2}$
ψ^5_y	$1/\sqrt{2}$	0	0	$1/\sqrt{2}$

	$u^6_{-1/2}v^7_{-1/2}$	$u^6_{-1/2}v^7_{1/2}$	$u^6_{1/2}v^7_{-1/2}$	$u^6_{1/2}v^7_{1/2}$
ψ_3	0	$-1/\sqrt{2}$	$1/\sqrt{2}$	0
ψ_4	0	$-i/\sqrt{2}$	$-i/\sqrt{2}$	0
ψ^5_x	$1/\sqrt{2}$	0	0	$1/\sqrt{2}$
ψ^5_y	$-i/\sqrt{2}$	0	0	$i/\sqrt{2}$

	$u^7_{-1/2}v^7_{-1/2}$	$u^7_{-1/2}v^7_{1/2}$	$u^7_{1/2}v^7_{-1/2}$	$u^7_{1/2}v^7_{1/2}$
ψ_1	0	$-1/\sqrt{2}$	$1/\sqrt{2}$	0
ψ_2	0	$-i/\sqrt{2}$	$-i/\sqrt{2}$	0
ψ^5_x	$-i/\sqrt{2}$	0	0	$i/\sqrt{2}$
ψ^5_y	$1/\sqrt{2}$	0	0	$1/\sqrt{2}$

D_{6h} is a hexagonal group. Referring to Fig. 5.2, the symmetry operators are

E identity
$2C_6$ rotations of $\pm 60°$ about the z axis
$2C_3$ rotations of $\pm 120°$ about the z axis
C_2 rotation of $180°$ about the z axis
$3C_2'$ rotations of $180°$ about axes such as 2–5
$3C_2''$ rotations of $180°$ about axes such as AB
i inversion in origin
σ_n reflection in xy plane
$2S_3$ $2C_3\sigma_h$
$2S_6$ $2C_6\sigma_h$
$3\sigma_d$ reflections in planes such as CD
$3\sigma_v$ reflection in planes such as 1–4

The character table for D_{6h} is included in Table 5.2.

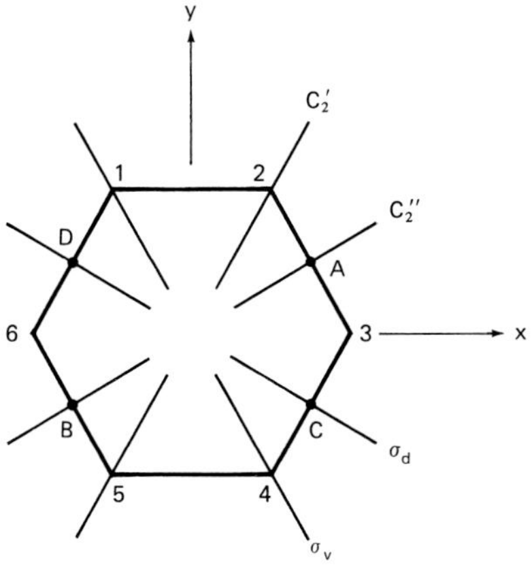

FIG. 5.2 Coordinate system and symmetry axes for the group D_{6h}

5.4 Permutation Groups S_n—Young Diagrams

A permutation (also called symmetric) group S_n consists of the $n!$ permutations on a set of n symbols. We devote this section to a summary of selected properties of permutation groups that have relevance to quantum mechanical applications.

Let

$$P = \begin{pmatrix} a_1 & a_2 & a_3 \\ a_2 & a_3 & a_1 \end{pmatrix} \equiv \begin{pmatrix} 1 & 2 & 3 \\ 2 & 3 & 1 \end{pmatrix} \equiv (1 \quad 2 \quad 3) \tag{5.4-1}$$

represent a permutation in which the sequence $a_1 a_2 a_3$ is permuted into the sequence $a_2 a_3 a_1$. Specifically we interpret the symbol in (5.4-1) to mean

$$\begin{gathered} 1 \text{ is replaced by } 2 \text{ or } 1 \rightarrow 2, \\ 2 \text{ is replaced by } 3 \text{ or } 2 \rightarrow 3, \\ 3 \text{ is replaced by } 1 \text{ or } 3 \rightarrow 1. \end{gathered}$$

It is also useful to think of $(1 \quad 2 \quad 3)$ as an operator:

$$(1 \quad 2 \quad 3)1 = 2, \qquad (1 \quad 2 \quad 3)2 = 3, \qquad (1 \quad 2 \quad 3)3 = 1. \tag{5.4-2}$$

In the form $(1 \quad 2 \quad 3)$, P is represented by a *cycle* which may also be written as $(2 \quad 3 \quad 1)$ or $(3 \quad 1 \quad 2)$. Hence the order in which the columns are written in the double-rowed parentheses is immaterial. Thus

$$\begin{pmatrix} 1 & 2 & 3 \\ 2 & 3 & 1 \end{pmatrix} = \begin{pmatrix} 2 & 3 & 1 \\ 3 & 1 & 2 \end{pmatrix} = \begin{pmatrix} 3 & 1 & 2 \\ 1 & 2 & 3 \end{pmatrix}.$$

Every permutation of a finite set of elements can be written as a product of cycles with no common elements. For example,

$$\begin{pmatrix} 1 & 2 & 3 & 4 \\ 2 & 3 & 1 & 4 \end{pmatrix} = (1 \quad 2 \quad 3)(4) = (4)(1 \quad 2 \quad 3) = (1 \quad 2 \quad 3),$$

$$\begin{pmatrix} 1 & 2 & 3 & 4 & 5 \\ 2 & 4 & 5 & 1 & 3 \end{pmatrix} = (1 \quad 2 \quad 4)(3 \quad 5) = (3 \quad 5)(1 \quad 2 \quad 4),$$

where a product like $(1 \quad 2 \quad 4)(3 \quad 5)$ has the meaning

$$\begin{pmatrix} 1 & 2 & 4 \\ 2 & 4 & 1 \end{pmatrix}\begin{pmatrix} 3 & 5 \\ 5 & 3 \end{pmatrix} = \begin{pmatrix} 1 & 2 & 4 & 3 & 5 \\ 2 & 4 & 1 & 5 & 3 \end{pmatrix} = \begin{pmatrix} 1 & 2 & 3 & 4 & 5 \\ 2 & 4 & 5 & 1 & 3 \end{pmatrix}.$$

Let

$$P_a = \begin{pmatrix} 1 & 2 & 3 & 4 \\ 2 & 4 & 3 & 1 \end{pmatrix}, \qquad P_b = \begin{pmatrix} 1 & 2 & 3 & 4 \\ 1 & 4 & 3 & 2 \end{pmatrix}.$$

The product $P_a P_b = P_c$ is given by

$$\begin{aligned} P_c &= \begin{pmatrix} 1 & 2 & 3 & 4 \\ 2 & 4 & 3 & 1 \end{pmatrix}\begin{pmatrix} 1 & 2 & 3 & 4 \\ 1 & 4 & 3 & 2 \end{pmatrix} \\ &= \begin{pmatrix} 1 & 4 & 3 & 2 \\ 2 & 1 & 3 & 4 \end{pmatrix}\begin{pmatrix} 1 & 2 & 3 & 4 \\ 1 & 4 & 3 & 2 \end{pmatrix} = \begin{pmatrix} 1 & 2 & 3 & 4 \\ 2 & 1 & 3 & 4 \end{pmatrix}. \end{aligned} \tag{5.4-3}$$

It is important to note that $P_a P_b$ means that the permutation P_a is performed *after* P_b in agreement with the standard convention for operator products. This means that the permutations proceed from right to left as, for example, in (5.4-3), where 2 is replaced by 4 (in P_b) and 4 is replaced by 1 (in P_a) giving a net change whereby 2 is replaced by 1 as in P_c.

A cycle containing two symbols is called a *transposition*. Any cycle can be written as a product of transpositions

$$(1 \quad 2 \cdots n) = (1 \quad n) \cdots (1 \quad 3)(1 \quad 2) \tag{5.4-4a}$$

or, alternatively,

$$(1 \quad 2 \cdots n) = (1 \quad 2)(2 \quad 3) \cdots (n-1 \quad n). \tag{5.4-4b}$$

Thus

$$\begin{aligned} (1 \quad 2 \quad 3) &= (1 \quad 3)(1 \quad 2), \\ &= (1 \quad 2)(2 \quad 3), \end{aligned}$$

which means (remembering to go from right to left) that

$$\begin{pmatrix} 1 & 2 & 3 \\ 2 & 3 & 1 \end{pmatrix} = \begin{pmatrix} 1 & 3 \\ 3 & 1 \end{pmatrix}\begin{pmatrix} 1 & 2 \\ 2 & 1 \end{pmatrix}.$$

Cycles without common elements commute; if they have an element in common, they generally do not commute as is evident from (5.4-4).

Since permutations can be decomposed into cycles and the latter into transpositions, any permutation can be written as a product of transpositions. If the number of transpositions is even (odd), the *parity p* of the permutation P is said to be even (odd).

Consider the group S_3 with six elements labeled as:

$$E = \begin{pmatrix} 1 & 2 & 3 \\ 1 & 2 & 3 \end{pmatrix}, \quad A = \begin{pmatrix} 1 & 2 & 3 \\ 1 & 3 & 2 \end{pmatrix}, \quad B = \begin{pmatrix} 1 & 2 & 3 \\ 3 & 2 & 1 \end{pmatrix},$$

$$C = \begin{pmatrix} 1 & 2 & 3 \\ 2 & 1 & 3 \end{pmatrix}, \quad D = \begin{pmatrix} 1 & 2 & 3 \\ 2 & 3 & 1 \end{pmatrix}, \quad F = \begin{pmatrix} 1 & 2 & 3 \\ 3 & 1 & 2 \end{pmatrix}. \tag{5.4-5}$$

The product BC, for example, is given by

$$BC = \begin{pmatrix} 1 & 2 & 3 \\ 3 & 2 & 1 \end{pmatrix}\begin{pmatrix} 1 & 2 & 3 \\ 2 & 1 & 3 \end{pmatrix} = \begin{pmatrix} 1 & 2 & 3 \\ 2 & 3 & 1 \end{pmatrix} = D. \tag{5.4-6}$$

If we take all possible pairs of permutations and organize them into a table, the result is just Table 3.1 which, we saw in Section 3.1, was the multiplication table for the group D_3. The set of permutations (5.4-5) therefore constitutes a group—specifically, the group S_3 which is isomorphic to D_3. In the same fashion it may be shown that S_2 is isomorphic to C_i and S_4 to O, etc. These are just a few examples of *Cayley's Theorem* which states that every finite group of order n is isomorphic with a subgroup of S_n.

If an integer n is split up into a sum of integers according to the scheme

$$\lambda_1 + \lambda_2 + \cdots + \lambda_h = n, \quad \lambda_1 \geqslant \lambda_2 \geqslant \cdots \geqslant \lambda_h, \tag{5.4-7}$$

we call this a *partition* $[\lambda_1, \lambda_2, \ldots, \lambda_h]$. The partitions of $n = 2, 3, 4$ are

$n = 2$: $[2]$, $[1, 1] \equiv [1^2]$,

$n = 3$: $[3]$, $[2, 1]$, $[1, 1, 1] \equiv [1^3]$,

$n = 4$: $[4]$, $[3, 1]$, $[2, 2] \equiv [2^2]$, $[2, 1, 1] \equiv [2, 1^2]$, $[1, 1, 1, 1] = [1^4]$.
$$\tag{5.4-8}$$

Every partition may be represented by a *Young diagram* which consists of an arrangement of n cells in h rows; each row begins with the same vertical line and the number of cells in successive rows is $\lambda_1, \lambda_2, \ldots, \lambda_h$. For $n = 2, 3$, and 4 the Young diagrams are:

$n = 2$:

$$[2] \,\square\square \qquad [1^2] \,\begin{matrix}\square\\\square\end{matrix} \tag{5.4-9}$$

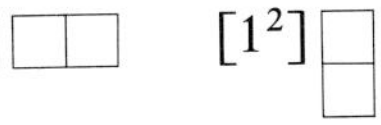

$n = 3$:

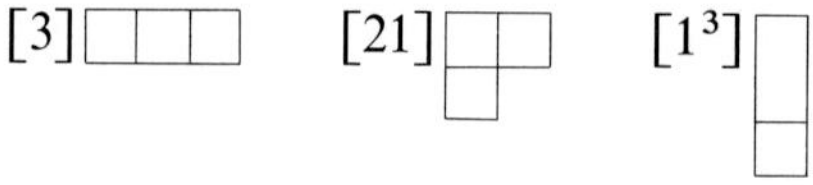

(5.4-10)

$n = 4$:

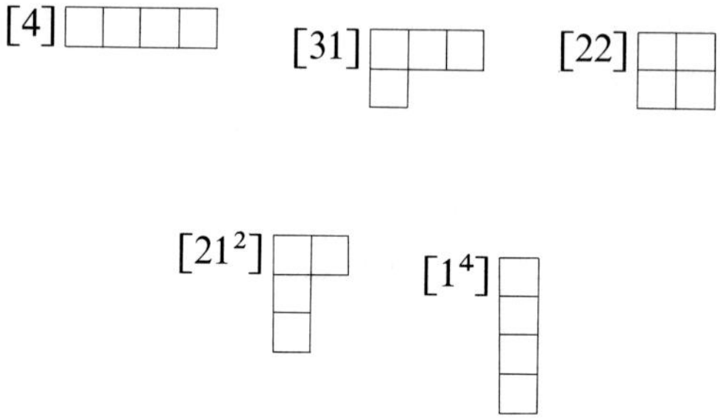

(5.4-11)

Each partition defines a class of S_n. Therefore the number of irreducible representations of S_n is equal to the number of partitions (or Young diagrams). Thus S_2, S_3, and S_4 have 2, 3, and 5 irreducible representations, respectively.

We may also obtain the dimensions of the irreducible representations by the following method: When the squares are numbered with the n integers we call the resulting pattern a *Young tableau.* If the numbers increase on going down a column and also from left to right along rows, the tableau is called a standard tableau. For S_2, S_3, and S_4 the standard tableaux are

S_2:

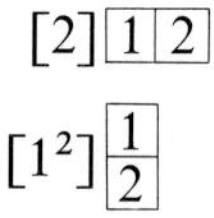

(5.4-12)

S_3:

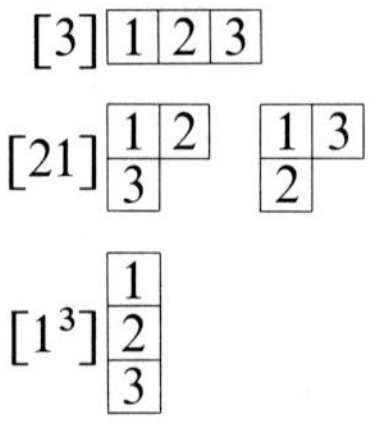

(5.4-13)

S_4:

$$
\begin{array}{ll}
[4] & \begin{array}{|c|c|c|c|}\hline 1&2&3&4\\\hline\end{array} \\[2ex]
[31] & \begin{array}{|c|c|c|}\hline 1&2&3\\\hline 4\\\cline{1-1}\end{array} \quad \begin{array}{|c|c|c|}\hline 1&3&4\\\hline 2\\\cline{1-1}\end{array} \quad \begin{array}{|c|c|c|}\hline 1&2&4\\\hline 3\\\cline{1-1}\end{array} \\[2ex]
[2^2] & \begin{array}{|c|c|}\hline 1&2\\\hline 3&4\\\hline\end{array} \quad \begin{array}{|c|c|}\hline 1&3\\\hline 2&4\\\hline\end{array} \\[2ex]
[21^2] & \begin{array}{|c|c|}\hline 1&2\\\hline 3\\\cline{1-1} 4\\\cline{1-1}\end{array} \quad \begin{array}{|c|c|}\hline 1&3\\\hline 2\\\cline{1-1} 4\\\cline{1-1}\end{array} \quad \begin{array}{|c|c|}\hline 1&4\\\hline 2\\\cline{1-1} 3\\\cline{1-1}\end{array} \\[2ex]
[1^4] & \begin{array}{|c|}\hline 1\\\hline 2\\\hline 3\\\hline 4\\\hline\end{array}
\end{array}
\qquad (5.4\text{-}14)
$$

Tableaux that are related to one another by an interchange of rows and columns are said to be *adjoint* as, for example, the pair

$$
\begin{array}{|c|c|}\hline 1&2\\\hline 3\\\cline{1-1}\end{array} \quad \begin{array}{|c|c|}\hline 1&3\\\hline 2\\\cline{1-1}\end{array}
$$

or

$$
\begin{array}{|c|c|c|}\hline 1&2&3\\\hline 4\\\cline{1-1}\end{array} \quad \begin{array}{|c|c|}\hline 1&4\\\hline 2\\\cline{1-1} 3\\\cline{1-1}\end{array}
\qquad (5.4\text{-}15)
$$

The importance of the standard tableaux lies in the fact that the dimension of an irreducible representation of S_n is equal to the number of standard tableaux that can be constructed from the corresponding partition. Hence we may now draw the following conclusions concerning the dimensions of the irreducible representations in S_2, S_3, and S_4:

S_2 two one-dimensional representations associated with $[2]$ and $[1^2]$, respectively,

S_3 two one-dimensional representations associated with $[3]$ and $[1^3]$, respectively,
one two-dimensional representation associated with $[21]$,

S_4 two one-dimensional representations associated with $[4]$ and $[1^4]$, respectively,
one two-dimensional representation associated with $[2^2]$,
two three-dimensional representations associated with $[31]$ and $[21^2]$, respectively.

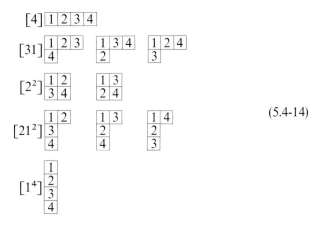

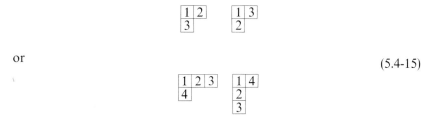

Each of these groups has two one-dimensional representations. This is a general property of the groups S_n in which the partitions $[n]$ and $[1^n]$ are associated with one-dimensional representations. All other irreducible representations of S_n are of dimension greater than one.

Basis functions for the irreducible representations of S_n can be constructed with the help of certain operators which we now define. Corresponding to any standard tableau let P be a permutation which interchanges numbers in one row. Such a permutation will be called a *horizontal permutation*. Similarly, a *vertical permutation* Q interchanges numbers in one column. Also, let

$$S = \sum_P P, \tag{5.4-16}$$

$$A = \sum_Q (-1)^q Q. \tag{5.4-17}$$

The sum in (5.4-16) is over all horizontal permutations; S is then the *symmetrizing* operator with respect to the rows of a Young tableau. With q the parity of the permutation Q, A is the *antisymmetrizing* operator with respect to the columns. The product

$$Y = AS \tag{5.4-18}$$

is known as a *Young operator*.

To construct basis functions for an irreducible representation with which a particular partition is associated it is first necessary to construct Young operators corresponding to all the standard tableaux for that partition. Each Young operator acting on a function $f(\chi_1, \chi_2, \ldots, \chi_n)$ projects a basis function for the irreducible representation associated with that particular tableau. For the partitions $[n]$ and $[1^n]$ corresponding to one-dimensional representations the Young operator Y reduces to S and A, respectively. Therefore, in the former case, the basis function is totally symmetric (remains unaltered on interchange of any two indices) and the representation is said to be totally symmetric; in the latter case the basis function is totally antisymmetric (changes sign upon an interchange of any two indices) and the representation is said to be antisymmetric.

A few examples will help to clarify the method. Let $n = 2$ and let

$$f(\chi_1, \chi_2) = a(\chi_1)b(\chi_2) \equiv a_1 b_2.$$

For the tableau $\boxed{1\,|\,2}$ there is only one row; we therefore expect to find a symmetric basis function

$$f_S^{(2)} = Ya_1b_2 = ASa_1b_2 = Sa_1b_2 = \sum_P Pa_1b_2 = a_1b_2 + a_2b_1 \tag{5.4-19}$$

and for the tableau $\begin{array}{|c|}\hline 1\\\hline 2\\\hline\end{array}$ with just one column there is an antisymmetric basis function

$$f_A^{(2)} = Ya_1b_2 = ASa_1b_2 = Aa_1b_2 = \sum_Q (-1)^q Qa_1b_2 = a_1b_2 - a_2b_1 = \begin{vmatrix} a_1 & b_1 \\ a_2 & b_2 \end{vmatrix}. \tag{5.4-20}$$

Now let $n = 3$ and let

$$f(\chi_1, \chi_2, \chi_3) = a(\chi_1)b(\chi_2)c(\chi_3) \equiv a_1b_2c_3. \tag{5.4-21}$$

For the one-dimensional representations we have

$$\begin{array}{|c|c|c|}\hline 1&2&3\\\hline\end{array}: \quad f_S^{(3)} = Ya_1b_2c_3 = Sa_1b_2c_3 = \sum_P Pa_1b_2c_3$$
$$= a_1b_2c_3 + a_1b_3c_2 + a_2b_1c_3 + a_2b_3c_1 + a_3b_1c_2 + a_3b_2c_1. \tag{5.4-22}$$

$$\begin{array}{|c|}\hline 1\\\hline 2\\\hline 3\\\hline\end{array}: \quad f_A^{(3)} = Ya_1b_2c_3 = Aa_1b_2c_3 = \sum_Q (-1)^q Qa_1b_2c_3$$
$$= a_1b_2c_3 - a_1b_3c_2 - a_2b_1c_3 + a_2b_3c_1 + a_3b_1c_2 - a_3b_2c_1$$
$$= \begin{vmatrix} a_1 & b_1 & c_1 \\ a_2 & b_2 & c_2 \\ a_3 & b_3 & c_3 \end{vmatrix}. \tag{5.4-23}$$

Proceeding to the two-dimensional representation we consider the tableau

$$\begin{array}{|c|c|}\hline 1&2\\\hline 3\\\cline{1-1}\end{array}$$

In this case

$$Ya_1b_2c_3 = ASa_1b_2c_3$$

where S is the symmetrization operator with respect to indices 1, 2 and A the antisymmetrization operator with respect to indices 1, 3. This gives

$$\begin{aligned} Sa_1b_2c_3 &= a_1b_2c_3 + a_2b_1c_3, \\ ASa_1b_2c_3 &= A(a_1b_2c_3 + a_2b_1c_3) \\ &= a_1b_2c_3 - a_3b_2c_1 + a_2b_1c_3 - a_2b_3c_1 \equiv f_1^{(3)}. \end{aligned} \tag{5.4-24}$$

Finally we have the tableau

1	3
2	

for which

$$\begin{aligned} Ya_1b_2c_3 &= ASa_1b_2c_3 = A(a_1b_2c_3 + a_3b_2c_1) \\ &= a_1b_2c_3 - a_2b_1c_3 + a_3b_2c_1 - a_3b_1c_2 \equiv f_2^{(3)}. \end{aligned} \tag{5.4-25}$$

The two functions $f_1^{(3)}$ and $f_2^{(3)}$ are basis functions for the two-dimensional irreducible representation of S_3.

CHAPTER 6

TENSORS

6.1 Irreducible Tensor Operators

An irreducible tensor operator $\mathbf{T}^{(k)}$ of rank k is a quantity with $2k+1$ components $T_q^{(k)}$ which, under a coordinate rotation $\mathbf{r}' = R\mathbf{r}$, satisfy

$$P_R T_q^{(k)} P_R^{-1} = \sum_{q'} T_{q'}^{(k)} D_{q'q}^{(k)}(R) \quad k = 0, 1, 2, \ldots,$$
$$q, q' = k, k-1, \ldots, -k \tag{6.1-1}$$

where $D_{q'q}^{(k)}(R)$ are matrix elements associated with the irreducible representation $D^{(k)}$ of $O^+(3)$. It will be noticed that (6.1-1) is identical in form to (2.3-28) thereby establishing the similarity in transformation properties between irreducible tensors and spherical harmonics. The latter may therefore be regarded as irreducible tensors. Furthermore, from the standpoint of group theory, the transformation law (6.1-1) permits the interpretation that the $T_q^{(k)}$ form a (tensor) basis for an irreducible representation $D^{(k)}$ as do the spherical harmonics Y_{kq}.

The definition of an irreducible tensor may be put in another form. In (4.5-3b) it was shown that

$$P_R|jm\rangle = \sum_{m'} |jm'\rangle D_{m'm}^{(j)}(R). \tag{6.1-2}$$

With the orthonormality property of the $|jm\rangle$ we then have

$$D_{m'm}^{(j)}(R) = \langle jm'|P_R|jm\rangle, \tag{6.1-3}$$

and, letting $j = k$, $m = q$, $m' = q'$,

$$D_{q'q}^{(k)}(R) = \langle kq'|P_R|kq\rangle. \tag{6.1-4}$$

From (2.2-8), to first order,

$$P_R(\delta\alpha, z) = 1 + i\,\delta\alpha\, J_z; \tag{6.1-5}$$

therefore

$$\begin{aligned} P_R(\delta\alpha, z) T_q^{(k)} P_R^{-1}(\delta\alpha, z) &= (1 + i\,\delta\alpha\, J_z) T_q^{(k)} (1 - i\,\delta\alpha\, J_z) \\ &= T_q^{(k)} + i\,\delta\alpha\,[J_z, T_q^{(k)}] \end{aligned} \tag{6.1-6}$$

to first order, and

$$D_{q'q}^{(k)}(\delta\alpha, z) = \langle kq'|1 + i\,\delta\alpha\, J_z|kq\rangle. \tag{6.1-7}$$

Substituting (6.1-6) and (6.1-7) into (6.1-1) we obtain

$$[J_z, T_q^{(k)}] = \sum_{q'} T_{q'}^{(k)} \langle kq'|J_z|kq\rangle = qT_q^{(k)}, \tag{6.1-8}$$

since $\langle kq'|kq\rangle = \delta_{q'q}$ and $J_z|kq\rangle = q|kq\rangle$. The same procedure can be applied to infinitesimal rotations about x and y to give

$$[J_x, T_q^{(k)}] = \sum_{q'} T_{q'}^{(k)} \langle kq'|J_x|kq\rangle, \tag{6.1-9}$$

$$[J_y, T_q^{(k)}] = \sum_{q'} T_{q'}^{(k)} \langle kq'|J_y|kq\rangle, \tag{6.1-10}$$

which may be combined into the single expression

$$[J_{\pm 1}, T_q^{(k)}] = \sum_{q'} T_{q'}^{(k)} \langle kq'|J_{\pm 1}|kq\rangle. \tag{6.1-11}$$

But, from (1.7-11) and (1.7-12),

$$\langle kq'|J_{\pm 1}|kq\rangle = \mp \sqrt{\tfrac{1}{2}[k(k+1) - q(q \pm 1)]}\, \delta_{q'\, q\pm 1}. \tag{6.1-12}$$

Therefore (6.1-11) becomes

$$[J_{\pm 1}, T_q^{(k)}] = \mp \sqrt{\tfrac{1}{2}[k(k+1) - q(q \pm 1)]}\, T_{q\pm 1}^{(k)}, \tag{6.1-13}$$

which, together with

$$[J_0, T_q^{(k)}] = qT_q^{(k)}, \tag{6.1-14}$$

specify all the commutators between components of $\mathbf{J}$ and components of $\mathbf{T}^{(k)}$.

It may also be shown that the transformation law (6.1-1) is derivable from the commutation relations (6.1-13) and (6.1-14); the latter therefore serve as an alternative but completely equivalent definition of an irreducible tensor. Such a formulation was employed by Racah (1942a, 1942b, 1943, 1949) in his pioneer work on the application of irreducible tensors to atomic structure.

The lowest rank tensor $\mathbf{T}^{(0)}$ has but one component $T_0^{(0)}$ which, according to (6.1-13) and (6.1-14), commutes with all the components of $\mathbf{J}$. In view of the discussion in Section 2.2, $T_0^{(0)}$ is invariant under a rotation and is therefore a scalar.

When $k = 1$ the commutation relations are

$$[J_{+1}, T_{-1}^{(1)}] = -T_0, \tag{6.1-15}$$

$$[J_0, T_{\pm 1}^{(1)}] = \pm T_{\pm 1}. \tag{6.1-16}$$

These are precisely of the same form as (1.7-4); we may therefore regard angular momentum operators, which are vector operators, as irreducible tensors of rank 1. Alternatively, it may be said that vector operators satisfy commutation relations with angular momentum operators analogous to those of the angular momentum operators among themselves.

The scalar product of two tensor operators, $\mathbf{T}^{(k)}$ and $\mathbf{U}^{(k)}$, is defined by

$$\mathbf{T}^{(k)} \cdot \mathbf{U}^{(k)} = \sum_q (-1)^q T_q^{(k)} U_{-q}^{(k)}. \tag{6.1-17}$$

For vector operators, $k = 1$; letting $\mathbf{T}^{(1)} = \mathbf{A}$ and $\mathbf{U}^{(k)} = \mathbf{B}$,

$$\begin{aligned} \mathbf{A} \cdot \mathbf{B} &= -A_{+1}B_{-1} + A_0 B_0 - A_{-1}B_{+1} \\ &= \sum_q (-1)^q A_q B_{-q} \qquad (q = 1, 0, -1) \\ &= \sum_q A_q{}^* B_q = A_x B_x + A_y B_y + A_z B_z \end{aligned} \tag{6.1-18}$$

where

$$\begin{aligned} A_{\pm 1} &= \mp \frac{1}{\sqrt{2}}(A_x \pm iA_y), \qquad A_0 = A_z, \\ B_{\pm 1} &= \mp \frac{1}{\sqrt{2}}(B_x \pm iB_y), \qquad B_0 = B_z. \end{aligned} \tag{6.1-19}$$

An arbitrary vector $\mathbf{A}$ can be expressed in terms of a set of unit vectors along the three coordinate axes. Let these unit vectors be $\hat{\mathbf{e}}_x$, $\hat{\mathbf{e}}_y$, and $\hat{\mathbf{e}}_z$; they can also be written as column matrices:

$$\hat{\mathbf{e}}_x = \begin{pmatrix} 1 \\ 0 \\ 0 \end{pmatrix}, \qquad \hat{\mathbf{e}}_y = \begin{pmatrix} 0 \\ 1 \\ 0 \end{pmatrix}, \qquad \hat{\mathbf{e}}_z = \begin{pmatrix} 0 \\ 0 \\ 1 \end{pmatrix}. \tag{6.1-20}$$

Thus

$$\mathbf{A} = A_x \hat{\mathbf{e}}_x + A_y \hat{\mathbf{e}}_y + A_z \hat{\mathbf{e}}_z \tag{6.1-21}$$

where A_x, A_y, and A_z are the rectangular components of the vector. The unit vectors may also be written in the spherical basis:

$$\begin{aligned}\hat{\mathbf{e}}_{+1} &= -\frac{1}{\sqrt{2}}(\hat{\mathbf{e}}_x + i\hat{\mathbf{e}}_y) = -\frac{1}{\sqrt{2}}\begin{pmatrix}1\\ i\\ 0\end{pmatrix},\\ \hat{\mathbf{e}}_0 &= \hat{\mathbf{e}}_z = \begin{pmatrix}0\\ 0\\ 1\end{pmatrix},\\ e_{-1} &= \frac{1}{\sqrt{2}}(\hat{\mathbf{e}}_x - i\hat{\mathbf{e}}_y) = \frac{1}{\sqrt{2}}\begin{pmatrix}1\\ -i\\ 0\end{pmatrix},\end{aligned} \tag{6.1-22}$$

with the inverse relations

$$\hat{\mathbf{e}}_x = -\frac{1}{\sqrt{2}}(\hat{\mathbf{e}}_{+1} + \hat{\mathbf{e}}_{-1}), \qquad \hat{\mathbf{e}}_y = \frac{i}{\sqrt{2}}(\hat{\mathbf{e}}_{+1} - \hat{\mathbf{e}}_{-1}), \qquad \hat{\mathbf{e}}_z = \hat{\mathbf{e}}_0. \tag{6.1-23}$$

Therefore the vector **A** (6.1-21) in the spherical basis assumes the form

$$\begin{aligned}\mathbf{A} &= -A_{+1}\hat{\mathbf{e}}_{-1} + A_0\hat{\mathbf{e}}_0 - A_{-1}\hat{\mathbf{e}}_{+1}\\ &= \sum_q (-1)^q A_q \hat{\mathbf{e}}_{-q} \qquad (q = 1, 0, -1)\end{aligned} \tag{6.1-24}$$

where A_{+1}, A_0, and A_{-1} are the spherical components (6.1-19).

In view of the orthogonality of the unit vectors in the rectangular basis

$$\hat{\mathbf{e}}_i \cdot \hat{\mathbf{e}}_j = \delta_{ij} \qquad (i, j = x, y, z), \tag{6.1-25}$$

there is also an orthogonality relation in the spherical basis

$$\hat{\mathbf{e}}_p{}^* \cdot \hat{\mathbf{e}}_q = (-1)^p \hat{\mathbf{e}}_{-p} \cdot \hat{\mathbf{e}}_q = \delta_{pq} \qquad (p, q = 1, 0, -1). \tag{6.1-26}$$

To verify (6.1-26), $\hat{\mathbf{e}}_p$ and $\hat{\mathbf{e}}_q$ may be converted to rectangular components according to (6.1-22) followed by the application of the orthogonality relation (6.1-25). Thus, for example,

$$\begin{aligned}\hat{\mathbf{e}}_{+1}^* \cdot \hat{\mathbf{e}}_{-1} &= -\frac{1}{\sqrt{2}}(\hat{\mathbf{e}}_x + i\hat{\mathbf{e}}_y)^* \cdot \frac{1}{\sqrt{2}}(\hat{\mathbf{e}}_x - i\hat{\mathbf{e}}_y)\\ &= \frac{1}{2}(\hat{\mathbf{e}}_x - i\hat{\mathbf{e}}_y) \cdot (\hat{\mathbf{e}}_x - i\hat{\mathbf{e}}_y) = 0.\end{aligned}$$

If one were to use the column vectors in (6.1-22) to represent $\hat{\mathbf{e}}_p$ and $\hat{\mathbf{e}}_q$ it is necessary to keep in mind that $\hat{\mathbf{e}}_p{}^*$ is to be interpreted as the Hermitian conjugate of $\hat{\mathbf{e}}_p$. Thus

$$\hat{\mathbf{e}}^*_{+1} \cdot \hat{\mathbf{e}}_{-1} = -\frac{1}{\sqrt{2}}(1, -i, 0)\frac{1}{\sqrt{2}}\begin{pmatrix} 1 \\ -i \\ 0 \end{pmatrix} = -\frac{1}{2}[1 \; -1] = 0.$$

6.2 Tensor Products

Let $T^{(k)}_q$ and $U^{(k')}_{q'}$ be the components of two irreducible tensor operators of rank k and k', respectively. The *tensor product* of the two tensor operators is defined by the relation

$$V^{(K)}_Q = \sum_{qq'} T^{(k)}_q U^{(k')}_{q'} \langle kk'qq' | kk'KQ \rangle$$

$$= \sqrt{2k+1} \sum_{qq'} (-1)^{-k+k'-Q} \begin{pmatrix} k & k' & K \\ q & q' & -Q \end{pmatrix} T^{(k)}_q U^{(k')}_{q'} \tag{6.2-1}$$

in which $\langle kkqq' | kk'KQ \rangle$ is a CG coefficient, $\left(\begin{smallmatrix} k & k' & K \\ q & q' & -Q \end{smallmatrix}\right)$ a $3j$ symbol, and the values of K and Q are restricted to

$$\begin{aligned} K &= k + k', k + k' - 1, \ldots, |k - k'|, \\ Q &= q + q' = K, K - 1, \ldots, -K. \end{aligned} \tag{6.2-2}$$

In an abbreviated notation we may write

$$V^{(K)}_Q = [\mathbf{T}^{(k)}\mathbf{U}^{(k')}]^{(K)}_Q. \tag{6.2-3}$$

Since $\mathbf{T}^{(k)}$ and $\mathbf{U}^{(k')}$ satisfy the commutation relations (6.1-13) and (6.1-14),

$$[J_0, V^{(K)}_Q] = QV^{(K)}_Q, \tag{6.2-4}$$

$$[J_{\pm 1}, V^{(K)}_Q] = \mp\sqrt{\tfrac{1}{2}[K(K+1) - Q(Q \pm 1)]}\, V^{(K)}_{Q \pm 1}, \tag{6.2-5}$$

so that $\mathbf{V}^{(K)}$ is an irreducible tensor.

The tensor product of $\mathbf{T}^{(k)}$ and $\mathbf{U}^{(k')}$ may also be given a group-theoretical interpretation based on the fact that, according to the fundamental definition (6.1-1), the components of $\mathbf{T}^{(k)}$ and $\mathbf{U}^{(k')}$ are bases for the irreducible representations $D^{(k)}$ and $D^{(k')}$ of $O^+(3)$, respectively. The products $T^{(k)}_q U^{(k')}_{q'}$ therefore form a basis for the product representation $D^{(k)} \times D^{(k')}$ which is reducible into

$$D^{(k)} \times D^{(k')} = \sum_{k=|k-k'|}^{K=k+k'} D^{(K)}. \tag{6.2-6}$$

Carrying the group-theoretical interpretation further, it is seen that rule (6.2-1) for the construction of tensor products is identical to rule (4.5-12) for the construction of basis functions of $D^{(j)}$ which appears in the decomposition of the product representation $D^{(j_1)} \times D^{(j_2)}$. Also, it is recalled that the same rule is employed for the construction of eigenfunctions in the (angular momentum) coupled representation. Therefore the tensor components $V_Q^{(K)}$ form a basis for $D^{(K)}$ which is an irreducible representation of $O^+(3)$ contained in the decomposition of the product representation $D^{(k)} \times D^{(k')}$.

A few special cases are of interest. Let $K = 0$, $Q = 0$. In order to satisfy (6.2-2), k and k' must be equal; also $Q = q + q' = 0$. We then have

$$V_0^{(0)} = [T_q^{(k)} U_q^{(k)}]_0^{(0)}$$

$$= \sum_{qq'} T_q^{(k)} U_{q'}^{(k)} \langle kkqq' | kk00 \rangle \tag{6.2-7}$$

$$= \sum_q T_q^{(k)} U_{-q}^{(k)} \langle kkq-q | kk00 \rangle. \tag{6.2-8}$$

Expressed in terms of $3j$ symbols as in (6.2-1),

$$V_0^{(0)} = \sum_q \begin{pmatrix} k & k & 0 \\ q & -q & 0 \end{pmatrix} T_q^{(k)} U_{-q}^{(k)}, \tag{6.2-9}$$

but from Table 1.5,

$$\begin{pmatrix} k & k & 0 \\ q & -q & 0 \end{pmatrix} = (-1)^{k-q} \frac{1}{\sqrt{2k+1}};$$

therefore

$$V_0^{(0)} = \frac{(-1)^k}{\sqrt{2k+1}} \sum_q (-1)^q T_q^{(k)} U_{-q}^{(k)} = \frac{(-1)^k}{\sqrt{2k+1}} \mathbf{T}^{(k)} \cdot \mathbf{U}^{(k)}. \tag{6.2-10}$$

The tensor product of two vector operators $\mathbf{T}^{(1)}$ and $\mathbf{U}^{(1)}$ is

$$V_Q^{(K)} = [T_q^{(1)} U_{q'}^{(1)}]_Q^{(K)}$$

with $K = 0, 1, 2$. From (6.2-1) and Table 1.4 or 1.6,

$$V_0^{(0)} = -\frac{1}{\sqrt{3}} (-T_{+1}^{(1)} U_{-1}^{(1)} + T_0^{(1)} U_0^{(1)} - T_{-1}^{(1)} U_{+1}^{(1)})$$

$$= -\frac{1}{\sqrt{3}} \mathbf{T}^{(1)} \cdot \mathbf{U}^{(1)}, \tag{6.2-11}$$

$$V_Q^{(1)} = \sqrt{3} \sum_{qq'} (-1)^Q \begin{pmatrix} 1 & 1 & 1 \\ q & q' & -Q \end{pmatrix} T_q^{(1)} U_{q'}^{(1)}, \tag{6.2-12}$$

from which we obtain

$$V_{+1}^{(1)} = \frac{1}{\sqrt{2}}(T_{+1}^{(1)}U_0^{(1)} - T_0^{(1)}U_{+1}^{(1)}),$$

$$V_0^{(1)} = \frac{1}{\sqrt{2}}(T_{+1}^{(1)}U_{-1}^{(1)} - T_{-1}^{(1)}U_{+1}^{(1)}), \tag{6.2-13}$$

$$V_{-1}^{(1)} = \frac{1}{\sqrt{2}}(T_0^{(1)}U_{-1}^{(1)} - T_{-1}^{(1)}U_0^{(1)}).$$

For $K = 2$,

$$V_Q^{(2)} = \sqrt{5}\sum_{qq'}(-1)^Q\begin{pmatrix}1 & 1 & 2\\ q & q' & -Q\end{pmatrix}T_q^{(1)}U_{q'}^{(1)} \tag{6.2-14}$$

with components

$$V_{+2}^{(2)} = T_{+1}^{(1)}U_{+1}^{(1)}, \qquad V_{+1}^{(2)} = \frac{1}{\sqrt{2}}(T_{+1}^{(1)}U_0^{(1)} + T_0^{(1)}U_{+1}^{(1)}),$$

$$\begin{aligned} V_0^{(2)} &= \frac{1}{\sqrt{6}}T_{+1}^{(1)}U_{-1}^{(1)} + \sqrt{\frac{2}{3}}\,T_0^{(1)}U_0^{(1)} + \frac{1}{\sqrt{6}}T_{-1}^{(1)}U_{+1}^{(1)} \\ &= \frac{1}{\sqrt{6}}[3T_0^{(1)}U_0^{(1)} - \mathbf{T}^{(1)}\cdot\mathbf{U}^{(1)}], \end{aligned} \tag{6.2-15}$$

$$V_{-1}^{(2)} = \frac{1}{\sqrt{2}}(T_0^{(1)}U_{-1}^{(1)} + T_{-1}^{(1)}U_0^{(1)}), \qquad V_{-2}^{(2)} = T_{-1}^{(1)}U_{-1}^{(1)}.$$

If we let $\mathbf{T}^{(1)} = \mathbf{A}$ and $\mathbf{U}^{(1)} = \mathbf{B}$, the expressions that are often found to be useful are:

$$V_0^{(0)} = -\frac{1}{\sqrt{3}}\mathbf{A}\cdot\mathbf{B}, \tag{6.2-16}$$

$$\begin{aligned} -\frac{1}{\sqrt{2}}(V_{+1}^{(1)} - V_{-1}^{(1)}) &= \frac{i}{\sqrt{2}}(A_yB_z - A_zB_y) = \frac{i}{\sqrt{2}}(\mathbf{A}\times\mathbf{B})_x, \\ \frac{i}{\sqrt{2}}(V_{+1}^{(1)} + V_{-1}^{(1)}) &= \frac{i}{\sqrt{2}}(A_zB_x - A_xB_z) = \frac{i}{\sqrt{2}}(\mathbf{A}\times\mathbf{B})_y, \\ V_{\pm 1}^{(1)} &= \frac{1}{2}[A_zB_x - A_xB_z \pm i(A_zB_y - A_yB_z)], \\ V_0^{(1)} &= \frac{i}{\sqrt{2}}(A_xB_y - A_yB_x) = \frac{i}{\sqrt{2}}(\mathbf{A}\times\mathbf{B})_z, \end{aligned} \tag{6.2-17}$$

$$\frac{1}{\sqrt{2}}(V_{+2}^{(2)} + V_{-2}^{(2)}) = \frac{1}{\sqrt{2}}(A_x B_x - A_y B_y),$$

$$\frac{i}{\sqrt{2}}(-V_{+2}^{(2)} + V_{-2}^{(2)}) = \frac{1}{\sqrt{2}}(A_x B_y + A_y B_x),$$

$$V_{\pm 2}^{(2)} = \frac{1}{2}[A_x B_x - A_y B_y \pm i(A_x B_y + A_y B_x)],$$

$$\frac{i}{\sqrt{2}}(V_{+1}^{(2)} + V_{-1}^{(2)}) = \frac{1}{\sqrt{2}}(A_y B_z + A_z B_y), \tag{6.2-18}$$

$$\frac{1}{\sqrt{2}}(-V_{+1}^{(2)} + V_{-1}^{(2)}) = \frac{1}{\sqrt{2}}(A_z B_x + A_x B_z),$$

$$V_{\pm 1}^{(2)} = \mp\frac{1}{2}[A_z B_x + A_x B_z \pm i(A_y B_z + A_z B_y)],$$

$$V_0^{(2)} = \frac{1}{\sqrt{6}}(3A_z B_z - \mathbf{A}\cdot\mathbf{B}).$$

In the special case where $\mathbf{T}^{(1)} = \mathbf{U}^{(1)} = \mathbf{J}$,

$$V_0^{(0)} = -\frac{1}{\sqrt{3}}[-J_{+1}J_{-1} + J_0{}^2 - J_{-1}J_{+1}] = -\frac{1}{\sqrt{3}}J^2, \tag{6.2-19}$$

$$V_{+1}^{(1)} = \frac{1}{\sqrt{2}}[J_{+1}J_0 - J_0 J_{+1}] = \frac{1}{\sqrt{2}}[J_{+1}, J_0] = -\frac{1}{\sqrt{2}}J_{+1},$$

$$V_0^{(1)} = \frac{1}{\sqrt{2}}[J_{+1}J_{-1} - J_{-1}J_{+1}] = \frac{1}{\sqrt{2}}[J_{+1}, J_{-1}] = -\frac{1}{\sqrt{2}}J_0, \tag{6.2-20}$$

$$V_{-1}^{(1)} = \frac{1}{\sqrt{2}}[J_0 J_{-1} - J_{-1}J_0] = \frac{1}{\sqrt{2}}[J_0, J_{-1}] = -\frac{1}{\sqrt{2}}J_{-1},$$

$$V_{+2}^{(2)} = J_{+1}^2,$$

$$V_{+1}^{(2)} = \frac{1}{\sqrt{2}}(J_{+1}J_0 + J_0 J_{+1}) = \frac{1}{\sqrt{2}}J_{+1}(2J_0 + 1),$$

$$V_0^{(2)} = \frac{1}{\sqrt{6}}J_{+1}J_{-1} + \sqrt{\frac{2}{3}}J_0{}^2 + \frac{1}{\sqrt{6}}J_{-1}J_{+1}$$

$$= \frac{1}{\sqrt{6}}(J_{+1}J_{-1} - J_0{}^2 + J_{-1}J_{+1} + 3J_0{}^2) = \frac{1}{\sqrt{6}}(3J_0{}^2 - J^2) \tag{6.2-21}$$

$$V_{-1}^{(2)} = \frac{1}{\sqrt{2}}(J_{-1}J_0 + J_0J_{-1}) = \frac{1}{\sqrt{2}}J_{-1}(2J_0 - 1),$$

$$V_{-2}^{(2)} = J_{-1}^2.$$

6.3 Wigner–Eckart Theorem

The chief reason for the introduction of irreducible tensors stems from an important theorem known as the Wigner–Eckart theorem:

$$\langle \alpha jm|T_q^{(k)}|\alpha' j'm'\rangle = \frac{1}{\sqrt{2j+1}}\langle j'km'q|j'kjm\rangle\langle \alpha j|\,|\mathbf{T}^{(k)}|\,|\alpha' j'\rangle \quad (6.3\text{-}1a)$$

$$= (-1)^{j-m}\begin{pmatrix} j & k & j' \\ -m & q & m' \end{pmatrix}\langle \alpha j|\,|\mathbf{T}^{(k)}|\,|\alpha' j'\rangle. \quad (6.3\text{-}1b)$$

The left side is a matrix element of an irreducible tensor operator $T_q^{(k)}$ taken between two states which are labeled by the quantum numbers α, j, m and α', j', m', respectively; j, m and j', m' are angular momentum quantum numbers, while α and α' are additional quantum numbers which are required in order that the states be completely specified. In the case of atomic states, for example, α and α' might be quantum numbers associated with the radial part of the wave function. The right side of (6.3-1) contains a coupling coefficient (*CG* coefficient or $3j$ symbol) and another quantity, $\langle \alpha j|\,|\mathbf{T}^{(k)}|\,|\alpha' j'\rangle$, called a *reduced matrix element* whose significance will emerge from the proof of the theorem.

Suppose we form the matrix element of $[J_{+1}, T_q^{(k)}]$ between two angular momentum eigenfunctions:

$$\langle jm|[J_{+1}, T_q^{(k)}]|j'm'\rangle \equiv \langle jm|J_{+1}T_q^{(k)}|j'm'\rangle - \langle jm|T_q^{(k)}J_{+1}|j'm'\rangle. \quad (6.3\text{-}2)$$

For the first term on the right we have

$$\begin{aligned}\langle jm|J_{+1}T_q^{(k)}|j'm'\rangle &= \langle J_{+1}^{\dagger}jm|T_q^{(k)}|j'm'\rangle \\ &= -\langle J_{-1}jm|T_q^{(k)}|j'm'\rangle \\ &= -\sqrt{\tfrac{1}{2}[j(j+1) - m(m-1)]}\langle j\,m-1|T_q^{(k)}|j'm'\rangle, \end{aligned}\quad (6.3\text{-}3)$$

and for the second term

$$\langle jm|T_q^{(k)}J_{+1}|j'm'\rangle = -\sqrt{\tfrac{1}{2}[j'(j'+1) - m'(m'+1)]}\langle jm|T_q^{(k)}|j'\,m'+1\rangle. \quad (6.3\text{-}4)$$

Also, according to (6.1-13),

$$\langle jm|[J_{+1}, T_q^{(k)}]|j'm'\rangle = -\sqrt{\tfrac{1}{2}[k(k+1) - q(q+1)]}\langle jm|T_{q+1}^{(k)}|j'm'\rangle. \quad (6.3\text{-}5)$$

Combining (6.3-3)–(6.3-5),

$$\begin{aligned}\sqrt{j(j+1)-m(m-1)}\langle j\,m-1|T_q^{(k)}|j'm'\rangle\\ =\sqrt{j'(j'+1)-m'(m'+1)}\langle jm|T_q^{(k)}|j'\,m'+1\rangle\\ +\sqrt{k(k+1)-q(q+1)}\langle jm|T_{q+1}^{(k)}|j'm'\rangle.\end{aligned} \tag{6.3-6}$$

This expression may be compared with the recursion relation (1.5-45b) for the *CG* coefficients. To facilitate the comparison let us establish a correspondence between j, j', k, m', q, and m in (6.3-6) with j, j_1, j_2, m_1, m_2, and m in (1.5-45b). Both expressions then have the same dependence on their respective projection quantum numbers, namely, m', q, and m in (6.3-6) and m_1, m_2, and m in (1.5-45b). In the latter the three coupling coefficients vanish unless $m_1 + m_2 + 1 = m$; it may be verified that the matrix elements in (6.3-6) behave in similar fashion. Thus

$$\begin{aligned}\langle jm|[J_0, T_q^{(k)}]|j'm'\rangle &\equiv \langle jm|J_0 T_q^{(k)}|j'm'\rangle - \langle jm|T_q^{(k)}J_0|j'm'\rangle\\ &= (m-m')\langle jm|T_q^{(k)}|j'm'\rangle.\end{aligned} \tag{6.3-7}$$

Also, from (6.1-14),

$$\langle jm|[J_0, T_q^{(k)}]|j'm'\rangle = q\langle jm|T_q^{(k)}|j'm'\rangle. \tag{6.3-8}$$

Therefore

$$\langle jm|T_q^{(k)}|j'm'\rangle = 0 \qquad \text{unless} \quad m = q + m',$$

and the matrix elements in (6.3-6) will vanish unless $m = q + m' + 1$. We may therefore conclude that the dependence of $\langle jm|T_q^{(k)}|j'm'\rangle$ on the projection quantum numbers m, q, and m' is entirely contained in the coupling coefficient. It is therefore permissible to express $\langle jm|T_q^{(k)}|j'm'\rangle$ as the product of a coupling coefficient which contains the total dependence on m, q, and m' and another factor which contains the dependence on everything else. It is seen that (6.3-1) is of this form and that the second factor corresponds to the reduced matrix element which is independent of m, q, and m'. The constants that appear in the theorem are inserted for reasons of convenience.

We observe that the Wigner–Eckart theorem separates those features of a physical process which depend on geometry or symmetry properties (contained in the coupling coefficients) from those which depend on the detailed physical interactions (contained in the reduced matrix elements).

The general selection rules, which emerge from (6.3-1) and which are a direct consequence of the properties of the coupling coefficients, are

$$m = m' + q, \qquad \triangle(jkj'); \tag{6.3-9}$$

that is, $\langle\alpha jm|T_q^{(k)}|\alpha'j'm'\rangle = 0$ unless (6.3-9) is satisfied. We list a few special cases:

$$\langle\alpha jm|T_0^{(0)}|\alpha'j'm'\rangle = 0 \qquad \text{unless} \quad \begin{cases} j' = j, \\ m' = m, \end{cases} \tag{6.3-10}$$

$$\langle\alpha jm|T_q^{(1)}|\alpha'j'm'\rangle = 0 \qquad \text{unless} \quad \begin{cases} \quad j' = j+1, j, j-1 \\ \qquad \text{or} \quad \Delta j = 0, \pm 1, \\ \quad m' = m - q \quad (q = 1, 0, -1) \\ \qquad \text{or} \quad \Delta m = 0, \pm 1, \\ j + j' \geqslant 1, \end{cases} \tag{6.3-11}$$

$$\langle\alpha jm|T_q^{(2)}|\alpha'j'm'\rangle = 0 \qquad \text{unless} \quad \begin{cases} \quad j' = j+2, j+1, j, j-1, j-2 \\ \qquad \text{or} \quad \Delta j = 0, \pm 1, \pm 2, \\ \quad m' = m - q \quad (q = 2, 1, 0, -1, -2) \\ \qquad \text{or} \quad \Delta m = 0, \pm 1, \pm 2, \\ j + j' \geqslant 2. \end{cases} \tag{6.3-12}$$

The restriction $j + j' \geqslant 1$ which arises from the triangle conditions implies that the matrix elements of $\mathbf{T}^{(1)}$ vanish when $j = j' = 0$ despite the fact that $\Delta j = 0$ is satisfied. Similarly $j + j' \geqslant 2$ means that the matrix elements of $\mathbf{T}^{(2)}$ vanish when $j = 0, j' = 0, 1$ or when $j = \frac{1}{2}, j' = \frac{1}{2}$.

The Wigner–Eckart theorem also establishes a proportionality between matrix elements of irreducible tensor operators with the same values of k and q. Thus suppose $T_q^{(k)}$ and $U_q^{(k)}$ are two such operators. The coupling coefficients in the evaluation of $\langle\alpha jm|T_q^{(k)}|\alpha'j'm'\rangle$ and $\langle\alpha jm|U_q^{(k)}|\alpha'j'm'\rangle$ are the same. Therefore

$$\frac{\langle\alpha jm|T_q^{(k)}|\alpha'j'm'\rangle}{\langle\alpha jm|U_q^{(k)}|\alpha'j'm'\rangle} = \frac{\langle\alpha j\|\mathbf{T}^{(k)}\|\alpha'j'\rangle}{\langle\alpha j\|\mathbf{U}^{(k)}\|\alpha'j'\rangle} \equiv C. \tag{6.3-13}$$

An important special case occurs when $k = 1$. Let $\mathbf{T}^{(1)} = \mathbf{A}$ and $\mathbf{U}^{(1)} = \mathbf{J}$ where both $\mathbf{A}$ and $\mathbf{J}$ are vector operators (irreducible tensors of rank 1). $\mathbf{A}$ is assumed to be an arbitrary operator but $\mathbf{J}$ is, specifically, an angular momentum operator. From (6.3-13) we have

$$\langle\alpha jm|\mathbf{A}|\alpha jm'\rangle = C\langle\alpha jm|\mathbf{J}|\alpha jm'\rangle. \tag{6.3-14}$$

The constant C may be evaluated as:

$$\langle\alpha jm|\mathbf{A}\cdot\mathbf{J}|\alpha jm\rangle = \sum_{m'}\langle\alpha jm|\mathbf{A}|\alpha jm'\rangle\cdot\langle\alpha jm'|\mathbf{J}|\alpha jm\rangle. \tag{6.3-15}$$

Substituting (6.3-14),

$$\langle \alpha jm|\mathbf{A}\cdot\mathbf{J}|\alpha jm\rangle = C\sum_{m'}\langle \alpha jm|\mathbf{J}|\alpha jm'\rangle\cdot\langle \alpha jm'|\mathbf{J}|\alpha jm\rangle$$

$$= C\langle \alpha jm|J^2|\alpha jm\rangle \tag{6.3-16}$$

$$= Cj(j+1), \tag{6.3-17}$$

or

$$C = \frac{\langle \alpha jm|\mathbf{A}\cdot\mathbf{J}|\alpha jm\rangle}{j(j+1)}. \tag{6.3-18}$$

Therefore

$$\langle \alpha jm|\mathbf{A}|\alpha jm'\rangle = \frac{\langle \alpha jm|\mathbf{A}\cdot\mathbf{J}|\alpha jm\rangle}{j(j+1)}\langle \alpha jm|\mathbf{J}|\alpha jm'\rangle. \tag{6.3-19}$$

This is a special form of the Wigner–Eckart theorem known as the *Landé formula*.

In some cases it is possible to give an explicit evaluation of the reduced matrix element. We illustrate with several examples. From (1.2-29),

$$\langle l'm'|Y_{LM}|lm\rangle = (-1)^{m'}\sqrt{\frac{(2l'+1)(2L+1)(2l+1)}{4\pi}}\begin{pmatrix} l' & L & l\\ -m' & M & m\end{pmatrix}\begin{pmatrix} l' & L & l\\ 0 & 0 & 0\end{pmatrix}, \tag{6.3-20}$$

and from the Wigner–Eckart theorem (6.3-1b),

$$\langle l'm'|Y_{LM}|lm\rangle = (-1)^{l'-m'}\begin{pmatrix} l' & L & l\\ -m' & M & m\end{pmatrix}\langle l'\|\mathbf{Y}^{(L)}\|l\rangle. \tag{6.3-21}$$

Therefore

$$\langle l'\|\mathbf{Y}^{(L)}\|l\rangle = (-1)^{l'}\sqrt{\frac{(2l'+1)(2L+1)(2l+1)}{4\pi}}\begin{pmatrix} l' & L & l\\ 0 & 0 & 0\end{pmatrix}. \tag{6.3-22}$$

Interchanging l and l' in (6.3-22) and noting that

$$\begin{pmatrix} l & L & l'\\ 0 & 0 & 0\end{pmatrix} = \begin{pmatrix} l' & L & l\\ 0 & 0 & 0\end{pmatrix},$$

we obtain

$$\langle l\|\mathbf{Y}^{(L)}\|l'\rangle = (-1)^{l-l'}\langle l'\|\mathbf{Y}^{(L)}\|l\rangle. \tag{6.3-23}$$

For a scalar operator $T_0^{(0)} = 1$ the Wigner–Eckart theorem (6.3-1) takes the form

$$\langle jm|1|j'm'\rangle = (-1)^{j-m}\begin{pmatrix} j & 0 & j' \\ -m & 0 & m' \end{pmatrix}\langle j\|1\|j'\rangle.$$

But from Table 1.5,

$$\begin{pmatrix} j & 0 & j' \\ -m & 0 & m' \end{pmatrix} = (-1)^{j-m}\frac{1}{\sqrt{2j+1}}\,\delta_{jj'}\,\delta_{mm'}.$$

Since

$$\langle jm|1|j'm'\rangle = \delta_{jj'}\,\delta_{mm'},$$

we have

$$\langle j\|1\|j'\rangle = \sqrt{2j+1}\,\delta_{jj'}. \tag{6.3-24}$$

In general, the evaluation of reduced matrix elements of $\mathbf{T}^{(k)}$ depends on our ability to compute the complete matrix element of at least one component of $\mathbf{T}^{(k)}$. Thus suppose $T_0^{(1)} = J_0$; then

$$\begin{aligned}\langle jm|J_0|j'm'\rangle &= (-1)^{j-m}\begin{pmatrix} j & 1 & j' \\ -m & 0 & m' \end{pmatrix}\langle j\|\mathbf{J}\|j'\rangle \\ &= m\,\delta_{jj'}\,\delta_{mm'}.\end{aligned}$$

Again from Table 1.5,

$$\begin{pmatrix} j & 1 & j \\ -m & 0 & m' \end{pmatrix} = (-1)^{j-m}\frac{m}{\sqrt{(2j+1)(j+1)j}}\,\delta_{mm'};$$

therefore

$$\langle j\|\mathbf{J}\|j'\rangle = \sqrt{(2j+1)(j+1)j}\,\delta_{jj'}. \tag{6.3-25}$$

Since the reduced matrix element in (6.3-1) is a function of k but not of q, (6.3-25) is valid for any component of $\mathbf{J}$ and we have the general expression

$$\langle jm|J_q|j'm'\rangle = (-1)^{j-m}\begin{pmatrix} j & 1 & j' \\ -m & q & m' \end{pmatrix}\sqrt{(2j+1)(j+1)j}\,\delta_{jj'} \tag{6.3-26}$$

which is nothing more than (1.7-10)–(1.7-12) expressed as a single formula.

The coupling coefficients are real; hence

$$\langle \alpha jm|T_q^{(k)}|\alpha'j'm'\rangle^* = (-1)^{j-m}\begin{pmatrix} j & k & j' \\ -m & q & m' \end{pmatrix}\langle \alpha j\|\mathbf{T}^{(k)}\|\alpha'j'\rangle^*,$$

and, since reduced matrix elements are independent of the projection quantum numbers,

$$\sum_{mm'} |\langle\alpha jm|T_q^{(k)}|\alpha' j'm'\rangle|^2 = |\langle\alpha j\|\mathbf{T}^{(k)}\|\alpha' j'\rangle|^2 \sum_{mm'} \begin{pmatrix} j & k & j' \\ -m & q & m' \end{pmatrix}^2. \tag{6.3-27}$$

The orthogonality property (1.5-50) may now be used to evaluate the sum over the $3j$ symbols with the result that

$$\sum_{mm'} |\langle\alpha jm|T_q^{(k)}|\alpha' j'm'\rangle|^2 = \frac{1}{2k+1} |\langle\alpha j\|\mathbf{T}^{(k)}\|\alpha' j'\rangle|^2. \tag{6.3-28}$$

Since the right side is independent of q and there are just $2k + 1$ values of q, the sum over q is

$$\sum_q \sum_{mm'} |\langle\alpha jm|T_q^{(k)}|\alpha' j'm'\rangle|^2 = |\langle\alpha j\|\mathbf{T}^{(k)}\|\alpha' j'\rangle|^2. \tag{6.3-29}$$

The matrix element of the scalar product of two tensor operators $\mathbf{T}^{(k)}$ and $\mathbf{U}^{(k)}$ may be written as

$$\begin{aligned} &\langle\alpha jm|\mathbf{T}^{(k)} \cdot \mathbf{U}^{(k)}|\alpha' j'm'\rangle \\ &= \sum_q (-1)^q \langle\alpha jm|T_q^{(k)} U_{-q}^{(k)}|\alpha' j'm'\rangle \\ &= \sum_{\alpha'' j'' m''} \sum_q (-1)^q \langle\alpha jm|T_q^{(k)}|\alpha'' j'' m''\rangle\langle\alpha'' j'' m''|U_{-q}^{(k)}|\alpha' j'm'\rangle. \end{aligned} \tag{6.3-30}$$

The individual matrix elements in (6.3-30) may be expressed in terms of the Wigner–Eckart theorem so that

$$\begin{aligned} &\langle\alpha jm|\mathbf{T}^{(k)} \cdot \mathbf{U}^{(k)}|\alpha' j'm'\rangle \\ &= \sum_{\alpha'' j'' m''} \sum_q (-1)^{q+j+j''-m-m''} \\ &\times \langle\alpha j\|\mathbf{T}^{(k)}\|\alpha'' j''\rangle\langle\alpha'' j''\|\mathbf{U}^{(k)}\|\alpha' j'\rangle \begin{pmatrix} j & k & j'' \\ -m & q & m'' \end{pmatrix} \begin{pmatrix} j'' & k & j' \\ -m'' & -q & m' \end{pmatrix}. \end{aligned} \tag{6.3-31}$$

But

$$-m + q + m'' = 0$$

from the first $3j$ symbol. Therefore the phase factor becomes

$$(-1)^{j+j''-2m''} = (-1)^{j+2j''-2m''-j''} = (-1)^{j-j''} \tag{6.3-32}$$

in which the last equality is a result of the fact that $j'' - m''$ is an integer. The $3j$ symbols can be manipulated in order to put them into a form suitable for the application of the orthogonality theorem (1.5-50):

$$\begin{pmatrix} j & k & j'' \\ -m & q & m'' \end{pmatrix} = \begin{pmatrix} k & j'' & j \\ q & m'' & -m \end{pmatrix},$$

$$\begin{pmatrix} j'' & k & j' \\ -m'' & -q & m' \end{pmatrix} = (-1)^{j''+k+j'} \begin{pmatrix} j'' & k & j' \\ m'' & q & -m' \end{pmatrix}$$

$$= \begin{pmatrix} k & j'' & j' \\ q & m'' & -m' \end{pmatrix} (-1)^{2(j''+k+j')}.$$

Since $j'' + k + j'$ is an integer,

$$(-1)^{2(j''+k+j')} = 1.$$

We then have, from (1.5-50),

$$\sum_{qm''} \begin{pmatrix} j & k & j'' \\ -m & q & m'' \end{pmatrix} \begin{pmatrix} j'' & k & j' \\ -m'' & -q & m' \end{pmatrix}$$

$$= \sum_{qm''} \begin{pmatrix} k & j'' & j \\ q & m'' & -m \end{pmatrix} \begin{pmatrix} k & j'' & j' \\ q & m'' & -m' \end{pmatrix} = \frac{\delta_{jj'}\,\delta_{mm'}}{2j+1}. \quad (6.3\text{-}33)$$

Substituting (6.3-32) and (6.3-33) into (6.3-31) we obtain

$$\langle \alpha jm | \mathbf{T}^{(k)} \cdot \mathbf{U}^{(k)} | \alpha' j' m' \rangle$$

$$= \sum_{\alpha'' j''} (-1)^{j-j''} \langle \alpha j \| \mathbf{T}^{(k)} \| \alpha'' j'' \rangle \langle \alpha'' j'' \| \mathbf{U}^{(k)} \| \alpha' j' \rangle \frac{\delta_{jj'}\,\delta_{mm'}}{2j+1}. \quad (6.3\text{-}34)$$

Suppose $\mathbf{T}^{(k)}$ and $\mathbf{U}^{(l)}$ are two irreducible tensor operators that act on two different noninteracting systems with angular momenta $\mathbf{J}_1$ and $\mathbf{J}_2$, respectively. For example, $\mathbf{T}^{(k)}$ might act on spatial coordinates and $\mathbf{U}^{(l)}$ on spin coordinates or $\mathbf{T}^{(k)}$ might act on the coordinates of one particle while $\mathbf{U}^{(l)}$ acts on the coordinates of another particle. It will therefore be assumed that

$$[J_{1\,\pm 1}, T_q^{(k)}] = \mp\sqrt{\tfrac{1}{2}[k(k+1) - q(q \pm 1)]}\,T_{q\pm 1}^{(k)}, \qquad [J_{1\,0}, T_q^{(k)}] = qT_q^{(k)} \quad (6.3\text{-}35)$$

and

$$[J_{2\pm 1}, U_p^{(l)}] = \mp\sqrt{\tfrac{1}{2}[l(l+1 - p(p \pm 1)]}\,U_{p\pm 1}^{(l)}, \qquad [J_{2\,0}, U_p^{(l)}] = pU_p^{(l)}, \quad (6.3\text{-}36)$$

but that all other commutators among the components of $\mathbf{T}^{(k)}$, $\mathbf{U}^{(l)}$, $\mathbf{J}_1$, and $\mathbf{J}_2$ vanish. Now suppose that the two angular momenta are coupled, i.e., $\mathbf{J} = \mathbf{J}_1 + \mathbf{J}_2$. The matrix elements of the scalar product $\mathbf{T}^{(k)} \cdot \mathbf{U}^{(k)}$ in the coupled representation is given by (Edmonds, 1960; Judd, 1963)

$$\langle \alpha j_1 j_2 jm | \mathbf{T}^{(k)} \cdot \mathbf{U}^{(k)} | \alpha' j_1' j_2' j' m' \rangle = (-1)^{j+j_2+j_1'} \delta_{jj'} \delta_{mm'} \begin{Bmatrix} j_1' & j_2' & j \\ j_2 & j_1 & k \end{Bmatrix} \times \sum_{\alpha''} \langle \alpha j_1 \| \mathbf{T}^{(k)} \| \alpha'' j_1' \rangle \langle \alpha'' j_2 \| \mathbf{U}^{(k)} \| \alpha' j_2' \rangle. \tag{6.3-37}$$

The quantity in the curly brackets is a $6j$ symbol (see Section 1.6). For $\mathbf{T}^{(k)}$ which acts on system 1, the matrix element is

$$\langle \alpha j_1 j_2 jm | T_q^{(k)} | \alpha' j_1' j_2' j' m' \rangle = (-1)^{j-m} \begin{pmatrix} j & k & j' \\ -m & q & m' \end{pmatrix} \langle \alpha j_1 j_2 j \| \mathbf{T}^{(k)} \| \alpha' j_1' j_2' j' \rangle. \tag{6.3-38}$$

The reduced matrix element in (6.3-38) is in the coupled representation; it may be expressed in terms of the reduced matrix element in the uncoupled representation (Judd, 1963):

$$\langle \alpha j_1 j_2 j \| \mathbf{T}^{(k)} \| \alpha' j_1' j_2' j' \rangle = (-1)^{j_1+j_2+j'+k} \delta_{j_2 j_2'} \times \sqrt{(2j+1)(2j'+1)} \begin{Bmatrix} j & k & j' \\ j_1' & j_2 & j_1 \end{Bmatrix} \langle \alpha j_1 \| \mathbf{T}^{(k)} \| \alpha' j_1' \rangle. \tag{6.3-39}$$

The parallel expressions for $\mathbf{U}^{(l)}$ which acts on system 2 are

$$\langle \alpha j_1 j_2 jm | U_p^{(l)} | \alpha' j_1' j_2' j' m' \rangle = (-1)^{j-m} \begin{pmatrix} j & l & j' \\ -m & p & m' \end{pmatrix} \langle \alpha j_1 j_2 j \| \mathbf{U}^{(l)} \| \alpha' j_1' j_2' j' \rangle, \tag{6.3-40}$$

$$\langle \alpha j_1 j_2 j \| \mathbf{U}^{(l)} \| \alpha' j_1' j_2' j' \rangle = (-1)^{j_1+j_2'+j+l} \delta_{j_1 j_1'} \times \sqrt{(2j+1)(2j'+1)} \begin{Bmatrix} j & l & j' \\ j_2' & j_1 & j_2 \end{Bmatrix} \langle \alpha j_2 \| \mathbf{U}^{(l)} \| \alpha' j_2' \rangle. \tag{6.3-41}$$

The tensor product of two irreducible tensor operators $\mathbf{T}^{(k)}$ and $\mathbf{U}^{(l)}$ was defined by (6.2-1). We now define the *direct product* of $\mathbf{T}^{(k)}$ and $\mathbf{U}^{(l)}$ as the set of $(2k+1)(2l+1)$ operators $T_q^{(k)} U_p^{(l)}$ obtained by multiplying in all possible ways the components of $\mathbf{T}^{(k)}$ with the components of $\mathbf{U}^{(l)}$. Under the previous assumption that $\mathbf{T}^{(k)}$ and $\mathbf{U}^{(l)}$ act on two noninteracting systems 1 and 2, respectively, the products

$$R_{qp}^{(kl)} \equiv T_q^{(k)} U_p^{(l)} \tag{6.3-42}$$

are components of an irreducible tensor $\mathbf{R}^{(kl)}$ which behaves as an irreducible tensor of rank k with respect to system 1 and as an irreducible tensor of rank l with respect to system 2. Specifically $\mathbf{R}^{(kl)}$ satisfies

$$\begin{aligned}[J_{1\,\pm1}, R_{qp}^{(kl)}] &= \mp\sqrt{\tfrac{1}{2}[k(k+1) - q(q\pm1)]}R_{q\pm1,p}^{(kl)},\\ [J_{1\,0}, R_{qp}^{(kl)}] &= qR_{qp}^{(kl)},\end{aligned} \tag{6.3-43a}$$

$$\begin{aligned}[J_{2\,\pm1}, R_{qp}^{(kl)}] &= \mp\sqrt{\tfrac{1}{2}[l(l+1) - p(p\pm1)]}R_{qp\pm1}^{(kl)},\\ [J_{2\,0}, R_{qp}^{(kl)}] &= pR_{qp}^{(kl)}.\end{aligned} \tag{6.3-43b}$$

The Wigner–Eckart theorem may be applied to $\mathbf{R}^{(kl)}$:

$$\begin{aligned}&\langle j_1 j_2 m_1 m_2 | R_{qp}^{(kl)} | j_1' j_2' m_1' m_2' \rangle\\ &\quad = (-1)^{j_1+j_2-m_1-m_2}\langle j_1 \| \mathbf{T}^{(k)} \| j_1' \rangle \langle j_2 \| \mathbf{U}^{(l)} \| j_2' \rangle\\ &\quad\quad \times \begin{pmatrix} j_1 & k & j_1' \\ -m_1 & q & m_1' \end{pmatrix}\begin{pmatrix} j_2 & l & j_2' \\ -m_2 & p & m_2' \end{pmatrix}.\end{aligned} \tag{6.3-44}$$

The scalar product of $\mathbf{R}^{(kl)}$ and $\mathbf{Q}^{(kl)}$ is defined by

$$\mathbf{R}^{(kl)} \cdot \mathbf{Q}^{(kl)} = \sum_{qp} (-1)^{q+p} R_{qp}^{(kl)} Q_{-q\,-p}^{(kl)}. \tag{6.3-45}$$

If $\mathbf{R}^{(kl)}$ satisfies the commutation rules (6.3-43a) with respect to L_1, S_1 and commutes with L_2, S_2 while $\mathbf{Q}^{(kl)}$ satisfies the commutation rules (6.3-43b) with respect to L_2, S_2 and commutes with L_1, S_1, the matrix element of (6.3-45) is given by

$$\begin{aligned}&\langle \alpha L_1 S_1 L_2 S_2 LSM_L M_S | \mathbf{R}^{(kl)} \cdot \mathbf{Q}^{(kl)} | \alpha' L_1' S_1' L_2' S_2' LSM_L M_S \rangle\\ &\quad = (-1)^{L_1'+S_1'+L_2+S_2+L+S} \sum_{\alpha''} \langle \alpha L_1 S_1 \| R^{(kl)} \| \alpha'' L_1' S_1' \rangle\\ &\quad\quad \times \langle \alpha'' L_2 S_2 \| Q^{(kl)} \| \alpha' L_2' S_2' \rangle \begin{Bmatrix} L_1 & L_2 & L \\ L_2' & L_1' & k \end{Bmatrix}\begin{Bmatrix} S_1 & S_2 & S \\ S_2' & S_1' & l \end{Bmatrix}.\end{aligned} \tag{6.3-46}$$

6.4 Cartesian Tensors

We consider an orthogonal transformation of coordinates

$$\mathbf{r}' = R\mathbf{r} \tag{6.4-1}$$

in a three-dimensional space with

$$R = \begin{pmatrix} a_{11} & a_{12} & a_{13} \\ a_{21} & a_{22} & a_{23} \\ a_{31} & a_{32} & a_{33} \end{pmatrix} \tag{6.4-2}$$

and

$$\sum_i a_{ij} a_{ik} = \sum_i a_{ji} a_{ki} = \delta_{jk}. \tag{6.4-3}$$

A tensor of rank r with 3^r components $T_{i_1 i_2 \cdots i_r}$ where

$$i_1, i_2, \ldots, i_r = 1, 2, 3 \tag{6.4-4}$$

is defined by the transformation law

$$T'_{i_1 i_2 \cdots i_r} = \sum_{j_1, j_2, \ldots, j_r} a_{i_1 j_1} a_{i_2 j_2} \cdots a_{i_r j_r} T_{j_1 j_2 \cdots j_r}, \tag{6.4-5}$$

which relates tensor components with respect to the coordinate system **r** to those with respect to **r**$'$ where **r** and **r**$'$ transform according to (6.4-1). The definition of a tensor is easily extended to spaces of more than three dimensions as well as to more general coordinate transformations, but we shall postpone such considerations until the next section.

A scalar quantity is a tensor of rank zero and its transformation law is simply

$$T' = T; \tag{6.4-6}$$

a vector is a tensor of rank 1 whose components satisfy

$$T_i' = \sum_j a_{ij} T_j; \tag{6.4-7}$$

and a tensor of rank 2 (or dyadic) transforms according to

$$T'_{ij} = \sum_{kl} a_{ik} a_{jl} T_{kl}. \tag{6.4-8}$$

A physical example which is describable in terms of a second rank tensor is the dipole–dipole interaction. If **P** and **Q** are two dipoles and the position of **Q** relative to **P** is given by **r** (Fig. 6.1), the interaction energy U may be

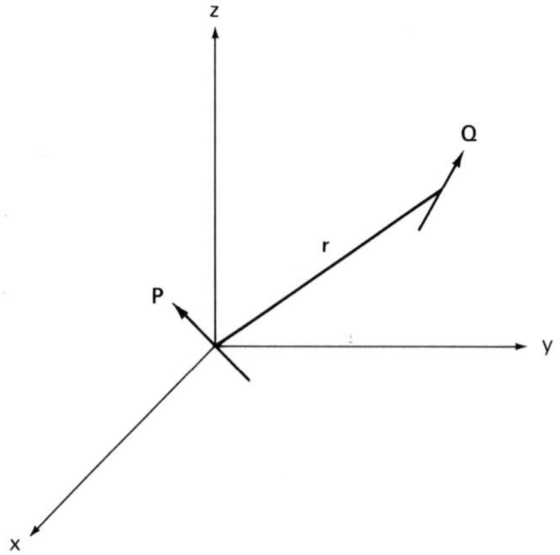

FIG. 6.1 Coordinate system for the interaction of two dipoles.

written as

$$U = \frac{1}{r^3}\left[\mathbf{P}\cdot\mathbf{Q} - \frac{3(\mathbf{P}\cdot\mathbf{r})(\mathbf{Q}\cdot\mathbf{r})}{r^2}\right] \tag{6.4-9}$$

or, in terms of Cartesian components

$$\begin{aligned} U = \frac{1}{r^3}\Bigg[&\left(1-\frac{3x^2}{r^2}\right)P_xQ_x - \frac{3xy}{r^2}P_xQ_y - \frac{3xz}{r^2}P_xQ_z \\ &- \frac{3yx}{r^2}P_yQ_x + \left(1-\frac{3y^2}{r^2}\right)P_yQ_y - \frac{3yz}{r^2}P_yQ_z \\ &- \frac{3zx}{r^2}P_zQ_x - \frac{3zy}{r^2}P_zQ_y + \left(1-\frac{3z^2}{r^2}\right)P_zQ_z\Bigg] \end{aligned} \tag{6.4-10}$$

$$= \frac{1}{r^3}(P_xP_yP_z)\begin{pmatrix} 1-\dfrac{3x^2}{r^2} & -\dfrac{3xy}{r^2} & -\dfrac{3xz}{r^2} \\ -\dfrac{3yx}{r^2} & 1-\dfrac{3y^2}{r^2} & -\dfrac{3yz}{r^2} \\ -\dfrac{3zx}{r^2} & -\dfrac{3zy}{r^2} & 1-\dfrac{3z^2}{r^2} \end{pmatrix}\begin{pmatrix} Q_x \\ Q_y \\ Q_z \end{pmatrix} \tag{6.4-11}$$

$$= \frac{1}{r^3}\mathbf{P}\cdot\mathbf{T}\cdot\mathbf{Q} \tag{6.4-12}$$

in which $\mathbf{T}$ the *dipolar coupling tensor* is symmetric and has zero trace. In dyadic notation

$$\begin{aligned} \mathbf{T} = &\left(1-\frac{3x^2}{r^2}\right)\hat{\mathbf{i}}\hat{\mathbf{i}} - \frac{3xy}{r^2}\hat{\mathbf{i}}\hat{\mathbf{j}} - \frac{3xz}{r^2}\hat{\mathbf{i}}\hat{\mathbf{k}} - \frac{3yx}{r^2}\hat{\mathbf{j}}\hat{\mathbf{i}} + \left(1-\frac{3y^2}{r^2}\right)\hat{\mathbf{j}}\hat{\mathbf{j}} \\ &- \frac{3yz}{r^2}\hat{\mathbf{j}}\hat{\mathbf{k}} - \frac{3zx}{r^2}\hat{\mathbf{k}}\hat{\mathbf{i}} - \frac{3zy}{r^2}\hat{\mathbf{k}}\hat{\mathbf{j}} + \left(1-\frac{3z^2}{r^2}\right)\hat{\mathbf{k}}\hat{\mathbf{k}} \end{aligned} \tag{6.4-13a}$$

in which the general component may be written

$$T_{ik} = \delta_{ik} - \frac{3r_ir_k}{r^2} \qquad (r_i, r_k = x, y, z). \tag{6.4-13b}$$

If $\mathbf{P}$ and $\mathbf{Q}$ are written in spherical components

$$P_{\pm 1} = \mp\frac{P_x \pm iP_y}{\sqrt{2}}, \quad P_0 = P_z, \qquad Q_{\pm 1} = \mp\frac{Q_x \pm iQ_y}{\sqrt{2}}, \quad Q_0 = Q_z,$$

the interaction energy U takes the form

$$U = \frac{1}{r^3}\sqrt{\frac{4\pi}{5}}\,[-\sqrt{6}Y_{2\,-2}P_{+1}Q_{+1} + \sqrt{3}Y_{2\,-1}P_{+1}Q_0 - Y_{2\,0}P_{+1}Q_{-1}$$
$$+ \sqrt{3}Y_{2\,-1}P_0Q_{+1} - 2Y_{2\,0}P_0Q_0 + \sqrt{3}Y_{2\,1}P_0Q_{-1}$$
$$- Y_{2\,0}P_{-1}Q_{+1} + \sqrt{3}Y_{2\,1}P_{-1}Q_0 - \sqrt{6}Y_{2\,2}P_{-1}Q_{-1}] \tag{6.4-14a}$$

$$= \frac{1}{r^3}\Bigg[\frac{1}{2}(3P_0Q_0 - \mathbf{P}\cdot\mathbf{Q})(1 - 3\cos^2\theta)$$
$$+ \frac{3}{\sqrt{2}}(P_{+1}Q_0 + P_0Q_{+1})\sin\theta\cos\theta\, e^{-i\varphi}$$
$$- \frac{3}{\sqrt{2}}(P_{-1}Q_0 + P_0Q_{-1})\sin\theta\cos\theta\, e^{i\varphi}$$
$$- \frac{3}{2}P_{+1}Q_{+1}\sin^2\theta\, e^{-2i\varphi} - \frac{3}{2}P_{-1}Q_{-1}\sin^2\theta\, e^{2i\varphi}\Bigg]. \tag{6.4-14b}$$

The interaction energy may also be expressed very compactly in terms of irreducible tensors of rank 2. By analogy with (6.2-15) we define

$$V^{(2)}_{\pm 2} = P_{\pm 1}Q_{\pm 1},$$
$$V^{(2)}_{\pm 1} = \frac{1}{\sqrt{2}}[P_{\pm 1}Q_0 + P_0Q_{\pm 1}], \tag{6.4-15}$$
$$V^{(2)}_0 = \frac{1}{\sqrt{6}}[3P_0Q_0 - \mathbf{P}\cdot\mathbf{Q}],$$

which then permits the conversion of (6.4-14) to

$$U = -\frac{1}{r^3}\sqrt{\frac{24\pi}{5}}\sum_m(-1)^m Y_{2m}V^{(2)}_{-m} = -\frac{1}{r^3}\sqrt{\frac{24\pi}{5}}\,\mathbf{Y}^{(2)}\cdot\mathbf{V}^{(2)}. \tag{6.4-16}$$

We now return to the general second rank tensor defined by the transformation law (6.4-8). In the first place it is observed that as far as the transformation properties are concerned, the second rank tensor may be regarded as a direct (Kronecker) product of two vectors. Thus let

$$T_{ij} = P_iQ_j \tag{6.4-17}$$

where P_i and Q_j are components of the vectors **P** and **Q**, respectively. The vector transformation law (6.4-7) applied to the right side of (6.4-17) gives

$$P_i'Q_j' = \left(\sum_k a_{ik}P_k\right)\left(\sum_l a_{jl}Q_l\right) = \sum_{kl} a_{ik}a_{jl}P_kQ_l = \sum_{kl} a_{ik}a_{jl}T_{kl} = T'_{ij}, \quad (6.4\text{-}18)$$

which proves that T_{ij} is, indeed, a tensor of rank 2.

Some tensors have special properties. For example, an *antisymmetric tensor* A_{ij} is one whose components satisfy

$$A_{ij} = -A_{ji}. \quad (6.4\text{-}19)$$

The diagonal components must then all vanish:

$$A_{11} = A_{22} = A_{33} = 0 \quad (6.4\text{-}20)$$

and of the remaining six off-diagonal components only three can be independent. A *symmetric tensor* B_{ij} on the other hand, satisfies

$$B_{ij} = B_{ji} \quad (6.4\text{-}21)$$

and contains six independent components—three diagonal and three off-diagonal.

Any tensor T_{ij} may be written as a sum of an antisymmetric tensor A_{ij} and a symmetric tensor B_{ij}. Thus if

$$A_{ij} = \tfrac{1}{2}(T_{ij} - T_{ji}) = -A_{ji}, \qquad B_{ij} = \tfrac{1}{2}(T_{ij} + T_{ji}) = B_{ji}, \quad (6.4\text{-}22)$$

we have the identity

$$T_{ij} = A_{ij} + B_{ij}. \quad (6.4\text{-}23)$$

Now let us apply the transformation law (6.4-8) to A_{ij} and B_{ij}, separately:

$$A'_{ij} = \frac{1}{2}(T'_{ij} - T'_{ji}) = \frac{1}{2}\left[\sum_{kl} a_{ik}a_{jl}T_{kl} - \sum_{kl} a_{jl}a_{ik}T_{lk}\right]$$

$$= \frac{1}{2}\sum_{kl} a_{ik}a_{jl}(T_{kl} - T_{lk}) = \sum_{kl} a_{ik}a_{jl}A_{kl}, \quad (6.4\text{-}24a)$$

$$B'_{ij} = \frac{1}{2}(T'_{ij} + T'_{ji}) = \frac{1}{2}\sum_{kl} a_{ik}a_{jl}(T_{kl} + T_{lk}) = \sum_{kl} a_{ik}a_{jl}B_{kl}, \quad (6.4\text{-}24b)$$

The important property embodied by (6.4-24a) and (6.4-24b) is that an arbitrary tensor T_{ij} may be written as the sum of two parts A_{ij} and B_{ij} which do not mix under the transformation (6.4-8). The components A_{ij} transform among themselves as do the components B_{ij}. It is further noted that this conclusion has been reached without invoking the orthogonality condition (6.4-3). Indeed, all that is required is that (6.4-1) be a linear transformation of coordinates with a nonsingular matrix R.

If the orthogonality condition is imposed, however, the decomposition of T_{ij} may be carried a step further. Upon forming the trace which, in tensor terminology, is known as *contraction*, it is found that

$$\operatorname{Tr} B' = \sum_i B'_{ii} = \sum_i \sum_{kl} a_{ik} a_{il} B_{kl}.$$

But in view of (6.4-3) we have

$$\sum_{kl} \sum_i a_{ik} a_{il} B_{kl} = \sum_{kl} \delta_{kl} B_{kl} = \sum_k B_{kk}$$

or

$$\operatorname{Tr} B' = \operatorname{Tr} B = T_{11} + T_{22} + T_{33}. \tag{6.4-25}$$

Thus Tr B is an invariant under the orthogonal transformation of coordinates (6.4-1). We may therefore express the components B_{ij} as a sum of parts which tranform among themselves:

$$B_{ij} = S_{ij} + X_{ij}, \tag{6.4-26}$$

where

$$S_{ij} = \tfrac{1}{2}(T_{ij} + T_{ji}) - X_{ij}, \qquad X_{ij} = \tfrac{1}{3}(T_{11} + T_{22} + T_{33})\delta_{ij}. \tag{6.4-27}$$

We note that S_{ij} is a traceless symmetric tensor with five independent components.

The three independent components of the antisymmetric tensor A_{ij} may be given another interpretation which is facilitated by writing the tensor components as in (6.4-17). Thus, let

$$\begin{aligned} A_{12} &= \tfrac{1}{2}(T_{12} - T_{21}) = \tfrac{1}{2}(P_1 Q_2 - P_2 Q_1) = \tfrac{1}{2}(\mathbf{P} \times \mathbf{Q})_3 \\ A_{23} &= \tfrac{1}{2}(T_{23} - T_{32}) = \tfrac{1}{2}(P_2 Q_3 - P_3 Q_2) = \tfrac{1}{2}(\mathbf{P} \times \mathbf{Q})_1 \\ A_{31} &= \tfrac{1}{2}(T_{31} - T_{13}) = \tfrac{1}{2}(P_3 Q_1 - P_1 Q_3) = \tfrac{1}{2}(\mathbf{P} \times \mathbf{Q})_2. \end{aligned} \tag{6.4-28}$$

It appears then that the components A_{ij} may be associated with the components of a vector $\tfrac{1}{2}(\mathbf{P} \times \mathbf{Q})$. The latter, however, is an *axial* (or *pseudo*-) vector which is to be distinguished from a *polar* vector. The essential difference is that under an improper rotation (rotation inversion), the components of polar vectors change sign but those of axial vectors do not. Under proper rotations axial and polar vectors are indistinguishable.

The decomposition of the general second rank tensor with components T_{ij} may now be described as follows: Under a linear, but not necessarily orthogonal, transformation of coordinates $\mathbf{r}' = R\mathbf{r}$, the components T_{ij} may be decomposed into symmetrical and antisymmetrical parts

$$T_{ij} = A_{ij} + B_{ij} \qquad \text{(linear transformation)} \tag{6.4-29}$$

where A_{ij} is an antisymmetric second rank tensor with three independent components which may be put into one-to-one correspondence with an axial vector and B_{ij} is a symmetric second rank tensor. The components A_{ij} transform among themselves; the B_{ij} do likewise and as long as the coordinate transformation remains unrestricted (apart from linearity) it is not possible to carry the decomposition of T_{ij} into sets of components which transform among themselves beyond that represented by (6.4-29). If, however, the coordinate transformation is orthogonal, it is possible to decompose B_{ij} into a traceless symmetric tensor S_{ij} with five independent components and a tensor X_{ij} with one independent component. The latter, clearly, must be equivalent to a scalar. Hence, under an orthogonal transformation R with $\det R = \pm 1$,

$$T_{ij} = X_{ij} + A_{ij} + S_{ij} \qquad \text{(orthogonal transformation)} \tag{6.4-30}$$

in which X_{ij}, A_{ij}, and S_{ij} are second rank tensors whose components transform only among themselves.

The decomposition of a (Cartesian) tensor in a three-dimensional space may be discussed from the stand point of the rotation group. We saw previously (6.4-17) that if P_i and Q_j are the components of two vectors each of which transforms according to (6.4-7), the nine products P_iQ_j transform according to (6.4-8). Hence the transformation law of a tensor of rank 2 is identical with the transformation law of the direct product of two vectors. Now, a vector, as far as its transformation properties are concerned, is also an *irreducible* tensor of rank 1. This becomes apparent when, for example, the components of a vector **r** are written as

$$r_{\pm 1} = \mp \frac{1}{\sqrt{2}}(x + iy), \qquad r_0 = z. \tag{6.4-31}$$

The right sides of the expressions in (6.4-31) are recognized as the components of the spherical harmonic Y_{1m} (within a constant factor). But, as shown in Section 6.1, the transformation properties of $T_q^{(1)}$ and Y_{1m} are identical. Therefore $r_{\pm 1}, r_0$ qualify as the components of a first rank irreducible tensor. At the same time we also know, from (6.4-7), that a vector is a Cartesian tensor of rank 1. Hence any vector **T** may be classified either as a Cartesian tensor of rank 1 with components T_1, T_2, and T_3 or as an irreducible tensor of rank 1 with components T_{+1}, T_0, and T_{-1}.

To say that a vector transforms as an irreducible tensor of rank 1 is equivalent to the statement that the (spherical) components of a vector constitute a basis for the irreducible representation $D^{(1)}$ of $O^{+(3)}$. Since a tensor of rank 2 transforms as the direct product of two vectors, the components of a tensor of rank 2 must then form a basis for the direct product representation $D^{(1)} \times D^{(1)}$. But the latter is reducible (Section 4.5):

$$D^{(1)} \times D^{(1)} = D^{(0)} + D^{(1)} + D^{(2)}. \tag{6.4-32}$$

We therefore conclude that under proper rotations ($\det R = +1$) the tensor representation whose basis consists of the Cartesian components of a tensor of rank 2 is reducible into the irreducible representation $D^{(0)}, D^{(1)}$, and $D^{(2)}$. Indeed, by means of the coupling coefficients it is possible to construct bases for $D^{(0)}, D^{(1)}$, and $D^{(2)}$ expressed as linear combinations of the components T_{ij}. Upon replacing x, y, z by $1, 2, 3$; $A_i B_j$ by T_{ij}; and V by T in (6.2-16)–(6.2-18) the relations between Cartesian tensors of rank 2 and irreducible tensors are given by

$$\begin{aligned}
T^{(0)} &= -\frac{1}{\sqrt{3}}[T_{11} + T_{22} + T_{33}], \\
T^{(1)}_{\pm 1} &= \frac{1}{2}[T_{31} - T_{13} \pm i(T_{32} - T_{23})], \\
T^{(1)}_{0} &= \frac{i}{\sqrt{2}}(T_{12} - T_{21}), \\
T^{(2)}_{\pm 2} &= \frac{1}{2}[T_{11} - T_{22} \pm i(T_{12} + T_{21})], \\
T^{(2)}_{\pm 1} &= \mp\frac{1}{2}[T_{31} + T_{13} \pm i(T_{23} + T_{32})], \\
T^{(2)}_{0} &= \frac{1}{\sqrt{6}}[3T_{33} - (T_{11} + T_{22} + T_{33})].
\end{aligned} \tag{6.4-33}$$

It is also instructive to reexamine the terms in (6.4-30) from the present viewpoint. X_{ij}, with only one independent component, must be equivalent to an irreducible tensor of rank 0. This is verified by comparing X_{ij} in (6.4-27) with $T^{(0)}_0$ in (6.4-33); they are the same to within a constant factor. The three independent components of A_{ij} transform like the components of a vector as shown in (6.4.28). From (6.4-33) or (6.2-17) it is seen that

$$\begin{aligned}
A_{12} &= \frac{1}{2}(T_{12} - T_{21}) = -\frac{i}{\sqrt{2}} T^{(1)}_0, \\
A_{23} &= \frac{1}{2}(T_{23} - T_{32}) = \frac{i}{2}(T^{(1)}_{+1} - T^{(1)}_{-1}), \\
A_{31} &= \frac{1}{2}(T_{31} - T_{13}) = \frac{1}{2}(T^{(1)}_{+1} + T^{(1)}_{-1}).
\end{aligned} \tag{6.4-34}$$

Hence A_{12}, A_{23}, and A_{31} transform like the components of an irreducible tensor of rank 1. The five independent components of the traceless symmetric tensor S_{ij} may be taken as:

$$S_{12} = S_{21} = \frac{1}{2}(T_{12} + T_{21}) = \frac{i}{2}(-T^{(2)}_{+2} + T^{(2)}_{-2}),$$

$$S_{13} = S_{31} = \frac{1}{2}(T_{13} + T_{31}) = \frac{1}{2}(-T^{(2)}_{+1} + T^{(2)}_{-1}),$$

$$S_{23} = S_{32} = \frac{1}{2}(T_{23} + T_{32}) = \frac{i}{2}(T^{(2)}_{+1} + T^{(2)}_{-1}), \qquad (6.4\text{-}35)$$

$$S_{33} = T_{33} - \frac{1}{3}(T_{11} + T_{22} + T_{33}) = \sqrt{\frac{2}{3}}\, T^{(2)}_{0},$$

$$S_{11} - S_{22} = T_{11} - T_{22} = T^{(2)}_{+2} + T^{(2)}_{-2}.$$

Thus the components of S transform like the components of an irreducible tensor of rank 2.

In summary, a general second rank tensor in a space of three dimensions has nine independent components T_{ij}. It is always possible to organize the T_{ij} into linear combinations to form a new set of nine independent components. If the linear combinations are constructed as in (6.4-22), then, under a general linear transformation of coordinates which may include orthogonal transformations, the three independent components A_{ij} transform among themselves and the six independent components B_{ij} transform among themselves. We then say that T_{ij} has been *reduced* to A_{ij} and B_{ij}. Furthermore, the A_{ij} and B_{ij} are irreducible in the sense that there does not exist any smaller set of linear combinations of T_{ij} which will transform among themselves under a general linear transformation. If, however, the coordinate transformations are restricted to those that are orthogonal, it is found that B_{ij} can be further reduced as in (6.4-26). In that case the five independent components S_{ij} transform among themselves and the single component X_{ij} is an invariant. Finally, X_{ij}, A_{ij}, and S_{ij} are shown to be bases for the irreducible representations $D^{(0)}$, $D^{(1)}$, and $D^{(2)}$, respectively, of $O^{+(3)}$.

6.5 Tensors, Permutation Groups, Continuous Groups

We now wish to enlarge on the ideas presented in the last section concerning the decomposition of tensors into sets of components which transform irreducibly among themselves. An important ingredient in this discussion is the connection between permutation groups and tensors.

Referring to Section 5.4, it was shown that the permutation group S_2 has two irreducible (one-dimensional) representations characterized by the diagrams

$$[2]\ \begin{array}{|c|c|}\hline i & j \\ \hline \end{array} \qquad [1^2]\ \begin{array}{|c|}\hline i \\ \hline j \\ \hline \end{array} \tag{6.5-1}$$

For each partition there is but one tableau into which the indices i, j have been inserted in anticipation of constructing a basis for each tableau from the components T_{ij} of a second rank tensor. Such bases are constructed by means of the Young operator (5.4-18). Thus for the partition $[2]$ it is necessary to symmetrize with respect to i and j, and for $[1^2]$ it is necessary to antisymmetrize with respect to the two indices. This may be indicated by writing (Hamermesh, 1962)

$$T_{[2]} = T_{ij} + T_{ji}, \qquad T_{[1^2]} = T_{ij} - T_{ji} \tag{6.5-2}$$

or

$$T_{ij} = \tfrac{1}{2}(T_{[2]} + T_{[1^2]}), \tag{6.5-3}$$

which is nothing more than (6.4-23) in another notation with

$$T_{[1^2]} = 2A_{ij}, \qquad T_{[2]} = 2B_{ij}. \tag{6.5-4}$$

It is seen that the tensor components A_{ij} and B_{ij} now serve as bases for the irreducible representations of S_2.

Let us now go through the same procedure with S_3 and tensors of rank 3. The Young operator will project out the necessary linear combinations of tensor components which will form bases for the irreducible representations of S_3. Since the typical expressions are already available in (5.4-22)–(5.4-25), we now have

$$[3]\ \begin{array}{|c|c|c|}\hline i & j & k \\ \hline \end{array}$$

$$T_{[3]} = T_{ijk} + T_{jki} + T_{kij} + T_{jik} + T_{kji} + T_{ikj}, \tag{6.5-5}$$

$$[1^3]\ \begin{array}{|c|}\hline i \\ \hline j \\ \hline k \\ \hline \end{array}$$

$$T_{[1^3]} = T_{ijk} + T_{jki} + T_{kij} - T_{jik} - T_{kji} - T_{ikj}, \tag{6.5-6}$$

$$[21a]\ \begin{array}{|c|c|}\hline i & j \\ \hline k & \multicolumn{1}{c}{} \\ \cline{1-1} \end{array}$$

$$T_{[21a]} = T_{ijk} + T_{jik} - T_{kji} - T_{jki}, \tag{6.5-7}$$

$$[21b]\begin{array}{|c|c|}\hline i & k \\ \hline j & \\ \cline{1-1}\end{array}$$

$$T_{[21b]} = T_{ijk} + T_{kji} - T_{jik} - T_{kij}. \tag{6.5-8}$$

Combining (6.5-7) and (6.5-8),

$$T_{[21]} \equiv T_{[21a]} + T_{[21b]} = 2T_{ijk} - T_{jki} - T_{kij}; \tag{6.5-9}$$

hence the analog to (6.5-3) for third rank tensors is

$$T_{ijk} = \tfrac{1}{6}(T_{[3]} + T_{[1^3]} + 2T_{[21]}). \tag{6.5-10}$$

These examples may be generalized. Consider a space of n dimensions and a linear (not necessarily orthogonal) transformation of coordinates given by

$$\begin{aligned} x_1' &= a_{11}x_1 + a_{12}x_2 + \cdots + a_{1n}x_n, \\ x_2' &= a_{21}x_1 + a_{22}x_2 + \cdots + a_{2n}x_n, \\ &\vdots \\ x_n' &= a_{n1}x_1 + a_{n2}x_2 + \cdots + a_{nn}x_n \end{aligned} \tag{6.5-11}$$

or

$$\mathbf{r}' = R\mathbf{r} \tag{6.5-12}$$

where $\mathbf{r}$ and $\mathbf{r}'$ are n-dimensional vectors and the transformation matrix R is

$$R = \begin{pmatrix} a_{11} & a_{12} & \cdots & a_{1n} \\ a_{21} & a_{22} & \cdots & a_{2n} \\ \vdots & \vdots & \cdots & \vdots \\ a_{n1} & a_{n2} & \cdots & a_{nn} \end{pmatrix} \tag{6.5-13}$$

in which the numerical coefficients a_{ij} are, is general, complex and $\det R \neq 0$. The matrices R are elements of the general linear group in n dimensions $GL(n)$. An rth rank tensor in this space is defined as a quantity with n^r components, which, under the coordinate transformation R, satisfy

$$T'_{i_1 i_2 \cdots i_r} = \sum_{j_1, j_2, \ldots, j_r} a_{i_1 j_1} a_{i_2 j_2} \cdots a_{i_r j_r} T_{j_1 j_2 \cdots j_r}. \tag{6.5-14}$$

This definition is of the same form as (6.4-5) but, in contrast to the latter, the transformation coefficients $a_{i_r j_r}$ in (6.5-14) refer to a general linear transformation in n-dimensional space (i.e., without restrictions such as orthogonality).

As previously illustrated, components of a tensor of rank r may be organized into linear combinations which form basis sets for the irreducible representations of the permutation group S_r. The dimension of the space is arbitrary; but if r exceeds n, the linear combinations of tensor components for Young diagrams containing more than n rows become identically zero. For example,

$$T_{[1^3]} \equiv 0 \qquad \text{for} \quad n = 2.$$

Permutation operators which act on the indices of the tensor components commute with the linear transformation (6.5-14). That is to say, if one permutes the indices of a tensor, the new tensor obeys the same transformation law as the original tensor. As a consequence, tensor components may be organized into linear combinations which form bases for irreducible representations of the permutation groups and such combinations will still satisfy (6.5-14), or, in other words they will transform among themselves. An alternative statement is that the entire space of rth rank tensors is reducible into subspaces consisting of tensors which belong to the irreducible representations of the permutation group S_r. Thus the construction of combinations of tensor components of rank r which transform among themselves under the operations of the linear group $GL(n)$ is equivalent to the construction of bases for irreducible representations of S_r. The latter procedure is accomplished by means of the Young operator (5.4-18) associated with each Young tableau and acting on the indices of $T_{i_1 i_2 \cdots i_r}$.

The tensor transformation law may be given a group theoretical interpretation in a manner analogous to that employed in the previous section in connection with second rank tensors. That is, under a general linear transformation of coordinates, a component of an rth rank tensor referred to the new coordinate system can be expressed, according to (6.5-14), as a linear combination of components of the rth rank tensor referred to the old coordinate system. This property automatically identifies the tensor components as a basis for a representation of the group $GL(n)$. However, the representation is reducible. When we construct tensors which belong to the various irreducible representations of the permutation group, such tensors will transform among themselves under the transformation (6.5-14) or, what is the same thing, under the operations of $GL(n)$. Hence these same tensors are bases for irreducible representations of the (continuous) general linear group in n dimensions. In short, the rth rank tensors of a given (permutation) symmetry form a basis for an irreducible representation of $GL(n)$.

In the previous section we saw that under a general linear coordinate transformation, second rank tensors decompose into symmetric and antisymmetric parts that are irreducible, provided no additional restrictions are imposed on the transformations but that further reduction is possible if the coordinate transformation is orthogonal. Analogous features may be expected in the general case when $GL(n)$ is restricted in various ways, i.e., when the matrices R (6.5-13) which describe the possible coordinate transformations are elements of a *subgroup* of $GL(n)$. A few examples of such subgroups are:

$SL(n)$	same as $GL(n)$ but with $\det R = +1$,
$U(n)$	$RR^\dagger = 1, \quad \lvert\det R\rvert = 1$
$SU(n)$	same as $U(n)$ but with $\det R = +1$,
$O(n)$	$R\tilde{R} = 1 \quad (a_{ij}$ real), $\det R = \pm 1$,
$O^+(n)$	same as $O(n)$ but with $\det R = +1$.

It may be shown (Hamermesh, 1962) that the irreducible representations of $GL(n)$ are also irreducible representations with respect to the subgroups $SL(n)$, $U(n)$, or $SU(n)$, but not with respect to $O(n)$ (or $O^+(n)$). In the latter cases, irreducible representations of $GL(n)$ may generally be decomposed with respect to the irreducible representations of $O(n)$ with still further reduction possible when we go to $O^+(n)$. This situation is reminiscent of the discussion in Section 5.3 where it was shown that the irreducible representations of $O^+(3)$ are generally reducible with respect to the irreducible representations of the subgroups of $O^+(3)$ as, for example, the finite point group O.

We now wish to connect this discussion with the angular momentum eigenfunctions $|jm\rangle$ which, for convenience, will be written as $\psi_m{}^j$. Recalling the discussion in Section 4.5, it was shown that under a coordinate rotation in three-dimensional space the $\psi_m{}^j$ transformed according to

$$P_R\psi_m{}^j \equiv \psi_m'^j = \sum_{m'} \psi_{m'}^j D_{m'm}^{(j)} \tag{6.5-15}$$

where $D^{(j)}$ is an irreducible representation of $O^+(3)$. Now the structure of (6.5-15) is essentially the same as that of (6.4-7), the latter being the transformation law for vectors. It is therefore tempting to regard $\psi_m{}^j$ as a vector. Just as in (6.4-7) the dimension of the space is given by the range of values over which the summation index is summed, so in (6.5-15) the dimension of the space is given by the range of values of m', namely $2j + 1$. Thus $\psi_m{}^j$ may be interpreted as a vector in an abstract space of dimension $2j + 1$. Furthermore, since the $D^{(j)}(R)$ are unitary and unimodular ($\det D^{(j)} = 1$), they are elements of the group $SU(2j + 1)$ which is a subgroup of the general linear group $GL(2j + 1)$. In other words the $\psi_m{}^j$ form a basis for a representation of $SU(2j + 1)$ and, because angular momentum eigenfunctions belonging to a particular value of j transform among themselves, the representation of $SU(2j + 1)$ whose basis consists of the functions $\psi_m{}^j$ is irreducible. This is yet another interpretation of the functions $\psi_m{}^j$ which have previously appeared as eigenfunctions of J^2 and J_z and as basis functions for the irreducible representations of $O^+(3)$. Using another language, the $\psi_m{}^j$ form a bridge between the group $O^+(3)$ and the group $SU(2j + 1)$ in the sense that a rotation in ordinary three-dimensional space induces a rotation in a unitary function space of dimension $2j + 1$.

The utility of introducing the group $SU(2j + 1)$ emerges when we consider systems of more than one particle. If $\psi^j_{m1}(1)$ is a wave function occupied by particle 1 and $\psi^j_{m_2}(2)$ is another wave function (but with the same value of j) occupied by an identical particle 2, a tensor of rank 2 can be constructed with components consisting of all possible products $\psi^j_{m_1}(1)\psi^j_{m_2}(2)$. More generally, if we have r identical particles, it is possible to construct a tensor of rank r with components

$$\psi^j_{m_1}(1)\psi^j_{m_2}(2) \cdots \psi^j_{m_r}(r). \tag{6.5-16}$$

From the previous discussion such (Cartesian) tensors are bases for representations of the group $GL(2j + 1)$ or $SU(2j + 1)$. Such tensor representations are, however, in general, reducible and in order to perform the reduction and to find the linear combinations of tensor components which transform irreducibly among themselves under the transformations $GL(2j + 1)$, we apply the Young operator to the appropriate tableaux as discussed above. Since irreducible representations of $GL(2j + 1)$ remain irreducible when the transformations are restricted to the subgroup $SU(2j + 1)$ we have a method of decomposing tensor representations of $SU(2j + 1)$ into its irreducible components. It should be clear that this procedure is entirely analogous to the decomposition of product representations as discussed in Sections 3.6 and 4.5.

An important subgroup of $SU(2j + 1)$ is $O^+(2j + 1)$. For future applications, our main interest is in the case with $j = l = 0, 1, 2, \ldots$. It may be shown (Hamermesh, 1962) that the irreducible representations of $O^+(2l + 1)$ are labeled by a set of nonnegative integers

$$(\mu_1, \mu_2, \ldots, \mu_l) \tag{6.5-17}$$

in which

$$\mu_1 \geqslant \mu_2 \geqslant \cdots \geqslant \mu_l, \qquad \mu_1 + \mu_2 + \cdots + \mu_l = r \tag{6.5-18}$$

where r is the rank of the tensor which serves as a basis for the irreducible representations of $O^+(2l + 1)$. When $l = 1$, the irreducible representations of $O^+(3)$ are specified by one parameter μ_1 which may assume the values $0, 1, 2, \ldots$. This is nothing more than a statement that irreducible tensors $\mathbf{T}^{(0)}, \mathbf{T}^{(1)}, \mathbf{T}^{(2)}, \ldots$ of rank $0, 1, 2, \ldots$ are bases for irreducible representations $D^{(0)}, D^{(1)}, D^{(2)}, \ldots$ of $O^+(3)$. When $l = 2$, the irreducible representations of $O^+(5)$ are specified by (μ_1, μ_2). Typical examples are (0 0), (1 0), (2 0), (1 1), (2 2), ... and the rank of the tensor that belongs to each of these irreducible representations is $0, 1, 2, 2, 4, \ldots$ respectively.

Methods for the decomposition of irreducible representations of $SU(2l+1)$ into irreducible representations of $O^+(2l+1)$ are described by Hamermesh (1962) and Judd (1963). Examples of such reductions are shown in Table 6.1. The basic feature is the invariance of the trace so that going from $SU(2l+1)$ to $O^+(2l+1)$ is equivalent to a contraction.

Finally, the irreducible representations of $O^+(2l+1)$ may be decomposed into irreducible representations of $O^+(3)$. These, too, are shown in Table 6.1.

TABLE 6.1

Decomposition of Selected Irreducible Representations of $SU(2l+1)$ into Irreducible Representations of $O^+(2l+1)$ and the Further Decomposition into Irreducible Representations of $O^+(3)$[a]

l	r	$SU(2l+1)$	$O^+(2l+1)$	$O^+(3)$
1	0	[0]	(0)	0
	1	[1]	(1)	1
	2	[2]	(0), (2)	0, 2
			(1)	1
	3	[21]	(1), (2)	1, 2
		[1 1 1]	(0)	0
2	0	[0]	(0 0)	0
	1	[1]	(1 0)	2
	2	[2]	(0 0)	0
			(2 0)	2, 4
		[1 1]	(1 1)	1, 3
	3	[2 1]	(1 0)	2
			(2 1)	1, 2, 3, 4, 5
		[1 1 1]	(1 1)	1, 3
2	4	[2 2]	(0 0)	0
			(2 0)	2, 4
			(2 2)	0, 2, 3, 4, 6
		[2 1 1]	(1 1)	1, 3
			(2 1)	1, 2, 3, 4, 5
		[1 1 1 1]	(1 0)	2
	5	[2 2 1]	(1 0)	2
			(2 1)	1, 2, 3, 4, 5
			(2 2)	0, 2, 3, 4, 6
		[2 1 1 1]	(1 1)	1, 3
			(2 0)	2, 4
		[1 1 1 1 1]	(0 0)	0

[a] The irreducible representations of $O^+(3)$ are indicated by the value of L in $D^{(L)}$. Only those representations of $SU(2l+1)$ are shown that are of interest for many-electron atomic states (Section 20.4).

CHAPTER 7

VECTOR FIELDS

7.1 Rotational Properties

Consider a vector field described by

$$\mathbf{V}(\mathbf{r}) = (x + 2y)\hat{\mathbf{i}} + x\hat{\mathbf{j}}. \tag{7.1-1}$$

Now let the coordinates be rotated by 45° in the positive sense about the z axis (Fig. 7.1). The two coordinate systems are related by

$$x' = \frac{1}{\sqrt{2}}(x + y), \qquad y' = \frac{1}{\sqrt{2}}(-x + y), \tag{7.1-2a}$$

or

$$\mathbf{r}' = R\mathbf{r} \tag{7.1-2b}$$

with

$$R = \frac{1}{\sqrt{2}}\begin{pmatrix} 1 & 1 \\ -1 & 1 \end{pmatrix}. \tag{7.1-2c}$$

The unit vectors also transform according to (7.1-2b) and (7.1-2c) so that

$$\hat{\mathbf{i}}' = \frac{1}{\sqrt{2}}(\hat{\mathbf{i}} + \hat{\mathbf{j}}), \qquad \hat{\mathbf{j}}' = \frac{1}{\sqrt{2}}(-\hat{\mathbf{i}} + \hat{\mathbf{j}}). \tag{7.1-3}$$

The inverse transformations corresponding to (7.1-2) and (7.1-3) are

$$\begin{aligned} x &= \frac{1}{\sqrt{2}}(x' - y'), \qquad & \hat{\mathbf{i}} &= \frac{1}{\sqrt{2}}(\hat{\mathbf{i}}' - \hat{\mathbf{j}}'), \\ y &= \frac{1}{\sqrt{2}}(x' + y'), \qquad & \hat{\mathbf{j}} &= \frac{1}{\sqrt{2}}(\hat{\mathbf{i}}' + \hat{\mathbf{j}}'). \end{aligned} \tag{7.1-4}$$

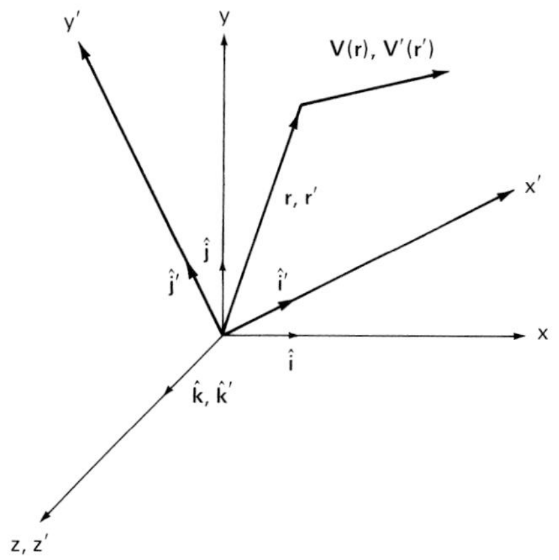

FIG. 7.1 Coordinate transformations for a vector field.

Substituting (7.1-4) into (7.1-1) we obtain

$$\mathbf{V}'(\mathbf{r}') = 2x'\hat{\mathbf{i}}' - (x' + y')\hat{\mathbf{j}}' \tag{7.1-5}$$

where $\mathbf{V}'(\mathbf{r}')$ is the description of the vector field (7.1-1) relative to the new coordinate system.

As in Section 2.1, we are interested in constructing an operator P_R that will transform (7.1-1) into (7.1-5). Such an operator is defined by

$$P_R\mathbf{V}(\mathbf{r}) = R\mathbf{V}(R^{-1}\mathbf{r}). \tag{7.1-6}$$

To see how this works in the example just cited we have

$$R = \frac{1}{\sqrt{2}}\begin{pmatrix} 1 & 1 \\ -1 & 1 \end{pmatrix}, \qquad R^{-1} = \frac{1}{\sqrt{2}}\begin{pmatrix} 1 & -1 \\ 1 & 1 \end{pmatrix},$$

$$R^{-1}\mathbf{r} = \frac{1}{\sqrt{2}}\begin{pmatrix} 1 & -1 \\ 1 & 1 \end{pmatrix}\begin{pmatrix} x \\ y \end{pmatrix} = \frac{1}{\sqrt{2}}\begin{pmatrix} x - y \\ x + y \end{pmatrix},$$

$$V_x(\mathbf{r}) = x + 2y,$$

$$V_x(R^{-1}\mathbf{r}) = \frac{1}{\sqrt{2}}(x - y) + \sqrt{2}(x + y) = \frac{1}{\sqrt{2}}(3x + y),$$

$$V_y(\mathbf{r}) = x,$$

$$V_y(R^{-1}\mathbf{r}) = \frac{1}{\sqrt{2}}(x - y),$$

$$R\mathbf{V} = \frac{1}{\sqrt{2}}\begin{pmatrix} 1 & 1 \\ -1 & 1 \end{pmatrix}\begin{pmatrix} V_x \\ V_y \end{pmatrix} = \frac{1}{\sqrt{2}}\begin{pmatrix} V_x + V_y \\ -V_x + V_y \end{pmatrix}.$$

Therefore

$$P_R V_x(\mathbf{r}) = R V_x(R^{-1}\mathbf{r}) = \frac{1}{\sqrt{2}}[V_x(R^{-1}\mathbf{r}) + V_y(R^{-1}\mathbf{r})]$$

$$= \frac{1}{\sqrt{2}}\left[\frac{1}{\sqrt{2}}(3x + y) + \frac{1}{\sqrt{2}}(x - y)\right] = 2x, \qquad (7.1\text{-}7a)$$

$$P_R V_y(\mathbf{r}) = R V_y(R^{-1}\mathbf{r}) = \frac{1}{\sqrt{2}}[-V_x(R^{-1}\mathbf{r}) + V_y(R^{-1}\mathbf{r})]$$

$$= \frac{1}{\sqrt{2}}\left[-\frac{1}{\sqrt{2}}(3x + y) + \frac{1}{\sqrt{2}}(x - y)\right] = -(x + y). \qquad (7.1\text{-}7b)$$

Equations (7.1-7a) and (7.1-7b) provide the form for the components of the transformed vector (7.1-5); thus the operator defined in (7.1-6) is capable of generating the desired transformation.

When the coordinate system is rotated through an infinitesimal angle $\delta\alpha$ about the z axis,

$$R(\delta\alpha, z) = \begin{pmatrix} 1 & \delta\alpha & 0 \\ -\delta\alpha & 1 & 0 \\ 0 & 0 & 1 \end{pmatrix}, \qquad (7.1\text{-}8)$$

$$R^{-1}(\delta\alpha, z)\mathbf{r} = \begin{pmatrix} 1 & -\delta\alpha & 0 \\ \delta\alpha & 1 & 0 \\ 0 & 0 & 1 \end{pmatrix}\begin{pmatrix} x \\ y \\ z \end{pmatrix} = \begin{pmatrix} x - y\,\delta\alpha \\ x\,\delta\alpha + y \\ z \end{pmatrix}, \qquad (7.1\text{-}9)$$

and

$$R(\delta\alpha, z)V = \begin{pmatrix} 1 & \delta\alpha & 0 \\ -\delta\alpha & 1 & 0 \\ 0 & 0 & 1 \end{pmatrix}\begin{pmatrix} V_x \\ V_y \\ V_z \end{pmatrix} = \begin{pmatrix} V_x + V_y\,\delta\alpha \\ -V_x\,\delta\alpha + V_y \\ V_z \end{pmatrix}. \qquad (7.1\text{-}10)$$

Therefore

$$\begin{aligned} P_R(\delta\alpha, z)V_x(\mathbf{r}) &= R V_x(R^{-1}\mathbf{r}) \\ &= V_x(R^{-1}\mathbf{r}) + \delta\alpha\, V_y(R^{-1}\mathbf{r}) \\ &= V_x(x - y\,\delta\alpha, x\,\delta\alpha + y, z) + \delta\alpha\, V_y(x - y\,\delta\alpha, x\,\delta\alpha + y, z). \end{aligned} \qquad (7.1\text{-}11)$$

If one writes to first order,

$$P_R V_x(\mathbf{r}) = V_x(\mathbf{r}) + \frac{\partial V_x}{\partial x}\,dx + \frac{\partial V_x}{\partial y}\,dy + \frac{\partial V_x}{\partial z}\,dz, \qquad (7.1\text{-}12)$$

a comparison of (7.1-11) with (7.1-12) yields

$$\begin{aligned} P_R(\delta\alpha, z)V_x(\mathbf{r}) &= V_x(\mathbf{r}) + \frac{\partial V_x}{\partial x}(-y\,\delta\alpha) + \frac{\partial V_x}{\partial y}(x\,\delta\alpha) + V_y\,\delta\alpha \\ &= V_x(\mathbf{r}) + \delta\alpha\left(x\frac{\partial}{\partial y} - y\frac{\partial}{\partial x}\right)V_x + V_y\,\delta\alpha \\ &= (1 + i\,\delta\alpha\,L_z)V_x + V_y\,\delta\alpha. \end{aligned} \tag{7.1-13a}$$

Similarly,

$$P_R(\delta\alpha, z)V_y(\mathbf{r}) = (1 + i\,\delta\alpha\,L_z)V_y - \delta\alpha\,V_x, \tag{7.1-13b}$$

$$P_R(\delta\alpha, z)V_z(\mathbf{r}) = (1 + i\,\delta\alpha\,L_z)V_z. \tag{7.1-13c}$$

The three expressions in (7.1-13) may be combined into

$$P_R(\delta\alpha, z)\mathbf{V}(\mathbf{r}) = (1 + i\,\delta\alpha\,L_z)\mathbf{V}(\mathbf{r}) + \delta\alpha\begin{pmatrix} 0 & 1 & 0 \\ -1 & 0 & 0 \\ 0 & 0 & 0 \end{pmatrix}\begin{pmatrix} V_x(\mathbf{r}) \\ V_y(\mathbf{r}) \\ V_z(\mathbf{r}) \end{pmatrix} \tag{7.1-14}$$

and, on defining

$$S_z = \begin{pmatrix} 0 & -i & 0 \\ i & 0 & 0 \\ 0 & 0 & 0 \end{pmatrix} \tag{7.1-15}$$

and

$$J_z = L_z + S_z, \tag{7.1-16}$$

Eq. (7.1-14) takes the form

$$P_R(\delta\alpha, z)\mathbf{V}(\mathbf{r}) = (1 + i\,\delta\alpha\,J_z)\mathbf{V}(\mathbf{r}). \tag{7.1-17}$$

It is instructive to compare (7.1-17) with (2.2-8); the two expressions are very similar in form. The transformation of the scalar function is governed by the operator L_z whereas the transformation of the vector function is governed by the operator J_z which is the sum of L_z and the new operator S_z whose properties remain to be explored.

By considering rotations about x and y, matrices analogous to S_z are generated. They are

$$S_x = \begin{pmatrix} 0 & 0 & 0 \\ 0 & 0 & -i \\ 0 & i & 0 \end{pmatrix}, \qquad S_y = \begin{pmatrix} 0 & 0 & i \\ 0 & 0 & 0 \\ -i & 0 & 0 \end{pmatrix}. \tag{7.1-18a}$$

Similarly, corresponding to (7.1-16) one obtains

$$J_x = L_x + S_x, \qquad J_y = L_y + S_y \tag{7.1-18b}$$

The three matrices S_x, S_y, and S_z are manifestly Hermitian; they also satisfy the commutation rules

$$\mathbf{S} \times \mathbf{S} = i\mathbf{S} \tag{7.1-19}$$

from which it may be inferred that the components of **S** qualify as angular momentum operators. The eigenvalues of S_z are readily determined from the secular equation

$$\begin{vmatrix} -\lambda & -i & 0 \\ i & -\lambda & 0 \\ 0 & 0 & -\lambda \end{vmatrix} = 0 \tag{7.1-20}$$

which yields $\lambda = 1, 0, -1$. For the eigenvalue $\lambda = 1$, the eigenvalue equation

$$\begin{pmatrix} 0 & -i & 0 \\ i & 0 & 0 \\ 0 & 0 & 0 \end{pmatrix} \begin{pmatrix} a_1 \\ a_2 \\ a_3 \end{pmatrix} = \begin{pmatrix} a_1 \\ a_2 \\ a_3 \end{pmatrix}$$

provides the normalized eigenvector

$$\hat{\mathbf{e}}_{+1} = -\frac{1}{\sqrt{2}} \begin{pmatrix} 1 \\ i \\ 0 \end{pmatrix} \qquad (\lambda = 1). \tag{7.1-21}$$

In similar fashion it is found that

$$\hat{\mathbf{e}}_0 = \begin{pmatrix} 0 \\ 0 \\ 1 \end{pmatrix} \quad (\lambda = 0); \qquad \hat{\mathbf{e}}_{-1} = \frac{1}{\sqrt{2}} \begin{pmatrix} 1 \\ -i \\ 0 \end{pmatrix} \quad (\lambda = -1). \tag{7.1-22}$$

These are just the unit vectors in the spherical basis (6.1-22) and are related to the unit vectors $\hat{\mathbf{e}}_x$, $\hat{\mathbf{e}}_y$, and $\hat{\mathbf{e}}_z$ along the coordinate axes by (6.1-23). Other relations among the unit vectors are given in Appendix 4.

One may also construct the operator S^2 defined by

$$S^2 = S_x^{\,2} + S_y^{\,2} + S_z^{\,2}. \tag{7.1-23}$$

Combining (7.1-15) and (7.1-18)

$$S^2 = \begin{pmatrix} 2 & 0 & 0 \\ 0 & 2 & 0 \\ 0 & 0 & 2 \end{pmatrix} \tag{7.1-24}$$

and, from the general properties of angular momentum operators, it is expected that $\hat{\mathbf{e}}_{+1}$, $\hat{\mathbf{e}}_0$, and $\hat{\mathbf{e}}_{-1}$ will also be eigenvectors of S^2. Thus, from

(7.1-21)–(7.1-23), we have

$$S^2\hat{\mathbf{e}}_q = 2\hat{\mathbf{e}}_q = 1(1+1)\hat{\mathbf{e}}_q \qquad (q = 1, 0, -1). \tag{7.1-25}$$

In an alternative notation one may write

$$|SM_s\rangle = \begin{cases} |1\ 1\rangle = \hat{\mathbf{e}}_{+1} \\ |1\ 0\rangle = \hat{\mathbf{e}}_0 \\ |1\ -1\rangle = \hat{\mathbf{e}}_{-1} \end{cases} \tag{7.1-26}$$

in which case

$$\begin{aligned} S_z|SM_s\rangle &= M_s|SM_s\rangle \qquad (M_s = 1, 0, -1), \\ S^2|SM_s\rangle &= S(S+1)|SM_s\rangle = 1(1+1)|SM_s\rangle. \end{aligned} \tag{7.1-27}$$

In summary, the transformation properties of a vector field are governed by an orbital angular momentum operator **L**, a spin angular momentum operator **S**, and that the two operators are coupled according to the relation

$$\mathbf{J} = \mathbf{L} + \mathbf{S}. \tag{7.1-28}$$

On the basis of (7.1-27) it is seen that a vector field has unit spin ($S = 1$). On the other hand, as was shown in Section 2.2, the transformation of a scalar field is describable entirely on the basis of an orbital angular momentum operator **L**, or in other words, the spin of a scalar field is zero. Considerations such as these retain their significance even when the fields are quantized as, for example, in the case of photons arising from the quantization of an electromagnetic field.

7.2 Vector Spherical Harmonics

It was shown in the previous section that the transformation of a vector field is characterized by an angular momentum operator **J** which arises from the coupling of an orbital angular momentum operator **L** and a spin angular momentum operator **S**. The simultaneous eigenfunctions of L^2 and L_z are the spherical harmonics $Y_{lm}(\theta, \varphi)$ and the simultaneous eigenfunctions of S^2 and S_z are $|SM_s\rangle$ with $S = 1$ and $M_s = 1, 0, -1$. It is then possible to construct simultaneous eigenfunctions of J^2 and J_z according to the general rules for the coupling of angular momentum eigenfunctions (1.7-23). Adopting the notation

$$\begin{aligned} j_1 &= l & m_1 &= m \\ j_2 &= S = 1 & m_2 &= M_s = 1, 0, -1 \\ j &= J & m &= M \\ |j_1 m_1\rangle &= Y_{lm}(\theta, \varphi) \\ |j_2 m_2\rangle &= |SM_s\rangle = \hat{\mathbf{e}}_q = \hat{\mathbf{e}}_{+1}, \hat{\mathbf{e}}_0, \hat{\mathbf{e}}_{-1} \end{aligned} \tag{7.2-1}$$

one obtains

$$|JM\rangle \equiv |l1\ JM\rangle = \sum_{mq} Y_{lm}\hat{\mathbf{e}}_q \langle l1\ mq|l1\ JM\rangle$$

$$= (-1)^{1-l-M}\sqrt{2J+1}\sum_{mq}\begin{pmatrix} l & 1 & J \\ m & q & -M \end{pmatrix} Y_{lm}\hat{\mathbf{e}}_q. \quad (7.2\text{-}2)$$

Equation (7.2-2) is the definition of a *vector spherical harmonic* usually designated $\mathbf{Y}_{JlM}$; that is,

$$\mathbf{Y}_{JlM} \equiv |JM\rangle.$$

Since $M = m + q$, the double sum in (7.2-2) may be replaced by the single sum

$$\mathbf{Y}_{JlM} = (-1)^{1-l-M}\sqrt{2J+1}\sum_{q}\begin{pmatrix} l & 1 & J \\ M-q & q & -M \end{pmatrix} Y_{l\,M-q}\hat{\mathbf{e}}_q \quad (7.2\text{-}3)$$

with the following properties:

$$J = l+1, l, l-1, \qquad M = m + q = J, J-1, \ldots, -J, \quad (7.2\text{-}4)$$

$$L^2\mathbf{Y}_{JlM} = l(l+1)\mathbf{Y}_{JlM}, \qquad S^2\mathbf{Y}_{JlM} = 1(1+1)\mathbf{Y}_{JlM} = 2\mathbf{Y}_{JlM},$$
$$J^2\mathbf{Y}_{JlM} = J(J+1)\mathbf{Y}_{JlM}, \qquad J_z\mathbf{Y}_{JlM} = M\mathbf{Y}_{JlM}. \quad (7.2\text{-}5)$$

The vector spherical harmonic $\mathbf{Y}_{llm}$ with $J = l$ and $m = M = l, l-1, \ldots, -l$ is a particularly important one in electromagnetic theory. For this case the $3j$ symbol will vanish unless the triangle condition $\triangle(l1l)$ is satisfied; the possible values of l are then $l = 1, 2, 3, \ldots$ but not $l = 0$. Using the formulas in Table 1.5 one may write

$$\mathbf{Y}_{llm} = -\sqrt{\frac{l(l+1) - m(m-1)}{2l(l+1)}}\, Y_{l\,m-1}\hat{\mathbf{e}}_{+1} + \frac{m}{\sqrt{l(l+1)}} Y_{lm}\hat{\mathbf{e}}_0$$
$$+ \sqrt{\frac{l(l+1) - m(m+1)}{2l(l+1)}}\, Y_{l\,m+1}\hat{\mathbf{e}}_{-1}. \quad (7.2\text{-}6)$$

But

$$L_{\pm 1}Y_{lm} = \mp\frac{1}{\sqrt{2}}\sqrt{l(l+1) - m(m\pm 1)}\,Y_{l\,m\pm 1},$$

$$L_0 Y_{lm} = mY_{lm}, \qquad \mathbf{L} = -L_{+1}\hat{\mathbf{e}}_{-1} + L_0\hat{\mathbf{e}}_0 - L_{-1}\hat{\mathbf{e}}_{+1}.$$

Therefore

$$\mathbf{Y}_{llm} = \frac{1}{\sqrt{l(l+1)}}[-L_{-1}Y_{lm}\hat{\mathbf{e}}_{+1} + L_0 Y_{lm}\hat{\mathbf{e}}_0 - L_{+1}Y_{lm}\hat{\mathbf{e}}_{-1}]$$
$$= \frac{1}{\sqrt{l(l+1)}}\sum_q (-1)^q L_q Y_{lm}\hat{\mathbf{e}}_{-q} = \frac{\mathbf{L}Y_{lm}}{\sqrt{l(l+1)}}. \quad (7.2\text{-}7)$$

It is observed that $\mathbf{Y}_{llm}$ is orthogonal to the radius vector $\mathbf{r}$ since

$$\mathbf{r}\cdot\mathbf{Y}_{llm} = \mathbf{r}\cdot\mathbf{L}\frac{Y_{lm}}{\sqrt{l(l+1)}} = \mathbf{r}\cdot\mathbf{r}\times\mathbf{p}\frac{Y_{lm}}{\hbar\sqrt{l(l+1)}} = 0. \tag{7.2-8}$$

A number of vector spherical harmonics are listed in Table 7.1. For example,

$$\mathbf{Y}_{1\,1\,0} = \sqrt{\frac{3}{8\pi}}\, i\sin\theta(-\sin\varphi\,\hat{\mathbf{e}}_x + \cos\varphi\,\hat{\mathbf{e}}_y). \tag{7.2-9a}$$

If $\mathbf{Y}_{1\,1\,0}$ were to acquire the significance of an electric field, we would say that the point of observation lies on a radius vector whose direction is specified by (θ, φ) and the electric field at the point of observation is specified by the components in the direction of the unit vectors $\hat{\mathbf{e}}_x$, $\hat{\mathbf{e}}_y$, $\hat{\mathbf{e}}_z$. Thus, for $\mathbf{Y}_{1\,1\,0}$, it is evident from (7.2-9a) that the electric field whose direction is that of the polarization vector, is perpendicular to both the z axis and the radius vector. Let us now examine $Y_{1\,1\,\pm1}$ which, from Table 7.1, may be written as

$$\mathbf{Y}_{1\,1\,\pm1} = \mp\sqrt{\frac{3}{8\pi}}\left(\frac{\pm\sin\theta}{\sqrt{2}}\,e^{\pm i\varphi}\,\hat{\mathbf{e}}_0 + \cos\theta\,\hat{\mathbf{e}}_{\pm1}\right). \tag{7.2-9b}$$

Suppose $\theta = 90^\circ$ so that the radius vector lies in the xy plane. The field is parallel to $\hat{\mathbf{e}}_0$; hence we have a linearly polarized field along the z axis, perpendicular to the radius vector. When, however $\theta = 0$, the field direction is given by $\hat{\mathbf{e}}_{\pm1}$. The interpretation of such a field is based on the property

$$\hat{\mathbf{e}}_q{}^* \times \hat{\mathbf{e}}_{q'} = iq\hat{\mathbf{e}}_0\,\delta_{qq'} \qquad (q, q' = 1, 0, -1) \tag{7.2-10}$$

where $\hat{\mathbf{e}}_0$ is the unit vector in the z direction. With $q = +1$ the cross product is a vector parallel to the z axis whereas the cross product with $q = -1$ is antiparallel to the z axis. For this reason, fields specified by $\hat{\mathbf{e}}_{+1}$ or $\hat{\mathbf{e}}_{-1}$ are said to have positive or negative helicity, respectively. In the context of wave propagation $\hat{\mathbf{e}}_{+1}$ corresponds to a counter clockwise rotation of the polarization vector when the observer is facing the oncoming beam. In optical terminology this is a left circularly polarized wave. Hence positive and negative helicity correspond to left and right circular polarization (see also Section 22.1). Finally, when $0 < \theta < 90^\circ$, the field in elliptically polarized.

One often needs the scalar product of a vector spherical harmonic with itself, i.e., the absolute square. To compute this quantity, it is recalled that, according to (6.1-18),

$$\begin{aligned}\mathbf{A}\cdot\mathbf{A} &= \sum_q (-1)^q A_q A_{-q}. \\ &= \sum_q A_q{}^* A_q \qquad (q = 1, 0, -1) \\ &= |A|^2.\end{aligned} \tag{7.2-11}$$

TABLE 7.1

Vector Spherical Harmonics[a]

$$\mathbf{Y}_{010} = \sqrt{\frac{1}{3}}\, Y_{1\,-1}\hat{\mathbf{e}}_{+1} - \sqrt{\frac{1}{3}}\, Y_{1\,0}\hat{\mathbf{e}}_0 + \sqrt{\frac{1}{3}}\, Y_{1\,1}\hat{\mathbf{e}}_{-1}$$

$$\mathbf{Y}_{101} = Y_{0\,0}\hat{\mathbf{e}}_{+1}, \qquad \mathbf{Y}_{100} = Y_{0\,0}\hat{\mathbf{e}}_0, \qquad \mathbf{Y}_{10\,-1} = Y_{0\,0}\hat{\mathbf{e}}_{-1}$$

$$\mathbf{Y}_{111} = -\sqrt{\frac{1}{2}}\, Y_{1\,0}\hat{\mathbf{e}}_{+1} + \sqrt{\frac{1}{2}}\, Y_{1\,1}\hat{\mathbf{e}}_0$$

$$= -\sqrt{\frac{3}{8\pi}}\left(\frac{\sin\theta}{\sqrt{2}}\, e^{i\varphi}\hat{\mathbf{e}}_0 + \cos\theta\,\hat{\mathbf{e}}_{+1}\right)$$

$$= \sqrt{\frac{3}{8\pi}}\left(-\frac{\sin\theta}{\sqrt{2}}\, e^{i\varphi}\hat{\mathbf{e}}_z + \cos\theta\frac{\hat{\mathbf{e}}_x + i\hat{\mathbf{e}}_y}{\sqrt{2}}\right)$$

$$= -\frac{1}{4}\sqrt{\frac{3}{\pi}}\, e^{i\varphi}\,(\hat{\imath}_2 - i\cos\theta\,\hat{\imath}_3)$$

$$\mathbf{Y}_{110} = -\sqrt{\frac{1}{2}}\, Y_{1\,-1}\hat{\mathbf{e}}_{+1} + \sqrt{\frac{1}{2}}\, Y_{1\,1}\hat{\mathbf{e}}_{-1}$$

$$= -\frac{1}{4}\sqrt{\frac{3}{\pi}}\,[\sin\theta(e^{i\varphi}\hat{\mathbf{e}}_{-1} + e^{i\varphi}\hat{\mathbf{e}}_{+1})$$

$$= \sqrt{\frac{3}{8\pi}}\, i\sin\theta(-\sin\varphi\,\hat{\mathbf{e}}_x + \cos\varphi\,\hat{\mathbf{e}}_y)$$

$$= \sqrt{\frac{3}{8\pi}}\, i\sin\theta\,\hat{\imath}_3$$

$$\mathbf{Y}_{11\,-1} = -\sqrt{\frac{1}{2}}\, Y_{1\,-1}\hat{\mathbf{e}}_0 + \sqrt{\frac{1}{2}}\, Y_{1\,0}\hat{\mathbf{e}}_{-1}$$

$$= \sqrt{\frac{3}{8\pi}}\left(-\frac{\sin\theta}{\sqrt{2}}\, e^{-i\varphi}\hat{\mathbf{e}}_0 + \cos\theta\,\hat{\mathbf{e}}_{-1}\right)$$

$$= \sqrt{\frac{3}{8\pi}}\left(-\frac{\sin\theta}{\sqrt{2}}\, e^{-i\varphi}\,\hat{\mathbf{e}}_z + \cos\theta\,\frac{\hat{\mathbf{e}}_x - i\hat{\mathbf{e}}_y}{\sqrt{2}}\right)$$

$$= -\frac{1}{4}\sqrt{\frac{3}{\pi}}\, e^{-i\varphi}\,(\hat{\imath}_2 + i\cos\theta\,\hat{\imath}_3)$$

$$\mathbf{Y}_{121} = \sqrt{\frac{1}{10}}\, Y_{2\,0}\hat{\mathbf{e}}_{+1} - \sqrt{\frac{3}{10}}\, Y_{2\,1}\hat{\mathbf{e}}_0 + \sqrt{\frac{3}{5}}\, Y_{2\,2}\hat{\mathbf{e}}_{-1}$$

$$= \frac{1}{4}\sqrt{\frac{1}{2\pi}}\,(3\cos^2\theta - 1)\,\hat{\mathbf{e}}_{+1} + \frac{3}{4}\sqrt{\frac{1}{\pi}}\cos\theta\sin\theta\, e^{i\varphi}\hat{\mathbf{e}}_0 + \frac{3}{4}\sqrt{\frac{1}{2\pi}}\sin^2\theta\, e^{2i\varphi}\,\hat{\mathbf{e}}_{-1}$$

$$= \frac{1}{8}\sqrt{\frac{1}{\pi}}\,(3\sin^2\theta\, e^{2i\varphi} - 3\cos^2\theta + 1)\hat{\mathbf{e}}_x$$

$$- \frac{i}{8}\sqrt{\frac{1}{\pi}}\,(3\sin^2\theta\, e^{2i\varphi} + 3\cos^2\theta - 1)\hat{\mathbf{e}}_y + \frac{3}{4}\sqrt{\frac{1}{\pi}}\cos\theta\sin\theta\, e^{i\varphi}\hat{\mathbf{e}}_z$$

[a] The unit vectors $\hat{\imath}_1, \hat{\imath}_2, \hat{\imath}_3$ are shown in Fig. A4.1.

TABLE 7.1 (continued)

$$\mathbf{Y}_{1\,2\,0} = \sqrt{\frac{3}{10}}\, Y_{2\,-}\hat{\mathbf{e}}_{+1} - \sqrt{\frac{2}{5}}\, Y_{2\,0}\hat{\mathbf{e}}_0 + \sqrt{\frac{3}{10}}\, Y_{2\,1}\hat{\mathbf{e}}_{-1}$$

$$= \frac{3}{4}\sqrt{\frac{1}{2\pi}} \cos\theta \sin\theta\, e^{-i\varphi}\, \hat{\mathbf{e}}_{+1} - \frac{1}{4}\sqrt{\frac{2}{\pi}}\,(3\cos^2\theta - 1)\hat{\mathbf{e}}_0 - \frac{3}{4}\sqrt{\frac{1}{2\pi}} \cos\theta \sin\theta\, e^{i\varphi}\hat{\mathbf{e}}_{-1}$$

$$= \frac{3}{4}\sqrt{\frac{1}{\pi}} \cos\theta \sin\theta(\hat{\mathbf{e}}_x \cos\varphi - \hat{\mathbf{e}}_y \sin\varphi) - \sqrt{\frac{1}{8\pi}}\,(3\cos^2\theta - 1)\hat{\mathbf{e}}_z$$

$$\mathbf{Y}_{1\,2\,-1} = \sqrt{\frac{3}{5}}\, Y_{2\,-2}\hat{\mathbf{e}}_{+1} - \sqrt{\frac{3}{10}}\, Y_{2\,-1}\hat{\mathbf{e}}_0 + \sqrt{\frac{1}{10}}\, Y_{2\,0}\hat{\mathbf{e}}_{-1}$$

$$= \frac{3}{4}\sqrt{\frac{1}{2\pi}} \sin^2\theta\, e^{-2i\varphi}\hat{\mathbf{e}}_{+1} - \frac{3}{4}\sqrt{\frac{1}{\pi}} \cos\theta \sin\theta\, e^{-i\varphi}\hat{\mathbf{e}}_0 + \frac{1}{4}\sqrt{\frac{1}{2\pi}}\,(3\cos^2\theta - 1)\hat{\mathbf{e}}_{-1}$$

$$= -\frac{1}{8}\sqrt{\frac{1}{\pi}}\,(3\sin^2\theta\, e^{-2i\varphi} - 3\cos^2\theta + 1)\hat{\mathbf{e}}_x$$

$$-\frac{i}{8}\sqrt{\frac{1}{\pi}}\,(3\sin^2\theta\, e^{-2i\varphi} + 3\cos^2\theta - 1)\hat{\mathbf{e}}_y - \frac{3}{4}\sqrt{\frac{1}{\pi}} \cos\theta \sin\theta\, e^{-i\varphi}\hat{\mathbf{e}}_z$$

$$\mathbf{Y}_{2\,1\,2} = Y_{1\,1}\hat{\mathbf{e}}_{+1}$$

$$\mathbf{Y}_{2\,1\,1} = \sqrt{\frac{1}{2}}\,[Y_{1\,0}\hat{\mathbf{e}}_{+1} + Y_{1\,1}\hat{\mathbf{e}}_0]$$

$$\mathbf{Y}_{2\,1\,0} = \sqrt{\frac{1}{6}}\, Y_{1\,-1}\hat{\mathbf{e}}_{+1} + \sqrt{\frac{2}{3}}\, Y_{1\,0}\hat{\mathbf{e}}_0 + \sqrt{\frac{1}{6}}\, Y_{1\,1}\hat{\mathbf{e}}_{-1}$$

$$\mathbf{Y}_{2\,1\,-1} = \sqrt{\frac{1}{2}}\,[Y_{1-1}\hat{\mathbf{e}}_0 + Y_{1\,0}\hat{\mathbf{e}}_{-1}]$$

$$\mathbf{Y}_{2\,1\,-2} = Y_{1\,-1}\hat{\mathbf{e}}_{-1}$$

$$\mathbf{Y}_{2\,2\,2} = -\sqrt{\frac{1}{3}}\, Y_{2\,1}\hat{\mathbf{e}}_{+1} + \sqrt{\frac{2}{3}}\, Y_{2\,2}\hat{\mathbf{e}}_0$$

$$\mathbf{Y}_{2\,2\,1} = -\sqrt{\frac{1}{2}}\, Y_{2\,0}\hat{\mathbf{e}}_{+1} + \sqrt{\frac{1}{6}}\, Y_{2\,1}\hat{\mathbf{e}}_0 + \sqrt{\frac{1}{3}}\, Y_{2\,2}\hat{\mathbf{e}}_{-1}$$

$$\mathbf{Y}_{2\,2\,0} = -\sqrt{\frac{1}{2}}\, Y_{2\,-1}\hat{\mathbf{e}}_{+1} + \sqrt{\frac{1}{2}}\, Y_{2\,1}\hat{\mathbf{e}}_{-1}$$

$$\mathbf{Y}_{2\,2\,-1} = \sqrt{\frac{1}{3}}\, Y_{2\,-2}\hat{\mathbf{e}}_{+1} - \sqrt{\frac{1}{6}}\, Y_{2-1}\hat{\mathbf{e}}_0 + \sqrt{\frac{1}{2}}\, Y_{2\,0}\hat{\mathbf{e}}_{-1}$$

$$\mathbf{Y}_{2\,2\,-2} = -\sqrt{\frac{2}{3}}\, Y_{2\,-2}\hat{\mathbf{e}}_0 + \sqrt{\frac{1}{3}}\, Y_{2\,-1}\hat{e}_{-1}$$

$$\mathbf{Y}_{3\,2\,\pm 3} = Y_{2\,\pm 2}e_{\pm 1}$$

$$\mathbf{Y}_{3\,2\,\pm 2} = \sqrt{\frac{2}{3}}\, Y_{2\,\pm 1}\hat{\mathbf{e}}_{\pm 1} + \sqrt{\frac{1}{3}}\, Y_{2\,\pm 2}\hat{\mathbf{e}}_0$$

$$\mathbf{Y}_{3\,2\,\pm 1} = \sqrt{\frac{2}{5}}\, Y_{2\,0}\hat{\mathbf{e}}_{\pm 1} + \sqrt{\frac{8}{15}}\, Y_{2\,\pm 1}\hat{\mathbf{e}}_0 + \sqrt{\frac{1}{15}}\, Y_{2\,\pm 2}\hat{\mathbf{e}}_{\mp 1}$$

$$\mathbf{Y}_{3\,2\,0} = \sqrt{\frac{1}{5}}\, Y_{2\,-1}\hat{\mathbf{e}}_{+1} + \sqrt{\frac{3}{5}}\, Y_{2\,0}\hat{\mathbf{e}}_0 + \sqrt{\frac{1}{5}}\, Y_{2\,1}\hat{\mathbf{e}}_{-1}$$

Therefore using (7.2-6) together with the orthogonality property of the unit vectors $\hat{\mathbf{e}}_q$ as given by (6.1-26) one obtains

$$\mathbf{Y}_{llm} \cdot \mathbf{Y}_{llm} = \frac{l(l+1) - m(m-1)}{2l(l+1)} |Y_{l\,m-1}(\theta, \varphi)|^2 + \frac{m^2}{l(l+1)} |Y_{lm}(\theta, \varphi)|^2 + \frac{l(l+1) - m(m+1)}{2l(l+1)} |Y_{l\,m+1}(\theta, \varphi)|^2 \tag{7.2-12}$$

which is seen to be independent of the azimuthal angle φ. For illustration and for future reference one may cite:

$$\mathbf{Y}_{1\,1\,\pm1} \cdot \mathbf{Y}_{1\,1\,\pm1} = \frac{1}{2}(|Y_{1\,0}|^2 + |Y_{1\,\pm1}|^2) = \frac{3}{16\pi}(\cos^2\theta + 1), \tag{7.2-13}$$

$$\mathbf{Y}_{1\,1\,0} \cdot \mathbf{Y}_{1\,1\,0} = \frac{1}{2}(|Y_{1\,-1}|^2 + |Y_{1\,+1}|^2) = \frac{3}{8\pi}\sin^2\theta, \tag{7.2-14}$$

$$\mathbf{Y}_{2\,2\,\pm2} \cdot \mathbf{Y}_{2\,2\,\pm2} = \frac{1}{3}|Y_{2\,\pm1}|^2 + \frac{2}{3}|Y_{2\,\pm2}|^2 = \frac{5}{16\pi}(1 - \cos^4\theta), \tag{7.2-15}$$

$$\mathbf{Y}_{2\,2\,\pm1} \cdot \mathbf{Y}_{2\,2\,\pm1} = \frac{1}{2}|Y_{2\,0}|^2 + \frac{1}{6}|Y_{2\,\pm1}|^2 + \frac{1}{3}|Y_{2\,\pm2}|^2 = \frac{5}{16\pi}(1 - 3\cos^2\theta + 4\cos^4\theta), \tag{7.2-16}$$

$$\mathbf{Y}_{2\,2\,0} \cdot \mathbf{Y}_{2\,2\,0} = \frac{1}{2}|Y_{2\,-1}|^2 + \frac{1}{2}|Y_{2\,+1}|^2 = \frac{15}{8\pi}\sin^2\theta\cos^2\theta. \tag{7.2-17}$$

These angular distributions are shown in Fig. 7.2. Those with $l = 1$ are known as dipole fields and those with $l = 2$ are quadrupole fields.

When $\theta = 0$, the spherical harmonics vanish except for Y_{l0}, which has the value

$$Y_{l0} = \sqrt{\frac{2l+1}{4\pi}}. \tag{7.2-18}$$

In that case, (7.2-12) simplifies to

$$\mathbf{Y}_{llm} \cdot \mathbf{Y}_{llm} = \mathbf{Y}_{l\,l\,\pm1} \cdot \mathbf{Y}_{l\,l\,\pm1} = \frac{2l+1}{8\pi}; \tag{7.2-19}$$

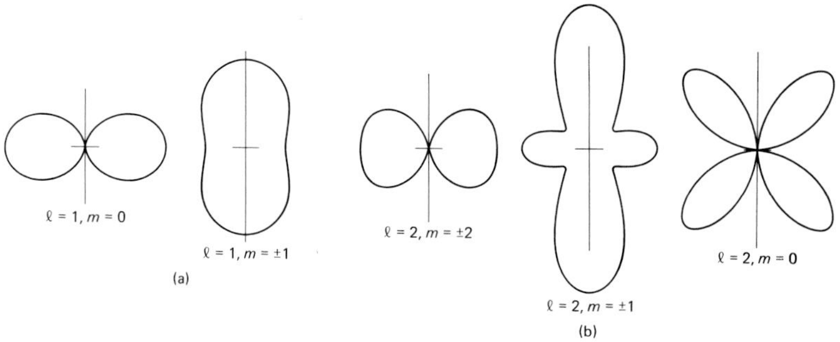

FIG. 7.2 Angular distribution in (a) dipole and (b) quadrupole fields. (Jackson, 1962.)

that is, when $\theta = 0$ $\mathbf{Y}_{llm} \cdot \mathbf{Y}_{llm}$ has a nonvanishing value only when $m = \pm 1$.

Referring to (7.2-3) it is seen that the only spherical harmonics that appear in the definition of $\mathbf{Y}_{JlM}$ are those of order l. $\mathbf{Y}_{JlM}$ therefore has odd parity when l is odd and even parity when l is even. Also, as a consequence of the orthogonality property of the spherical harmonics and the unit vectors, the vector spherical harmonics satisfy the orthogonality relation

$$\int \mathbf{Y}_{JlM} \cdot \mathbf{Y}_{J'l'M'} \sin\theta \, d\theta \, d\varphi = \delta_{JJ'} \delta_{ll'} \delta_{MM'} \tag{7.2-20}$$

with

$$\mathbf{Y}^*_{JlM} = (-1)^{J+l+1+M} \mathbf{Y}_{Jl\,-M} \tag{7.2-21}$$

7.3 Plane Wave Expansion

An important application of vector spherical harmonics is in the expansion of plane waves. According to (1.2-28),

$$e^{i\mathbf{k}\cdot\mathbf{r}} = 4\pi \sum_{l=0}^{\infty} \sum_{m=-l}^{l} i^l j_l(kr) Y_{lm}(\theta_r, \varphi_r) Y^*_{lm}(\theta_k, \varphi_k)$$

where $j_l(kr)$ is a spherical Bessel function (Appendix 5) and (θ_r, φ_r), (θ_k, φ_k) are the angular coordinates of the position and wave vectors, respectively (Fig. 1.4). If it is assumed that the propagation is along the z direction, $\theta_k = 0$ in which case only Y^*_{l0} is nonzero and is given by (7.2-18). The sum over m then consists of only one term, namely the one with $m = 0$. Therefore

$$e^{ikz} = \sum_{l=0}^{\infty} i^l \sqrt{4\pi(2l+1)}\, j_l(kr) Y_{l0}(\theta) \tag{7.3-1}$$

where $\theta \equiv \theta_r$. Now let

$$\mathbf{A} = \hat{\mathbf{e}}_x e^{ikz}. \tag{7.3-2}$$

Apart from an exponential time factor, (7.3-2) represents a plane wave propagating along the z axis and polarized along the x axis. By writing

$$\mathbf{A} = -\frac{1}{\sqrt{2}}(\mathbf{A}_{+1} - \mathbf{A}_{-1}) \equiv -\frac{1}{\sqrt{2}}(\hat{\mathbf{e}}_{+1} - \hat{\mathbf{e}}_{-1})e^{ikz}$$

we have

$$\mathbf{A}_{\pm 1} = \hat{\mathbf{e}}_{\pm 1}e^{ikz}, \tag{7.3-3}$$

which, we saw, represents right and left circularly polarized waves. Using (7.3-1),

$$\mathbf{A}_{\pm 1} = \sum_l i^l \sqrt{4\pi(2l+1)}\, j_l(kr) Y_{l0}(\theta)\hat{\mathbf{e}}_{\pm 1}. \tag{7.3-4}$$

We shall now express $Y_{l0}(\theta)\hat{\mathbf{e}}_{\pm 1}$ in terms of vector spherical harmonics. From (1.5-54),

$$|j_1 j_2 m_1 m_2\rangle = \sum_{JM} (-1)^{j_2 - j_1 - M}\sqrt{2J+1}\begin{pmatrix} j_1 & j_2 & J \\ m_1 & m_2 & -M \end{pmatrix} |j_1 j_2 JM\rangle$$

with $M = m_1 + m_2$. In the present case

$$|j_1 m_1\rangle = Y_{l0}, \qquad |j_2 m_2\rangle = |1q\rangle = \hat{\mathbf{e}}_q = \hat{\mathbf{e}}_{\pm 1}, \qquad |j_1 j_2 JM\rangle = \mathbf{Y}_{JlM} = \mathbf{Y}_{Jl\,\pm 1}.$$

Therefore

$$Y_{l0}\hat{\mathbf{e}}_{+1} = \sum_J (-1)^{-l}\sqrt{2J+1}\begin{pmatrix} l & 1 & J \\ 0 & 1 & -1 \end{pmatrix}\mathbf{Y}_{Jl\,+1}. \tag{7.3-5}$$

Since the possible values of J are $l-1, l, l+1$,

$$\begin{aligned} Y_{l0}\hat{\mathbf{e}}_{+1} = (-1)^{-l}\Bigg[&\sqrt{2l-1}\begin{pmatrix} l & 1 & l-1 \\ 0 & 1 & -1 \end{pmatrix}\mathbf{Y}_{l-1\,l\,1} \\ &+ \sqrt{2l+1}\begin{pmatrix} l & 1 & l \\ 0 & 1 & -1 \end{pmatrix}\mathbf{Y}_{ll1} + \sqrt{2l+3}\begin{pmatrix} l & 1 & l+1 \\ 0 & 1 & -1 \end{pmatrix}\mathbf{Y}_{l+1\,l\,1}\Bigg]. \end{aligned} \tag{7.3-6}$$

From Table 1.5,

$$\begin{aligned} \begin{pmatrix} l & 1 & l-1 \\ 0 & 1 & -1 \end{pmatrix} &= (-1)^l \sqrt{\frac{l-1}{2(2l+1)(2l-1)}}, \\ \begin{pmatrix} l & 1 & l \\ 0 & 1 & -1 \end{pmatrix} &= (-1)^{l+1} \sqrt{\frac{1}{2(2l+1)}}, \\ \begin{pmatrix} l & 1 & l+1 \\ 0 & 1 & -1 \end{pmatrix} &= (-1)^l \sqrt{\frac{l+2}{2(2l+1)(2l+3)}}. \end{aligned} \tag{7.3-7}$$

Therefore

$$Y_{l0}\hat{\mathbf{e}}_{+1} = \sqrt{\frac{(l-1)}{2(2l+1)}}\mathbf{Y}_{l-1\,l1} - \frac{1}{\sqrt{2}}\mathbf{Y}_{ll1} + \sqrt{\frac{(l+2)}{2(2l+1)}}\mathbf{Y}_{l+1\,l1}, \quad (7.3\text{-}8)$$

and in the same fashion

$$Y_{l0}\hat{\mathbf{e}}_{-1} = \sqrt{\frac{(l-1)}{2(2l+1)}}\mathbf{Y}_{l-1\,l\,-1} + \frac{1}{\sqrt{2}}\mathbf{Y}_{ll\,-1} + \sqrt{\frac{(l+2)}{2(2l+1)}}\mathbf{Y}_{l+1\,l\,-1}. \quad (7.3\text{-}9)$$

Note that when $\mathbf{Y}_{JlM} = \mathbf{Y}_{l-1\,l1}$ the value of M is 1; therefore J cannot be smaller than 1. But since $J = l - 1$ we must have $l \geqslant 2$. Similarly, when $\mathbf{Y}_{JlM} = \mathbf{Y}_{ll1}$, it is necessary that $l \geqslant 1$, but when $\mathbf{Y}_{JlM} = \mathbf{Y}_{l+1\,l1}$, the smallest value of l can be zero. With (7.3-8) and (7.3-9) substituted into (7.3-4) we obtain the plane wave expansion in terms of vector spherical harmonics:

$$\begin{aligned}\mathbf{A}_{\pm 1} = \hat{\mathbf{e}}_{\pm 1}e^{ikz} &= \sum_{l=0}^{\infty} i^l\sqrt{2\pi(l+2)}\,j_l(kr)\mathbf{Y}_{l+1\,l\,\pm 1} \\ &\mp \sum_{l=1}^{\infty} i^l\sqrt{2\pi(2l+1)}\,j_l(kr)\mathbf{Y}_{ll\,\pm 1} \\ &+ \sum_{l=2}^{\infty} i^l\sqrt{2\pi(l-1)}\,j_l(kr)\mathbf{Y}_{l-1\,l\,\pm 1}. \end{aligned} \quad (7.3\text{-}10)$$

For physical applications one is usually interested in the first few terms of expansion (7.3-10), particularly under conditions where $kr \ll 1$. The leading term is the one containing $j_0(kr)$, that is,

$$\mathbf{A}^{(1)}_{\pm 1} = \sqrt{4\pi}\,j_0(kr)\mathbf{Y}_{1\,0\,\pm 1}. \quad (7.3\text{-}11)$$

From Table 7.1 and the asymptotic values of the spherical Bessel functions ((A5-17) in Appendix 5)

$$\mathbf{Y}_{1\,0\,\pm 1} = Y_{00}\hat{\mathbf{e}}_{\pm 1} = \frac{1}{\sqrt{4\pi}}\hat{\mathbf{e}}_{\pm 1},$$

$$j_0(kr) = 1 \qquad (kr \ll 1),$$

so that (7.3-11) becomes

$$\mathbf{A}^{(1)}_{\pm 1} = \hat{\mathbf{e}}_{\pm 1}. \quad (7.3\text{-}12)$$

Hence the approximation of $\mathbf{A}_{\pm 1}$ by $\mathbf{A}^{(1)}_{\pm 1}$ is equivalent to setting $e^{ikz} = 1$ in (7.3-3).

The next higher-order terms are those containing $j_1(kr)$. These are

$$\mathbf{A}^{(2)}_{\pm 1} = i\sqrt{6\pi}\,j_1(kr)\mathbf{Y}_{2\,1\,\pm 1} \mp i\sqrt{6\pi}\,j_1(kr)\mathbf{Y}_{1\,1\,\pm 1}. \quad (7.3\text{-}13)$$

Again using Table 7.1 and (A5-17)

$$\mathbf{Y}_{2\,1\,\pm1} = \sqrt{\tfrac{1}{2}}[Y_{10}\hat{\mathbf{e}}_{\pm1} + Y_{1\,\pm1}\hat{\mathbf{e}}_0],$$
$$\mathbf{Y}_{1\,1\,\pm1} = \sqrt{\tfrac{1}{2}}[\mp Y_{10}\hat{\mathbf{e}}_{\pm1} \pm Y_{1\,\pm1}\hat{\mathbf{e}}_0],$$
$$j_1(kr) = \tfrac{1}{3}kr \qquad (kr \ll 1).$$

Therefore

$$\mathbf{A}^{(2)}_{\pm1} = i\sqrt{\frac{4\pi}{3}}\,krY_{10}\hat{\mathbf{e}}_{\pm1} = i\sqrt{\frac{4\pi}{3}}\,kr\sqrt{\frac{3}{4\pi}}\cos\theta\,\hat{\mathbf{e}}_{\pm1}$$
$$= ikr\cos\theta\,\hat{\mathbf{e}}_{\pm1} = ikz\hat{\mathbf{e}}_{\pm1} \tag{7.3-14}$$

and

$$\mathbf{A}^{(1)}_{\pm1} + \mathbf{A}^{(2)}_{\pm1} = (1 + ikz)\hat{\mathbf{e}}_{\pm1} \tag{7.3-15}$$

which corresponds to the first two terms in the power series expansion of the exponential in (7.3-3).

7.4 Multipole Expansion of the Electromagnetic Field

The free-field Maxwell equations are

$$\begin{aligned} \nabla\cdot\mathbf{B} &= 0, & \nabla\cdot\mathbf{E} &= 0,\\ \nabla\times\mathbf{B} &= \frac{1}{c}\frac{\partial\mathbf{E}}{\partial t}, & \nabla\times\mathbf{E} &= -\frac{1}{c}\frac{\partial\mathbf{B}}{\partial t} \end{aligned} \tag{7.4-1}$$

with

$$\mathbf{B} = \nabla\times\mathbf{A}, \qquad \mathbf{E} = -\frac{1}{c}\frac{\partial\mathbf{A}}{\partial t} \tag{7.4-2}$$

where **A** is the vector potential. The time dependence of the field vectors may be taken to be of the form $e^{-i\omega t}$ in which case (7.4-1) and (7.4-2) become

$$\begin{aligned} \nabla\cdot\mathbf{B} &= 0, & \nabla\cdot\mathbf{E} &= 0,\\ \nabla\times\mathbf{B} &= -ik\mathbf{E}, & \nabla\times\mathbf{E} &= ik\mathbf{B}, \end{aligned} \tag{7.4-3}$$

$$\mathbf{B} = \nabla\times\mathbf{A}, \qquad \mathbf{E} = ik\mathbf{A}, \tag{7.4-4}$$

where $k = \omega/c$. From (7.4-3) and (7.4-4) and the assumption that $\nabla\cdot\mathbf{A} = 0$ (Coulomb gauge) one obtains the vector Helmholtz equations:

$$(\nabla^2 + \mathrm{k}^2)\mathbf{B}(\mathbf{r}) = 0, \tag{7.4-5a}$$

$$(\nabla^2 + k^2)\mathbf{E}(\mathbf{r}) = 0, \tag{7.4-5b}$$

$$(\nabla^2 + k^2)\mathbf{A}(\mathbf{r}) = 0. \tag{7.4-5c}$$

As a step toward the solution of (7.4-5) we consider first the scalar Helmholtz equation

$$(\nabla^2 + k^2)\psi(\mathbf{r}) = 0. \tag{7.4-6}$$

Equation (7.4-6) when written in spherical coordinates is separable into two equations one of which depends on r only and the other on θ and φ. Solutions to the angular equation are expressible in terms of the spherical harmonics. We therefore write

$$\psi(r, \theta, \varphi) = \sum_{lm} f_l(r) Y_{lm}(\theta, \varphi), \tag{7.4-7}$$

and on substituting (7.4-7) into (7.4-6), one obtains the radial equation

$$\left[\frac{d^2}{dr^2} + \frac{1}{r}\frac{d}{dr} + k^2 - \frac{(l + \frac{1}{2})^2}{r^2}\right] u_l(r) = 0 \tag{7.4-8}$$

where

$$u_l(r) = \sqrt{r} f_l(r).$$

Equation (7.4-8) is a form of the Bessel equation (Appendix 5). Therefore the solutions to (7.4-6) may be written

$$\psi(\mathbf{r}) = \sum_{lm} \left[A_{lm}^{(1)} h_l^{(1)}(kr) + A_{lm}^{(2)} h_l^{(2)}(kr)\right] Y_{lm}(\theta, \varphi) \tag{7.4-9}$$

where $h_l^{(1)}(kr)$ and $h_l^{(2)}(kr)$ are spherical Hankel functions of the first and second kind, respectively, and the coefficients $A_{lm}^{(1)}$ and $A_{lm}^{(2)}$ are determined from the boundary conditions. In place of the Hankel functions one may write $\psi(\mathbf{r})$ in (7.4-9) in terms of $j_l(kr)$ and $n_l(kr)$ since

$$h_l^{(1)}(kr) = j_l(kr) + i n_l(kr), \qquad h_l^{(2)}(kr) = j_l(kr) - i n_l(kr). \tag{7.4-10}$$

Thus any product of the form $f_l(kr) Y_{lm}(\theta, \varphi)$, where $f_l(kr)$ stands for any one of the spherical Bessel functions $h_l^{(1)}(kr), h_l^{(2)}(kr), j_l(kr)$, or $n_l(kr)$ will satisfy (7.4-6), i.e.,

$$(\nabla^2 + k^2) f_l(kr) Y_{lm}(\theta, \varphi) = 0. \tag{7.4-11}$$

Now consider the commutator $[\mathbf{L}, \nabla^2]$. In spherical coordinates

$$\nabla^2 = \frac{\partial^2}{\partial r^2} + \frac{2}{r}\frac{\partial}{\partial r} - \frac{L^2}{r^2},$$

and since the components of $\mathbf{L}$ operate only on angular coordinates and L^2 commutes with every component of $\mathbf{L}$, we have

$$[\mathbf{L}, \nabla^2] = 0. \tag{7.4-12}$$

Therefore

$$\mathbf{L}(\nabla^2 + k^2) f_l(kr) Y_{lm}(\theta, \varphi) = (\nabla^2 + k^2) f_l(kr) \mathbf{L} Y_{lm}(\theta, \varphi) = 0, \quad (7.4\text{-}13)$$

from which we conclude that $f_l(kr)\mathbf{L}Y_{lm}(\theta, \varphi)$ satisfies the vector Helmholtz equation (7.4-5). Using (7.2-7), a solution to (7.4-5a) may be written as

$$(\nabla^2 + k^2)\mathbf{B}_E(l, m, \mathbf{r}) = 0 \quad (7.4\text{-}14)$$

where

$$\mathbf{B}_E(l, m, \mathbf{r}) = f_l(kr)\mathbf{L}Y_{lm}(\theta, \varphi) = \sqrt{l(l+1)} f_l(kr)\mathbf{Y}_{llm}, \quad (7.4\text{-}15)$$

and the corresponding electric field, according to the Maxwell equations, is

$$\mathbf{E}_E(l, m, \mathbf{r}) = \frac{i}{k} \nabla \times \mathbf{B}_E(l, m, \mathbf{r}). \quad (7.4\text{-}16)$$

The transversality relation (7.2-8) implies that

$$\mathbf{r} \cdot \mathbf{B}_E(l, m, \mathbf{r}) = 0; \quad (7.4\text{-}17)$$

therefore an electromagnetic field characterized by (7.4-15)–(7.4-17) is known as a TM (transverse magnetic) field of order l. However, the transversality condition does not apply to $\mathbf{E}_E$ which may then have non-vanishing radial components. Such a field is therefore also designated as an *electric multipole* of order l. Since $f_l(kr)\mathbf{L}Y_{lm}(\theta, \varphi)$ is a solution to the vector Helmholtz equation it could equally well serve as a solution to (7.4-5b). We therefore have another pair of solutions

$$\mathbf{E}_M(l, m, \mathbf{r}) = f_l(kr)\mathbf{L}Y_{lm}(\theta, \varphi), \quad (7.4\text{-}18)$$

$$\mathbf{B}_M(l, m, \mathbf{r}) = \frac{-i}{k} \nabla \times \mathbf{E}_M(l, m, \mathbf{r}), \quad (7.4\text{-}19)$$

$$\mathbf{r} \cdot \mathbf{E}_M(l, m, \mathbf{r}) = 0, \quad (7.4\text{-}20)$$

and such a field is known as a *magnetic multipole* or a TE (transverse electric) field of order l. In this case $\mathbf{B}_M$ may have radial components.

The general solution to the vector Helmholtz equations for the electromagnetic field may now be expressed as a linear superposition of electric and magnetic multipoles of all orders:

$$\mathbf{B}(\mathbf{r}) = \sum_{lm} [a_E(l, m)\mathbf{B}_E(l, m, \mathbf{r}) + a_M(l, m)\mathbf{B}_M(l, m, \mathbf{r})], \quad (7.4\text{-}21)$$

$$\mathbf{E}(\mathbf{r}) = \sum_{lm} [a_E(l, m)\dot{\mathbf{E}}_E(l, m, \mathbf{r}) + a_M(l, m)\mathbf{E}_E(l, m, \mathbf{r})], \quad (7.4\text{-}22)$$

in which the coefficients depend on the sources and the boundary conditions.

The two types of multipoles are distinguished by the fact that the roles of the electric and magnetic fields in the two cases are interchanged. Thus from the fact that the parity of $\mathbf{Y}_{JlM}$ is $(-1)^l$ we see that the parities of $\mathbf{B}_\mathrm{E}(l, m, \mathbf{r})$ and $\mathbf{E}_\mathrm{M}(l, m, \mathbf{r})$, which we designate by $\pi[B_\mathrm{E}(l, m, \mathbf{r})]$ and $\pi[E_\mathrm{M}(l, m, \mathbf{r})]$, respectively, are both $(-1)^l$. Since $\mathbf{E}_\mathrm{E}$ is proportional to $\nabla \times \mathbf{B}_\mathrm{E}$ according to (7.4-3), the parity of $\mathbf{E}_\mathrm{E}(l, m, \mathbf{r})$ must be opposite to that of $\mathbf{B}_\mathrm{E}(l, m, \mathbf{r})$, or $(-1)^{l+1}$ because the curl operation transforms a polar vector into an axial vector and vice versa. The same applies to $\mathbf{B}_\mathrm{M}(l, m, \mathbf{r})$ whose parity is also $(-1)^{l+1}$. Also in view of the proportionality between $\mathbf{A}$ and $\mathbf{E}$ [Eq. (7.4-4)], the parity of $\mathbf{A}$ is the same as the parity of $\mathbf{E}$. Thus we have

$$\begin{aligned} \pi[\mathbf{B}_\mathrm{E}(l, m, \mathbf{r})] &= \pi[\mathbf{E}_\mathrm{M}(l, m, \mathbf{r})] = \pi[\mathbf{A}_\mathrm{M}(l, m, \mathbf{r})] = (-1)^l, \\ \pi[\mathbf{E}_\mathrm{E}(l, m, \mathbf{r})] &= \pi[\mathbf{B}_\mathrm{M}(l, m, \mathbf{r})] = \pi[\mathbf{A}_\mathrm{E}(l, m, \mathbf{r})] = (-1)^{l+1}. \end{aligned} \tag{7.4-23}$$

Since $\mathbf{Y}_{llm} = 0$ for $l = 0$ (Section 7.2), multipoles of order zero vanish identically. This ensures the transversality of the fields to the direction of propagation in an electromagnetic wave in free space.

It is customary to employ the following conventions and notations:

(1) The parity of an electromagnetic field π_em is defined as the parity of $\mathbf{B}$, i.e., $\pi_\mathrm{em} = \pi(\mathbf{B})$.

(2) The polarization of an electromagnetic field is in the direction of $\mathbf{E}$.

(3) The multipoles are named according to the values of 2^l; thus dipole ($l = 1, 2^l = 2$), quadrupole ($l = 2, 2^l = 4$), octapole ($l = 3, 2^l = 8$), etc. Additional designations are El for electric multipole of order l and Ml for magnetic multipole of order l. Thus E1, E2, . . . are electric dipole, quadrupole, . . . and M1, M2, . . . are magnetic dipole, quadrupole, ...

The angular distribution of a multiple may be calculated from the fact that the energy density of a radiation field is proportional to $\mathbf{E} \cdot \mathbf{E}$ or $\mathbf{B} \cdot \mathbf{B}$. Therefore for magnetic multipoles the angular distribution may be obtained from $\mathbf{E}_\mathrm{M}(l, m, \mathbf{r}) \cdot \mathbf{E}_\mathrm{M}(l, m, \mathbf{r})$ which is proportional to $\mathbf{Y}_{llm} \cdot \mathbf{Y}_{llm}$ whose general form is given by (7.2-12). For electric multipoles the angular distribution can be obtained from $\mathbf{B}_\mathrm{E}(l, m, \mathbf{r}) \cdot \mathbf{B}_\mathrm{E}(l, m, \mathbf{r})$ which again is proportional to $\mathbf{Y}_{llm} \cdot \mathbf{Y}_{llm}$. Thus the angular distribution is the same for electric and magnetic multipoles of the same order. For dipole and quadrupole radiation the patterns are shown in Fig. 7.2. One should note, however, that since $\mathbf{E}_\mathrm{M}(l, m, \mathbf{r})$ has the same form as $\mathbf{B}_\mathrm{E}(l, m, \mathbf{r})$ but $\mathbf{E}_\mathrm{E}(l, m, \mathbf{r})$ is proportional to $\nabla \times \mathbf{B}_\mathrm{E}(l, m, \mathbf{r})$, the polarizations of electric and magnetic multipoles of the same order are perpendicular to one another.

Let the vector potential be given by

$$\mathbf{A}_{\pm 1} = \hat{\mathbf{e}}_{\pm 1} e^{ikz} \tag{7.4-24}$$

as in (7.3-3). The expansion of $\mathbf{A}_{\pm 1}$ was given by (7.3-10). Our objective now is to examine the multipolarity of the electromagnetic fields associated with the first few terms in the expansion of $\mathbf{A}_{\pm 1}$.

According to (7.3-11), the leading term ($kr \ll 1$) is

$$\mathbf{A}_{\pm 1}^{(1)} = \sqrt{4\pi} j_0(kr)\mathbf{Y}_{1\,0\,\pm 1} = \hat{\mathbf{e}}_{\pm 1} \tag{7.4-25}$$

and

$$\mathbf{B}_{\pm 1}^{(1)} = \nabla \times \mathbf{A}_{\pm 1}^{(1)}.$$

To evaluate the curl of quantities such as those on the right side of (7.4-25) the following special formulas (Edmonds, 1960) are useful:

$$\nabla \times f_l(kr)\mathbf{Y}_{llm} = ik\left[-\sqrt{\frac{l}{2l+1}}\, f_{l+1}(kr)\mathbf{Y}_{l\,l+1\,m} + \sqrt{\frac{l+1}{2l+1}}\, f_{l-1}(kr)\mathbf{Y}_{l\,l-1\,m}\right] \tag{7.4-26a}$$

$$\nabla \times f_{l-1}(kr)\mathbf{Y}_{l\,l-1\,m} = -ikf_l(kr)\sqrt{\frac{l+1}{2l+1}}\, Y_{llm}. \tag{7.4-26b}$$

Thus

$$\mathbf{B}_{\pm 1}^{(1)} = -i\sqrt{\frac{8\pi}{3}}\, kj_1(kr)\mathbf{Y}_{1\,1\,\pm 1}. \tag{7.4-26c}$$

Since $l = 1$ the field is a dipole field. Also, since the parity of $\mathbf{Y}_{JlM}$ is $(-1)^l$, the parity of $\mathbf{B}_{\pm 1}^{(1)}$ is $(-1)^l = -1$. From Table 7.2 the electromagnetic field is further identified as E1, that is, electric dipole. We conclude that $\mathbf{A}_{\pm 1}^{(1)}$ in (7.4-25) is the vector potential for an electric dipole field. Following the

TABLE 7.2

Multipole Relations[a]

Multipole	Field vector	Proportional to	Parity	Electromagnetic field or photon parity
El	$\mathbf{B}_E(l, m, \mathbf{r})$	$f_l(kr)\mathbf{Y}_{llm}$	$(-1)^l$	
	$\mathbf{E}_E(l, m, \mathbf{r})$	$\nabla \times f_l(kr)\mathbf{Y}_{llm}$	$(-1)^{l+1}$	$(-1)^l$
	$\mathbf{A}_E(l, m, \mathbf{r})$	$\nabla \times f_l(kr)\mathbf{Y}_{llm}$	$(-1)^{l+1}$	
Ml	$\mathbf{B}_M(l, m, \mathbf{r})$	$\nabla \times f_l(kr)\mathbf{Y}_{llm}$	$(-1)^{l+1}$	
	$\mathbf{E}_M(l, m, \mathbf{r})$	$f_l(kr)\mathbf{Y}_{llm}$	$(-1)^l$	$(-1)^{l+1}$
	$\mathbf{A}_M(l, m, \mathbf{r})$	$f_l(kr)\mathbf{Y}_{llm}$	$(-1)^l$	

[a] The parity of an electromagnetic field is defined by the parity of **B**.

argument in Section 7.3 it is seen that the approximation whereby

$$\mathbf{A}_{\pm 1} = \hat{\mathbf{e}}_{\pm 1} e^{ikz}$$

is replaced by

$$\mathbf{A}^{(1)}_{\pm 1} = \hat{\mathbf{e}}_{\pm 1},$$

that is, $e^{ikz} = 1$ corresponds to the approximation whereby a plane wave is replaced by an electric dipole field.

The next term in expansion (7.3-10) is

$$\mathbf{A}^{(2)}_{\pm 1} = i\sqrt{6\pi} j_1(kr)[\mathbf{Y}_{2\,1\,\pm 1} \mp \mathbf{Y}_{1\,1\,\pm 1}] = ikz\hat{\mathbf{e}}_{\pm 1} \qquad (7.4\text{-}27)$$

as shown in (7.3-13) and (7.3-14). The term containing $\mathbf{Y}_{1\,1\,\pm 1}$ has a parity of -1. Since the vector potential $\mathbf{A}$ is proportional to the electric field $\mathbf{E}$, the parity of $\mathbf{E}$ is -1. With $l = 1$ such a field is identified (Table 7.2) as a magnetic dipole (M1) field. To identify the term containing $\mathbf{Y}_{2\,1\,\pm 1}$ we note from (7.4-26b) that

$$\nabla \times j_1(kr)\mathbf{Y}_{2\,1\,\pm 1} = -ikj_2(kr)\mathbf{Y}_{2\,2\,\pm 2}. \qquad (7.4\text{-}28)$$

Therefore $\nabla \times \mathbf{A}^{(2)}_{\pm 1}$ contains $\mathbf{Y}_{2\,2\,\pm 2}$ which means that there is a $\mathbf{B}$ field with $l = 2$ or $(-1)^l = +1$. From Table 7.2 such a field is identified as electric quadrupole (E2). Thus we have the result that $\mathbf{A}^{(2)}_{\pm 1}$ in (7.4-27) represents a combination of a magnetic dipole (M1) and an electric quadrupole (E2) field.

The results we have obtained may be summarized by writing

$$e^{ikz} = 1(\text{E1}) + ikz(\text{E2} + \text{M1}) + \cdots . \qquad (7.4\text{-}29)$$

In further detail the M1 part of the vector potential is

$$(\mathbf{A}^{(2)}_{\pm 1})_{\text{M1}} = \mp i\sqrt{6\pi} j_1(kr)\mathbf{Y}_{1\,1\,\pm 1} = i\sqrt{\frac{\pi}{3}}\, kr(Y_{10}\hat{\mathbf{e}}_{\pm 1} - Y_{1\,\pm 1}\hat{\mathbf{e}}_0) \quad (7.4\text{-}30)$$

in which the approximation

$$j_1(kr) = \tfrac{1}{3}kr \qquad (kr \ll 1)$$

has been made. The rectangular components are

$$(\mathbf{A}^{(2)}_x)_{\text{M1}} = i\frac{k}{2}(z\hat{\mathbf{e}}_x - x\hat{\mathbf{e}}_z), \qquad (\mathbf{A}^{(2)}_y)_{\text{M1}} = i\frac{k}{2}(z\hat{\mathbf{e}}_y - y\hat{\mathbf{e}}_z). \quad (7.4\text{-}31)$$

The E2 part of the vector potential is

$$(\mathbf{A}^{(2)}_{\pm 1})_{\text{E2}} = i\sqrt{6\pi} j_1(kr)\mathbf{Y}_{2\,1\,\pm 1} = i\sqrt{\frac{\pi}{3}}\, kr(Y_{10}\hat{\mathbf{e}}_{\pm 1} + Y_{1\,\pm 1}\hat{\mathbf{e}}_0), \quad (7.4\text{-}32)$$

and the rectangular components are

$$(\mathbf{A}_x^{(2)})_{\mathrm{E2}} = i\frac{k}{2}(z\hat{\mathbf{e}}_x + x\hat{\mathbf{e}}_z), \qquad (\mathbf{A}_y^{(2)})_{\mathrm{E2}} = i\frac{k}{2}(z\hat{\mathbf{e}}_y + y\hat{\mathbf{e}}_z). \tag{7.4-33}$$

We also note that

$$\mathbf{A}_x^{(2)} = (\mathbf{A}_x^{(2)})_{\mathrm{M1}} + (\mathbf{A}_x^{(2)})_{\mathrm{E2}} = ikz\hat{\mathbf{e}}_x, \qquad \mathbf{A}_y^{(2)} = (\mathbf{A}_y^{(2)})_{\mathrm{M1}} + (\mathbf{A}^{(2)})_{\mathrm{E2}} = ikz\hat{\mathbf{e}}_y. \tag{7.4-34}$$

PART II

Quantum-Mechanical Background

CHAPTER 8

SYMMETRY ELEMENTS OF THE HAMILTONIAN

8.1 Connection between Group Theory and Quantum Mechanics

In Section 2.1 we saw that if an operator $\mathscr{H}(\mathbf{r})$ is invariant under a coordinate transformation R, then

$$[P_R, \mathscr{H}(\mathbf{r})] = 0. \tag{8.1-1}$$

Let us now examine the consequences of this commutation relation for the solutions of the Schrödinger equation

$$\mathscr{H}(\mathbf{r})\psi_\nu(\mathbf{r}) = E\psi_\nu(\mathbf{r}) \qquad (\nu = 1, 2, \ldots, n) \tag{8.1-2}$$

where $\psi_\nu(\mathbf{r})$ is a member of an n-fold degenerate set of independent eigenfunctions all of which belong to the same eigenvalue E. If we operate on both sides of (8.1-2) with P_R and make use of (8.1-1),

$$P_R\mathscr{H}(\mathbf{r})\psi_\nu(\mathbf{r}) = \mathscr{H}(\mathbf{r})P_R\psi_\nu(\mathbf{r}) = EP_R\psi_\nu(\mathbf{r}). \tag{8.1-3}$$

Thus, if $\psi_\nu(\mathbf{r})$ is an eigenfunction of $\mathscr{H}(\mathbf{r})$ with eigenvalue E, then $P_R\psi_\nu(\mathbf{r})$ is also an eigenfunction of $\mathscr{H}(\mathbf{r})$ with the same eigenvalue E. Since there are just n independent eigenfunctions with eigenvalue E, $P_R\psi_\nu(\mathbf{r})$ can only be some linear combination of the n independent eigenfunctions which means that

$$P_R\psi_\nu(\mathbf{r}) = \sum_\mu \psi_\mu(\mathbf{r})\Gamma_{\mu\nu}(R) \tag{8.1-4}$$

where $\Gamma_{\mu\nu}(R)$ are constants. A relation of the type of (8.1-4) will be obtained when P_R acts on each of the n independent eigenfunctions.

It will be recognized that (8.1-4) is of the same form as (3.4-9). Therefore, if the coordinate transformations under which the Hamiltonian is invariant constitute a group G, which may be called the symmetry group of the Hamiltonian, the interpretation of (8.1-4) is that the n independent eigenfunctions of the Hamiltonian, associated with the eigenvalue E, are basis functions for the n-dimensional representation Γ of the group G.

The representation Γ may be assumed to be irreducible. This would not be so if, for some reason, the eigenfunctions belonging to the eigenvalue E were to divide themselves into sets and within each set the eigenfunctions were to transform only among themselves. Such a situation, although possible, is highly improbable; irreducibility will therefore be assumed unless it is stated otherwise.

These considerations may be fruitfully employed to deduce useful information concerning the solutions to the Schrödinger equation from the transformation properties of the Hamiltonian. The general statement is that if a Hamiltonian is invariant under the operations of a group G, the eigenfunctions of the Hamiltonian may be classified according to the irreducible representations of G and the degeneracy of the eigenfunctions is equal to the dimensionality of the irreducible representation to which they belong. Thus each eigenvalue of the Hamiltonian is associated with some irreducible representation of G. Such symmetry classifications are of considerable practical importance because they exert a profound influence on the matrix elements of various operators that may represent physical interactions. Selection rules, relative intensities of transitions, mixing of states under perturbations, and removal of degeneracies are all affected by the symmetry properties of the pertinent states.

Group-theoretical deductions are usually quite easy to perform and the information so obtained concerning the solutions, although not complete, often contains the essential physics. On the other hand, detailed solutions of the Schrödinger equation usually require laborious procedures. In any event, group-theoretical considerations, when applicable, will always reduce the labor involved in obtaining detailed solutions. In the next few sections we shall examine several types of symmetry and the ensuing consequences.

8.2 Geometrical Symmetries

To illustrate the foregoing, suppose we have a hydrogen atom surrounded by three unit charges located at positions 1, 2, and 3 which form an equilateral triangle of side a; the proton is at the origin O and the electron at $\mathbf{r}$ (Fig. 8.1). The Hamiltonian of the electron is

$$\mathscr{H} = \mathscr{H}_0 + V \tag{8.2-1}$$

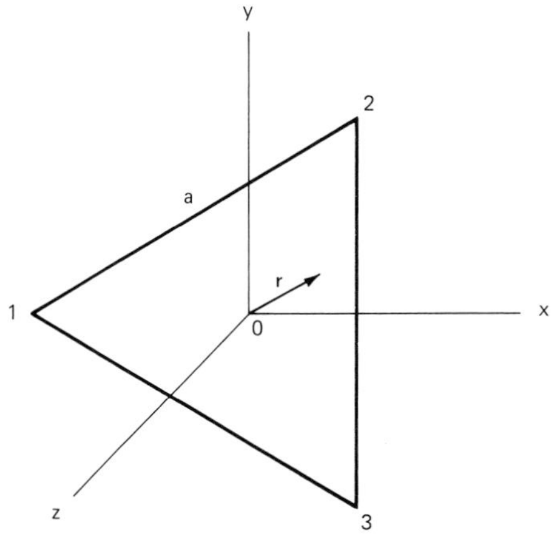

FIG. 8.1 Hydrogen atom surrounded by three charges located at the corners of an equilateral triangle.

with

$$\mathscr{H}_0 = -\frac{\hbar^2}{2m}\nabla^2 - \frac{Ze^2}{r},$$
$$V = e^2\left[\frac{1}{|\mathbf{r}-\mathbf{r}_1|} + \frac{1}{|\mathbf{r}-\mathbf{r}_2|} + \frac{1}{|\mathbf{r}-\mathbf{r}_3|}\right], \tag{8.2-2}$$

in which $|\mathbf{r}_1|$, $|\mathbf{r}_2|$, and $|\mathbf{r}_3|$ are the distances O1, O2, and O3, respectively, and

$$|\mathbf{r}-\mathbf{r}_2|^2 = \left(x - \frac{a}{2\sqrt{3}}\right)^2 + \left(y - \frac{a}{2}\right)^2 + z^2 \equiv f_2(\mathbf{r}), \tag{8.2-3a}$$

$$|\mathbf{r}-\mathbf{r}_3|^2 = \left(x - \frac{a}{2\sqrt{3}}\right)^2 + \left(y + \frac{a}{2}\right)^2 + z^2 \equiv f_3(\mathbf{r}), \tag{8.2-3b}$$

$$|\mathbf{r}-\mathbf{r}_1| = \left(x + \frac{a}{\sqrt{3}}\right)^2 + y^2 + z^2 \equiv f_1(\mathbf{r}). \tag{8.2-3c}$$

In (8.2-1), $\mathscr{H}_0$ is invariant under all three-dimensional rotations; hence it is invariant under the specific coordinate transformations which are the elements of the group D_3. The potential V remains to be investigated. Let us take, for example, the coordinate transformation D in (3.1-6) which is a positive rotation through 120°. We have

$$P_D f(\mathbf{r}) = f(D^{-1}\mathbf{r}),$$
$$P_D f(x, y, z) = f\left(-\frac{1}{2}x - \frac{\sqrt{3}}{2}y, \frac{\sqrt{3}}{2}x - \frac{1}{2}y, z\right). \tag{8.2-4}$$

Therefore

$$P_D f_2(\mathbf{r}) = \left(-\frac{x}{2} - \frac{\sqrt{3}}{2}y - \frac{a}{2\sqrt{3}}\right)^2 + \left(\frac{\sqrt{3}}{2}x - \frac{1}{2}y - \frac{a}{2}\right)^2 + z^2 = f_3(\mathbf{r}) \tag{8.2-5}$$

and

$$P_D f_3(\mathbf{r}) = f_1(\mathbf{r}), \tag{8.2-6}$$

$$P_D f_1(\mathbf{r}) = f_2(\mathbf{r}), \tag{8.2-7}$$

so that

$$P_D[f_1(\mathbf{r}) + f_2(\mathbf{r}) + f_3(\mathbf{r})] = f_1(\mathbf{r}) + f_2(\mathbf{r}) + f_3(\mathbf{r})$$

and

$$P_D V P_D^{-1} = V(D^{-1}\mathbf{r}) = V. \tag{8.2-8}$$

In other words, V in (8.2-2) is invariant under the coordinate transformation D. In similar fashion it may be verified that V is invariant under all the coordinate transformations that belong to the group D_3. From this information alone and a knowledge of the dimensionalities of the irreducible representations of D_3—which we obtain from a character table (Table 5.2)—it may immediately be established that the eigenvalues of $\mathscr{H}$ are either nondegenerate or, at most, twofold degenerate.

This example illustrates another important point. We have seen that $\mathscr{H}_0$ in (8.2-2) is invariant under all three-dimensional rotations. Therefore the eigenfunctions of $\mathscr{H}_0$ (exclusive of spin) are of the form $f(r)Y_{lm}(\theta, \varphi)$ which are basis functions for the irreducible representations $D^{(l)}$ of $O^+(3)$. The dimensionality of $D^{(l)}$ is $(2l + 1)$ which is also the degeneracy of $f(r)Y_{lm}(\theta, \varphi)$. The potential V is invariant only under a smaller set of rotations, namely, those which are elements of the group D_3. Therefore the symmetry group of $\mathscr{H}$ $(=\mathscr{H}_0 + V)$ is D_3 and, as we have seen, the eigenvalues of $\mathscr{H}$ are confined to being either nondegenerate or twofold degenerate. Thus the effect of adding V to $\mathscr{H}_0$ has been to reduce the symmetry of the total Hamiltonian and the consequence of the reduction in symmetry has been to remove degeneracies in whole or in part. This feature is also contained in the more formal statement that $D^{(l)}$, an irreducible representation of $O^+(3)$, may generally be decomposed into irreducible representations of a subgroup of $O^+(3)$. Thus if $l = 2$, for example, we saw (5.1-3) that

$$D^{(2)} = \Gamma^{(1)} + 2\Gamma^{(3)}$$

where $\Gamma^{(1)}$ and $\Gamma^{(3)}$ are, respectively, one- and two-dimensional irreducible representations of D_3.

Another important geometrical symmetry operation is inversion. Let i be an operator defined by

$$\mathbf{r}' = i\mathbf{r} \tag{8.2-9}$$

where

$$i = \begin{pmatrix} -1 & 0 & 0 \\ 0 & -1 & 0 \\ 0 & 0 & -1 \end{pmatrix}. \tag{8.2-10}$$

Under the coordinate transformation (8.2-9),

$$P_i f(x, y, z) = f(-x, -y, -z). \tag{8.2-11}$$

The inversion operator, together with the identity operator, are the sole elements of the group C_i which therefore has two one-dimensional representations (Table 5.2) known as the even and odd representations.

If a Hamiltonian is invariant with respect to inversion,

$$[\mathscr{H}, P_i] = 0. \tag{8.2-12}$$

This means that $\mathscr{H}$ is invariant under the operations of C_i and, as a consequence, the eigenfunctions of $\mathscr{H}$ must belong to the irreducible representations of C_i, i.e.,

$$P_i\psi = \pm\psi. \tag{8.2-13}$$

Thus all the eigenfunctions of a Hamiltonian that satisfies (8.2-12) may be classified as either even or odd under inversion. One also refers to such eigenfunctions as having either even or odd parity. We note that P_i commutes with the angular momentum operators:

$$[P_i, L_x] = [P_i, L_y] = [P_i, L_z] = [P_i, L^2] = 0 \tag{8.2-14}$$

8.3 Time Reversal and Kramers' Theorem

We define an operator K, called the *time reversal* or *Kramers'* operator, by

$$Kf(\mathbf{r}, t) = f^*(\mathbf{r}, -t). \tag{8.3-1}$$

If the Hamiltonian operator $\mathscr{H}(\mathbf{r})$ is a real function of $\mathbf{r}$,

$$K\mathscr{H}(\mathbf{r})\psi(\mathbf{r}, t) = \mathscr{H}^*(\mathbf{r})\psi^*(\mathbf{r}, -t) = \mathscr{H}(\mathbf{r})\psi^*(\mathbf{r}, -t) = \mathscr{H}(\mathbf{r})K\psi(\mathbf{r}, t). \tag{8.3-2}$$

Since

$$K\mathscr{H}(\mathbf{r})\psi(\mathbf{r}, t) = K\mathscr{H}(\mathbf{r})K^{-1}K\psi(\mathbf{r}, t) \tag{8.3-3}$$

we have

$$K\mathscr{H}K^{-1} = \mathscr{H} \qquad \text{or} \qquad [K, \mathscr{H}] = 0. \tag{8.3-4}$$

Thus a real Hamiltonian commutes with the Kramers' operator, or alternatively such a Hamiltonian is said to be invariant under time reversal. Provided $\mathscr{H}$ is real, the statement holds even if $\mathscr{H}$ depends on t (as well as $\mathbf{r}$), i.e., $\mathscr{H} = \mathscr{H}(\mathbf{r}, t)$.

If $\psi(\mathbf{r}, t)$ satisfies the time-dependent Schrödinger equation,

$$\mathscr{H}(\mathbf{r})\psi(\mathbf{r}, t) = i\hbar \frac{\partial}{\partial t} \psi(\mathbf{r}, t), \tag{8.3-5}$$

then

$$\begin{aligned} K\mathscr{H}(\mathbf{r})\psi(\mathbf{r}, t) &= Ki\hbar \frac{\partial}{\partial t} \psi(\mathbf{r}, t) \\ &= -i\hbar K \frac{\partial}{\partial t} \psi(\mathbf{r}, t) = -i\hbar \frac{\partial}{\partial(-t)} K\psi(\mathbf{r}, t) = i\hbar \frac{\partial}{\partial t} K\psi(\mathbf{r}, t), \end{aligned} \tag{8.3-6}$$

which shows that if $\psi(\mathbf{r}, t)$ is a solution to the time-dependent Schrödinger equation, then $K\psi(\mathbf{r}, t)$ is also a solution. We should therefore expect to encounter degeneracies associated with the invariance of the Hamiltonian under time reversal. This is another type of symmetry, distinct from any spatial symmetries.

For stationary states $K\psi(\mathbf{r}) = \psi^*(\mathbf{r})$; hence if ψ satisfies $\mathscr{H}\psi = E\psi$, then $\mathscr{H}\psi^* = E\psi^*$. From these two relations it is seen that if $\psi = u + iv$, then u and v separately satisfy the Schrödinger equation provided, of course, that $\mathscr{H}$ is real.

The components of the orbital angular momentum operator $\mathbf{L}$ (1.1-3) are pure imaginary quantities. Therefore $KL_z Y_{lm} = L_z^* Y_{lm}^* = -L_z K Y_{lm}$. But $KL_z Y_{lm} \equiv KL_z K^{-1} K Y_{lm}$; hence $KL_z K^{-1} = -L_z$ with similar relations for the other components. We then have

$$K\mathbf{L}K^{-1} = -\mathbf{L} \qquad \text{or} \qquad K\mathbf{L} + \mathbf{L}K \equiv \{K, \mathbf{L}\} = 0 \tag{8.3-7}$$

in which the braces are used to designate the *anticommutator* $K\mathbf{L} + \mathbf{L}K$. To maintain consistency, other angular momentum operators will be required to satisfy similar relations, i.e.,

$$K\mathbf{S}K^{-1} = -\mathbf{S}, \tag{8.3-8a}$$

$$K\mathbf{J}K^{-1} = -\mathbf{J}. \tag{8.3-8b}$$

If an *external* magnetic field $\mathbf{B}$ is present, the Hamiltonian will contain terms such as $\mathbf{L} \cdot \mathbf{B}$, $\mathbf{S} \cdot \mathbf{B}$, $\mathbf{J} \cdot \mathbf{B}$. In that event, as a result of (8.3-7) and (8.3-8), K

and $\mathscr{H}$ do not commute. It should be noted, however, that this does not apply to internal magnetic fields generated by moving electrons within the system as, for example, in the case of spin–orbit coupling or hyperfine interactions. Such interactions contain bilinear combinations of angular momentum operators so that K and $\mathscr{H}$ again commute.

The effect of K on spin functions may be deduced from (8.3-8a). Thus

$$S_z\alpha = \tfrac{1}{2}\alpha, \qquad KS_z\alpha = \tfrac{1}{2}K\alpha. \tag{8.3-9}$$

But, from (8.3-8a),

$$KS_z\alpha = -S_zK\alpha. \tag{8.3-10}$$

Comparing (8.3-9) with (8.3-10), it is seen that

$$S_zK\alpha = -\tfrac{1}{2}K\alpha. \tag{8.3-11}$$

Thus $K\alpha$ is an eigenfunction of S_z with the eigenvalue $-\frac{1}{2}$; in similar fashion

$$S_zK\beta = \tfrac{1}{2}K\beta \tag{8.3-12}$$

indicating that $K\beta$ is an eigenfunction of S_z with the eigenvalue $\frac{1}{2}$. It is seen that $K\alpha$ behaves like β and $K\beta$ like α. We may therefore satisfy (8.3-11) and (8.3-12) by choosing

$$K\alpha = c_1\beta, \qquad K\beta = c_2\alpha, \tag{8.3-13}$$

where c_1 and c_2 are arbitrary constants. Further restrictions on the constants are obtained by considering the effect of K on other spin operators. Thus

$$S_x\alpha = \tfrac{1}{2}\beta, \qquad S_x\beta = \tfrac{1}{2}\alpha,$$

and

$$KS_x\alpha = \tfrac{1}{2}K\beta = \tfrac{1}{2}c_2\alpha.$$

But

$$KS_x\alpha = -S_xK\alpha = -S_xc_1\beta = -\tfrac{1}{2}c_1\alpha.$$

Therefore $c_1 = -c_2$ and in place of (8.3-13) one may write

$$K\alpha = c\beta, \qquad K\beta = -c\alpha \tag{8.3-14}$$

with the consequence that

$$K^2\alpha = Kc\beta = c^*K\beta = -c^*c\alpha. \tag{8.3-15}$$

It is desirable to preserve the normalization of the spin functions under the time reversal operation. We therefore set $c^*c = 1$ so that c may be any complex number of modulus unity. In that case

$$K^2 = -1. \tag{8.3-16}$$

A common choice for the constant is $c = i$ so that

$$K\alpha = i\beta, \qquad K\beta = -i\alpha \tag{8.3-17}$$

or, written more compactly,

$$K|m_s\rangle = (-1)^{m_s}|-m_s\rangle. \tag{8.3-18}$$

With $|lm_l\rangle \equiv Y_{lm_l}$ we have, from (8.3-1), that

$$KY_{lm_l} = Y^*_{lm_l} = (-1)^{m_l} Y_{l\,-m_l} \tag{8.3-19a}$$

or

$$K|lm_l\rangle = (-1)^{m_l}|l\,-m_l\rangle \tag{8.3-19b}$$

A typical one-electron wave function is a product of a radial function, a spherical harmonic, and a spin function. It may be symbolized by $\psi = |nlm_lm_s\rangle$ in which the principal quantum number n characterizes the radial part, l and m_l, the spherical harmonic, and m_s the spin function. Combining (8.3-18) and (8.3-19),

$$K\psi = K|nlm_lm_s\rangle = (-1)^{m_l+m_s}|nl - m_l - m_s\rangle \tag{8.3-20}$$

and

$$K^2\psi = -\psi. \tag{8.3-21}$$

We must now investigate more closely the relationship between ψ and $K\psi$. Are they linearly independent? Suppose they are not. Then it is possible to write

$$K\psi = a\psi \tag{8.3-22}$$

where a is a constant. Repeating the operation with K,

$$K^2\psi = K(K\psi) = K(a\psi) = a^*K\psi = a^*a\psi. \tag{8.3-23}$$

But, from (8.3-16), $K^2 = -1$ which would then produce the nonsensical result that $a^*a = -1$. We conclude that ψ and $K\psi$ are linearly independent and the state described by ψ is therefore twofold degenerate.

Now suppose we have a 2-electron wave function, say $\psi_1\psi_2$. For this case

$$K^2\psi_1\psi_2 = \psi_1\psi_2 \tag{8.3-24}$$

and there is no problem in satisfying a relation of the type

$$K\psi_1\psi_2 = a\psi_1\psi_2 \tag{8.3-25}$$

because then

$$K^2\psi_1\psi_2 = |a|^2\psi_1\psi_2 = \psi_1\psi_2 \tag{8.3-26}$$

or $|a|^2 = 1$. Hence $\psi_1\psi_2$ and $K\psi_1\psi_2$ are not linearly independent. More generally,

$$K^2\psi = (-1)^n\psi \tag{8.3-27}$$

where ψ is a wave function for an n-electron system. If n is odd, ψ and $K\psi$ are linearly independent and there is, at least, a twofold degeneracy. If n is even, ψ and $K\psi$ are not independent and the time reversal operator has no effect on degeneracies.

The results we have just obtained do not hold when an external magnetic field is present because then $\mathscr{H}$ does not commute with K and $K\psi$ need not be a solution of the Schrödinger equation when ψ is a solution.

Kramers' theorem may now be stated as follows: With *no external magnetic field*, the Hamiltonian is invariant under time reversal, i.e., $[K, \mathscr{H}] = 0$. For an *odd* number of electrons, $K^2 = -1$ and the energy levels may be n-fold degenerate but n must be even. No combination of electric fields can remove the degeneracies completely; at least twofold degeneracies will remain. For an even number of electrons $K^2 = 1$; hence there are no additional degeneracies of the system from time reversal symmetry. This case also corresponds to the Schrödinger theory without spin. When *an external magnetic field is present*, the Hamiltonian is not invariant under time reversal, i.e., $[K, \mathscr{H}] \neq 0$. For an odd or an even number of electrons, degeneracies can be completely removed.

The statement of Kramers' theorem may require modification if other symmetries are present. Let us suppose that

$$[K, \mathscr{H}] = 0 \tag{8.3-28}$$

so that if

$$\mathscr{H}\psi = \mathrm{E}\psi, \tag{8.3.29}$$

then $K\psi$ is also an eigenfunction of $\mathscr{H}$ with the same eigenvalue E. In addition, it will be assumed that $\mathscr{H}$ is invariant not only under time reversal but also under the operations of a group whose irreducible representations Γ belong to one of the three types a, b, or c described in Section 5.2. We consider these cases separately.

Case a Γ is real. Kramers' theorem remains as stated above.

Case b Γ and Γ^* are not equivalent. If ψ is a basis function for Γ, ψ^* is a basis function for Γ^* but since Γ and Γ^* are inequivalent, ψ and ψ^* are linearly independent. Therefore $K\psi(\equiv\psi^*)$ and ψ are linearly independent.

Since they both belong to the same eigenvalue we must have at least a twofold degeneracy *regardless of the number of electrons*. This means that when time reversal symmetry is present, the two inequivalent complex conjugate representations Γ and Γ^* belong to the same eigenvalue. This is the reason for the common practice of combining the complex characters of Γ and Γ^* into a set of real characters which belong to a representation whose dimension is twice that of Γ or Γ^*.

Case c Γ and Γ^* are equivalent but cannot be transformed to real form. It may be shown (Wigner, 1959; Heine, 1960) that this case is the inverse of case a, that is, for an even number of electrons there is an additional degeneracy due to time reversal but not for an odd number. These results are summarized in Table 8.1.

TABLE 8.1

Degeneracies Due to Time Reversal Symmetry

Case	Even number of electrons $K^2 = 1$	Odd number of electrons $K^2 = -1$
a	$K\psi$ and ψ not linearly independent No additional degeneracy	$K\psi$ and ψ linearly independent $2n$ degeneracy
b	$K\psi$ and ψ linearly independent $2n$ degeneracy	$K\psi$ and ψ linearly independent $2n$ degeneracy
c	$K\psi$ and ψ linearly independent $2n$ degeneracy	$K\psi$ and ψ not linearly independent No additional degeneracy

As an application of the effect of time reversal symmetry, let

$$\mathscr{H} = \mathscr{H}_0 + V \tag{8.3-30}$$

in which $\mathscr{H}_0$ is invariant under all three-dimensional rotations and V is invariant under the coordinate transformations of D_3. Assuming that orbital and spin angular momenta have been coupled, the eigenstates of $\mathscr{H}_0$ will belong to $D^{(j)}$. Suppose $j = \frac{3}{2}$. Such a state can only arise in a system with an odd number of electrons. When $V \neq 0$ the symmetry of $\mathscr{H}$ is reduced to D_3 and the state with $j = \frac{3}{2}$ is split according to

$$D^{(3/2)} = \Gamma^{(4)} + \Gamma^{(5)} + \Gamma^{(6)} \tag{8.3-31}$$

as in (5.2-3). This means that the original fourfold degeneracy of the state with $j = \frac{3}{2}$ has been split into a twofold degenerate state $\Gamma^{(4)}$ and two nondegenerate states $\Gamma^{(5)}$ and $\Gamma^{(6)}$, that is, a doublet and two singlets (Table 5.3). However, time reversal symmetry will modify this result because $\Gamma^{(5)}$ and

$\Gamma^{(6)}$ are type b representations (Section 5.2) which must have even degeneracies. Therefore the states corresponding to $\Gamma^{(5)}$ and $\Gamma^{(6)}$ must have the same energy. $\Gamma^{(4)}$ is of type c for which there are no additional degeneracies for an odd number electrons.

8.4 Indistinguishability of Particles

An N-electron wave function will be written as

$$\psi(\lambda_1, \lambda_2, \ldots, \lambda_N) \tag{8.4-1}$$

with λ_i standing for the three spatial coordinates r_i, θ_i, φ_i (or x_i, y_i, z_i) and the spin coordinate m_{si}.

Let us now consider two electrons with two different sets of space and spin coordinates represented by λ_1 and λ_2. The wave function for the two-electron system will be written as $\psi(\lambda_1, \lambda_2)$. The principle of indistinguishability states that upon interchange of the two electrons, that is, when λ_1 and λ_2 are interchanged

$$|\psi(\lambda_2, \lambda_1)|^2 = |\psi(\lambda_1, \lambda_2)|^2. \tag{8.4-2}$$

If (8.4-2) is satisfied, $\psi(\lambda_2, \lambda_1)$ can differ from $\psi(\lambda_1, \lambda_2)$ by, at most, a phase factor, i.e.,

$$\psi(\lambda_2, \lambda_1) = e^{i\alpha}\psi(\lambda_1, \lambda_2) \tag{8.4-3}$$

which implies that no experiment will be capable of distinguishing between the states $\psi(\lambda_1, \lambda_2)$ and $\psi(\lambda_2, \lambda_1)$.

Let P_{12} be an operator that interchanges λ_1 and λ_2:

$$P_{12}\psi(\lambda_1, \lambda_2) = \psi(\lambda_2, \lambda_1). \tag{8.4-4}$$

Now let P_{12} operate on both sides of (8.4-3):

$$P_{12}\psi(\lambda_2, \lambda_1) = e^{i\alpha}P_{12}\psi(\lambda_1, \lambda_2) \tag{8.4-5}$$

$$= e^{i\alpha}\psi(\lambda_2, \lambda_1)$$

$$= e^{2i\alpha}\psi(\lambda_1, \lambda_2). \tag{8.4-6}$$

But

$$P_{12}\psi(\lambda_2, \lambda_1) = \psi(\lambda_1, \lambda_2). \tag{8.4-7}$$

Therefore

$$e^{2i\alpha} = 1$$

or

$$e^{i\alpha} = \pm 1. \tag{8.4-8}$$

We conclude from (8.4-4) and (8.4-8) that

$$\psi(\lambda_2, \lambda_1) = \pm\psi(\lambda_1, \lambda_2). \tag{8.4-9}$$

When the plus sign prevails, $\psi(\lambda_1, \lambda_2)$ is said to be a symmetric function with respect to the interchange of two electrons. When the minus sign holds, $\psi(\lambda_1, \lambda_2)$ is antisymmetric. The same considerations may be extended to any number of electrons. Thus the principle of indistinguishability forces the multielectron wave function to be either symmetric or antisymmetric. Since most functions are neither symmetric nor antisymmetric, it is seen that the principle of indistinguishability severely restricts the form of the wave function.

Electrons are not the only particles that are indistinguishable—so are photons, protons, neutrons, pions, and many others and in all such cases, an equation of form (8.4-9) is applicable. However, there is an important distinction. Symmetric wave functions describe particles that obey Bose–Einstein statistics and are therefore called *bosons*. These are particles with integral values of the spin (e.g., photons, pions, deuterons, α-particles). On the other hand, antisymmetric wave functions describe particles that obey Fermi–Dirac statistics. Such particles, called *fermions* have half-integral values of the spin (e.g., protons, neutrons, electrons).

It is quite easy to construct symmetric wave functions. Thus for two particles a symmetric wave function would be

$$\Psi_S(\lambda_1, \lambda_2) = \frac{1}{\sqrt{2}}[\psi_1(\lambda_1)\psi_2(\lambda_2) + \psi_2(\lambda_1)\psi_1(\lambda_2)] \tag{8.4-10}$$

in which ψ_1 and ψ_2 are one-particle functions, i.e., functions that depend on the space and spin coordinates of a single particle. Such a function is usually called a *spin orbital*. For systems with more than two electrons

$$\Psi_S(\lambda_1, \lambda_2, \ldots, \lambda_N) = \frac{1}{\sqrt{N!}}\sum_P P\psi_1(\lambda_1)\psi_2(\lambda_2)\cdots\psi_N(\lambda_N) \tag{8.4-11}$$

where the sum is taken over all permutations of the N variables $\lambda_1, \lambda_2, \ldots, \lambda_N$, that is, all possible permutations of the N particles; $1/\sqrt{N!}$ is a normalizing constant. Any interchange of the λs in (8.4-11), which is equivalent to an interchange of particles, leaves Ψ_s unaltered. Hence Ψ_s satisfies (8.4-9) with the plus sign.

A two-electron antisymmetric wave function is also fairly simple, namely

$$\Psi_A(\lambda_1, \lambda_2) = \frac{1}{\sqrt{2}}[\psi_1(\lambda_1)\psi_2(\lambda_2) - \psi_2(\lambda_1)\psi_1(\lambda_2)]. \tag{8.4-12}$$

Upon interchange of λ_1 and λ_2, Ψ_A changes sign thereby satisfying (8.4-9) with the minus sign. The most convenient way to extend (8.4-12) to more than two electrons is by means of a *Slater determinant*:

$$\Psi_A(\lambda_1, \lambda_2, \ldots, \lambda_N) = \frac{1}{\sqrt{N!}} \begin{vmatrix} \psi_1(\lambda_1) & \psi_2(\lambda_1) & \cdots & \psi_N(\lambda_1) \\ \psi_1(\lambda_2) & \psi_2(\lambda_2) & \cdots & \psi_N(\lambda_2) \\ \vdots & \vdots & \cdots & \vdots \\ \psi_1(\lambda_N) & \psi_2(\lambda_N) & \cdots & \psi_N(\lambda_N) \end{vmatrix} \tag{8.4-13a}$$

$$= \frac{1}{\sqrt{N!}} \det\{\psi_1, \psi_2, \ldots, \psi_N\} \tag{8.4-13b}$$

$$= \frac{1}{\sqrt{N!}} \sum_p (-1)^p P \psi_1(\lambda_1)\psi_2(\lambda_2) \cdots \psi_N(\lambda_N) \tag{8.4-13c}$$

with p the parity of the permutation P (see also Section 20.4). That a function of form (8.4-13) is antisymmetric is seen from the fact that an interchange of two particles corresponds to an interchange of two rows in the determinant as a result of which the determinant changes sign. Furthermore, if any two orbitals, say ψ_i and ψ_j are the same $(i = j)$, then two columns of the determinant will be the same and the determinant will vanish. Thus no orbital can be occupied by more than one electron. This is the Pauli exclusion principle.

CHAPTER 9

TIME DEVELOPMENT OF A QUANTUM SYSTEM

9.1 Schrödinger Representation

The time-dependent Schrödinger equation

$$i\hbar\frac{\partial\psi(\mathbf{r},t)}{\partial t} = \mathscr{H}\psi(\mathbf{r},t) \tag{9.1-1}$$

governs the evolution of a quantum system in time. $\mathscr{H}$, the Hamiltonian of the system, may or may not itself be time-dependent. In a purely formal manner one may write

$$\psi(t) = U(t,t_0)\psi(t_0) \tag{9.1-2}$$

with

$$U(t_0,t_0) = 1 \tag{9.1-3}$$

in which $U(t,t_0)$ is known as an *evolution operator*. The dependence of the wave function on spatial coordinates is unimportant in the present discussion and may therefore be suppressed. Clearly, to obtain $\psi(t)$ one may attempt to solve the time-dependent Schrödinger equation or one may seek the detailed form of the evolution operator since both methods will give $\psi(t)$ at an arbitrary time t once $\psi(t_0)$ at an initial time t_0 is known. A priori, there is no reason to suppose that one method is superior to the other.

A useful interpretation of $U(t,t_0)$ may be obtained as follows: Suppose a system is known to be in the state ψ_a at an initial time t_0. At a later time t the system has evolved into the state $U(t,t_0)\psi_a$. The probability amplitude that $U(t,t_0)\psi_a$ is a particular state ψ_b is given by the overlap integral $\langle\psi_b|U(t,t_0)|\psi_a\rangle$ which is the projection of ψ_b on $U(t,t_0)\psi_a$. Hence the probability of finding a system in a state ψ_{b} at a time t when the system is known to have been in the state ψ_{a} at $t = t_0$ is

$$w_{\mathrm{ba}} = |\langle\psi_{\mathrm{b}}|U(t, t_0)|\psi_{\mathrm{a}}\rangle|^2. \tag{9.1-4}$$

From (9.1-2)

$$\psi(t_2) = U(t_2, t_0)\psi(t_0)$$

which may be written as

$$\psi(t_2) = U(t_2, t_1)\psi(t_1) = U(t_2, t_1)U(t_1, t_0)\psi(t_0).$$

Therefore

$$U(t_2, t_0) = U(t_2, t_1)U(t_1, t_0). \tag{9.1-5}$$

If t and t_0 are interchanged in (9.1-2), we have

$$\psi(t_0) = U(t_0, t)\psi(t)$$

which is equivalent to multiplying both sides of (9.1-2) by $U^{-1}(t, t_0)$. Therefore

$$U^{-1}(t, t_0) = U(t_0, t). \tag{9.1-6}$$

We now show that $U(t, t_0)$ is a unitary operator. Substitution of (9.1-2) into (9.1-1) leads to the operator equation

$$i\hbar\frac{\partial U(t, t_0)}{\partial t} = \mathscr{H}U(t, t_0) \tag{9.1-7}$$

and its Hermitian conjugate

$$-i\hbar\frac{\partial U^\dagger(t, t_0)}{\partial t} = U^\dagger(t, t_0)\mathscr{H} \tag{9.1-8}$$

with the assumption that $\mathscr{H}^\dagger = \mathscr{H}$. If (9.1-7) is premultiplied by $U^\dagger$ and (9.1-8) is postmultiplied by U, the result is

$$i\hbar\left[U^\dagger\frac{\partial U}{\partial t} + \frac{\partial U^\dagger}{\partial t}U\right] = i\hbar\frac{\partial}{\partial t}U^\dagger U = 0, \tag{9.1-9}$$

or $U^\dagger U$ is a constant which must be set equal to unity in order to satisfy (9.1-3). Thus

$$U^\dagger U = 1, \tag{9.1-10}$$

proving that $U(t, t_0)$ is a unitary operator. If $\psi(t_0)$ is visualized as a vector pointing in a certain direction in a Hilbert space, the effect of $U(t, t_0)$ on $\psi(t_0)$, from (9.1-2), is to rotate this vector in a continuous fashion as time progresses.

If it is stipulated that the Hamiltonian is independent of time, as will henceforth be assumed, a formal solution to the Schrödinger equation (9.1-1) is

$$\psi(\mathbf{r}, t) = e^{-i\mathscr{H}(t-t_0)/\hbar}\psi(\mathbf{r}, t_0) \tag{9.1-11}$$

in which the exponential operator is interpreted as

$$e^{-i\mathscr{H}(t-t_0)/\hbar} = 1 + \frac{1}{i\hbar}\mathscr{H}(t - t_0) + \frac{1}{2!}\left(\frac{1}{i\hbar}\right)^2 \mathscr{H}^2(t - t_0)^2 + \cdots . \tag{9.1-12}$$

Comparing (9.1-2) with (9.1-11), it is seen that an explicit form for the evolution operator is

$$U(t, t_0) = e^{-i\mathscr{H}(t-t_0)/\hbar}. \tag{9.1-13}$$

Another formulation for the evolution operator is achieved by converting (9.1-7) and condition (9.1-3) into the integral equation

$$U(t, t_0) = 1 - \frac{i}{\hbar}\int_{t_0}^{t} \mathscr{H}U(t', t_0)\, dt'. \tag{9.1-14}$$

It is often desirable to express the solution to the time-dependent Schrödinger equation in terms of a complete set of eigenstates $\psi_k(\mathbf{r})$ which satisfy the time-independent Schrödinger equation

$$\mathscr{H}\psi_k(\mathbf{r}) = E_k\psi_k(\mathbf{r}). \tag{9.1-15}$$

Since

$$\sum_k |\psi_k(r)\rangle\langle\psi_k(r)| = 1 \tag{9.1-16}$$

we have

$$\begin{aligned} \psi(\mathbf{r}, t) &= U(t, t_0)\psi(\mathbf{r}, t_0) \\ &= e^{-i\mathscr{H}(t-t_0)/\hbar}\psi(\mathbf{r}, t_0) \\ &= \sum_k e^{-i\mathscr{H}(t-t_0)/\hbar}|\psi_k(\mathbf{r})\rangle\langle\psi_k(\mathbf{r})|\psi(\mathbf{r}, t_0)\rangle \\ &= \sum_k e^{-iE_k(t-t_0)/\hbar}\langle\psi_k(\mathbf{r})|\psi(\mathbf{r}, t_0)\rangle\psi_k(\mathbf{r}) \\ &= \sum_k a_k(t)\psi_k(\mathbf{r}) \end{aligned} \tag{9.1-17}$$

with

$$a_k(t) = \langle\psi_k(\mathbf{r})|\psi(\mathbf{r}, t_0)\rangle e^{-iE_k(t-t_0)/\hbar}.$$

Suppose A is a time-independent operator. Then

$$
\begin{aligned}
i\hbar \frac{\partial}{\partial t}\langle A\rangle &\equiv i\hbar \frac{\partial}{\partial t}\langle\psi(t)|A|\psi(t)\rangle \\
&\equiv i\hbar\left[\left\langle\frac{\partial\psi(t)}{\partial t}|A|\psi(t)\right\rangle+\left\langle\psi(t)|A|\frac{\partial\psi(t)}{\partial t}\right\rangle\right] \\
&= -\left\langle i\hbar\frac{\partial\psi(t)}{\partial t}|A|\psi(t)\right\rangle+\left\langle\psi(t)|A|i\hbar\frac{\partial\psi(t)}{\partial t}\right\rangle \\
&= -\langle\mathscr{H}\psi(t)|A|\psi(t)\rangle+\langle\psi(t)|A|\mathscr{H}\psi(t)\rangle \\
&= -\langle\psi(t)|\mathscr{H}A|\psi(t)\rangle+\langle\psi(t)|A\mathscr{H}|\psi(t)\rangle \\
&= \langle\psi(t)|[A,\mathscr{H}]|\psi(t)\rangle \equiv \langle[A,\mathscr{H}]\rangle.
\end{aligned}
\tag{9.1-18}
$$

It follows at once that the expectation value of an operator is constant in time if the operator commutes with the Hamiltonian. In that event, the operator and the physical observable associated with it are known as *constants of the motion.* As examples, if $\mathscr{H}$ commutes with a component of angular momentum, say J_x, then the x component of angular momentum is conserved; similarly parity is conserved if $[P_i, \mathscr{H}] = 0$ where i is the inversion operator.

If

$$\psi(\mathbf{r}, t) = O\varphi(\mathbf{r}, t) \tag{9.1-19}$$

where O is a unitary operator which may depend on time and $\psi(\mathbf{r}, t)$ satisfies (9.1-1), we have

$$i\hbar\frac{\partial\psi}{\partial t} = i\hbar\left[O\frac{\partial\varphi}{\partial t} + \varphi\frac{\partial O}{\partial t}\right], \qquad \mathscr{H}\psi = \mathscr{H}O\varphi.$$

Therefore

$$i\hbar\left[O\frac{\partial\varphi}{\partial t} + \frac{\partial O}{\partial t}\varphi\right] = \mathscr{H}O\varphi$$

and upon premultiplying by $O^\dagger$ we obtain the Schrödinger equation for $\varphi(\mathbf{r}, t)$:

$$i\hbar\frac{\partial\varphi(\mathbf{r}, t)}{\partial t} = \mathscr{H}'\varphi(\mathbf{r}, t) \tag{9.1-20}$$

with the transformed Hamiltonian

$$\mathscr{H}' = O^\dagger\mathscr{H}O - i\hbar O^\dagger\frac{\partial O}{\partial t}. \tag{9.1-21}$$

The formulation to this point was based on the Schrödinger equation (9.1-1). This is called the *Schrödinger representation* (or picture) although it should be noted that "representation" in the present context is completely unrelated to the group-theoretical representations discussed previously.

9.2 Heisenberg Representation

The *Heisenberg representation* is one in which a new wave function ψ_{H} is defined by

$$\psi_{\mathrm{H}} = e^{i\mathscr{H}t/\hbar}\psi(t) \tag{9.2-1}$$

where $\psi(t)$ (without a subscript) is a Schrödinger wave function, that is, a function which satisfies the Schrödinger equation (9.1-1). Taking the time derivative of (9.2-1),

$$\begin{aligned}\frac{\partial\psi_{\mathrm{H}}}{\partial t} &= \frac{i}{\hbar}\mathscr{H}e^{i\mathscr{H}t/\hbar}\psi(t) + e^{i\mathscr{H}t/\hbar}\frac{\partial\psi(t)}{\partial t} \\ &= \frac{i}{\hbar}\mathscr{H}e^{i\mathscr{H}t/\hbar}\psi(t) + e^{i\mathscr{H}t/\hbar}\frac{1}{i\hbar}\mathscr{H}\psi(t) = 0.\end{aligned} \tag{9.2-2}$$

Thus, from (9.2-1),

$$\psi_{\mathrm{H}} = \psi(0)$$

for all time.

The expectation value of an operator A in the Schrödinger representation

$$\langle A\rangle \equiv \langle\psi(t)|A|\psi(t)\rangle$$

may be converted to the Heisenberg representation by means of (9.2-1):

$$\langle A\rangle = \langle e^{-i\mathscr{H}t/\hbar}\psi_{\mathrm{H}}|A|e^{-i\mathscr{H}t/\hbar}\psi_{\mathrm{H}}\rangle$$

which, using the turn-over rule is

$$\langle A\rangle = \langle\psi_{\mathrm{H}}|e^{i\mathscr{H}t/\hbar}Ae^{-i\mathscr{H}t/\hbar}|\psi_{\mathrm{H}}\rangle. \tag{9.2-3}$$

This suggests that it might be useful to define an operator in the Heisenberg representation by the relation

$$A_{\mathrm{H}}(t) = e^{i\mathscr{H}t/\hbar}Ae^{-i\mathscr{H}t/\hbar} \tag{9.2-4}$$

so that the expectation value of A, written in the Heisenberg representation becomes

$$\langle A\rangle = \langle A_{\mathrm{H}}(t)\rangle = \langle\psi_{\mathrm{H}}|A_{\mathrm{H}}(t)|\psi_{\mathrm{H}}\rangle. \tag{9.2-5}$$

Clearly, the expectation value of an operator which is a pure number must be the same whether it is calculated in one representation or another. We also note that since $\mathscr{H}$ commutes with $e^{i\mathscr{H}t/\hbar}$,

$$\mathscr{H}_{\mathrm{H}} = \mathscr{H}. \tag{9.2-6}$$

The time dependence of $A_{\mathrm{H}}(t)$ is obtained directly from (9.2-4). Assuming again that A has no intrinsic time dependence

$$i\hbar\frac{\partial A_{\mathrm{H}}(t)}{\partial t} = e^{i\mathscr{H}t/\hbar}A\mathscr{H}e^{-i\mathscr{H}t/\hbar} - e^{i\mathscr{H}t/\hbar}\mathscr{H}Ae^{-i\mathscr{H}t/\hbar} = [A_{\mathrm{H}}(t), \mathscr{H}]. \tag{9.2-7}$$

Equation (9.2-7) is the equation of motion for the operator $A_{\mathrm{H}}(t)$ and is known as the *Heisenberg equation.*

A further point to note is that commutation relations are preserved in the passage from one representation to the other. Thus let

$$[A, B] = C; \tag{9.2-8}$$

upon converting each of the operators to the Heisenberg representation by means of (9.2-4) we find

$$[A_{\mathrm{H}}(t), B_{\mathrm{H}}(t)] = C_{\mathrm{H}}(t). \tag{9.2-9}$$

Thus if $[A, \mathscr{H}] = 0$, then $[A_{\mathrm{H}}(t), \mathscr{H}] = 0$ and, according to (9.2-7), $A_{\mathrm{H}}(t)$ is a constant of the motion. We see then that the physical content of the Heisenberg equation is identical with the analogous expression (9.1-18) in the Schrödinger representation. The basic distinction between the two representations resides in the fact that in the Schrödinger representation the time development of the system is contained in $\psi(t)$ while in the Heisenberg representation it is contained in $A_{\mathrm{H}}(t)$, the time-dependent operator which represents a physical observable.

9.3 Interaction Representation

In addition to the Schrödinger and Heisenberg representations there is yet another one known as the *interaction representation* which is particularly appropriate for certain formulations of time-dependent perturbation theory. Let the time-independent Hamiltonian in the Schrödinger representation be written as the sum of two parts:

$$\mathscr{H} = \mathscr{H}_0 + V. \tag{9.3-1}$$

A wave function $\psi_{\mathrm{I}}(t)$ and an operator $A_{\mathrm{I}}(t)$ in the interaction representation are defined by

$$\psi_{\mathrm{I}}(t) = e^{i\mathscr{H}_0 t/\hbar}\psi(t) \tag{9.3-2}$$

and

$$A_{\rm I}(t) = e^{i\mathscr{H}_0 t/\hbar} A e^{-i\mathscr{H}_0 t/\hbar}. \tag{9.3-3}$$

Again we note that $\mathscr{H}_0$ in the interaction representation is identical with $\mathscr{H}_0$ in the Schrödinger representation. Also

$$\begin{aligned}\langle A\rangle &= \langle\psi(t)|A|\psi(t)\rangle \\ &= \langle e^{-i\mathscr{H}_0 t/\hbar}\psi_{\rm I}(t)|A|e^{-i\mathscr{H}_0 t/\hbar}\psi_{\rm I}(t)\rangle \\ &= \langle\psi_{\rm I}(t)|e^{i\mathscr{H}_0 t/\hbar}Ae^{-i\mathscr{H}_0 t/\hbar}|\psi_{\rm I}(t)\rangle = \langle\psi_{\rm I}(t)|A_{\rm I}(t)|\psi_{\rm I}(t)\rangle\end{aligned} \tag{9.3-4}$$

which indicates, as we have already seen for the Heisenberg representation, that the expectation value of an operator is independent of the representation.

The time derivative of $\psi_{\rm I}(t)$ in (9.3-2) is

$$\frac{\partial\psi_{\rm I}(t)}{\partial t} = \frac{i}{\hbar}\mathscr{H}_0 e^{i\mathscr{H}_0 t/\hbar}\psi(t) + e^{i\mathscr{H}_0 t/\hbar}\frac{\partial\psi(t)}{\partial t}$$

and upon substituting

$$i\hbar\frac{\partial\psi}{\partial t} = \mathscr{H}\psi = (\mathscr{H}_0 + V)\psi$$

we find

$$\begin{aligned}i\hbar\frac{\partial\psi_{\rm I}(t)}{\partial t} &= e^{i\mathscr{H}_0 t/\hbar}V\psi(t) \\ &= e^{i\mathscr{H}_0 t/\hbar}Ve^{-i\mathscr{H}_0 t/\hbar}e^{i\mathscr{H}_0 t/\hbar}\psi(t) = V_{\rm I}(t)\psi_{\rm I}(t).\end{aligned} \tag{9.3-5}$$

Similarly, if (9.3-3) is differentiated with respect to time, the result is

$$\begin{aligned}i\hbar\frac{\partial A_{\rm I}(t)}{\partial t} &= -\mathscr{H}_0 e^{i\mathscr{H}_0 t/\hbar}Ae^{-i\mathscr{H}_0 t/\hbar} + e^{i\mathscr{H}_0 t/\hbar}A\mathscr{H}_0 e^{-i\mathscr{H}_0 t/\hbar} \\ &= -\mathscr{H}_0 A_{\rm I}(t) + A_{\rm I}(t)\mathscr{H}_0 = [A_{\rm I}(t), \mathscr{H}_0].\end{aligned} \tag{9.3-6}$$

Equations (9.3-5) and (9.3-6) are the equations of motion for wave functions and operators, respectively, in the interactions representation.

The time development of a Schrödinger wave function was described in terms of an evolution operator by (9.1-2). An analogous expression can be established for a wave function in the interaction representation. For this purpose we define an operator $U_{\rm I}(t, t_0)$ by

$$\psi_{\rm I}(t) = U_{\rm I}(t, t_0)\psi_{\rm I}(t_0) \tag{9.3-7}$$

with

$$U_{\rm I}(t_0, t_0) = 1. \tag{9.3-8}$$

Upon combining (9.1-2) and (9.3-2) one finds

$$\begin{aligned}\psi_{\mathrm{I}}(t) &= e^{i\mathscr{H}_0 t/\hbar}\psi(t)\\ &= e^{i\mathscr{H}_0 t/\hbar}U(t,t_0)\psi(t_0) = e^{i\mathscr{H}_0 t/\hbar}U(t,t_0)e^{-i\mathscr{H}_0 t_0/\hbar}\psi_{\mathrm{I}}(t_0),\end{aligned} \tag{9.3-9}$$

from which it is concluded that

$$U_{\mathrm{I}}(t,t_0) = e^{i\mathscr{H}_0 t/\hbar}U(t,t_0)e^{-i\mathscr{H}_0 t_0/\hbar} \tag{9.3-10}$$

$$= e^{i\mathscr{H}_0 t/\hbar}e^{-i\mathscr{H}(t-t_0)/\hbar}e^{-i\mathscr{H}_0 t_0/\hbar}, \tag{9.3-11}$$

with the second equality based on (9.1-13). The unitarity property of $U(t, t_0)$ confers unitarity to $U_{\mathrm{I}}(t, t_0)$ so that

$$U_{\mathrm{I}}^{\dagger}(t,t_0) = U_{\mathrm{I}}^{-1}(t,t_0). \tag{9.3-12}$$

We may also find the differential equation obeyed by $U_{\mathrm{I}}(t, t_0)$ by substituting (9.3-7) into (9.3-5). Since $\psi_{\mathrm{I}}(t_0)$ is constant,

$$i\hbar\,\frac{\partial U_{\mathrm{I}}(t,t_0)}{\partial t} = V_{\mathrm{I}}(t)U_{\mathrm{I}}(t,t_0). \tag{9.3-13a}$$

An equivalent equation may be obtained by interchanging t and t_0 and taking the Hermitian conjugate:

$$-i\hbar\,\frac{\partial U_{\mathrm{I}}^{\dagger}(t_0,t)}{\partial t_0} = U_{\mathrm{I}}^{\dagger}(t_0,t)V_{\mathrm{I}}^{\dagger}(t_0).$$

But

$$U_{\mathrm{I}}^{\dagger}(t_0,t) = U_{\mathrm{I}}(t,t_0), \qquad V_{\mathrm{I}}^{\dagger}(t_0) = V_{\mathrm{I}}(t_0),$$

hence

$$i\hbar\,\frac{\partial U_{\mathrm{I}}(t,t_0)}{\partial t_0} = -U_{\mathrm{I}}(t,t_0)V_{\mathrm{I}}(t_0). \tag{9.3-13b}$$

In place of differential equations for $U_{\mathrm{I}}(t, t_0)$ it is often convenient to have integral equations. From (9.3-13a) and the initial condition (9.3-8) one immediately obtains the integral equation

$$U_{\mathrm{I}}(t,t_0) = 1 - \frac{i}{\hbar}\int_{t_0}^{t} V_{\mathrm{I}}(t')U_{\mathrm{I}}(t',t_0)\,dt' \tag{9.3-14a}$$

whereas (9.3-13b) and the initial condition yield

$$U_{\mathrm{I}}(t,t_0) = 1 - \frac{i}{\hbar}\int_{t_0}^{t} U_{\mathrm{I}}(t,t')V_{\mathrm{I}}(t')\,dt'. \tag{9.3-14b}$$

Although the two integral equations are entirely equivalent, (9.3-14a) is the more common form and we shall use it to develop a series solution for

$U_{\mathrm{I}}(t, t_0)$ by means of successive iterations. Thus, starting with

$$U_{\mathrm{I}}^{(0)} = 1$$

and substituting in the integral of (9.3-14a), the next approximation is

$$U_{\mathrm{I}}^{(1)} = 1 - \frac{i}{\hbar} \int_{t_0}^{t} V_{\mathrm{I}}(t_1)\, dt_1 .$$

Continuing in this fashion we obtain

$$\begin{aligned} U_{\mathrm{I}}(t, t_0) &= U_{\mathrm{I}}^{(0)}(t, t_0) + U_{\mathrm{I}}^{(1)}(t, t_0) + \cdots \\ &= 1 - \frac{i}{\hbar} \int_{t_0}^{t} V_{\mathrm{I}}(t_1)\, dt_1 + \left(\frac{-i}{\hbar}\right)^2 \int_{t_0}^{t} V_{\mathrm{I}}(t_1)\, dt_1 \int_{t_0}^{t_1} V_{\mathrm{I}}(t_2)\, dt_2 + \cdots \\ &= 1 + \sum_{n=1}^{\infty} \left(\frac{-i}{\hbar}\right)^n \int_{t_0}^{t} dt_1 \cdots \int_{t_0}^{t_{n-1}} dt_n\, V_{\mathrm{I}}(t_1) \cdots V_{\mathrm{I}}(t_n), \end{aligned} \tag{9.3-15}$$

where

$$t_0 < t_n < t_{n-1} < \cdots < t_2 < t_1 < t. \tag{9.3-16}$$

It is possible to cast (9.3-15) into another form by means of the *Dyson chronological operator* or *time-ordering* P which is defined by

$$\begin{aligned} P\{A(t_1)B(t_2)\} &= P\{B(t_2)A(t_1)\} \\ &= \begin{cases} A(t_1)B(t_2) & \text{if} \quad t_1 > t_2, \\ B(t_2)A(t_1) & \text{if} \quad t_2 > t_1. \end{cases} \end{aligned} \tag{9.3-17}$$

When there are more than two operators they are ordered so that time increases from right to left. It will now be shown that (9.3-15) may be written as

$$\begin{aligned} U_{\mathrm{I}}(t, t_0) &= P\left\{\exp\left[-\frac{i}{\hbar} \int_{t_0}^{t} V_{\mathrm{I}}(t')\, dt'\right]\right\} = P\left\{\sum_{n=0}^{\infty} \left(-\frac{i}{\hbar}\right)^n \frac{1}{n!} \left(\int_{t_0}^{t} V_{\mathrm{I}}(t')\, dt'\right)^n\right\} \end{aligned} \tag{9.3-18}$$

$$= P\left\{1 + \sum_{n=1}^{\infty} \left(\frac{-i}{\hbar}\right)^n \frac{1}{n!} \int_{t_0}^{t} dt_1 \int_{t_0}^{t} dt_2 \cdots \int_{t_0}^{t} dt_n\, V_{\mathrm{I}}(t_1) V_{\mathrm{I}}(t_2) \cdots V_{\mathrm{I}}(t_n)\right\}. \tag{9.3-19}$$

Consider the term

$$I = \left(\frac{-i}{\hbar}\right)^2 \frac{1}{2!} \int_{t_0}^{t} dt_1 \int_{t_0}^{t} dt_2\, P\{V_{\mathrm{I}}(t_1) V_{\mathrm{I}}(t_2)\}. \tag{9.3-20}$$

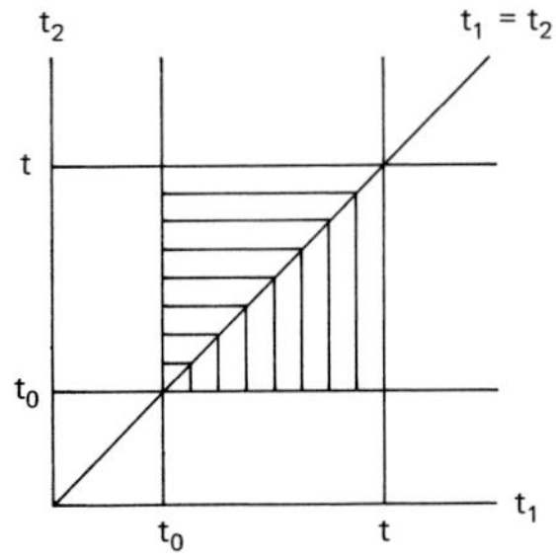

FIG. 9.1 Region of integration for Eq. (9.3-21).

The region of integration extends over the hatched area in Fig. 9.1; hence the integral may be divided into two parts:

$$I = \left(\frac{-i}{\hbar}\right)^2 \frac{1}{2!}\left[\int_{t_0}^{t} dt_1 \int_{t_0}^{t_1} dt_2\, P\{V_I(t_1)V_I(t_2)\} + \int_{t_0}^{t} dt_2 \int_{t_0}^{t_2} dt_1\, P\{V_I(t_1)V_I(t_2)\}\right] \tag{9.3-21}$$

in which the first integral is over the lower triangle where $t_1 > t_2$ and the second is over the upper triangle where $t_1 < t_2$. Therefore

$$P\{V_I(t_1)V_I(t_2)\} = \begin{cases} V_I(t_1)V_I(t_2) & \text{if} \quad t_1 > t_2, \\ V_I(t_2)V_I(t_1) & \text{if} \quad t_1 < t_2, \end{cases}$$

and the integral (9.3-21) becomes

$$I = \left(\frac{-i}{\hbar}\right)^2 \frac{1}{2!}\left[\int_{t_0}^{t} dt_1 \int_{t_0}^{t_1} dt_2\, V_I(t_1)V_I(t_2) + \int_{t_0}^{t} dt_2 \int_{t_0}^{t_2} dt_1\, V_I(t_2)V_I(t_1)\right].$$

However, by interchanging the variables in the second integral it becomes identical to the first. Hence

$$I = \left(\frac{-i}{\hbar}\right)^2 \int_{t_0}^{t} dt_1 \int_{t_0}^{t_1} dt_2\, V_I(t_1)V_I(t_2)$$

which corresponds to the $n = 2$ term in (9.3-15). The same type of argument can be applied to other terms of the series, that is, the general proof can be obtained by mathematical induction to show the equality between (9.3-15) and (9.3-18).

The expansion of $U_I(t, t_0)$ provides an expansion of the expectation value of an operator A. In the interaction representation

$$\begin{aligned} \langle A_I \rangle &= \langle \psi_I(t)|A_I(t)|\psi_I(t)\rangle \\ &= \langle U_I(t, t_0)\psi(t_0)|A_I(t)|U_I(t, t_0)\psi(t_0)\rangle \\ &= \langle \psi_I(t_0)|U_I^\dagger(t, t_0)A_I(t)U_I(t, t_0)|\psi(t_0)\rangle. \end{aligned} \tag{9.3-22}$$

TABLE 9.1

Time Development of a Quantum System According to the Schrödinger, Heisenberg, and Interaction Representations[a]

Schrödinger	Heisenberg	Interaction
Schrödinger equation	Wave function	Hamiltonian
$i\hbar \dfrac{\partial\psi(t)}{\partial t} = \mathscr{H}\psi(t)$	$\psi_{\mathrm{H}} = e^{i\mathscr{H}t/\hbar}\psi(t) = \psi(0)$	$\mathscr{H} = \mathscr{H}_0 + V$
Evolution operator	Operators	Wave function
$\psi(t) = U(t, t_0)\psi(t_0)$	$A_{\mathrm{H}}(t) = e^{i\mathscr{H}t/\hbar} A e^{-i\mathscr{H}t/\hbar}$	$\psi_{\mathrm{I}}(t) = e^{i\mathscr{H}_0 t/\hbar}\psi(t)$
$i\hbar \dfrac{\partial U(t, t_0)}{\partial t} = \mathscr{H} U(t, t_0)$	$\langle A_{\mathrm{H}}(t)\rangle = \langle \psi_{\mathrm{H}} \vert A_{\mathrm{H}}(t) \vert \psi_{\mathrm{H}}\rangle = \langle A\rangle$	$i\hbar \dfrac{\partial\psi_{\mathrm{I}}(t)}{\partial t} = V_{\mathrm{I}}(t)\psi_{\mathrm{I}}(t)$
$U(t_0, t_0) = 1$	Heisenberg equation	Operators
$U^{\dagger}(t, t_0) = U^{-1}(t, t_0)$	$i\hbar \dfrac{\partial A_{\mathrm{H}}(t)}{\partial t} = [A_{\mathrm{H}}(t), \mathscr{H}_{\mathrm{H}}]$	$A_{\mathrm{I}}(t) = e^{i\mathscr{H}_0 t/\hbar} A e^{-i\mathscr{H}_0 t/\hbar}$
Operators	$\mathscr{H}_{\mathrm{H}} = \mathscr{H}$	$\langle A_{\mathrm{I}}(t)\rangle = \langle \psi_{\mathrm{I}}(t) \vert A_{\mathrm{I}}(t) \vert \psi_{\mathrm{I}}(t)\rangle = \langle A\rangle$
$\langle A\rangle = \langle \psi(t) \vert A \vert \psi(t)\rangle$	Assumption	$i\hbar \dfrac{\partial A_{\mathrm{I}}(t)}{\partial t} = [A_{\mathrm{I}}(t), \mathscr{H}_0]$
$i\hbar \dfrac{\partial}{\partial t}\langle A\rangle = \langle [A, \mathscr{H}]\rangle$	$\mathscr{H}$ independent of time	Evolution operator
If $\mathscr{H}$ is independent of time,		$\psi_{\mathrm{I}}(t) = U_{\mathrm{I}}(t, t_0)\psi_{\mathrm{I}}(t_0)$
$U(t, t_0) = e^{-i\mathscr{H}(t - t_0)/\hbar}$		$U_{\mathrm{I}}(t_0, t_0) = 1$
		$U_{\mathrm{I}}^{\dagger}(t, t_0) = U_{\mathrm{I}}^{-1}(t, t_0)$
		$U_{\mathrm{I}}(t, t_0) = e^{i\mathscr{H}_0 t/\hbar} U(t, t_0) e^{-i\mathscr{H}_0 t_0/\hbar}$
		$i\hbar \dfrac{\partial U_{\mathrm{I}}(t, t_0)}{\partial t} = V_{\mathrm{I}}(t) U_{\mathrm{I}}(t, t_0)$
		$U_{\mathrm{I}}(t, t_0) = 1 - \dfrac{i}{\hbar}\int_{t_0}^{t} V_{\mathrm{I}}(t') U_{\mathrm{I}}(t', t_0)\, dt'$
		$= 1 - \dfrac{i}{\hbar}\int_{t_0}^{t} V_{\mathrm{I}}(t_1)\, dt_1 + \left(\dfrac{-i}{\hbar}\right)^2 \int_{t_0}^{t} V_{\mathrm{I}}(t_1)\, dt_1 \int_{t_0}^{t_1} V_{\mathrm{I}}(t_2)\, dt_2 + \cdots$

[a] Schrödinger: $\psi(t)$ carries time dependence; A is independent of time. Heisenberg: ψ_{H} is independent of time; $A_{\mathrm{H}}(t)$ carries time dependence. Interaction: $\psi_{\mathrm{I}}(t)$ carries time dependence associated with V; $A_{\mathrm{I}}(t)$ carries time dependence associated with $\mathscr{H}_0$.

When (9.3-15) is incorporated into (9.3-22) one finds

$$\begin{aligned}\langle A_{\mathrm{I}}\rangle &= \langle A\rangle \\ &= \langle\psi_{\mathrm{I}}(t_0)|A_{\mathrm{I}}(t)|\psi_{\mathrm{I}}(t_0)\rangle \\ &\quad + \left(\frac{-i}{\hbar}\right)\int_{t_0}^{t} dt_1\, \langle\psi_{\mathrm{I}}(t_0)|[A_{\mathrm{I}}(t), V_{\mathrm{I}}(t_1)]|\psi_{\mathrm{I}}(t_0)\rangle \\ &\quad + \left(\frac{-i}{\hbar}\right)^2 \int_{t_0}^{t} dt_1 \int_{t_0}^{t} dt_2\, \langle\psi_{\mathrm{I}}(t_0)|[[A_{\mathrm{I}}(t), V_{\mathrm{I}}(t_1)], V_{\mathrm{I}}(t_2)]|\psi(t_0)\rangle + \cdots .\end{aligned} \tag{9.3-23}$$

The differences among the three representations may now be summarized.

Schrödinger representation. $\psi(t)$ obeys the Schrödinger equation and is the source of information on the time development of the system. The operators are independent of time (excluding the possibility that they may have an explicit time dependence), but the expectation value of an operator is *not* independent of time unless the operator commutes with the Hamiltonian.

Heisenberg representation. The state vector ψ_{H} is independent of time, but operators obey the Heisenberg equation. The time-dependent operators provide the information on the time-development of the system.

Interaction representation. Both state vectors and operators are time-dependent. When $\mathscr{H} = \mathscr{H}_0 + V$, the state vectors carry the time dependence due to V and the operators carry the time dependence due to $\mathscr{H}_0$.

A summary of the pertinent information concerning the three representations is presented in Table 9.1.

9.4 Infinite Limits

We return to the integral equation (9.3-14a):

$$U_{\mathrm{I}}(t, t_0) = 1 - \frac{i}{\hbar}\int_{t_0}^{t} V_{\mathrm{I}}(t')U_{\mathrm{I}}(t', t_0)\, dt' \tag{9.4-1}$$

satisfied by $U_{\mathrm{I}}(t, t_0)$, the evolution operator in the interaction representation. The present objective is to investigate the behavior of $U_{\mathrm{I}}(t, t_0)$ as one or both limits in the integral are taken to infinity. In order to ensure the proper treatment of convergence properties it is convenient to incorporate a factor

$$e^{-\varepsilon|t|/\hbar}, \qquad \varepsilon > 0 \tag{9.4-2}$$

into the integral; after the integration is performed (with infinite limits), another limiting procedure is carried out in which $\varepsilon \to 0$. Thus

$$U_{\mathrm{I}}(\infty, t_0) = 1 - \frac{i}{\hbar} \lim_{\varepsilon \to 0} \int_{t_0}^{\infty} e^{-\varepsilon|t'|/\hbar} V_{\mathrm{I}}(t') U_{\mathrm{I}}(t', t_0)\, dt' \tag{9.4-3}$$

and

$$U_{\mathrm{I}}(\infty, -\infty) = 1 - \frac{i}{\hbar} \lim_{\varepsilon \to 0} \int_{-\infty}^{\infty} e^{-\varepsilon|t'|/\hbar} V_{\mathrm{I}}(t') U_{\mathrm{I}}(t', -\infty)\, dt'. \tag{9.4-4}$$

The operator $U_{\mathrm{I}}(\infty, -\infty)$ is also known as the *scattering operator* S because of its role in the description of scattering processes.

Another expression for $U_{\mathrm{I}}(\infty, -\infty)$ may be derived by taking the Hermitian conjugate of (9.4-1):

$$\begin{aligned} U_{\mathrm{I}}^{\dagger}(t, t_0) &= 1 + \frac{i}{\hbar} \int_{t_0}^{t} [V_{\mathrm{I}}(t') U_{\mathrm{I}}(t', t_0)]^{\dagger}\, dt' \\ &= 1 + \frac{i}{\hbar} \int_{t_0}^{t} U_{\mathrm{I}}^{\dagger}(t', t_0) V_{\mathrm{I}}^{\dagger}(t')\, dt'. \end{aligned}$$

Since $V_{\mathrm{I}}(t)$ is Hermitian and $U_{\mathrm{I}}(t, t_0)$ is unitary

$$U_{\mathrm{I}}^{\dagger}(t, t_0) = U_{\mathrm{I}}^{-1}(t, t_0) = U_{\mathrm{I}}(t_0, t) = 1 - \frac{i}{\hbar} \int_{t}^{t_0} U_{\mathrm{I}}(t_0, t') V_{\mathrm{I}}(t')\, dt'.$$

Incorporating the convergence factor (9.4-2) we now have

$$U_{\mathrm{I}}(t_0, -\infty) = 1 - \frac{i}{\hbar} \lim_{\varepsilon \to 0} \int_{-\infty}^{t_0} e^{-\varepsilon|t'|/\hbar} U_{\mathrm{I}}(t_0, t') V_{\mathrm{I}}(t')\, dt',$$

which may be written more generally as

$$U_{\mathrm{I}}(t, -\infty) = 1 - \frac{i}{\hbar} \lim_{\varepsilon \to 0} \int_{-\infty}^{t} e^{-\varepsilon|t'|/\hbar} U_{\mathrm{I}}(t, t') V_{\mathrm{I}}(t')\, dt', \tag{9.4-5}$$

and upon passing to the limit $t \to \infty$,

$$U_{\mathrm{I}}(\infty, -\infty) = 1 - \frac{i}{\hbar} \lim_{\varepsilon \to 0} \int_{-\infty}^{\infty} e^{-\varepsilon|t'|/\hbar} U_{\mathrm{I}}(\infty, t') V_{\mathrm{I}}(t')\, dt'. \tag{9.4-6}$$

We adopt the notation

$$\psi_{\mathrm{I}}(-\infty) = \varphi_{\mathrm{a}}; \qquad \psi_{\mathrm{I}}(0) = \psi_{\mathrm{a}} \tag{9.4-7}$$

with the assumptions that

$$V_{\mathrm{I}}(-\infty) = 0 \tag{9.4-8}$$

$$\mathscr{H}_0 \varphi_{\mathrm{a}} = E_{\mathrm{a}} \varphi_{\mathrm{a}}. \tag{9.4-9}$$

When (9.4-7) is substituted into (9.3-7) and $U_{\rm I}(0, -\infty)$ is written in accordance with (9.4-5),

$$\begin{aligned}\psi_{\rm I}(0) = \psi_{\rm a} &= U_{\rm I}(0, -\infty)\psi_{\rm I}(-\infty)\\ &= U_{\rm I}(0, -\infty)\varphi_{\rm a}\\ &= \left[1 - \frac{i}{\hbar}\lim_{\varepsilon\to 0}\int_{-\infty}^{0} dt'\, e^{-\varepsilon|t'|/\hbar} U_{\rm I}(0, t')V_{\rm I}(t')\right]\varphi_{\rm a}.\end{aligned} \tag{9.4-10}$$

But from (9.3-3) and (9.3-11)

$$V_{\rm I}(t') = e^{i\mathscr{H}_0 t'/\hbar} V e^{-i\mathscr{H}_0 t'/\hbar}, \qquad U_{\rm I}(0, t') = e^{i\mathscr{H}t'/\hbar} e^{-i\mathscr{H}_0 t'/\hbar},$$

so that

$$U_{\rm I}(0, t')V_{\rm I}(t') = e^{i\mathscr{H}t'/\hbar} V e^{-i\mathscr{H}_0 t'/\hbar}. \tag{9.4-11}$$

Therefore

$$\psi_{\rm a} = \varphi_{\rm a} - \frac{i}{\hbar}\lim_{\varepsilon\to 0}\int_{-\infty}^{0} dt'\, e^{-\varepsilon|t'|/\hbar} e^{i(\mathscr{H} - E_{\rm a})t'/\hbar} V\varphi_{\rm a}$$

which may be integrated to give

$$\psi_{\rm a} = \varphi_{\rm a} + \lim_{\varepsilon\to 0}\frac{1}{E_{\rm a} - \mathscr{H} + i\varepsilon} V\varphi_{\rm a}. \tag{9.4-12}$$

This permits us to identify

$$U_{\rm I}(0, -\infty) = 1 + \lim_{\varepsilon\to 0}\frac{1}{E_{\rm a} - \mathscr{H} + i\varepsilon} V. \tag{9.4-13}$$

There is yet another useful expression for $\psi_{\rm a}$. The differential equation (9.3-5)

$$i\hbar\frac{\partial\psi_{\rm I}(t)}{\partial t} = V_{\rm I}(t)\psi_{\rm I}(t)$$

may be converted to the integral equation

$$\psi_{\rm I}(t) = \psi_{\rm I}(t_0) - \frac{i}{\hbar}\int_{t_0}^{t} V_{\rm I}(t')\psi_{\rm I}(t')\, dt'. \tag{9.4-14}$$

On setting $t_0 = -\infty$, $t = 0$, and inserting the convergence factor (9.4-2) we have, in the notation of (9.4-7),

$$\begin{aligned}\psi_{\rm a} &= \varphi_{\rm a} - \frac{i}{\hbar}\lim_{\varepsilon\to 0}\int_{-\infty}^{0} dt'\, e^{-\varepsilon|t'|\hbar} V_{\rm I}(t')\psi_{\rm I}(t')\\ &= \varphi_{\rm a} - \frac{i}{\hbar}\lim_{\varepsilon\to 0}\int_{-\infty}^{0} dt'\, e^{-\varepsilon|t'|\hbar} e^{i\mathscr{H}_0 t'/\hbar} V e^{-i\mathscr{H}_0 t'/\hbar}\psi_{\rm I}(t').\end{aligned} \tag{9.4-15}$$

This equation may be solved by iteration as was done to obtain (9.3-15). Thus

$$\psi_a^{(0)} = \varphi_a$$

which corresponds to the initial value at $-\infty$. For the next approximation $\psi_I(t')$ in (9.4-15) is replaced by φ_a giving

$$\begin{aligned}\psi_a^{(1)} &= \varphi_a - \frac{i}{\hbar}\lim_{\varepsilon\to 0}\int_{-\infty}^{0} dt'\, e^{-\varepsilon|t'|/\hbar} e^{i\mathscr{H}_0 t'/\hbar} V e^{-i\mathscr{H}_0 t'/\hbar}\varphi_a \\ &= \varphi_a + \lim_{\varepsilon\to 0}\frac{1}{E_a - \mathscr{H}_0 + i\varepsilon} V\varphi_a. \end{aligned} \tag{9.4-16}$$

The process may be continued indefinitely to obtain

$$\begin{aligned}\psi_a &= \varphi_a + \lim_{\varepsilon\to 0}\frac{1}{E_a - \mathscr{H}_0 + i\varepsilon} V\varphi_a \\ &\quad + \lim_{\varepsilon\to 0}\frac{1}{E_a - \mathscr{H}_0 + i\varepsilon} V \frac{1}{E_a - \mathscr{H}_0 + i\varepsilon} V\varphi_a + \cdots \end{aligned} \tag{9.4-17}$$

which is just the iterative solution to the *Lippmann–Schwinger* equation

$$\psi_a = \varphi_a + \lim_{\varepsilon\to 0}\frac{1}{E_a - \mathscr{H}_0 + i\varepsilon} V\psi_a. \tag{9.4-18}$$

The similarity as well as the difference between expressions (9.4-12) and (9.4-18) should be noted. In the former case, ψ_a is expressed in terms of the total Hamiltonian $\mathscr{H} = \mathscr{H}_0 + V$, whereas the latter expression is based on $\mathscr{H}_0$ and represents an infinite series. In either case we may verify that ψ_a is an eigenfunction of $\mathscr{H}$ with eigenvalue E_i. Thus in the limit $\varepsilon \to 0$, and bearing in mind that φ_a is an eigenfunction of $\mathscr{H}_0$, we obtain from (9.4-12):

$$(E_a - \mathscr{H})\psi_a = (E_a - \mathscr{H})\varphi_a + V\varphi_a = (E_a - \mathscr{H}_0 - V)\varphi_a + V\varphi_a = 0$$

and from (9.4-18):

$$(E_a - \mathscr{H}_0)\psi_a = (E_a - \mathscr{H}_0)\varphi_a + V\psi_a = V\psi_a,$$

or

$$(E_a - \mathscr{H})\psi_a = 0.$$

Finally we note that operators such as $(E_a - \mathscr{H} + i\varepsilon)^{-1}$ and $(E_a - \mathscr{H}_0 + i\varepsilon)^{-1}$ appear in scattering theory and are usually known as Green's functions.

One often needs matrix elements of the scattering operator S in a basis set φ_k which are eigenfunctions of an unperturbed Hamiltonian $\mathscr{H}_0$ with eigenvalues E_k. Using (9.4-4),

$$S_{\mathrm{ba}} = \langle \varphi_{\mathrm{b}} | U_{\mathrm{I}}(\infty, -\infty) | \varphi_{\mathrm{a}} \rangle$$

$$= \left\langle \varphi_{\mathrm{b}} \left| 1 - \frac{i}{\hbar} \lim_{\varepsilon \to 0} \int_{-\infty}^{\infty} e^{-\varepsilon |t'|/\hbar} V_{\mathrm{I}}(t') U_{\mathrm{I}}(t', -\infty)\, dt' \right| \varphi_{\mathrm{a}} \right\rangle. \quad (9.4\text{-}19)$$

From (9.1-5) and (9.3-11),

$$U_{\mathrm{I}}(t', -\infty) = U_{\mathrm{I}}(t', 0) U_{\mathrm{I}}(0, -\infty) = e^{i\mathscr{H}_0 t'/\hbar} e^{-i\mathscr{H} t'/\hbar} U_{\mathrm{I}}(0, -\infty).$$

Therefore

$$V_{\mathrm{I}}(t') U_{\mathrm{I}}(t', -\infty) = e^{i\mathscr{H}_0 t'/\hbar} V e^{-i\mathscr{H}_0 t'/\hbar} U_{\mathrm{I}}(t', -\infty)$$
$$= e^{i\mathscr{H}_0 t'/\hbar} V e^{-i\mathscr{H} t'/\hbar} U_{\mathrm{I}}(0, -\infty)$$

and

$$S_{\mathrm{ba}} = \delta_{\mathrm{ba}} - \frac{i}{\hbar} \lim_{\varepsilon \to 0} \int_{-\infty}^{\infty} dt'\, e^{-\varepsilon |t'|/\hbar} \langle \varphi_{\mathrm{b}} | e^{i\mathscr{H}_0 t'/\hbar} V e^{-i\mathscr{H} t'/\hbar} U_{\mathrm{I}}(0, -\infty) | \varphi_{\mathrm{a}} \rangle. \quad (9.4\text{-}20)$$

Also, it has been shown that the function

$$\psi_{\mathrm{a}} = U_{\mathrm{I}}(0, -\infty) \varphi_{\mathrm{a}}$$

is an eigenfunction of $\mathscr{H}$ with eigenvalue E_{a}. We then have

$$\langle \varphi_{\mathrm{b}} | e^{i\mathscr{H}_0 t'/\hbar} V e^{-i\mathscr{H} t'/\hbar} U_{\mathrm{I}}(0, -\infty) | \varphi_{\mathrm{a}} \rangle = \langle \varphi_{\mathrm{b}} | V | \psi_{\mathrm{a}} \rangle e^{i(E_{\mathrm{b}} - E_{\mathrm{a}}) t'/\hbar}$$

which, when substituted into (9.4-20), permits the evaluation of the integral. This is facilitated by writing

$$\alpha = \varepsilon/\hbar, \qquad \beta = (E_{\mathrm{b}} - E_{\mathrm{a}})/\hbar$$

in which case

$$\lim_{\varepsilon \to 0} \int_{-\infty}^{\infty} dt'\, e^{-\varepsilon |t'|/\hbar} \langle \varphi_{\mathrm{b}} | V | \psi_{\mathrm{a}} \rangle e^{i(E_{\mathrm{b}} - E_{\mathrm{a}}) t'/\hbar}$$

$$= \lim_{\alpha \to 0} \langle \varphi_{\mathrm{b}} | V | \psi_{\mathrm{a}} \rangle \int_{-\infty}^{\infty} dt'\, e^{-\alpha |t'|} e^{i\beta t'}$$

$$= \lim_{\alpha \to 0} \langle \varphi_{\mathrm{b}} | V | \psi_{\mathrm{a}} \rangle \left[\int_{-\infty}^{0} dt'\, e^{\alpha t'} e^{i\beta t'} + \int_{0}^{\infty} dt'\, e^{-\alpha t'} e^{i\beta t'} \right]$$

$$= \lim_{\alpha \to 0} \langle \varphi_{\mathrm{b}} | V | \psi_{\mathrm{a}} \rangle \frac{2\alpha}{\alpha^2 + \beta^2} = 2\pi \langle \varphi_{\mathrm{b}} | V | \psi_{\mathrm{a}} \rangle\, \delta(\beta)$$

$$= 2\pi \langle \varphi_{\mathrm{b}} | V | \psi_{\mathrm{a}} \rangle\, \delta\left(\frac{E_{\mathrm{b}} - E_{\mathrm{a}}}{\hbar} \right).$$

Hence the S matrix (9.4-19) becomes

$$S_{\mathrm{ba}} = \delta_{\mathrm{ba}} - 2\pi i \langle \varphi_{\mathrm{b}} | V | \psi_{\mathrm{a}} \rangle\, \delta(E_{\mathrm{b}} - E_{\mathrm{a}}) \tag{9.4-21a}$$

$$= \delta_{\mathrm{ba}} - 2\pi i R_{\mathrm{ba}}\, \delta(E_{\mathrm{b}} - E_{\mathrm{a}}) \tag{9.4-21b}$$

where

$$R_{\mathrm{ba}} \equiv \langle \varphi_{\mathrm{b}} | V | \psi_{\mathrm{a}} \rangle$$

is known as the *reaction matrix.*

A more useful form for the reaction matrix may be obtained by replacing ψ_{a} by (9.4-18):

$$\begin{aligned}
R_{\mathrm{ba}} &= \langle \varphi_{\mathrm{b}} | V | \varphi_{\mathrm{a}} \rangle + \lim_{\varepsilon \to 0} \left\langle \varphi_{\mathrm{b}} \left| V \frac{1}{E_{\mathrm{a}} - \mathscr{H}_0 + i\varepsilon} V \right| \psi_{\mathrm{a}} \right\rangle \\
&= \langle \varphi_{\mathrm{b}} | V | \varphi_{\mathrm{a}} \rangle + \lim_{\varepsilon \to 0} \sum_c \left\langle \varphi_{\mathrm{b}} \left| V \frac{1}{E_{\mathrm{a}} - \mathscr{H}_0 + i\varepsilon} \right| \varphi_c \right\rangle \langle \varphi_c | V | \psi_{\mathrm{a}} \rangle \\
&= \langle \varphi_{\mathrm{b}} | V | \varphi_{\mathrm{a}} \rangle + \lim_{\varepsilon \to 0} \sum_c \frac{\langle \varphi_{\mathrm{b}} | V | \varphi_c \rangle}{E_{\mathrm{a}} - E_c + i\varepsilon} \langle \varphi_c | V | \psi_{\mathrm{a}} \rangle \\
&= V_{\mathrm{ba}} + \lim_{\varepsilon \to 0} \sum_c \frac{V_{\mathrm{b}c} R_{c\mathrm{a}}}{E_{\mathrm{a}} - E_c + i\varepsilon} \\
&= V_{\mathrm{ba}} + \lim_{\varepsilon \to 0} \sum_c \frac{V_{\mathrm{b}c} V_{c\mathrm{a}}}{E_{\mathrm{a}} - E_c + i\varepsilon} + \text{higher-order terms.}
\end{aligned} \tag{9.4-22}$$

CHAPTER 10

HARMONIC OSCILLATOR

10.1 Schrödinger Solutions

For a one-dimensional harmonic oscillator of mass m and frequency ω the Hamiltonian is

$$\mathscr{H} = \frac{p^2}{2m} + \frac{m}{2}\,\omega^2 q^2 \tag{10.1-1}$$

where p is the momentum and q the displacement. With

$$p = -i\hbar\,\frac{d}{dq}$$

the Schrödinger equation corresponding to (10.1-1) is

$$\frac{d^2\psi(q)}{dq^2} + \frac{2m}{\hbar^2}\left(E - \frac{1}{2}\,m\omega^2 q^2\right)\psi(q) = 0. \tag{10.1-2}$$

The eigenfunctions of (10.1-2) are

$$\psi_n(q) = \frac{1}{2^{n/2}\sqrt{n!}}\left(\frac{m\omega}{\hbar\pi}\right)^{1/4}\exp\left(-\frac{m\omega}{2\hbar}\,q^2\right)H_n\left(\sqrt{\frac{m\omega}{\hbar}}\,q\right) \tag{10.1-3}$$

in which $H_n(x)$ is a Hermite polynomial of degree n (Appendix 7) and the eigenvalues are

$$E_n = (n + \tfrac{1}{2})\hbar\omega, \qquad n = 0, 1, 2, \ldots . \tag{10.1-4}$$

A more compact form of (10.1-3) is achieved by changing the variables to

$$\xi = \sqrt{\frac{m\omega}{\hbar}}\,q, \qquad u_n = \sqrt{\frac{\hbar}{m\omega}}\,\psi_n. \tag{10.1-5}$$

We then have

$$u_n(\xi) = \frac{1}{(\sqrt{\pi}2^n n!)^{1/2}} e^{-\xi^2/2} H_n(\xi) \tag{10.1-6}$$

which has the properties:

$$\int_{-\infty}^{\infty} u_n u_p \, d\xi = \delta_{np}, \tag{10.1-7}$$

$$\frac{1}{\sqrt{2}}\left(\xi + \frac{d}{d\xi}\right) u_n = \sqrt{n} u_{n-1}, \tag{10.1-8}$$

$$\frac{1}{\sqrt{2}}\left(\xi - \frac{d}{d\xi}\right) u_n = \sqrt{n+1} u_{n+1}, \tag{10.1-9}$$

$$\xi u_n = \sqrt{\frac{n+1}{2}} u_{n+1} + \sqrt{\frac{n}{2}} u_{n-1}, \tag{10.1-10}$$

$$u_n(-\xi) = (-1)^n u_n(\xi). \tag{10.1-11}$$

10.2 Matrix Formulation

We shall now investigate the harmonic oscillator by means of a matrix formulation. For convenience, let

$$P^2 = p^2/m, \qquad Q^2 = mq^2,$$

so that the Hamiltonian (10.1-1) becomes

$$\mathscr{H} = \tfrac{1}{2}(P^2 + \omega^2 Q^2). \tag{10.2-1}$$

The transition to quantum mechanics is made by reinterpreting P and Q as Hermitian operators which obey the commutation relation

$$[Q, P] = i\hbar \tag{10.2-2}$$

which is equivalent to the replacement of P by $-i\hbar\, \partial/\partial Q$. It is important to note that whereas the canonical variables Q and P in the classical Hamiltonian are functions of time, the quantum-mechanical operators Q and P are independent of time (Schrödinger representation).

We now construct the linear combinations

$$a = \frac{1}{\sqrt{2\hbar\omega}}(\omega Q + iP), \qquad a^\dagger = \frac{1}{\sqrt{2\hbar\omega}}(\omega Q - iP); \tag{10.2-3}$$

or

$$Q = \sqrt{\frac{\hbar}{2\omega}}(a + a^\dagger), \qquad P = i\sqrt{\frac{\hbar\omega}{2}}(a^\dagger - a). \tag{10.2-4}$$

Since Q and P are in the Schrödinger representation, a and $a^\dagger$ must also be in the same representation which means then that they, too, are independent of time. In terms of a and $a^\dagger$, (10.2-2) becomes

$$[a, a^\dagger] = 1 \tag{10.2-5}$$

and the Hamiltonian (10.2-1) takes the form

$$\mathscr{H} = \tfrac{1}{2}\hbar\omega(a^\dagger a + aa^\dagger) = \hbar\omega(a^\dagger a + \tfrac{1}{2}) \tag{10.2-6}$$

$$= \hbar\omega(N + \tfrac{1}{2}) \tag{10.2-7}$$

where

$$N = a^\dagger a \tag{10.2-8}$$

is called the *number operator*. Although a and $a^\dagger$ are not Hermitian, as is evident from definitions (10.2-3), N is nevertheless Hermitian. Also, from (10.2-5),

$$Na = a^\dagger aa = (aa^\dagger - 1)a = a(a^\dagger a - 1) = a(N - 1), \tag{10.2-9}$$

$$Na^\dagger = a^\dagger aa^\dagger = a^\dagger(a^\dagger a + 1) = a^\dagger(N + 1). \tag{10.2-10}$$

Now let $|n\rangle$ be an eigenstate of N with eigenvalue n, i.e., let

$$N|n\rangle = n|n\rangle. \tag{10.2-11}$$

Since N is Hermitian the eigenvalue n is real. Moreover it is a nondegenerate eigenvalue because the symmetry group of the Hamiltonian (10.1-1) is C_i, that is, $\mathscr{H}$ is invariant under the identity and inversion ($q \to -q$) operations. The irreducible representations of C_i are one-dimensional (Table 5.2); hence the eigenvalues of the Hamiltonian and therefore the eigenvalues of N must be nondegenerate. The Hermitian property of N also require that

$$\langle n'|n\rangle = \delta_{n'n}. \tag{10.2-12}$$

From (10.2-9) and (10.2-11),

$$Na|n\rangle = a(N - 1)|n\rangle = aN|n\rangle - a|n\rangle = an|n\rangle - a|n\rangle = (n - 1)a|n\rangle \tag{10.2-13}$$

and from (10.2-10) and (10.2-11),

$$Na^\dagger|n\rangle = a^\dagger(N + 1)|n\rangle = a^\dagger N|n\rangle + a^\dagger|n\rangle = a^\dagger n|n\rangle + a^\dagger|n\rangle = (n + 1)a^\dagger|n\rangle. \tag{10.2-14}$$

Thus we have the result that if $|n\rangle$ is an eigenstate of N with eigenvalue n, then $a|n\rangle$ is also an eigenstate of N with eigenvalue $n - 1$. Similarly, $a^\dagger|n\rangle$ is also an eigenstate of N with eigenvalue $n + 1$. The eigenvalues of N and the corresponding eigenstates may be displayed in the form of a ladder (Fig. 10.1), and in view of (10.2-7), the energy levels of the harmonic oscillator must have a constant spacing with an energy $\hbar\omega$ between adjacent levels.

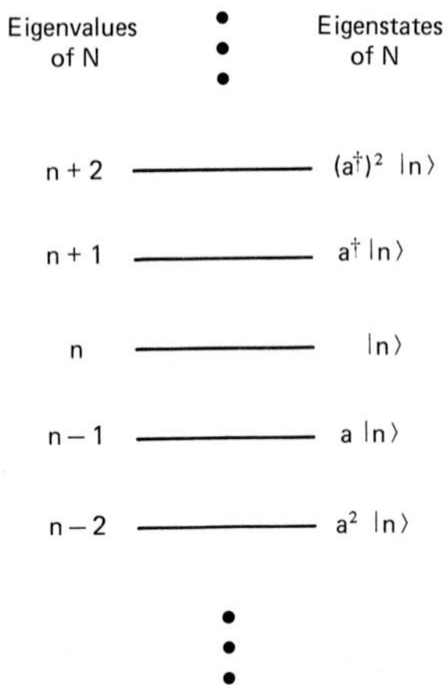

FIG. 10.1 Eigenvalues and eigenstates of the number operator N.

However, the ladder has a lower bound because

$$\langle n|N|n\rangle = \langle n|a^\dagger a|n\rangle = \langle an|an\rangle \geqslant 0,$$

and $\langle n|N|n\rangle = n$ so that

$$\langle n|N|n\rangle = \langle an|an\rangle = n \geqslant 0. \tag{10.2-15}$$

From (10.2-11) and (10.2-13),

$$N|n-1\rangle = (n-1)|n-1\rangle, \qquad Na|n\rangle = (n-1)a|n\rangle.$$

Since there are no degeneracies

$$a|n\rangle = c_n|n-1\rangle \tag{10.2-16}$$

where c_n is a constant of proportionality. Since

$$\langle an|an\rangle = c_n{}^2\langle n-1|n-1\rangle = c_n{}^2,$$

and from (10.2-15), where

$$\langle an|an\rangle = n,$$

we have

$$c_n = \sqrt{n} \tag{10.2-17}$$

where the positive square root is in accord with convention. Thus

$$a|n\rangle = \sqrt{n}|n-1\rangle. \tag{10.2-18}$$

In the same way we find

$$a^\dagger|n\rangle = \sqrt{n+1}|n+1\rangle. \tag{10.2-19}$$

Because of (10.2-18) and (10.2-19), the operators a and $a^\dagger$ have been given the names *annihilation* and *creation* operators, respectively.

Is n an integer? Assume that it is not; then (10.2-18) will ultimately lead to a negative value of n which would then violate (10.2-15). On the other hand, when n is an integer, (10.2-18) leads to

$$a|0\rangle = 0 \tag{10.2-20}$$

which is the lower limit of the sequence and is consistent with (10.2-15).

It is now possible to construct matrices to represent the various operators:

$$\langle n-1|a|n\rangle = \sqrt{n},$$

$$a = \begin{pmatrix} 0 & \sqrt{1} & 0 & 0 & \cdots \\ 0 & 0 & \sqrt{2} & 0 & \cdots \\ 0 & 0 & 0 & \sqrt{3} & \cdots \\ \vdots & \vdots & \vdots & \vdots & \cdots \end{pmatrix}, \tag{10.2-21}$$

$$\langle n+1|a^\dagger|n\rangle = \sqrt{n+1},$$

$$a^\dagger = \begin{pmatrix} 0 & 0 & 0 & \cdots \\ \sqrt{1} & 0 & 0 & \cdots \\ 0 & \sqrt{2} & 0 & \cdots \\ 0 & 0 & \sqrt{3} & \cdots \\ \vdots & \vdots & \vdots & \cdots \end{pmatrix}, \tag{10.2-22}$$

$$\langle n|a^\dagger a|n\rangle \equiv \langle n|N|n\rangle = n,$$

$$N = a^\dagger a = \begin{pmatrix} 0 & 0 & 0 & 0 & \cdots \\ 0 & 1 & 0 & 0 & \cdots \\ 0 & 0 & 2 & 0 & \cdots \\ 0 & 0 & 0 & 3 & \cdots \\ \vdots & \vdots & \vdots & \vdots & \cdots \end{pmatrix}, \tag{10.2-23}$$

$$\langle n|aa^\dagger|n\rangle = \langle n|a^\dagger a|n\rangle + \langle n|n\rangle = n+1,$$

$$aa^\dagger = \begin{pmatrix} 1 & 0 & 0 & 0 & \cdots \\ 0 & 2 & 0 & 0 & \cdots \\ 0 & 0 & 3 & 0 & \cdots \\ \vdots & \vdots & \vdots & \vdots & \cdots \end{pmatrix}. \tag{10.2-24}$$

The eigenfunctions $|n\rangle$ may also be given a matrix representation:

$$|0\rangle = \begin{pmatrix} 1 \\ 0 \\ 0 \\ \vdots \end{pmatrix}, \quad |1\rangle = \begin{pmatrix} 0 \\ 1 \\ 0 \\ \vdots \end{pmatrix}, \quad |2\rangle = \begin{pmatrix} 0 \\ 0 \\ 1 \\ \vdots \end{pmatrix}, \quad \text{etc.} \tag{10.2-25}$$

Q and P are related to a and $a^\dagger$ by (10.2-4). Hence

$$Q = \sqrt{\frac{\hbar}{2\omega}} \begin{pmatrix} 0 & \sqrt{1} & 0 & 0 & 0 & \cdots \\ \sqrt{1} & 0 & \sqrt{2} & 0 & 0 & \cdots \\ 0 & \sqrt{2} & 0 & \sqrt{3} & 0 & \cdots \\ 0 & 0 & \sqrt{3} & 0 & \sqrt{4} & \cdots \\ \vdots & \vdots & \vdots & \vdots & \vdots & \cdots \end{pmatrix}, \tag{10.2-26}$$

$$P = i\sqrt{\frac{\hbar\omega}{2}} \begin{pmatrix} 0 & -\sqrt{1} & 0 & 0 & 0 & \cdots \\ \sqrt{1} & 0 & -\sqrt{2} & 0 & 0 & \cdots \\ 0 & \sqrt{2} & 0 & -\sqrt{3} & 0 & \cdots \\ 0 & 0 & \sqrt{3} & 0 & -\sqrt{4} & \cdots \\ \vdots & \vdots & \vdots & \vdots & \vdots & \cdots \end{pmatrix}. \tag{10.2-27}$$

Finally, the Hamiltonian (10.2-7) is

$$\mathscr{H} = \hbar\omega \begin{pmatrix} \frac{1}{2} & 0 & 0 & 0 & \cdots \\ 0 & \frac{3}{2} & 0 & 0 & \cdots \\ 0 & 0 & \frac{5}{2} & 0 & \cdots \\ 0 & 0 & 0 & \frac{7}{2} & \cdots \\ \vdots & \vdots & \vdots & \vdots & \cdots \end{pmatrix}. \tag{10.2-28}$$

In diagramatic fashion the eigenstates and eigenvalues of the harmonic oscillator may be represented as shown in Fig. 10.2.

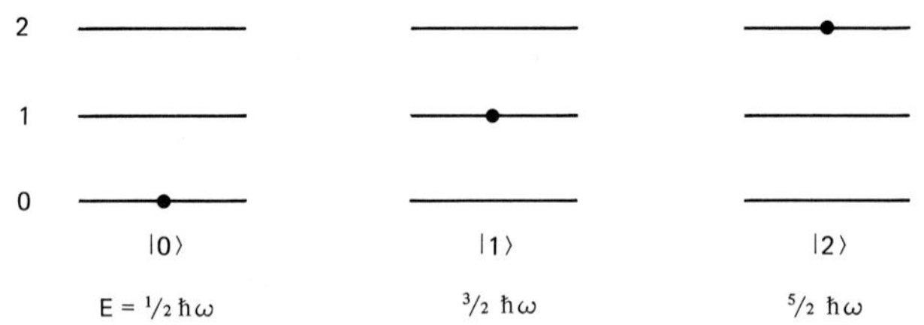

FIG. 10.2 Eigenvalues and eigenstates of a single harmonic oscillator.

10.3 Heisenberg Representation

It is recalled (Section 9.2) that a Heisenberg operator $A_H(t)$ is related to a Schrödinger operator A by

$$A_H(t) = e^{i\mathscr{H}t/\hbar} A e^{-i\mathscr{H}t/\hbar} \tag{10.3-1}$$

and satisfies the Heisenberg equation

$$i\hbar(\partial A_H(t)/\partial t) = [A_H(t), \mathscr{H}_H]. \tag{10.3-2}$$

For the annihilation operator a we then have

$$i\hbar(\partial a_H(t)/\partial t) = [a_H(t), \mathscr{H}_H]. \tag{10.3-3}$$

In the Schrödinger representation

$$\begin{aligned}[a, \mathscr{H}] &= [a, \hbar\omega(N + \tfrac{1}{2})] = \hbar\omega[a, N] \\ &= \hbar\omega(aN - Na) = \hbar\omega(aN - aN + a) = \hbar\omega a.\end{aligned} \tag{10.3-4}$$

The same relationship must hold in the Heisenberg representation, i.e.,

$$[a_H(t), \mathscr{H}_H] = \hbar\omega a_H$$

and the Heisenberg equation becomes

$$\partial a_H(t)/\partial t = -i\omega a_H. \tag{10.3-5}$$

In like fashion the creation operator $a^\dagger$ satisfies

$$\partial a_H^\dagger(t)/\partial t = i\omega a_H^\dagger. \tag{10.3-6}$$

The two differential equations may be integrated to give

$$a_H(t) = a_H(0)e^{-i\omega t}, \qquad a_H^\dagger(t) = a_H^\dagger(0)e^{i\omega t} \tag{10.3-7}$$

and since Heisenberg and Schrödinger operators are the same at $t = 0$,

$$a_H(0) = a, \qquad a_H^\dagger(0) = a^\dagger,$$

so that

$$a_H(t) = ae^{-i\omega t}, \qquad a_H^\dagger(t) = a^\dagger e^{i\omega t}. \tag{10.3-8}$$

CHAPTER 11

SLATER DETERMINANTS

11.1 Matrix Elements—General

In Section 8.4 we saw that multielectron wave functions $\psi(\lambda_1, \lambda_2, \ldots, \lambda_N)$ must be antisymmetric with respect to an interchange of the (space and spin) coordinates of any two electrons. Antisymmetry can be ensured by expressing the wave function in terms of Slater determinants as in (8.4-13). To facilitate the calculation of various physical quantities, we shall need expressions for matrix elements of operators when the wave functions are written in determinantal form.

Consider a two-electron system and let

$$\Psi_1 = \frac{1}{\sqrt{2}} \begin{vmatrix} \psi_j(\lambda_1) & \psi_k(\lambda_1) \\ \psi_j(\lambda_2) & \psi_k(\lambda_2) \end{vmatrix}, \tag{11.1-1}$$

$$\Psi_2 = \frac{1}{\sqrt{2}} \begin{vmatrix} \psi_j(\lambda_1) & \psi_l(\lambda_1) \\ \psi_j(\lambda_2) & \psi_l(\lambda_2) \end{vmatrix}, \tag{11.1-2}$$

$$\Psi_3 = \frac{1}{\sqrt{2}} \begin{vmatrix} \psi_l(\lambda_1) & \psi_m(\lambda_1) \\ \psi_l(\lambda_2) & \psi_m(\lambda_2) \end{vmatrix}. \tag{11.1-3}$$

In $\psi_k(\lambda_i)$, k is a label that identifies a particular spin orbital, i.e., a one-electron function that depends on both space and spin coordinates; the index i is an electron label. The notation may be shortened by writing

$$\psi_k(\lambda_i) \equiv \psi_k(i). \tag{11.1-4}$$

It will also be assumed that for any two spin orbitals such as ψ_k and ψ_l

$$\langle \psi_k(i) | \psi_l(i) \rangle = \delta_{kl}. \tag{11.1-5}$$

This has the immediate consequence that

$$\langle \Psi_i | \Psi_j \rangle = \delta_{ij} \tag{11.1-6}$$

where Ψ_i and Ψ_j are any of the determinantal functions (11.1-1)–(11.1-3).

Let us now suppose that we have a sum of one-electron operators

$$F = f_1 + f_2$$

where f_1 and f_2 have the same functional dependence but f_1 operates only on the spin orbital occupied by electron 1, namely $\psi(\lambda_1)$, and f_2 operates only on $\psi(\lambda_2)$. Since variables of integration are dummy variables we may write

$$\langle \psi_k(1) | f_1 | \psi_l(1) \rangle = \langle \psi_k(2) | f_2 | \psi_l(2) \rangle, = \langle \psi_k | f | \psi_l \rangle. \tag{11.1-7}$$

Therefore, in view of the orthonormality relation (11.1-5),

$$\langle \Psi_1 | F | \Psi_1 \rangle = \langle \psi_j | f | \psi_j \rangle + \langle \psi_k | f | \psi_k \rangle \tag{11.1-8}$$

with analogous expressions for $\langle \Psi_2 | F | \Psi_2 \rangle$ and $\langle \Psi_3 | F | \Psi_3 \rangle$. For the off-diagonal elements

$$\langle \Psi_1 | F | \Psi_2 \rangle = \langle \psi_k | f | \psi_l \rangle, \tag{11.1-9a}$$

$$\langle \Psi_1 | F | \Psi_3 \rangle = 0, \tag{11.1-9b}$$

$$\langle \Psi_2 | F | \Psi_3 \rangle = -\langle \psi_j | f | \psi_m \rangle. \tag{11.1-9c}$$

A two-electron operator g_{12} operates on both $\psi(\lambda_1)$ and $\psi(\lambda_2)$, as, for example, in the case of the electronic Coulomb repulsion operator e^2/r_{12}. For a typical diagonal element

$$\langle \Psi_1 | g_{12} | \Psi_1 \rangle = \langle \psi_j(1)\psi_k(2) | g_{12} | \psi_j(1)\psi_k(2) \rangle - \langle \psi_j(1)\psi_k(2) | g_{12} | \psi_k(1)\psi_j(2) \rangle, \tag{11.1-10}$$

and for off-diagonal elements

$$\langle \Psi_1 | g_{12} | \Psi_2 \rangle = \langle \psi_j(1)\psi_k(2) | g_{12} | \psi_j(1)\psi_l(2) \rangle - \langle \psi_j(1)\psi_k(2) | g_{12} | \psi_l(1)\psi_j(2) \rangle, \tag{11.1-11a}$$

$$\langle \Psi_1 | g_{12} | \Psi_3 \rangle = \langle \psi_j(1)\psi_k(2) | g_{12} | \psi_l(1)\psi_m(2) \rangle - \langle \psi_j(1)\psi_k(2) | g_{12} | \psi_m(1)\psi_l(2) \rangle, \tag{11.1-11b}$$

$$\langle \Psi_2 | g_{12} | \Psi_3 \rangle = -\langle \psi_j(1)\psi_l(2) | g_{12} | \psi_m(1)\psi_l(2) \rangle + \langle \psi_j(1)\psi_l(2) | g_{12} | \psi_l(1)\psi_m(2) \rangle. \tag{11.1-11c}$$

These results for the special case of the Slater determinants (11.1-1)–(11.1-3) may be generalized to determinants of arbitrary dimension. Thus, let

$$A = \frac{1}{\sqrt{N!}} \begin{vmatrix} a_1(1) & a_2(1) & \cdots & a_N(1) \\ a_1(2) & a_2(2) & \cdots & a_N(2) \\ \vdots & \vdots & \cdots & \vdots \\ a_1(N) & a_2(N) & \cdots & a_N(N) \end{vmatrix}, \tag{11.1-12}$$

$$B = \frac{1}{\sqrt{N!}} \begin{vmatrix} b_1(1) & b_2(1) & \cdots & b_N(1) \\ b_1(2) & b_2(2) & \cdots & b_N(2) \\ \vdots & \vdots & \cdots & \vdots \\ b_1(N) & b_2(N) & \cdots & b_N(N) \end{vmatrix}, \tag{11.1-13}$$

$$F = \sum_{i=1}^{N} f_i, \tag{11.1-14}$$

$$G = \sum_{i<j} g_{ij} = \tfrac{1}{2} \sum_{i \neq j} g_{ij}, \tag{11.1-15}$$

in which $a_k(i), b_k(i)$ are spin orbitals; f_i and g_{ij} are one- and two-electron operators, respectively.

We must also take note of the order in which the orbitals appear in (11.1-12) and (11.1-13) because an interchange of two columns (or rows) will change the sign of the determinantal wave function. As previously written the order is

$$A: \quad a_1, a_2, \ldots, a_k, a_{k+1}, \ldots, a_l, a_{l+1}, \ldots, a_N,$$
$$B: \quad b_1, b_2, \ldots, b_k, b_{k+1}, \ldots, b_l, b_{l+1}, \ldots, b_N.$$

For the diagonal matrix element of F,

$$\langle A|F|A\rangle = \sum_{k=1}^{N} \langle a_k|f|a_k\rangle \tag{11.1-16}$$

in which the argument of a_k and the subscript on f have been omitted, as they will be henceforth, since they are arbitrary (see, for example, (11.1-7)). The matrix element $\langle B|F|B\rangle$ has the same form with respect to the b orbitals. For an off-diagonal matrix element

$$\langle A|F|B\rangle = 0 \tag{11.1-17}$$

if A and B differ by more than one pair of orbitals, and

$$\langle A|F|B\rangle = \pm\langle a_k|f|b_l\rangle \tag{11.1-18}$$

if $a_k \neq b_l$, but the rest of the orbitals in B are the same as those in A. The plus sign occurs when an even number of interchanges are required to move the b_l orbital into the kth position or, in other words, when the parity of the permutation is even; the minus sign appears as a result of an odd-parity permutation. Examples of (11.1-18) are provided by (11.1-9a) and (11.1-9c). It may also be remarked that for one-electron operators such as (11.1-14), simple product functions, and determinantal functions give the same matrix elements.

The diagonal matrix elements of G are

$$\begin{aligned}\langle A|G|A\rangle = \sum_{k<t} [&\langle a_k(1)a_t(2)|g_{12}|a_k(1)a_t(2)\rangle \\ &- \langle a_k(1)a_t(2)|g_{12}|a_t(1)a_k(2)\rangle]\end{aligned} \tag{11.1-19}$$

and for off-diagonal elements we have the cases:

(1) If A and B differ by more than *two* pairs of spin orbitals,

$$\langle A|G|B\rangle = 0. \tag{11.1-20}$$

(2) If A and B differ by two pairs of orbitals, i.e., a_k, a_l in A are not the same as b_m, b_n in B but all other orbitals in A are the same as orbitals in B,

$$\begin{aligned}\langle A|G|B\rangle = \pm[&\langle a_k(1)a_l(2)|g_{12}|b_m(1)b_n(2)\rangle \\ &- \langle a_k(1)a_l(2)|g_{12}|b_n(1)b_m(2)\rangle].\end{aligned} \tag{11.1-21}$$

(3) If A and B differ by one pair of orbitals, e.g., $a_k \neq b_l$,

$$\begin{aligned}\langle A|G|B\rangle = \pm\sum_{t\neq k} [&\langle a_k(1)a_t(2)|g_{12}|b_l(1)a_t(2)\rangle \\ &- \langle a_k(1)a_t(2)|g_{12}|a_t(1)b_l(2)\rangle].\end{aligned} \tag{11.1-22}$$

The same rule as in (11.1-18) applies to the $\pm$ signs in (11.1-21) and (11.1-22). Examples of diagonal and off-diagonal elements are given by (11.1-10) and (11.1-11).

It will now be assumed that the general spin orbital $a(\lambda_i)$ consists of a product of a spatial function $\varphi_a(\mathbf{r}_i)$ and a spin function $\xi_i^a(m_s)$. The latter is always either an α or a β spin function depending on whether m_s is $+\frac{1}{2}$ or $-\frac{1}{2}$. Thus

$$a(\lambda_i) = \varphi_a(\mathbf{r}_i)\xi_i^a(m_s) \qquad \text{or} \qquad a(i) = \varphi_a(i)\xi_i^a(m_s). \tag{11.1-23}$$

Therefore

$$\langle a|f|b\rangle = \langle \varphi_a|f|\varphi_b\rangle\langle \xi^a(m_s)|\xi^b(m_s)\rangle = \langle \varphi_a|f|\varphi_b\rangle\, \delta(m_s{}^a, m_s{}^b) \tag{11.1-24}$$

in which the orthonormality of the spin functions has been inserted.

If a, b, c, and d are spin orbitals of form (11.1-23), the general matrix element of a two-electron operator becomes

$$\begin{aligned}\langle a(1)b(2)|g_{12}|c(1)d(2)\rangle &= \langle \varphi_a(1)\varphi_b(2)|g_{12}|\varphi_c(1)\varphi_d(2)\rangle \\ &\quad \times \langle \xi_1{}^a(m_s)|\xi_1{}^c(m_s)\rangle\langle \xi_2{}^b(m_s)|\xi_2{}^d(m_s)\rangle \\ &= \langle \varphi_a(1)\varphi_b(2)|g_{12}|\varphi_c(1)\varphi_d(2)\rangle\, \delta(m_s{}^a, m_s{}^c)\,\delta(m_s{}^b, m_s{}^d).\end{aligned} \tag{11.1-25}$$

11.2 Matrix Elements—Special Cases

We shall now specialize the discussion in two ways. It will be assumed that the spin–orbital $a(i)$ in (11.1-23) is given by

$$a(i) = R_{nl}^a(r_i)Y_{lm_l}^a(\mathbf{\Omega}_i)\xi_i{}^a(m_s) = \frac{1}{r_i}P_{nl}^a(r_i)Y_{lm_l}^a(\mathbf{\Omega}_i)\xi_i{}^a(m_s), \tag{11.2-1}$$

that is, the spatial part of $a(i)$ is a product of a radial function and a spherical harmonic; and that

$$g_{12} = e_2/r_{12} \tag{11.2-2}$$

in which $r_{12} = |\mathbf{r}_1 - \mathbf{r}_2|$ is the distance between electrons 1 and 2. With these assumptions, integrals of the type that occur in (11.1-24) and (11.1-25) can be given a more explicit form.

As a first step, e^2/r_{12} is expanded in spherical harmonics as in (1.2-22):

$$\frac{e^2}{r_{12}} = 4\pi e^2 \sum_{k=0}^{\infty}\sum_{m=-k}^{k} \frac{1}{2k+1}\frac{r_<{}^k}{r_>^{k+1}} Y_{km}^*(\Omega_1)Y_{km}(\Omega_2). \tag{11.2-3}$$

Then

$$\begin{aligned}&\langle a(1)b(2)|g_{12}|c(1)d(2)\rangle \\ &= \left\langle a(1)b(2)\left|\frac{e^2}{r_{12}}\right|c(1)d(2)\right\rangle \\ &= \left\langle \frac{1}{r_1}P_{nl}^a(r_1)Y_{lm_l}^a(\Omega_1)\frac{1}{r_2}P_{nl}^b(r_2)Y_{lm_l}^b(\Omega_2)\left|\frac{e^2}{r_{12}}\right|\frac{1}{r_1}P_{nl}^c(r_1)Y_{lm_l}^c(\Omega_1)\right. \\ &\quad \times \left.\frac{1}{r_2}P_{nl}^d(r_2)Y_{lm_l}^d(\Omega_2)\right\rangle \delta(m_s{}^a, m_s{}^c)\,\delta(m_s{}^b, m_s{}^d).\end{aligned} \tag{11.2-4}$$

When e^2/r_{12} is replaced by (11.2-3), the radial part of (11.2-4) is a sum over quantities such as

$$R^k(abcd) \equiv e^2\int_0^\infty\int_0^\infty \frac{r_<{}^k}{r_>^{k+1}} P_{nl}^a(r_1)P_{nl}^b(r_2)P_{nl}^c(r_1)P_{nl}^d(r_2)\,dr_1\,dr_2. \tag{11.2-5}$$

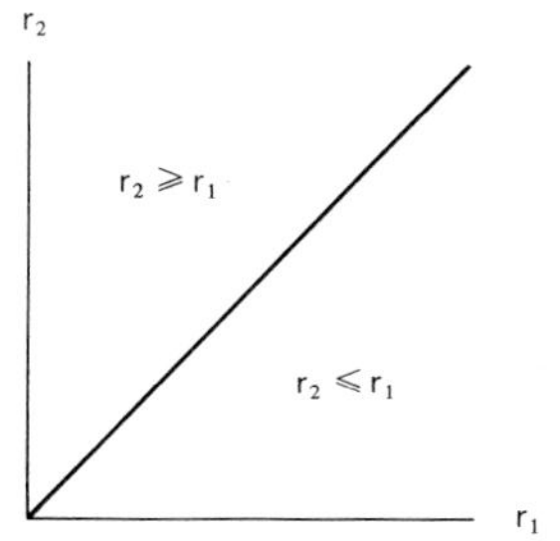

FIG. 11.1 Regions of integration for Eq. (11.2-7).

Integrals of this type may now be evaluated: Let

$$I = \int_0^\infty \int_0^\infty \frac{r_<^{\,k}}{r_>^{\,k}} f(r_1)g(r_2)\,dr_1\,dr_2 \tag{11.2-6}$$

where it is understood that $r_<$ means the smaller of the quantities r_1 and r_2 while $r_>$ means the greater of the two. If the integration space is subdivided into two regions as shown in Fig. 11.1, then

$$I = \int_0^\infty dr_1 \int_0^{r_1} \frac{r_2^{\,k}}{r_1^{k+1}} f(r_1)g(r_2)\,dr_2 + \int_0^\infty dr_2 \int_0^{r_2} \frac{r_1^{\,k}}{r_2^{k+1}} f(r_1)g(r_2)\,dr_1. \tag{11.2-7}$$

In order to perform the radial integration we require the radial wave functions which, for a multielectron atom, necessitate rather elaborate computations (see Chapter 19).

The angular integrals in (11.2-4) are of the form

$$\frac{4\pi}{2k+1}\langle Y_{lm_l}^a(\Omega_1)|Y_{km}^*(\Omega_1)|Y_{lm_l}^c(\Omega_1)\rangle\langle Y_{lm_l}^b(\Omega_2)|Y_{km}(\Omega_2)|Y_{lm_l}^d(\Omega_2)\rangle.$$

These are readily evaluated by means of the general formula (1.2-29), namely:

$$\langle l'm'|Y_{LM}|lm\rangle = (-1)^{m'}\sqrt{\frac{(2l'+1)(2L+1)(2l+1)}{4\pi}} \times \begin{pmatrix} l' & L & l \\ -m' & M & m \end{pmatrix}\begin{pmatrix} l' & L & l \\ 0 & 0 & 0 \end{pmatrix} \tag{11.2-8}$$

which is nonvanishing only when

$$-m' + M + m = 0, \qquad l' + L + l \text{ is even}, \qquad \triangle(l'\,L\,l). \tag{11.2-9}$$

It is customary to define a quantity $c^k(lm, l'm')$ as

$$c^k(lm, l'm') = \sqrt{\frac{4\pi}{2k+1}}\langle Y_{lm}|Y_{k\,m-m'}|Y_{l'm'}\rangle = (-1)^{m-m'}c^k(l'm', lm). \tag{11.2-10}$$

These integrals have been evaluated for special cases and are tabulated in Table 11.1. Remembering that $Y^*_{km} = (-1)^m Y_{k\,-m}$ and adopting the notation in which $m_l^a, m_l^b, \ldots$ are the projection quantum numbers associated with $Y^a_{lm_l}, Y^b_{lm_l}, \ldots,$ we have, from (11.2-9), the conditions on the projection quantum numbers:

$$-m_l^a - m + m_l^c = 0, \qquad -m_l^b + m + m_l^d = 0,$$

or

$$m_l^a + m_l^b = m_l^c + m_l^d. \tag{11.2-11}$$

TABLE 11.1

$c^k(lm, l'm')$ *for* s, p, *and* d Electrons[a]

	m	m'	k = 0	1	2	3	4
ss	0	0	1	0	0	0	0
sp	0	±1	0	$-\sqrt{1/3}$	0	0	0
	0	0	0	$\sqrt{1/3}$	0	0	0
pp	±1	±1	1	0	$-\sqrt{1/25}$	0	0
	±1	0	0	0	$\sqrt{3/25}$	0	0
	±1	∓1	0	0	$-\sqrt{6/25}$	0	0
	0	0	1	0	$\sqrt{4/25}$	0	0
sd	0	±2	0	0	$\sqrt{1/5}$	0	0
	0	±1	0	0	$-\sqrt{1/5}$	0	0
	0	0	0	0	$\sqrt{1/5}$	0	0
pd	±1	±2	0	$-\sqrt{6/15}$	0	$\sqrt{3/245}$	0
	±1	±1	0	$\sqrt{3/15}$	0	$-\sqrt{9/245}$	0
	±1	0	0	$-\sqrt{1/15}$	0	$\sqrt{18/245}$	0
	±1	∓1	0	0	0	$-\sqrt{30/245}$	0
	±1	∓2	0	0	0	$\sqrt{45/245}$	0
	0	±2	0	0	0	$\sqrt{15/245}$	0
	0	±1	0	$-\sqrt{3/15}$	0	$-\sqrt{24/245}$	0
	0	0	0	$\sqrt{4/15}$	0	$\sqrt{27/245}$	0
dd	±2	±2	1	0	$-\sqrt{4/49}$	0	$\sqrt{1/441}$
	±2	±1	0	0	$\sqrt{6/49}$	0	$-\sqrt{5/441}$
	±2	0	0	0	$-\sqrt{4/49}$	0	$\sqrt{15/441}$
	±2	∓1	0	0	0	0	$-\sqrt{35/441}$
	±2	∓2	0	0	0	0	$\sqrt{70/441}$
	±1	±1	1	0	$\sqrt{1/49}$	0	$-\sqrt{16/441}$
	±1	0	0	0	$\sqrt{1/49}$	0	$\sqrt{30/441}$
	±1	∓1	0	0	$-\sqrt{6/49}$	0	$-\sqrt{40/441}$
	0	0	1	0	$\sqrt{4/49}$	0	$\sqrt{36/441}$

[a] (Slater, 1960)

Combining (11.2-3)–(11.2-5), (11.2-10), and (11.2-11),

$$\left\langle a(1)b(2)\left|\frac{e^2}{r_{12}}\right| c(1)d(2)\right\rangle$$

$$= \delta(m_s{}^a, m_s{}^c)\,\delta(m_s{}^b, m_s{}^d)\,\delta(m_l{}^a + m_l{}^b, m_l{}^c + m_l{}^d)$$

$$\times \sum_{k=0}^{\infty} c^k(l^a m_l{}^a, l^c m_l{}^c) c^k(l^d m_l{}^d, l^b m_l{}^b) R^k(abcd). \qquad (11.2\text{-}12)$$

A special case of (11.2-12) is one in which

$$a = c, \qquad b = d. \qquad (11.2\text{-}13)$$

The resulting integral $J(a, b)$ is known as a *Coulomb* or *direct* integral and is given by

$$J(a,b) = \left\langle a(1)b(2)\left|\frac{e^2}{r_{12}}\right| a(1)b(2)\right\rangle \qquad (11.2\text{-}14a)$$

$$= \left\langle \varphi_a(1)\varphi_b(2)\left|\frac{e^2}{r_{12}}\right| \varphi_a(1)\varphi_b(2)\right\rangle \qquad (11.2\text{-}14b)$$

$$= \sum_{k=0}^{\infty} c^k(l^a m_l{}^a, l^a m_l{}^a) c^k(l^b m_l{}^b, l^b m_l{}^b) R^k(abab), \qquad (11.2\text{-}14c)$$

or defining

$$a^k(l^a m_l{}^a, l^b m_l{}^b) \equiv c^k(l^a m_l{}^a, l^a m_l{}^a) c^k(l^b m_l{}^b, l^b m_l{}^b)$$

$$= \frac{4\pi}{2k+1} \langle l^a m_l{}^a | Y_{k0} | l^a m_l{}^a \rangle \langle l^b m_l{}^b | Y_{k0} | l^b m_l{}^b \rangle, \qquad (11.2\text{-}15)$$

$$F^k(n^a l^a, n^b l^b) = R^k(abab),$$

$$= e^2 \int_0^\infty \int_0^\infty \frac{r_<{}^k}{r_>^{k+1}} [P_{nl}^a(r_1)]^2 [P_{nl}^b(r_2)]^2\, dr_1\, dr_2. \qquad (11.2\text{-}16)$$

we have

$$J(a,b) = \sum_{k=0}^{\infty} a^k F^k = J(b,a). \qquad (11.2\text{-}17)$$

Numerical values of $a^k(lm, l'm') = a^k(l'm', lm)$ are tabulated in Table 11.2.

A second special case occurs when

$$a = d, \qquad b = c.$$

TABLE 11.2

$a^k(lm, l'm')$ for s, p, *and* d Electrons[a]

	m	m	k = 0	2	4
ss	0	0	1	0	0
sp	0	±1	1	0	0
	0	0	1	0	0
pp	±1	±1	1	1/25	0
	±1	0	1	−2/25	0
	0	0	1	4/25	0
sd	0	±2	1	0	0
	0	±1	1	0	0
	0	0	1	0	0
pd	±1	±2	1	2/35	0
	±1	±1	1	−1/35	0
	±1	0	1	−2/35	0
	0	±2	1	−4/35	0
	0	±1	1	2/35	0
	0	0	1	4/35	0
dd	±2	±2	1	4/49	1/441
	±2	±1	1	−2/49	−4/441
	±2	0	1	−4/49	6/441
	±1	±1	1	1/49	16/441
	±1	0	1	2/49	−24/441
	0	0	1	4/49	36/441

[a] (Slater, 1960)

This produces another integral $K(a, b)$ known as an *exchange* integral:

$$K(a, b) = \left\langle a(1)b(2) \left| \frac{e^2}{r_{12}} \right| b(1)a(2) \right\rangle \tag{11.2-18a}$$

$$= \delta(m_s{}^a, m_s{}^b) \left\langle \varphi_a(1)\varphi_b(2) \left| \frac{e^2}{r_{12}} \right| \varphi_b(1)\varphi_a(2) \right\rangle \tag{11.2-18b}$$

$$= \delta(m_s{}^a, m_s{}^b) \sum_{k=0}^{\infty} c^k(l^a m_l{}^a, l^b m_l{}^b) c^k(l^a m_l{}^a, l^b m_l{}^b) R^k(abba). \tag{11.2-18c}$$

We define

$$b^k(l^a m_l{}^a, l^b m_l{}^b) \equiv [c^k(l^a m_l{}^a, l^b m_l{}^b)]^2 \tag{11.2-19}$$

$$G^k(n^a l^a, n^b l^b) = R^k(abba),$$

$$= e^2 \int_0^\infty \int_0^\infty \frac{r_<{}^k}{r_>^{k+1}} P_{nl}^a(r_1) P_{nl}^b(r_2) P_{nl}^b(r_1) P_{nl}^a(r_2)\, dr_1\, dr_2. \tag{11.2-20}$$

Then

$$K(a,b) = \delta(m_s{}^a, m_s{}^b) \sum_{k=0}^{\infty} b^k G^k = K(b,a). \tag{11.2-21}$$

The radial integrals F^k and G^k are both positive and are known as Slater–Condon parameters. Often it is convenient to define related parameters as shown in Table 11.3.

TABLE 11.3

Slater-Condon and Racah Parameters

	Slater-Condon	Racah
s	$F_0 = F^0$	
p	$F_0 = F^0$ $F_2 = F^2/25$	
d	$F_0 = F^0$ $F_2 = F^2/49$ $F_4 = F^4/441$	$A = F_0 - 49F_4$ $B = F_2 - 5F_4$ $C = 35F_4$
f	$F_0 = F^0$ $F_2 = F^2/225$ $F_4 = F^4/1089$ $F_6 = F^6/184041$	

Two electrons with the same values of n and l are said to be *equivalent*. In that case the radial parts of the two wave functions are the same, i.e.,

$$P_{nl}^a(r_i) = P_{nl}^b(r_i)$$

or

$$n^a = n^b, \qquad l^a = l^b$$

and the integrals in (11.2-16) and (11.2-20) become identical so that

$$F^k = G^k. \tag{11.2-22}$$

The integrals $J(a,b)$ and $K(a,b)$ may be expressed in terms of two operators J_b and K_b known as *Coulomb* or *direct* and *exchange* operators, respectively, where

$$\begin{aligned} J_b(1)a(1) &= \left[\int b^*(2)\frac{e^2}{r_{12}} b(2)\,d\tau_2\right]a(1), \\ K_b(1)a(1) &= \left[\int b^*(2)\frac{e^2}{r_{12}} a(2)\,d\tau_2\right]b(1), \end{aligned} \tag{11.2-23}$$

or, in terms of spatial orbitals alone,

$$J_b(1)\varphi_a(1) = \left[\int \varphi_b^*(2)\,\frac{e^2}{r_{12}}\,\varphi_b(2)\,d\mathbf{r}_2\right]\varphi_a(1),$$

$$K_b(1)\varphi_a(1) = \left[\delta(m_s^{\,a}, m_s^{\,b}) \int \varphi_b^*(2)\,\frac{e^2}{r_{12}}\,\varphi_a(2)\,d\mathbf{r}_2\right]\varphi_b(1). \tag{11.2-24}$$

The Coulomb and exchange integrals may now be written as

$$J(a,b) = \langle a(1)|J_b(1)|a(1)\rangle = \left\langle a(1)b(2)\left|\frac{e^2}{r_{12}}\right| a(1)b(2)\right\rangle, \tag{11.2-25}$$

$$K(a,b) = \langle a(1)|K_b(1)|a(1)\rangle = \left\langle a(1)b(2)\left|\frac{e^2}{r_{12}}\right| b(1)a(2)\right\rangle, \tag{11.2-26}$$

as in (11.2-14) and (11.2-18). It is also observed that if

$$G = \sum_{i<j} e^2/r_{ij},$$

it is possible to express (11.1-19) in the form

$$\langle A|G|A\rangle = \sum_{k<t} [J(k,t) - K(k,t)] \tag{11.2-27}$$

where A is a determinantal wave function such as (11.1-12).

Let us now consider the Hamiltonian

$$\mathscr{H} = \sum_i^N \mathscr{H}_0(i) + \sum_{i<j} e^2/r_{ij} \tag{11.2-28}$$

where

$$\mathscr{H}_0(i) = \frac{p_i^{\,2}}{2m} - \frac{Ze^2}{r_i}$$

and a wave function $\psi(\lambda_1, \lambda_2, \ldots, \lambda_N)$ expressed in the form of a single Slater determinant

$$\Psi = \frac{1}{\sqrt{N!}}\begin{vmatrix} \psi_1(\lambda_1) & \psi_2(\lambda_1) & \cdots & \psi_N(\lambda_1) \\ \psi_1(\lambda_2) & \psi_2(\lambda_2) & \cdots & \psi_N(\lambda_2) \\ \vdots & \vdots & \cdots & \vdots \\ \psi_1(\lambda_N) & \psi_2(\lambda_N) & \cdots & \psi_N(\lambda_N) \end{vmatrix} \tag{11.2-29}$$

in which $\psi_k(\lambda_i) \equiv \psi_k(i)$ is a spin orbital and

$$\langle \psi_k(i)|\psi_j(i)\rangle = \delta_{kj}.$$

We shall be interested in the total energy

$$E = \langle \Psi | \mathcal{H} | \Psi \rangle$$

which may be written, with the aid of (11.1-16) and (11.1-19) as

$$E = \sum_{i=1}^{N} \langle \psi_i(1) | \mathcal{H}_0(1) | \psi_i(1) \rangle + \sum_{i<j} \left[\left\langle \psi_i(1)\psi_j(2) \left| \frac{e^2}{r_{12}} \right| \psi_i(1)\psi_j(2) \right\rangle - \left\langle \psi_i(1)\psi_j(2) \left| \frac{e^2}{r_{12}} \right| \psi_j(1)\psi_i(2) \right\rangle \right], \tag{11.2-30}$$

or, in terms of the spatial orbitals, using (11.1-24), (11.2-14), and (11.2-18),

$$E = \sum_{i}^{N} I_i + \sum_{i<j} \left[\left\langle \varphi_i(1)\varphi_j(2) \left| \frac{e^2}{r_{12}} \right| \varphi_i(1)\varphi_j(2) \right\rangle - \delta(m_s^{\,i}, m_s^{\,j}) \left\langle \varphi_i(1)\varphi_j(2) \left| \frac{e^2}{r_{12}} \right| \varphi_j(1)\varphi_i(2) \right\rangle \right] \tag{11.2-31}$$

where

$$I_i \equiv \langle \psi_i | \mathcal{H}_0 | \psi_i \rangle = \langle \varphi_i | \mathcal{H}_0 | \varphi_i \rangle. \tag{11.2-32}$$

An important special case is one in which the determinantal wave function represents a system of $2N$ electrons distributed among N spatial orbitals each of which is occupied by two electrons with opposite spin. Such a system is known as a *closed shell*. We consider a specific example. Let

$$\Psi = \frac{1}{\sqrt{4!}} \begin{vmatrix} \varphi_1(1)\alpha(1) & \varphi_1(1)\beta(1) & \varphi_2(1)\alpha(1) & \varphi_2(1)\beta(1) \\ \varphi_1(2)\alpha(2) & \vdots & \vdots & \vdots \\ \varphi_1(3)\alpha(3) & & & \\ \varphi_1(4)\alpha(4) & & & \varphi_2(4)\beta(4) \end{vmatrix} \tag{11.2-33}$$

be the Slater determinant for a closed shell containing four electrons. To calculate the energy we may use (11.2-30) and (11.2-31) and for this purpose it is convenient to define

$$\psi_a = \varphi_1\alpha, \qquad \psi_b = \varphi_1\beta, \qquad \psi_c = \varphi_2\alpha, \qquad \psi_d = \varphi_2\beta. \tag{11.2-34}$$

We then have

$$\langle \psi_a | \mathcal{H}_0 | \psi_a \rangle = \langle \psi_b | \mathcal{H}_0 | \psi_b \rangle = \langle \varphi_1 | \mathcal{H}_0 | \varphi_1 \rangle = I_1,$$
$$\langle \psi_c | \mathcal{H}_0 | \psi_c \rangle = \langle \psi_b | \mathcal{H}_0 | \psi_d \rangle = \langle \varphi_2 | \mathcal{H}_0 | \varphi_2 \rangle = I_2,$$

and the contribution to the energy from the one-electron integrals is

$$2(I_1 + I_2). \tag{11.2-35}$$

The six possible Coulomb integrals are

$$\left\langle \psi_a(1)\psi_b(2) \left| \frac{e^2}{r_{12}} \right| \psi_a(1)\psi_b(2) \right\rangle = \left\langle \varphi_1(1)\varphi_1(2) \left| \frac{e^2}{r_{12}} \right| \varphi_1(1)\varphi_1(2) \right\rangle = J(1,1)$$

$$\left.\begin{aligned} &\left\langle \psi_a(1)\psi_c(2) \left| \frac{e^2}{r_{12}} \right| \psi_a(1)\psi_c(2) \right\rangle \\ &\left\langle \psi_a(1)\psi_d(2) \left| \frac{e^2}{r_{12}} \right| \psi_a(1)\psi_d(2) \right\rangle \\ &\left\langle \psi_b(1)\psi_c(2) \left| \frac{e^2}{r_{12}} \right| \psi_b(1)\psi_c(2) \right\rangle \\ &\left\langle \psi_b(1)\psi_d(2) \left| \frac{e^2}{r_{12}} \right| \psi_b(1)\psi_d(2) \right\rangle \end{aligned}\right\} = \left\langle \varphi_1(1)\varphi_2(2) \left| \frac{e^2}{r_{12}} \right| \varphi_1(1)\varphi_2(2) \right\rangle = J(1,2)$$

$$\left\langle \psi_c(1)\psi_d(2) \left| \frac{e^2}{r_{12}} \right| \psi_c(1)\psi_d(2) \right\rangle = \left\langle \varphi_2(1)\varphi_2(2) \left| \frac{e^2}{r_{12}} \right| \varphi_2(1)\varphi_2(2) \right\rangle = J(2,2)$$

and their contribution to the energy is

$$J(1,1) + J(2,2) + 4J(1,2).$$

The orthogonality of the spin functions causes all but two of the six exchange integrals to vanish. The ones that remain are

$$\begin{aligned} \left\langle \psi_a(1)\psi_c(2) \left| \frac{e^2}{r_{12}} \right| \psi_c(1)\psi_a(2) \right\rangle &= \left\langle \psi_b(1)\psi_d(2) \left| \frac{e^2}{r_{12}} \right| \psi_d(1)\psi_b(2) \right\rangle \\ &= \left\langle \varphi_1(1)\varphi_2(2) \left| \frac{e^2}{r_{12}} \right| \varphi_2(1)\varphi_1(2) \right\rangle = K(1,2) \end{aligned}$$

with a contribution of $-2K(1,2)$ to the energy. The total energy is the sum of the contributions from the three kinds of integrals:

$$E = 2(I_1 + I_2) + J(1,1) + J(2,2) + 4J(1,2) - 2K(1,2).$$

It is a simple matter to extend this formalism to a system of $2N$ electrons in a closed shell configuration. The Slater determinant is

$$\Psi = \frac{1}{\sqrt{(2N)!}} \begin{vmatrix} \varphi_1(1)\alpha(1) & \varphi_1(1)\beta(1) & \cdots & \varphi_N(1)\alpha(1) & \varphi_N(1)\beta(1) \\ \vdots & \vdots & \vdots & \vdots & \vdots \\ \varphi_1(2N)\alpha(2N) & \varphi_1(2N)\beta(2N) & \cdots & \varphi_N(2N)\alpha(2N) & \varphi_N(2N)\beta(2N) \end{vmatrix} \tag{11.2-36}$$

and the total energy may be written

$$E = 2 \sum_{k=1}^{N} I_k + \sum_{k,l}^{N} [2J(k,l) - K(k,l)] \tag{11.2-37}$$

provided one sets

$$J(k,k) = K(k,k) = \left\langle \varphi_k(1)\varphi_k(2) \left| \frac{e^2}{r_{12}} \right| \varphi_k(1)\varphi_k(2) \right\rangle \tag{11.2-38}$$

The sums in (11.2-37) are taken over N spatial orbitals which correspond to $2N$ spin orbitals (or $2N$ electrons). It should also be noted that the relation $J(k,k) = K(k,k)$ is a necessary requirement for the validity of the energy expression (11.2-37) but is not true in general.

CHAPTER 12

SECOND QUANTIZATION

12.1 Creation and Annihilation Operators

Suppose we have an orthonormal set of (one-electron) spin orbitals ψ_1, $\psi_2, \ldots, \psi_N$ some of which are occupied by electrons and others are vacant. There are several ways in which such a system may be described. We give a few examples from which the general picture will emerge:

(a) All orbitals are vacant. This is generally called the *vacuum state*. Such a state, designated by $|0\rangle$ or Ψ_0 could be described by writing

$$\Psi_0 = |0\rangle = |0_1, 0_2, \ldots, 0, \ldots\rangle \tag{12.1-1}$$

in which the subscripts in the state vector refer to the spin orbitals. Thus 0_1 means that ψ_1 is vacant (zero electrons), 0_2 means that ψ_2 is vacant, etc.

(b) ψ_m is occupied; all other orbitals are vacant. Here we might write

$$\Psi_m = |0_1, 0_2, \ldots, 1_m, \ldots, 0, \ldots\rangle = \psi_m(1) \tag{12.1-2}$$

in which 1_m means that ψ_m is occupied by one electron. In view of the restriction from the Pauli principle, the maximum number of electrons that can reside in ψ_m (or any other orbital) is just one.

(c) Two orbitals ψ_l and ψ_m each contain one electron; all other orbitals are vacant. In the state vector notation this would be

$$\Psi_{lm} = |0_1, 0_2, \ldots, 1_l, \ldots, 1_m, \ldots, 0, \ldots\rangle. \tag{12.1-3}$$

But Ψ_{lm} is a 2-electron wave function which must be antisymmetric with respect to an interchange of the two electrons. It is therefore expressible as a Slater determinant

$$\Psi_{lm} = \frac{1}{\sqrt{2}} \begin{vmatrix} \psi_l(1) & \psi_m(1) \\ \psi_l(2) & \psi_m(2) \end{vmatrix} \tag{12.1-4}$$

with

$$\Psi_{ml} = -\Psi_{lm}. \tag{12.1-5}$$

Therefore

$$|0_1, 0_2, \ldots, 1_m, \ldots, 1_l, \ldots, 0_N\rangle = -|0_1, 0_2, \ldots, 1_l, \ldots, 1_m, \ldots, 0, \ldots\rangle. \tag{12.1-6}$$

(d) Three orbitals, ψ_k, ψ_l, ψ_m, are occupied and the rest are vacant. Then

$$\Psi_{klm} = |0_1, 0_2, \ldots, 1_k, \ldots, 1_l, \ldots, 1_m, \ldots, 0, \ldots\rangle \tag{12.1-7}$$

and the corresponding Slater determinant is

$$\Psi_{klm} = \frac{1}{\sqrt{3!}} \begin{vmatrix} \psi_k(1) & \psi_l(1) & \psi_m(1) \\ \psi_k(2) & \psi_l(2) & \psi_m(2) \\ \psi_k(3) & \psi_l(3) & \psi_m(3) \end{vmatrix}. \tag{12.1-8}$$

This is sufficient to illustrate the general pattern. The description in terms of state vectors as in (12.1-1)–(12.1-3) and (12.1-7) is called the *occupation number* or *Fock space* description; it lists all the orbitals and tells which are occupied and which are vacant. A Slater determinant lists only the occupied orbitals. A general state vector or occupation number wave function for a system of electrons can be written as

$$|n_1, n_2, \ldots, n_k, \ldots\rangle \tag{12.1-9}$$

with $n_k = 0, 1$. In order for the state vectors to be in a one-to-one correspondence with Slater determinants it is necessary to arrange the one-electron orbitals in a certain "standard" order and then to employ the same order in the corresponding Slater determinant. Since the latter are orthonormal it follows that

$$\langle n_1', n_2', \ldots, n_k', \ldots | n_1, n_2, \ldots, n_k, \ldots\rangle = \delta_{n_1', n_1} \delta_{n_2', n_2} \cdots \delta_{n_k', n_k}. \tag{12.1-10}$$

An *annihilation operator* c_k is defined as an operator that removes or annihilates an electron in the kth orbital provided that the orbital in question initially contained an electron. Thus

$$c_k|0_1, 0_2, \ldots, 1_k, \ldots, 1_l, \ldots, 1_m, \ldots, 0, \ldots\rangle$$
$$= |0_1, 0_2, \ldots, 0_k, \ldots, 1_l, \ldots, 1_m, \ldots, 0, \ldots\rangle$$

or

$$c_k \Psi_{klm} = \Psi_{lm} \tag{12.1-11}$$

where Ψ_{lm} and Ψ_{klm} are understood to be either the state vectors (12.1-3) and (12.1-7) or the Slater determinants (12.1-4) and (12.1-8). Also, as a consequence of the antisymmetry requirement,

$$c_l \Psi_{klm} = -c_l \Psi_{lkm} = -\Psi_{km}. \tag{12.1-12}$$

If the annihilation operator operates on a vacant orbital, the result is zero; for example,

$$c_i |0_1, 0_2, \ldots, 0_i, \ldots, 1_k, \ldots, 1_l, \ldots, 1_m, \ldots\rangle = c_i \Psi_{klm} = 0. \tag{12.1-13}$$

In particular, an annihilation operator acting on the vacuum state is identically zero:

$$c_i |0\rangle = 0. \tag{12.1-14}$$

The distinction between the vacuum state $|0\rangle$ in which all orbitals are vacant and the number 0 should be carefully noted.

Two successive operations with annihilation operators are easily interpreted. Thus

$$c_l c_k \Psi_{klm} = c_l \Psi_{lm} = \Psi_m, \qquad c_k c_l \Psi_{klm} = -c_k \Psi_{km} = -\Psi_m. \tag{12.1-15}$$

Therefore c_k, c_l satisfy

$$c_k c_l + c_l c_k \equiv \{c_k, c_l\} = 0 \tag{12.1-16}$$

in which $\{c_k, c_l\}$ is known as an *anticommutator*. An immediate consequence of (12.1-16) is that

$$c_k c_k = 0 \tag{12.1-17}$$

which simply says that an electron in the kth orbital can only be annihilated once—the second operator finds a vacant orbital.

Having defined an annihilation operator it must be possible to define a *creation operator* $c_k^\dagger$ which creates a particle in the kth orbital provided the orbital is initially vacant. Thus

$$c_m^\dagger \Psi_0 = \Psi_m, \qquad c_l^\dagger \Psi_m = \Psi_{lm}, \qquad c_k^\dagger \Psi_{lm} = \Psi_{klm}. \tag{12.1-18}$$

However, because of the Pauli principle,

$$c_l^\dagger \Psi_{lm} = 0. \tag{12.1-19}$$

In the occupation number description

$$\begin{aligned} c_m^\dagger|0_1,0_2,\ldots,0,\ldots\rangle &= |0_1,0_2,\ldots,1_m,\ldots,0,\ldots\rangle, \\ c_l^\dagger|0_1,0_2,\ldots,1_m,\ldots\rangle &= |0_1,0_2,\ldots,1_l,\ldots,1_m,\ldots,0,\ldots\rangle \\ c_k^\dagger|0_1,0_2,\ldots,1_l,\ldots,1_m,\ldots,0,\ldots\rangle &= |0_1,0_2,\ldots,1_k,\ldots,1_l,\ldots,1_m,\ldots,0,\ldots\rangle \end{aligned} \tag{12.1-20}$$

or

$$c_k^\dagger c_l^\dagger c_m^\dagger|0\rangle = |0,0,\ldots,1_k,\ldots,1_l,\ldots,1_m,\ldots,0,\ldots\rangle \tag{12.1-21}$$

indicating that any state can be generated from the vacuum state by an appropriate set of creation operators. But

$$c_l^\dagger|0,0,\ldots,1_l,\ldots,1_m,\ldots,0,\ldots\rangle = 0, \tag{12.1-22}$$

that is, a creation operator acting on an occupied orbital produces a wave function that is identically zero.

If ψ_k, ψ_l are two vacant orbitals,

$$c_l^\dagger c_k^\dagger \Psi_m = \Psi_{lkm} = -\Psi_{klm}, \tag{12.1-23}$$

but

$$c_k^\dagger c_l^\dagger \Psi_m = \Psi_{klm}. \tag{12.1-24}$$

Therefore

$$\{c_k^\dagger, c_l^\dagger\} = 0 \tag{12.1-25a}$$

and

$$(c_k^\dagger)^2 = 0. \tag{12.1-25b}$$

The latter is merely a statement of the Pauli principle which states that for particles obeying Fermi statistics an orbital cannot be occupied by more than one particle. For this reason c_k and $c_k^\dagger$ are known as *fermion* operators in Fock space.

Next we consider mixed products of c and $c^\dagger$. An example would be

$$c_l^\dagger c_k \Psi_k = \Psi_l. \tag{12.1-26}$$

$$c_k c_l^\dagger \Psi_k = c_k \Psi_{lk} = -\Psi_l \tag{12.1-27}$$

so that

$$\{c_k, c_l^\dagger\} = 0 \qquad (k \neq l). \tag{12.1-28}$$

When $k = l$ we have

$$\begin{aligned} c_k^\dagger c_k \Psi_0 &= 0, \qquad & c_k c_k^\dagger \Psi_0 &= \Psi_0, \\ c_k^\dagger c_k \Psi_k &= \Psi_k, \qquad & c_k c_k^\dagger \Psi_k &= 0. \end{aligned} \tag{12.1-29}$$

We note that the eigenvalue of $c_k^\dagger c_k$ is the occupation number (0 or 1) in the kth orbital. One may then define a *number operator for the kth orbital* N_k where

$$N_k = c_k^\dagger c_k = N_k^\dagger \tag{12.1-30}$$

which also implies that $N_k^2 = N_k$. From (12.1-29),

$$(c_k c_k^\dagger + c_k^\dagger c_k)\Psi_0 = \Psi_0, \qquad (c_k c_k^\dagger + c_k^\dagger c_k)\Psi_k = \Psi_k. \tag{12.1-31}$$

In either case, $\{c_k, c_k^\dagger\} = 1$.

In summary, the fermion operators obey the commutation rules

$$\{c_k, c_l\} = \{c_k^\dagger, c_l^\dagger\} = 0, \qquad \{c_k, c_l^\dagger\} = \delta_{kl}. \tag{12.1-32}$$

Also, the generalization of expressions such as (12.1-11) and (12.1-20) may be written as

$$\begin{aligned} c_k |\cdots n_k \cdots\rangle &= (-1)^{p_k}\sqrt{n_k}\,|\cdots 0_k \cdots\rangle, \\ c_k |\cdots n_k \cdots\rangle &= (-1)^{p_k}\sqrt{1-n_k}\,|\cdots 1_k \cdots\rangle \end{aligned} \tag{12.1-33}$$

where $n_k = 0$ or 1, p_k is the number of occupied orbitals to the left of k, and $(-1)^{p_k}$ is a parity factor which takes into account the antisymmetry of the N-electron wave function.

We have seen previously (Section 8.4) that the wave function of a system of indistinguishable particles must be either symmetric (for bosons) or antisymmetric (for fermions) with respect to an interchange of any two particles. The occupation number formalism described so far pertains to fermions. We shall now describe the analogous formalism appropriate for bosons.

Bosons, in contrast to fermions, are not restricted by the Pauli principle; hence a many-boson wave function expressed in terms of occupation numbers would be written as

$$|n_1, n_2, \ldots, n_k, \ldots\rangle$$

where n_k is any positive integer or zero. In this connection it may be helpful to picture an assembly of harmonic oscillators with each oscillator in an arbitrary state of excitation. Although such a physical system may not be of any great interest per se, it does provide an excellent model for other physical systems such as a quantized electromagnetic field in a cavity. Now, the occupation number formalism for harmonic oscillators has already been worked out in Section 10.2 where it was shown that an oscillator in an eigenstate $|n\rangle$ has an energy $\hbar\omega(n+\frac{1}{2})$. Furthermore it was shown that one may define annihilation and creation operators a and $a^\dagger$ which satisfy

$$a|n\rangle = \sqrt{n}|n-1\rangle, \qquad a^\dagger|n\rangle = \sqrt{n+1}|n+1\rangle, \qquad a|0\rangle = 0,$$
$$aa^\dagger - a^\dagger a = [a, a^\dagger] = 1, \qquad \langle n'|n\rangle = \delta_{n'n}.$$

The extension to a many-boson system therefore leads to the expressions:

$$\langle n_1', n_2', \ldots, n_k', \ldots | n_1, n_2, \ldots, n_k, \ldots \rangle = \delta_{n_1'n_1}\delta_{n_2'n_2} \cdots \delta_{n_k'n_k} \cdots, \tag{12.1-34}$$

$$\begin{aligned} a_k|\cdots n_k \cdots\rangle &= \sqrt{n_k}|\cdots n_k - 1 \cdots\rangle, \\ a_k^\dagger|\cdots n_k \cdots\rangle &= \sqrt{n_k+1}|\cdots n_k + 1 \cdots\rangle. \end{aligned} \tag{12.1-35}$$

The commutation rules for $a_k, a_k^\dagger$ follow directly from these relations; they are

$$[a_k, a_l] = [a_k^\dagger, a_l^\dagger] = 0, \qquad [a_k, a_l^\dagger] = \delta_{kl}. \tag{12.1-36}$$

These relations look remarkably similar to those in (12.1-32) for fermions, but the similarity is entirely superficial because the distinction between commutators and anticommutators is as profound as the distinction between fermions and bosons.

12.2 Matrix Elements of Operators

One- and two-electron operators may be expressed in terms of the creation and annihilation operators. For example, suppose

$$\Psi_{lm} = \frac{1}{\sqrt{2}}\begin{vmatrix} \psi_l(1) & \psi_m(1) \\ \psi_l(2) & \psi_m(2) \end{vmatrix}, \tag{12.2-1}$$

$$F = f_1 + f_2 \tag{12.2-2}$$

where ψ_l, ψ_m are one-electron orbitals and f_1, f_2 are one-electron operators that operate on electrons 1 and 2, respectively. According to (11.1-16),

$$\langle\Psi_{lm}|F|\Psi_{lm}\rangle = \langle l|f|l\rangle + \langle m|f|m\rangle \tag{12.2-3}$$

where we have abbreviated $\langle\psi_l|f|\psi_l\rangle$ to $\langle l|f|l\rangle$, etc. Now let

$$F = \langle l|f|l\rangle c_l^\dagger c_l + \langle l|f|m\rangle c_l^\dagger c_m + \langle m|f|l\rangle c_m^\dagger c_l + \langle m|f|m\rangle c_m^\dagger c_m \tag{12.2-4}$$

$$\Psi_{lm} = |0_1, 0_2, \ldots, 1_l 1_m, \ldots, 0, \ldots\rangle. \tag{12.2-5}$$

Then

$$\begin{aligned}\langle \Psi_{lm}|F|\Psi_{lm}\rangle = {} & \langle l|f|l\rangle\langle 1_l 1_m|c_l{}^\dagger c_l|1_l 1_m\rangle \\ & + \langle l|f|m\rangle\langle 1_l 1_m|c_l{}^\dagger c_m|1_l 1_m\rangle \\ & + \langle m|f|l\rangle\langle 1_l 1_m|c_m{}^\dagger c_l|1_l 1_m\rangle \\ & + \langle m|f|m\rangle\langle 1_l 1_m|c_m{}^\dagger c_m|1_l 1_m\rangle\end{aligned}$$

which reduces, with the application of (12.1-33), to

$$\langle \Psi_{lm}|F|\Psi_{lm}\rangle = \langle l|f|l\rangle + \langle m|f|m\rangle. \tag{12.2-6}$$

We may therefore infer that

$$F = \sum_i f_i = \sum_{kl} \langle k|f|l\rangle c_k{}^\dagger c_l \tag{12.2-7}$$

with

$$\langle k|f|l\rangle \equiv \langle \psi_k(i)|f_i|\psi_l(i)\rangle \tag{12.2-8}$$

and with k, l summed over all occupied orbitals. Because of the single occupancy of the spin orbitals, such a summation is equivalent to a sum over all the electrons. Written as in (12.2-7), F is an operator in Fock space, that is, it must operate on a state vector written in terms of the occupation numbers. Such operators are equivalent but not identical to operators in configuration space.

A special case of (12.2-7) is one in which each $f_i = 1$ so that F is an operator representing the total number of electrons. Calling this operator N, the *number operator*, we have

$$N = \sum_{kl} \langle k|l\rangle c_k{}^\dagger c_l = \sum_{kl} \delta_{kl} c_k{}^\dagger c_l = \sum_k c_k{}^\dagger c_k = \sum_k N_k. \tag{12.2-9}$$

Another special case of importance is one in which $F = \mathscr{H}_0$ where

$$\mathscr{H}_0 = \sum_i \mathscr{H}_0(i) \tag{12.2-10}$$

and $\mathscr{H}_0(i)$ is a one-particle Hamiltonian (e.g., kinetic energy operator) which operates only on the coordinates of the ith particle. If $\psi_k(i)$, $\psi_l(i)$, . . . are eigenfunctions of $\mathscr{H}_0(i)$ with eigenvalues $E_k, E_l, \ldots$, then, according to (12.2-7),

$$\begin{aligned}\mathscr{H}_0 &= \sum_{kl} \langle \psi_k(i)|\mathscr{H}_0(i)|\psi_l(i)\rangle c_k{}^\dagger c_l \\ &= \sum_{kl} E_l \langle \psi_k(i)|\psi_l(i)\rangle c_k{}^\dagger c_l \\ &= \sum_{kl} E_l\, \delta_{kl} c_k{}^\dagger c_l = \sum_k E_k c_k{}^\dagger c_k = \sum_k N_k E_k.\end{aligned} \tag{12.2-11}$$

We may now use the anticommutator relations (12.1-32) to establish the relations

$$[c_l, \mathscr{H}_0] = E_l c_l, \qquad [c_l^\dagger, \mathscr{H}_0] = -E_l c_l^\dagger. \tag{12.2-12}$$

In the interaction representation

$$c_l^{\mathrm{I}}(t) = e^{i\mathscr{H}_0 t/\hbar} c_l e^{-i\mathscr{H}_0 t/\hbar}.$$

On differentiating with respect to time we obtain the differential equation for $c_l^{\mathrm{I}}(t)$:

$$i\hbar\, \partial c_l^{\mathrm{I}}(t)/\partial t = [c_l^{\mathrm{I}}(t), \mathscr{H}_0] = E_l c_l^{\mathrm{I}}(t).$$

Therefore

$$c_l^{\mathrm{I}}(t) = c_l(0) e^{-iE_l t/\hbar} = c_l e^{-iE_l t/\hbar}. \tag{12.2-13a}$$

The same procedure applied to $c_l^\dagger(t)$ gives

$$c_l^{\mathrm{I}\dagger}(t) = c_l^\dagger e^{iE_l t/\hbar}. \tag{12.2-13b}$$

The expressions in (12.2-13) provide the relationships that connect the Schrödinger and interaction representations for the annihilation and creation operators. We note the analogy between these expressions and those of (10.3-8).

For two-electron operators the prescription is

$$G = \tfrac{1}{2} \sum_{i \neq j} g_{ij} = \tfrac{1}{2} \sum_{kl,mn} \langle kl|g_{12}|mn\rangle c_k^\dagger c_l^\dagger c_n c_m \tag{12.2-14}$$

where

$$\langle kl|g_{12}|mn\rangle = \langle \psi_k(1)\psi_l(2)|g_{12}|\psi_m(1)\psi_n(2)\rangle. \tag{12.2-15}$$

We may test this expression by calculating several matrix elements and comparing them with expressions for matrix elements between Slater determinants given in Section 11.1. Let

$$\Psi_g = |1_k 1_l 1_m \cdots 0_n \cdots\rangle \equiv |1_k 1_l 1_m\rangle \tag{12.2-16}$$

be a ground state configuration in which orbitals ψ_k, ψ_l, and ψ_m are occupied and all the rest are vacant. We may think of these orbitals as having the lowest energies in the entire set but that is not a requirement for the present purpose. In evaluating the matrix element $\langle\Psi_g|G|\Psi_g\rangle$ there will be terms such as

$$\langle kl|g_{12}|kl\rangle\langle 1_k 1_l 1_m|c_k^\dagger c_l^\dagger c_l c_k|1_k 1_l 1_m\rangle = \langle kl|g_{12}|kl\rangle$$

in which rules (12.1-33) have been applied to evaluate the matrix element of the product of creation and annihilation operators. In the same way

$$\langle kl|g_{12}|lk\rangle\langle 1_k 1_l 1_m|c_k^\dagger c_l^\dagger c_k c_l|1_k 1_l 1_m\rangle = -\langle kl|g_{12}|lk\rangle$$

but

$$\langle kl|g_{12}|lm\rangle\langle 1_k1_l1_m|c_k{}^{\dagger}c_l{}^{\dagger}c_mc_l|1_k1_l1_m\rangle = 0$$

and all other terms of this type vanish. All together there are 12 nonvanishing matrix elements which combine to give

$$\begin{aligned}\langle\Psi_g|G|\Psi_g\rangle = \langle kl|g_{12}|kl\rangle - \langle kl|g_{12}|lk\rangle + \langle km|g_{12}|km\rangle \\ - \langle km|g_{12}|mk\rangle + \langle lm|g_{12}|lm\rangle - \langle lm|g_{12}|ml\rangle \quad (12.2\text{-}17)\end{aligned}$$

in agreement with the result obtained from (11.1-19).

Let us now examine off-diagonal matrix elements between the ground state Ψ_g and excited states. A singly excited state would be of the form

$$\Psi_1 = |1_k0_l1_m \cdots 1_p \cdots 0_n \cdots\rangle \equiv |1_k1_m1_p\rangle$$

in which an electron has been excited from ψ_l to ψ_p. The matrix element

$$\langle\Psi_1|G|\Psi_g\rangle = \langle 1_k1_m1_p|G|1_k1_l1_m\rangle$$

contains nonvanishing contributions such as

$$\langle pk|g_{12}|lk\rangle\langle 1_k1_m1_p|c_p{}^{\dagger}c_k{}^{\dagger}c_kc_l|1_k1_l1_m\rangle = \langle pk|g_{12}|lk\rangle$$

in which we have used

$$c_p{}^{\dagger}c_k{}^{\dagger}c_kc_l|1_k1_l1_m\rangle = -|1_p1_k0_l1_m\rangle = |1_p1_k1_m\rangle.$$

Similarly,

$$\langle pk|g_{12}|kl\rangle\langle 1_k1_m1_p|c_p{}^{\dagger}c_k{}^{\dagger}c_lc_k|1_k1_l1_m\rangle = -\langle pk|g_{12}|kl\rangle.$$

The net result is

$$\langle\Psi_1|G|\Psi_g\rangle = \pm[\langle pk|g_{12}|lk\rangle - \langle pk|g_{12}|kl\rangle + \langle pm|g_{12}|lm\rangle - \langle pm|g_{12}|ml\rangle] \quad (12.2\text{-}18)$$

which is what one obtains from (11.1-22).

A doubly excited state might be written as

$$\Psi_2 = |0_k1_l0_m \cdots 1_p \cdots 1_q \cdots 0_n \cdots\rangle \equiv |1_l1_p1_q\rangle.$$

The matrix element

$$\langle\Psi_2|G|\Psi_g\rangle = \langle 1_l1_m1_p|G|1_k1_l1_m\rangle$$

is the sum of the two contributions

$$\begin{aligned}\langle pq|g_{12}|km\rangle\langle 1_l1_p1_q|c_p{}^{\dagger}c_q{}^{\dagger}c_mc_k|1_k1_l1_m\rangle = \langle pq|g_{12}|km\rangle, \\ \langle pq|g_{12}|mk\rangle\langle 1_l1_p1_q|c_p{}^{\dagger}c_q{}^{\dagger}c_kc_m|1_k1_l1_m\rangle = -\langle pq|g_{12}|mk\rangle,\end{aligned} \quad (12.2\text{-}19)$$

again in agreement with the Slater determinant computation (11.1-21). If more than two orbitals are excited, the resulting state has no matrix elements with the ground state.

For bosons

$$F = \sum_i f_i = \sum_{kl} \langle k|f|l\rangle a_k^{\dagger} a_l, \qquad (12.2\text{-}20)$$

$$G = \tfrac{1}{2} \sum_{i \neq j} g_{ij} = \tfrac{1}{2} \sum_{kl,mn} \langle kl|g_{12}|mn\rangle a_k^{\dagger} a_l^{\dagger} a_n a_m \qquad (12.2\text{-}21)$$

$$N = \sum_k a_k^{\dagger} a_k. \qquad (12.2\text{-}22)$$

12.3 Diagrams

The various matrix elements for fermion operators that were obtained in the previous section may be schematized by means of diagrams according to the following rules:

1. A broken line (---) represents the two-particle interaction g_{12}.
2. At each end of the broken line draw two solid lines with arrows pointing toward and away from the broken line.
3. For a matrix element $\langle ab|g_{12}|cd\rangle$ we follow the convention that a is the line out of the left vertex, b, the line out of the right vertex, c, the line into the left vertex, and d, the line into the right vertex.

We may illustrate such diagrams with some of the matrix elements in (12.2-17)–(12.2-19):

$\langle kl|g_{12}|kl\rangle$

$\langle kl|g_{12}|lk\rangle$

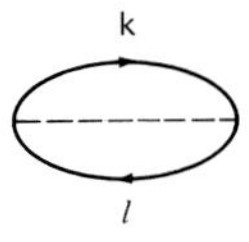

$\langle pk|g_{12}|lk\rangle$

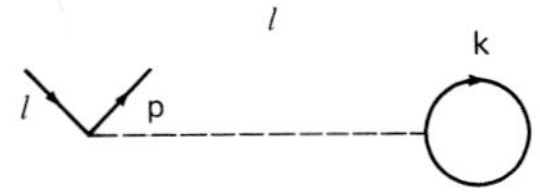

$\langle pk|g_{12}|kl\rangle$

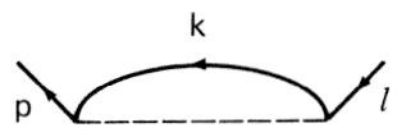

$\langle pq|g_{12}|km\rangle$

$\langle pq|g_{12}|mk\rangle$

Such matrix elements and their diagrams may be interpreted in terms of creation and annihilation of particles or in terms of scattering (or excitation) from one state to another. For example, in the matrix element $\langle pk|g_{12}|lk\rangle$ one may picture the interaction as annihilating a particle in ψ_l (or creating a hole in ψ_l) and creating a particle in ψ_p (or annihilating a hole in ψ_p). Alternatively, one may say that a particle has been scattered (or excited) from ψ_l into ψ_p. The second particle is annihilated in ψ_k and created in ψ_k which means in effect that nothing has happened to the particle in ψ_k. In the exchange integral $\langle kl|g_{12}|lk\rangle$ particle 1 is annihilated in ψ_l and created in ψ_k while particle 2 is annihilated in ψ_k and created in ψ_l thereby exchanging the orbitals for the two particles. The other matrix elements and diagrams are analogously interpreted.

We shall now examine the expansion of the time development operator $U_{\mathrm{I}}(t, t_0)$ [Eq. (9.3-15)] when the perturbing potential $V = G$ has the form (12.2-14). The first few terms in the series for $U_{\mathrm{I}}(t, t_0)$ are

$$U_{\mathrm{I}}(t, t_0) = 1 - \frac{i}{\hbar}\int_{t_0}^{t} V_{\mathrm{I}}(t_1)\,dt_1 + \frac{1}{2}\left(\frac{-i}{\hbar}\right)^2 \int_{t_0}^{t} dt_1 \int_{t_0}^{t} dt_2\, P\{V_{\mathrm{I}}(t_1)V_{\mathrm{I}}(t_2)\} + \cdots$$

where P is the Dyson time-ordering operator (9.3-17). The matrix element

$$\langle \Psi_{\mathrm{g}}|U_{\mathrm{I}}(t, t_0)|\Psi_{\mathrm{g}}\rangle$$

will contain contributions from the various terms in the series. Writing

$$V = \tfrac{1}{2}\sum_{kl,mn} \langle kl|g_{12}|mn\rangle c_k^{\dagger} c_l^{\dagger} c_n c_m$$

and using (12.2-13),

$$V_{\mathrm{I}}(t) = \tfrac{1}{2}\sum_{kl,mn} \langle kl|g_{12}|mn\rangle c_k^{\dagger} c_l^{\dagger} c_n c_m \exp[i(E_k + E_l - E_n - E_m)t/\hbar]. \quad (12.3\text{-}1)$$

Hence the matrix elements are

$$\langle \Psi_{\mathrm{g}}|V|\Psi_{\mathrm{g}}\rangle$$

in first order and, assuming $t_2 > t_1$,

$$\langle \Psi_{\mathrm{g}}|V(t_2)V(t_1)|\Psi_{\mathrm{g}}\rangle$$

in second order. All quantities in these matrix elements are in the Schrödinger representation.

The first-order matrix elements and their diagrams have already been exhibited. We shall therefore concentrate on the second-order terms which will be written as

$$\langle \Psi_{\mathrm{g}}|V(t_2)V(t_1)|\Psi_{\mathrm{g}}\rangle = \sum_n \langle \Psi_{\mathrm{g}}|V(t_2)|\Psi_n\rangle\langle \Psi_n|V(t_1)|\Psi_{\mathrm{g}}\rangle.$$

But, as we have seen, there will be nonvanishing contributions only when Ψ_n corresponds to zero-, one- and two-particle excitations of the type previously calculated. Here, too, we may use diagrams to represent the various types of contributions. To accomplish this it is necessary to add a time scale and to draw two interaction lines one at t_1 and the other at t_2. We shall now indicate typical products and their diagrams.

$\langle\Psi_g|V(t_2)|\Psi_g\rangle\langle\Psi_g|V(t_1)|\Psi_g\rangle$:

$\langle k'l'|g_{12}|k'l'\rangle\langle kl|g_{12}|kl\rangle$

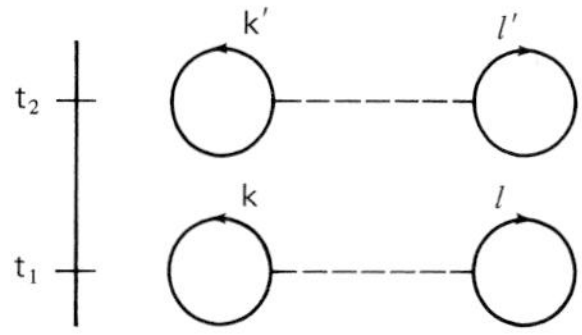

$\langle k'l'|g_{12}|k'l'\rangle\langle kl|g_{12}|lk\rangle$

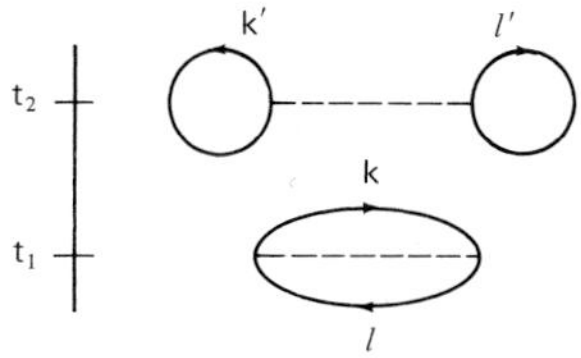

$\langle\Psi_g|V(t_2)|\Psi_1\rangle\langle\Psi_1|V(t_1)|\Psi_g\rangle$:

$\langle lk'|g_{12}|pk'\rangle\langle pk|g_{12}|lk\rangle$

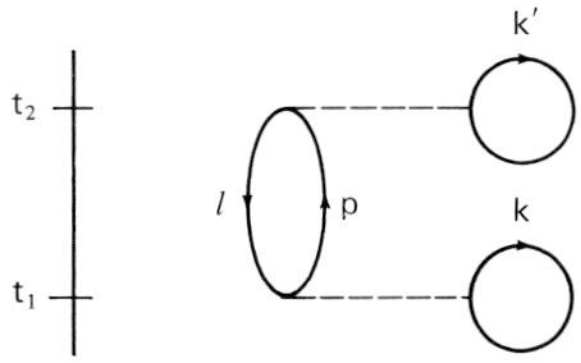

$\langle\Psi_g|V(t_2)|\Psi_2\rangle\langle\Psi_2|V(t_1)|\Psi_g\rangle$:

$\langle pq|g_{12}|km\rangle\langle km|g_{12}|pq\rangle$

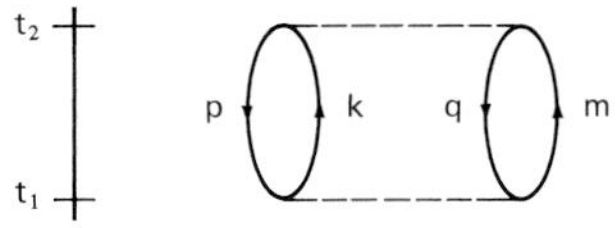

$\langle pq|g_{12}|km\rangle\langle mk|g_{12}|pq\rangle$

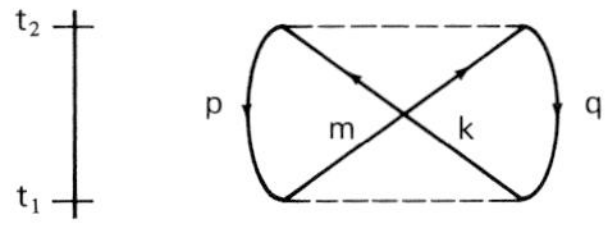

12.4 Field Operators

For fermions we define the *field operators* $\psi_F(\mathbf{r})$ and $\psi_F^\dagger(\mathbf{r})$ by

$$\psi_F(\mathbf{r}) = \sum_k \varphi_k(\mathbf{r})c_k, \qquad \psi_F^\dagger(\mathbf{r}) = \sum_k \varphi_k^*(\mathbf{r})c_k^\dagger \tag{12.4-1}$$

in which the φ_k are a complete set of orthonormal wave functions but otherwise arbitrary. In some instances it may be desirable to employ a complete set of orthonormal spin orbitals in which case $\varphi_k(\mathbf{r})$ would be replaced by $\psi_k(\lambda)$ as in Section 11.1 where λ represents the totality of spatial and spin coordinates.

The commutation rules pertaining to ψ_{F} and $\psi_{\mathrm{F}}^\dagger$ are directly derivable from those pertaining to c_k and $c_k^\dagger$ [Eq. (12.1-32)]. Thus

$$\{\psi_{\mathrm{F}}(\mathbf{r}), \psi_{\mathrm{F}}(\mathbf{r}')\} = \sum_{kl} \{c_k, c_l\}\varphi_k(\mathbf{r})\varphi_l(\mathbf{r}') = 0, \tag{12.4-2}$$

$$\{\psi_{\mathrm{F}}^\dagger(\mathbf{r}), \psi_{\mathrm{F}}^\dagger(\mathbf{r}')\} = \sum_{kl} \{c_k^\dagger, c_l^\dagger\}\varphi_k^*(\mathbf{r})\varphi_l^*(\mathbf{r}') = 0, \tag{12.4-3}$$

$$\begin{aligned}\{\psi_{\mathrm{F}}(\mathbf{r}), \psi_{\mathrm{F}}^\dagger(\mathbf{r}')\} &= \sum_{kl} \{c_k, c_l^\dagger\}\varphi_k(\mathbf{r})\varphi_l^*(\mathbf{r}') \\ &= \sum_{kl} \delta_{kl}\varphi_k(\mathbf{r})\varphi_l^*(\mathbf{r}') = \sum_k \varphi_k(\mathbf{r})\varphi_k^*(\mathbf{r}') = \delta(\mathbf{r} - \mathbf{r}').\end{aligned} \tag{12.4-4}$$

The δ-function in (12.4-4) is a result of the closure property of the orthonormal set φ_k (A8-14). The field operators therefore satisfy

$$\begin{aligned}\{\psi_{\mathrm{F}}(\mathbf{r}), \psi_{\mathrm{F}}(\mathbf{r}')\} &= \{\psi_{\mathrm{F}}^\dagger(\mathbf{r}), \psi_{\mathrm{F}}^\dagger(\mathbf{r}')\} = 0, \\ \{\psi_{\mathrm{F}}(\mathbf{r}), \psi_{\mathrm{F}}^\dagger(\mathbf{r}')\} &= \delta(\mathbf{r} - \mathbf{r}').\end{aligned} \tag{12.4-5}$$

Sums of one-particle and two-particle operators such as F (12.2-7) and G (12.2-14) may be expressed in terms of the field operators. Thus

$$\begin{aligned}\int \psi_{\mathrm{F}}^\dagger(\mathbf{r}) f \psi_{\mathrm{F}}(\mathbf{r})\, d\mathbf{r} &= \sum_{kl} c_k^\dagger c_l \int \varphi_k^*(\mathbf{r}) f \varphi_l(\mathbf{r})\, d\mathbf{r} \\ &= \sum_{kl} \langle \varphi_k(\mathbf{r}) | f | \varphi_l(\mathbf{r}) \rangle c_k^\dagger c_l,\end{aligned} \tag{12.4-6}$$

which, upon comparison with (12.2-7), leads to the interpretation that

$$F = \int \psi_{\mathrm{F}}^\dagger(\mathbf{r}) f \psi_{\mathrm{F}}(\mathbf{r})\, d\mathbf{r}. \tag{12.4-7}$$

If $f = 1$,

$$F = \int \psi_{\mathrm{F}}^\dagger(\mathbf{r})\psi_{\mathrm{F}}(\mathbf{r})\, d\mathbf{r} = \sum_k c_k^\dagger c_k = \sum_k N_k = N. \tag{12.4-8}$$

This suggests that we may define a (number) *density operator* $\rho(\mathbf{r})$ as

$$\rho(\mathbf{r}) = \psi_{\mathrm{F}}^\dagger(\mathbf{r})\psi_{\mathrm{F}}(\mathbf{r}). \tag{12.4-9}$$

A sum of two-particle operators assumes the form

$$G = \tfrac{1}{2} \int \psi_{\mathrm{F}}^\dagger(\mathbf{r}_1)\psi_{\mathrm{F}}^\dagger(\mathbf{r}_2) g_{12} \psi_{\mathrm{F}}(\mathbf{r}_2)\psi_{\mathrm{F}}(\mathbf{r}_1)\, d\mathbf{r}_1\, d\mathbf{r}_2 \tag{12.4-10}$$

since the right-hand side is

$$\tfrac{1}{2} \sum_{kl,mn} c_k{}^\dagger c_l{}^\dagger c_n c_m \int \varphi_k{}^*(\mathbf{r}_1)\varphi_l{}^*(\mathbf{r}_2) g_{12} \varphi_m(\mathbf{r}_1)\varphi_n(\mathbf{r}_2)\, d\mathbf{r}_1\, d\mathbf{r}_2$$

$$= \tfrac{1}{2} \sum_{kl,mn} c_k{}^\dagger c_l{}^\dagger c_n c_m \langle kl|g_{12}|mn\rangle \tag{12.4-11}$$

which has the same structure as (12.2-14).

As examples, the kinetic energy operator for an arbitrary number of electrons may be written as

$$T = \frac{1}{2m} \sum_{kl} \langle k|p^2|l\rangle c_k{}^\dagger c_l \tag{12.4-12a}$$

$$= \frac{1}{2m} \int \psi_{\mathrm{F}}{}^\dagger(\mathbf{r}) p^2 \psi_{\mathrm{F}}(\mathbf{r})\, d\mathbf{r} \tag{12.4-12b}$$

and the Coulomb repulsion energy as

$$G = \frac{1}{2} \sum_{kl,mn} \left\langle kl \left| \frac{e^2}{r_{12}} \right| mn \right\rangle c_k{}^\dagger c_l{}^\dagger c_n c_m \tag{12.4-13a}$$

$$= \frac{1}{2} \int \psi_{\mathrm{F}}{}^\dagger(\mathbf{r}_1)\psi_{\mathrm{F}}(\mathbf{r}_2) \frac{e^2}{r_{12}} \psi_{\mathrm{F}}(\mathbf{r}_2)\psi_{\mathrm{F}}(\mathbf{r}_1)\, d\mathbf{r}_1\, d\mathbf{r}_2. \tag{12.4-13b}$$

For bosons the formalism is much the same with commutators replacing anticommutators. In place of the fermion field operators ψ_{F} (12.4-1) we now have

$$\psi_{\mathrm{B}}(\mathbf{r}) = \sum_i \psi_i(\mathbf{r}) a_i, \qquad \psi_{\mathrm{B}}{}^\dagger(\mathbf{r}) = \sum_i \psi_i{}^*(\mathbf{r}) a_i{}^\dagger \tag{12.4-14}$$

which satisfy

$$\begin{aligned} [\psi_{\mathrm{B}}(\mathbf{r}), \psi_{\mathrm{B}}(\mathbf{r}')] &= [\psi_{\mathrm{B}}{}^\dagger(\mathbf{r}), \psi_{\mathrm{B}}{}^\dagger(\mathbf{r}')] = 0, \\ [\psi_{\mathrm{B}}(\mathbf{r}), \psi_{\mathrm{B}}{}^\dagger(\mathbf{r}')] &= \delta(\mathbf{r} - \mathbf{r}'). \end{aligned} \tag{12.4-15}$$

The operators F, G, and N may also be written as

$$F = \int \psi_{\mathrm{B}}{}^\dagger(\mathbf{r}) f \psi_{\mathrm{B}}(\mathbf{r})\, d\mathbf{r}, \tag{12.4-16}$$

$$G = \tfrac{1}{2} \int \psi_{\mathrm{B}}{}^\dagger(\mathbf{r}_1)\psi_{\mathrm{B}}{}^\dagger(\mathbf{r}_2) g_{12} \psi_{\mathrm{B}}(\mathbf{r}_2)\psi_{\mathrm{B}}(\mathbf{r}_1)\, d\mathbf{r}_1\, d\mathbf{r}_2, \tag{12.4-17}$$

$$N = \int \psi_{\mathrm{B}}{}^\dagger(\mathbf{r}) \psi_{\mathrm{B}}(\mathbf{r})\, d\mathbf{r}.$$

CHAPTER 13

DENSITY MATRICES

13.1 General Properties

A quantum–mechanical wave function, when it exists, conveys the maximum amount of information permitted by quantum mechanics concerning the properties of a physical system in the state described by the wave function. The wave function may be an eigenfunction of an observable (Hermitian operator which represents a physical quantity) or a linear superposition of eigenfunctions of an observable. Thus, if ψ_1 and ψ_2 are two possible eigenfunctions, the principle of superposition tells us that the normalizable function

$$\psi = c_1\psi_1 + c_2\psi_2$$

also represents a possible state.

Situations in which we have accurate wave functions for a physical system are actually quite rare. More often, the complexity of the system owing to its many degrees of freedom precludes the possibility of constructing a wave function. It is then necessary to resort to statistical methods, the same as in classical physics. Even in relatively simple systems for which wave functions exist, it may be impossible to construct a wave function pertaining to a *component* of the system. Consider, for example, the wave function

$$\psi = \frac{1}{\sqrt{2}}[\alpha(1)\beta(2) + \beta(1)\alpha(2)] \tag{13.1-1}$$

which describes a spin state of a two-electron system. For this case there is no way to construct wave functions for the individual electrons. Physically, this means that no experiment gives a unique result predictable with certainty on the spins of the individual electrons.

Quantum–mechanical states describable by wave functions are said to be *pure*; those that are not are said to be *mixed*. Thus, in the example just cited, the spin state of the two-electron system is pure, but the spin states of the individual electrons are mixed. It is primarily for the purpose of treating mixed states that the density matrix formalism was developed although it is also applicable to pure states.

We define a *density operator*

$$\rho = \sum_i |\psi^i\rangle p_i \langle\psi^i| \tag{13.1-2}$$

in which p_i is the probability of the system being in the normalized state $|\psi^i\rangle$ and the sum is taken over all states that are accessible to the system. The probability p_i evidently satisfies

$$0 \leqslant p_i \leqslant 1, \qquad \sum_i p_i = 1, \qquad \sum_i p_i^{\,2} \leqslant 1. \tag{13.1-3}$$

For a pure state there is just one p_i which is equal to unity and all the rest are zero. In that case

$$\rho = |\psi\rangle\langle\psi| \qquad \text{(pure state).} \tag{13.1-4}$$

It is convenient to introduce a complete orthonormal set $|\varphi_n\rangle$ to serve as a basis set for the expansion of $|\psi^i\rangle$ and for the computation of matrix elements. With the notation $|\varphi_n\rangle \equiv |n\rangle$,

$$|\psi^i\rangle = \sum_n c_n^{\,i} |n\rangle \tag{13.1-5a}$$

and from the orthonormality property of the φ_n,

$$c_k^{\,i} = \langle k|\psi^i\rangle$$

so that (13.1-5a) can be written as

$$|\psi^i\rangle = \sum_n |n\rangle\langle n|\psi^i\rangle. \tag{13.1-5b}$$

This gives rise to the operator relation

$$\sum_n |n\rangle\langle n| = 1. \tag{13.1-6}$$

We may now construct the density matrix which consists of the matrix elements of the density operator. In the basis set φ_n, using (13.1-2),

$$\langle n|\rho|n'\rangle = \sum_i \langle n|\psi^i\rangle p_i \langle\psi^i|n'\rangle = \sum_i p_i c_n^{\,i} c_{n'}^{i*} \tag{13.1-7}$$

which characterizes ρ as a Hermitian operator since

$$\langle n|\rho|n'\rangle = \langle n'|\rho|n\rangle^*.$$

The trace of ρ, i.e., the sum of the diagonal elements in (13.1-7) is

$$\operatorname{Tr}\rho = \sum_n \langle n|\rho|n\rangle = \sum_i \sum_n \langle\psi^i|n\rangle\langle n|\psi^i\rangle p_i$$
$$= \sum_i p_i \langle\psi^i|\psi^i\rangle = \sum_i p_i = 1. \qquad (13.1\text{-}8)$$

Since ρ is Hermitian, the diagonal elements $\langle n|\rho|n\rangle$ must be real and from (13.1-7) it follows that

$$0 \leqslant \langle n|\rho|n\rangle \leqslant 1. \qquad (13.1\text{-}9)$$

For a pure state $\langle n|\rho|n\rangle = c_n c_n^*$ which is simply the probability of finding the system in the state φ_n.

Consider now the matrix elements of ρ^2; again in the basis set φ_n,

$$\langle n|\rho^2|n''\rangle = \sum_{n'} \langle n|\rho|n'\rangle\langle n'|\rho|n''\rangle.$$

Substituting (13.1-2) for ρ,

$$\langle n|\rho^2|n''\rangle = \sum_j \sum_i \sum_{n'} \langle n|\psi^i\rangle p_i \langle\psi^i|n'\rangle\langle n'|\psi^j\rangle p_j \langle\psi^j|n''\rangle$$

and

$$\operatorname{Tr}\rho^2 = \sum_i \sum_{n'} \sum_n \langle n|\psi^i\rangle p_i \langle\psi^i|n'\rangle\langle n'|\psi^i\rangle p_i \langle\psi^i|n\rangle$$
$$= \sum_i \sum_n \langle n|\psi^i\rangle p_i^2 \langle\psi^i|n\rangle = \sum_i p_i^2. \qquad (13.1\text{-}10)$$

But since $\sum_i p_i^2 \leqslant 1$, we have

$$\operatorname{Tr}\rho^2 \leqslant 1. \qquad (13.1\text{-}11)$$

For a pure state there is only one p_i and it is equal to unity. Therefore

$$\operatorname{Tr}\rho^2 = 1 \qquad \text{(pure state)} \qquad (13.1\text{-}12)$$

and

$$\rho^2 = |\psi\rangle\langle\psi|\psi\rangle\langle\psi| = |\psi\rangle\langle\psi|$$
$$= \rho \qquad \text{(pure state)}. \qquad (13.1\text{-}13)$$

Thus whether a state is pure or not can be established by testing whether (13.1-12) or (13.1-13) is satisfied or not.

So far we have focused on the formal properties of the density operator ρ. Its utility arises from the following considerations. Let

$$\langle A\rangle_i = \langle\psi^i|A|\psi^i\rangle, \qquad (13.1\text{-}14a)$$

$$\langle A\rangle = \sum_i p_i \langle\psi^i|A|\psi^i\rangle. \qquad (13.1\text{-}14b)$$

The distinction between $\langle A\rangle_i$ and $\langle A\rangle$ is that the former is a quantum-mechanical average or the expectation value of an operator A when the system is definitely in the state ψ^i. On the other hand, $\langle A\rangle$ is a statistical or ensemble average which, from (13.1-14b), is seen to be a weighted average of $\langle A\rangle_i$ taken over all the states that the system may occupy. Again for a pure state

$$\langle A\rangle = \langle A\rangle_i \qquad \text{(pure state)}. \tag{13.1-15}$$

We now construct the operator ρA. From (13.1-2),

$$\rho A = \sum_i |\psi^i\rangle p_i \langle\psi^i|A$$

and the matrix element of ρA in the basis set φ_n is

$$\langle n|\rho A|n'\rangle = \sum_i \langle n|\psi^i\rangle p_i \langle\psi^i|A|n'\rangle.$$

Taking the trace of ρA,

$$\operatorname{Tr}\rho A = \sum_n \langle n|\rho A|n\rangle = \sum_i\sum_n \langle n|\psi^i\rangle p_i\langle\psi^i|A|n\rangle = \sum_i p_i\langle\psi^i|A|\psi^i\rangle$$

which, according to (13.1-14b), is

$$\operatorname{Tr} A\rho = \operatorname{Tr}\rho A = \langle A\rangle. \tag{13.1-16}$$

Thus the average value of an operator A is known as soon as the density operator ρ is known.

13.2 Spin States

Suppose the spin state of an electron is given by

$$|\psi\rangle = |\alpha\rangle \tag{13.2-1}$$

so that the density operator is

$$\rho = |\alpha\rangle\langle\alpha|.$$

In the basis set $\varphi_1 = \alpha$, $\varphi_2 = \beta$, the density matrix is

$$\rho = \begin{pmatrix} \langle\alpha|\alpha\rangle\langle\alpha|\alpha\rangle & \langle\alpha|\alpha\rangle\langle\alpha|\beta\rangle \\ \langle\beta|\alpha\rangle\langle\alpha|\alpha\rangle & \langle\beta|\alpha\rangle\langle\alpha|\beta\rangle \end{pmatrix} = \begin{pmatrix} 1 & 0 \\ 0 & 0 \end{pmatrix}. \tag{13.2-2}$$

For the spin operators S_x, S_y, and S_z whose matrix representations are given in (1.7-16),

$$\langle S_x\rangle = \operatorname{Tr}\rho S_x = \operatorname{Tr}\tfrac{1}{2}\begin{pmatrix} 1 & 0 \\ 0 & 0 \end{pmatrix}\begin{pmatrix} 0 & 1 \\ 1 & 0 \end{pmatrix} = \operatorname{Tr}\tfrac{1}{2}\begin{pmatrix} 0 & 1 \\ 0 & 0 \end{pmatrix} = 0, \tag{13.2-3a}$$

$$\langle S_y \rangle = \operatorname{Tr} \rho S_y = \operatorname{Tr} \tfrac{1}{2}\begin{pmatrix} 1 & 0 \\ 0 & 0 \end{pmatrix}\begin{pmatrix} 0 & -i \\ i & 0 \end{pmatrix} = \operatorname{Tr} \tfrac{1}{2}\begin{pmatrix} 0 & -i \\ 0 & 0 \end{pmatrix} = 0, \tag{13.2-3b}$$

$$\langle S_z \rangle = \operatorname{Tr} \rho S_z = \operatorname{Tr} \tfrac{1}{2}\begin{pmatrix} 1 & 0 \\ 0 & 0 \end{pmatrix}\begin{pmatrix} 1 & 0 \\ 0 & -1 \end{pmatrix} = \operatorname{Tr} \tfrac{1}{2}\begin{pmatrix} 1 & 0 \\ 0 & 0 \end{pmatrix} = \tfrac{1}{2}, \tag{13.2-3c}$$

as one would expect. More generally, suppose the electron is in a state described by

$$|\psi\rangle = a_1|\alpha\rangle + a_2|\beta\rangle \tag{13.2-4}$$

with

$$|a_1|^2 + |a_2|^2 = 1.$$

The density operator in this case is

$$\rho = (a_1|\alpha\rangle + a_2|\beta\rangle)(a^*\langle\alpha| + a_2{}^*\langle\beta|) \tag{13.2-5}$$

and the density matrix is

$$\rho = \begin{pmatrix} |a_1|^2 & a_1 a_2{}^* \\ a_2 a_1{}^* & |a_2|^2 \end{pmatrix} \tag{13.2-6}$$

which indicates that the diagonal elements $|a_1|^2$ and $|a_2|^2$ are just the probabilities that the electron is in the state α or β, respectively.

Another useful form for the density matrix of spin $\frac{1}{2}$ particles is obtained by writing

$$\rho = c_0 I + c_1 S_x + c_2 S_y + c_3 S_z \tag{13.2-7}$$

where I is the unit 2×2 matrix. The density matrix then becomes

$$\rho = \begin{pmatrix} \langle\alpha|\rho|\alpha\rangle & \langle\alpha|\rho|\beta\rangle \\ \langle\beta|\rho|\alpha\rangle & \langle\beta|\rho|\beta\rangle \end{pmatrix} = \begin{pmatrix} c_0 + \frac{1}{2}c_3 & \frac{1}{2}(c_1 - ic_2) \\ \frac{1}{2}(c_1 + ic_2) & c_0 - \frac{1}{2}c_3 \end{pmatrix} \tag{13.2-8}$$

where (1.7-16) has been used to evaluate the matrix elements. Now,

$$\operatorname{Tr} \rho = 1 = 2c_0, \tag{13.2-9a}$$

$$\begin{aligned} \operatorname{Tr} \rho S_x &= \langle S_x \rangle \\ &= \operatorname{Tr}\begin{pmatrix} c_0 + \frac{1}{2}c_3 & \frac{1}{2}(c_1 - ic_3) \\ \frac{1}{2}(c_1 + ic_2) & c_0 - \frac{1}{2}c_3 \end{pmatrix}\begin{pmatrix} 0 & \frac{1}{2} \\ \frac{1}{2} & 0 \end{pmatrix} \\ &= \operatorname{Tr}\begin{pmatrix} \frac{1}{4}(c_1 - ic_2) & c_0 + \frac{1}{2}c_3 \\ c_0 - \frac{1}{2}c_3 & \frac{1}{4}(c_1 + ic_2) \end{pmatrix} = \tfrac{1}{2}c_1. \end{aligned} \tag{13.2-9b}$$

Similarly,

$$\operatorname{Tr} \rho S_y = \langle S_y \rangle = \tfrac{1}{2}c_2, \tag{13.2-9c}$$

$$\operatorname{Tr} \rho S_z = \langle S_z \rangle = \tfrac{1}{2}c_3. \tag{13.2-9d}$$

Substituting (13.2-9) in (13.2-8),

$$\rho = \begin{pmatrix} \frac{1}{2} + \langle S_z \rangle & \langle S_x \rangle - i\langle S_y \rangle \\ \langle S_x \rangle + i\langle S_y \rangle & \frac{1}{2} - \langle S_z \rangle \end{pmatrix} \tag{13.2-10}$$

or, written more compactly,

$$\rho = \tfrac{1}{2}I + 2\langle \mathbf{S} \rangle \cdot \mathbf{S} = \tfrac{1}{2}[I + \langle \boldsymbol{\sigma} \rangle \cdot \boldsymbol{\sigma}]. \tag{13.2-11}$$

13.3 Reduced Density Matrices

Let us now consider a two-electron system and let the basis set consist of the functions

$$\varphi_{11} = \alpha(1)\alpha(2), \qquad \varphi_{12} = \alpha(1)\beta(2), \qquad \varphi_{21} = \beta(1)\alpha(2), \qquad \varphi_{22} = \beta(1)\beta(2). \tag{13.3-1}$$

A typical element of the 4×4 density matrix will then be

$$\langle \varphi_{kl} | \rho | \varphi_{k'l'} \rangle \equiv \langle kl | \rho | k'l' \rangle. \tag{13.3-2}$$

Here the first index (k, k') refers to the spin state of electron 1 and the second (l, l') to electron 2. Suppose now that we have an operator $F(1)$ which acts only on electron 1. In the basis set (13.3-1) a typical matrix element of $F(1)$ will be

$$\langle \varphi_{kl} | F(1) | \varphi_{k'l'} \rangle \equiv \langle kl | F(1) | k'l' \rangle = \langle k | F(1) | k' \rangle \, \delta_{ll'} \tag{13.3-3}$$

in which the last form on the right takes into account the orthonormality of the one-electron spin functions. It is now possible to express the trace of $\rho F(1)$ in terms of (13.3-2) and (13.3-3). Thus

$$\begin{aligned} \operatorname{Tr} \rho F(1) &= \sum_{kl} \langle kl | \rho F(1) | kl \rangle \\ &= \sum_{kl} \sum_{k'l'} \langle kl | \rho | k'l' \rangle \langle k'l' | F(1) | kl \rangle \\ &= \sum_{kl} \sum_{k'l'} \langle kl | \rho | k'l' \rangle \langle k' | F(1) | k \rangle \, \delta_{l'l} \\ &= \sum_{kk'l} \langle kl | \rho | k'l \rangle \langle k' | F(1) | k \rangle. \end{aligned} \tag{13.3-4}$$

Now let us define the *reduced density matrix*:

$$\langle k | \rho(1) | k' \rangle = \sum_{l} \langle kl | \rho | k'l \rangle. \tag{13.3-5}$$

We then have

$$\begin{aligned} \operatorname{Tr} \rho F(1) &= \sum_{k,k'} \langle k | \rho(1) | k' \rangle \langle k' | F(1) | k \rangle \\ &= \sum_{k} \langle k | \rho(1) F(1) | k \rangle = \operatorname{Tr} \rho(1) F(1) = \langle F(1) \rangle. \end{aligned} \tag{13.3-6}$$

Equation (13.3-6) tells us that the average value of a one-electron operator $F(1)$ may be computed with the help of the (reduced) density operator $\rho(1)$ whose matrix elements are derivable from those of the density operator ρ. The important distinction is that ρ refers to the system as a whole while $\rho(1)$ refers to a component of the system, namely electron 1. The generalization of (13.3-5) is that if A and B are components of a composite system denoted by AB, that is, a system in which A and B interact, then

$$\rho_A = \mathrm{Tr}_B\, \rho_{AB}, \qquad \rho_B = \mathrm{Tr}_A\, \rho_{AB} \tag{13.3-7}$$

where ρ_A and ρ_B are the reduced density operators for components A and B belonging to the composite system AB whose density operator is ρ_{AB}. Tr_A and Tr_B denote the sum over the indices associated with components A and B, respectively. This means that if we are interested only in the development of system A, for example, the pertinent density operator is ρ_A which contains no reference to indices associated with B. In this way unessential indices (or coordinates) are eliminated.

Let us now take an example where the state of the system is represented by the wave function

$$\psi = \frac{1}{\sqrt{2}}[\alpha(1)\beta(2) + \beta(1)\alpha(2)] = \frac{1}{\sqrt{2}}[\varphi_{12} + \varphi_{21}] \tag{13.3-8}$$

as in (13.1-1). Since the system, as a whole, is in a pure state,

$$\rho = |\psi\rangle\langle\psi|; \tag{13.3-9}$$

the density matrix in the basis set (13.3-1) is

	φ_{11}	φ_{12}	φ_{21}	φ_{22}
φ_{11}	0	0	0	0
φ_{12}	0	$\frac{1}{2}$	$\frac{1}{2}$	0
φ_{21}	0	$\frac{1}{2}$	$\frac{1}{2}$	0
φ_{22}	0	0	0	0

(13.3-10)

and from (13.3-5)

$$\rho(1) = \begin{pmatrix} \frac{1}{2} & 0 \\ 0 & \frac{1}{2} \end{pmatrix}. \tag{13.3-11}$$

It is now possible to calculate $\langle F(1)\rangle$ by three different methods:

(1) Substituting (13.3-8) for ψ one obtains

$$\langle F(1)\rangle = \langle\psi|F(1)|\psi\rangle = \tfrac{1}{2}[\langle\alpha(1)|F(1)|\alpha(1)\rangle + \langle\beta(1)|F(1)|\beta(1)\rangle]. \tag{13.3-12}$$

(2) The matrix representation of $F(1)$ in the basis set (13.3-1) is

$$F(1) = \begin{pmatrix} \langle\varphi_{11}|F(1)|\varphi_{11}\rangle & \cdots & \langle\varphi_{11}|F(1)|\varphi_{22}\rangle \\ \vdots & \vdots & \vdots \\ \langle\varphi_{22}|F(1)|\varphi_{11}\rangle & \cdots & \langle\varphi_{22}|F(1)|\varphi_{22}\rangle \end{pmatrix}.$$

With the density matrix (13.3-10),

$$\begin{aligned} \operatorname{Tr} \rho F(1) &= \tfrac{1}{2}[\langle\varphi_{12}|F(1)|\varphi_{12}\rangle + \langle\varphi_{12}|F(1)|\varphi_{21}\rangle \\ &\quad + \langle\varphi_{21}|F(1)|\varphi_{12}\rangle + \langle\varphi_{21}|F(1)|\varphi_{21}\rangle] \\ &= \tfrac{1}{2}[\langle\varphi_{12}|F(1)|\varphi_{12}\rangle + \langle\varphi_{21}|F(1)|\varphi_{21}\rangle] \\ &= \tfrac{1}{2}[\alpha(1)|F(1)|\alpha(1)\rangle + \langle\beta(1)|F(1)|\beta(1)\rangle]. \end{aligned} \tag{13.3-13}$$

(3) Using (13.3-11) for $\rho(1)$ and basis set $\alpha(1)$, $\beta(1)$ which is the appropriate basis set for a single electron, we obtain

$$\begin{aligned} \langle F(1)\rangle &= \operatorname{Tr} \rho(1)F(1) \\ &= \operatorname{Tr}\begin{pmatrix} \langle\alpha(1)|F(1)|\alpha(1)\rangle & \langle\alpha(1)|F(1)|\beta(1)\rangle \\ \langle\beta(1)|F(1)|\alpha(1)\rangle & \langle\beta(1)|F(1)|\beta(1)\rangle \end{pmatrix}\begin{pmatrix} \frac{1}{2} & 0 \\ 0 & \frac{1}{2} \end{pmatrix} \\ &= \tfrac{1}{2}[\langle\alpha(1)|F(1)|\alpha(1)\rangle + \langle\beta(1)|F(1)|\beta(1)\rangle]. \end{aligned} \tag{13.3-14}$$

A few comments at this stage are appropriate. The two-electron system is described by the wave function ψ (13.3-8); for the density matrix (13.3-10),

$$\operatorname{Tr} \rho = 1, \qquad \rho^2 = \rho, \qquad \operatorname{Tr} \rho^2 = 1. \tag{13.3-15}$$

These are properties of a pure state. For electron 1, the density matrix $\rho(1)$ is given by (13.3-11) for which

$$\operatorname{Tr} \rho(1) = 1, \qquad \rho^2(1) = \begin{pmatrix} \frac{1}{4} & 0 \\ 0 & \frac{1}{4} \end{pmatrix} \neq \rho(1), \qquad \operatorname{Tr} \rho^2(1) = \tfrac{1}{2}. \tag{13.3-16}$$

Electron 1 is therefore in a mixed state. For comparison let us suppose that we have a single electron whose wave function is

$$\psi' = \frac{1}{\sqrt{2}}(\alpha + \beta). \tag{13.3-17}$$

This is a special case of (13.2-4); therefore the density matrix according to (13.2-6) is

$$\rho' = \begin{pmatrix} \frac{1}{2} & \frac{1}{2} \\ \frac{1}{2} & \frac{1}{2} \end{pmatrix} \tag{13.3-18}$$

and

$$\operatorname{Tr} \rho' = 1, \qquad \rho'^2 = \rho', \qquad \operatorname{Tr} \rho'^2 = 1 \tag{13.3-19}$$

as we expect for a pure state. Let us now compare $\rho(1)$ in (13.3-11) and ρ' in (13.3-18). In both cases the diagonal elements are the same so that in both cases the probability of the electron being in an α and β state is $\frac{1}{2}$. However, for the electron in the state ψ' given by (13.3-17), the superposition of the α and β states is coherent, whereas for electron 1, residing in a mixed state, the superposition of α and β states is incoherent. What this means is that for an operator F acting on the electron in ψ':

$$\begin{aligned}\langle F\rangle = \langle\psi|F|\psi\rangle &= \tfrac{1}{2}\langle\alpha+\beta|F|\alpha+\beta\rangle \\ &= \tfrac{1}{2}[\langle\alpha|F|\alpha\rangle + \langle\beta|F|\beta\rangle + 2\langle\alpha|F|\beta\rangle]. \qquad (13.3\text{-}20)\end{aligned}$$

It is the presence of the cross term $\langle\alpha|F|\beta\rangle$ in (13.3-20) for $\langle F\rangle$ and the absence of any cross term in $\langle F(1)\rangle$ as it was calculated in (13.3-12)–(13.3-14) which distinguishes the coherent from the incoherent case.

13.4 Thermal Equilibrium

Let φ_n be a complete set of orthonormal functions that satisfy

$$\mathscr{H}\varphi_n = E_n\varphi_n \qquad (13.4\text{-}1)$$

and let the states be occupied according to the Boltzmann distribution

$$p_n = Ne^{-E_n/kT} \qquad (13.4\text{-}2)$$

where p_n is the probability of finding the system in the eigenstate φ_n with energy E_n, k is the Boltzmann constant, T the absolute temperature, and N a normalizing constant chosen to ensure that

$$\sum_n p_n = 1. \qquad (13.4\text{-}3)$$

Condition (13.4-2) defines thermal equilibrium and the corresponding density operator ρ_{T}, according to the basic definition (13.1-2), becomes

$$\rho_{\mathrm{T}} = N\sum_n e^{-E_n/kT}|\varphi_n\rangle\langle\varphi_n|. \qquad (13.4\text{-}4)$$

Now consider the operator $e^{-\mathscr{H}/kT}$ defined by its power series expansion. In view of (13.4-1),

$$e^{-\mathscr{H}/kT}|\varphi_n\rangle = e^{-E_n/kT}|\varphi_n\rangle \qquad (13.4\text{-}5)$$

which enables us to write

$$\rho_{\mathrm{T}}|\psi\rangle = N\sum_n e^{-\mathscr{H}/kT}|\varphi_n\rangle\langle\varphi_n|\psi\rangle = Ne^{-\mathscr{H}/kT}|\psi\rangle$$

where ψ is an arbitrary function expandable in terms of the φ_n. Thus

$$\rho_{\mathrm{T}} = Ne^{-\mathscr{H}/kT}. \qquad (13.4\text{-}6)$$

To find N we note that

$$\mathrm{Tr}\,\rho_{\mathrm{T}} = N\,\mathrm{Tr}\,e^{-\mathscr{H}/kT} = 1$$

or

$$N = 1/\mathrm{Tr}\,e^{-\mathscr{H}/kT}.$$

Hence the density operator under thermal equilibrium is

$$\rho_{\mathrm{T}} = \frac{e^{-\mathscr{H}/kT}}{\mathrm{Tr}\,e^{-\mathscr{H}/kT}} = \frac{1}{Z}\,e^{-\mathscr{H}/kT} \tag{13.4-7}$$

where

$$Z = \mathrm{Tr}\,e^{-\mathscr{H}/kT} \tag{13.4-8}$$

is known as the *partition function.*

As an application of (13.4-7), consider an ensemble of electrons with angular momenta J subjected to a constant magnetic field B which may be assumed to be oriented in the z direction. Under thermal equilibrium the magnetic substates M_J will be populated according to the Boltzmann distribution. We wish to find the average magnetization (magnetic moment per unit volume) in a sample containing N electrons per unit volume. The general relation between the magnetic moment operator $\boldsymbol{\mu}_J$ and the angular momentum operator $\mathbf{J}$ is

$$\boldsymbol{\mu}_J = -\beta g_J \mathbf{J} \tag{13.4-9}$$

and the Hamiltonian for the interaction of the electron with the magnetic field is

$$\mathscr{H} = -\boldsymbol{\mu}_J \cdot \mathbf{B} = \beta g_J \mathbf{J} \cdot \mathbf{B}. \tag{13.4-10}$$

Here β is a constant known as the Bohr magneton and g_J is another constant expressible in terms of the orbital and spin angular momenta of the electrons (see also Section 17.1). The average for the z component of the magnetization $\langle M_z \rangle$ is defined as $N\langle \mu_{Jz} \rangle$. Thus

$$\begin{aligned}\langle M_z \rangle &= -N\beta g_J \langle J_z \rangle \\ &= -N\beta g_J\,\mathrm{Tr}\,J_z\,\rho_{\mathrm{T}} = -\frac{N\beta g_J}{Z}\,\mathrm{Tr}\,J_z e^{-\mathscr{H}/kT}.\end{aligned} \tag{13.4-11}$$

The partition function Z is

$$\begin{aligned}Z = \mathrm{Tr}\,e^{-\mathscr{H}/kT} &= \mathrm{Tr}\,e^{-\beta g_J B J_z/kT} \\ &= \sum_{M_J=-J}^{J} \langle M_J | e^{-\beta g_J B J_z/kT} | M_J \rangle \\ &= \sum_{M_J=-J}^{J} e^{-\beta g_J B M_J/kT}.\end{aligned} \tag{13.4-12}$$

We shall assume that the temperature is sufficiently high to justify setting the exponentials equal to unity. In that case

$$Z = 2J + 1. \tag{13.4-13}$$

For the trace of $J_z e^{-\mathscr{H}/kT}$ we confine ourselves to linear terms in the magnetic field so that

$$\operatorname{Tr} J_z e^{-\mathscr{H}/kT} = \operatorname{Tr}\left[J_z - \frac{J_z \mathscr{H}}{kT}\right]. \tag{13.4-14}$$

Evaluating the two terms separately,

$$\operatorname{Tr} J_z = \sum_{M_J=-J}^{J} \langle M_J|J_z|M_J\rangle = 0, \tag{13.4-15}$$

$$-\operatorname{Tr}\frac{J_z \mathscr{H}}{kT} = -\frac{\beta g_J B}{kT} \sum_{M_J=-J}^{J} \langle M_J|J_z^2|M_J\rangle. \tag{13.4-16}$$

The summation on the right side of (13.4-16) may be shown to be

$$\tfrac{1}{3}J(J+1)(2J+1).$$

Therefore

$$\operatorname{Tr} J_z e^{-\mathscr{H}/kT} = -\frac{\beta g_J B}{3kT} J(J+1)(2J+1)$$

and the final expression for the magnetization in this approximation is

$$\langle M_z\rangle = N\beta^2 g_J{}^2 BJ(J+1)/3kT. \tag{13.4-17}$$

13.5 Equation of Motion

The density operator ρ has been defined by (13.1-2). If ψ^i is a function of time, as it is in the Schrödinger representation, then ρ is also a function of time:

$$\rho(t) = \sum_i |\psi^i(t)\rangle p_i \langle \psi^i(t)|. \tag{13.5-1}$$

The time dependence of $\psi^i(t)$ may be expressed in terms of the evolution operator $U(t, t_0)$:

$$\psi^i(t) = U(t, t_0)\psi^i(t_0);$$

therefore

$$\begin{aligned}\rho(t) &= \sum_i |U(t,t_0)\psi^i(t_0)\rangle p_i \langle U(t,t_0)\psi^i(t_0)| \\ &= \sum_i U(t,t_0)|\psi^i(t_0)\rangle p_i \langle\psi^i(t_0)|U^\dagger(t,t_0) \\ &= U(t,t_0)\rho(t_0)U^\dagger(t,t_0)\end{aligned} \tag{13.5-2}$$

where

$$\rho(t_0) = \sum_i |\psi^i(t_0)\rangle p_i \langle \psi^i(t_0)|. \tag{13.5-3}$$

Upon differentiating (13.5-2) with respect to t and replacing the time derivatives of $U(t, t_0)$ and $U^\dagger(t, t_0)$ by (9.1-7) and (9.1-8) we obtain

$$\begin{aligned} i\hbar\dot{\rho}(t) &= i\hbar\left[U\rho(t_0)\frac{\partial U^\dagger}{\partial t} + \frac{\partial U}{\partial t}\rho(t_0)U^\dagger\right] \\ &= -U\rho(t_0)U^\dagger\mathscr{H} + \mathscr{H}U\rho(t_0)U^\dagger = -\rho(t)\mathscr{H} + \mathscr{H}\rho(t) = [\mathscr{H}, \rho(t)]. \end{aligned} \tag{13.5-4}$$

This is the equation of motion for the density operator $\rho(t)$. It is important to note that this development has been carried out in the Schrödinger representation, although (13.5-4) differs from the Heisenberg equation (9.2-7) only in the sign.

Suppose A is a Schrödinger operator (with no intrinsic time dependence). From (13.1-16),

$$i\hbar\frac{\partial}{\partial t}\langle A\rangle = i\hbar\frac{\partial}{\partial t}\operatorname{Tr}\rho A = i\hbar\operatorname{Tr}\frac{\partial\rho}{\partial t}A.$$

Using (13.5-4),

$$i\hbar\frac{\partial}{\partial t}\langle A\rangle = i\hbar\operatorname{Tr}[\mathscr{H}, \rho]A = i\hbar\operatorname{Tr}(\mathscr{H}\rho A - \rho\mathscr{H}A). \tag{13.5-5}$$

But since $\operatorname{Tr}ABC = \operatorname{Tr}BCA = \operatorname{Tr}CAB$,

$$\operatorname{Tr}\mathscr{H}\rho A = \operatorname{Tr}\rho A\mathscr{H} = \langle A\mathscr{H}\rangle, \qquad \operatorname{Tr}\rho\mathscr{H}A = \langle\mathscr{H}A\rangle,$$

so that

$$i\hbar\frac{\partial}{\partial t}\langle A\rangle = \langle[A, \mathscr{H}]\rangle \tag{13.5-6}$$

the same as in (9.1-18).

We return to the equation of motion for the density operator with the assumption that

$$\mathscr{H} = \mathscr{H}_0 + V.$$

If φ_k and φ_l are eigenstates of $\mathscr{H}_0$ with eigenvalues E_k and E_l,

$$\begin{aligned} i\hbar\frac{\partial}{\partial t}\langle\varphi_k|\rho(t)|\varphi_l\rangle &= \langle\varphi_k|[\mathscr{H}_0, \rho(t)]|\varphi_l\rangle + \langle\varphi_k|[V, \rho(t)]|\varphi_l\rangle \\ &= (E_k - E_l)\langle\varphi_k|\rho(t)|\varphi_l\rangle + \langle\varphi_k|[V, \rho(t)]|\varphi_l\rangle. \end{aligned} \tag{13.5-7}$$

It is often more convenient to express the equation of motion in the interaction representation. Thus

$$\rho_{\rm I}(t) = e^{i\mathscr{H}_0 t/\hbar}\rho(t)e^{-i\mathscr{H}_0 t/\hbar}$$

$$\begin{aligned} i\hbar\dot{\rho}_{\rm I}(t) &= i\hbar\left[\frac{i}{\hbar}\mathscr{H}_0 e^{i\mathscr{H}_0 t/\hbar}\rho(t)e^{-i\mathscr{H}_0 t/\hbar} + e^{i\mathscr{H}_0 t/\hbar}\frac{\partial\rho}{\partial t}e^{-i\mathscr{H}_0 t/\hbar}\right. \\ &\quad \left. - e^{i\mathscr{H}_0 t/\hbar}\rho(t)\frac{i}{\hbar}\mathscr{H}_0 e^{-i\mathscr{H}_0 t/\hbar}\right] \\ &= [-\mathscr{H}_0 e^{i\mathscr{H}_0 t/\hbar}\rho(t)e^{-i\mathscr{H}_0 t/\hbar} + e^{i\mathscr{H}_0 t/\hbar}[\mathscr{H}_0 + V, \rho(t)]e^{-i\mathscr{H}_0 t/\hbar} \\ &\quad + e^{i\mathscr{H}_0 t/\hbar}\rho(t)\mathscr{H}_0 e^{-i\mathscr{H}_0 t/\hbar}] \\ &= e^{i\mathscr{H}_0 t/\hbar}[V, \rho(t)]e^{-i\mathscr{H}_0 t/\hbar} = [V_{\rm I}(t), \rho_{\rm I}(t)]. \end{aligned} \tag{13.5-8}$$

A formal solution to (13.5-8) may be written as

$$\rho_{\rm I}(t) = \rho_{\rm I}(t_0) - \frac{i}{\hbar}\int_{t_0}^{t} dt_1\,[V_{\rm I}(t_1), \rho_{\rm I}(t_1)]$$

which can be iterated to give

$$\begin{aligned} \rho_{\rm I}(t) &= \rho_{\rm I}(t_0) - \frac{i}{\hbar}\int_{t_0}^{t} dt_1\,[V_{\rm I}(t_1), \rho_{\rm I}(t_0)] \\ &\quad + \left(-\frac{i}{\hbar}\right)^2 \int_{t_0}^{t} dt_1 \int_{t_0}^{t_1} dt_2\,[V_{\rm I}(t_1), [V_{\rm I}(t_2), \rho_{\rm I}(t_0)]] + \cdots. \end{aligned} \tag{13.5-9}$$

13.6 Multielectron Systems

Suppose the normalized wave function of a particular state of a one-particle system is $\psi(\lambda)$ where λ symbolizes the aggregate of space and spin coordinates. The density matrix, according to (13.1-4), is

$$\rho(\lambda) = |\psi(\lambda)\rangle\langle\psi(\lambda)| \equiv \psi(\lambda)\psi^*(\lambda). \tag{13.6-1}$$

By a trivial extension we may define

$$\rho(\lambda;\lambda') = \psi(\lambda)\psi^*(\lambda') \tag{13.6-2}$$

which then reduces to (13.6-1) as soon as one sets $\lambda' = \lambda$. It is also possible to define the trace of $\rho(\lambda;\lambda')$ so as to be consistent with previous results:

$$\operatorname{Tr}\rho(\lambda;\lambda') = \int [\psi(\lambda)\psi^*(\lambda')]_{\lambda'=\lambda}\,d\lambda \tag{13.6-3}$$

in which the notation is intended to convey that before the integration is performed, λ' is set equal to λ in the integrand. Performing the indicated operations and taking note of the normalization of $\psi(\lambda)$, it is seen that

$$\operatorname{Tr} \rho(\lambda;\lambda') = 1$$

which is consistent with (13.1-8). Thus $\rho(\lambda;\lambda')$ is analogous to a matrix element; the continuous variables λ and λ' perform the same function as the two discrete indices which identify an ordinary matrix element. In this sense it is permissible to refer to $\rho(\lambda;\lambda)$ as a diagonal matrix element.

The expectation value of an operator may also be expressed in terms of $\rho(\lambda;\lambda')$. To see this, let $A(\lambda)$ be an operator which operates on functions of λ. Then

$$\begin{aligned}\langle A(\lambda)\rangle &= \langle\psi(\lambda)|A(\lambda)|\psi(\lambda)\rangle \\ &= \int \psi^*(\lambda)A(\lambda)\psi(\lambda)\,d\lambda = \int A(\lambda)[\psi(\lambda)\psi^*(\lambda')]_{\lambda'=\lambda}\,d\lambda\end{aligned}$$

where it shall be understood henceforth that the operations are carried out in the order: (1) let $A(\lambda)$ act on the (unprimed) $\psi(\lambda)$, (2) set $\lambda' = \lambda$, and (3) integrate. On inserting (13.6-2) and dropping the subscript $\lambda' = \lambda$,

$$\langle A(\lambda)\rangle = \int A(\lambda)\rho(\lambda;\lambda')\,d\lambda = \operatorname{Tr} A(\lambda)\rho(\lambda;\lambda') \tag{13.6-4}$$

which is consistent with (13.1-16).

In a multielectron system the wave function depends on the coordinates and spins of all the electrons. Let $\psi(\lambda_1, \lambda_2)$ be a two-electron wave function in which λ_1 stands for the set of space and spin coordinates of electron 1 and λ_2 conveys the corresponding information for electron 2. In this case we define

$$\rho(\lambda_1;\lambda_1') = 2\int \psi(\lambda_1,\lambda_2)\psi^*(\lambda_1',\lambda_2)\,d\lambda_2. \tag{13.6-5}$$

If $A(\lambda_1)$ is an operator that operates on functions of λ_1 (but not on functions of λ_2), then

$$\begin{aligned}\langle A(\lambda_1)\rangle &= \langle\psi(\lambda_1,\lambda_2)|A(\lambda_1)|\psi(\lambda_1,\lambda_2)\rangle \\ &= \int \psi^*(\lambda_1,\lambda_2)A(\lambda_1)\psi(\lambda_1,\lambda_2)\,d\lambda_1\,d\lambda_2 \\ &= \int A(\lambda_1)\psi(\lambda_1,\lambda_2)\psi^*(\lambda_1',\lambda_2)\,d\lambda_1\,d\lambda_2 \qquad (13.6\text{-}6a) \\ &= \tfrac{1}{2}\int A(\lambda_1)\rho(\lambda_1;\lambda_1')\,d\lambda_1. \qquad (13.6\text{-}6b)\end{aligned}$$

The expectation value of $A(\lambda_2)$ would similarly be given by

$$\begin{aligned}\langle A(\lambda_2)\rangle &= \int A(\lambda_2)\psi(\lambda_1,\lambda_2)\psi^*(\lambda_1,\lambda_2')\,d\lambda_1\,d\lambda_2 \qquad (13.6\text{-}7a) \\ &= \tfrac{1}{2}\int A(\lambda_2)\rho(\lambda_2;\lambda_2')\,d\lambda_2. \qquad (13.6\text{-}7b)\end{aligned}$$

Since we are dealing with identical particles (electrons) the wave function $\psi(\lambda_1, \lambda_2)$ must be antisymmetric with respect to an interchange of the two particles. In that case

$$\langle A(\lambda_1)\rangle = \langle A(\lambda_2)\rangle$$

as is immediately apparent from the definition of the expectation value. The same result holds for identical particles described by symmetric wave functions. If, now

$$A = A(\lambda_1) + A(\lambda_2), \tag{13.6-8}$$

we have the result

$$\langle A\rangle = \int A(\lambda_1)\rho(\lambda_1; \lambda_1')\, d\lambda_1. \tag{13.6-9}$$

With these examples, we may now generalize. For an N-electron wave function $\psi(\lambda_1, \lambda_2, \ldots, \lambda_N)$, the *one-particle density matrix* is defined by

$$\rho_1(\lambda_1, \lambda_1') = N \int \psi(\lambda_1, \lambda_2, \ldots, \lambda_N)\psi^*(\lambda_1', \lambda_2, \ldots, \lambda_N)\, d\lambda_2\, d\lambda_3 \cdots d\lambda_N, \tag{13.6-10}$$

and *the two-particle density matrix* by

$$\rho_2(\lambda_1, \lambda_2; \lambda_1', \lambda_2') = N(N-1) \int \psi(\lambda_1, \lambda_2, \ldots, \lambda_N)\psi^*(\lambda_1', \lambda_2' \cdots \lambda_N)\, d\lambda_3 \cdots d\lambda_N. \tag{13.6-11}$$

These definitions may be readily extended to N-particle density matrices but such extensions are generally not required. To clarify the meaning of the designations one-particle, two-particle, etc., when referring to density matrices, we consider the expectation value of an operator

$$A = \sum_{i=1}^{N} A(\lambda_i) \tag{13.6-12}$$

where $A(\lambda_i)$ operates exclusively on functions of λ_i, the coordinates of the ith electron. In terms of (13.6-10)

$$\langle A\rangle = \int A(\lambda_1)\rho_1(\lambda_1; \lambda_1')\, d\lambda_1. \tag{13.6-13}$$

Similarly, for a sum of two-electron operators of the form

$$B = \tfrac{1}{2} \sum_{i \neq j}^{N} B(\lambda_i, \lambda_j) = \sum_{i<j} B(\lambda_i, \lambda_j) \tag{13.6-14}$$

the expectation value of B is given by

$$\langle B\rangle = \tfrac{1}{2} \int B(\lambda_1, \lambda_2)\rho_2(\lambda_1, \lambda_2; \lambda_1', \lambda_2')\, d\lambda_1\, d\lambda_2. \tag{13.6-15}$$

It is seen that the expectation value of a sum of one-electron operators is expressible in terms of the one-particle density matrix and for the expectation value of a sum of two-electron operators we require the two-particle density matrix. This is as far as we need to go because there are no known three-body, or higher, interactions in atomic or molecular physics. It is further observed that the one- and two-particle density matrices are related by

$$(N-1)\rho_1(\lambda_1;\lambda_1') = \int \rho_2(\lambda_1,\lambda_2;\lambda_1',\lambda_2')\,d\lambda_2. \tag{13.6-16}$$

To give a specific example, let

$$\mathscr{H} = \mathscr{H}_0 + \sum_{i<j} \frac{e^2}{r_{ij}} \tag{13.6-17}$$

with

$$\mathscr{H}_0 = \sum_i \left(\frac{p_i^2}{2m} - \frac{Ze^2}{r_i}\right).$$

The expectation value of $\mathscr{H}$, i.e., the average energy, written in terms of density matrices, takes the form

$$\begin{aligned} E &= \langle \mathscr{H} \rangle \\ &= \int \mathscr{H}_0(1)\rho_1(\lambda_1;\lambda_1')\,d\lambda_1 + \frac{e^2}{2}\int \frac{\rho_2(\lambda_1,\lambda_2;\lambda_1',\lambda_2')}{r_{12}}\,d\lambda_1\,d\lambda_2. \end{aligned} \tag{13.6-18}$$

Written this way, the energy is independent of the choice of one-electron orbitals used in the construction of the N-electron wave function.

It is instructive to compare the present formalism with the density matrix of a component of a system as given by (13.3-5) or more generally by (13.3-7). The N-electron system is described by the wave function $\psi(\lambda_1, \lambda_2, \ldots, \lambda_N)$. A component of the system consisting of fewer than N electrons cannot, in general, be described by a wave function. Therefore, the component is in a mixed state which can only be described by a density matrix expressed in terms of the density matrix of the entire system. Equations (13.6-10) and (13.6-11) provide the specific relationships and are precisely analogous to (13.3-7). In this sense one may regard $\rho_1(\lambda_1;\lambda_1')$ and $\rho_2(\lambda_1,\lambda_2;\lambda_1',\lambda_2')$ as *reduced* density matrices.

13.7 Fock–Dirac Density Matrices

The imposition of the antisymmetry requirement of a system of particles results in certain simplifications in the density matrix formalism. Consider by way of an example

$$\Psi(\lambda_1,\lambda_2) = \frac{1}{\sqrt{2}}\begin{vmatrix} \psi_a(\lambda_1) & \psi_b(\lambda_1) \\ \psi_a(\lambda_2) & \psi_b(\lambda_2) \end{vmatrix} \tag{13.7-1}$$

with

$$\langle \psi_a | \psi_b \rangle = \delta_{ab}.$$

The one-particle density matrix (13.6-10) then becomes

$$\rho_1(\lambda_1; \lambda_1') = 2 \int \Psi(\lambda_1, \lambda_2) \Psi^*(\lambda_1', \lambda_2)\, d\lambda_2$$

which, in view of the orthonormality of the spin orbitals, may be written as

$$\rho_1(\lambda_1; \lambda_1') = \psi_a(\lambda_1)\psi_a^*(\lambda_1') + \psi_b(\lambda_1)\psi_b(\lambda_1').$$

This is easily extended to the N-particle antisymmetric wave function $\Psi(\lambda_1, \lambda_2, \ldots, \lambda_N)$ to give

$$\rho_1(\lambda_1; \lambda_1') = \sum_{i=1}^{N} \psi_i(\lambda_1)\psi_i^*(\lambda_1')$$

or, in a slightly more general form,

$$\rho_1(\lambda_1; \lambda_2') = \sum_{i=1}^{N} \psi_i(\lambda_1)\psi_i^*(\lambda_2'). \tag{13.7-2}$$

Equation (13.7-2) defines the *Fock–Dirac density matrix*.

To obtain the form of the two-particle density matrix we again consider $\psi(\lambda_1, \lambda_2)$ given by (13.7-1). From definition (13.6-11),

$$\begin{aligned}
\rho_2(\lambda_1, \lambda_2; \lambda_1', \lambda_2') &= 2\psi(\lambda_1, \lambda_2)\psi^*(\lambda_1', \lambda_2') \\
&= \psi_a(\lambda_1)\psi_b(\lambda_2)\psi_a^*(\lambda_1')\psi_b^*(\lambda_2') \\
&\quad - \psi_a(\lambda_1)\psi_b(\lambda_2)\psi_b^*(\lambda_1')\psi_a^*(\lambda_2') \\
&\quad - \psi_b(\lambda_1)\psi_a(\lambda_2)\psi_a^*(\lambda_1')\psi_b^*(\lambda_2') \\
&\quad + \psi_b(\lambda_1)\psi_a(\lambda_2)\psi_b^*(\lambda_1')\psi_a^*(\lambda_2').
\end{aligned}$$

However, on the basis of (13.7-2), we find

$$\rho_1(\lambda_1; \lambda_1')\rho_1(\lambda_2; \lambda_2') - \rho_1(\lambda_1; \lambda_2')\rho_1(\lambda_2; \lambda_1') = \rho_2(\lambda_1, \lambda_2; \lambda_1', \lambda_2') \tag{13.7-3a}$$

or

$$\rho_2(\lambda_1, \lambda_2; \lambda_1', \lambda_2') = \begin{vmatrix} \rho_1(\lambda_1; \lambda_1') & \rho_1(\lambda_1; \lambda_2') \\ \rho_1(\lambda_2; \lambda_1') & \rho_1(\lambda_2; \lambda_2') \end{vmatrix}. \tag{13.7-3b}$$

Relations (13.7-3a) or (13.7-3b) are quite general and are valid for N-particle antisymmetric wave functions. The expression for the energy in (13.6-18) may now be written entirely in terms of one-particle density matrices:

$$\begin{aligned}
E = \langle \mathscr{H} \rangle &= \langle \psi(\lambda_1, \lambda_2, \ldots, \lambda_N) | \mathscr{H} | \psi(\lambda_1, \lambda_2, \ldots, \lambda_N) \rangle \\
&= \int \mathscr{H}_0(1)\rho_1(\lambda_1; \lambda_1')\, d\lambda_1 + \frac{e^2}{2} \int \frac{1}{r_{12}} [\rho_1(\lambda_1; \lambda_1')\rho_1(\lambda_2; \lambda_2') \\
&\quad - \rho_1(\lambda_1; \lambda_2')\rho_1(\lambda_2; \lambda_1')]\, d\lambda_1\, d\lambda_2.
\end{aligned} \tag{13.7-4}$$

We note that $1/r_{12}$ is merely a multiplicative operator which does not alter the form of the wave function (in contrast to, say, the Laplacian operator). Hence there is no need to concern ourselves with primes (which are inserted only for the purpose of protecting certain variables from the effects of an operator). Similarly, $1/r_1$ is also a multiplicative operator. Since

$$\mathscr{H}_0(1) = -\frac{\hbar^2}{2m}\nabla_1{}^2 - \frac{Ze^2}{r_1},$$

Eq. (13.7-4) may be simplified to

$$E = \langle \mathscr{H} \rangle = -\frac{\hbar^2}{2m}\int \nabla_1{}^2 \rho_1(\lambda_1;\lambda_1')\,d\lambda_1 - Ze^2\int \frac{\rho_1(\lambda_1)}{r_1}\,d\lambda_1 + \frac{e^2}{2}\int \frac{\rho_2(\lambda_1,\lambda_2)}{r_{12}}\,d\lambda_1\,d\lambda_2 \tag{13.7-5a}$$

$$= -\frac{\hbar^2}{2m}\int \nabla_1{}^2 \rho_1(\lambda_1;\lambda_1')\,d\lambda_1 - Ze^2\int \frac{\rho_1(\lambda_1)}{r_1}\,d\lambda_1 + \frac{e^2}{2}\int \frac{1}{r_{12}}[\rho_1(\lambda_1)\rho_1(\lambda_2) - \rho_1(\lambda_1;\lambda_2)\rho_1(\lambda_2;\lambda_1)]\,d\lambda_1\,d\lambda_2 \tag{13.7-5b}$$

where the notation has been abbreviated to

$$\rho_1(\lambda_1) \equiv \rho_1(\lambda_1;\lambda_1), \qquad \rho_1(\lambda_2) \equiv \rho_1(\lambda_2;\lambda_2), \qquad \rho_2(\lambda_1,\lambda_2) \equiv \rho_2(\lambda_1,\lambda_2;\lambda_1,\lambda_2). \tag{13.7-6}$$

13.8 Spinless Density Matrices

The density matrices (13.6-10) and (13.6-11) are defined in terms of wave functions $\Psi(\lambda_1, \lambda_2, \ldots, \lambda_N)$ which depend on both space and spin coordinates. It is sometimes desirable to have density matrices that are expressed in terms of functions which depend only on the space coordinates. Such *spinless* density matrices (McWeeny and Sutcliffe, 1969) may be achieved by integrating over the spins. We therefore define

$$P_1(\mathbf{r}_1;\mathbf{r}_1') = \int \rho_1(\lambda_1;\lambda_1')\,ds_1, \tag{13.8-1}$$

$$P_2(\mathbf{r}_1,\mathbf{r}_2;\mathbf{r}_1',\mathbf{r}_2') = \int \rho_2(\lambda_1,\lambda_2;\lambda_1',\lambda_2')\,ds_1\,ds_2. \tag{13.8-2}$$

For antisymmetric wave functions $\Psi(\lambda_1, \lambda_2, \ldots, \lambda_N)$ $P_1(\mathbf{r}_1;\mathbf{r}_1')$ assumes the form

$$P_1(\mathbf{r}_1;\mathbf{r}_1') = P_1^{\alpha\alpha}(\mathbf{r}_1,\mathbf{r}_1') + P_1^{\beta\beta}(\mathbf{r}_1,\mathbf{r}_1') \tag{13.8-3}$$

where $P_1^{\alpha\alpha}(\mathbf{r}_1;\mathbf{r}_1')$ consists of all the terms in the Fock–Dirac density matrix with spin α and $P_1^{\beta\beta}(\mathbf{r}_1,\mathbf{r}_1')$ is the corresponding quantity for spin β. Thus, for example, let

$$\psi_a(\lambda_1) = \varphi_a(\mathbf{r}_1)\alpha(1), \qquad \psi_b(\lambda_1) = \varphi_b(\mathbf{r}_1)\beta(1) \tag{13.8-4}$$

and

$$\Psi(\lambda_1,\lambda_2) = \frac{1}{\sqrt{2}}\begin{vmatrix}\psi_a(\lambda_1) & \psi_b(\lambda_1)\\ \psi_a(\lambda_2) & \psi_b(\lambda_2)\end{vmatrix}. \tag{13.8-5}$$

The Fock–Dirac density matrix (13.7-2) for this case is

$$\begin{aligned}\rho_1(\lambda_1;\lambda_1') &= \psi_a(\lambda_1)\psi_a^*(\lambda_1') + \psi_b(\lambda_1)\psi_b^*(\lambda_1')\\ &= \varphi_a(\mathbf{r}_1)\varphi_a^*(\mathbf{r}_1')\alpha(1)\alpha^*(1') + \varphi_b(\mathbf{r}_1)\varphi_b^*(\mathbf{r}_1')\beta(1)\beta^*(1')\end{aligned} \tag{13.8-6}$$

in which the notation α^* and β^* is to be understood as $\langle\alpha|$ and $\langle\beta|$, respectively. Following the usual rules that the primes are to be removed before integration

$$\begin{aligned}P_1(\mathbf{r}_1;\mathbf{r}_1') &= \int \rho_1(\lambda_1;\lambda_1')\,ds_1,\\ &= \varphi_a(\mathbf{r}_1)\varphi_a^*(\mathbf{r}_1')\int \alpha(1)\alpha^*(1')\,ds_1\\ &\quad + \varphi_b(\mathbf{r}_1)\varphi_b^*(\mathbf{r}_1')\int \beta(1)\beta^*(1')\,ds_1,\\ &= \varphi_a(\mathbf{r}_1)\varphi_a^*(\mathbf{r}_1')\langle\alpha(1)|\alpha(1)\rangle\\ &\quad + \varphi_b(\mathbf{r}_1)\varphi_b(\mathbf{r}_1')\langle\beta(1)|\beta(1)\rangle\\ &= \varphi_a(\mathbf{r}_1)\varphi_a^*(\mathbf{r}_1') + \varphi_b(\mathbf{r}_1)\varphi_b(\mathbf{r}_1')\end{aligned} \tag{13.8-7}$$

so that, for this particular example

$$P_1^{\alpha\alpha}(\mathbf{r}_1,\mathbf{r}_1') = \varphi_a(\mathbf{r}_1)\varphi_a^*(\mathbf{r}_1'), \qquad P_1^{\beta\beta}(\mathbf{r}_1,\mathbf{r}_1') = \varphi_b(\mathbf{r}_1)\varphi_b^*(\mathbf{r}_1'). \tag{13.8-8}$$

For the diagonal element, i.e. when $\mathbf{r}_1 = \mathbf{r}_1'$, and with the notation $P_1{}^\alpha(\mathbf{r}_1) \equiv P_1^{\alpha\alpha}(\mathbf{r}_1,\mathbf{r}_1)$, $P_1{}^\beta(\mathbf{r}_1) \equiv P_1^{\beta\beta}(\mathbf{r}_1,\mathbf{r}_1)$, the spinless density matrix (13.8-3) becomes

$$P_1(\mathbf{r}_1) = P_1{}^\alpha(\mathbf{r}_1) + P_1{}^\beta(\mathbf{r}_1). \tag{13.8-9}$$

These quantities lend themselves to a direct physical interpretation. Thus $P^\alpha{}_1(\mathbf{r}_1)\,d\mathbf{r}_1$ is the probability of finding an electron with α-spin in the volume element $d\mathbf{r}_1$; $P_1{}^\beta(\mathbf{r}_1)\,d\mathbf{r}_1$ is the same for an electron with β-spin; and $P_1(\mathbf{r}_1)$ is the ordinary electron density function, i.e., $P_1(\mathbf{r}_1)\,d\mathbf{r}_1$ is the probability of finding an electron in the volume element $d\mathbf{r}_1$ regardless of spin.

Another useful parameter is the *spin density* defined as

$$Q_1(\mathbf{r}_1) = P_1{}^{\alpha}(\mathbf{r}_1) - P_1{}^{\beta}(\mathbf{r}_1) \tag{13.8-10}$$

which is a measure of the excess of α-spins over β-spins. If Q_1 is integrated over all space, we obtain a quantity proportional to the number of unpaired spins $N_\alpha - N_\beta$.

The analogous development for the spinless two-particle density matrix yields

$$\begin{aligned} P_2(\mathbf{r}_1,\mathbf{r}_2;\mathbf{r}_1',\mathbf{r}_2') = {} & P_2^{\alpha\alpha,\alpha\alpha}(\mathbf{r}_1,\mathbf{r}_2;\mathbf{r}_1',\mathbf{r}_2') + P_2^{\alpha\beta,\alpha\beta}(\mathbf{r}_1,\mathbf{r}_2;\mathbf{r}_1',\mathbf{r}_2') \\ & + P_2^{\beta\alpha,\beta\alpha}(\mathbf{r}_1,\mathbf{r}_2;\mathbf{r}_1',\mathbf{r}_2') + P_2^{\beta\beta,\beta\beta}(\mathbf{r}_1,\mathbf{r}_2;\mathbf{r}_1',\mathbf{r}_2') \end{aligned} \tag{13.8-11}$$

where, owing to the antisymmetry of the N-particle wave function

$$\begin{aligned} P_2^{\alpha\alpha,\alpha\alpha}(\mathbf{r}_1,\mathbf{r}_2;\mathbf{r}_1',\mathbf{r}_2') &= P_1^{\alpha\alpha}(\mathbf{r}_1;\mathbf{r}_1')P_1^{\alpha\alpha}(\mathbf{r}_2;\mathbf{r}_2') - P_1^{\alpha\alpha}(\mathbf{r}_1;\mathbf{r}_2')P_1^{\alpha\alpha}(\mathbf{r}_2;\mathbf{r}_1'), \\ P_2^{\alpha\beta,\alpha\beta}(\mathbf{r}_1,\mathbf{r}_2;\mathbf{r}_1',\mathbf{r}_2') &= P_1^{\alpha\alpha}(\mathbf{r}_1;\mathbf{r}_1')P_1^{\beta\beta}(\mathbf{r}_2;\mathbf{r}_2'), \\ P_2^{\beta\alpha,\beta\alpha}(\mathbf{r}_1,\mathbf{r}_2;\mathbf{r}_1',\mathbf{r}_2') &= P_1^{\beta\beta}(\mathbf{r}_1;\mathbf{r}_1')P_1^{\alpha\alpha}(\mathbf{r}_2;\mathbf{r}_2'), \\ P_2^{\beta\beta,\beta\beta}(\mathbf{r}_1,\mathbf{r}_2;\mathbf{r}_1',\mathbf{r}_2') &= P_1^{\beta\beta}(\mathbf{r}_1;\mathbf{r}_1')P_1^{\beta\beta}(\mathbf{r}_2;\mathbf{r}_2') - P_1^{\beta\beta}(\mathbf{r}_1;\mathbf{r}_2')P_1^{\beta\beta}(\mathbf{r}_2;\mathbf{r}_1'). \end{aligned} \tag{13.8-12}$$

For the diagonal elements, using the contracted notation,

$$\begin{aligned} P_2^{\alpha\alpha}(\mathbf{r}_1,\mathbf{r}_2) &\equiv P_2^{\alpha\alpha,\alpha\alpha}(\mathbf{r}_1,\mathbf{r}_2;\mathbf{r}_1,\mathbf{r}_2), \\ P_1{}^{\alpha}(\mathbf{r}_1) &\equiv P_1^{\alpha\alpha}(\mathbf{r}_1;\mathbf{r}_1), \end{aligned} \tag{13.8-13}$$

with similar abbreviations for the other terms, we have

$$P_2(\mathbf{r}_1,\mathbf{r}_2) = P_2^{\alpha\alpha}(\mathbf{r}_1,\mathbf{r}_2) + P_2^{\alpha\beta}(\mathbf{r}_1,\mathbf{r}_2) + P_2^{\beta\alpha}(\mathbf{r}_1,\mathbf{r}_2) + P_2^{\beta\beta}(\mathbf{r}_1,\mathbf{r}_2) \tag{13.8-14}$$

where

$$\begin{aligned} P_2^{\alpha\alpha}(\mathbf{r}_1,\mathbf{r}_2) &= P_1{}^{\alpha}(\mathbf{r}_1)P_1{}^{\alpha}(\mathbf{r}_2) - P_1^{\alpha\alpha}(\mathbf{r}_1;\mathbf{r}_2)P_1^{\alpha\alpha}(\mathbf{r}_2;\mathbf{r}_1), \\ P_2^{\alpha\beta}(\mathbf{r}_1,\mathbf{r}_2) &= P_1{}^{\alpha}(\mathbf{r}_1)P_1{}^{\beta}(\mathbf{r}_2), \\ P_2^{\beta\alpha}(\mathbf{r}_1,\mathbf{r}_2) &= P_1{}^{\beta}(\mathbf{r}_1)P_1{}^{\alpha}(\mathbf{r}_2) \\ P_2^{\beta\beta}(\mathbf{r}_1,\mathbf{r}_2) &= P_1{}^{\beta}(\mathbf{r}_1)P_1{}^{\beta}(\mathbf{r}_2) - P_1^{\beta\beta}(\mathbf{r}_1;\mathbf{r}_2)P_1^{\beta\beta}(\mathbf{r}_2;\mathbf{r}_1). \end{aligned} \tag{13.8-15}$$

A convenient expression for $P_2(\mathbf{r}_1,\mathbf{r}_2)$ based on (13.8-9) and (13.8-15) is

$$\begin{aligned} P_2(\mathbf{r}_1,\mathbf{r}_2) = {} & P_1(\mathbf{r}_1)P_1(\mathbf{r}_2) - [P_1^{\alpha\alpha}(\mathbf{r}_1;\mathbf{r}_2)P_1^{\alpha\alpha}(\mathbf{r}_2;\mathbf{r}_1) \\ & + P_1^{\beta\beta}(\mathbf{r}_1;\mathbf{r}_2)P_1^{\beta\beta}(\mathbf{r}_2;\mathbf{r}_1)]. \end{aligned} \tag{13.8-16}$$

The physical interpretation, in analogy with that given for the one-particle spinless (diagonal) density matrix elements, is that $P_2^{\alpha\alpha}(\mathbf{r}_1,\mathbf{r}_2)$ is the probability density for finding an electron with α-spin at $\mathbf{r}_1$ and another electron with α-spin at $\mathbf{r}_2$; $P_2^{\alpha\beta}(\mathbf{r}_1,\mathbf{r}_2)$ is the same for an electron with α-spin at $\mathbf{r}_1$ and another with β-spin at $\mathbf{r}_2$. The other terms have analogous meanings and their sum, which is $P_2(\mathbf{r}_1, \mathbf{r}_2)$ as shown in (13.8-14), is the probability density for an electron at $\mathbf{r}_1$ and another at $\mathbf{r}_2$ irrespective of their spins. In view of this interpretation $P_2(\mathbf{r}_1, \mathbf{r}_2)$ is known as a *pair function.*

Several interesting conclusions may be drawn from the expressions in (13.8-15). Suppose $\mathbf{r}_1 = \mathbf{r}_2$. In that case both $P_2^{\alpha\alpha}(\mathbf{r}_1; \mathbf{r}_2)$ and $P_2^{\beta\beta}(\mathbf{r}_1, \mathbf{r}_2)$ vanish but $P_2^{\alpha\beta}(\mathbf{r}_1, \mathbf{r}_2)$ and $P_2^{\beta\alpha}(\mathbf{r}_1, \mathbf{r}_2)$ do not. This is simply another version of the Pauli principle which prohibits two electrons with parallel spin from occupying the same point in space but does not impose any restrictions on two electrons with opposite spin. Another way to look at this is to note that $P_2^{\alpha\beta}(\mathbf{r}_1, \mathbf{r}_2)$ and $P_2^{\beta\alpha}(\mathbf{r}_1, \mathbf{r}_2)$ are simple products of probability densities for an electron with α-spin and another with β-spin. There is therefore no correlation between the positions of electrons of opposite spin. On the other hand, $P_2^{\alpha\alpha}(\mathbf{r}_1, \mathbf{r}_2)$ and $P_2^{\beta\beta}(\mathbf{r}_1, \mathbf{r}_2)$ may be said to be correlated functions in the sense that an electron at one position, say $\mathbf{r}_1$, influences the probability of finding an electron at $\mathbf{r}_2$ when the two electrons have the same spin. All of these effects, it must be remembered, are a direct consequence of the antisymmetry of the N-particle wave function.

The correlation in $P_2^{\alpha\alpha}(\mathbf{r}_1, \mathbf{r}_2)$ and $P_2^{\beta\beta}(\mathbf{r}_1, \mathbf{r}_2)$ is often called *Fermi correlation* because of the association of Fermi particles with antisymmetric wave functions. However, one might also expect a correlation between two electrons, regardless of their spin, because of their mutual Coulomb repulsion. This feature is not included in the present formalism.

We may now rewrite the energy of the multielectron system in terms of the spinless density matrices. Referring to (13.7-5a), we suppose that the spin integrations (summations) have been performed. Then

$$E = \langle \mathscr{H} \rangle = -\frac{\hbar^2}{2m}\int \nabla_1{}^2 P_1(\mathbf{r}_1, \mathbf{r}_1')\,d\mathbf{r}_1 - Ze^2 \int \frac{P_1(\mathbf{r}_1)}{r_1}\,dr_1 + \frac{e^2}{2}\int \frac{P_2(\mathbf{r}_1, \mathbf{r}_2)}{r_{12}}\,d\mathbf{r}_1\,d\mathbf{r}_2 \tag{13.8-17}$$

or, upon replacing $P_2(\mathbf{r}_1, \mathbf{r}_2)$ by (13.8-16),

$$\begin{aligned} E = \langle \mathscr{H} \rangle = &-\frac{\hbar^2}{2m}\int \nabla_1{}^2 P_1(\mathbf{r}_1, \mathbf{r}_1')\,d\mathbf{r}_1 - Ze^2 \int \frac{P_1(\mathbf{r}_1)}{r_1}\,d\mathbf{r}_1 \\ &+ \frac{e^2}{2}\int \frac{P_1(\mathbf{r}_1)P_1(\mathbf{r}_2)}{r_{12}}\,d\mathbf{r}_1\,d\mathbf{r}_2 \\ &- \frac{e^2}{2}\int \frac{P_1^{\alpha\alpha}(\mathbf{r}_1; \mathbf{r}_2)P_1^{\alpha\alpha}(\mathbf{r}_2; \mathbf{r}_1) + P_1^{\beta\beta}(\mathbf{r}_1; \mathbf{r}_2)P_1^{\beta\beta}(\mathbf{r}_2; \mathbf{r}_1)}{r_{12}}\,d\mathbf{r}_1\,d\mathbf{r}_2. \end{aligned} \tag{13.8-18}$$

The terms on the right side of (13.8-18) in the order of their appearance are kinetic energy, potential energy due to the nuclear field, the direct or Coulomb term, and the exchange term. It is particularly noteworthy that the direct term contains no reference to the spins of the particles whereas the exchange term, which owes its existence to the antisymmetry of the wave function, requires that the electronic spins be parallel.

CHAPTER 14

APPROXIMATIONS

14.1 Variational Methods

We begin by showing that the Schrödinger equation may be derived from a variational principle. Let

$$I = \langle\psi|\mathscr{H}|\psi\rangle \tag{14.1-1}$$

and let ψ be a function such that

$$\delta I = 0 \tag{14.1-2}$$

subject to the condition

$$\langle\psi|\psi\rangle = 1. \tag{14.1-3}$$

In other words we seek the Euler equation which ψ must satisfy in order that I have a stationary value as ψ is varied arbitrarily.

According to the general theorems of variational calculus condition (14.1-3) is taken into account by requiring that

$$\delta[\langle\psi|\mathscr{H}|\psi\rangle - \lambda\langle\psi|\psi\rangle] = 0 \tag{14.1-4}$$

where λ is a Lagrange multiplier. Writing (14.1-4) as

$$\delta\langle\psi|\mathscr{H} - \lambda|\psi\rangle = 0 \tag{14.1-5}$$

we have

$$\langle\delta\psi|\mathscr{H} - \lambda|\psi\rangle + \langle\psi|\mathscr{H} - \lambda|\delta\psi\rangle = 0. \tag{14.1-6}$$

It is tempting to set the individual terms in (14.1-6) equal to zero but $\langle\delta\psi|$ and $|\delta\psi\rangle$ are not independent so it would appear that such a procedure is not permissible. Nevertheless, it will be shown that the individual terms in

(14.1-6) may, in fact, be set equal to zero. Let

$$\delta\psi = \delta u + i\,\delta v \tag{14.1-7}$$

where δu and δv are arbitrary real, independent, variations. Then

$$\begin{aligned} &\langle\delta\psi|\mathscr{H} - \lambda|\psi\rangle + \langle\psi|\mathscr{H} - \lambda|\delta\psi\rangle \\ &\quad = \langle\delta u|\mathscr{H} - \lambda|\psi\rangle + \langle\psi|\mathscr{H} - \lambda|\delta u\rangle \\ &\qquad + \langle i\,\delta v|\mathscr{H} - \lambda|\psi\rangle + \langle\psi|\mathscr{H} - \lambda|i\,\delta v\rangle \\ &\quad = \langle\delta u|\mathscr{H} - \lambda|\psi + \psi^*\rangle - i\langle\delta v|\mathscr{H} - \lambda|\psi - \psi^*\rangle = 0 \end{aligned} \tag{14.1-8}$$

for all δu and δv. Therefore the Euler equations for the variational problem are

$$(\mathscr{H} - \lambda)|\psi + \psi^*\rangle = 0, \qquad (\mathscr{H} - \lambda)|\psi - \psi^*\rangle = 0. \tag{14.1-9}$$

By adding and subtracting the two equations we obtain

$$(\mathscr{H} - \lambda)|\psi\rangle = 0, \qquad (\mathscr{H} - \lambda)|\psi^*\rangle = \langle\psi|(\mathscr{H} - \lambda) = 0, \tag{14.1-10}$$

which is tantamount to having set each term in (14.1-6), separately, equal to zero.

The two equations (14.1-10) are equivalent forms of the Schrödinger equations

$$(\mathscr{H} - E)\psi = 0 \tag{14.1-11}$$

where, it is seen, the Lagrange multiplier λ acquires the significance of an energy eigenvalue. Thus the variational equation (14.1-2) with the subsidiary condition (14.1-3) leads directly to the Schrödinger equation (14.1-11).

The variational method achieves its greatest utility in the calculation of approximate solutions to the Schrödinger equation. We will suppose that ψ is an arbitrary trial function which is intended to serve as an approximate solution to the Schrödinger equation. Let E be an energy defined by

$$E = \frac{\langle\psi|\mathscr{H}|\psi\rangle}{\langle\psi|\psi\rangle} \tag{14.1-12}$$

and let E_k and ψ_k be an eigenvalue and the corresponding eigenstate of $\mathscr{H}$. We may then write

$$\begin{aligned} E - E_k &= \frac{\langle\psi|\mathscr{H}|\psi\rangle}{\langle\psi|\psi\rangle} - E_k \\ &= \frac{\langle\psi|\mathscr{H}|\psi\rangle - \langle\psi|\psi\rangle E_k}{\langle\psi|\psi\rangle} = \frac{\langle\psi|\mathscr{H} - E_k|\psi\rangle}{\langle\psi|\psi\rangle}. \end{aligned} \tag{14.1-13}$$

If ψ and ψ_k differ by $\delta\psi$, i.e., if

$$\psi = \psi_k + \delta\psi, \tag{14.1-14}$$

the substitution of (14.1-14) into (14.1-13), together with the Hermitian property of $\mathscr{H}$, gives

$$E - E_k = \frac{\langle\delta\psi|\mathscr{H} - E_k|\delta\psi\rangle}{\langle\psi|\psi\rangle} \tag{14.1-15}$$

which indicates that the difference between E and the true eigenvalue E_k, when the trial function ψ differs from the true eigenstate ψ_k by $\delta\psi$, varies quadratically with $\delta\psi$. This means then that the lowest nonvanishing order of the variation in E is of second order and therefore the first-order variation in E must vanish, or

$$\delta E = 0. \tag{14.1-16}$$

This is the *variation principle*. It says that the best approximation to the eigenvalue E_k is obtained by varying ψ, or the parameters on which it may depend, so that (14.1-16) is satisfied.

In terms of a complete set of orthonormal functions ψ_k that satisfy

$$\mathscr{H}\psi_k = E_k\psi_k, \qquad \langle\psi_k|\psi_l\rangle = \delta_{kl}, \tag{14.1-17}$$

the normalized trial function ψ is

$$\psi = \sum_k a_k\psi_k.$$

We then obtain

$$E = \langle\psi|\mathscr{H}|\psi\rangle = \sum_{kl} a_k{}^*a_l\langle\psi_k|\mathscr{H}|\psi_l\rangle = \sum_{kl} a_k{}^*a_lE_l\langle\psi_k|\psi_l\rangle = \sum_k |a_k|^2E_k. \tag{14.1-18}$$

Also, because ψ is normalized,

$$\langle\psi|\psi\rangle = \sum_{kl} a_k{}^*a_l\langle\psi_k|\psi_l\rangle = \sum_{kl} a_k{}^*a_l\,\delta_{kl} = \sum_k |a_k|^2 = 1. \tag{14.1-19}$$

Now let E_0 be the lowest energy among the eigenvalues E_k. Then

$$E = \sum_k |a_k|^2E_k \geqslant E_0\sum_k |a_k|^2 \geqslant E_0. \tag{14.1-20}$$

In other words, the energy E based on the trial function ψ can never lie below the ground state energy E_0. The closer ψ is to the true ground state wave function ψ_0, the smaller will be the difference between E and E_0, but $E - E_0$ will always be equal to or greater than zero. Of two trial functions, the one that gives the lower value of E is closer to the correct ground state wave function and this property may be used to assess the relative merits of the two trial functions. It should be noted, however, that a trial function that has been optimized with respect to the total energy is not necessarily the best function with which to calculate properties other than the energy.

We now illustrate the method with a more specific example. Suppose the trial function ψ is expressed as a linear combination of some other set of functions:

$$\psi = \sum_i a_i \varphi_i. \tag{14.1-21}$$

With an eye toward future applications it is assumed that the φ_i are not necessarily orthogonal but rather that

$$\langle \varphi_i | \varphi_j \rangle = S_{ij} \tag{14.1-22}$$

where S_{ij} is known as the *overlap integral.*

Substituting (14.1-21) into (14.1-12),

$$E \sum_{ij} a_i{}^* a_j S_{ij} = \sum_{ij} a_i{}^* a_j H_{ij} \tag{14.1-23}$$

where

$$H_{ij} = \langle \varphi_i | \mathscr{H} | \varphi_j \rangle. \tag{14.1-24}$$

When both sides of (14.1-23) are differentiated with respect to one of the coefficients, say $a_k{}^*$, one obtains

$$\partial E / \partial a_k{}^* \sum_{ij} a_i{}^* a_j S_{ij} + E \sum_j a_j S_{kj} = \sum_j a_j H_{kj},$$

from which a stationary value of the energy is obtained by setting

$$\partial E / \partial a_k{}^* = 0 \tag{14.1-25}$$

or

$$\sum_j a_j (H_{kj} - E S_{kj}) = 0. \tag{14.1-26}$$

When the process is repeated for other coefficients, additional equations of the type (14.1-26) are obtained and the number of such equations is equal to the number of functions φ_i in (14.1-21). Solutions to a set of homogeneous equations of type (14.1-26) exist only when

$$\begin{vmatrix} H_{11} - ES_{11} & H_{12} - ES_{12} & \cdots \\ H_{21} - ES_{21} & H_{22} - ES_{22} & \cdots \\ \vdots & \vdots & \vdots \end{vmatrix} = 0. \tag{14.1-27}$$

This is known as the *secular equation* and its solution provides a set of energies with the lowest one being an upper bound to the true ground state energy. The remaining solutions are often interpreted as approximations to the energies of excited states but such an interpretation must be used cautiously and with due regard to the orthogonality requirements associated with states of different energies.

We have here an example of the variational principle imposed on a trial function constrained to be of the form of (14.1-21). In this particular case the embodiment of the variational principle resides in the homogeneous equations (14.1-26) and the secular equation (14.1-27).

14.2 Time-Independent Perturbations

It is assumed that the time-independent Hamiltonian for the system can be written as the sum of two terms

$$\mathscr{H} = \mathscr{H}_0 + V \tag{14.2-1}$$

in which effects on the system due to $\mathscr{H}_0$ are the dominant ones while those due to V (the perturbation) are relatively weaker. We write the eigenvalues and eigenfunctions of $\mathscr{H}$ in the form

$$E_n = E_n^{(0)} + E_n^{(1)} + E_n^{(2)} + \cdots, \qquad \psi_n = \psi_n^{(0)} + \psi_n^{(1)} + \psi_n^{(2)} + \cdots \tag{14.2-2}$$

where $E_n^{(0)}$ and $\psi_n^{(0)}$ satisfy the Schrödinger equation with the unperturbed Hamiltonian $\mathscr{H}_0$:

$$\mathscr{H}_0\psi_n^{(0)} = E_n^{(0)}\psi_n^{(0)} \tag{14.2-3}$$

and $E_n^{(1)}$, $\psi_n^{(1)}$ are first-order corrections to $E_n^{(0)}$, $\psi_n^{(0)}$, respectively; $E_n^{(2)}$, $\psi_n^{(2)}$ are second-order corrections, etc.

The starting point of the perturbation approach consists of the set of eigenvalues and eigenfunctions $E_n^{(0)}$ and $\psi_n^{(0)}$ (zero order). These are presumed to be known. Corrections to any given order can be expressed in terms of quantities of next lower order or alternatively, all corrections may be based on zero-order quantities. We distinguish between nondegenerate and degenerate cases. For the nondegenerate case, in which there is just one eigenfunction $\psi_n^{(0)}$ associated with the eigenvalue $E_n^{(0)}$, one obtains for the first few terms in (14.2-2) (see, for example, Merzbacher, 1970; Messiah, 1962; Ziman, 1969):

$$E_n^{(1)} = \langle\psi_n^{(0)}|V|\psi_n^{(0)}\rangle, \tag{14.2-4}$$

$$E_n^{(2)} = \sum_{k\neq n} \frac{\langle\psi_n^{(0)}|V|\psi_k^{(0)}\rangle\langle\psi_k^{(0)}|V|\psi_n^{(0)}\rangle}{E_n^{(0)} - E_k^{(0)}}, \tag{14.2-5}$$

$$|\psi_n^{(1)}\rangle = \sum_{k\neq n} \frac{\langle\psi_k^{(0)}|V|\psi_n^{(0)}\rangle}{E_n^{(0)} - E_k^{(0)}} |\psi_k^{(0)}\rangle. \tag{14.2-6}$$

An alternative form for $E_n^{(2)}$ is

$$E_n^{(2)} = \langle \psi_n^{(0)} | V | \psi_n^{(1)} \rangle, \tag{14.2-7}$$

as may be verified by substituting (14.2-6) into (14.2-7).

The perturbation expressions (14.2-4)–(14.2-6) as well as higher-order corrections may be written in terms of operators P_k defined by

$$P_k = |\psi_k^{(0)}\rangle\langle\psi_k^{(0)}|. \tag{14.2-8}$$

Since $\mathscr{H}_0$ is an observable, the eigenfunctions $\psi_k^{(0)}$ form a complete orthonormal set so that

$$\sum_k P_k = \sum_k |\psi_k^{(0)}\rangle\langle\psi_k^{(0)}| = 1 \tag{14.2-9}$$

and

$$\begin{aligned} 1 - P_n &= \sum_k |\psi_k^{(0)}\rangle\langle\psi_k^{(0)}| - |\psi_n^{(0)}\rangle\langle\psi_n^{(0)}| \\ &= \sum_{k \neq n} |\psi_k^{(0)}\rangle\langle\psi_k^{(0)}|. \end{aligned} \tag{14.2-10}$$

From (14.2-3) we have

$$(E_n^{(0)} - \mathscr{H}_0)|\psi_k^{(0)}\rangle = (E_n^{(0)} - E_k^{(0)})|\psi_k^{(0)}\rangle$$

which, formally, may be written as

$$(E_n^{(0)} - \mathscr{H}_0)^{-1}|\psi_k^{(0)}\rangle = \frac{1}{E_n^{(0)} - E_k^{(0)}}|\psi_k^{(0)}\rangle. \tag{14.2-11}$$

Equation (14.2-11) defines the operator $(E_n^{(0)} - \mathscr{H}_0)^{-1}$. We then have

$$\begin{aligned} (E_n^{(0)} - \mathscr{H}_0)^{-1}(1 - P_n)V|\psi_n^{(0)}\rangle &= (E_n^{(0)} - \mathscr{H}_0)^{-1} \sum_{k \neq n} |\psi_k^{(0)}\rangle\langle\psi_k^{(0)}|V|\psi_n^{(0)}\rangle \\ &= \sum_{k \neq n} \frac{\langle\psi_k^{(0)}|V|\psi_n^{(0)}\rangle}{E_n^{(0)} - E_k^{(0)}}|\psi_k^{(0)}\rangle \end{aligned} \tag{14.2-12}$$

which is identical with $|\psi_n^{(1)}\rangle$ in (14.2-6). In the same way

$$\begin{aligned} &\langle\psi_n^{(0)}|V(E_n^{(0)} - \mathscr{H}_0)^{-1}(1 - P_n)V|\psi_n^{(0)}\rangle \\ &\quad = \langle\psi_n^{(0)}|V(E_n^{(0)} - \mathscr{H}_0)^{-1} \sum_{k \neq n} |\psi_k^{(0)}\rangle\langle\psi_k^{(0)}|V|\psi_n^{(0)}\rangle \\ &\quad = \sum_{k \neq n} \frac{\langle\psi_n^{(0)}|V|\psi_k^{(0)}\rangle\langle\psi_k^{(0)}|V|\psi_n^{(0)}\rangle}{E_n^{(0)} - E_k^{(0)}} \end{aligned} \tag{14.2-13}$$

which is just $E_n^{(2)}$ in (14.2-5). These expressions may be generalized to

$$E_n = E_n^{(0)} + \langle\psi_n^{(0)}|V|\psi_n^{(0)}\rangle + \langle\psi_n^{(0)}|V(E_n^{(0)} - \mathscr{H}_0)^{-1}(1 - P_n)V|\psi_n^{(0)}\rangle$$
$$+ \langle\psi_n^{(0)}|V(E_n^{(0)} - \mathscr{H}_0)^{-1}(1 - P_n)V(E_n^{(0)} - \mathscr{H}_0)^{-1}(1 - P_n)V|\psi_n^{(0)}\rangle + \cdots, \qquad (14.2\text{-}14)$$

$$\psi_n = \psi_n^{(0)} + (E_n^{(0)} - \mathscr{H}_0)^{-1}(1 - P_n)V|\psi_n^{(0)}\rangle$$
$$+ (E_n^{(0)} - \mathscr{H}_0)^{-1}(1 - P_n)V(E_n^{(0)} - \mathscr{H}_0)^{-1}(1 - P_n)V|\psi_n^{(0)}\rangle + \cdots. \qquad (14.2\text{-}15)$$

Equations (14.2-14) and (14.2-15), which are the formal perturbation expansions in the nondegenerate case, are known as the *Rayleigh–Schrödinger* expansions. There exist similar expansions but with $E_n^{(0)}$ replaced by E_n; these are known as *Brillouin–Wigner* expansions.

It is sometimes useful to have a general form for the energy correction

$$\Delta E_n \equiv E_n - E_n^{(0)}. \qquad (14.2\text{-}16)$$

Since

$$\mathscr{H}_0\psi_n^{(0)} = E_n^{(0)}\psi_n^{(0)}, \qquad (\mathscr{H}_0 + V)\psi_n = E_n\psi_n \qquad (14.2\text{-}17)$$

we have

$$\langle\psi_n^{(0)}|\mathscr{H}_0|\psi_n\rangle = E_n^{(0)}\langle\psi_n^{(0)}|\psi_n\rangle$$
$$\langle\psi_n^{(0)}|\mathscr{H}_0 + V|\psi_n\rangle = (E_n^{(0)} + \Delta E_n)\langle\psi_n^{(0)}|\psi_n\rangle. \qquad (14.2\text{-}18)$$

By subtraction

$$\Delta E_n = \langle\psi_n^{(0)}|V|\psi_n\rangle / \langle\psi_n^{(0)}|\psi_n\rangle. \qquad (14.2\text{-}19)$$

This result, which expresses the correction of ΔE_n to the eigenvalue E_n due to the perturbation V, is of interest mainly in a formal sense since we do not have the eigenfunctions of the complete Hamiltonian $\mathscr{H} = \mathscr{H}_0 + V$.

In the degenerate case there exists more than one eigenfunction $\psi_n^{(0)}$ belonging to the eigenvalue $E_n^{(0)}$. Let

$$\begin{aligned} \mathscr{H}_0\psi_{n1}^{(0)} &= E_n^{(0)}\psi_{n1}^{(0)}, \\ \mathscr{H}_0\psi_{n2}^{(0)} &= E_n^{(0)}\psi_{n2}^{(0)}, \\ &\vdots \\ \mathscr{H}_0\psi_{np}^{(0)} &= E_n^{(0)}\psi_{np}^{(0)}. \end{aligned} \qquad (14.2\text{-}20)$$

In this case it may be shown that $E_n^{(1)}$ is obtained by solving the secular equation

$$\begin{vmatrix} V_{11} - E_n^{(1)} & V_{12} & \cdots & V_{1p} \\ V_{21} & V_{22} - E_n^{(1)} & \cdots & V_{2p} \\ \vdots & \vdots & \vdots & \vdots \\ V_{p1} & V_{p2} & \cdots & V_{pp} - E_n^{(1)} \end{vmatrix} = 0 \qquad (14.2\text{-}21)$$

or, more compactly

$$|V_{ij} - \delta_{ij}E_n^{(1)}| = 0 \tag{14.2-22}$$

with

$$V_{ij} = \langle\psi_{ni}^{(0)}|V|\psi_{nj}^{(0)}\rangle. \tag{14.2-23}$$

14.3 Time-Dependent Perturbations—Harmonic Potential

Starting with the same Hamiltonian

$$\mathscr{H} = \mathscr{H}_0 + V \tag{14.3-1}$$

as in (14.2-1) with $\mathscr{H}_0$ the time-independent Hamiltonian of the unperturbed system, the condition on the perturbation V is now relaxed to permit it to vary in time. We shall seek solutions to the time-dependent Schrödinger equation

$$i\hbar\frac{\partial\psi(\mathbf{r},t)}{\partial t} = \mathscr{H}\psi(\mathbf{r},t) = (\mathscr{H}_0 + V)\psi(\mathbf{r},t) \tag{14.3-2}$$

in terms of the orthonormal set $\varphi_k(\mathbf{r})$ which satisfies

$$\mathscr{H}_0\varphi_k(\mathbf{r}) = E_k\varphi_k(\mathbf{r}). \tag{14.3-3}$$

Let

$$\psi(\mathbf{r},t) = \sum_k c_k(t)\varphi_k(\mathbf{r})e^{-iE_kt/\hbar}, \tag{14.3-4}$$

in which case

$$i\hbar\frac{\partial\psi(\mathbf{r},t)}{\partial t} = \sum_k E_kc_k(t)\varphi_k(\mathbf{r})e^{-iE_kt/\hbar} + i\hbar\sum_k\frac{\partial c_k}{\partial t}\varphi_k(\mathbf{r})e^{-iE_kt/h},$$

$$(\mathscr{H}_0 + V)\psi(\mathbf{r},t) = \mathscr{H}_0\psi(\mathbf{r},t) + \sum_k Vc_k(t)\varphi_k(\mathbf{r})e^{-iE_kt/\hbar},$$

and

$$i\hbar\sum_k\frac{\partial c_k(t)}{\partial t}\varphi_k(\mathbf{r})e^{-iE_kt/\hbar} = \sum_k Vc_k(t)\varphi_k(\mathbf{r})e^{-iE_kt/\hbar}. \tag{14.3-5}$$

If (14.3-5) is multiplied on the left by $\varphi_l{}^*(\mathbf{r})$ and integrated over $\mathbf{r}$ one obtains, with the orthonormality of the $\varphi_k(\mathbf{r})$,

$$i\hbar\frac{\partial c_l(t)}{\partial t} = \sum_k\langle\varphi_l|V|\varphi_k\rangle c_k(t)e^{i(E_l-E_k)t/\hbar}. \tag{14.3-6}$$

This is the differential equation whose solution provides the coefficients in expansion (14.3-4). The quantity $|c_l(t)|^2$ is the probability of finding the system in the state φ_l at the time t.

To proceed further it is necessary to specify the perturbation V and the initial conditions. Let

$$V = Ay(t) \tag{14.3-7}$$

in which A is independent of t but may depend on the coordinates. We shall focus attention on two states, say φ_a and φ_b and assume that initially ($t = 0$)

$$c_a(0) = 1, \qquad c_b(0) = 0, \tag{14.3-8}$$

that is, at $t = 0$, φ_a is occupied and φ_b is vacant. With these assumptions, the differential equation (14.3-6) becomes

$$i\hbar \frac{\partial c_b(t)}{\partial t} = \langle \varphi_b | A | \varphi_a \rangle y(t) c_a(t) e^{i(E_b - E_a)t/\hbar} \tag{14.3-9}$$

from which we may infer that $c_b(t)$ will not remain zero for all time but rather, as time progresses, will acquire nonzero values as a result of the perturbation. In other words, there will be a nonvanishing probability for the occurrence of a transition from φ_a to φ_b governed by the probability amplitude $c_b(t)$ which in this case may be written more explicitly as $c_{ba}(t)$ to indicate both the initial and final states.

To obtain a first-order approximation we may set $c_a(t)$ in (14.3-9) equal to its initial value, namely, unity. Hence

$$c_{ba}(t) = -\frac{i}{\hbar} \langle \varphi_b | A | \varphi_a \rangle \int_0^t y(t_1) e^{i\omega_{ba} t_1} \, dt_1 \tag{14.3-10}$$

where

$$\omega_{ba} = (E_b - E_a)/\hbar.$$

Now suppose that

$$y(t) = 2 \cos \omega t. \tag{14.3-11}$$

Direct integration of (14.3-10) gives

$$c_{ba}(t) = -\frac{1}{\hbar} \langle \varphi_b | A | \varphi_a \rangle \left[\frac{e^{i(\omega_{ba} + \omega)t} - 1}{\omega_{ba} + \omega} + \frac{e^{i(\omega_{ba} - \omega)t} - 1}{\omega_{ba} - \omega} \right]. \tag{14.3-12}$$

The probability of finding the system in the state φ_b is appreciable only when one or the other of the two terms is close to zero. Assuming that $E_b > E_a$, ω_{ba} is positive and the probability per unit time W_{ba} for a transition from φ_a to φ_b is

$$W_{ba} = \frac{1}{t}|c_{ba}(t)|^2 = \frac{2\pi}{\hbar^2}|\langle\varphi_b|A|\varphi_a\rangle|^2 \frac{\sin^2\frac{1}{2}(\omega_{ba}-\omega)t}{2\pi t[\frac{1}{2}(\omega_{ba}-\omega)]^2}$$

$$= \frac{2\pi}{\hbar^2}|\langle\varphi_b|A|\varphi_a\rangle|^2\,\delta(\omega_{ba}-\omega) \qquad (14.3\text{-}13a)$$

$$= \frac{2\pi}{\hbar}|\langle\varphi_b|A|\varphi_a\rangle|^2\,\delta(E_b - E_a - \hbar\omega). \qquad (14.3\text{-}13b)$$

Equation (14.3-13) is a special case of Fermi's golden rule, the more general form of which will be developed in the next section.

14.4 Fermi's Golden Rule

In Section 9.1 it was shown that if a system is known to be in a state ψ_a at $t = t_0$, the probability for the system to be in the state ψ_b at a time t is given by (9.1-4):

$$w_{ba} = |\langle\psi_b|U(t,t_0)|\psi_a\rangle|^2.$$

We shall now let

$$H = \mathscr{H}_0 + V, \qquad \mathscr{H}_0\varphi_a = E_a\varphi_a, \qquad \mathscr{H}_0\varphi_b = E_b\varphi_b,$$

and if we specialize to the case of a transition between the stationary states φ_a and φ_b, the transition probability is

$$w_{ba} = |\langle\varphi_b|U(t,t_0)|\varphi_a\rangle|^2 \qquad (14.4\text{-}1a)$$

$$= |\langle\varphi_b|U_I(t,t_0)|\varphi_a\rangle|^2. \qquad (14.4\text{-}1b)$$

The equality between the two forms in (14.4-1) arises as a result of (9.3-10) which merely introduces a phase factor into the matrix element when $U(t,t_0)$ is replaced by $U_I(t,t_0)$. Since the probability is determined by the absolute square of the matrix element, such phase factors are of no consequence. We shall use the form (14.4-1b) to evaluate the probability per unit time:

$$W_{ba} = \frac{d}{dt}|\langle\varphi_b|U_I(t,t_0)|\varphi_a\rangle|^2 \qquad (14.4\text{-}2a)$$

$$= \frac{d}{dt}[\langle\varphi_b|U_I(t,t_0)|\varphi_a\rangle\langle\varphi_b|U_I(t,t_0)|\varphi_a\rangle^*|. \qquad (14.4\text{-}2b)$$

Since the time dependence of $\langle\varphi_b|U_I(t,t_0)|\varphi_a\rangle$ is contained entirely in the operator $U_I(t,t_0)$ which satisfies (9.3-13a),

$$i\hbar\frac{dU_I(t,t_0)}{dt} = V_I(t)U_I(t,t_0),$$

we have

$$\begin{aligned}W_{ba} &= -\frac{i}{\hbar}[\langle\varphi_b|V_I(t)U_I(t,t_0)|\varphi_a\rangle\langle\varphi_b|U_I(t,t_0)|\varphi_a\rangle^* \\ &\quad - \langle\varphi_b|U_I(t,t_0)|\varphi_a\rangle\langle\varphi_b|V_I(t)U_I(t,t_0)|\varphi_a\rangle^*] \\ &= -\frac{i}{\hbar}[\langle\varphi_b|V_I(t)U_I(t,t_0)|\varphi_a\rangle\langle\varphi_b|U_I(t,t_0)|\varphi_a\rangle^* \\ &\quad - \text{complex conjugate}] \\ &= \frac{2}{\hbar}\,\text{Im}[\langle\varphi_b|V_I(t)U_I(t,t_0)|\varphi_a\rangle\langle\varphi_b|U_I(t,t_0)|\varphi_a\rangle^*]. \end{aligned} \tag{14.4-3}$$

We shall now insert, using (9.3-3) and (9.3-11),

$$U_I(t,t_0) = U_I(t,0)U_I(0,t_0) = e^{i\mathcal{H}_0 t/\hbar}e^{-i\mathcal{H}t/\hbar}U_I(0,t_0),$$
$$V_I(t) = e^{i\mathcal{H}_0 t/\hbar}Ve^{-\mathcal{H}_0 t/\hbar}$$

into (14.4-3) to obtain

$$\begin{aligned}W_{ba} &= \frac{2}{\hbar}\,\text{Im}[\langle\varphi_b|e^{i\mathcal{H}_0 t/\hbar}Ve^{-i\mathcal{H}t/\hbar}U_I(0,t_0)|\varphi_a\rangle \\ &\quad \times \langle\varphi_b|e^{i\mathcal{H}_0 t/\hbar}e^{-i\mathcal{H}t/\hbar}U_I(0,t_0)|\varphi_a\rangle^*].\end{aligned} \tag{14.4-4}$$

Up to this point t_0 was arbitrary; it will now be assumed that $t_0 \to -\infty$. This will allow us to apply the perturbation V adiabatically by which it is meant that V is multiplied by the convergence factor (9.4-2). This is a convenient mathematical device and has the effect of causing the perturbation to vanish at very early and very late times but to remain unaffected at $t = 0$. In Section 9.4 it was shown that

$$\psi_a = U_I(0,-\infty)\varphi_a$$

is an eigenfunction of $\mathcal{H}$ with eigenvalue E_a. Therefore

$$\begin{aligned}W_{ba} &= \frac{2}{\hbar}\,\text{Im}[e^{iE_b t/\hbar}\langle\varphi_b|V|\psi_a\rangle e^{-iE_a t/\hbar}e^{-iE_b t/\hbar}\langle\varphi_b|\psi_a\rangle^* e^{iE_a t/\hbar}] \\ &= \frac{2}{\hbar}\,\text{Im}[\langle\varphi_b|V|\psi_a\rangle\langle\varphi_b|\psi_a\rangle^* \\ &= \frac{2}{\hbar}\,\text{Im}[R_{ba}\langle\varphi_b|\psi_a\rangle^*]\end{aligned} \tag{14.4-5}$$

where

$$R_{\mathrm{ba}} = \langle \varphi_{\mathrm{b}} | V | \psi_{\mathrm{a}} \rangle \tag{14.4-6a}$$

$$= V_{\mathrm{ba}} + \sum_c \frac{V_{\mathrm{bc}} V_{\mathrm{ca}}}{E_{\mathrm{a}} - E_c} + \cdots \tag{14.4-6b}$$

with the second equality coming from (9.4-22). We shall now replace ψ_{a} in (14.4-5) by (9.4-18),

$$\psi_{\mathrm{a}} = \varphi_{\mathrm{a}} + \lim_{\varepsilon \to 0} \frac{1}{E_{\mathrm{a}} - \mathscr{H}_0 + i\varepsilon} V \psi_{\mathrm{a}}$$

giving

$$W_{\mathrm{ba}} = \frac{2}{\hbar} \operatorname{Im} \left[R_{\mathrm{ba}} \langle \varphi_{\mathrm{b}} | \varphi_{\mathrm{a}} \rangle^* + \lim_{\varepsilon \to 0} R_{\mathrm{ba}} \left\langle \varphi_{\mathrm{b}} \middle| \frac{1}{E_{\mathrm{a}} - \mathscr{H}_0 + i\varepsilon} V \middle| \psi_{\mathrm{a}} \right\rangle^* \right].$$

Assuming $\varphi_{\mathrm{a}} \neq \varphi_{\mathrm{b}}$, the first term in the brackets vanishes and

$$W_{\mathrm{ba}} = \frac{2}{\hbar} \operatorname{Im} \left[R_{\mathrm{ba}} \lim_{\varepsilon \to 0} \frac{\langle \varphi_{\mathrm{b}} | V | \psi_{\mathrm{a}} \rangle^*}{E_{\mathrm{a}} - E_{\mathrm{b}} - i\varepsilon} \right]$$

$$= \frac{2}{\hbar} \operatorname{Im} \lim_{\varepsilon \to 0} \frac{|R_{\mathrm{ba}}|^2}{E_{\mathrm{a}} - E_{\mathrm{b}} - i\varepsilon}.$$

Now

$$\operatorname{Im} \frac{1}{E_{\mathrm{a}} - E_{\mathrm{b}} - i\varepsilon} = \frac{\varepsilon}{(E_{\mathrm{a}} - E_{\mathrm{b}})^2 + \varepsilon^2},$$

$$\lim_{\varepsilon \to 0} \frac{\varepsilon}{(E_{\mathrm{a}} - E_{\mathrm{b}})^2 + \varepsilon^2} = \pi \, \delta(E_{\mathrm{a}} - E_{\mathrm{b}}),$$

so that, finally

$$W_{\mathrm{ba}} = \frac{2\pi}{\hbar} |R_{\mathrm{ba}}|^2 \, \delta(E_{\mathrm{a}} - E_{\mathrm{b}}) \tag{14.4-7}$$

which is the most general form of Fermi's golden rule.

14.5 Density Matrices—Random Perturbations

Transition probabilities may also be expressed in terms of density matrices. This is based on the interpretation of the diagonal matrix element $\langle \varphi_{\mathrm{b}} | \rho | \varphi_{\mathrm{b}} \rangle$ as the probability of finding the system in the state φ_{b}, as shown in Section 13.1. Hence, if at $t = 0$, the state φ_{a} is occupied while all other states accessible to the system, including φ_{b}, are vacant, the probability per unit time for a transition from φ_{a} to φ_{b} is

$$\frac{d}{dt} \langle \varphi_{\mathrm{b}} | \rho | \varphi_{\mathrm{b}} \rangle. \tag{14.5-1}$$

It will be convenient to work in the interaction representation where, according to the development in Section 13.5,

$$\rho_{\mathrm{I}}(t) = e^{i\mathscr{H}_0 t}\rho e^{-i\mathscr{H}_0 t}, \tag{14.5-2}$$

$$i\hbar \frac{\partial \rho_{\mathrm{I}}(t)}{\partial t} = [V_{\mathrm{I}}(t), \rho_{\mathrm{I}}(t)]. \tag{14.5-3}$$

Here $V_{\mathrm{I}}(t)$ is the perturbation—in the interaction representation—which drives the transition between one state and another. Using (13.5-9) and differentiating once with respect to the time we obtain

$$\frac{d\rho_{\mathrm{I}}(t)}{dt} = -\frac{i}{\hbar}[V_{\mathrm{I}}(t), \rho_{\mathrm{I}}(0)] + \left(\frac{-i}{\hbar}\right)^2 \int_0^t dt' [V_{\mathrm{I}}(t), [V_{\mathrm{I}}(t'), \rho_{\mathrm{I}}(0)] + \cdots$$

and on expanding and keeping terms up to second order in the perturbation,

$$\begin{aligned}\frac{d\rho_{\mathrm{I}}(t)}{dt} = {} & -\frac{i}{\hbar}[V_{\mathrm{I}}(t)\rho_{\mathrm{I}}(0) - \rho_{\mathrm{I}}(0)V_{\mathrm{I}}(t)] + \left(-\frac{i}{\hbar}\right)^2\left[\int_0^t dt'\, V_{\mathrm{I}}(t)V_{\mathrm{I}}(t')\rho_{\mathrm{I}}(0)\right. \\ & - \int_0^t dt'\, V_{\mathrm{I}}(t)\rho_{\mathrm{I}}(0)V_{\mathrm{I}}(t') - \int_0^t dt'\, V_{\mathrm{I}}(t')\rho_{\mathrm{I}}(0)V_{\mathrm{I}}(t) \\ & \left. + \int_0^t dt'\, \rho_{\mathrm{I}}(0)V_{\mathrm{I}}(t')V_{\mathrm{I}}(t)\right].\end{aligned} \tag{14.5-4}$$

The assumption concerning the occupancy of the states at $t = 0$ implies that for any state $\varphi_k \neq \varphi_{\mathrm{a}}$

$$\langle\varphi_k|\rho(0)|\varphi_{\mathrm{a}}\rangle = \langle\varphi_k|\rho_{\mathrm{I}}(0)|\varphi_{\mathrm{a}}\rangle = \delta_{k\mathrm{a}} \tag{14.5-5}$$

in which orthonormality of the φs has been assumed. Therefore in computing the diagonal matrix element

$$\left\langle \varphi_{\mathrm{b}} \left| \frac{d\rho_{\mathrm{I}}(t)}{dt} \right| \varphi_{\mathrm{b}} \right\rangle$$

we have, for $\varphi_{\mathrm{b}} \neq \varphi_{\mathrm{a}}$,

$$\langle\varphi_{\mathrm{b}}|V_{\mathrm{I}}(t)\rho_{\mathrm{I}}(0)|\varphi_{\mathrm{b}}\rangle = \sum_k \langle\varphi_{\mathrm{b}}|V_{\mathrm{I}}(t)|\varphi_k\rangle\langle\varphi_k|\rho_{\mathrm{I}}(0)|\varphi_{\mathrm{b}}\rangle = 0.$$

Similarly

$$\begin{aligned}\langle\varphi_{\mathrm{b}}|\rho_{\mathrm{I}}(0)V_{\mathrm{I}}(t)|\varphi_{\mathrm{b}}\rangle &= 0, \\ \langle\varphi_{\mathrm{b}}|V_{\mathrm{I}}(t)V_{\mathrm{I}}(t')\rho_{\mathrm{I}}(0)|\varphi_{\mathrm{b}}\rangle &= 0, \\ \langle\varphi_{\mathrm{b}}|\rho_{\mathrm{I}}(0)V_{\mathrm{I}}(t')V_{\mathrm{I}}(t)|\varphi_{\mathrm{b}}\rangle &= 0,\end{aligned}$$

so the remaining terms are

$$\left\langle \varphi_b \left| \frac{d\rho_I(t)}{dt} \right| \varphi_b \right\rangle = \frac{1}{\hbar^2}\left[\int_0^t dt' \langle\varphi_b|V_I(t)\rho_I(0)V_I(t')|\varphi_b\rangle\right.$$

$$= \left. + \int_0^t dt' \langle\varphi_b|V_I(t')\rho_I(0)V_I(t)|\varphi_b\rangle\right]$$

$$= \frac{1}{\hbar^2}\left[\int_0^t dt' \langle\varphi_b|V_I(t)|\varphi_a\rangle\langle\varphi_a|\rho_I(0)|\varphi_a\rangle\langle\varphi_a|V_I(t')|\varphi_b\rangle\right.$$

$$\left. + \int_0^t dt' \langle\varphi_b|V_I(t')|\varphi_a\rangle\langle\varphi_a|\rho_I(0)|\varphi_a\rangle\langle\varphi_a|V_I(t)|\varphi_b\rangle\right]$$

$$= \frac{1}{\hbar^2}\left[\int_0^t dt' \langle\varphi_b|V_I(t)|\varphi_a\rangle\langle\varphi_a|V_I(t')|\varphi_b\rangle\right.$$

$$\left. + \int_0^t dt' \langle\varphi_b|V_I(t')|\varphi_a\rangle\langle\varphi_a|V_I(t)|\varphi_b\rangle\right]. \tag{14.5-6}$$

On converting to the Schrödinger representation and assuming the φs are eigenfunctions of the unperturbed Hamiltonian $\mathscr{H}_0$,

$$\left\langle \varphi_b \left| \frac{d\rho_I(t)}{dt} \right| \varphi_b \right\rangle$$

$$= \frac{\partial}{\partial t}\langle\varphi_b|\rho_I(t)|\varphi_b\rangle = \frac{\partial}{\partial t}\langle\varphi_b|\rho|\varphi_b\rangle$$

$$= \frac{1}{\hbar^2}\left[\int_0^t dt' \langle\varphi_b|V(t)|\varphi_a\rangle\langle\varphi_a|V(t')|\varphi_b\rangle \exp\frac{i(E_b - E_a)(t - t')}{\hbar}\right.$$

$$\left. + \int_0^t dt' \langle\varphi_b|V(t')|\varphi_a\rangle\langle\varphi_a|V(t)|\varphi_b\rangle \exp\frac{i(E_b - E_a)(t' - t)}{\hbar}\right]. \tag{14.5-7}$$

This is the general expression for the probability per unit time for a transition from a stationary state φ_a to another stationary state φ_b to second order in the perturbation.

We shall apply (14.5-7) to the case where $V(t)$ is a random function of time, that is, at any instant t, the value of V is not fixed but is subject to a law of probability. Assuming $V(t)$ is of the form

$$V(t) = Ay(t)$$

where the total time dependence of V is contained in $y(t)$ which may be taken to be a real function of the time, the statistical average of V is

$$\overline{V(t)} = A\overline{y(t)}.$$

It will further be assumed that the random function $V(t)$ is stationary, that is invariant under a change in the origin of time. In that case $\overline{y(t)\,y(t')}$ does not depend on t and t' separately but on their difference

$$\tau = t - t'.$$

The quantity

$$G(\tau) = \overline{y(t)\,y(t-\tau)} \equiv \lim_{T\to\infty} \int_{-T/2}^{T/2} y(t)\,y(t-\tau)\,dt = G(-\tau) \qquad (14.5\text{-}8)$$

is known as the *autocorrelation* function. Hence

$$\begin{aligned}\frac{\partial}{\partial t}\overline{\langle\varphi_{\rm b}|\rho|\varphi_{\rm b}\rangle} &= \frac{|\langle\varphi_{\rm b}|A|\varphi_{\rm a}\rangle|^2}{\hbar^2}\left[\int_0^t G(\tau)\exp\frac{-i(E_{\rm b}-E_{\rm a})\tau}{\hbar}\,d\tau\right.\\ &\qquad\left.+\int_0^t G(\tau)\exp\frac{i(E_{\rm b}-E_{\rm a})\tau}{\hbar}\,d\tau\right]\\ &= \frac{|\langle\varphi_{\rm b}|A|\varphi_{\rm a}\rangle|^2}{\hbar^2}\int_{-t}^t G(\tau)\exp\frac{i(E_{\rm b}-E_{\rm a})\tau}{\hbar}\,d\tau. \qquad (14.5\text{-}9)\end{aligned}$$

The autocorrelation function is a measure of the persistence of the fluctuations. As the time interval τ increases $G(\tau)$ goes to zero and frequently $G(\tau)$ decays exponentially. We shall assume the latter to be the case, i.e.,

$$G(\tau) = e^{-|\tau|/\tau_{\rm c}} \qquad (14.5\text{-}10)$$

where $\tau_{\rm c}$ is a characteristic time for the correlation between fluctuations at t and t' and is therefore known as a *correlation* time. Then

$$\frac{\partial}{\partial t}\overline{\langle\varphi_{\rm b}|\rho|\varphi_{\rm b}\rangle} = \frac{|\langle\varphi_{\rm b}|A|\varphi_{\rm a}\rangle|^2}{\hbar^2}\int_{-t}^t e^{-|\tau|/\tau_{\rm c}}e^{i\omega_{\rm ba}\tau}\,d\tau$$

with

$$\omega_{\rm ba} = \frac{E_{\rm b}-E_{\rm a}}{\hbar}.$$

If t is long compared to the correlation time $\tau_{\rm c}$, the correlation function will have decayed sufficiently so that the limits may be extended to $\pm\infty$ without introducing an appreciable error. In that case, bearing in mind that $e^{-|\tau|/\tau_{\rm c}}$ is an even function of τ,

$$\begin{aligned}\int_{-\infty}^{\infty} e^{-|\tau|/\tau_{\rm c}}e^{i\omega_{ba}\tau}\,d\tau &= \int_{-\infty}^{\infty} e^{-|\tau|/\tau_{\rm c}}\cos\omega_{\rm ba}\tau\,d\tau\\ &= 2\int_0^{\infty} e^{-|\tau|/\tau_{\rm c}}\cos\omega_{\rm ba}\tau\,d\tau = \frac{2\tau_{\rm c}}{\tau_{\rm c}{}^2\omega_{\rm ba}^2+1}\end{aligned}$$

and

$$\frac{\partial}{\partial t}\overline{\langle\varphi_b|\rho|\varphi_b\rangle} = \frac{|\langle\varphi_b|A|\varphi_a\rangle|^2}{\hbar^2}\frac{2\tau_c}{1+\omega_{ba}^2\tau_c^{\,2}}. \tag{14.5-11}$$

14.6 Response Function; Susceptibility

Suppose an external influence f acting on a physical system elicits a response y. For example, we may think of an electric field inducing an electric dipole moment in an atom or molecule. Typically, the relation between y and f is linear and $y(t)$ depends only on the past history of the system and on the external influence up to but not later than t. The second condition is a causality requirement which prohibits the system from responding before the influence has been imposed.

The general relation that satisfies the two assumptions is

$$y(t) = \int_{-\infty}^{t} K(t-t')f(t')\,dt' \tag{14.6-1}$$

where $K(t-t')$ is known as the *response function* and is characteristic of the physical system. By writing $K(t-t')$ rather than $K(t,t')$ we have implicitly assumed that the system is invariant under a translation in time which is usually the case. With a change of variable

$$\tau = t - t', \tag{14.6-2}$$

Eq. (14.6-1) acquires the more convenient form

$$y(t) = \int_{0}^{\infty} K(\tau)f(t-\tau)\,d\tau. \tag{14.6-3}$$

We shall now add the assumptions that (1) $K(\tau)$ is real and (2) $K(\tau)=0$ for $\tau<0$. The first assumption ensures that $y(t)$ is real when $f(t)$ is real and the second permits the extension of the lower limit to $-\infty$:

$$y(t) = \int_{-\infty}^{\infty} K(\tau)f(t-\tau)\,d\tau \tag{14.6-4}$$

without violating the causality requirement. It will be observed that in this form $y(t)$ is expressed as a convolution integral.

An interpretation of the response function may be obtained by letting $f(t-\tau)$ become $\delta(t-\tau)$. In that case $y(t)=K(t)$, indicating that $K(t)$ is the response of the physical system due to the application of a δ-function external influence.

Let us now express $f(t)$ in terms of its Fourier components:

$$f(t) = \int_{-\infty}^{\infty} f(\omega)e^{-i\omega t}\,d\omega \tag{14.6-5}$$

where

$$f(\omega) = \frac{1}{2\pi} \int_{-\infty}^{\infty} f(t) e^{i\omega t}\, dt. \tag{14.6-6}$$

On substituting (14.6-5) into (14.6-4),

$$\begin{aligned} y(t) &= \int_{-\infty}^{\infty} d\tau\, K(\tau) \int_{-\infty}^{\infty} f(\omega) e^{-i\omega(t-\tau)}\, d\omega \\ &= \int_{-\infty}^{\infty} \chi(\omega) f(\omega) e^{-i\omega t}\, d\omega \end{aligned} \tag{14.6-7}$$

where

$$\chi(\omega) = \int_{-\infty}^{\infty} K(\tau) e^{i\omega\tau}\, d\tau \tag{14.6-8}$$

is known as the *generalized susceptibility* and is defined as the Fourier transform of the response function $K(\tau)$. Since (14.6-4) is a convolution integral the Fourier transforms are related by

$$y(\omega) = \chi(\omega) f(\omega), \tag{14.6-9}$$

and since $K(\tau)$ is real,

$$\chi(-\omega) = \chi^*(\omega). \tag{14.6-10}$$

If χ is written in terms of real and imaginary parts

$$\chi = \chi' + i\chi'', \tag{14.6-11}$$

then

$$\chi'(-\omega) = \chi'(\omega), \qquad \chi''(-\omega) = -\chi''(\omega), \tag{14.6-12}$$

that is, χ' is an even function and χ'' an odd function of ω.

An important feature of the separation of χ into real and imaginary components is the association of χ'' with absorption or the dissipation of energy, as will now be shown. We may identify the energy associated with $f(t)$ as

$$U = -yf(t). \tag{14.6-13}$$

This becomes somewhat more transparent if, as before, we think of y as an induced dipole moment $\boldsymbol{\mu}$ and $f(t)$ as an external electric field $\mathbf{E}(t)$, in which case $U = -\boldsymbol{\mu} \cdot \mathbf{E}(t)$. When

$$f(t) = f_0 \cos \omega t \qquad (f_0 \text{ real}) \tag{14.6-14}$$

and y is intrinsically independent of time,

$$dU/dt = \omega y f_0 \sin \omega t. \tag{14.6-15}$$

But, from (14.6-4),

$$
\begin{aligned}
y(t) &= \int_{-\infty}^{\infty} K(\tau) f_0 \cos \omega(t-\tau)\, d\tau \\
&= \int_{-\infty}^{\infty} K(\tau) f_0 \operatorname{Re}[e^{-i\omega(t-\tau)}]\, d\tau \\
&= f_0 \operatorname{Re}\left[e^{-i\omega t} \int_{-\infty}^{\infty} K(\tau) e^{i\omega\tau}\, d\tau \right] \\
&= f_0 \operatorname{Re}[\chi(\omega) e^{-i\omega t}] \\
&= \tfrac{1}{2} f_0 [\chi(\omega) e^{-i\omega t} + \chi^*(\omega) e^{i\omega t}].
\end{aligned} \tag{14.6-16}
$$

Substituting in (14.6-15),

$$
\begin{aligned}
\frac{dU}{dt} &= \frac{1}{2} \omega f_0^2 \sin \omega t [\chi(\omega) e^{-i\omega t} + \chi^*(\omega) e^{i\omega t}] \\
&= \frac{1}{4i} \omega f_0^2 (e^{i\omega t} - e^{-i\omega t}) [\chi(\omega) e^{-i\omega t} + \chi^*(\omega) e^{i\omega t}] \\
&= \frac{1}{4i} \omega f_0^2 [\chi(\omega) + \chi^*(\omega) e^{2i\omega t} - \chi(\omega) e^{-2i\omega t} - \chi^*(\omega)].
\end{aligned}
$$

Averaged over a cycle, the terms containing $e^{\pm 2i\omega t}$ vanish and we have

$$
\left\langle \frac{dU}{dt} \right\rangle = \frac{i\omega}{4} f_0^2 [\chi^*(\omega) - \chi(\omega)] = \frac{\omega}{2} f_0^2 \chi'' \tag{14.6-17}
$$

where χ'' is defined in (14.6-11). We see then that the imaginary part of the susceptibility determines the rate of energy dissipation in the medium.

Normally when Fourier transforms are computed ω is taken to be real. However, under suitable restrictions it is possible to allow ω to be complex and thereby provide additional mathematical flexibility from which useful physical information can be extracted. Writing

$$
\omega' = \omega_r + i\omega_i, \tag{14.6-18}
$$

the susceptibility (14.6-8) becomes

$$
\chi(\omega') = \int_{-\infty}^{\infty} d\tau\, K(\tau) e^{i\omega_r \tau} e^{-\omega_i \tau}, \tag{14.6-19}
$$

which now modifies (14.6-10) to

$$
\chi^*(\omega') = \chi(-\omega'^*). \tag{14.6-20}
$$

The integral in (14.6-19) is zero when $\tau < 0$ since $K(\tau) = 0$ but when $\tau > 0$ it is necessary to restrict ω_i to positive values in order to ensure convergence of the integral; that is, ω' is confined to the upper half of the complex plane (Fig. 14.1).

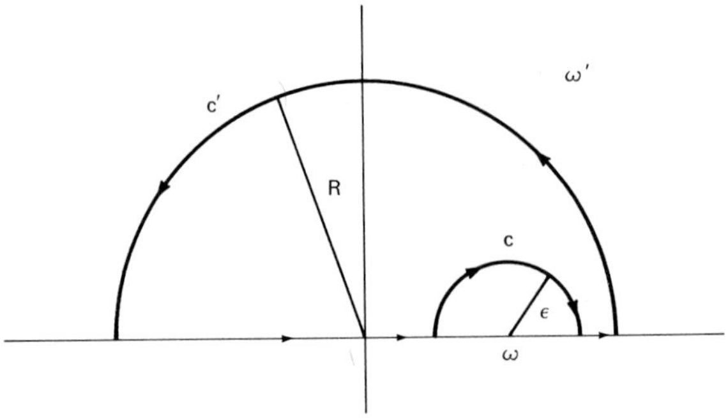

FIG. 14.1 Integration paths for the Kramers–Kronig relations.

There exists a general relation between the real and imaginary parts of the susceptibility. Assuming $\chi(\omega') = \chi'(\omega') + i\chi''(\omega')$ has no singularities in the upper half-plane the integral of the function $\chi(\omega')/(\omega' - \omega)$ (ω real) over the contour indicated in Fig. 14.1 equals zero since there are no singularities with the contour. Thus

$$\oint \frac{\chi(\omega')}{\omega' - \omega}\, d\omega' = \int_{C'} \frac{\chi(\omega')}{\omega' - \omega}\, d\omega' + \int_{-R}^{\omega-\varepsilon} \frac{\chi(\omega')}{\omega' - \omega}\, d\omega' + \int_{\omega+\varepsilon}^{R} \frac{\chi(\omega')}{\omega' - \omega}\, d\omega' + \int_{C} \frac{\chi(\omega')}{\omega' - \omega}\, d\omega' = 0.$$

Assuming further that $|\chi(\omega')|$ goes to zero faster than $1/|\omega'|$ as $\omega' \to \infty$ we also have

$$\lim_{\substack{R\to\infty \\ \varepsilon\to 0}} \int_{C'} \frac{\chi(\omega')}{\omega' - \omega}\, d\omega' = 0.$$

The integral over C may be evaluated by writing

$$\omega' = \omega + \varepsilon e^{i\varphi}, \qquad d\omega' = i\varepsilon e^{i\varphi}\, d\varphi, \qquad \omega' - \omega = \varepsilon e^{i\varphi}.$$

Thus

$$\lim_{\varepsilon\to 0} \int_{c} \frac{\chi(\omega')}{\omega' - \omega}\, d\omega' = \lim_{\varepsilon\to 0} \int_{\pi}^{0} i\, \frac{\chi(\omega + \varepsilon e^{i\varphi})}{\varepsilon e^{i\varphi}}\, \varepsilon e^{i\varphi}\, d\varphi = -i\pi\chi(\omega)$$

and the remaining two integrals constitute the principal value, i.e.,

$$\lim_{\substack{R\to\infty \\ \varepsilon\to 0}} \left[\int_{-R}^{\omega-\varepsilon} \frac{\chi(\omega')}{\omega' - \omega}\, d\omega' + \int_{\omega+\varepsilon}^{R} \frac{\chi(\omega')}{\omega' - \omega}\, d\omega' \right] \equiv ⨍_{-\infty}^{\infty} \frac{\chi(\omega')}{\omega' - \omega}\, d\omega'. \quad (14.6\text{-}21)$$

Therefore

$$⨍_{-\infty}^{\infty} \frac{\chi(\omega')}{\omega' - \omega}\, d\omega' = i\pi\chi(\omega), \qquad (14.6\text{-}22)$$

and upon separation into real and imaginary components we obtain the *Kramers–Kronig relations* also known as *Hilbert transforms* or *dispersion relations*:

$$\chi'(\omega) = \frac{1}{\pi} \int_{-\infty}^{\infty} \frac{\chi''(\omega')}{\omega' - \omega} \, d\omega',$$

$$\chi''(\omega) = -\frac{1}{\pi} \int_{-\infty}^{\infty} \frac{\chi'(\omega')}{\omega' - \omega} \, d\omega'. \tag{14.6-23}$$

It is important to note that these relations stem directly from the causality requirement and involve only real frequencies.

The discussion until now has been based on a linear relation between the response and the external influence. There are, however, situations in which the linear relation is not valid as, for example, in various optical applications employing high-intensity laser sources. In such cases the polarization $\mathbf{P}(t)$ of a medium may depend on higher powers of the applied electric field $\mathbf{E}(t)$. We therefore write

$$\mathbf{P}(t) = \mathbf{P}^{(1)}(t) + \mathbf{P}^{(2)}(t) + \cdots + \mathbf{P}^{(r)}(t) + \cdots \tag{14.6-24}$$

where $\mathbf{P}^{(1)}(t)$ is linear in $\mathbf{E}(t)$, $\mathbf{P}^{(2)}(t)$ is quadratic, etc. The development to this point considered only $\mathbf{P}^{(1)}(t)$ and in the present context (14.6-4) translates into

$$\mathbf{P}^{(1)}(t) = \int_{-\infty}^{\infty} K^{(1)}(\tau)\mathbf{E}(t - \tau)\, d\tau \tag{14.6-25}$$

where, in general, $K^{(1)}(\tau)$ is a tensor of rank 2 since it cannot be presumed that $\mathbf{P}^{(1)}$ is colinear with $\mathbf{E}$ unless the medium has special characteristics. The tensor $K^{(1)}(\tau)$ is the *linear polarization response function* and, as before, we may interpret $K^{(1)}(\tau)$ as the polarization of the medium due to a δ-function electric field. Also, as was done previously, $K^{(1)}(\tau)$ is taken to be real and, in order to satisfy the causality requirement, we set $K^{(1)}(\tau) = 0$ when $\tau < 0$.

Extending (14.6-25) to $\mathbf{P}^{(2)}(t)$ gives

$$\mathbf{P}^{(2)}(t) = \int_{-\infty}^{\infty} d\tau_1 \int_{-\infty}^{\infty} d\tau_2\, K^{(2)}(\tau_1, \tau_2)\mathbf{E}(t - \tau_1)\mathbf{E}(t - \tau_2). \tag{14.6-26}$$

$K^{(2)}(\tau_1, \tau_2)$ is the *quadratic polarization response function* and is a tensor of rank 3. The general expression for $\mathbf{P}^{(r)}$ is

$$\mathbf{P}^{(r)}(t) = \int_{-\infty}^{\infty} d\tau_1 \cdots \int_{-\infty}^{\infty} d\tau_r\, K^{(r)}(\tau_1, \ldots, \tau_r)\mathbf{E}(t - \tau_1) \cdots \mathbf{E}(t - \tau_r). \tag{14.6-27}$$

By analogy with (14.6-8) the susceptibility is

$$\chi^{(1)}(\omega) = \int_{-\infty}^{\infty} K^{(1)}(\tau) e^{i\omega\tau}\, d\tau \tag{14.6-28}$$

in terms of which the polarization is written as

$$\mathbf{P}^{(1)}(t) = \int_{-\infty}^{\infty} \chi^{(1)}(\omega)\mathbf{E}(\omega)e^{-i\omega t}\, d\omega \tag{14.6-29}$$

as in (14.6-7). All the previous results follow. With $\omega = \omega' + i\omega''$, ω is restricted to the upper half of the complex plane; the real and imaginary parts of $\chi^{(1)}(\omega)$ satisfy the Kramers–Kronig relations and

$$[\chi^{(1)}(\omega)]^* = \chi^{(1)}(-\omega^*). \tag{14.6-30}$$

Note that $\chi^{(1)}(\omega)$ is a tensor of rank 2 the same as $K^{(1)}(\tau)$.

For the second and higher orders,

$$\chi^{(2)}(\omega_1, \omega_2) = \int_{-\infty}^{\infty} d\tau_1 \int_{-\infty}^{\infty} d\tau_2\, K^{(2)}(\tau_1, \tau_2) \exp i(\omega_1\tau_1 + \omega_2\tau_2), \tag{14.6-31}$$

$$\mathbf{P}^{(2)}(t) = \int_{-\infty}^{\infty} d\omega_1 \int_{-\infty}^{\infty} d\omega_2\, \chi^{(2)}(\omega_1, \omega_2)\mathbf{E}(\omega_1)\mathbf{E}(\omega_2) \exp -i(\omega_1 + \omega_2)t, \tag{14.6-32}$$

$$[\chi^{(2)}(\omega_1, \omega_2)]^* = \chi^{(2)}(-\omega_1^*, -\omega_2^*), \tag{14.6-33}$$

$$\chi^{(r)}(\omega_1, \ldots, \omega_r) = \int_{-\infty}^{\infty} d\tau_1 \cdots \int_{-\infty}^{\infty} d\tau_r\, K^{(r)}(\tau_1, \ldots, \tau_r) \times \exp i(\omega_1\tau_1 + \omega_2\tau_2 + \cdots + \omega_r\tau_r), \tag{14.6-34}$$

$$\mathbf{P}^{(r)}(t) = \int_{-\infty}^{\infty} d\omega_1 \cdots \int_{-\infty}^{\infty} d\omega_r\, \chi^{(r)}(\omega_1, \ldots, \omega_r)\mathbf{E}(\omega_1) \cdots \mathbf{E}(\omega_r) \times \exp -i(\omega_1 + \cdots + \omega_r)t, \tag{14.6-35}$$

$$[\chi^{(r)}(\omega_1, \ldots, \omega_r)]^* = \chi^{(r)}(-\omega_1^*, \ldots, -\omega_r^*). \tag{14.6-36}$$

Response function may also be formulated quantum mechanically. For this purpose we refer to the discussion of the interaction representation in Section 9.3. It was shown there that for a Hamiltonian $\mathscr{H} = \mathscr{H}_0 + V$ the expectation value of an operator A can be developed into a series

$$\langle A_{\mathrm{I}} \rangle = \langle \psi_{\mathrm{I}}(t_0) | A_{\mathrm{I}}(t) | \psi_{\mathrm{I}}(t_0) \rangle - \frac{i}{\hbar} \int_{t_0}^{t} dt' \langle \psi_{\mathrm{I}}(t_0) | [A_{\mathrm{I}}(t), V_{\mathrm{I}}(t')] | \psi_{\mathrm{I}}(t_0) \rangle + \cdots. \tag{14.6-37}$$

The quantity

$$\Delta\langle A_{\mathrm{I}} \rangle \equiv \langle A_{\mathrm{I}} \rangle - \langle \psi_{\mathrm{I}}(t_0) | A_{\mathrm{I}}(t) | \psi_{\mathrm{I}}(t_0) \rangle = -\frac{i}{\hbar} \int_{t_0}^{t} dt' \langle \psi_{\mathrm{I}}(t_0) | [A_{\mathrm{I}}(t), V_{\mathrm{I}}(t')] | \psi_{\mathrm{I}}(t_0) \rangle \tag{14.6-38}$$

may be regarded as the first-order response to an external perturbation V.

We shall assume that the perturbation can be written as

$$V = -yf(t) \tag{14.6-39}$$

which is of the same form as (14.6-13) except that y is now interpreted as an operator; $f(t)$ represents an external field. In the interaction representation

$$V_{\mathrm{I}} = -y_{\mathrm{I}} f(t). \tag{14.6-40}$$

With this perturbation the response is

$$\begin{aligned} \Delta\langle A_{\mathrm{I}}\rangle &= \frac{i}{\hbar}\int_{t_0}^{t} dt' \, \langle \psi_{\mathrm{I}}(t_0)|[A_{\mathrm{I}}(t), y_{\mathrm{I}}(t')] f(t')|\psi_{\mathrm{I}}(t_0)\rangle \\ &= \int_{t_0}^{t} K(t, t') f(t')\, dt' \end{aligned} \tag{14.6-41}$$

where

$$K(t, t') = \frac{i}{\hbar} \langle \psi_{\mathrm{I}}(t_0)|[A_{\mathrm{I}}(t), y_{\mathrm{I}}(t')]|\psi_{\mathrm{I}}(t_0)\rangle. \tag{14.6-42}$$

Assuming $K(t, t') = K(t - t') = K(\tau)$ and $t_0 = -\infty$,

$$\Delta\langle A_{\mathrm{I}}\rangle = \int_{0}^{\infty} K(\tau) f(t - \tau)\, d\tau. \tag{14.6-43}$$

The lower limit may be extended to $-\infty$ since $K(\tau) = 0$ for $\tau < 0$. We then have

$$\Delta\langle A_{\mathrm{I}}\rangle = \int_{-\infty}^{\infty} K(\tau) f(t - \tau)\, d\tau \tag{14.6-44}$$

with

$$\begin{aligned} K(\tau) &= \frac{i}{\hbar} \langle \psi_{\mathrm{I}}(-\infty)|[A_{\mathrm{I}}(t), y_{\mathrm{I}}(t - \tau)]|\psi_{\mathrm{I}}(-\infty)\rangle \\ &\equiv \frac{i}{\hbar} \langle [A_{\mathrm{I}}(t), y_{\mathrm{I}}(t - \tau)]\rangle. \end{aligned} \tag{14.6-45}$$

$K(\tau)$ is the linear response function. As before, the susceptibility is defined as the Fourier transform of the response function

$$\chi(\omega) = \int_{-\infty}^{\infty} K(\tau) e^{i\omega\tau}\, d\tau. \tag{14.6-46}$$

It is often more appropriate to consider the perturbing potential V as composed of a set of terms

$$V = -\sum_{a} y_a f_a(t) \tag{14.6-47}$$

or in the interaction representation

$$V_{\mathrm{I}} = -\sum_a y_a{}^{\mathrm{I}} f_a(t). \tag{14.6-48}$$

We may wish to know the response associated with the change in expectation value of one of the operators y^{I}, say $y_b{}^{\mathrm{I}}$. From (14.6-44) it follows that

$$\Delta\langle y_b{}^{\mathrm{I}}\rangle = \sum_a \int_{-\infty}^{\infty} K_{ba}(\tau) f_a(t-\tau)\, d\tau \tag{14.6-49}$$

where

$$K_{ba}(\tau) = \frac{i}{\hbar}\langle[y_b{}^{\mathrm{I}}(t), y_a{}^{\mathrm{I}}(t-\tau)]\rangle \tag{14.6-50}$$

or in terms of a density operator

$$K_{ba}(\tau) = \frac{i}{\hbar}\,\mathrm{Tr}\{[y_b{}^{\mathrm{I}}(t), y_a{}^{\mathrm{I}}(t-\tau)]\rho\}. \tag{14.6-51}$$

Similarly the susceptibility is

$$\begin{aligned} \chi_{ba}(\omega) &= \int_{-\infty}^{\infty} K_{ba}(\tau) e^{i\omega\tau}\, d\tau \\ &= \frac{i}{\hbar}\int_{-\infty}^{\infty} \mathrm{Tr}\{[y_b{}^{\mathrm{I}}(t), y_a{}^{\mathrm{I}}(t-\tau)]\rho\} e^{i\omega\tau}\, d\tau. \end{aligned} \tag{14.6-52}$$

An application to optical susceptibility is discussed in Section 24.4.

PART III

One-Electron Atoms

CHAPTER 15

DIRAC EQUATION

The phenomena occurring in atoms and molecules are describable, for the most part, on the basis of nonrelativistic quantum mechanics. It is nevertheless advantageous to start with the relativistic equation of Dirac and to proceed to its nonrelativistic approximation. In doing so one obtains expressions for the various interaction terms that appear in the Schrödinger Hamiltonian and no further derivations are required. Strictly speaking, this approach is rigorous only for the one-electron case because the Dirac equation applies only to a single particle. Still, much of the information obtained by this route is also applicable to a many-electron system.

The Dirac equation for a free particle serves to introduce the formalism. Electromagnetic couplings are then added and the approximation to order v^2/c^2 provides the desired results.

15.1 Free Particle Equation

The time-independent Dirac equation for a free particle is

$$(c\boldsymbol{\alpha} \cdot \mathbf{p} + \beta mc^2)\psi = E\psi \tag{15.1-1}$$

in which

$$\alpha_x = \begin{pmatrix} 0 & 0 & 0 & 1 \\ 0 & 0 & 1 & 0 \\ 0 & 1 & 0 & 0 \\ 1 & 0 & 0 & 0 \end{pmatrix}, \qquad \alpha_y = \begin{pmatrix} 0 & 0 & 0 & -i \\ 0 & 0 & i & 0 \\ 0 & -i & 0 & 0 \\ i & 0 & 0 & 0 \end{pmatrix}, \tag{15.1-2}$$

$$\alpha_z = \begin{pmatrix} 0 & 0 & 1 & 0 \\ 0 & 0 & 0 & -1 \\ 1 & 0 & 0 & 0 \\ 0 & -1 & 0 & 0 \end{pmatrix}, \qquad \beta = \begin{pmatrix} 1 & 0 & 0 & 0 \\ 0 & 1 & 0 & 0 \\ 0 & 0 & -1 & 0 \\ 0 & 0 & 0 & -1 \end{pmatrix},$$

$$\mathbf{p} = -i\hbar\nabla, \tag{15.1-3}$$

$$\boldsymbol{\alpha}\cdot\mathbf{p} = \alpha_x p_x + \alpha_y p_y + \alpha_z p_z, \tag{15.1-4}$$

$$\alpha_x p_x = \begin{pmatrix} 0 & 0 & 0 & p_x \\ 0 & 0 & p_x & 0 \\ 0 & p_x & 0 & 0 \\ p_x & 0 & 0 & 0 \end{pmatrix}, \quad \text{etc,}$$

and m is the particle rest mass, c the velocity of light, E the relativistic energy (including rest mass energy), and ψ a four-component (spinor) function

$$\psi = \begin{pmatrix} \psi_1 \\ \psi_2 \\ \psi_3 \\ \psi_4 \end{pmatrix}. \tag{15.1-5}$$

In component form (15.1-1) becomes a set of four equations:

$$\begin{aligned} c(p_x - ip_y)\psi_4 + cp_z\psi_3 + (mc^2 - E)\psi_1 &= 0, \\ c(p_x + ip_y)\psi_3 - cp_z\psi_4 + (mc^2 - E)\psi_2 &= 0, \\ c(p_x - ip_y)\psi_2 + cp_z\psi_1 - (mc^2 + E)\psi_3 &= 0, \\ c(p_x + ip_y)\psi_1 - cp_z\psi_2 - (mc^2 + E)\psi_4 &= 0. \end{aligned} \tag{15.1-6}$$

The Dirac equation (15.1-1) may also be written in two-component form by means of the Pauli spin matrices

$$\sigma_x = \begin{pmatrix} 0 & 1 \\ 1 & 0 \end{pmatrix}, \quad \sigma_y = \begin{pmatrix} 0 & -i \\ i & 0 \end{pmatrix}, \quad \sigma_z = \begin{pmatrix} 1 & 0 \\ 0 & -1 \end{pmatrix}, \tag{15.1-7}$$

and the 2×2 identity matrix

$$I = \begin{pmatrix} 1 & 0 \\ 0 & 1 \end{pmatrix}.$$

With $\boldsymbol{\sigma} = (\sigma_x, \sigma_y, \sigma_z)$ it is seen that

$$\boldsymbol{\alpha} = \begin{pmatrix} 0 & \boldsymbol{\sigma} \\ \boldsymbol{\sigma} & 0 \end{pmatrix}, \qquad \beta = \begin{pmatrix} I & 0 \\ 0 & -I \end{pmatrix}, \tag{15.1-8}$$

so that in place of (15.1-6) we have

$$c\boldsymbol{\sigma}\cdot\mathbf{p}\psi_{\rm v} + (mc^2 - E)\psi_{\rm u} = 0, \tag{15.1-9a}$$

$$c\boldsymbol{\sigma}\cdot\mathbf{p}\psi_{\rm u} - (mc^2 + E)\psi_{\rm v} = 0, \tag{15.1-9b}$$

where

$$\psi_{\rm u} = \begin{pmatrix}\psi_1\\ \psi_2\end{pmatrix}, \qquad \psi_{\rm v} = \begin{pmatrix}\psi_3\\ \psi_4\end{pmatrix}. \tag{15.1-10}$$

Equations (15.1-9) provide a useful starting point for the nonrelativistic approximation. From (15.1-9a),

$$\psi_{\rm u} = \frac{c\boldsymbol{\sigma}\cdot\mathbf{p}}{E - mc^2}\psi_{\rm v}, \tag{15.1-11}$$

which enables us to make an estimate of the relative magnitude of $\psi_{\rm u}$ and $\psi_{\rm v}$. Approximating $\boldsymbol{\sigma}\cdot\mathbf{p} \approx mv$, $E - mc^2 \approx \frac{1}{2}mv^2$,

$$\psi_{\rm u}/\psi_{\rm v} \approx cmv/\tfrac{1}{2}mv^2 = 2c/v. \tag{15.1-12}$$

Consequently, for $v/c \ll 1$, the component of interest is $\psi_{\rm u}$, also known as the "large" component. Eliminating $\psi_{\rm v}$ between (15.1-9a) and (15.1-9b), $\psi_{\rm u}$ satisfies

$$c^2(\boldsymbol{\sigma}\cdot\mathbf{p})^2\psi_{\rm u} = (E - mc^2)(E + mc^2)\psi_{\rm u}. \tag{15.1-13}$$

At nonrelativistic energies

$$E' \equiv E - mc^2 \ll mc^2, \tag{15.1-14}$$

or

$$(E - mc^2)(E + mc^2) = 2mc^2E'\left(1 + \frac{E'}{2mc^2}\right) \approx 2mc^2E'. \tag{15.1-15}$$

Equation (15.1-13) then becomes

$$E'\psi_{\rm u} = \frac{(\boldsymbol{\sigma}\cdot\mathbf{p})(\boldsymbol{\sigma}\cdot\mathbf{p})}{2m}\psi_{\rm u}. \tag{15.1-16}$$

Further reduction of (15.1-16) is possible with the aid of the identity

$$(\boldsymbol{\sigma}\cdot\mathbf{A})(\boldsymbol{\sigma}\cdot\mathbf{B}) = (\mathbf{A}\cdot\mathbf{B}) + i\boldsymbol{\sigma}\cdot(\mathbf{A}\times\mathbf{B}) \tag{15.1-17}$$

which is easily verified from the definitions of the Pauli spin matrices (15.1-7). Therefore

$$(\boldsymbol{\sigma}\cdot\mathbf{p})(\boldsymbol{\sigma}\cdot\mathbf{p}) = p^2, \tag{15.1-18}$$

and, substituting in (15.1-16),

$$E'\psi_u = \frac{p^2}{2m}\psi_u \tag{15.1-19}$$

or

$$E'\psi_u = -\frac{\hbar^2}{2m}\nabla^2\psi_u. \tag{15.1-20}$$

Equation (15.1-20) is still in two-component form since ψ_u is a two-component function. But each component of ψ_u satisfies (15.1-20) so that we may drop all subscripts and write

$$E'\psi = -\frac{\hbar^2}{2m}\nabla^2\psi \tag{15.1-21}$$

which is the time-independent Schrödinger equation for a free particle.

15.2 Dirac Equation with Electromagnetic Coupling

The free-particle Dirac equation (15.1-1) must now be modified to include effects due to external fields. Classical considerations suggest how this may be accomplished. In the presence of external fields a possible Lagrangian for the system is

$$L = \frac{1}{2}mv^2 + \frac{q}{c}\mathbf{v}\cdot\mathbf{A} - q\varphi \tag{15.2-1}$$

in which v is the velocity of a particle with positive charge q and mass m, $\mathbf{A}$ and φ are the vector and scalar potentials, respectively. The fields are then given by

$$\mathbf{B} = \nabla\times\mathbf{A}, \qquad \mathbf{E} = -\frac{1}{c}\frac{\partial\mathbf{A}}{\partial t} - \nabla\varphi. \tag{15.2-2}$$

To verify that a Lagrangian of form (15.2-1) leads to correct physical results it is sufficient to demonstrate the derivation of the Lorentz force law

$$\mathbf{F} = q\left(\mathbf{E} + \frac{1}{c}\mathbf{v}\times\mathbf{B}\right) \tag{15.2-3}$$

from (15.2-1). Thus the Lagrange equations

$$\frac{d}{dt}\left(\frac{\partial L}{\partial \dot{x}_i}\right) = \frac{\partial L}{\partial x_i} \qquad (x_i = x, y, z), \tag{15.2-4}$$

with the Lagrangian as in (15.2-1), become

$$\frac{\partial L}{\partial x_i} = m\ddot{x}_i + \frac{q}{c}\frac{dA_i}{dt} \tag{15.2-5}$$

where, for example,

$$\frac{dA_x}{dt} = \frac{\partial A_x}{\partial x}\dot{x} + \frac{\partial A_x}{\partial y}\dot{y} + \frac{\partial A_x}{\partial z}\dot{z} + \frac{\partial A_x}{\partial t} \tag{15.2-6}$$

is the total time derivative of A_x and the last term on the right represents the intrinsic time dependence, if any, of A_x. Since

$$\frac{\partial L}{\partial x} = \frac{q}{c}\left(\frac{\partial A_x}{\partial x}\dot{x} + \frac{\partial A_y}{\partial x}\dot{y} + \frac{\partial A_z}{\partial x}\dot{z}\right) - q\frac{\partial \varphi}{\partial x}$$

the Lagrange equation (15.2-5) for the x component is

$$\begin{aligned} m\ddot{x} &= q\left[\frac{1}{c}\left(\frac{\partial A_y}{\partial x} - \frac{\partial A_x}{\partial y}\right)\dot{y} - \frac{1}{c}\left(\frac{\partial A_x}{\partial z} - \frac{\partial A_z}{\partial x}\right)\dot{z} - \frac{1}{c}\frac{\partial A_x}{\partial t} - \frac{\partial \varphi}{\partial x}\right] \\ &= q\left[E_x + \frac{1}{c}(\mathbf{v} \times \mathbf{B})_x\right] = F_x. \end{aligned} \tag{15.2-7}$$

The other components are obtained in similar fashion thus verifying the derivation of the Lorentz force law from (15.2-1).

On this basis one may now proceed to define the canonical momentum

$$p_i = \frac{\partial L}{\partial \dot{x}_i} = m\dot{x}_i + \frac{q}{c}A_i \tag{15.2-8}$$

and the Hamiltonian

$$\mathscr{H} = \sum_i \frac{\partial L}{\partial \dot{x}_i}\dot{x}_i - L = \mathbf{p}\cdot\mathbf{v} - L = \frac{1}{2m}\left(\mathbf{p} - \frac{q}{c}A\right)^2 + q\varphi. \tag{15.2-9}$$

Equations (15.2-8) and (15.2-9) indicate the modifications in the canonical momenta and the Hamiltonian brought about by the presence of the external fields.

It will now be assumed that the same modifications can be introduced into the free particle Dirac equation (15.1-1) so that the proper equation for a particle of (positive) charge q and rest mass m in a field with vector potential $\mathbf{A}$ and scalar potential φ is

$$[\boldsymbol{\alpha}\cdot(c\mathbf{p} - q\mathbf{A}) + \beta mc^2 + q\varphi]\psi = E\psi. \tag{15.2-10}$$

Needless to say this classical development is merely suggestive; the Dirac equation is to be regarded as a fundamental equation whose validity must be investigated—ultimately by resort to experiment—independently of any classical arguments. In two-component form, by analogy with (15.1-9),

$$\boldsymbol{\sigma}\cdot(c\mathbf{p} - q\mathbf{A})\psi_{\mathrm{v}} + (mc^2 + q\varphi)\psi_{\mathrm{u}} = E\psi_{\mathrm{u}}, \tag{15.2-11a}$$

$$\boldsymbol{\sigma}\cdot(c\mathbf{p} - q\mathbf{A})\psi_{\mathrm{u}} - (mc^2 - q\varphi)\psi_{\mathrm{v}} = E\psi_{\mathrm{v}}. \tag{15.2-11b}$$

As in the free particle case we shall be interested mainly in the "large" component ψ_u. From (15.2-11),

$$(E' - q\varphi)\psi_u = \frac{1}{2m}(\boldsymbol{\sigma} \cdot \boldsymbol{\pi})K(\boldsymbol{\sigma} \cdot \boldsymbol{\pi})\psi_u \tag{15.2-12}$$

where

$$E' = E - mc^2,$$

$$K = \frac{2mc^2}{E' + 2mc^2 - q\varphi} = \frac{1}{1 + [(E' - q\varphi)/2mc^2]}, \tag{15.2-13}$$

$$\boldsymbol{\pi} = \mathbf{p} - \frac{q}{c}\mathbf{A}. \tag{15.2-14}$$

Our objective now is to obtain an approximation to (15.2-12) to order v^2/c^2.

Identity (15.1-17) gives

$$(\boldsymbol{\sigma} \cdot \boldsymbol{\pi})^2 = \pi^2 + i\boldsymbol{\sigma} \cdot \boldsymbol{\pi} \times \boldsymbol{\pi}$$
$$= \left(\mathbf{p} - \frac{q}{c}\mathbf{A}\right)^2 - i\frac{q}{c}\sigma \cdot (\mathbf{p} \times \mathbf{A} + \mathbf{A} \times \mathbf{p}).$$

But

$$\mathbf{p} \times \mathbf{A} = -i\hbar \nabla \times \mathbf{A} - \mathbf{A} \times \mathbf{p}; \tag{15.2-15}$$

therefore

$$(\boldsymbol{\sigma} \cdot \boldsymbol{\pi})^2 = \left(\mathbf{p} - \frac{q}{c}\mathbf{A}\right)^2 - \frac{q\hbar}{c}\boldsymbol{\sigma} \cdot \nabla \times \mathbf{A} \tag{15.2-16}$$

and the approximation to (15.2-12) with $K = 1$ is

$$(E' - q\varphi)\psi_u = \left[\frac{1}{2m}\left(\mathbf{p} - \frac{q}{c}\mathbf{A}\right)^2 - \frac{q\hbar}{2mc}\boldsymbol{\sigma} \cdot \nabla \times \mathbf{A}\right]\psi_u. \tag{15.2-17}$$

The next higher approximation is obtained by setting

$$K = 1 - \frac{E' - q\varphi}{2mc^2} \tag{15.2-18}$$

so that (15.2-12) now becomes

$$(E' - q\varphi)\psi_u = \frac{1}{2m}\left[(\boldsymbol{\sigma} \cdot \boldsymbol{\pi})^2 - (\boldsymbol{\sigma} \cdot \boldsymbol{\pi})\frac{E' - q\varphi}{2mc^2}(\boldsymbol{\sigma} \cdot \boldsymbol{\pi})\right]\psi_u. \tag{15.2-19}$$

It might be supposed that a straightforward reduction of (15.2-19) by the same methods that were used to arrive at (15.2-17) would lead to the desired results. It turns out that this is not so; various difficulties develop (Foldy and Wouthuysen, 1958; Feynman, 1961; Sakurai, 1967), and to circumvent them it is necessary to follow a different route. Let

$$\psi = \Omega\psi_{\mathrm{u}} \tag{15.2-20}$$

where‡

$$\Omega = 1 + \frac{(\boldsymbol{\sigma}\cdot\boldsymbol{\pi})^2}{8m^2c^2} \tag{15.2-21}$$

and, to order v^2/c^2,

$$\Omega^{-1} = 1 - \frac{(\boldsymbol{\sigma}\cdot\boldsymbol{\pi})^2}{8m^2c^2}. \tag{15.2-22}$$

Multiplying (15.2-19) on the left by Ω^{-1} and replacing ψ_{u} by $\Omega^{-1}\psi$ one obtains

$$\Omega^{-1}(E' - q\varphi)\Omega^{-1}\psi = \frac{1}{2m}\Omega^{-1}\left[(\boldsymbol{\sigma}\cdot\boldsymbol{\pi})^2 - (\sigma\cdot\pi)\frac{E' - q\varphi}{2mc^2}(\boldsymbol{\sigma}\cdot\boldsymbol{\pi})\right]\Omega^{-1}\psi. \tag{15.2-23}$$

To order v^2/c^2,

$$\Omega^{-1}(E' - q\varphi)\Omega^{-1}\psi = (E' - q\varphi)\psi - \frac{(\boldsymbol{\sigma}\cdot\boldsymbol{\pi})^2}{8m^2c^2}(E' - q\varphi)\psi - (E' - q\varphi)\frac{(\boldsymbol{\sigma}\cdot\boldsymbol{\pi})^2}{8m^2c^2}\psi,$$

$$\frac{1}{2m}\Omega^{-1}(\boldsymbol{\sigma}\cdot\boldsymbol{\pi})^2\Omega^{-1}\psi = \frac{1}{2m}(\boldsymbol{\sigma}\cdot\boldsymbol{\pi})^2\psi - \frac{(\sigma\cdot\pi)^4}{8m^3c^2}\psi,$$

$$\frac{1}{2m}\Omega^{-1}(\boldsymbol{\sigma}\cdot\boldsymbol{\pi})\frac{E' - q\varphi}{2mc^2}(\boldsymbol{\sigma}\cdot\boldsymbol{\pi})\Omega^{-1}\psi = (\boldsymbol{\sigma}\cdot\boldsymbol{\pi})\frac{E' - q\varphi}{4m^2c^2}(\boldsymbol{\sigma}\cdot\boldsymbol{\pi})\psi.$$

Hence the transformed equation (15.2-23), to order v^2/c^2, becomes

$$(E' - q\varphi)\psi = \left[\frac{1}{2m}(\boldsymbol{\sigma}\cdot\boldsymbol{\pi})^2 - \frac{1}{8m^3c^2}(\boldsymbol{\sigma}\cdot\boldsymbol{\pi})^4 + \frac{(\boldsymbol{\sigma}\cdot\boldsymbol{\pi})^2}{8m^2c^2}(E' - q\varphi) - \frac{1}{4m^2c^2}(\boldsymbol{\sigma}\cdot\boldsymbol{\pi})(E' - q\varphi)(\boldsymbol{\sigma}\cdot\boldsymbol{\pi}) + (E' - q\varphi)\frac{(\boldsymbol{\sigma}\cdot\boldsymbol{\pi})^2}{8m^2c^2}\right]\psi. \tag{15.2-24}$$

A number of simplifications are possible with the aid of:

$$(\boldsymbol{\sigma}\cdot\boldsymbol{\pi})(E' - q\varphi) - (E' - q\varphi)(\boldsymbol{\sigma}\cdot\boldsymbol{\pi}) = q\varphi\boldsymbol{\sigma}\cdot\mathbf{p} - q\boldsymbol{\sigma}\cdot\mathbf{p}\varphi = i\hbar q(\boldsymbol{\sigma}\cdot\nabla\varphi), \tag{15.2-25}$$

$$(\boldsymbol{\sigma}\cdot\boldsymbol{\pi})(\boldsymbol{\sigma}\cdot\nabla\varphi) - (\boldsymbol{\sigma}\cdot\nabla\varphi)(\boldsymbol{\sigma}\cdot\boldsymbol{\pi}) = -i\hbar\nabla\cdot\nabla\varphi + 2i\boldsymbol{\sigma}\cdot\boldsymbol{\pi}\times\nabla\varphi. \tag{15.2-26}$$

‡ ψ is normalized to order v^2/c^2 and therefore contains a contribution from the "small" component.

Therefore

$$\begin{aligned}(\boldsymbol{\sigma}\cdot\boldsymbol{\pi})^2(E'-q\varphi)+(E'-q\varphi)(\boldsymbol{\sigma}\cdot\boldsymbol{\pi})^2 \\ &\equiv (\boldsymbol{\sigma}\cdot\boldsymbol{\pi})[(\boldsymbol{\sigma}\cdot\boldsymbol{\pi})(E'-q\varphi)-(E'-q\varphi)(\boldsymbol{\sigma}\cdot\boldsymbol{\pi})] \\ &\quad -[(\boldsymbol{\sigma}\cdot\boldsymbol{\pi})(E'-q\varphi)-(E'-q\varphi)(\boldsymbol{\sigma}\cdot\boldsymbol{\pi})](\boldsymbol{\sigma}\cdot\boldsymbol{\pi}) \\ &\quad +2(\boldsymbol{\sigma}\cdot\boldsymbol{\pi})(E'-q\varphi)(\boldsymbol{\sigma}\cdot\boldsymbol{\pi}) \\ &= q\hbar^2\,\nabla\cdot\nabla\varphi-2q\hbar\boldsymbol{\sigma}\cdot\boldsymbol{\pi}\times\nabla\varphi \\ &\quad +2(\boldsymbol{\sigma}\cdot\boldsymbol{\pi})(E'-q\varphi)(\boldsymbol{\sigma}\cdot\boldsymbol{\pi}).\end{aligned} \tag{15.2-27}$$

Substitution of (15.2-27) into (15.2-24) gives

$$\begin{aligned}(E'-q\varphi)\psi = \Bigg[&\frac{1}{2m}(\boldsymbol{\sigma}\cdot\boldsymbol{\pi})^2-\frac{1}{8m^3c^2}(\boldsymbol{\sigma}\cdot\boldsymbol{\pi})^4 \\ &+\frac{q\hbar^2}{8m^2c^2}\nabla\cdot\nabla\varphi-\frac{q\hbar}{4m^2c^2}\boldsymbol{\sigma}\cdot\boldsymbol{\pi}\times\nabla\varphi\Bigg]\psi.\end{aligned} \tag{15.2-28}$$

Also, from (15.2-16),

$$\frac{1}{2m}(\boldsymbol{\sigma}\cdot\boldsymbol{\pi})^2=\frac{1}{2m}\left(\mathbf{p}-\frac{q}{c}\mathbf{A}\right)^2-\frac{q\hbar}{2mc}\boldsymbol{\sigma}\cdot\nabla\times\mathbf{A} \tag{15.2-29}$$

and, to order v^2/c^2,

$$\frac{1}{c^2}(\boldsymbol{\sigma}\cdot\boldsymbol{\pi})^4=\frac{p^4}{c^2}, \tag{15.2-30}$$

$$\begin{aligned}\frac{1}{c^2}\boldsymbol{\pi}\times\nabla\varphi=\frac{1}{c^2}\mathbf{p}\times\nabla\varphi&=\frac{1}{c^2}(-i\hbar\nabla\times\nabla\varphi-\nabla\varphi\times\mathbf{p}) \\ &=-\frac{1}{c^2}\nabla\varphi\times\mathbf{p}.\end{aligned} \tag{15.2-31}$$

We shall now replace q by $-e$, where e is the magnitude (absolute value) of the electronic charge, to obtain the Dirac equation for an electron to order v^2/c^2:

$$\begin{aligned}(E'+e\varphi)\psi=\Bigg[&\frac{1}{2m}\left(\mathbf{p}+\frac{e}{c}\mathbf{A}\right)^2+\frac{e\hbar}{2mc}\boldsymbol{\sigma}\cdot\nabla\times\mathbf{A} \\ &-\frac{p^4}{8m^3c^2}-\frac{e\hbar^2}{8m^2c^2}\nabla\cdot\nabla\varphi-\frac{e\hbar}{4m^2c^2}\boldsymbol{\sigma}\cdot\nabla\varphi\times\mathbf{p}\Bigg]\psi.\end{aligned} \tag{15.2-32}$$

Equation (15.2-32), which may also be regarded as the Schrödinger equation for an electron interacting with fields describable by the potentials $\mathbf{A}$ and φ, is the starting point for discussions of atomic and molecular properties. The significance of the various terms and their energies, indicated to within an order of magnitude, are:

$e\varphi$ scalar potential energy ($10^5\ \text{cm}^{-1}$).

$(1/2m)(\mathbf{p} + (e/c)\mathbf{A})^2$ contains kinetic energy and interaction terms with a field represented by a vector potential $\mathbf{A}$ ($10^5\ \text{cm}^{-1}$). The interaction terms are responsible or contribute to numerous physical processes among which are absorption, emission and scattering of electromagnetic waves, diamagnetism, and the Zeeman effect.

$(e\hbar/2mc)\boldsymbol{\sigma} \cdot \nabla \times \mathbf{A}$ interaction of the spin magnetic moment with a magnetic field $\mathbf{B} = \nabla \times \mathbf{A}$ ($1\ \text{cm}^{-1}$).

$p^4/8m^3c^2$ this term appears in the expansion of the relativistic energy

$$\sqrt{(mc^2)^2 + p^2c^2} = mc^2 + \frac{p^2}{2m} - \frac{p^4}{8m^3c^2} + \cdots .$$

It is therefore a relativistic correction to the kinetic energy ($0.1\ \text{cm}^{-1}$).

$-(e\hbar^2/8m^2c^2)\nabla \cdot \nabla\varphi$ produces an energy shift in s-states and is known as the Darwin term ($<0.1\ \text{cm}^{-1}$).

$-(e\hbar/4m^2c^2)\boldsymbol{\sigma} \cdot \nabla\varphi \times \mathbf{p}$ spin-orbit interaction (10–$10^3\ \text{cm}^{-1}$).

It is necessary to add a word of caution concerning the validity of (15.2-32) which contains the approximation

$$K = \frac{2mc^2}{E' + 2mc^2 - q\varphi} \approx 1 - \frac{E' - q\varphi}{2mc^2}.$$

Clearly, this will not be legitimate if φ becomes singular as it does, for example, in certain hyperfine interactions. In such cases a separate treatment is required (see Section 18.1).

CHAPTER 16

HYDROGEN ATOM

16.1 Schrödinger Equation

For an electron in a static field whose potential is φ, the Schrödinger equation (15.2-32), without the higher-order relativistic corrections, simplifies to

$$(E' + e\varphi)\psi = \frac{p^2}{2m}\psi. \tag{16.1-1}$$

Replacing E' by E (nonrelativistic energy), $e\varphi$ by $-V$, and $\mathbf{p}$ by $-i\hbar\boldsymbol{\nabla}$, Eq. (16.1-1) becomes

$$\mathscr{H}\psi = \left(-\frac{\hbar^2}{2m}\nabla^2 + V\right)\psi = E\psi \tag{16.1-2}$$

or, in spherical coordinates,

$$r^2\left(\frac{\partial^2}{\partial r^2} + \frac{2}{r}\frac{\partial}{\partial r}\right)\psi + \frac{2mr^2}{\hbar^2}(E - V)\psi = L^2\psi \tag{16.1-3}$$

in which

$$L^2 = -\left[\frac{1}{\sin\theta}\frac{\partial}{\partial\theta}\left(\sin\theta\frac{\partial}{\partial\theta}\right) + \frac{1}{\sin^2\theta}\frac{\partial^2}{\partial\varphi^2}\right]. \tag{16.1-4}$$

Assuming $V = V(r)$ and

$$\psi(r, \theta, \varphi) = \frac{1}{r}P(r)Y(\theta, \varphi) = R(r)Y(\theta, \varphi), \tag{16.1-5}$$

Eq. (16.1-3) separates into

$$\frac{d^2P(r)}{dr^2} + \frac{2m}{\hbar^2}[E - V(r)]P(r) = \frac{\lambda}{r^2}P(r), \tag{16.1-6}$$

$$L^2 Y(\theta, \varphi) = \lambda Y(\theta, \varphi), \tag{16.1-7}$$

with λ as a separation constant. Further progress requires that boundary conditions be specified. On physical grounds it is assumed that ψ and its first derivatives are everywhere continuous, single-valued, and finite. The consequences of imposing these conditions are that

$$\lambda = l(l+1) \qquad (l = 0, 1, 2, \ldots) \tag{16.1-8}$$

and that the functions $Y(\theta, \varphi)$ are the spherical harmonics $Y_{lm}(\theta, \varphi)$ with‡

$$m = l, l-1, \ldots, -l. \tag{16.1-9}$$

Substituting (16.1-8) into (16.1-6),

$$\left\{\frac{d^2}{dr^2} + \frac{2m}{\hbar^2}[E - V(r)] - \frac{l(l+1)}{r^2}\right\}P(r) = 0 \tag{16.1-10}$$

on which we impose the requirement

$$V(r) \to 0 \quad \text{as} \quad r \to \infty. \tag{16.1-11}$$

Solutions to (16.1-10) may be obtained by first considering the behavior at large r. In the asymptotic region $r \to \infty$,

$$\frac{d^2P(r)}{dr^2} + \frac{2m}{\hbar^2}EP(r) = 0 \tag{16.1-12}$$

has solutions

$$P(r) = e^{\pm ar} \tag{16.1-13}$$

where

$$a = \sqrt{-(2m/\hbar^2)E}. \tag{16.1-14}$$

If $E < 0, e^{ar} \to \infty$ as $r \to \infty$. Since this violates the conditions that the wave function must be finite everywhere, it is not an acceptable solution. On the other hand $e^{-ar} \to 0$ as $r \to \infty$; it is therefore a possible solution. If $E > 0$,

‡ Note that m appears both as a symbol to represent the mass of the electron and as a label for the spherical harmonics. Such confusion and others like it are unavoidable if we are not to depart too drastically from conventional notation.

either sign in the exponent will satisfy the boundary conditions. We concentrate on the case $E < 0$, that is, the bound states of the atom. The asymptotic behavior suggests that solutions to (16.1-10) be sought in the form

$$P(r) = e^{-ar} f(r) \tag{16.1-15}$$

where $f(r)$ is a function to be determined by the radial equation (16.1-10) and the boundary conditions.

To proceed further it is necessary to specify the form of the potential $V(r)$. Let us now assume that the physical system consists of an electron of mass m‡ interacting with a nucleus of charge Ze via the Coulomb interaction

$$V = -Ze^2/r. \tag{16.1-16}$$

With $Z = 1$ we have hydrogen, while hydrogen-like atoms or ions are represented by assigning an appropriate value to Z. The radial equation (16.1-10) now becomes

$$\frac{d^2 P(r)}{dr^2} + \frac{2m}{\hbar^2}\left[E + \frac{Ze^2}{r} - \frac{l(l+1)}{r^2}\right] P(r) = 0 \tag{16.1-17a}$$

or, in Rydberg units (Table 16.1),

$$\frac{d^2 P(r)}{dr^2} + \left[E + \frac{2Z}{r} - \frac{l(l+1)}{r^2}\right] P(r) = 0. \tag{16.1-17b}$$

The substitution of (16.1-15) into (16.1-17b) yields

$$\frac{d^2 f}{dr^2} - 2a\frac{df}{dr} + \left[\frac{2Z}{r} - \frac{l(l+1)}{r^2}\right] f = 0 \tag{16.1-18}$$

whose solution may be expressed as a power series

$$f = r^s[A_0 + A_1 r + A_2 r^2 + \cdots]. \tag{16.1-19}$$

To ensure that f remains finite as $r \to 0$ it is necessary for s to be positive. When (16.1-19) is substituted into (16.1-18), it is found that

$$s = l + 1 > 0 \tag{16.1-20}$$

and that the coefficients obey the recursion relation

$$A_k/A_{k-1} = 2[(l+k)a - Z]/[k^2 + (2l+1)k]. \tag{16.1-21}$$

As $k \to \infty$,

$$A_k/A_{k-1} = 2a/k. \tag{16.1-22}$$

‡ To take into account the finite mass of the nucleus, m would be replaced by the reduced mass $\mu = mM/(m + M)$ where M is the mass of the nucleus (see also Section 23.8).

TABLE 16.1

Relations among Atomic Units[a]

	Rydberg	Hartree
Length r	$r_R = \dfrac{r}{a_0}$	$r_H = \dfrac{r}{a_0}$
Energy E	$E_R = \dfrac{E}{R_\infty}$	$E_H = \dfrac{E}{2R_\infty}$
∇^2	$\nabla_R{}^2 = a_0{}^2\nabla^2$	$\nabla_H{}^2 = a_0{}^2\nabla^2$
$\left(-\dfrac{\hbar^2}{2m}\nabla^2 - \dfrac{Ze^2}{r}\right)\psi = E\psi$	$\left(-\nabla_R{}^2 - \dfrac{2Z}{r_R}\right)\psi = E_R\psi$	$\left(-\dfrac{1}{2}\nabla_H{}^2 - \dfrac{Z}{r_H}\right)\psi = E_H\psi$

[a] a_0 = Bohr radius $= \hbar^2/me^2 = 0.5292$ Å.
R_∞ = Rydberg constant $= \frac{1}{2}mc^2(e^2/\hbar c)^2 = 13.605$ eV $= 109{,}737.311$ cm^{-1}.

We note that

$$e^{2ar} = 1 + 2ar + \frac{(2ar)^2}{2!} + \cdots \tag{16.1-23}$$

and

$$\frac{(2a)^k/k!}{(2a)^{k-1}/(k-1)!} = \frac{2a}{k} \tag{16.1-24}$$

which indicates that f behaves like e^{2ar} for large k. Therefore

$$P(r) = e^{-ar}f \approx e^{ar} \tag{16.1-25}$$

for large k. Since this would mean that $P(r) \to \infty$ as $r \to \infty$, it is necessary that the series (16.1-19) be terminated at some finite value of k. We therefore set

$$(l + k)a - Z = (l + k)\sqrt{-E} - Z = 0 \tag{16.1-26}$$

or

$$E = -Z^2/(l + k)^2 \equiv -Z^2/n^2 \quad \text{(Ry)}. \tag{16.1-27}$$

It is seen that the boundary conditions have forced the bound states to be discrete.

The series solution for $P(r)$ may be expressed in terms of Laguerre polynomials (Appendix 6)

$$L_{n+l}^{2l+1}(x) = B_0 + B_1x + B_2x^2 + \cdots + B_{n-l-1}x^{n-l-1} \tag{16.1-28}$$

in which

$$B_{n-l-1} = (-1)^{n-l}[(n+l)!/(n-1-1)!]. \tag{16.1-29}$$

In order that B_{n-l-1} remain finite, $n - l - 1$ must be zero or a positive integer. Hence

$$n = l+1, l+2, \ldots . \tag{16.1-30}$$

The complete solution for the bound states of hydrogen may now be written as

$$|nlm\rangle \equiv \psi_{nlm}(r,\theta,\varphi) = \frac{1}{r} P_{nl}(r) Y_{lm}(\theta,\varphi) = R_{nl}(r) Y_{lm}(\theta,\varphi) \tag{16.1-31}$$

TABLE 16.2

Radial Wave Functions in Hydrogen

n	l	$P_{nl}(r)$
1	0	$\left(\frac{Z}{a_0}\right)^{3/2} 2re^{-Zr/a_0}$
2	0	$\left(\frac{Z}{a_0}\right)^{3/2} \frac{1}{\sqrt{2}} r\left(1 - \frac{Zr}{2a_0}\right) e^{-Zr/2a_0}$
2	1	$\left(\frac{Z}{a_0}\right)^{3/2} \frac{1}{2\sqrt{6}} \frac{Zr^2}{a_0} e^{-Zr/2a_0}$
3	0	$\left(\frac{Z}{a_0}\right)^{3/2} \frac{2}{3\sqrt{3}} r\left[1 - \frac{2Zr}{3a_0} + \frac{2}{27}\left(\frac{Zr}{a_0}\right)^2\right] e^{-Zr/3a_0}$
3	1	$\left(\frac{Z}{a_0}\right)^{3/2} \frac{8}{27\sqrt{6}} \frac{Zr^2}{a_0} \left(1 - \frac{Zr}{6a_0}\right) e^{-Zr/3a_0}$
3	2	$\left(\frac{Z}{a_0}\right)^{3/2} \frac{4}{81\sqrt{30}} \frac{Z^2r^3}{a_0^{\,2}} e^{-Zr/3a_0}$
4	0	$\left(\frac{Z}{a_0}\right)^{3/2} \frac{r}{4}\left[1 - \frac{3Zr}{4a_0} + \frac{1}{8}\left(\frac{Zr}{a_0}\right)^2 - \frac{1}{192}\left(\frac{Zr}{a_0}\right)^3\right] e^{-Zr/4a_0}$
4	1	$\left(\frac{Z}{a_0}\right)^{3/2} \frac{1}{16}\sqrt{\frac{5}{3}} \frac{Zr^2}{a_0}\left[1 - \frac{Zr}{4a_0} + \frac{1}{80}\left(\frac{Zr}{a_0}\right)^2\right] e^{-Zr/4a_0}$
4	2	$\left(\frac{Z}{a_0}\right)^{3/2} \frac{1}{64\sqrt{5}} \frac{Z^2r^3}{a_0^{\,2}} \left(1 - \frac{Zr}{12a_0}\right) e^{-Zr/4a_0}$
4	3	$\left(\frac{Z}{a_0}\right)^{3/2} \frac{1}{768\sqrt{35}} \frac{Z^3r^4}{a_0^{\,3}} e^{-Zr/4a_0}$

with

$$P_{nl}(r) = \sqrt{\frac{(n-l-1)!Z}{n^2[(n+l)!]^3 a_0}} \left(\frac{2Zr}{na_0}\right)^{l+1} e^{-Zr/na_0} L_{n+l}^{2l+1}\left(\frac{2Zr}{na_0}\right) \quad (16.1\text{-}32)$$

and

$$\int_0^\infty P_{nl}(r) P_{n'l}\, dr = \delta_{nn'}. \quad (16.1\text{-}33)$$

Explicit forms of $P_{nl}(r)$ for several values of n and l are given in Table 16.2. In view of the orthogonality properties of $P_{nl}(r)$ and $Y_{lm}(\theta, \varphi)$ the eigenfunctions (16.1-31) must also obey orthogonality relations

$$\begin{aligned} \langle nlm|n'l'm'\rangle &= \int \psi^*_{nlm}(r, \theta, \varphi)\psi_{n'l'm'}(r, \theta, \varphi)\, d\tau \\ &= \delta_{nn'}\, \delta_{ll'}\, \delta_{mm'}. \end{aligned} \quad (16.1\text{-}34)$$

To summarize, we have three quantum numbers n, l, and m, where n is the principal quantum number with possible values $1, 2, 3, \ldots$; l the orbital angular momentum quantum number (orbital quantum number, for short) with possible values $0, 1, 2, \ldots, n-1$; and, m the magnetic (or projection) quantum number whose values are restricted to $l, l-1, \ldots, -l$. In spectroscopic notation, s, p, d, f, g, ... correspond to $l = 0, 1, 2, 3, 4, \ldots$, respectively.

The bound states of hydrogen may now be described. Table 16.3 lists the quantum numbers, spectroscopic notation, wave functions, energy, and the degeneracy (not including spin). The latter gives the number of eigenfunctions associated with a particular energy and is equal to n^2. This is somewhat unexpected because the Hamiltonian in (16.1-2) with $V = V(r)$ is invariant under all three-dimensional rotations. Based on the discussion in Section 8.1 we expect the eigenfunctions of the Hamiltonian to be of the form $R(r)Y_{lm}(\theta, \varphi)$ as indeed was shown by the detailed calculation culminating in (16.1-31). In that case the degeneracy ought to be $2l + 1$ with $l = 0, 1, 2, \ldots$. The fact that in hydrogen the degeneracy is actually n^2 is due to the special property of the Coulomb field. It has been shown (Fock, 1935) that the Schrödinger equation for the motion of an electron in a Coulomb field, is invariant under transformations of the four-dimensional rotation group of which $O^+(3)$ is a subgroup. The classification of states with respect to the irreducible representations of the four-dimensional group then leads to the n^2 degeneracy.‡

The eigenfunctions in Table 16.3 are written with the phase factors all equal to $+1$ despite the fact that the phase factors depend on the quantum numbers in accordance with the definitions of $P_{nl}(r)$ [Eq. (16.1-32)] and

‡ The n^2 degeneracy in hydrogen is often said to be accidental.

TABLE 16.3

Hydrogen Atom Wave Functions[a]

n	l	m	Notation	Wave function	Energy (Ry)	Degeneracy (n^2)
1	0	0	1s	$\left(\frac{Z}{a_0}\right)^{3/2}\frac{1}{\sqrt{\pi}}e^{-Zr/a_0}$	$-Z^2$	1
2	0	0	2s	$\left(\frac{Z}{a_0}\right)^{3/2}\frac{1}{\sqrt{32\pi}}\left(2-\frac{Zr}{a_0}\right)e^{-Zr/2a_0}$	$-\frac{Z^2}{4}$	4
	1	0	$2p_0$	$\left(\frac{Z}{a_0}\right)^{3/2}\frac{1}{\sqrt{32\pi}}\frac{Zr}{a_0}e^{-Zr/2a_0}\cos\theta$		
	1	± 1	$2p_{\pm 1}$	$\left(\frac{Z}{a_0}\right)^{3/2}\frac{1}{\sqrt{64\pi}}\frac{Zr}{a_0}e^{-Zr/2a_0}\sin\theta\, e^{\pm i\varphi}$		
3	0	0	3s	$\left(\frac{Z}{a_0}\right)^{3/2}\frac{1}{81\sqrt{3\pi}}\left[27-18\frac{Zr}{a_0}+2\left(\frac{Zr}{a_0}\right)^2\right]e^{-Zr/3a_0}$	$-\frac{Z^2}{9}$	9
	1	0	$3p_0$	$\left(\frac{Z}{a_0}\right)^{3/2}\frac{1}{81}\sqrt{\frac{2}{\pi}}\left[6\frac{Zr}{a_0}-\left(\frac{Zr}{a_0}\right)^2\right]e^{-Zr/3a_0}\cos\theta$		
	1	± 1	$3p_{\pm 1}$	$\left(\frac{Z}{a_0}\right)^{3/2}\frac{1}{81\sqrt{\pi}}\left[6\frac{Zr}{a_0}-\left(\frac{Zr}{a_0}\right)^2\right]e^{-Zr/3a_0}\sin\theta\, e^{\pm i\varphi}$		
	2	0	$3d_0$	$\left(\frac{Z}{a_0}\right)^{3/2}\frac{1}{81\sqrt{6\pi}}\left(\frac{Zr}{a_0}\right)^2 e^{-Zr/3a_0}(3\cos^2\theta-1)$		
	2	± 1	$3d_{\pm 1}$	$\left(\frac{Z}{a_0}\right)^{3/2}\frac{1}{81\sqrt{\pi}}\left(\frac{Zr}{a_0}\right)^2 e^{-Zr/3a_0}\sin\theta\cos\theta\, e^{\pm i\varphi}$		
	2	± 2	$3d_{\pm 2}$	$\left(\frac{Z}{a_0}\right)\frac{1}{162\sqrt{\pi}}\left(\frac{Zr}{a_0}\right)^2 e^{-Zr/3a_0}\sin^2\theta\, e^{\pm 2i\varphi}$		

4	0	0	4s	$\left(\frac{Z}{a_0}\right)^{3/2}\frac{1}{1536\sqrt{\pi}}\left[192-144\frac{Zr}{a_0}+24\left(\frac{Zr}{a_0}\right)^2-\left(\frac{Zr}{a_0}\right)^3\right]e^{-Zr/4a_0}$	$-\frac{Z^2}{16}$	16
	1	0	$4p_0$	$\left(\frac{Z}{a_0}\right)^{3/2}\frac{1}{2560}\sqrt{\frac{5}{\pi}}\left[80\frac{Zr}{a_0}-20\left(\frac{Zr}{a_0}\right)^2+\left(\frac{Zr}{a_0}\right)^3\right]e^{-Zr/4a_0}\cos\theta$		
	1	±1	$4p_{\pm1}$	$\left(\frac{Z}{a_0}\right)^{3/2}\frac{1}{2560}\sqrt{\frac{5}{2\pi}}\left[80\frac{Zr}{a_0}-20\left(\frac{Zr}{a_0}\right)^2+\left(\frac{Zr}{a_0}\right)^3\right]e^{-Zr/4a_0}\sin\theta\, e^{\pm i\varphi}$		
	2	0	$4d_0$	$\left(\frac{Z}{a_0}\right)^{3/2}\frac{1}{3062\sqrt{\pi}}\left[12\left(\frac{Zr}{a_0}\right)^2-\left(\frac{Zr}{a_0}\right)^3\right]e^{-Zr/4a_0}(3\cos^2\theta-1)$		
	2	±1	$4d_{\pm1}$	$\left(\frac{Z}{a_0}\right)^{3/2}\frac{1}{1536}\sqrt{\frac{3}{2\pi}}\left[12\left(\frac{Zr}{a_0}\right)^2-\left(\frac{Zr}{a_0}\right)^3\right]e^{-Zr/4a_0}\sin\theta\cos\theta\, e^{\pm i\varphi}$		
	2	±2	$4d_{\pm2}$	$\left(\frac{Z}{a_0}\right)^{3/2}\frac{1}{3072}\sqrt{\frac{3}{2\pi}}\left[12\left(\frac{Zr}{a_0}\right)^2-\left(\frac{Zr}{a_0}\right)^3\right]e^{-Zr/4a_0}\sin^2\theta\, e^{\pm 2i\varphi}$		
	3	0	$4f_0$	$\left(\frac{Z}{a_0}\right)^{3/2}\frac{1}{3072\sqrt{5\pi}}\left(\frac{Zr}{a_0}\right)^3 e^{-Zr/4a_0}(5\cos^3\theta-3\cos\theta)$		
	3	±1	$4f_{\pm1}$	$\left(\frac{Z}{a_0}\right)^{3/2}\frac{1}{6144}\sqrt{\frac{3}{5\pi}}\left(\frac{Zr}{a_0}\right)^3 e^{-Zr/4a_0}\sin\theta(5\cos^2\theta-1)e^{\pm i\varphi}$		
	3	±2	$4f_{\pm2}$	$\left(\frac{Z}{a_0}\right)^{3/2}\frac{1}{3072}\sqrt{\frac{3}{2\pi}}\left(\frac{Zr}{a_0}\right)^3 e^{-Zr/4a_0}\sin^2\theta\cos\theta\, e^{\pm 2i\varphi}$		
	3	±3	$4f_{\pm3}$	$\left(\frac{Z}{a_0}\right)^{3/2}\frac{1}{6144\sqrt{\pi}}\left(\frac{Zr}{a_0}\right)^3 e^{-Zr/4a_0}\sin^3\theta\, e^{\pm 3i\varphi}$		

[a] All the wave functions are written with a positive phase factor.

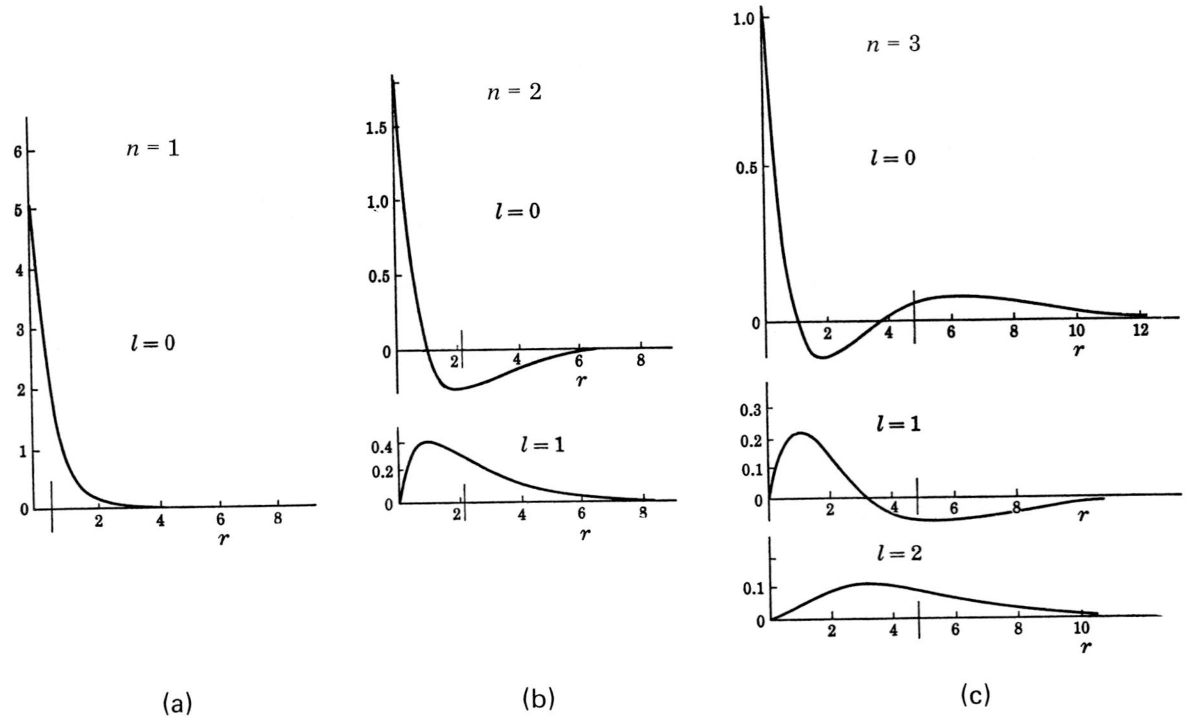

FIG. 16.1 Relative magnitude of the radial part $R_{nl}(r)$ of the hydrogen wave function $\psi_{nlm}(r, \theta, \varphi) = R_{nl}(r) Y_{lm}(\theta, \varphi)$ as a function of r (in angstroms). The vertical line on the horizontal axis is the radius of the corresponding Bohr orbit. (a) $n = 1$; (b) $n = 2$; (c) $n = 3$. (From Gerhard Herzberg "Atomic Spectra and Atomic Structure," Dover Publications, Inc., New York, 1944. Adapted through the permission of the publisher.)

$Y_{lm}(\theta, \varphi)$ [Eq. (1.2-1)]. Plots of $R_{nl}(r)$ $[=(1/r)P_{nl}(r)]$ are shown in Fig. 16.1. Note that the higher the value of n and therefore the energy, the greater is the number of nodes in $R_{nl}(r)$ for a given value of l. Energies are plotted in Fig. 16.2.

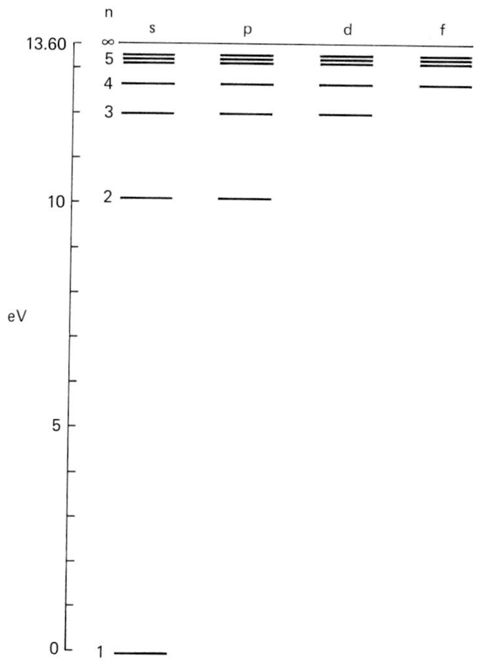

FIG. 16.2 Energy levels of the hydrogen atom, without fine structure.

The probability of finding an electron in an element of volume $d\tau$ is

$$\psi^*\psi \, d\tau = \frac{1}{r^2} P_{nl}^2(r) Y_{lm}^*(\theta, \varphi) Y_{lm}(\theta, \varphi) \, d\tau. \qquad (16.1\text{-}35)$$

If (16.1-35) is integrated over the surface of a sphere, we obtain the probability of finding an electron in a shell between two spheres of radii r and $r + dr$. Since the spherical harmonics are normalized to unity, the result is simply $P_{nl}^2(r)\,dr$. In the sense that $\psi^*\psi$ is a charge density, $P_{nl}^2(r)$ is a radial charge density; a plot is shown in Fig. 16.3. The probability of finding an electron between θ and $\theta + d\theta$ is proportional to

$$Y_{lm}^*(\theta, \varphi) Y_{lm}(\theta, \varphi) \sin\theta \, d\theta = [P_l^{|m|}(\cos\theta)]^2 \sin\theta \, d\theta. \qquad (16.1\text{-}36)$$

$[P_l^{|m|}(\cos\theta)]^2$ is shown in Fig. 16.4. Finally, the probability of finding an electron between φ and $\varphi + d\varphi$ is simply proportional to $d\varphi$.

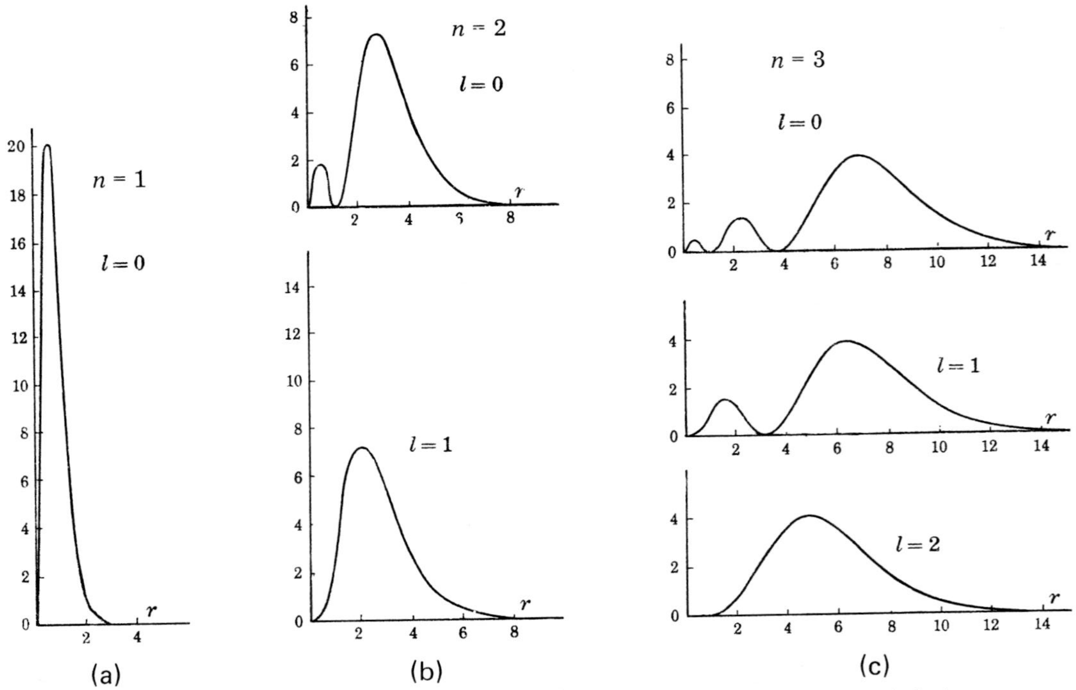

FIG. 16.3 Radial charge density $[P_{nl}(r)]^2$ for the hydrogen wave function $\psi_{nlm}(r, \theta, \varphi) = (1/r)\ P_{nl}(r) Y_{lm}(\theta, \varphi)$. (a) $n = 1$; (b) $n = 2$; (c) $n = 3$. (From Gerhard Herzberg, "Atomic Spectra and Atomic Structure," Dover Publications, Inc., New York, 1944. Adapted through the permission of the publisher.)

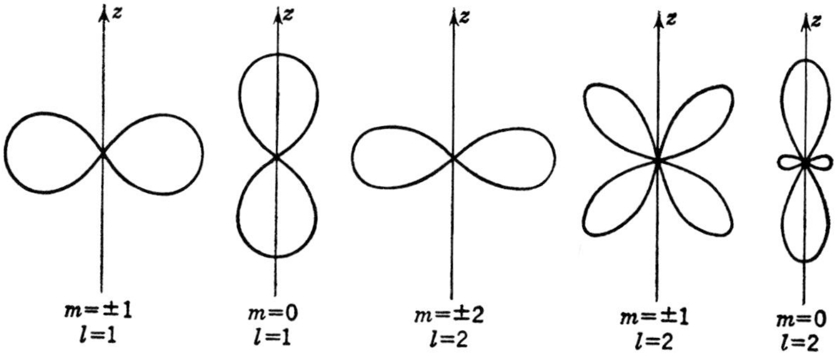

FIG. 16.4 Polar plots of $[P_l^{|m|}(\cos\theta)]^2$. (Slater, 1960)

Average values of various powers of r are often needed in a computation; these are defined by

$$\langle r^k \rangle = \int_0^\infty r^k P_{nl}^2(r)\, dr. \tag{16.1-37}$$

The results for several values of k are given in Table 16.4.

TABLE 16.4

Expectation Values of r^k [a]

k	$\langle r^k \rangle$
1	$\frac{a_0}{2Z}[3n^2 - l(l+1)]$
2	$\frac{a_0^2}{Z^2}\frac{n^2}{2}[5n^2 + 1 - 3l(l+1)]$
3	$\frac{a_0^3}{Z^3}\frac{n^2}{8}[35n^2(n^2-1) - 30n^2(l+2)(l-1) + 3(l+2)(l+1)l(l-1)]$
4	$\frac{a_0^4}{Z^4}\frac{n^4}{8}[63n^4 - 35n^2(2l^2+2l-3) + 5l(l+1)(3l^2+3l-10) + 12]$
−1	$\frac{Z}{a_0}\frac{1}{n^2}$
−2	$\frac{Z^2}{a_0^2}\frac{1}{n^3(l+\frac{1}{2})}$
−3	$\frac{Z^3}{a_0^3}\frac{1}{n^3(l+1)(l+\frac{1}{2})l}$
−4	$\frac{Z^4}{a_0^4}\frac{3n^2 - l(l+1)}{2n^5(l+\frac{3}{2})(l+1)(l+\frac{1}{2})l(l-\frac{1}{2})}$

[a] $\langle r^k \rangle = \int_0^\infty R_{nl}^2(r) r^{k+2}\, dr = \int_0^\infty P_{nl}^2(r) r^k\, dr.$

The states corresponding to $E > 0$, in contrast to those with $E < 0$, are not quantized; instead they form a continuum. This is evident from (16.1-13) which shows that, asymptotically, $P(r)$ has a sinusoidal behavior; it is therefore not necessary to terminate the series in (16.1-19). A more detailed development of the continuum states (Bethe and Salpeter, 1957) indicates that the radial solution is expressible in terms of the regular Coulomb function $F_l(\eta, kr)$ whose properties and numerical values are given by Abramowitz and Stegun (1965). It should be emphasized that, although the bound states of the hydrogen atom (16.1-31) form an infinite, discrete set, the set is not complete. If we wish to expand a function in terms of hydrogen atom solutions, it is necessary to include the continuum states as well as the bound states.

Previously it was shown (7.4-12) that $[\mathbf{L}, \nabla^2] = 0$. It then follows that the Hamiltonian

$$\mathscr{H} = -(\hbar^2/2m)\nabla^2 + V(r) \tag{16.1-38}$$

also commutes with $\mathbf{L}$ and L^2. However, the components of $\mathbf{L}$ do not commute with one another. We are therefore restricted to one component of $\mathbf{L}$ which will commute both with L^2 and $\mathscr{H}$. The usual choice is to pick L_z. Thus

$$[L_z, \mathscr{H}] = 0. \tag{16.1-39}$$

Since $\mathscr{H}$ does not contain any spin operators,

$$[S^2, \mathscr{H}] = [S_z, \mathscr{H}] = 0. \tag{16.1-40}$$

16.2 One-Electron Wave Functions

For the Hamiltonian

$$\mathscr{H} = \frac{p^2}{2m} - \frac{Ze^2}{r} \tag{16.2-1}$$

the complete solution to the Schrödinger equation $\mathscr{H}\psi = E\psi$ for the bound states consists of the orbitals $\psi_{nlm}(r, \theta, \varphi)$ [Eq. (16.1-31)] together with the energy eigenvalues E_n [Eq. (16.1-27)]. However, the Schrödinger equation (15.2-32) contains additional terms which will have an effect on the eigenfunctions and eigenvalues of the system. To deal with these effects we must inquire, in the first place, whether the $\psi_{nlm}(r, \theta, \varphi)$ are suitable as a starting point for a perturbation treatment, i.e., whether the $\psi_{nlm}(r, \theta, \varphi)$ can serve as a zero-order basis set.

Examination of (15.2-32) reveals the presence of terms containing the operator $\boldsymbol{\sigma}$ whose components are the Pauli matrices. According to the discussion in Section 1.4,

$$\boldsymbol{\sigma} = 2\mathbf{S}$$

where the rectangular components of **S** are angular momentum operators and

$$S_z\alpha = \tfrac{1}{2}\alpha, \qquad S_z\beta = -\tfrac{1}{2}\beta \tag{16.2-2a}$$

$$S^2\alpha = \tfrac{1}{2}(\tfrac{1}{2}+1)\alpha = \tfrac{3}{4}\alpha, \qquad S^2\beta = \tfrac{1}{2}(\tfrac{1}{2}+1)\beta = \tfrac{3}{4}\beta. \tag{16.2-2b}$$

Equations (16.2-2) identify the system as a particle with spin $s = \frac{1}{2}$. Thus the appearance of $\boldsymbol{\sigma}$ in the Hamiltonian indicates that the wave function of the system must include a spin eigenfunction. We adopt the notation:

$$\varphi(\mathbf{r}) \equiv \varphi_{nlm_l}(r, \theta, \varphi)$$

is a one-electron spatial orbital characterized by the quantum numbers n, l, m_l. This is identical to the function which had previously been written as $\psi_{nlm}(r, \theta, \varphi)$ as in (16.1-31).

$$\xi_{m_s} \equiv \xi(m_s) = \begin{cases} \alpha & \text{for } m_s = \frac{1}{2} \\ \beta & \text{for } m_s = -\frac{1}{2} \end{cases}$$

$$= \text{one-electron spin function,} \tag{16.2-3}$$

$$\psi(\lambda) = \psi(r, \theta, \varphi, m_s) = \varphi(\mathbf{r})\xi(m_s)$$

$$= \text{(one-electron) spin orbital.} \tag{16.2-4}$$

It shall be understood that an integral involving spin orbitals implies a spatial integration as well as a summation over spin coordinates.

In many instances the radial part of the wave function remains fixed so that the principal quantum number n may be suppressed without loss of information. Also, the Dirac notation is often the most convenient one. Thus $|lsm_lm_s\rangle$ identifies the angular and spin parts of the wave function. The subscript in m_l has been added to emphasize that m_l is the projection quantum number associated with l in distinction to m_s which is the projection quantum number associated with s. For one-electron wave functions it is customary to use lower case letters to denote the quantum numbers, although we continue to write the operators in upper case letters (for the present).

The spin orbitals (16.2-4) are, of course, eigenfunctions of $\mathscr{H}$ given by (16.2-1) and as long as we do not add any more terms to $\mathscr{H}$, the energy eigenvalues are the same as those associated with $\varphi_{nlm}(r, \theta, \varphi)$. The only difference is that we have an additional twofold spin degeneracy. As was shown in the previous section, the degeneracy of $\varphi_{nlm}(r, \theta, \varphi)$ is n^2. We now have a degeneracy of $2n^2$ associated with $\psi(\lambda) = \psi(r, \theta, \varphi, m_s)$. In summary, $|lsm_lm_s\rangle$ satisfies the relations:

$$\begin{aligned} L^2|lsm_lm_s\rangle &= l(l+1)|lsm_lm_s\rangle, \\ S^2|lsm_lm_s\rangle &= s(s+1)|lsm_lm_s\rangle = \tfrac{3}{4}|lsm_lm_s\rangle, \\ L_z|lsm_lm_s\rangle &= m_l|lsm_lm_s\rangle, \\ S_z|lsm_lm_s\rangle &= m_s|l_sm_lm_s\rangle = \pm\tfrac{1}{2}|lsm_lm_s\rangle. \end{aligned} \tag{16.2-5}$$

Degenerate eigenfunctions may be combined linearly to form other sets of degenerate eigenfunctions. Consider, for example, the linear combination

$$|lsjm\rangle = \sum_{m_lm_s} |lsm_lm_s\rangle\langle lsm_lm_s|lsjm\rangle \tag{16.2-6}$$

where $\langle lsm_lm_s|lsjm\rangle$ is a coupling coefficient of the kind discussed in Section 1.5. As shown there, the $|lsjm\rangle$ have the properties:

$$\begin{aligned} L^2|lsjm\rangle &= l(l+1)|lsjm\rangle, \\ S^2|lsjm\rangle &= s(s+1)|lsjm\rangle = \tfrac{3}{4}|lsjm\rangle, \\ J^2|lsjm\rangle &= j(j+1)|lsjm\rangle, \\ J_z|lsjm\rangle &= m|lsjm\rangle, \end{aligned} \tag{16.2-7}$$

in which

$$J^2 = J_x{}^2 + J_y{}^2 + J_z{}^2, \tag{16.2-8}$$

$$\mathbf{J} = \mathbf{L} + \mathbf{S}, \tag{16.2-9}$$

$$j = l+s, l+s-1, \ldots, |l-s| = l+\tfrac{1}{2}, l-\tfrac{1}{2}, \qquad m = m_l + m_s. \tag{16.2-10}$$

It should further be noted that the $|lsjm\rangle$ are not eigenfunctions of L_z or S_z. The eigenfunctions $|lsm_lm_s\rangle$ are said to be in the *uncoupled representation* while the $|lsjm\rangle$, in view of (16.2-9), are said to be in the *coupled representation.*

We present a summary of notations including those commonly employed in spectroscopy:

$$\begin{aligned} \alpha &= \tbinom{1}{0} = |\tfrac{1}{2}\tfrac{1}{2}\rangle = \xi(\tfrac{1}{2}) = \text{spin function with } s = \tfrac{1}{2}, \quad m_s = \tfrac{1}{2}, \\ \beta &= \tbinom{0}{1} = |\tfrac{1}{2}\ -\tfrac{1}{2}\rangle = \xi(-\tfrac{1}{2}) = \text{spin function with } s = \tfrac{1}{2}, \quad m_s = -\tfrac{1}{2}. \end{aligned}$$

In $|lsm_lm_s\rangle$ (uncoupled representation), s, p, d, f, ... stand for $l = 0, 1, 2, 3, \ldots$, with m_l written as a subscript. Example: $|1\ \tfrac{1}{2} - 1\ \tfrac{1}{2}\rangle = |\mathrm{p}_{-1}\alpha\rangle$ (with or without the ket). If it is necessary to specify the principal quantum number, we write $|nlsm_lm_s\rangle$. Thus, $|21\ \tfrac{1}{2} - 1\ \tfrac{1}{2}\rangle = 2\mathrm{p}_{-1}\alpha$. In $|lsjm\rangle$ (coupled representation—also called *terms*), S, P, D, F, ... stand for $l = 0, 1, 2, 3, \ldots$. The spin s is indicated by writing $2s + 1$ as a superscript to the left. Thus $^2\mathrm{D}$ means $l = 2$, $2s + 1 = 2$ or $s = \tfrac{1}{2}$, j is written as a subscript and m is indicated separately. Example: $|2\,\frac{1}{2}\,\frac{3}{2}\,-\frac{1}{2}\rangle = |^2\mathrm{D}_{3/2}\,-\frac{1}{2}\rangle$ (with or without the ket). It should be emphasized that the choice of representation is basically a matter of convenience in carrying out the computation and the same physics must emerge from any representation. Nevertheless, depending on the specific interactions contained in the Hamiltonian, it is often possible to choose a representation in which physical results are obtained in the most efficient manner.

It is also instructive to consider the two representations from the standpoint of group theory. The eigenfunctions $\varphi_{nlm}(r, \theta, \varphi)$ are basis functions for the irreducible representation $D^{(l)}$ of the three-dimensional rotation group $O^+(3)$. The two spin functions α, β are basis functions for the irreducible representation $D^{(1/2)}$ of $O^+(3)$. When we construct $|lsm_l m_s\rangle$, which are products of the angular part of $\varphi_{nlm}(r, \theta, \varphi)$ and the spin functions, we obtain basis functions for the product representation $D^{(l)} \times D^{(1/2)}$. But the product representation is reducible:

$$D^{(l)} \times D^{(1/2)} = D^{(l+1/2)} + D^{(l-1/2)} = \sum_{l-1/2}^{l+1/2} D^{(j)}. \tag{16.2-11}$$

The functions $|lsjm\rangle$ in the coupled representation constructed according to (16.2-6) are basis functions for the irreducible representations $D^{(j)}$.

16.3 Spin-Orbit Coupling

We turn next to the spin-orbit coupling term in the Schrödinger equation (15.2-32). The Hamiltonian, denoted by $\mathscr{H}_{\text{SO}}$, is given by

$$\mathscr{H}_{\text{SO}} = -\frac{e\hbar}{4m^2c^2}\,\boldsymbol{\sigma} \cdot \nabla\varphi \times \mathbf{p} \tag{16.3-1}$$

where φ is the electrostatic potential. If φ depends only on r,

$$\nabla\varphi = \frac{d\varphi}{dr}\frac{\mathbf{r}}{r}$$

and

$$\nabla\varphi \times \mathbf{p} = \frac{d\varphi}{dr}\frac{1}{r}\,\mathbf{r} \times \mathbf{p} = \frac{d\varphi}{dr}\frac{\hbar}{r}\,\mathbf{L} \tag{16.3-2}$$

in which $\mathbf{L}$ is the orbital angular momentum operator with components as in (1.1-3). When (16.3-2) is substituted in (16.3-1) and $\boldsymbol{\sigma}$ is replaced by $2\mathbf{S}$, we obtain

$$\mathscr{H}_{\text{SO}} = -\frac{e\hbar^2}{2m^2c^2}\frac{1}{r}\frac{d\varphi}{dr}\,\mathbf{L} \cdot \mathbf{S} = \xi(r)\mathbf{L} \cdot \mathbf{S}. \tag{16.3-3}$$

For hydrogen

$$\varphi = Ze/r$$

so that

$$\xi(r) = Ze^2\hbar^2/2m^2c^2r^3. \tag{16.3-4}$$

The one-electron Hamiltonian with spin-orbit coupling now becomes

$$\mathscr{H} = \mathscr{H}_0 + \mathscr{H}_{SO} \tag{16.3-5}$$

with

$$\mathscr{H}_0 = -\frac{\hbar^2}{2m}\nabla^2 - \frac{Ze^2}{r}, \qquad \mathscr{H}_{SO} = \xi(r)\mathbf{L}\cdot\mathbf{S}.$$

We regard $\mathscr{H}_{SO}$ as a perturbation, although the justification for this may not be apparent until a calculation of the magnitude of the effect has been carried out. This will be done at a later stage. Since $\mathscr{H}_0$ has degenerate eigenvalues, the first-order corrections to the energy will be obtained from the solutions of the secular equation

$$|H_{ij}^{SO} - E_n^{(1)}\delta_{ij}| = 0 \tag{16.3-6}$$

where

$$H_{ij}^{SO} = \langle\psi_{ni}^{(0)}|\mathscr{H}_{SO}|\psi_{nj}^{(0)}\rangle \tag{16.3-7}$$

and $\psi_{ni}^{(0)}, \psi_{nj}^{(0)}$ are degenerate eigenfunctions of $\mathscr{H}_0$. Remembering that linear combinations of the $\psi_{ni}^{(0)}$ are also eigenfunctions of $\mathscr{H}_0$, there is a certain latitude in the construction of basis functions with which to calculate the matrix elements. Examination of the commutators provides useful guidance. Thus

$$[\mathscr{H}_0, L^2] = [\mathscr{H}_0, S^2] = [\mathscr{H}_0, J^2] = [\mathscr{H}_0, L_\mu] = [\mathscr{H}_0, S_\mu] = [\mathscr{H}_0, J_\mu] = 0 \tag{16.3-8}$$

in which L_μ, S_μ, and J_μ refer to any component—rectangular or spherical—of **L**, **S**, and **J**, respectively. Also

$$[\mathbf{L}\cdot\mathbf{S}, L^2] = [\mathbf{L}\cdot\mathbf{S}, S^2] = [\mathbf{L}\cdot\mathbf{S}, J^2] = [\mathbf{L}\cdot\mathbf{S}, J_z] = 0 \tag{16.3-9}$$

but

$$[\mathbf{L}\cdot\mathbf{S}, L_z] \neq 0, \qquad [\mathbf{L}\cdot\mathbf{S}, S_z] \neq 0. \tag{16.3-10}$$

On the basis of (16.3-8)–(16.3-10) we may draw the following conclusions: Both $|lsm_lm_s\rangle$ and $|lsjm\rangle$ are eigenfunctions of $\mathscr{H}_0$ and it is immaterial which representation is used. However, $\mathbf{L}\cdot\mathbf{S}$ does not commute with L_z and S_z; this means that $|lsm_lm_s\rangle$ will not necessarily be an eigenfunction of $\mathbf{L}\cdot\mathbf{S}$. On the other hand, because of (16.3-9), $|lsjm\rangle$ is a simultaneous eigenfunction of $L^2 \cdot S^2, J^2, J_z$, and $\mathbf{L} \cdot \mathbf{S}$. We therefore expect the coupled representation to be the most convenient since only diagonal matrix elements will appear in the secular determinant.

To see how this works out, let

$$\mathbf{J} = \mathbf{L} + \mathbf{S}, \qquad J^2 = L^2 + S^2 + 2\mathbf{L} \cdot \mathbf{S},$$

or

$$\mathbf{L} \cdot \mathbf{S} = \tfrac{1}{2}(J^2 - L^2 - S^2). \tag{16.3-11}$$

Using the coupled representation

$$\begin{aligned}\langle l'sj'm'|\mathbf{L} \cdot \mathbf{S}|lsjm\rangle &= \tfrac{1}{2}\langle l'sj'm'|J^2 - L^2 - S^2|lsjm\rangle \\ &= \tfrac{1}{2}[j(j+1) - l(l+1) - s(s+1)]\,\delta_{j'j}\delta_{l'l}\delta_{m'm}.\end{aligned} \tag{16.3-12}$$

The Hamiltonian $\mathscr{H}_{\mathrm{SO}}$ in (16.3-3) also contains the radial function $\xi(r)$. To obtain the energy corrections it is also necessary to evaluate the expectation value of $\xi(r)$:

$$\begin{aligned}\langle\xi(r)\rangle = \xi_{nl} = \langle nl|\xi(r)|nl\rangle &= \int_0^\infty \xi(r)R_{nl}^2(r)r^2\,dr \\ &= \int_0^\infty \xi(r)P_{nl}^2(r)\,dr.\end{aligned} \tag{16.3-13}$$

In hydrogen $\xi(r)$ is given by (16.3-4) in which case

$$\xi_{nl} = \frac{Ze^2\hbar^2}{2m^2c^2}\left\langle nl\left|\frac{1}{r^3}\right|nl\right\rangle = \frac{Ze^2\hbar^2}{2m^2c^2}\left\langle\frac{1}{r^3}\right\rangle. \tag{16.3-14}$$

From Table 16.4,

$$\left\langle\frac{1}{r^3}\right\rangle = \frac{Z^3}{a_0{}^3}\frac{1}{n^3(l+1)(l+\frac{1}{2})l}. \tag{16.3-15}$$

On combining (16.3-12), (16.3-14), and (16.3-15), the spin-orbit interaction energy is

$$E_{\mathrm{SO}} = \frac{Z^4e^2\hbar^2}{4a_0{}^3m^2c^2}\frac{j(j+1) - l(l+1) - s(s+1)}{n^3(l+1)(l+\frac{1}{2})l} \tag{16.3-16a}$$

or, in Rydberg units,

$$E_{\mathrm{SO}} = \frac{Z^4\alpha^2}{n^3}\frac{j(j+1) - l(l+1) - s(s+1)}{2l(l+1)(l+\frac{1}{2})}\ \mathrm{Ry} \tag{16.3-16b}$$

with

$$\alpha = e^2/\hbar c, \qquad a_0 = \hbar^2/me^2.$$

E_{SO} becomes indeterminate when $l = 0$ (and $j = s = \frac{1}{2}$). The difficulty stems from the apparent singularity of $\langle 1/r^3 \rangle$ for $l = 0$. However, expression (16.3-15) for $\langle 1/r^3 \rangle$ contains the implicit assumption of the nonrelativistic Hamiltonian in (15.2-32) that $e\varphi \ll mc^2$. But in the present case the potential energy $e\varphi$ is the Coulomb energy Ze^2/r which becomes infinite at the origin where the wave function, for $l = 0$, is finite. As pointed out in Section 15.2, a separate treatment of such cases is required. The mechanics of carrying out a computation with a potential that becomes singular at the origin will be demonstrated in connection with the magnetic hyperfine interaction (Section 18.1). Here we simply quote the result, as shown, for example, by Bethe and Salpeter (1957), that $\langle 1/r^3 \rangle$ is indeed finite for $l = 0$. Since the matrix element of $\mathbf{L} \cdot \mathbf{S}$ vanishes for $l = 0$ we have $E_{SO} = 0$ for s states.

It is possible to rewrite (16.3-16) entirely in terms of l since j is confined to the values $l \pm \frac{1}{2}$ and $s = \frac{1}{2}$. We then have ($l \neq 0$)

$$E_{SO} = E_{l \pm 1/2} = \frac{Z^4 e^2 \hbar^2}{2m^2 c^2 a_0{}^3} \frac{1}{n^3 (l+1)(l+\frac{1}{2})l} \begin{cases} 1/2 \\ -\frac{1}{2}(l+1) \end{cases}, \tag{16.3-17}$$

$$\Delta E_{SO} = 5.84 \frac{Z^4}{n^3 l(l+1)} \quad \text{cm}^{-1} \tag{16.3-18}$$

and in Rydberg units

$$E_{SO} = E_{l+1/2} = \frac{Z^4 \alpha^2}{n^3} \frac{1}{2l(l+1)(l+\frac{1}{2})} \begin{cases} l \\ -(l+1) \end{cases} \quad \text{Ry} \tag{16.3-19}$$

$$\Delta E_{SO} = \frac{Z^4 \alpha^2}{n^3} \frac{1}{l(l+1)} \quad \text{Ry}. \tag{16.3-20}$$

For $Z = 1$ and $l = 1$, the splitting due to spin-orbit coupling is 0.36, 0.12, and 0.044 cm^{-1} for $n = 2, 3$, and 4, respectively. Since these energies are much smaller than the energies which separate states with different values of the principal quantum number n—about $10^4\ \text{cm}^{-1}$—the use of perturbation theory to first order is certainly appropriate. Nevertheless, one should observe the Z^4-dependence in E_{SO} which causes spin-orbit interaction energies to increase rapidly with atomic number.

From (16.3-12) it is seen that the matrix element vanishes unless

$$\Delta j = \Delta l = 0; \qquad \Delta m = \Delta(m_l + m_s) = 0. \tag{16.3-21}$$

These are the selection rules for spin-orbit coupling. It is also instructive to inquire whether there are any restrictions on m_l and m_s individually. For this purpose we must return to the $|lsm_l m_s\rangle$ basis set and examine the matrix element of $\mathbf{L} \cdot \mathbf{S}$ in this basis set with respect to dependence on m_l and m_s.

If **L** and **S** are expressed in terms of their spherical components,

$$\mathbf{L}\cdot\mathbf{S} = -L_{+1}S_{-1} + L_0S_0 - L_{-1}S_{+1} \tag{16.3-22}$$

the matrix element

$$\langle lsm_l'm_s'|-L_{+1}S_{-1} + L_0S_0 - L_{-1}S_{+1}|lsm_lm_s\rangle$$

will vanish unless

$$\Delta m_l = 0, \quad \pm 1, \qquad \Delta m_s = 0, \quad \pm 1. \tag{16.3-23}$$

These selection rules are a direct consequence of the properties of the raising and lowering operators as expressed by (1.7-11) and (1.7-12). In view of (16.3-21) it is seen that if m_l increases by one unit, m_s must decrease by one unit and vice versa. Values of $\langle 1\frac{1}{2}m_lm_s|\mathbf{L}\cdot\mathbf{S}|1\frac{1}{2}m_lm_s\rangle$ are shown in Table 16.5.

TABLE 16.5

Matrix Elements of $\mathbf{L}\cdot\mathbf{S}$ *for* p *States in Units of* $\xi_{nl} = \langle nl|\xi(r)|nl\rangle$[a]

	$\left\|1\frac{1}{2}\right\rangle$	$\left\|1\ -\frac{1}{2}\right\rangle$	$\left\|0\frac{1}{2}\right\rangle$	$\left\|0\ -\frac{1}{2}\right\rangle$	$\left\|-1\frac{1}{2}\right\rangle$	$\left\|-1\ -\frac{1}{2}\right\rangle$
$\left\langle 1\frac{1}{2}\right\|$	$\frac{1}{2}$					
$\left\langle 1\ -\frac{1}{2}\right\|$		$-\frac{1}{2}$	$\frac{1}{\sqrt{2}}$			
$\left\langle 0\frac{1}{2}\right\|$		$\frac{1}{\sqrt{2}}$	0			
$\left\langle 0\ -\frac{1}{2}\right\|$				0	$\frac{1}{\sqrt{2}}$	
$\left\langle -1\frac{1}{2}\right\|$				$\frac{1}{\sqrt{2}}$	$-\frac{1}{2}$	
$\left\langle -1\ -\frac{1}{2}\right\|$						$\frac{1}{2}$

[a] The notation for the states has been shortened to $|m_lm_s\rangle$ since $l = 1$, $s = \frac{1}{2}$ for all states.

16.4 Other Interactions

The term in the Schrödinger equation (15.2-32) that corresponds to the relativistic correction to the kinetic energy is

$$\mathscr{H}_{\mathrm{r}} = -\frac{p^4}{8m^3c^2} = -\frac{1}{2mc^2}\left(\frac{p^2}{2m}\right)^2 \tag{16.4-1}$$

or, with

$$\mathscr{H}_0 = \frac{p^2}{2m} - \frac{Ze^2}{r}, \tag{16.4-2}$$

$\mathscr{H}_r$ can be written as

$$\mathscr{H}_r = -\frac{1}{2mc^2}\left[\mathscr{H}_0 + \frac{Ze^2}{r}\right]^2 \tag{16.4-3}$$

$$= -\frac{1}{2mc^2}\left[\mathscr{H}_0^2 + Ze^2\left(\mathscr{H}_0\frac{1}{r} + \frac{1}{r}\mathscr{H}_0\right) + \frac{(Ze^2)^2}{r^2}\right]. \tag{16.4-4}$$

We shall be interested in the effects produced by $\mathscr{H}_r$ within a manifold of states specified by particular values of n, l, s, j as for example the $^2S_{1/2}$ or $^2P_{3/2}$ eigenstates of $\mathscr{H}_0$ belonging to a particular value of n. Therefore, treating $\mathscr{H}_r$ as a perturbation, the basis set is $|nlsjm\rangle$ and the relevant matrix elements are

$$\langle nlsjm'|\mathscr{H}_r|nlsjm\rangle$$
$$= -\frac{1}{2mc^2}\left[\langle nlsjm'|\mathscr{H}_0^2|nlsjm\rangle + Ze^2\left(\left\langle nlsjm'\left|\mathscr{H}_0\frac{1}{r}\right|nlsjm\right\rangle\right.\right.$$
$$\left.\left. + \left\langle nlsjm'\left|\frac{1}{r}\mathscr{H}_0\right|nlsjm\right\rangle\right) + (Ze^2)^2\left\langle nlsjm'\left|\frac{1}{r^2}\right|nlsjm\right\rangle\right]. \tag{16.4-5}$$

Equation (16.4-5) may be simplified. For the first term on the right we have

$$\langle nlsjm'|\mathscr{H}_0^2|nlsjm\rangle = (E_n^{(0)})^2\,\delta_{m'm} \tag{16.4-6}$$

where $E_n^{(0)}$ is an eigenvalue of $\mathscr{H}_0$. In view of the noncommutativity of $\mathbf{r}$ and $\mathbf{p}$, $\mathscr{H}_0$ and $1/r$ do not commute; nevertheless, the Hermitian property of $\mathscr{H}_0$ leads to the equality of the matrix elements

$$\left\langle nlsjm'\left|\mathscr{H}_0\frac{1}{r}\right|nlsjm\right\rangle = \left\langle nlsjm'\left|\frac{1}{r}\mathscr{H}_0\right|nlsjm\right\rangle$$
$$= E_n^{(0)}\left\langle nlsjm'\left|\frac{1}{r}\right|nlsjm\right\rangle = E_n^{(0)}\left\langle\frac{1}{r}\right\rangle\delta_{m'm} \tag{16.4-7}$$

in which $\langle 1/r\rangle$ depends only on the radial part of the wave function and is independent of m. Similarly

$$\left\langle nlsjm'\left|\frac{1}{r^2}\right|nlsjm\right\rangle = \left\langle\frac{1}{r^2}\right\rangle\delta_{m'm}. \tag{16.4-8}$$

The net result is that $\mathscr{H}_r$ has only diagonal matrix elements:

$$\langle \mathscr{H}_r \rangle = -\frac{(E_n^{(0)})^2}{2mc^2} - \frac{Ze^2}{mc^2} E_n^{(0)} \left\langle \frac{1}{r} \right\rangle - \frac{(Ze^2)^2}{2mc^2} \left\langle \frac{1}{r^2} \right\rangle. \tag{16.4-9}$$

For hydrogen, from (16.1-27)

$$E_n^{(0)} = -\frac{Z^2}{n^2}\frac{e^2}{2a_0} = -\frac{Z^2}{n^2}\frac{me^4}{2\hbar^2} \tag{16.4-10a}$$

$$= -\frac{Z^2}{n^2} \text{ Ry}. \tag{16.4-10b}$$

Also, from Table 16.4

$$\left\langle \frac{1}{r} \right\rangle = \frac{Z}{a_0}\frac{1}{n^2}, \tag{16.4-11}$$

$$\left\langle \frac{1}{r^2} \right\rangle = \frac{Z^2}{a_0{}^2}\frac{1}{n^3(l+\frac{1}{2})}. \tag{16.4-12}$$

Therefore

$$\langle \mathscr{H}_r \rangle = \frac{Z^2e^2E_n^{(0)}}{mc^2a_0}\left[\frac{1}{4n^2} - \frac{1}{n^2} + \frac{1}{n(l+\frac{1}{2})}\right] \tag{16.4-13a}$$

$$= E_n^{(0)}Z^2\alpha^2\left[-\frac{3}{4n^2} + \frac{1}{n(l+\frac{1}{2})}\right] \tag{16.4-13b}$$

with

$$\alpha^2 = \left(\frac{e^2}{\hbar c}\right)^2 = \frac{e^2}{mc^2a_0}, \qquad a_0 = \frac{\hbar^2}{me^2}.$$

In Rydberg units,

$$\langle \mathscr{H}_r \rangle = E_r = -\frac{Z^4}{n^3}\alpha^2\left[-\frac{3}{4n} + \frac{1}{l+\frac{1}{2}}\right] \text{ Ry}. \tag{16.4-14}$$

This expression holds for all values of l including $l = 0$.

Next we consider the Darwin term in (15.2-32):

$$\mathscr{H}_D = \frac{e\hbar^2}{8m^2c^2}\,\nabla \cdot \mathbf{E} \tag{16.4-15}$$

which is of the same order as $\mathscr{H}_r$. With $\mathbf{E} = -\nabla\varphi$ and $\varphi = Ze/r$,

$$\mathscr{H}_D = -\frac{Ze^2\hbar^2}{8m^2c^2}\nabla^2\left(\frac{1}{r}\right) = \frac{4\pi Ze^2\hbar^2}{8m^2c^2}\,\delta(\mathbf{r}) \tag{16.4-16}$$

where we have used the relationship

$$\nabla^2(1/r) = -4\pi\delta(\mathbf{r}). \tag{16.4-17}$$

Since only the radial part of the wave function will influence the matrix element of $\mathscr{H}_D$

$$\langle nl|\delta(\mathbf{r})|nl\rangle = |\psi(0)|^2, \tag{16.4-18}$$

and

$$\langle \mathscr{H}_D \rangle = \frac{\pi Z e^2 \hbar^2}{2m^2c^2}|\psi(0)|^2 = E_D. \tag{16.4-19}$$

Thus the matrix elements of $\mathscr{H}_D$ are nonzero only for s states. In hydrogen

$$|\psi(0)|^2 = Z^3/\pi n^3 a_0{}^3. \tag{16.4-20}$$

Therefore

$$\langle \mathscr{H}_D \rangle = E_D = Z^4 e^2 \hbar^2/2n^3 m^2 c^2 a_0{}^3 \quad (l = 0 \text{ only}) \tag{16.4-21}$$

$$= Z^4\alpha^2/n^3 \quad \text{Ry}. \tag{16.4-22}$$

It is now possible to combine the expressions for E_{SO} [Eq. (16.3-19)], E_r [Eq. (16.4-14)], and E_D [Eq. (16.4-22)] into a single expression which depends on n and j but not on l (or m):

$$E_{SO} + E_r + E_D = -\frac{Z^4\alpha^2}{n^3}\left[\frac{1}{j+\frac{1}{2}} - \frac{3}{4n}\right] \quad \text{Ry} \tag{16.4-23}$$

with $j = l \pm \frac{1}{2}$. When $l = 0$, $E_{SO} = 0$, and

$$E_r + E_D = -\frac{Z^4\alpha^2}{n^3}\left[1 - \frac{3}{4n}\right] \quad \text{Ry}. \tag{16.4-24}$$

When $l \neq 0$, $E_D = 0$ and

$$E_{SO} + E_r = -\frac{Z^4\alpha^2}{n^3}\left[\frac{1}{j+\frac{1}{2}} - \frac{3}{4n}\right] \quad \text{Ry}. \tag{16.4-25}$$

Therefore the combined effects of the spin-orbit coupling, the relativistic energy correction, and the Darwin term are all included in (16.4-23). It is of interest to note that on the basis of (16.4-23) the energies of $^2S_{1/2}$ and $^2P_{1/2}$ are identical for a given n. The same result is obtained from the exact solution of the Dirac equation for hydrogen. Experimentally, a small difference (0.035 cm^{-1} for $Z = 1$) between the energies of $^2S_{1/2}$ and $^2P_{1/2}$ has been observed. This is known as the *Lamb shift* and its explanation is based on higher-order radiative corrections.

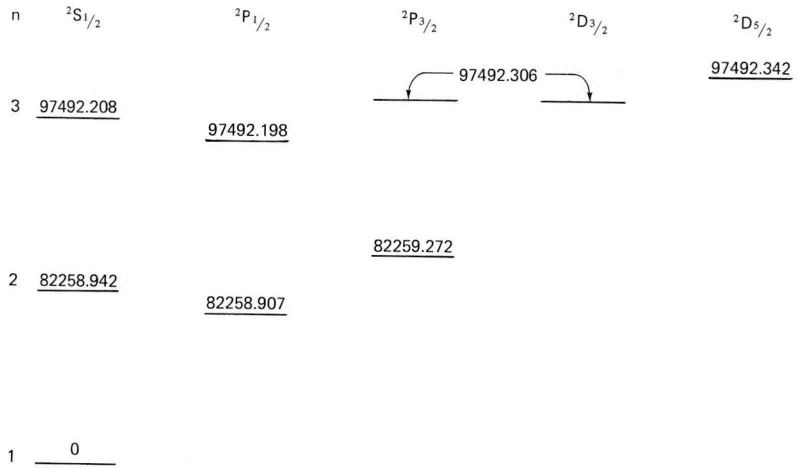

FIG. 16.5 Energy levels of the hydrogen atom for $n = 1, 2$, and 3. The energies are in reciprocal centimeters; the drawing is not to scale.

A partial energy level diagram of hydrogen is shown in Fig. 16.5. The splitting that arises within a manifold of states belonging to the same value of n is known as *fine structure* (see also Section 23.8).

CHAPTER 17

STATIC FIELDS

17.1 Magnetic Fields

It is assumed that the electron (or hydrogen atom) is placed in a constant magnetic field **B** with vector potential

$$\mathbf{A} = \tfrac{1}{2}\mathbf{B} \times \mathbf{r}. \tag{17.1-1}$$

Referring to the Schrödinger equation (15.2-32) the interaction terms that depend on the vector potential are

$$\frac{e}{2mc}(\mathbf{p}\cdot\mathbf{A} + \mathbf{A}\cdot\mathbf{p}) + \frac{e^2}{2mc^2}A^2 + \frac{e\hbar}{2mc}\boldsymbol{\sigma}\cdot\nabla\times\mathbf{A}. \tag{17.1-2}$$

When **A** has form (17.1-1), $\nabla\cdot\mathbf{A}$ is identically zero as a result of which

$$\mathbf{p}\cdot\mathbf{A} = \mathbf{A}\cdot\mathbf{p}. \tag{17.1-3}$$

We shall confine our attention, initially, to effects which are linear in $\mathbf{B} = \nabla\times\mathbf{A}$; hence the Hamiltonian describing the interaction with the field is

$$\mathscr{H}_{\mathrm{m}} = \frac{e}{mc}\mathbf{A}\cdot\mathbf{p} + \frac{e\hbar}{2mc}\boldsymbol{\sigma}\cdot\mathbf{B}. \tag{17.1-4}$$

But

$$\mathbf{A}\cdot\mathbf{p} = \tfrac{1}{2}\mathbf{B}\times\mathbf{r}\cdot\mathbf{p} = \tfrac{1}{2}\mathbf{B}\cdot\mathbf{r}\times\mathbf{p} = \tfrac{1}{2}\hbar\mathbf{B}\cdot\mathbf{L} \tag{17.1-5}$$

in which **L** is the orbital angular momentum operator. With the replacement of $\boldsymbol{\sigma}$ by 2**S** and substitution of (17.1-5) into (17.1-4) we have

$$\mathscr{H}_{\mathrm{m}} = \frac{e\hbar}{2mc}\mathbf{B}\cdot(\mathbf{L}+2\mathbf{S}) = \beta\mathbf{B}\cdot(\mathbf{L}+2\mathbf{S}). \tag{17.1-6}$$

The positive constant

$$\beta = \frac{e\hbar}{2mc} = 9.27 \times 10^{-21} \quad \text{erg/G}$$

$$= 9.27 \times 10^{-24}\ \text{J/T}. \tag{17.1-7}$$

is known as the *Bohr magneton.*

Equation (17.1-6) may also be written as

$$\mathscr{H}_{\text{m}} = -\boldsymbol{\mu}_L \cdot \mathbf{B} - \boldsymbol{\mu}_S \cdot \mathbf{B} \tag{17.1-8}$$

where

$$\boldsymbol{\mu}_L = -\beta\mathbf{L}; \qquad \boldsymbol{\mu}_S = -2\beta\mathbf{S}. \tag{17.1-9}$$

The resemblance of (17.1-8) to the classical expression for the energy of a magnetic dipole in a magnetic field suggests that $\boldsymbol{\mu}_L$ and $\boldsymbol{\mu}_S$ be interpreted as magnetic moment operators associated with **L** and **S**, respectively. The minus signs in (17.1-9) are due to the negative charge on the electron. Note that the factor of 2 appears in the relation between $\boldsymbol{\mu}_S$ and **S** and is absent in the relation between $\boldsymbol{\mu}_L$ and **L**. The latter has a classical analog but the former does not. Actually, the factor of 2 is slightly erroneous; higher order corrections show that

$$\boldsymbol{\mu}_S = -g_{\text{e}}\beta\mathbf{S} \tag{17.1-10}$$

with

$$g_{\text{e}} = 2.0023; \tag{17.1-11}$$

although in most cases it is sufficient to set $g_{\text{e}} = 2$.

It is important to distinguish between a magnetic moment operator $\boldsymbol{\mu}$ such as $\boldsymbol{\mu}_L$ or $\boldsymbol{\mu}_S$ defined by (17.1-9) from the quantity μ known as "the magnetic moment." The orbital magnetic moment μ_L is defined as

$$\mu_L = \langle l\ m_l = l|\mu_z{}^L|l\ m_l = l\rangle \tag{17.1-12}$$

where $\mu_z{}^L$ is the z component of $\boldsymbol{\mu}_L$. From (17.1-9),

$$\mu_L = -\beta\langle l\ m_l = l|L_z|l\ m_l = l\rangle = -\beta l. \tag{17.1-13}$$

Similarly, the spin magnetic moment μ_S is given by

$$\begin{aligned}\mu_S &= \langle s\ m_s = s|\mu_z{}^S|s\ m_s = s\rangle \\ &= -\beta g_{\text{e}}\langle s\ m_s = s|S_z|s\ m_s = s\rangle \\ &= -\beta g_{\text{e}} s = -\tfrac{1}{2}\beta g_{\text{e}} \approx -\beta.\end{aligned} \tag{17.1-14}$$

In other words, the absolute value of the spin magnetic moment of the electron is one Bohr magneton.

In place of (17.1-9) we may now write

$$\boldsymbol{\mu}_L = \frac{\mu_L}{l}\mathbf{L}; \qquad \boldsymbol{\mu}_S = \frac{\mu_S}{s}\mathbf{S} = 2\mu_S\mathbf{S}. \tag{17.1-15}$$

It is convenient, although not essential, to assume that the coordinate system has been chosen so that the z axis coincides with the direction of **B**. In that case (17.1-6) simplifies to

$$\mathscr{H}_{\mathrm{m}} = \beta B(L_z + 2S_z) \tag{17.1-16}$$

where $B = B_z$. We shall now divide the discussion of magnetic field effects into two parts: "weak" fields and "strong" fields. The scale is set by the spin-orbit interaction energy. If the changes in energy due to the application of a magnetic field are small compared with the spin-orbit coupling energy, the field is said to be "weak"; otherwise it is strong. The "weak" field case is the regime of the ordinary *Zeeman* effect while the "strong" field case corresponds to the *Paschen–Back* effect.

When the fields are "weak" it is presumed that the effects of spin-orbit coupling have already been taken into account so that the eigenstates are described in the coupled representation $|lsjm\rangle$. We shall therefore be interested in matrix elements of $\mathscr{H}_{\mathrm{m}}$ in this basis set. Furthermore, the assumption of a "weak" field implies that individual terms like ${}^2\mathrm{S}_{1/2}$, ${}^2\mathrm{P}_{1/2}$, ${}^2\mathrm{P}_{3/2}$, etc., are isolated so that attention can be confined to matrix elements which are diagonal in l, s, and j. To evaluate such matrix elements we apply the Landé formula (6.3-19):

$$\langle lsjm'|L_z + 2S_z|lsjm\rangle = \frac{\langle lsjm|(\mathbf{L} + 2\mathbf{S}) \cdot \mathbf{J}|lsjm\rangle}{j(j+1)} \langle lsjm'|J_z|lsjm\rangle. \tag{17.1-17}$$

But

$$\langle lsjm'|J_z|lsjm\rangle = m\,\delta_{m'm} \tag{17.1-18}$$

and

$$(\mathbf{L} + 2\mathbf{S}) \cdot \mathbf{J} = (\mathbf{J} + \mathbf{S}) \cdot \mathbf{J} = J^2 + \mathbf{S} \cdot \mathbf{J}.$$

Since

$$\mathbf{L} = \mathbf{J} - \mathbf{S}, \qquad L^2 = J^2 + S^2 - 2\mathbf{S} \cdot \mathbf{J},$$

and

$$\mathbf{S} \cdot \mathbf{J} = \tfrac{1}{2}(J^2 + S^2 - L^2),$$

we have

$$(\mathbf{L} + 2\mathbf{S}) \cdot \mathbf{J} = \tfrac{3}{2}J^2 + \tfrac{1}{2}(S^2 - L^2).$$

Therefore

$$\begin{aligned}\langle lsjm|(\mathbf{L}+2\mathbf{S})\cdot\mathbf{J}|lsjm\rangle &= \langle lsjm|\tfrac{3}{2}J^2+\tfrac{1}{2}(S^2-L^2)|lsjm\rangle\\ &= \tfrac{3}{2}j(j+1)+\tfrac{1}{2}s(s+1)-\tfrac{1}{2}l(l+1),\\ \frac{\langle lsjm|(\mathbf{L}+2\mathbf{S})\cdot\mathbf{J}|lsjm\rangle}{j(j+1)} &= 1+\frac{j(j+1)+s(s+1)-l(l+1)}{2j(j+1)}\\ &\equiv g_J = \text{Landé } g \text{ factor.}\end{aligned} \qquad (17.1\text{-}19)$$

Substituting (17.1-18) and (17.1-19) into (17.1-17) we obtain

$$\langle lsjm'|L_z+2S_z|lsjm\rangle = g_J m\,\delta_{m'm} \qquad (17.1\text{-}20)$$

which indicates that only diagonal elements are nonzero. Hence the energies in a "weak" magnetic field are given by

$$E_{\mathrm{m}} = \beta g_J B m. \qquad (17.1\text{-}21)$$

These are known as the *Zeeman levels* with energies proportional to the magnetic quantum number m. Thus, the effect of the magnetic field has been to remove the m-degeneracy.

The Landé g factor (17.1-19) may also be written as

$$g_J = 1+(g_{\mathrm{e}}-1)\frac{j(j+1)+s(s+1)-l(l+1)}{2j(j+1)} \qquad (17.1\text{-}22)$$

to permit the use of the more exact value of g_{e} given by (17.1-11).

It is now possible to define a total magnetic moment operator $\boldsymbol{\mu}_J$ by

$$\boldsymbol{\mu}_J = -\beta g_J \mathbf{J} \qquad (17.1\text{-}23)$$

which contains (17.1-9) as special cases. Corresponding to (17.1-12)–(17.1-14) we have, for the total magnetic moment

$$\mu_J = \langle j\,m=j|\mu_z{}^J|j\,m=j\rangle = -g_J\beta\langle j\,m=j|J_z|j\,m=j\rangle = -\beta g_J j. \quad (17.1\text{-}24)$$

Hence (17.1-23) may be written as

$$\boldsymbol{\mu}_J = (\mu_J/j)\mathbf{J},$$

and, in terms of $\boldsymbol{\mu}_J$, the magnetic Hamiltonian (17.1-6) is

$$\mathscr{H}_{\mathrm{m}} = -\boldsymbol{\mu}_J\cdot\mathbf{B} \qquad (17.1\text{-}25)$$

which then leads directly to (17.1-21). Also, on comparing (17.1-25) with (17.1-8), it is seen that

$$\boldsymbol{\mu}_J = \boldsymbol{\mu}_L + \boldsymbol{\mu}_S. \qquad (17.1\text{-}26)$$

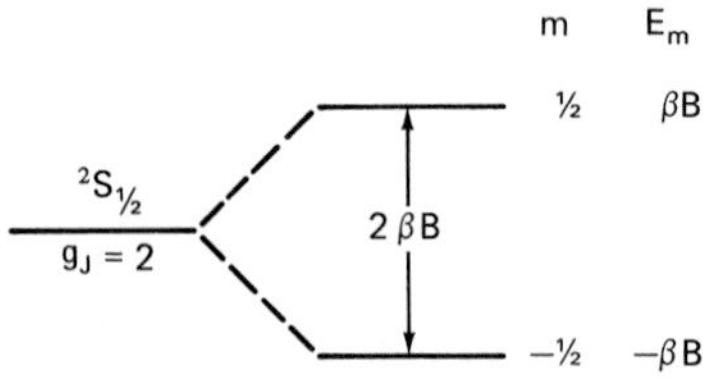

FIG. 17.1 Splitting of $^2S_{1/2}$ in a weak magnetic field.

For an electron in an s state ($^2S_{1/2}$), $l = 0$, $s = \frac{1}{2}$, $j = \frac{1}{2}$, $g_J = 2$ so that the energies, from (17.1-21) are

$$E_m = \pm\beta B \tag{17.1-27}$$

as shown in Fig. 17.1. The energy separations in $^2P_{1/2}$ and $^2P_{3/2}$ are

$$\begin{aligned} E_m(^2P_{1/2}) &= \pm\tfrac{1}{3}\beta B \\ E_m(^2P_{3/2}) &= \pm 2\beta B, \pm\tfrac{2}{3}\beta B \end{aligned} \tag{17.1-28}$$

as shown in Fig. 17.2.

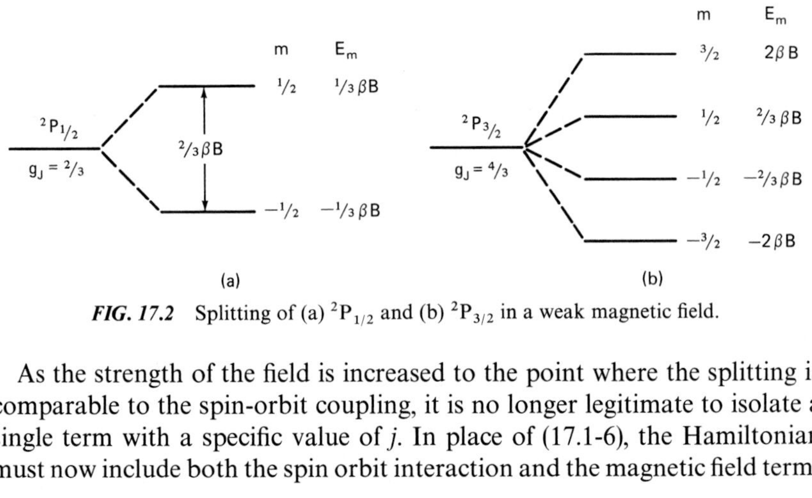

FIG. 17.2 Splitting of (a) $^2P_{1/2}$ and (b) $^2P_{3/2}$ in a weak magnetic field.

As the strength of the field is increased to the point where the splitting is comparable to the spin-orbit coupling, it is no longer legitimate to isolate a single term with a specific value of j. In place of (17.1-6), the Hamiltonian must now include both the spin orbit interaction and the magnetic field term:

$$\mathscr{H}_m' = \xi(r)\mathbf{L}\cdot\mathbf{S} + \beta B(L_z + 2S_z). \tag{17.1-29}$$

Matrix elements are most conveniently calculated in the $|lsm_lm_s\rangle$ representation, since then

$$\langle l'sm_l'm_s'|L_z + 2S_z|lsm_lm_s\rangle = (m_l + 2m_s)\,\delta_{l'l}\,\delta_{m_l'm_l}\,\delta_{m_s'm_s} \tag{17.1-30}$$

and

$$\langle l'sm_l'm_s'|\mathbf{L}\cdot\mathbf{S}|lsm_lm_s\rangle = \langle l'sm_l'm_s'| - L_{+1}S_{-1} + L_0S_0 - L_{-1}S_{+1}|lsm_lm_s\rangle. \tag{17.1-31}$$

TABLE 17.1

Matrix Elements of $\mathscr{H}_m'$*(17.1-29) for* p *States*

	$\left\|1\frac{1}{2}\right\rangle$	$\left\|1\ -\frac{1}{2}\right\rangle$	$\left\|0\frac{1}{2}\right\rangle$	$\left\|0\ -\frac{1}{2}\right\rangle$	$\left\|-1\ \frac{1}{2}\right\rangle$	$\left\|-1\ -\frac{1}{2}\right\rangle$
$\left\langle 1\frac{1}{2}\right\|$	$\frac{1}{2}\xi_{nl} + 2\beta B$					
$\left\langle 1\ -\frac{1}{2}\right\|$		$-\frac{1}{2}\xi_{nl}$	$\frac{1}{\sqrt{2}}\xi_{nl}$			
$\left\langle 0\frac{1}{2}\right\|$		$\frac{1}{\sqrt{2}}\xi_{nl}$	βB			
$\left\langle 0\ -\frac{1}{2}\right\|$				$-\beta B$	$\frac{1}{\sqrt{2}}\xi_{nl}$	
$\left\langle -1\ \frac{1}{2}\right\|$				$\frac{1}{\sqrt{2}}\xi_{nl}$	$-\frac{1}{2}\xi_{nl}$	
$\left\langle -1\ -\frac{1}{2}\right\|$						$\frac{1}{2}\xi_{nl} - 2\beta B$

For the 2p manifold, the matrix elements of $\mathbf{L}\cdot\mathbf{S}$ have already been calculated (Table 16.5). Adding the magnetic field interaction from (17.1-30), the results are those shown in Table 17.1. The secular equation, therefore, factors into

$$\left(2\beta B + \frac{1}{2}\xi_{nl} - E\right) = 0$$

$$\begin{vmatrix} -\frac{1}{2}\xi_{nl} - E & \frac{1}{\sqrt{2}}\xi_{nl} \\ \frac{1}{\sqrt{2}}\xi_{nl} & \beta B - E \end{vmatrix} = 0$$

$$\begin{vmatrix} -\beta B - E & \frac{1}{\sqrt{2}}\xi_{nl} \\ \frac{1}{\sqrt{2}}\xi_{nl} & -\frac{1}{2}\xi_{nl} - E \end{vmatrix} = 0 \tag{17.1-32}$$

$$\left(\frac{1}{2}\xi_{nl} - 2\beta B - E\right) = 0$$

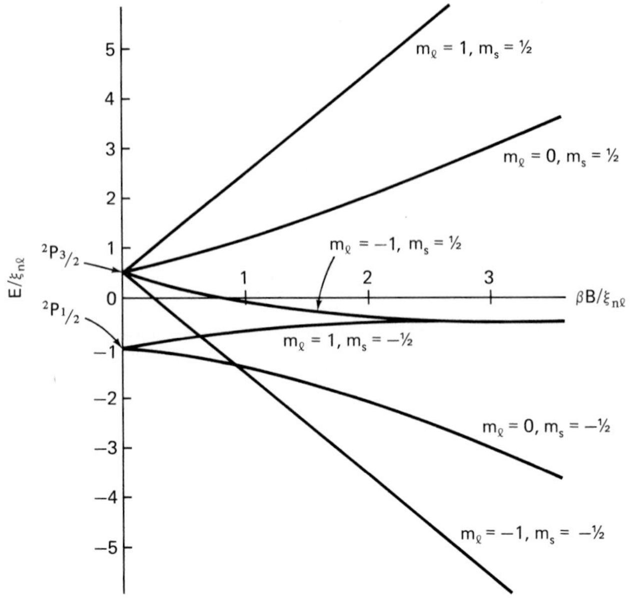

FIG. 17.3 Transition from a weak to a strong magnetic field for a ^{2}P term.

from which the energy eigenvalues are readily calculated. They are

$$\begin{aligned}
\frac{E_1}{\xi_{nl}} &= 2\frac{\beta B}{\xi_{nl}} + \frac{1}{2}, \\
\frac{E_{2,3}}{\xi_{nl}} &= \frac{1}{2}\left\{\left(\frac{\beta B}{\xi_{nl}} - \frac{1}{2}\right) \pm \left[\left(\frac{\beta B}{\xi_{nl}}\right)^2 + \left(\frac{\beta B}{\xi_{nl}}\right) + \frac{9}{4}\right]^{1/2}\right\}, \\
\frac{E_{4,5}}{\xi_{nl}} &= \frac{1}{2}\left\{-\left(\frac{\beta B}{\xi_{nl}} + \frac{1}{2}\right) \pm \left[\left(\frac{\beta B}{\xi_{nl}}\right)^2 - \left(\frac{\beta B}{\xi_{nl}}\right) + \frac{9}{4}\right]^{1/2}\right\}, \\
\frac{E_6}{\xi_{nl}} &= -2\frac{\beta B}{\xi_{nl}} + \frac{1}{2}.
\end{aligned} \tag{17.1-33}$$

These energies are plotted in Fig. 17.3. A few special cases are of interest.

$\beta B \gg \xi_{nl}$: This is the *Paschen–Back* region. In this approximation the energies conform to the expression

$$E_m = \beta B(m_l + 2m_s). \tag{17.1-34}$$

$\beta B \ll \xi_{nl}$: If we confine ourselves to linear terms in βB, the reduction of (17.1-33) gives

$$\begin{aligned}
&E_1 = \tfrac{1}{2}\xi_{nl} + 2\beta B, \qquad E_2 = \tfrac{1}{2}\xi_{nl} + \tfrac{2}{3}\beta B, \qquad E_3 = -\xi_{nl} + \tfrac{1}{3}\beta B, \\
&E_4 = \tfrac{1}{2}\xi_{nl} - \tfrac{2}{3}\beta B, \qquad E_5 = -\xi_{nl} - \tfrac{1}{3}\beta B, \qquad E_6 = \tfrac{1}{2}\xi_{nl} - 2\beta B.
\end{aligned} \tag{17.1-35}$$

These are precisely the energies obtained on the basis of (16.3-12) for the spin-orbit interaction with $l = 1$ and (17.1-21) for the splitting in a "weak" magnetic field. The correlations between the "weak" field and "strong" field levels are shown in Fig. 17.4. Note that states with the same value of $m\ (= m_l + m_s)$ do not cross. A further point is that the "weak" field case is best described in the coupled representation as evidenced by the fact that m is a "good" quantum number. In a classical sense this means that the orbital and spin angular momenta are coupled to produce a total angular momentum and it is the latter which precesses about the applied magnetic field. On the other hand, when the field is "strong," the "good" quantum numbers are m_l and m_s. This is in the uncoupled representation and corresponds, classically, to the orbital angular momentum and the spin angular momentum *individually* precessing about the field (Fig. 17.5).

A further point to note is that, when an atom is subjected to a magnetic field, the Hamiltonian is no longer invariant under all three-dimensional rotations but only under rotations about an axis parallel to the magnetic field. In other words, the symmetry has been reduced from $O^+(3)$ to C_∞. The consequence of this restriction in symmetry is that the Hamiltonian no longer commutes with J^2, although it commutes with J_z. Alternatively, it may be a said that, in a magnetic field, j is not a good quantum number but m is. When these features are fully realized, the field is regarded as "strong"

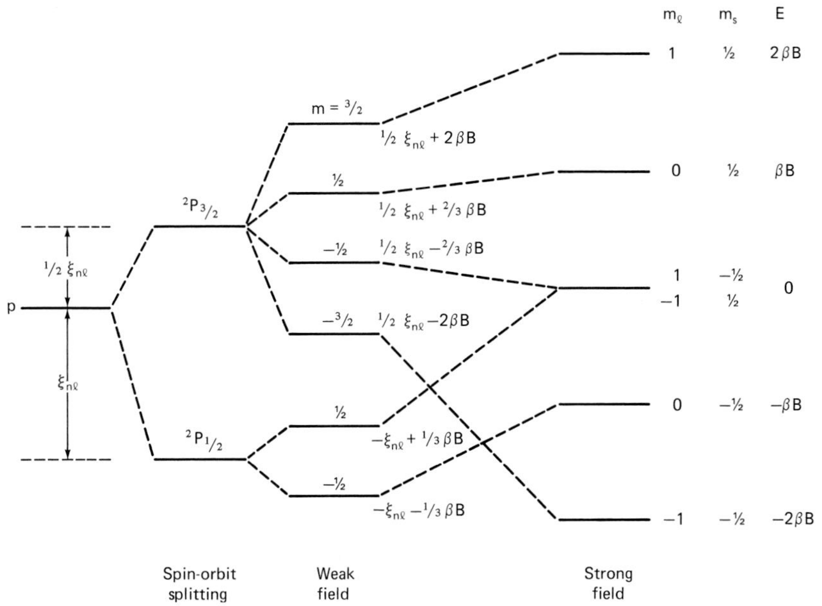

FIG. 17.4 Splitting of a p state under the influence of spin-orbit coupling and a magnetic field.

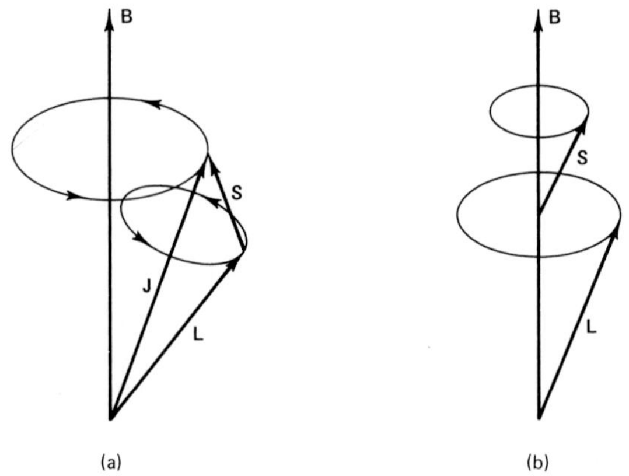

FIG. 17.5 Precession of **L**, **S**, **J** in (a) a weak and (b) a strong magnetic field.

and by the same token, a "weak" field is one for which, to a good approximation, j is still a good quantum number.

We note that the degeneracy in hydrogen between $2^2S_{1/2}$ and $2^2P_{1/2}$ has no effect on the computation performed above. Because of (17.1-29), matrix elements of $L_z + 2S_z$ between 2s and 2p vanish; the spin-orbit coupling matrix element (17.1-31) also vanishes unless $\Delta l = 0$.

Experiments in which photons are used to excite electronic transitions between magnetic substates are known as electron spin resonance (ESR) or electron paramagnetic resonance (EPR) experiments. For an s state the energies are given by (17.1-27); hence the photon energy must satisfy

$$\hbar\omega = 2\beta B \tag{17.1-36}$$

and upon inserting numerical values for the constants, the photon frequency ν is

$$\nu = \frac{\omega}{2\pi} = 2.80B\,\text{MHz}\ (s = \tfrac{1}{2},\ B \text{ in gauss}). \tag{17.1-37}$$

This relation is also applicable to a large number of free radicals.

The term in A^2 contained in (17.1-2) with **A** given by (17.1-1) is

$$\mathscr{H}_d = \frac{e^2}{2mc^2}\,\tfrac{1}{4}(\mathbf{B} \times \mathbf{r}) \cdot (\mathbf{B} \times \mathbf{r}) = \frac{e^2}{8mc^2} B^2 r^2 \sin^2\theta \tag{17.1-38}$$

where θ is the angle between **B** and **r**. For the ground state of hydrogen or any other spherically symmetric state the expectation value of (17.1-38) is

$$E_d = \frac{e^2}{8mc^2} B^2 \langle r^2 \rangle \overline{\sin^2\theta} = \frac{e^2}{12mc^2} B^2 \langle r^2 \rangle \tag{17.1-39}$$

in which $\langle r^2 \rangle$ is the expectation value of r^2, r is the distance of the electron from the nucleus and $\overline{\sin^2 \theta}$ is a spherical average. The ratio of E_d in (17.1-39) to E_m in (17.1-21) at fields of the order of 10^4 gauss is approximately equal to $\alpha^2 = (e^2/\hbar c)^2 = (1/137)^2$.

For the magnetic moment we may take

$$\mu = -\frac{\partial E_d}{\partial B} = -\frac{e^2}{6mc^2} B\langle r^2 \rangle \tag{17.1-40}$$

from which we obtain the *diamagnetic susceptibility*

$$\chi_d = \frac{\mu}{B} = -\frac{e^2}{6mc^2} \langle r^2 \rangle = -\tfrac{1}{6}\alpha^2 a_0 \langle r^2 \rangle \tag{17.1-41}$$

where a_0 is the Bohr radius.

Diamagnetism arises as a consequence of the orbital motion of the electrons; hence it is a general property of all atoms and molecules. Paramagnetism, on the other hand, has its origin in the alignments of magnetic moments associated with orbital and spin angular momenta in an external magnetic field. Quite often, an electronic system may possess both a diamagnetic and paramagnetic susceptibility.

17.2 Electric Fields

Electric fields may also have an effect on the states of an atom. This is known as the Stark effect. The discussion in this section will be confined to hydrogen which is somewhat unique in this respect.

If the coordinate system is chosen so that the z axis coincides with the direction of the electric field, the Hamiltonian for the interaction is

$$\mathscr{H}_s = eEz = eEr\cos\theta. \tag{17.2-1}$$

The situation of greatest physical interest is the one in which the splittings due to the Stark effect are large compared to the spin-orbit splittings. A numerical estimate of the required field will be given later. As in the case of high magnetic fields, the uncoupled representation is the appropriate one; we shall therefore calculate matrix elements of $\mathscr{H}_s$ in the $|nlsm_lm_s\rangle$ representation.

It is convenient to replace $\cos\theta$ in (17.2-1) by $\sqrt{4\pi/3}\,Y_{10}$ so that we may make use of the theorem (1.2-29):

$$\langle l'm_l'|Y_{LM}|lm_l\rangle = (-1)^{m_l'}\sqrt{(2l'+1)(2L+1)(2l+1)/4\pi} \times \begin{pmatrix} l' & L & l \\ -m_l' & M & m_l \end{pmatrix}\begin{pmatrix} l' & L & l \\ 0 & 0 & 0 \end{pmatrix}. \tag{17.2-2}$$

The matrix element vanishes when $l' + L + l$ is odd, that is, when the integrand is of odd parity. In the present case $Y_{LM} = Y_{10}$ which means that the expectation value of $\mathscr{H}_s$ must vanish for any state in which l is a good quantum number. From this we conclude that there is no first-order splitting due to an electric field.

Although this conclusion is generally correct, it does not hold for hydrogen. The reason is that the energy in hydrogen, in first approximation, depends only on the principal quantum number n and not on l. Consider, for example, the case $n = 2$. This consists of four degenerate states: $l = 0, m_l = 0; l = 1, m_l = 0, \pm 1$. Within this manifold of states there will be nonvanishing matrix elements of Y_{10} between the state with $l = 0$ and states with $l = 1$. A first-order Stark effect in hydrogen is therefore expected.

To proceed with the calculation it is noted that

$$\begin{pmatrix} l' & L & l \\ -m_l' & M & m_l \end{pmatrix} = 0$$

unless $-m_l' + M + m_l = 0$. Since $M = 0, m_l$ and m_l' are equal. Moreover $m_l' = m_l = 0$ because one of the l values is zero. This leaves two nonzero matrix elements in (17.2-2) which are

$$\langle 00|Y_{10}|10\rangle = \langle 10|Y_{10}|00\rangle = 1/\sqrt{4\pi}. \tag{17.2-3}$$

Also, the matrix element of r taken between the 2s function and the radial part of the $2p_0$ wave function is given by

$$\langle 2s|r|R(2p_0)\rangle = \langle R(2p_0)|r|2s\rangle = \frac{9}{\sqrt{3}}\frac{a_0}{Z}. \tag{17.2-4}$$

Combining the results from (17.2-3) and (17.2-4), the Hamiltonian matrix is the one shown in Table 17.2 with eigenvalues

$$E_s = 3eEa_0/Z, \quad 0, \quad 0, \quad -3eEa_0/Z. \tag{17.2-5}$$

TABLE 17.2

Matrix Elements for the Stark Effect in Hydrogen with $n = 2$, $s = \frac{1}{2}$, $m_s = m_s'$

	$\vert 0\ 0\rangle$	$\vert 1\ 0\rangle$	$\vert 1\ 1\rangle$	$\vert 1\ -1\rangle$
$\langle 0\ 0\vert$		$3eEa_0/Z$		
$\langle 1\ 0\vert$	$3eEa_0/Z$			
$\langle 1\ 1\vert$				
$\langle 1\ -1\vert$				

The two states with $m_l = 0$ are shifted up and down symmetrically while the states with $m_l = \pm 1$ are not affected by the electric field. Thus, the m degeneracy is only partially lifted.

In hydrogen, assuming $E = 10^4$ V cm^{-1} and $Z = 1$,

$$3eEa_0/Z = 1.3 \quad \text{cm}^{-1},$$

which is considerably larger than the fine structure splitting (Section 16.3 and Fig. 16.5).

It is seen that because of the degeneracy of states with different l and the same n there is a *linear* Stark effect in hydrogen. At very high field strengths a quadratic effect appears, superimposed upon the linear effect, and results in an asymmetric shift of energy levels.

CHAPTER 18

HYPERFINE INTERACTIONS

The most important interaction between a nucleus of charge Ze and an electron is, of course, the Coulumb interaction $-Ze^2/r$. All other interactions between a nucleus and the electrons of the same atom are classified as *hyperfine interactions*, and among these the most important are the ones that arise as a result of a nucleus possessing a magnetic dipole moment (associated with the nuclear spin) and an electric quadrupole moment (associated with a departure from a spherical charge distribution in the nucleus). In the former case the nuclear magnetic dipole moment interacts with the electronic magnetic dipole moments which are associated with the electronic orbital and/or spin angular momenta. In the latter, the interaction occurs when, at the position of the nucleus, the electronic charge distribution produces an electric field gradient with which the nuclear electric quadrupole moment can interact. We shall consider the two cases separately.

18.1 Hamiltonian for the Magnetic Hyperfine Interaction

To derive the form of this interaction we return to the Dirac equation (15.2-12):

$$E'\psi_{\mathrm{u}} = \left\{q\varphi + \frac{1}{2m}\left[\boldsymbol{\sigma}\cdot\left(\mathbf{p} - \frac{q}{c}\mathbf{A}\right)K\boldsymbol{\sigma}\cdot\left(\mathbf{p} - \frac{q}{c}\mathbf{A}\right)\right]\right\}\psi_{\mathrm{u}} \qquad (18.1\text{-}1)$$

with

$$K = \frac{2mc^2}{E' + 2mc^2 - q\varphi}, \qquad E' = E - mc^2. \qquad (18.1\text{-}2)$$

Because φ becomes infinite at the origin (nucleus), any approximation to K must be carried out very cautiously. For the time being, we shall leave K as is and make no approximation. The terms in (18.1-1) which are of interest for the magnetic hyperfine interaction are those that contain the vector potential **A**. Accordingly, it is advantageous to separate the Hamiltonian in (18.1-1) into two parts:

$$\begin{aligned}\mathscr{H} &= q\varphi + \frac{1}{2m}\left[\boldsymbol{\sigma}\cdot\left(\mathbf{p} - \frac{q}{c}\mathbf{A}\right)K\boldsymbol{\sigma}\cdot\left(\mathbf{p} - \frac{q}{c}\mathbf{A}\right)\right] \\ &= \mathscr{H}_0 + \mathscr{H}_1 \end{aligned} \tag{18.1-3}$$

where

$$\mathscr{H}_0 = q\varphi + \frac{1}{2m}(\boldsymbol{\sigma}\cdot\mathbf{p})K(\boldsymbol{\sigma}\cdot\mathbf{p}), \tag{18.1-4a}$$

$$\mathscr{H}_1 = -\frac{q}{2mc}[(\boldsymbol{\sigma}\cdot\mathbf{p})K(\boldsymbol{\sigma}\cdot\mathbf{A}) + (\boldsymbol{\sigma}\cdot\mathbf{A})K(\boldsymbol{\sigma}\cdot\mathbf{p})] + \frac{q^2}{2mc^2}(\boldsymbol{\sigma}\cdot\mathbf{A})K(\boldsymbol{\sigma}\cdot\mathbf{A}). \tag{18.1-4b}$$

We concentrate on $\mathscr{H}_1$ which contains the entire dependence on the vector potential. The last term on the right in (18.1-4b) is quadratic in **A**; it may therefore be neglected in a first approximation. The potential φ is assumed to be a function of r only,

$$\varphi = \varphi(r), \tag{18.1-5}$$

which means that K depends only on r,

$$K = K(r), \tag{18.1-6}$$

so that

$$\boldsymbol{\nabla}K = \frac{dK}{dr}\frac{\mathbf{r}}{r}. \tag{18.1-7}$$

Also, since

$$\mathbf{p}Kf(\mathbf{r}) = -i\hbar\boldsymbol{\nabla}Kf(\mathbf{r}) = -i\hbar[K\boldsymbol{\nabla}f + f\boldsymbol{\nabla}K] = K\mathbf{p}f - i\hbar\frac{dK}{dr}\frac{\mathbf{r}}{r}f, \tag{18.1-8}$$

we have

$$\mathbf{p}K = K\mathbf{p} - i\hbar\frac{dK}{dr}\frac{\mathbf{r}}{r}, \tag{18.1-9}$$

and (18.1-4b) becomes

$$\mathscr{H}_1 = -\frac{q}{2mc}\left[K(\boldsymbol{\sigma}\cdot\mathbf{p})(\boldsymbol{\sigma}\cdot\mathbf{A}) - i\hbar\frac{dK}{dr}\frac{1}{r}(\boldsymbol{\sigma}\cdot\mathbf{r})(\boldsymbol{\sigma}\cdot\mathbf{A}) + K(\boldsymbol{\sigma}\cdot\mathbf{A})(\boldsymbol{\sigma}\cdot\mathbf{p})\right]. \tag{18.1-10}$$

Using the theorem

$$(\boldsymbol{\sigma}\cdot\mathbf{A})(\boldsymbol{\sigma}\cdot\mathbf{B}) = \mathbf{A}\cdot\mathbf{B} + i\boldsymbol{\sigma}\cdot(\mathbf{A}\times\mathbf{B}), \tag{18.1-11}$$

$\mathscr{H}_1$ is converted to

$$\mathscr{H}_1 = -\frac{q}{2mc}K(\mathbf{p}\cdot\mathbf{A} + \mathbf{A}\cdot\mathbf{p}) + iK\boldsymbol{\sigma}\cdot(\mathbf{p}\times\mathbf{A} + \mathbf{A}\times\mathbf{p})$$
$$- i\hbar\frac{dK}{dr}\frac{1}{r}[\mathbf{r}\cdot\mathbf{A} + i\boldsymbol{\sigma}\cdot\mathbf{r}\times\mathbf{A}]. \tag{18.1-12}$$

It is now necessary to specify the vector potential **A**. The nucleus is presumed to be a point dipole with a magnetic dipole moment $\boldsymbol{\mu}$; hence the vector potential at **r** is

$$\mathbf{A} = (\boldsymbol{\mu}\times\mathbf{r})/r^3. \tag{18.1-13}$$

Availing ourselves of the vector identities

$$\begin{aligned}\nabla\cdot(\mathbf{a}\times\mathbf{b}) &= \mathbf{b}\cdot\nabla\times\mathbf{a} - \mathbf{a}\cdot\nabla\times\mathbf{b},\\ \nabla\times k\mathbf{a} &= \nabla k\times\mathbf{a} + k\nabla\times\mathbf{a},\end{aligned} \tag{18.1-14}$$

and assuming that $\boldsymbol{\mu} = 0$ outside of the origin so that $\nabla\times\boldsymbol{\mu} = 0$, it is found that

$$\nabla\cdot\mathbf{A} = 0 \tag{18.1-15}$$

and

$$\mathbf{r}\cdot\mathbf{A} = 0. \tag{18.1-16}$$

We also have

$$\begin{aligned}\mathbf{p}\cdot\mathbf{A} &= i\hbar\nabla\cdot\mathbf{A} = \mathbf{A}\cdot\mathbf{p},\\ \mathbf{p}\times\mathbf{A} &= -i\hbar\nabla\times\mathbf{A} - \mathbf{A}\times\mathbf{p}.\end{aligned} \tag{18.1-17}$$

Combining (18.1-15)–(18.1-17) and substituting in (18.1-12), the result, after replacing q by $-e$, is

$$\mathscr{H}_1 = \frac{e}{2mc}\left[2K\mathbf{A}\cdot\mathbf{p} + K\hbar\boldsymbol{\sigma}\cdot\nabla\times\mathbf{A} + \hbar\frac{dK}{dr}\frac{1}{r}\boldsymbol{\sigma}\cdot\mathbf{r}\times\mathbf{A}\right]. \tag{18.1-18}$$

This expression may be put into another form as follows: From (18.1-13)

$$\mathbf{A}\cdot\mathbf{p} = \frac{\boldsymbol{\mu}\times\mathbf{r}}{r^3}\cdot\mathbf{p} = \frac{1}{r^3}\boldsymbol{\mu}\cdot\mathbf{r}\times\mathbf{p} = \frac{\hbar}{r^3}\boldsymbol{\mu}\cdot\mathbf{L}. \tag{18.1-19}$$

Since $\boldsymbol{\mu}$ is constant

$$\nabla \times \left(\frac{\boldsymbol{\mu}}{r}\right) = \nabla\left(\frac{1}{r}\right) \times \boldsymbol{\mu} + \frac{1}{r} \nabla \times \boldsymbol{\mu}$$

$$= \nabla\left(\frac{1}{r}\right) \times \boldsymbol{\mu} = -\boldsymbol{\mu} \times \nabla\left(\frac{1}{r}\right) = \frac{\boldsymbol{\mu} \times \mathbf{r}}{r^3} = \mathbf{A}, \qquad (18.1\text{-}20)$$

$$\nabla^2\left(\frac{\boldsymbol{\mu}}{r}\right) = \boldsymbol{\mu}\, \nabla^2\left(\frac{1}{r}\right) = -4\pi\boldsymbol{\mu}\,\delta(\mathbf{r}), \qquad (18.1\text{-}21)$$

$$\nabla \cdot \left(\frac{\boldsymbol{\mu}}{r}\right) = \boldsymbol{\mu} \cdot \nabla\left(\frac{1}{r}\right) + \frac{1}{r} \nabla \cdot \boldsymbol{\mu} = \boldsymbol{\mu} \cdot \nabla\left(\frac{1}{r}\right)$$

$$= -\frac{\boldsymbol{\mu} \cdot \mathbf{r}}{r^3}, \qquad (18.1\text{-}22)$$

$$\nabla\nabla \cdot \left(\frac{\boldsymbol{\mu}}{r}\right) = -\nabla\left(\frac{\boldsymbol{\mu} \cdot \mathbf{r}}{r^3}\right)$$

$$= -\left[(\boldsymbol{\mu} \cdot \mathbf{r})\nabla\left(\frac{1}{r^3}\right) + \frac{1}{r^3} \nabla(\boldsymbol{\mu} \cdot \mathbf{r})\right]$$

$$= (\boldsymbol{\mu} \cdot \mathbf{r})\frac{3\mathbf{r}}{r^5} - \frac{1}{r^3}(\boldsymbol{\mu} \cdot \nabla)\mathbf{r}. \qquad (18.1\text{-}23)$$

But

$$(\boldsymbol{\mu} \cdot \nabla)\mathbf{r} = \left(\mu_x \frac{\partial}{\partial_x} + \mu_y \frac{\partial}{\partial_y} + \mu_z \frac{\partial}{\partial_z}\right)(x\hat{\mathbf{i}} + y\hat{\mathbf{j}} + z\hat{\mathbf{k}})$$

$$= \mu_x\hat{\mathbf{i}} + \mu_y\hat{\mathbf{j}} + \mu_z\hat{\mathbf{k}} = \boldsymbol{\mu}.$$

Therefore

$$\nabla\left(\nabla \cdot \frac{\boldsymbol{\mu}}{r}\right) = (\boldsymbol{\mu} \cdot \mathbf{r})\frac{3\mathbf{r}}{r^5} - \frac{\boldsymbol{\mu}}{r^3} \qquad (18.1\text{-}24)$$

With these relations

$$\nabla \times \mathbf{A} = \nabla \times \nabla \times \frac{\boldsymbol{\mu}}{r} = \nabla\nabla \cdot \left(\frac{\boldsymbol{\mu}}{r}\right) - \nabla^2\left(\frac{\boldsymbol{\mu}}{r}\right).$$

$$= (\boldsymbol{\mu} \cdot \mathbf{r})\frac{3\mathbf{r}}{r^5} - \frac{\boldsymbol{\mu}}{r^3} + 4\pi\boldsymbol{\mu}\,\delta(\mathbf{r}), \qquad (18.1\text{-}25)$$

Finally

$$\mathbf{r} \times \mathbf{A} = \frac{1}{r^3}[\mathbf{r} \times \boldsymbol{\mu} \times \mathbf{r}] = \frac{1}{r^3}[(\mathbf{r} \cdot \mathbf{r})\boldsymbol{\mu} - (\mathbf{r} \cdot \boldsymbol{\mu})\mathbf{r}]$$

$$= \frac{\boldsymbol{\mu}}{r} - \frac{(\mathbf{r} \cdot \boldsymbol{\mu})\mathbf{r}}{r^3}. \qquad (18.1\text{-}26)$$

When (18.1-19, 25 and 26) are inserted into (18.1-18) one obtains

$$\mathscr{H}_1 = 2\beta K\left[\frac{\boldsymbol{\mu}\cdot(\mathbf{L}-\mathbf{S})}{r^3} + \frac{3(\boldsymbol{\mu}\cdot\mathbf{r})(\mathbf{S}\cdot\mathbf{r})}{r^5} + 4\pi\,\delta(\mathbf{r})\boldsymbol{\mu}\cdot\mathbf{S}\right] + 2\beta\frac{dK}{dr}\left[\frac{\boldsymbol{\mu}\cdot\mathbf{S}}{r^2} - \frac{(\boldsymbol{\mu}\cdot\mathbf{r})(\mathbf{S}\cdot\mathbf{r})}{r^4}\right] \tag{18.1-27}$$

in which

$$\mathbf{S} = \frac{\boldsymbol{\sigma}}{2}, \qquad \beta = \frac{e\hbar}{2mc}.$$

We now examine K and dK/dr. If φ is replaced by Ze/r and q by $-e$ the expression for K in (18.1-2) becomes

$$K = 2mc^2\bigg/\left(E' + 2mc^2 + \frac{Ze^2}{r}\right). \tag{18.1-28}$$

In Section 15.2, the nonrelativistic approximation to the Dirac equation was made by assuming that $E' + Ze^2/r \ll 2mc^2$. Since an atom in an s state has a nonvanishing amplitude at $r = 0$, such an approximation is no longer valid. We follow the treatment of Blinder (1960), which is an adaptation of the original derivation by Fermi (1930). It will now be assumed that $E' = E - mc^2 \ll 2mc^2$, but that Ze^2/r and $2mc^2$ may be comparable in magnitude. Hence,

$$K = \frac{1}{1 + (Ze^2/2mc^2)(1/r)} \equiv \frac{1}{1 + (r_0/r)} \tag{18.1-29}$$

with

$$r_0 = Ze^2/2mc^2. \tag{18.1-30}$$

A plot of $K(r)$ is shown in Fig. 18.1. When $Z = 1$, $r_0 = 1.4 \times 10^{-13}$ cm; this is a nuclear dimension and is much smaller than the Bohr radius a_0 which is 0.52×10^{-8} cm. Thus

$$\frac{r_0}{a_0} = \frac{1}{2}\frac{e^2}{mc^2}\frac{me^2}{\hbar^2} = \frac{1}{2}\left(\frac{e^2}{\hbar c}\right) = \frac{1}{2}\alpha^2 = \frac{1}{2}\left(\frac{1}{137}\right)^2. \tag{18.1-31}$$

$K(r)$ rises from zero at $r = 0$ to almost unity in a distance of several units of r_0, i.e., in a distance very small compared to a_0. $K(r)$ may therefore be approximated by a step function

$$K(r) = \begin{cases} 0 & \text{when} \quad r = 0, \\ 1 & \text{when} \quad r \neq 0. \end{cases} \tag{18.1-32}$$

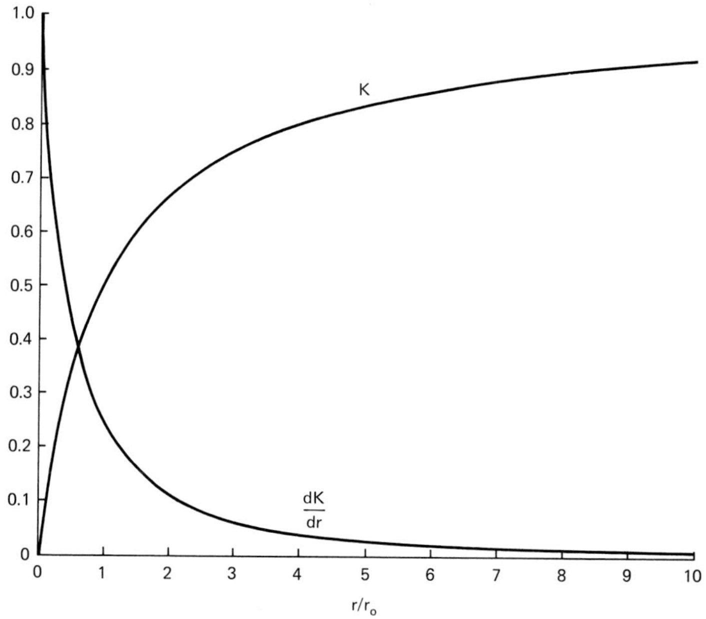

FIG. 18.1 Variation of K and K' with r.

With this approximation for K it may now be shown that the δ-function term in (18.1-27) does not contribute. Assume, first, that we are dealing with an s state. Then, since $K = 0$ at $r = 0$ and $\delta(\mathbf{r}) = 0$ at $r \neq 0$

$$\langle \psi_s | K\, \delta(\mathbf{r}) | \psi_s \rangle = K|\psi_s(0)|^2 = 0 \qquad (r = 0), \tag{18.1-33}$$

$$\langle \psi_s | \delta(\mathbf{r}) | \psi_s \rangle = 0 \qquad \text{at} \quad r \neq 0. \tag{18.1-34}$$

For a state with $l \neq 0$,

$$\langle \psi_{l\neq 0} | \delta(\mathbf{r}) | \psi_{l \neq 0} \rangle = 0 \qquad \text{at} \quad \begin{cases} r = 0, \\ r \neq 0, \end{cases} \tag{18.1-35}$$

because $\psi_{l \neq 0} = 0$ at $r = 0$, and $\delta(\mathbf{r}) = 0$ at $r \neq 0$. Therefore (18.1-27) may now be written without the δ function term:

$$\begin{aligned} \mathscr{H}_1 &= \mathscr{H}_a + \mathscr{H}_b, \\ \mathscr{H}_a &= 2\beta \left[\frac{\boldsymbol{\mu} \cdot (\mathbf{L} - \mathbf{S})}{r^3} + \frac{3(\boldsymbol{\mu} \cdot \mathbf{r})(\mathbf{S} \cdot \mathbf{r})}{r^5} \right], \\ \mathscr{H}_b &= 2\beta \frac{dK}{dr} \left[\frac{\boldsymbol{\mu} \cdot \mathbf{S}}{r^2} - \frac{(\boldsymbol{\mu} \cdot \mathbf{r})(\mathbf{S} \cdot \mathbf{r})}{r^4} \right]. \end{aligned} \tag{18.1-36}$$

The factor K that multiplies $\mathscr{H}_a$ has been set equal to one. This is permissible because, as will be shown in the next section, there is no contribution from $\mathscr{H}_a$ in s states which are the only states with nonvanishing amplitudes at $r = 0$ (where $K = 0$).

Next we examine dK/dr. Since K is approximated by a step function, $K' = dK/dr$ would be approximated by a δ function. To see this in greater detail, we have from (18.1-29),

$$K' = \frac{r_0/r^2}{(1 + r_0/r)^2}. \tag{18.1-37}$$

K' (in units of $1/r_0$) is shown in Fig. 18.1, from which it is evident that K' effectively decreases to zero in a distance equal to just a few nuclear radii so that, approximately

$$K' = 0 \qquad \text{for} \quad r \neq 0. \tag{18.1-38}$$

Also, within the approximation (18.1-32),

$$\int_0^\infty (dK/dr)\, dr = K(\infty) - K(0) = 1 - 0 = 1. \tag{18.1-39}$$

Since the behavior of K and K' at negative r has no physical meaning, we may suppose that $K = K' = 0$ for $r < 0$. Therefore (18.1-39) may be replaced by

$$\int_{-\infty}^{\infty} K'(r)\, dr = 1. \tag{18.1-40}$$

The relations expressed by (18.1-38 and 40) are those which characterize a δ function. Hence it is possible to express K' by

$$K' = 4\pi r^2\, \delta(\mathbf{r}) \tag{18.1-41}$$

since then

$$\int_{-\infty}^{\infty} (dK/dr)\, dr = 4\pi \int_{-\infty}^{\infty} \delta(\mathbf{r}) r^2\, dr = \int \delta(\mathbf{r})\, d\mathbf{r} = 1. \tag{18.1-42}$$

Let us now consider $\mathscr{H}_b$ in (18.1-36) and let

$$\mathscr{H}_b = -\boldsymbol{\mu} \cdot \mathbf{B} \tag{18.1-43a}$$

with

$$\mathbf{B} = -2\beta \frac{dK}{dr} \left[\frac{\mathbf{S}}{r^2} - \frac{\mathbf{r}(\mathbf{S} \cdot \mathbf{r})}{r^4} \right]. \tag{18.1-43b}$$

Applying the Landé formula (6.3-19),

$$\langle jm|\mathbf{T}|jm'\rangle = \frac{\langle jm|\mathbf{T} \cdot \mathbf{J}|jm\rangle}{j(j+1)} \langle jm|\mathbf{J}|jm'\rangle$$

and, with $\mathbf{T} = \mathbf{B}$, $\mathbf{J} = \mathbf{S}$, $j = s$, we obtain

$$\langle sm|\mathbf{B}|sm'\rangle = \frac{\langle sm|\mathbf{B}\cdot\mathbf{S}|sm\rangle}{s(s+1)}\langle sm|\mathbf{S}|sm'\rangle \tag{18.1-44}$$

with

$$\langle sm|\mathbf{B}\cdot\mathbf{S}|sm\rangle = -2\beta\left\langle sm\left|\frac{dK}{dr}\left[\frac{S^2}{r^2} - \frac{(\mathbf{S}\cdot\mathbf{r})^2}{r^4}\right]\right|sm\right\rangle \tag{18.1-45a}$$

and

$$(\mathbf{S}\cdot\mathbf{r})(\mathbf{S}\cdot\mathbf{r}) = \tfrac{1}{4}(\boldsymbol{\sigma}\cdot\mathbf{r})(\boldsymbol{\sigma}\cdot\mathbf{r}) = \tfrac{1}{4}r^2. \tag{18.1-45b}$$

Therefore

$$\begin{aligned}\langle sm|\mathbf{B}\cdot\mathbf{S}|sm\rangle &\equiv \langle\mathbf{B}\cdot\mathbf{S}\rangle\\ &= -2\beta\left\langle sm\left|\frac{dK}{dr}\left(\frac{S^2}{r^2} - \frac{1}{4r^2}\right)\right|sm\right\rangle\\ &= -2\beta\left\langle\frac{dK}{dr}\frac{1}{r^2}\right\rangle\left[s(s+1) - \frac{1}{4}\right]\\ &= -\beta\left\langle\frac{dK}{dr}\frac{1}{r^2}\right\rangle \qquad \text{for} \quad s = \frac{1}{2}\end{aligned} \tag{18.1-46}$$

where $\langle(dK/dr)(1/r^2)\rangle$ is an integral taken over the radial part of the wave function. But, from (18.1-41),

$$\left\langle\frac{dK}{dr}\frac{1}{r^2}\right\rangle = 4\pi\langle\delta(\mathbf{r})\rangle \tag{18.1-47}$$

or

$$\langle\mathbf{B}\cdot\mathbf{S}\rangle = -4\pi\beta\langle\delta(\mathbf{r})\rangle. \tag{18.1-48}$$

Substitution in (18.1-44) gives

$$\langle sm|\mathbf{B}|sm'\rangle = -\frac{16\pi}{3}\beta\langle\delta(\mathbf{r})\rangle\langle sm|\mathbf{S}|sm'\rangle. \tag{18.1-49}$$

As an operator relationship, we may write

$$\mathbf{B} = -\frac{16\pi}{3}\beta\,\delta(\mathbf{r})\mathbf{S} \tag{18.1-50}$$

so that $\mathscr{H}_b$ in (18.1-43a) becomes

$$\mathscr{H}_b = -\boldsymbol{\mu}\cdot\mathbf{B} = \frac{16\pi}{3}\beta\,\delta(\mathbf{r})\boldsymbol{\mu}\cdot\mathbf{S} = a\boldsymbol{\mu}\cdot\mathbf{S} \tag{18.1-51}$$

with

$$a = \frac{16\pi}{3} \beta\, \delta(\mathbf{r}). \tag{18.1-52}$$

At times it is convenient to regard $\mathbf{B} = -a\mathbf{S}$ as an effective magnetic field. Substituting (18.1-51) in (18.1-36), the magnetic hyperfine Hamiltonian acquires the form

$$\mathscr{H}_1 \equiv \mathscr{H}_h = 2\beta\left[\frac{\boldsymbol{\mu}\cdot(\mathbf{L}-\mathbf{S})}{r^3} + \frac{3(\boldsymbol{\mu}\cdot\mathbf{r})(\mathbf{S}\cdot\mathbf{r})}{r^5} + \frac{8\pi}{3}\delta(\mathbf{r})\boldsymbol{\mu}\cdot\mathbf{S}\right]. \tag{18.1-53}$$

The nuclear magnetic moment operator $\boldsymbol{\mu}$ is related to the nuclear spin operator $\mathbf{I}$ by an expression of the same form that relates electronic magnetic moments to their respective angular momenta (17.1-23), namely,

$$\boldsymbol{\mu} = g_N \beta_N \mathbf{I} \tag{18.1-54}$$

where

$$\begin{aligned} \beta_N = e\hbar/2Mc &= 5.0505 \times 10^{-24} \quad \text{erg/G} \\ &= 5.0505 \times 10^{-27} \quad \text{J/T} \end{aligned} \tag{18.1-55}$$

is the nuclear Bohr magneton. Here M is the mass of the proton and g_N a factor whose magnitude and sign are characteristic of each nucleus (Table 18.1). It is also customary to write (18.1-54) as

$$\boldsymbol{\mu} = \gamma\hbar\mathbf{I} \tag{18.1-56}$$

TABLE 18.1

Some Nuclear Properties[a]

	I	g_N	μ	γ (10^4 rad/SG)	Q (10^{-24} cm^2)
^{1}H	$\frac{1}{2}$	5.586	2.793	2.675	
^{2}H	1	0.857	0.857	0.411	2.77×10^{-3}
n	$\frac{1}{2}$	−3.826	−1.913	1.832	
^{12}C	0				
^{13}C	$\frac{1}{2}$	1.404	0.702	0.673	
^{14}N	1	0.404	0.404	0.193	7.1×10^{-2}
^{15}N	$\frac{1}{2}$	−0.566	−0.283	0.271	
^{16}O	0				
^{17}O	$\frac{5}{2}$	−0.757	−1.893	0.363	-4×10^{-3}
^{19}F	$\frac{1}{2}$	5.254	2.627	2.516	
^{35}Cl	$\frac{3}{2}$	0.548	0.821	0.262	-7.97×10^{-2}

[a] I is the spin, $\mu = g_N I$ the magnetic moment in units of the nuclear Bohr magneton ($\beta_N = e\hbar/2Mc = 5.05 \times 10^{-24}$ erg/G), $\gamma = g_N\beta_N/\hbar$ the gyromagnetic ratio, Q the electric quadrupole moment).

where $\gamma = g_N\beta_N/\hbar$ is known as the *gyromagnetic ratio.* Following the discussion in Section 17.1, the magnetic moment of the nucleus μ is defined as

$$\begin{aligned}\mu &= \langle I\ m_I = I|\mu_z|I\ m_I = I\rangle \\ &= g_N\beta_N\langle I\ m_I = I|I_z|I\ m_I = I\rangle = g_N\beta_N I = \gamma\hbar I \qquad (18.1\text{-}57)\end{aligned}$$

which permits the magnetic moment operator to be written as

$$\boldsymbol{\mu} = \mu\mathbf{I}/I. \qquad (18.1\text{-}58)$$

Substituting (18.1-56) into (18.1-53) we obtain

$$\mathscr{H}_h = 2\beta\gamma\hbar\left[\frac{\mathbf{I}\cdot(\mathbf{L}-\mathbf{S})}{r^3} + \frac{3(\mathbf{I}\cdot\mathbf{r})(\mathbf{S}\cdot\mathbf{r})}{r^5} + \frac{8\pi}{3}\delta(\mathbf{r})\mathbf{I}\cdot\mathbf{S}\right] \qquad (18.1\text{-}59)$$

in which the coefficient in front of the square brackets can be put in several equivalent forms:

$$2\beta\gamma\hbar = 2\beta g_N\beta_N \approx g_e\beta\gamma\hbar = g_e\beta g_N\beta_N. \qquad (18.1\text{-}60)$$

Equation (18.1-59) is the most common expression for the magnetic hyperfine interaction. The first two terms in the brackets taken together are known as the *dipole–dipole interaction* in analogy with the corresponding classical expression. The last term is the *Fermi contact interaction*; it has no classical analog.

In the absence of orbital angular momentum, (18.1-59) may be written as

$$\mathscr{H}_h = \mathbf{I}\cdot\mathbf{A}\cdot\mathbf{S} \qquad (18.1\text{-}61)$$

in which

$$\mathbf{A} = 2\beta\gamma\hbar\left[-\frac{1}{r^3} + \frac{3\mathbf{rr}}{r^5} + \frac{8\pi}{3}\delta(\mathbf{r})\right] \qquad (18.1\text{-}62)$$

is the magnetic *hyperfine coupling tensor.*

18.2 Magnetic Hyperfine Interaction in One-Electron Systems

An alternative expression for the magnetic hyperfine Hamiltonian (18.1-59) is obtained by transforming the dipole–dipole part by means of the Landé formula (6.3-19). Let

$$\mathbf{B}' = -2\beta\left[\frac{\mathbf{L}-\mathbf{S}}{r^3} + \frac{3\mathbf{r}(\mathbf{S}\cdot\mathbf{r})}{r^5}\right]. \qquad (18.2\text{-}1)$$

Then

$$\langle jm|\mathbf{B}'|jm'\rangle = \frac{\langle jm|\mathbf{B}'\cdot\mathbf{J}|jm\rangle}{j(j+1)}\langle jm|\mathbf{J}|jm'\rangle \qquad (18.2\text{-}2)$$

and

$$\langle jm|\mathbf{B}' \cdot \mathbf{J}|jm\rangle = -2\beta \left\langle jm \left| \frac{(\mathbf{L}-\mathbf{S})\cdot\mathbf{J}}{r^3} + \frac{3(\mathbf{r}\cdot\mathbf{J})(\mathbf{S}\cdot\mathbf{r})}{r^5} \right| jm \right\rangle. \quad (18.2\text{-}3)$$

But since

$$(\mathbf{L}-\mathbf{S})\cdot\mathbf{J} = (\mathbf{L}-\mathbf{S})\cdot(\mathbf{L}+\mathbf{S}) = L^2 - S^2$$

and recalling that $|jm\rangle \equiv |lsjm\rangle$ is an eigenfunction of L^2 and S^2 (as well as of J^2 and J_z),

$$\langle jm|(\mathbf{L}-\mathbf{S})\cdot\mathbf{J}|jm\rangle = \langle jm|L^2 - S^2|jm\rangle = l(l+1) - s(s+1). \quad (18.2\text{-}4)$$

Also,

$$\begin{aligned} \mathbf{r}\cdot\mathbf{J} = \mathbf{r}\cdot(\mathbf{L}+\mathbf{S}) &= \mathbf{r}\cdot\frac{\mathbf{r}\times\mathbf{p}}{\hbar} + \mathbf{r}\cdot\mathbf{S} \\ &= \mathbf{r}\cdot\mathbf{S} = \mathbf{S}\cdot\mathbf{r}. \end{aligned} \quad (18.2\text{-}5)$$

Therefore, as in (18.1-45b),

$$(\mathbf{r}\cdot\mathbf{J})(\mathbf{r}\cdot\mathbf{S}) = (\mathbf{r}\cdot\mathbf{S})(\mathbf{r}\cdot\mathbf{S}) = \tfrac{1}{4}(\boldsymbol{\sigma}\cdot\mathbf{r})(\boldsymbol{\sigma}\cdot\mathbf{r}) = \tfrac{1}{4}r^2.$$

Integrating over the radial part of the wave function we obtain

$$\begin{aligned} \langle jm|B'\cdot\mathbf{J}|jm\rangle &= -2\beta[l(l+1) - s(s+1) + \tfrac{3}{4}]\langle 1/r^3\rangle \\ &= -2\beta l(l+1)\langle 1/r^3\rangle \qquad (s=\tfrac{1}{2}) \end{aligned} \quad (18.2\text{-}6)$$

so that

$$\mathbf{B}' = -2\beta\frac{l(l+1)}{j(j+1)}\langle 1/r^3\rangle \qquad (s=\tfrac{1}{2}). \quad (18.2\text{-}7)$$

We may now replace (18.2-1) with (18.2-7) and substitute in (18.1-59). The Hamiltonian, for $s=\frac{1}{2}$, then takes the form

$$\begin{aligned} \mathscr{H}_h &= 2\beta\gamma\hbar\left[\frac{l(l+1)}{j(j+1)}\left\langle\frac{1}{r^3}\right\rangle \mathbf{I}\cdot\mathbf{J} + \frac{8\pi}{3}|\psi(0)|^2\mathbf{I}\cdot\mathbf{S}\right] \\ &= 2\beta\gamma\hbar\frac{l(l+1)}{j(j+1)}\left\langle\frac{1}{r^3}\right\rangle \mathbf{I}\cdot\mathbf{J} + A_F\mathbf{I}\cdot\mathbf{S} \end{aligned} \quad (18.2\text{-}8)$$

where

$$\begin{aligned} A_F &= \frac{16\pi}{3}\beta\gamma\hbar|\psi(0)|^2, \\ |\psi(0)|^2 &= \langle\delta(\mathbf{r})\rangle. \end{aligned} \quad (18.2\text{-}9)$$

The first (dipole) term in (18.2-8) is zero when $l=0$; on the other hand, the second (contact) term is zero when $l \neq 0$, which means, then, that the two parts of the magnetic hyperfine interaction do not overlap. For s states, we need only consider the contact interaction, whereas for non-s states, only the dipole part contributes.

As in the case of spin-orbit coupling, the form (18.2-8) suggests that it would be useful to couple the angular momenta **I** and **J** to form a new angular momentum operator

$$\mathbf{F} = \mathbf{I} + \mathbf{J} \tag{18.2-10}$$

with the notation $|jm\rangle, |Im_I\rangle$ and $|Fm_F\rangle$ to designate the eigenfunctions belonging to **J**, **I** and **F**, respectively.

For hydrogen in an s state only the contact term is effective. Since $l = 0$, **J** may be replaced by **S** so that

$$\mathbf{I} \cdot \mathbf{S} = \tfrac{1}{2}[F^2 - S^2 - I^2], \tag{18.2-11}$$

and, since the spin of the proton is $\frac{1}{2}$, we have

$$I = \tfrac{1}{2}, \qquad s = \tfrac{1}{2}, \qquad F = 0, 1. \tag{18.2-12}$$

Hence

$$\begin{aligned} \langle Fm_F|\mathbf{I} \cdot \mathbf{S}|Fm_F\rangle &= \langle Fm_F|\tfrac{1}{2}(F^2 - S^2 - I^2)|Fm_F\rangle \\ &= \tfrac{1}{2}[F(F+1) - s(s+1) - I(I+1)] \\ &= \tfrac{1}{2}[F(F+1) - \tfrac{3}{2}] \end{aligned} \tag{18.2-13}$$

which indicates that only diagonal elements are nonzero. For the two possible values of F the energies are

$$E = \begin{cases} \frac{4}{3}\pi\beta\gamma\hbar|\psi(0)|^2 = \dfrac{A_F}{4}, & F = 1, \\[2ex] -4\pi\beta\gamma\hbar|\psi(0)|^2 = \dfrac{-3A_F}{4}, & F = 0. \end{cases} \tag{18.2-14}$$

For hydrogen,

$$\begin{aligned} |\psi(0)|^2 &= 1/\pi a_0{}^3 \qquad (a_0 = 0.529 \times 10^{-8} \ \text{cm}), \\ \gamma\hbar = \beta_N g_N &= 5.05 \times 10^{-24} \ \text{erg/G} \times 5.586 \\ &= 2.82 \times 10^{-23} \ \text{erg/G}. \end{aligned}$$

With $\beta = 0.927 \times 10^{-20}$ erg/G, the difference in energy between the two levels in (18.2-14) is

$$\begin{aligned} \Delta E &= \frac{16\pi}{3} \beta\gamma\hbar|\psi(0)|^2 = A_F = \frac{16}{3}\frac{\beta\gamma\hbar}{a_0{}^3} \\ &= 0.04738 \ \text{cm}^{-1} = 5.9 \times 10^{-6} \ \text{eV}, \end{aligned} \tag{18.2-15}$$

as shown in Fig. 18.2. If we set $\Delta E = \hbar\nu$, the frequency ν is 1420.4058 MHz, which corresponds to a wavelength of 21 cm. In deuterium, the wavelength is 92 cm‡. It is seen that the energy separation ΔE is much smaller than the separation between atomic states, and, of course, much smaller than the separation between nuclear states (typically 10^5–10^6 eV). Hence, it is justifiable to assume that only the ground states of the atom and nucleus are affected by the magnetic hyperfine interaction.

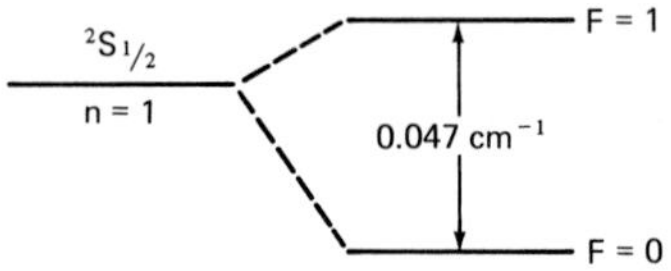

FIG. 18.2 Magnetic hyperfine splitting of the ground state of hydrogen. The entire splitting is due to the contact term.

It is also of interest to ascertain the magnitude of the effective magnetic fields at the positions of the electron and the proton. Considering first the electron, it is seen that $^2S_{1/2}$ is split into two states by the magnetic field produced by the proton. If this splitting is put equal to $g_e\beta B_{\text{eff}}^{\text{e}}$ as in Fig. 17.1, then, from (18.2-15)

$$\Delta E = g_e\beta B_{\text{eff}}^{\text{e}}$$

and $B_{\text{eff}}^{\text{e}}$ comes out to be 507 G. In the same way, the magnetic field at the proton is given by

$$\Delta E = g_N\beta_N B_{\text{eff}}^{\text{p}} = \gamma\hbar B_{\text{eff}}^{\text{p}}$$

where $g_N\beta_N = \gamma\hbar$ is the magnetic moment of the proton in analogy with $g_e\beta$ which is the magnetic moment of the electron. The value of $B_{\text{eff}}^{\text{p}}$ is 3.3×10^5 G. This is an example of the very high magnetic fields that can exist at the positions of the nuclei due to the relatively large magnetic moment of the electron.

Next we consider the dipole–dipole term in (18.2-8). For hydrogen (Table 16.4)

$$\left\langle \frac{1}{r^3} \right\rangle = \frac{Z^3}{a_0{}^3} \frac{1}{n^3(l+1)(l+\frac{1}{2})l}.$$

Also

$$\mathbf{I}\cdot\mathbf{J} = \tfrac{1}{2}[F^2 - I^2 - J^2],$$
$$\langle Fm_F|\mathbf{I}\cdot\mathbf{J}|Fm_F\rangle = \tfrac{1}{2}[F(F+1) - I(I+1) - j(j+1)].$$

‡ These wavelengths have been observed in astrophysical investigations.

Again only diagonal matrix elements are nonzero. Therefore the energies are given by

$$E = \beta\gamma\hbar \frac{Z^3}{a_0{}^3 n^3 j(j+1)(l+\frac{1}{2})}[F(F+1) - j(j+1) - I(I+1)]. \quad (18.2\text{-}16)$$

For $2\,^2P_{1/2}$, $Z = 1, j = \frac{1}{2}, I = \frac{1}{2}, n = 2, l = 1$,

$$E = \begin{cases} \frac{1}{18}(\beta\gamma\hbar/a_0{}^3), & F = 1, \\ -\frac{1}{6}(\beta\gamma\hbar/a_0{}^3), & F = 0, \end{cases} \quad (18.2\text{-}17a)$$

$$\Delta E = -\tfrac{2}{9}(\beta\gamma\hbar/9a_0{}^3) \quad (18.2\text{-}17b)$$

which is smaller, by a factor of 24, than the splitting due to the contact interaction. Similarly for $2\,^2P_{3/2}$ the energies are

$$E = \begin{cases} \frac{1}{30}(\beta\gamma\hbar/a_0{}^3), & F = 2, \\ -\frac{1}{18}(\beta\gamma\hbar/a_0{}^3), & F = 1, \end{cases} \quad (18.2\text{-}18a)$$

$$\Delta E = \tfrac{4}{45}(\beta\gamma\hbar/a_0{}^3). \quad (18.2\text{-}18b)$$

A case of practical importance arises when the magnetic hyperfine interaction occurs in the presence of an external magnetic field. This is the situation, for example, in a magnetic resonance experiment. We shall specialize to hydrogen in an s state so that only the contact term in (18.2-8) contributes. The Hamiltonian may then be written as

$$\mathscr{H} = 2\beta \mathbf{B} \cdot \mathbf{S} + A_F \mathbf{I} \cdot \mathbf{S} - \gamma\hbar \mathbf{B} \cdot \mathbf{I} \quad (18.2\text{-}19)$$

in which A_F is given by (18.2-9). The term proportional to $\mathbf{B} \cdot \mathbf{I}$ takes into account the interaction of the nuclear magnetic moment with the external field **B**. However, the nuclear magnetic moment is very small compared to the electronic magnetic moment so that the contribution from the term containing $\mathbf{B} \cdot \mathbf{I}$ is small and will be neglected here.

We have seen that the magnetic field felt by the s electron due to the nuclear magnetic moment is 507 G. If the external field is large compared with 507 G, the electronic and nuclear angular momenta will be decoupled. The argument is basically the same as in the case of "strong" versus "weak" fields in relation to spin-orbit coupling (Section 17.1). Since we wish to know the eigenvalues of $\mathscr{H}$ in (18.2-19) over a wide range of the external field, it will be more convenient to employ the uncoupled representation $|sIm_sm_I\rangle$. With s and I remaining equal to $\frac{1}{2}$, the notation could be shortened to $|m_sm_I\rangle$. The 4×4 matrix with m_s and m_I each taking on the possible values of $\pm\frac{1}{2}$ may now be calculated by rewriting (18.2-19), without the $\mathbf{B} \cdot \mathbf{I}$ term, as

$$\mathscr{H} = 2\beta BS_z + A_F[-I_{+1}S_{-1} + I_0S_0 - I_{-1}S_{+1}]. \quad (18.2\text{-}20)$$

The axes have been chosen, as we have before, so that the magnetic field is along the z axis and **B** has only one component $B_z = B$. With the aid of

TABLE 18.2

Matrix Elements of the Magnetic Hyperfine Interaction in the Hydrogen s State in the Presence of an External Magnetic Field[a]

	$\lvert\frac{1}{2}\frac{1}{2}\rangle$	$\lvert\frac{1}{2}\,-\frac{1}{2}\rangle$	$\lvert-\frac{1}{2}\,\frac{1}{2}\rangle$	$\lvert-\frac{1}{2}\,-\frac{1}{2}\rangle$
$\langle\frac{1}{2}\frac{1}{2}\rvert$	$\beta B + \frac{1}{4}A_F$			
$\langle\frac{1}{2}\,-\frac{1}{2}\rvert$		$\beta B - \frac{1}{4}A_F$	$\frac{1}{2}A_F$	
$\langle-\frac{1}{2}\,\frac{1}{2}\rvert$		$\frac{1}{2}A$	$-\beta B - \frac{1}{4}A_F$	
$\langle-\frac{1}{2}\,-\frac{1}{2}\rvert$				$-\beta B + \frac{1}{4}A_F$

[a] The bras and kets are of the form $\langle m_s m_I\rvert$ and $\lvert m_s m_I\rangle$, respectively; A_F is the hyperfine coupling constant.

Table 1.3, the Hamiltonian matrix acquires the form shown in Table 18.2 whose energy eigenvalues are then given by

$$E = \beta B + \tfrac{1}{4}A_F, \qquad -\beta B + \tfrac{1}{4}A_F \tag{18.2-21}$$

and

$$E = -\tfrac{1}{4}A_F \pm \tfrac{1}{2}A_F\sqrt{1 + (2\beta B/A_F)^2}; \tag{18.2-22}$$

the latter being associated with the 2 × 2 matrix. The energy eigenvalues are shown in Fig. 18.3. Several special cases are of interest:

$B = 0$: The eigenvalues are $\frac{1}{4}A_F$ (three-fold degenerate) and $-\frac{3}{4}A_F$. This result is identical with (18.2-14) and shows very clearly the development of the three-fold degeneracy associated with $F = 1$.

$\beta B \ll A_F$: For this case the approximate energies are

$$E = \begin{cases} \beta B + \frac{1}{4}A_F \\ \frac{1}{4}A_F + \beta^2B^2/A_F \\ -\frac{3}{4}A_F - \beta^2B^2/A_F \\ -\beta B + \frac{1}{4}A_F \end{cases} \tag{18.2-23}$$

Note that in this case the energies of two of the levels have a quadratic dependence on the field.

$\beta A \gg A$:

$$E = \begin{cases} \beta B + \frac{1}{4}A_F & \quad \frac{1}{2} \quad \frac{1}{2}, \\ \beta B - \frac{1}{4}A_F & \quad \frac{1}{2} \quad -\frac{1}{2}, \\ -\beta B - \frac{1}{4}A_F & \quad -\frac{1}{2} \quad \frac{1}{2}, \\ -\beta B + \frac{1}{4}A_F & \quad -\frac{1}{2} \quad -\frac{1}{2}. \end{cases} \qquad \begin{matrix} m_s & m_I \end{matrix} \tag{18.2-24}$$

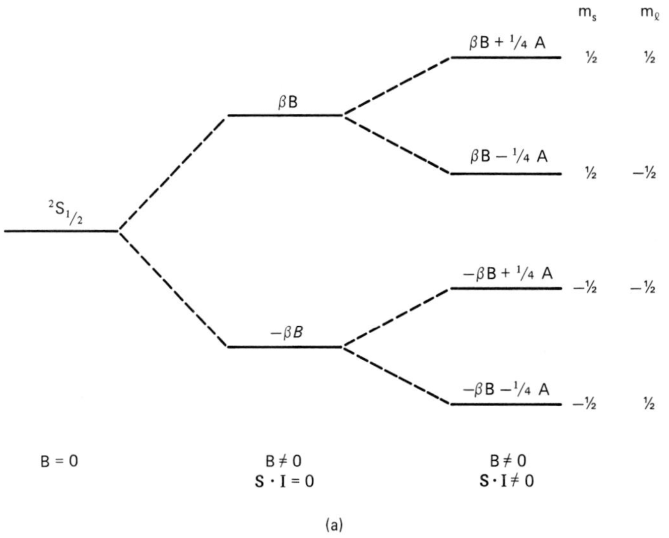

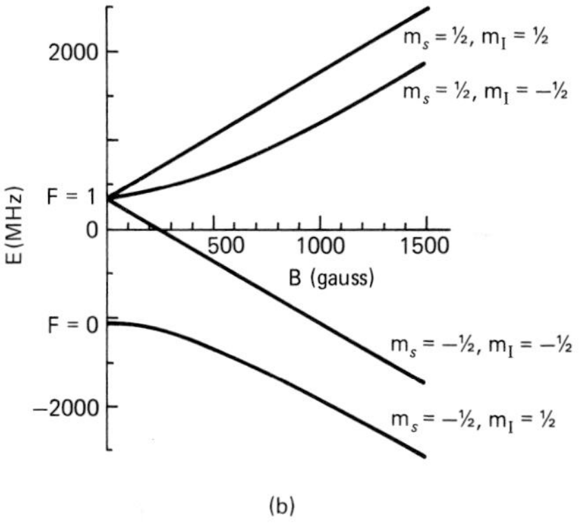

FIG. 18.3 Energy levels of hydrogen ground state in a magnetic field with hyperfine splitting: (a) first-order splitting and (b) variation with an external magnetic field.

These energies are obtained directly from (18.2-21) and (18.2-22); they also appear as the diagonal elements in the matrix of Table 18.2. It is from the latter that we obtain the identification of the levels as to the values of m_s and m_I. Note that when $\beta B \gg A$ it is permissible to replace the Hamiltonian (18.2-20) by

$$\mathscr{H} = 2\beta B S_z + A_F I_z S_z \tag{18.2-25}$$

since the eigenvalues of (18.2-25) are precisely the energies shown in (18.2-24).

This may be regarded as the Paschen–Back region for the magnetic hyperfine interaction.

All this is reminiscent of the situation in which the spin-orbit and magnetic field interactions are treated simultaneously. When the magnetic field is weak ($\ll 507$ G in hydrogen), the electronic and spin angular momenta are coupled to form states with $F = 0, 1$. The levels spread linearly with the field. At larger fields, the dependence is quadratic; at still higher fields ($\gg 507$ G) the electronic and spin angular momenta become uncoupled and the dependence on field becomes linear again. The so-called "good" quantum numbers at very low fields are s, I, F, m_F; at very high fields, the "good" quantum numbers are s, I, m_s, m_I. At intermediate fields the levels are mixed, and neither system of quantum numbers provides an adequate set of labels.

Figure 18.4 shows the level scheme for an s electron interacting with a nucleus with spin $I = 1$ through the magnetic hyperfine interaction; this occurs, for example, in ^{14}N.

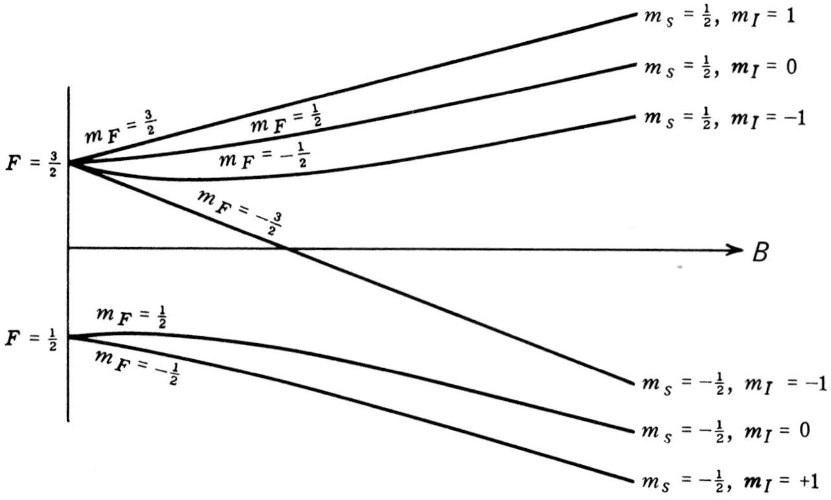

FIG. 18.4 Magnetic hyperfine levels for a nucleus with spin $I = 1$ and positive coupling constant as a function of magnetic field. (Reproduced from "Introduction to Electron Paramagnetic Resonance," written by Malcolm Bersohn and James C. Baird, with permission of publishers, Addison-Wesley/W. A. Benjamin, Inc., Advanced Book Program, Reading, Massachusetts.)

18.3 Electric Quadrupole Interaction

The quadrupole moment of a nucleus is a measure of the departure of the mean distribution of nuclear charge from spherical symmetry. It is positive for a distribution which is prolate ellipsoidal (football), negative for an oblate (door-knob) distribution, and zero for a spherically symmetric distribution. Some nuclei that possess quadrupole moments (Table 18.1) are ^{2}H, ^{14}N, ^{17}O, ^{35}Cl.

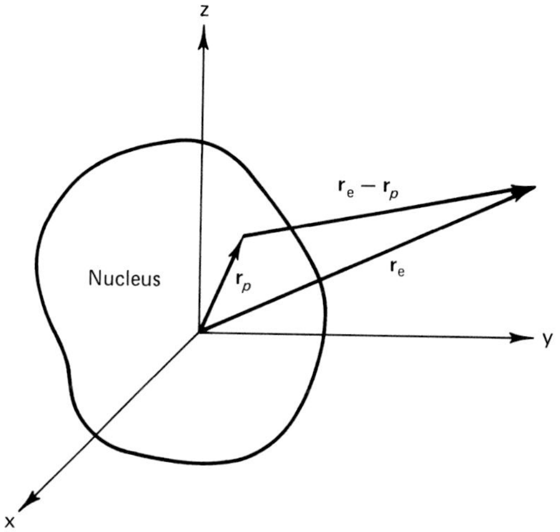

FIG. 18.5 Coordinate system for Eq. (18.3-1).

Referring to a coordinate system (Fig. 18.5) whose origin is located within the nucleus, the electrostatic interaction between a single electron and a nucleus containing Z protons is

$$W = -\sum_{p=1}^{Z} \frac{e^2}{|\mathbf{r}_e - \mathbf{r}_p|} \tag{18.3-1}$$

in which $\mathbf{r}_e$ is the position vector of the electron, $\mathbf{r}_p$ is the position vector of the pth proton, and the sum is taken over all Z protons. From (1.2-22),

$$\frac{1}{|\mathbf{r}_e - \mathbf{r}_p|} = 4\pi \sum_{l=0}^{\infty} \sum_{m=-l}^{l} \frac{1}{2l+1} \frac{r_<^{\,l}}{r_>^{l+1}} Y_{lm}^*(\theta_p, \varphi_p) Y_{lm}(\theta_e, \varphi_e) \tag{18.3-2}$$

and for electrons that do not penetrate the nucleus,‡ $r_e > r_p$, so that (18.3-1) becomes

$$W = -4\pi e^2 \sum_{p=1}^{Z} \sum_{l=0}^{\infty} \sum_{m=-l}^{l} \frac{1}{2l+1} \frac{r_p^{\,l}}{r_e^{l+1}} Y_{lm}^*(\theta_p, \varphi_p) Y_{lm}(\theta_e, \varphi_e). \tag{18.3-3}$$

‡ It is shown below that there is no quadrupole interaction in s states.

It is possible to express (18.3-3) more compactly by writing

$$\sum_{m=-l}^{l} Y_{lm}^* Y_{lm} = \sum_m (-1)^m Y_{l\,-m} Y_{lm} = \mathbf{Y}^{(l)}(\Omega) \cdot \mathbf{Y}^{(l)}(\Omega),$$

$$(\theta_p, \varphi_p) \equiv (\Omega_\mathrm{p}), \qquad (\theta_\mathrm{e}, \varphi_\mathrm{e}) \equiv (\Omega_\mathrm{e})$$

$$\mathbf{Q}^{(l)} \equiv e \sum_{p=1}^{Z} \sqrt{\frac{4\pi}{2l+1}}\, r_p{}^l \mathbf{Y}^{(l)}(\Omega_p)$$

$$\mathbf{U}^{(l)} \equiv -e \sqrt{\frac{4\pi}{2l+1}} \frac{1}{r_\mathrm{e}^{l+1}} \mathbf{Y}^{(l)}(\Omega_\mathrm{e}). \tag{18.3-4}$$

We then have

$$W = \sum_l \mathbf{Q}^{(l)} \cdot \mathbf{U}^{(l)}. \tag{18.3-5}$$

When $l = 0$, (18.3-5) becomes

$$W = -Ze^2/r_\mathrm{e} \qquad (l = 0), \tag{18.3-6}$$

which is just the ordinary Coulomb interaction with Z being the sum over the protons. For a nucleus of finite size there is a correction term to be added to (18.3-6):

$$\tfrac{2}{3}\pi e^2 Z |\psi(0)|^2 \langle R^2 \rangle \tag{18.3-7}$$

in which $\langle R^2 \rangle$ is the mean square charge radius of the nucleus and $e^2|\psi(0)|^2$ is the electronic charge density at the nucleus.

The term with $l = 1$ in (18.3-5) vanishes because it corresponds to the interaction between a nuclear electric dipole moment and the electric field established by the electrons. Nuclear states have well-defined parity (neglecting the "weak" interactions) so that the expectation value of the nuclear electric dipole moment—which is an odd operator—vanishes. The next term, with $l = 2$, is the electric quadrupole interaction

$$\mathscr{H}_Q = \mathbf{Q}^{(2)} \cdot \mathbf{U}^{(2)}. \tag{18.3-8}$$

We will now suppose that the electronic state is characterized by angular momentum quantum numbers j, m_j, the nuclear state by I, m_I and when the two angular momenta are coupled the quantum numbers are I, j, F, m_F. We shall compute the interaction energy associated with $\mathscr{H}_Q$ in the latter representation. The pertinent expression is (6.3-37) which, with the omission of quantum numbers other than those pertaining to angular momenta, is

$$\langle j_1 j_2 jm | \mathbf{T}^{(k)} \cdot \mathbf{U}^{(k)} | j_1' j_2' j' m' \rangle$$

$$= (-1)^{j+j_2+j_1'} \delta_{jj'} \delta_{mm'} \begin{Bmatrix} j_1' & j_2' & j \\ j_2 & j_1 & k \end{Bmatrix} \langle j_1 \| \mathbf{T}^{(k)} \| j_1' \rangle \langle j_2 \| \mathbf{U}^{(k)} \| j_2' \rangle. \tag{18.3-9}$$

In the present context this becomes

$$\langle IjFm_F|\mathbf{Q}^{(2)}\cdot\mathbf{U}^{(2)}|IjF'm_F'\rangle$$
$$= (-1)^{F+j+I}\delta_{FF'}\delta_{m_Fm_F'}\begin{Bmatrix} I & j & F \\ j & I & 2 \end{Bmatrix}\langle I\|\mathbf{Q}^{(2)}\|I\rangle\langle j\|\mathbf{U}^{(2)}\|j\rangle. \tag{(18.3-10)}$$

From Table 1.7,

$$\begin{Bmatrix} I & j & F \\ j & I & 2 \end{Bmatrix} = (-1)^{F+I+j}$$
$$\times \frac{2[3X(X-1)-4I(I+1)j(j+1)]}{\sqrt{(2I-1)2I(2I+1)(2I+2)(2I+3)(2j-1)2j(2j+1)(2j+2)(2j+3)}} \tag{18.3-11}$$

where

$$X = I(I+1) + j(j+1) - F(F+1).$$

The main problem then, is to evaluate the reduced matrix elements in (18.3-10).

First consider $\langle I\|\mathbf{Q}^{(2)}\|I\rangle$. This quantity can be related to the nuclear quadrupole moment Q which is defined by

$$Q = \left\langle I\ m_I = I\left|\sum_p (3z_p{}^2 - r_p{}^2)\right| I\ m_I = I\right\rangle. \tag{18.3-12}$$

Since

$$\sum_p (3z_p{}^2 - r_p{}^2) = 2\sum_p \sqrt{\tfrac{4}{5}\pi}\,r_p{}^2\, Y_{20}(\Omega_p) = (2/e)Q_0^{(2)} \tag{18.3-13}$$

where the second equality comes from (18.3-4), we have

$$\tfrac{1}{2}eQ = \langle I\ m_I = I|Q_0^{(2)}|I\ m_I = I\rangle. \tag{18.3-14}$$

It is now possible to invoke the Wigner–Eckart theorem

$$\langle I\ m_I|Q_0^{(2)}|I\ m_I\rangle = (-1)^{I-m}\begin{pmatrix} I & 2 & I \\ -m_I & 0 & m_I \end{pmatrix}\langle I\|\mathbf{Q}^{(2)}\|I\rangle, \tag{18.3-15}$$

and, on setting $m_I = I$ and evaluating the $3j$ symbol (Table 1.5),

$$\langle I\ m_I = I|Q_0^{(2)}|I\ m_I = I\rangle = \sqrt{I(2I-1)/(2I+1)(I+1)(2I+3)}\langle I\|\mathbf{Q}^{(2)}\|I\rangle \tag{18.3-16a}$$
$$= \tfrac{1}{2}eQ \tag{18.3-16b}$$

or

$$\langle I\|\mathbf{Q}^{(2)}\|I\rangle = \tfrac{1}{2}\sqrt{(2I+1)(I+1)(2I+3)/I(2I-1)}\,eQ. \tag{18.3-17}$$

Note that according to (18.3-16a) and (18.3-16b), nuclei with spins $I = 0$ or $\frac{1}{2}$ have no quadrupole moment.

The computation of $\langle j \| \mathbf{U}^{(2)} \| j \rangle$ proceeds in analogous fashion. This time we define a quantity eq as

$$\tfrac{1}{2}eq = \langle j\, m_j = j | U_0^{(2)} | j\, m_j = j \rangle \tag{18.3-18}$$

where, from (18.3-4),

$$U_0^{(2)} = -e\sqrt{\frac{4\pi}{5}}\frac{1}{r_e^{\,3}} Y_{20}(\Omega_e) = -\frac{1}{2}e\frac{3z_e^{\,2} - r_e^{\,2}}{r_e^{\,5}}. \tag{18.3-19}$$

We note that

$$\frac{\partial^2}{\partial z^2}\left(-\frac{e}{r}\right) = -e\frac{3z^2 - r^2}{r^5} \tag{18.3-20}$$

is the zz component of the electric field gradient tensor produced by an electron at a point whose coordinates with respect to the electron are (x, y, z). Since the origin of the coordinate system (Fig. 18.5) has been positioned at the nucleus $2U_0^{(2)}$ in (18.3-19) is the zz component of the electric field gradient tensor at the nucleus produced by an electron at $\mathbf{r}_e$, or

$$U_0^{(2)} = \tfrac{1}{2}(\partial^2 V/\partial z^2)_0 \equiv \tfrac{1}{2}V_{zz} \tag{18.3-21}$$

where V is the potential due to the electron and the second derivative is evaluated at the origin (nucleus). We then have

$$eq = \langle j\, m_j = j | V_{zz} | j\, m_j = j \rangle = \langle V_{zz} \rangle \tag{18.3-22}$$

which is the average or expectation value of V_{zz} taken over the electronic state $|jj\rangle$.

Again the use of the Wigner–Eckart theorem as in (18.3-15) leads to the result

$$\langle j \| \mathbf{U}^{(2)} \| j \rangle = \tfrac{1}{2}\sqrt{(2j+1)(j+1)(2j+3)/j(2j-1)}\,eq. \tag{18.3-23}$$

On substituting (18.3-11), (18.3-17), and (18.3-23) into (18.3-10) one obtains

$$\langle IjFm_F | \mathbf{Q}^{(2)} \cdot \mathbf{U}^{(2)} | IjFm_F \rangle = \frac{e^2qQ}{2I(2I-1)j(2j-1)}\left[\frac{3}{4}X(X-1) - I(I+1)j(j+1)\right] \tag{18.3-24}$$

in which e^2qQ is known as the *quadrupole coupling constant* and X is the same quantity as in (18.3-11). The quadrupole coupling constant may be positive or negative.

Equation (18.3-24) may be cast into another form. From

$$\mathbf{F} = \mathbf{I} + \mathbf{J} \qquad \text{and} \qquad -2\mathbf{I}\cdot\mathbf{J} = I^2 + J^2 - F^2,$$

$$\langle IjFm_F | -2\mathbf{I}\cdot\mathbf{J} | IjFm_F \rangle = I(I+1) + j(j+1) - F(F+1) = X. \tag{18.3-25}$$

Thus

$$\langle IjFm_F|3(\mathbf{I}\cdot\mathbf{J})^2 + \tfrac{3}{2}(\mathbf{I}\cdot\mathbf{J}) - I^2J^2|IjFm_F\rangle = \tfrac{3}{4}X(X-1) - I(I+1)j(j+1) \tag{18.3-26}$$

which, when compared with (18.3-24), gives

$$\mathscr{H}_Q = \mathbf{Q}^{(2)}\cdot\mathbf{U}^{(2)} = \frac{e^2qQ}{2I(2I-1)j(2j-1)}\left[3(\mathbf{I}\cdot\mathbf{J})^2 + \frac{3}{2}(\mathbf{I}\cdot\mathbf{J}) - I^2J^2\right]. \tag{18.3-27}$$

Another useful expression equivalent to (18.3-24) may now be obtained. Consider the tensor product of the nuclear angular momentum operator **I** with itself. According to (6.2-21) we can construct a tensor $\mathbf{T}^{(2)}$ with components

$$T^{(2)}_{\pm 2} = I^2_{\pm 1} \tag{18.3-28a}$$

$$T^{(2)}_{\pm 1} = \frac{1}{\sqrt{2}}[I_{\pm 1}I_0 + I_0I_{\pm 1}] \tag{18.3-28b}$$

$$T^{(2)}_0 = \frac{1}{\sqrt{6}}[3I_0{}^2 - I^2]. \tag{18.3-28c}$$

Once more the Wigner–Eckart theorem may be used in form (6.3-14) to express a proportionality between the matrix elements of $Q^{(2)}_0$ and $T^{(2)}_0$. If A is the proportionality constant

$$\langle II|Q^{(2)}_0|II\rangle = A\langle II|T^{(2)}_0|II\rangle \tag{18.3-29}$$

in which $|II\rangle$ is an abbreviation for $|I\ m_I = I\rangle$; then, using (18.3-16b) and (18.3-28c)

$$\frac{1}{2}eQ = A\frac{1}{\sqrt{6}}\langle II|3I_0{}^2 - I^2|II\rangle = \frac{1}{\sqrt{6}}[3I^2 - I(I+1)] \tag{18.3-30}$$

or

$$A = \sqrt{\tfrac{3}{2}}eQ/I(2I-1). \tag{18.3-31}$$

This permits the components of $\mathbf{Q}^{(2)}$ to be expressed in terms of the nuclear angular momentum operators:

$$Q^{(2)}_{\pm 2} = \sqrt{\frac{3}{2}}\frac{eQ}{I(2I-1)}I^2_{\pm 1}, \tag{18.3-32a}$$

$$Q^{(2)}_{\pm 1} = \sqrt{\frac{3}{2}}\frac{eQ}{I(2I-1)}[I_{\pm 1}I_0 + I_0I_{\pm 1}], \tag{18.3-32b}$$

$$Q^{(2)}_0 = \frac{1}{2}\frac{eQ}{I(2I-1)}[3I_0{}^2 - I^2]. \tag{18.3-32c}$$

Analogous expressions for the components of $\mathbf{U}^{(2)}$ may also be found. For this purpose we note that

$$\frac{\partial^2}{\partial r_i\,\partial r_j}\left(\frac{-e}{r}\right) = -e\left[\frac{3r_i r_j - r^2\,\delta_{ij}}{r^5}\right], \qquad (r_i, r_j = x, y, z)$$

$$= V_{r_i r_j} \equiv V_{ij} = V_{ji}. \tag{18.3-33}$$

Using (18.3-33) and Table 1.1 it is possible to establish a connection between the components of $\mathbf{Y}^{(2)}$ and V_{ij}. These are

$$Y_{2\,\pm 2} = -\frac{1}{e}\frac{1}{4}\sqrt{\frac{5}{6\pi}}\,r^3[V_{xx} - V_{yy} \pm 2iV_{xy}], \tag{18.3-34a}$$

$$Y_{2\,\pm 1} = \pm\frac{1}{e}\frac{1}{2}\sqrt{\frac{5}{6\pi}}\,r^3[V_{zx} \pm iV_{yz}], \tag{18.3-34b}$$

$$Y_{2\,0} = -\frac{1}{e}\frac{1}{4}\sqrt{\frac{5}{\pi}}\,r^3 V_{zz}. \tag{18.3-34c}$$

From (18.3-4) we have

$$\mathbf{U}^{(2)} = -e\,\frac{\mathbf{Y}^{(2)}}{r^3}\sqrt{\frac{4\pi}{5}}. \tag{18.3-35}$$

Therefore

$$U^{(2)}_{\pm 2} = \tfrac{1}{4}\sqrt{\tfrac{2}{3}}[V_{xx} - V_{yy} \pm 2iV_{xy}], \tag{18.3-36a}$$

$$U^{(2)}_{\pm 1} = \mp\tfrac{1}{2}\sqrt{\tfrac{2}{3}}[V_{zx} \pm iV_{yz}], \tag{18.3-36b}$$

$$U^{(2)}_{0} = \tfrac{1}{2}V_{zz}. \tag{18.3-36c}$$

Combining (18.3-32) and (18.3-36) one obtains an expression for the Hamiltonian (18.3-8):

$$\mathscr{H}_Q = \mathbf{Q}^{(2)} \cdot \mathbf{U}^{(2)}$$

$$= Q^{(2)}_{+2}U^{(2)}_{-2} - Q^{(2)}_{+1}U^{(2)}_{-1} + Q^{(2)}_{0}U^{(2)}_{0} - Q^{(2)}_{-1}U^{(2)}_{+1} + Q^{(2)}_{-2}U^{(2)}_{+2} \tag{18.3-37}$$

where

$$Q^{(2)}_{+2}U^{(2)}_{-2} = \frac{1}{4}\frac{eQ}{I(2I-1)}\,I^2_{+1}[V_{xx} - V_{yy} - 2iV_{xy}], \tag{18.3-38a}$$

$$Q^{(2)}_{+1}U^{(2)}_{-1} = \frac{1}{2\sqrt{2}}\frac{eQ}{I(2I-1)}\,[I_{+1}I_0 + I_0 I_{+1}][V_{zx} - iV_{yz}], \tag{18.3-38b}$$

$$Q^{(2)}_{0}U^{(2)}_{0} = \frac{1}{4}\frac{eQ}{I(2I-1)}\,[3I_0{}^2 - I^2]V_{zz}, \tag{18.3-38c}$$

$$Q^{(2)}_{-1}U^{(2)}_{+1} = -\frac{1}{2\sqrt{2}}\frac{eQ}{I(2I-1)}[I_{-1}I_0 + I_0 I_{-1}][V_{zx} + iV_{yz}], \quad (18.3\text{-}38d)$$

$$Q^{(2)}_{-2}U^{(2)}_{+2} = \frac{1}{4}\frac{eQ}{I(2I-1)}I^2_{-1}[V_{xx} - V_{yy} + 2iV_{xy}]. \quad (18.3\text{-}38e)$$

At this point let us choose a coordinate system such that $V_{ij} = 0$, $i \neq j$. This is possible since, according to (18.3-33), the V_{ij} are components of a symmetric tensor. Making use of this simplification, the Hamiltonian $\mathscr{H}_Q$ takes the form

$$\mathscr{H}_Q = \frac{eQ}{I(2I-1)}\left[\frac{1}{4}I^2_{+1}(V_{xx} - V_{yy}) + \frac{1}{4}(3I_0{}^2 - I^2)V_{zz} + \frac{1}{4}I^2_{-1}(V_{xx} - V_{yy})\right]$$

$$= \frac{eQ}{4I(2I-1)}[V_{zz}(3I_0{}^2 - I^2) + (V_{xx} - V_{yy})(I^2_{+1} + I^2_{-1})]. \quad (18.3\text{-}39)$$

Furthermore, since V is subject to the Laplace equation $\nabla^2 V = 0$, two parameters are sufficient to characterize the field gradient. It is also conventional to assume that the coordinate system has been so chosen that

$$|V_{zz}| \geqslant |V_{yy}| \geqslant |V_{xx}|.$$

Since, according to (18.3-22),

$$\langle V_{zz}\rangle = eq, \quad (18.3\text{-}40a)$$

it is also convenient to define

$$\langle V_{xx} - V_{yy}\rangle = \eta\langle V_{zz}\rangle = e\eta q \qquad (0 \leqslant \eta \leqslant 1). \quad (18.3\text{-}40b)$$

These may be inserted into (18.3-39) with the result, after integrating over electronic coordinates, that

$$\mathscr{H}_Q = \frac{e^2 qQ}{4I(2I-1)}[(3I_0{}^2 - I^2) + \eta(I^2_{+1} + I^2_{-1})]. \quad (18.3\text{-}41)$$

In this form $\mathscr{H}_Q$ consists entirely of nuclear operators. For s electrons the charge distribution is spherically symmetric; hence $V_{xx} = V_{yy} = V_{zz}$. But since $\nabla^2 V = 0$, the principal components are all zero and in particular $eq = \langle V_{zz}\rangle = 0$. It is therefore concluded that there can be no quadrupole interaction with s electrons.

A few examples will illustrate the characteristics of the quadrupole interaction. Assuming axial symmetry, $V_{xx} = V_{yy}$, we have

$$\mathscr{H}_Q = \frac{e^2 qQ}{4I(2I-1)}(3I_0{}^2 - I^2), \quad (18.3\text{-}42)$$

$$E_Q = \langle I\ m_I|\mathscr{H}_Q|I\ m_I\rangle = \frac{e^2 qQ}{4I(2I-1)}[3m_I{}^2 - I(I+1)]. \quad (18.3\text{-}43)$$

For this case there are only diagonal elements with twofold degeneracies in m_I, that is, states with $\pm m_I$ have the same energy. As has already been noted, the quadrupole interaction vanishes when $I = 0$ or $\frac{1}{2}$. When $I = 1$, the energies are (Fig. 18.6)

$$E_Q = \tfrac{1}{4}e^2qQ(3m_I^{\,2} - 2)$$
$$= \begin{cases} \tfrac{1}{4}e^2qQ, & m_I = \pm 1, \\ -\tfrac{1}{2}e^2qQ, & m_I = 0. \end{cases} \qquad (18.3\text{-}44a)$$

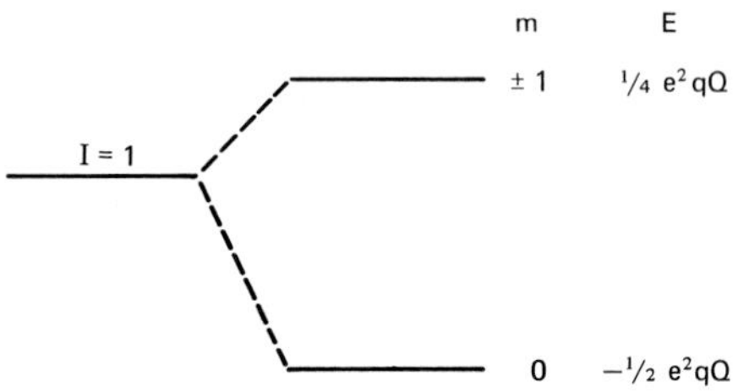

FIG. 18.6 Quadrupole splitting: $I = 1, e^2qQ > 0$.

For $I = \frac{3}{2}$ (Fig. 18.7a),

$$E_Q = \tfrac{1}{12}e^2qQ(3M_I^{\,2} - \tfrac{15}{4})$$
$$= \begin{cases} \tfrac{1}{4}e^2qQ, & m_I = \pm\tfrac{3}{2}, \\ -\tfrac{1}{4}e^2qQ, & m_I = \pm\tfrac{1}{2}. \end{cases} \qquad (18.3\text{-}44b)$$

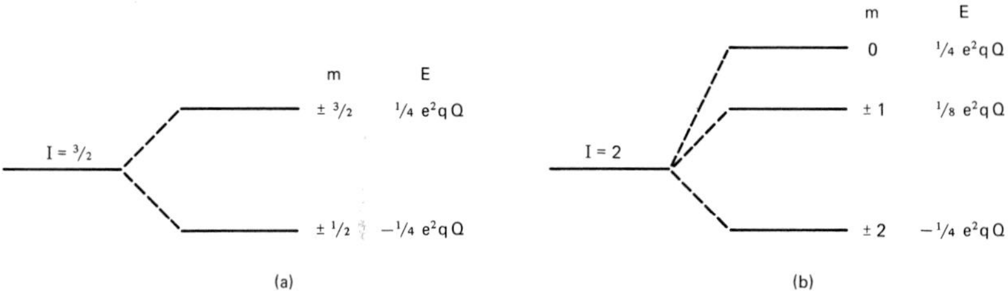

FIG. 18.7 Quadrupole splitting: (a) $I = \frac{3}{2}, e^2qQ > 0$; (b) $I = 2, e^2qQ < 0$.

Let us now assume that $V_{xx} \neq V_{yy}$, $I = \frac{3}{2}$. Then

$$\mathscr{H}_Q = \tfrac{1}{12}e^2qQ[(3I_0^{\,2} - I^2) + \eta(I_{+1}^2 + I_{-1}^2)]. \qquad (18.3\text{-}45)$$

The matrix elements of $\mathscr{H}_Q$ in the basis set $|I\ M_I\rangle$ can be evaluated with the aid of Table 1.3 and are listed, in units of $\frac{1}{4}e^2qQ$, in Table 18.3. The energies,

TABLE 18.3

Matrix Elements of the Electric Quadrupole Interaction in Units of $\frac{1}{4}e^2qQ$ for $I = \frac{3}{2}$

	$\lvert\frac{3}{2}\frac{3}{2}\rangle$	$\lvert\frac{3}{2}-\frac{1}{2}\rangle$	$\lvert\frac{3}{2}-\frac{3}{2}\rangle$	$\lvert\frac{3}{2}\frac{1}{2}\rangle$
$\langle\frac{3}{2}\frac{3}{2}\rvert$	1	$\frac{\eta}{\sqrt{3}}$		
$\langle\frac{3}{2}-\frac{1}{2}\rvert$	$\frac{\eta}{\sqrt{3}}$	-1		
$\langle\frac{3}{2}-\frac{3}{2}\rvert$			1	$\frac{\eta}{\sqrt{3}}$
$\langle\frac{3}{2}\frac{1}{2}\rvert$			$\frac{\eta}{\sqrt{3}}$	-1

obtained by diagonalizing the 2×2 matrix, are

$$E = \pm\tfrac{1}{4}e^2qQ\sqrt{1 + \tfrac{1}{3}\eta^2} \tag{18.3-46}$$

which indicates that states with $\pm m_I$ are degenerate even when $\eta \neq 0$. The difference in energy between the two states is

$$\Delta E = \tfrac{1}{2}e^2qQ\sqrt{1 + \tfrac{1}{3}\eta^2}. \tag{18.3-47}$$

The quadrupole splitting for $I = 2$ is shown in Fig. 18.7b.

We now compute the matrix elements of V_{zz} and $V_{xx} - V_{yy}$ with respect to the electronic wave functions. From (18.3-34c),

$$V_{zz} = -4\sqrt{\frac{\pi}{5}}\,eY_{20}\frac{1}{r^3} \tag{18.3-48}$$

and using the general theorem (1.2-29),

$$\langle Y_{l'm'}|Y_{20}|Y_{lm}\rangle = (-1)^{m'}\sqrt{\frac{5}{4\pi}}\,[2l' + 1)(2l + 1)]^{1/2} \times \begin{pmatrix} l' & 2 & l \\ -m' & 0 & m \end{pmatrix}\begin{pmatrix} l' & 2 & l \\ 0 & 0 & 0 \end{pmatrix}. \tag{18.3-49}$$

The first $3j$ symbol indicates that $m = m'$. For s electrons $l = l' = 0$ and the $3j$ symbols vanish. Hence

$$\langle\psi_s|V_{zz}|\psi_s\rangle = 0 \tag{18.3-50}$$

as we have already seen. For p electrons

$$\frac{1}{e}\langle\psi_\mathrm{p}|V_{zz}|\psi_\mathrm{p}\rangle = \begin{cases} \dfrac{2}{5}\left\langle\dfrac{1}{r^3}\right\rangle_{l=1}, & m_l = 1, \\ -\dfrac{4}{5}\left\langle\dfrac{1}{r^3}\right\rangle_{l=1}, & m_l = 0, \\ \dfrac{2}{5}\left\langle\dfrac{1}{r^3}\right\rangle_{l=1}, & m_l = -1. \end{cases} \tag{18.3-51}$$

For d electrons

$$\frac{1}{e}\langle\psi_\mathrm{d}|V_{zz}|\psi_\mathrm{d}\rangle = \begin{cases} \dfrac{4}{7}\left\langle\dfrac{1}{r^3}\right\rangle_{l=2}, & m_l = 2, \\ -\dfrac{2}{7}\left\langle\dfrac{1}{r^3}\right\rangle_{l=2}, & m_l = 1, \\ -\dfrac{4}{7}\left\langle\dfrac{1}{r^3}\right\rangle_{l=2}, & m_l = 0, \\ -\dfrac{2}{7}\left\langle\dfrac{1}{r^3}\right\rangle_{l=2}, & m_l = -1, \\ \dfrac{4}{7}\left\langle\dfrac{1}{r^3}\right\rangle_{l=2}, & m_l = -2. \end{cases} \tag{18.3-52}$$

Using (18.3-34a),

$$V_{xx} - V_{yy} = -2\sqrt{\frac{6\pi}{5}}\, e\,\frac{1}{r^3}(Y_{2\,2} + Y_{2\,-2}) \tag{18.3-53}$$

and, from (1.2-29),

$$\langle Y_{2m'}|Y_{2\,2} + Y_{2\,-2}|Y_{2m}\rangle$$
$$= -10\sqrt{\frac{5}{4\pi}}\sqrt{\frac{1}{70}}(-1)^{m'}\left[\begin{pmatrix} 2 & 2 & 2 \\ -m' & 2 & m \end{pmatrix} + \begin{pmatrix} 2 & 2 & 2 \\ -m' & -2 & m \end{pmatrix}\right]. \tag{18.3-54}$$

In this case the diagonal elements are zero. The nonzero elements are those for which

$$\begin{matrix} m' = 2 & 1 & 0 & -1 & -2 \\ m = 0 & -1 & \pm 2 & 1 & 0. \end{matrix} \tag{18.3-55}$$

PART IV

N–Electron Atoms

CHAPTER 19

HARTREE–FOCK FORMULATION

19.1 The Hamiltonian

The Schrödinger equation (15.2-32) is a one-electron equation since it originated from a nonrelativistic approximation to the Dirac equation (15.2-10) which is itself a one-electron equation. The extension to an N-electron atom requires two steps: The first is to sum the one-electron operators in (15.2-32) over all of the N electrons and the second is to add to the Hamiltonian all the relevant electron–electron interactions. Among the latter, the most important is the electron–electron Coulomb repulsion summed over all pairs of electrons. It is further assumed that it is not necessary to include operators involving more than two electrons.

Accordingly, the most important terms in the Hamiltonian for the N-electron atom, in the absence of external fields, are

$$\mathscr{H} = \mathscr{H}_0 + \mathscr{H}_1 + \mathscr{H}_2 \tag{19.1-1}$$

where

$$\mathscr{H}_0 = \sum_{i=1}^{N} [(p_i^2/2m) - (Ze^2/r_i)], \tag{19.1-2}$$

$$\mathscr{H}_1 = \sum_{\text{pairs}} e^2/r_{ij} = \sum_{i<j}^{N} e^2/r_{ij}, \tag{19.1-3}$$

$$\mathscr{H}_2 = \sum_{i=1}^{N} \xi(r_i)\mathbf{l}_i \cdot \mathbf{s}_i. \tag{19.1-4}$$

Here r_i is the distance from the nucleus of charge Ze to the ith electron with mass m, charge e, and momentum $\mathbf{p}_i$; r_{ij} the distance between the ith and jth electrons. The term $\mathscr{H}_0$ contains the kinetic energy and the potential energy of the electrons in the field of the nucleus, $\mathscr{H}_1$ represents the interelectron Coulomb repulsion, and $\mathscr{H}_2$ the spin-orbit interaction. Other effects such as those due to hyperfine interactions, relativistic corrections, etc., are treated as perturbations.

The relative effects of $\mathscr{H}_1$ and $\mathscr{H}_2$ depend on the atomic number. For low Z atoms, $\mathscr{H}_1$ is the dominant term compared to $\mathscr{H}_2$ which may therefore be treated as a perturbation, whereas at high Z it is $\mathscr{H}_2$ that is dominant with $\mathscr{H}_1$ acting as a perturbation. It is therefore useful to recognize two limiting cases—one for low Z with the Hamiltonian

$$\mathscr{H}_{\mathrm{I}} = \mathscr{H}_0 + \mathscr{H}_1 \tag{19.1-5}$$

and the other for high Z with the Hamiltonian

$$\mathscr{H}_{\mathrm{II}} = \mathscr{H}_0 + \mathscr{H}_2. \tag{19.1-6}$$

Despite the fact that these are limiting cases, the properties of most atoms can be systematized on the basis of one of the two Hamiltonians, especially $\mathscr{H}_{\mathrm{I}}$ which is often regarded as the fundamental Hamiltonian in atomic physics. Let us now examine the symmetries of the two Hamiltonians $\mathscr{H}_{\mathrm{I}}$ and $\mathscr{H}_{\mathrm{II}}$. In the first place, since we are dealing with indistinguishable particles, every Hamiltonian for an N-particle system must be invariant under an interchange in the coordinates (spatial and spin) of any two particles. For a system of fermions (e.g., electrons) this means, as shown in Section 8.4, that the total N-particle wave function must be antisymmetric with respect to an interchange of two particles.

Another general symmetry which applies to both $\mathscr{H}_{\mathrm{I}}$ and $\mathscr{H}_{\mathrm{II}}$ is inversion symmetry since the general Hamiltonian (19.1-1) is clearly invariant under inversion (reflection in the origin). As shown in Section 8.2, the direct consequence of this symmetry is that the eigenfunctions of $\mathscr{H}$ have a definite parity, or, as it is usually stated, parity is a good quantum number for atomic states.

Although $\mathscr{H}_{\mathrm{I}}$ does not commute with $\mathbf{l}_i$, the orbital angular momentum operator for the ith electron,‡ it will be shown that $\mathscr{H}_{\mathrm{I}}$ commutes with $\mathbf{L}$ where

$$\mathbf{L} = \sum_i \mathbf{l}_i. \tag{19.1-7}$$

Since

$$p_i^2 = -(\hbar^2/2m)\nabla_i^2$$

and the angular part of ∇_i^2 is proportional to l_i^2, we have

$$[p_i^2, \mathbf{l}_i] = 0.$$

‡ We shall now and henceforth employ the convention that quantities pertaining to a single electron (e.g., operators, quantum members, etc.) are designated by lower case letters whereas quantities pertaining to the entire electronic system are designated by upper case letters.

Also, $\mathbf{l}_i$ commutes with $1/r_i$ since, in spherical coordinates, the components of $\mathbf{l}_i$ act only on the angular variables. Therefore $\mathbf{L}$ commutes with $\mathscr{H}_0$. It then remains to be shown that $\mathbf{L}$ commutes with $\mathscr{H}_1$ [Eq. (19.1-3)]. We have

$$[\mathscr{H}_1, \mathbf{L}] = e^2 \sum_{i<j} \sum_k \frac{1}{r_{ij}} \mathbf{l}_k - e^2 \sum_{i<j} \sum_k \mathbf{l}_k \frac{1}{r_{ij}} \tag{19.1-8}$$

with

$$\mathbf{l}_k = -i\mathbf{r}_k \times \nabla_k.$$

Allowing the operators to act on an arbitrary function ψ,

$$\mathbf{l}_k \frac{1}{r_{ij}} \psi = \frac{1}{r_{ij}} \mathbf{l}_k \psi + \psi \mathbf{l}_k \frac{1}{r_{ij}}$$

or

$$\mathbf{l}_k \frac{1}{r_{ij}} = \frac{1}{r_{ij}} \mathbf{l}_k - i\mathbf{r}_k \times \nabla_k \left(\frac{1}{r_{ij}}\right). \tag{19.1-9}$$

Substituting (19.1-9) into (19.1-8),

$$[\mathscr{H}_1, \mathbf{L}] = ie^2 \sum_{i<j} \sum_k \mathbf{r}_k \times \nabla_k \left(\frac{1}{r_{ij}}\right).$$

But

$$\sum_k \mathbf{r}_k \times \nabla_k \left(\frac{1}{r_{ij}}\right) = \sum_k \left[-\mathbf{r}_k \times \frac{\mathbf{r}_i - \mathbf{r}_j}{r_{ij}^3} \delta_{ik} + \mathbf{r}_k \times \frac{\mathbf{r}_i - \mathbf{r}_j}{r_{ij}^3} \delta_{jk} \right]$$

$$= \frac{\mathbf{r}_i \times \mathbf{r}_j}{r_{ij}^3} + \frac{\mathbf{r}_j \times \mathbf{r}_i}{r_{ij}^3} = 0;$$

therefore

$$[\mathscr{H}_\mathrm{I}, \mathbf{L}] = 0. \tag{19.1-10}$$

The absence of spin operators in $\mathscr{H}_\mathrm{I}$ would lead one to conclude that

$$[\mathscr{H}_\mathrm{I}, \mathbf{S}] = 0 \tag{19.1-11}$$

where

$$\mathbf{S} = \sum_i \mathbf{s}_i \tag{19.1-12}$$

and $\mathbf{s}_i$ is the spin operator of the ith electron. However, the argument is not completely rigorous because the antisymmetry requirement on the wave function prevents the spin and orbital angular momenta from being completely independent (see, for example, Section 20.1 where it is shown that for two equivalent electrons $L + S$ is even). Nevertheless, as shown by

Bethe and Jackiw (1968) by means of a more elaborate derivation, the components of **S** still commute with $\mathscr{H}_{\mathrm{I}}$.

Since $\mathscr{H}_{\mathrm{I}}$ commutes with both **L** and **S**, it must also commute with $\mathbf{J} = \mathbf{L} + \mathbf{S}$, i.e.,

$$[\mathscr{H}_{\mathrm{I}}, \mathbf{J}] = 0. \tag{19.1-13}$$

We now turn to $\mathscr{H}_{\mathrm{II}}$. As shown in Section 16.3, $\mathbf{l} \cdot \mathbf{s}$ does not commute with **l** or **s** but does commute with $\mathbf{j} = \mathbf{l} + \mathbf{s}$. Therefore

$$[\mathscr{H}_{\mathrm{II}}, \mathbf{J}] = 0$$

with

$$\mathbf{J} = \sum_i \mathbf{j}_i \tag{19.1-14}$$

and

$$\mathbf{j}_i = \mathbf{l}_i + \mathbf{s}_i;$$

but $\mathscr{H}_{\mathrm{II}}$ does not commute with **L** or **S**.

Thus we have two coupling schemes:

$$\mathbf{L} = \sum_i \mathbf{l}_i, \qquad \mathbf{S} = \sum_i \mathbf{s}_i, \tag{19.1-15}$$

$$\mathbf{J} = \mathbf{L} + \mathbf{S}. \tag{19.1-16}$$

This is known as *LS coupling or Russell–Saunders coupling* and is used when $\mathscr{H}_{\mathrm{I}}$ is the appropriate Hamiltonian, that is, when the interelectron Coulomb repulsion energy is far greater than the spin-orbit interaction so that orbital and spin angular momenta are conserved separately. The coupling represented by (19.1-14) is known as *jj coupling* and is associated with the Hamiltonian $\mathscr{H}_{\mathrm{II}}$ when the spin-orbit interaction exceeds the electron repulsion energy.

The fact that $\mathscr{H}_{\mathrm{I}}$ commutes with **L**, **S**, and **J** is equivalent to saying that $\mathscr{H}_{\mathrm{I}}$ is invariant under a rotation of spatial and spin coordinates. As a further consequence the eigenfunctions of $\mathscr{H}_{\mathrm{I}}$ are also eigenfunctions of L^2, L_z, S^2, S_z. In ket form, such eigenfunctions may be written $|LSM_LM_S\rangle$—a notation that identifies the eigenvalues of L^2, S^2, L_z and S_z as $L(L+1)$, $S(S+1)$, M_L and M_S or, more briefly, that the good quantum numbers are L, S, M_L and M_S. One may also use the more complete form $|\alpha LSM_LM_S\rangle$ where α is an index to represent additional information required to specify the state unambiguously, such as the radial part of the wave function, the parity, the electronic configuration, etc. When **L** and **S** are coupled to form **J**, the eigenfunctions of L^2, S^2, J^2, and J_z are denoted by $|LSJM\rangle$. In this case, the good quantum numbers are L, S, J, and M, and here, too, one may add an index α, as needed. Whether one writes the eigenfunctions of $\mathscr{H}_{\mathrm{I}}$ as $|LSM_LM_S\rangle$ in the uncoupled representation, or as $|LSJM\rangle$ in the coupled representation is purely a matter of choice and convenience.

In the language of group theory, the invariance properties of the Hamiltonian $\mathscr{H}_{\mathrm{I}}$ imply that the $|LSM_LM_S\rangle$ eigenfunctions of $\mathscr{H}_{\mathrm{I}}$ are basis functions for the irreducible representations $D^{(L)}$ and $D^{(S)}$ of $O^+(3)$ while $|LSJM\rangle$ belongs to $D^{(J)}$ which is an irreducible representation contained in the reduction of the product representation $D^{(L)} \times D^{(S)}$.

We have seen that $\mathbf{L}$ and $\mathbf{S}$ are sums of one-electron operators. This implies that the construction of eigenfunctions of L^2 and L_z from eigenfunctions of l_i^2 and l_{iz} is implemented by the standard rules of angular momentum coupling. Group theoretically, a one-electron eigenfunction of l_i^2 and l_{iz} belongs to the irreducible representation $D^{(l_i)}$. A product of N such orbitals belongs to the product representation $D^{(l_1)} \times D^{(l_2)} \times \cdots \times D^{(l_N)}$ and the latter is reducible into $\sum_L D^{(L)}$. By means of the coupling coefficients the products may be combined into basis functions for the irreducible representations $D^{(L)}$. The same methods apply to the construction of eigenfunctions of S^2 and S_z from eigenfunctions of s_i^2 and s_{iz} and, by extension, to the construction of eigenfunctions of J^2 and J_z.

A further property of $\mathscr{H}_{\mathrm{I}}$ is that the energy eigenvalues are independent of M_L and M_S. This may be seen qualitatively as a consequence of the fact that the energy of a system described by $\mathscr{H}_{\mathrm{I}}$ is independent of external coordinates and the relative orientations of orbital and spin angular momenta. A more formal demonstration is given in Section 21.1.

19.2 Central Field Approximation

In view of its wider applicability we shall concentrate our attention on the Hamiltonian $\mathscr{H}_{\mathrm{I}}$ [Eq. (19.1-5)]. The basic difficulty in solving the Schrödinger equation with this Hamiltonian stems from the fact that the interelectron repulsion is too large an effect to be treated as a perturbation. There is, however, a saving feature to the problem, although it only becomes apparent at a later stage. It is that the interelectron repulsion contains a large spherically symmetric component. This motivates us to imagine that it is possible to construct a potential energy function $U(r_i)$ which is a spherically symmetric, one-electron operator and is a good approximation to the actual potential energy of the ith electron in the field of the nucleus and the other $N-1$ electrons. Assuming that $U(r_i)$ can be constructed to the required accuracy, the Hamiltonian (19.1-5) may be written as (dropping the subscript on $\mathscr{H}_{\mathrm{I}}$):

$$\mathscr{H} = \mathscr{H}_0' + \mathscr{H}_1' \tag{19.2-1}$$

where

$$\mathscr{H}_0' = \sum_{i=1}^{N} \left(\frac{p_i^2}{2m} + U(r_i) \right), \tag{19.2-2a}$$

$$\sum_i U(r_i) = -\sum_{i=1}^{N} \frac{Ze^2}{r_i} + \left\langle \sum_{i<j}^{N} \frac{e^2}{r_{ij}} \right\rangle \tag{19.2-2b}$$

$$\mathscr{H}_1' = -\sum_{i=1}^{N} \frac{Ze^2}{r_i} + \sum_{i<j}^{N} \frac{e^2}{r_{ij}} - \sum_{i=1}^{N} U(r_i)$$

$$= \sum_{i<j}^{N} \frac{e^2}{r_{ij}} - \left\langle \sum_{i<j}^{N} \frac{e^2}{r_{ij}} \right\rangle \tag{19.2-2c}$$

in which $\langle \sum_{i<j}^{N} (e^2/r_{ij}) \rangle$ is the average over a sphere of the electron repulsion. It is therefore independent of the angular coordinates. Then $\mathscr{H}_1'$ becomes the Hamiltonian which contains the nonspherical part of the electronic repulsions whereas $\mathscr{H}_0'$ contains the kinetic energy, potential energy in the field of the nucleus, and the (spherical) average electron repulsion energy:

$$\mathscr{H}_0' = \sum_{i}^{N} \left(\frac{p_i^2}{2m} - \frac{Ze^2}{r_i} \right) + \left\langle \sum_{i<j}^{N} \frac{e^2}{r_{ij}} \right\rangle. \tag{19.2-3}$$

This is the *central field approximation*. The advantages of this approach may now be perceived—the main features of the electronic motion are described by the Hamiltonian $\mathscr{H}_0'$. Since it is assumed that $\mathscr{H}_0'$ contains most of the interelectron repulsion, the remaining term $\mathscr{H}_1'$ is small enough to be treated as a perturbation. Thus, conceptually, the system of interacting electrons has been replaced by a system which, in the zeroth approximation consists of independent electrons moving in a spherically symmetric field produced by the nucleus and all the other electrons. Higher approximations must take into account the departure from spherical symmetry in the effective field seen by the electrons.

Since $U(r_i)$ is a one-electron operator, the Schrödinger equation

$$\mathscr{H}_0'\Phi = \left\{ \sum_{i=1}^{N} \left[\frac{p_i^2}{2m} + U(r_i) \right] \right\} \Phi = E\Phi \tag{19.2-4}$$

for the N-electron spatial wave function Φ is separable into N equations of the form

$$[(p^2/2m) + U(r)]\varphi(\mathbf{r}) = \varepsilon\varphi(\mathbf{r}). \tag{19.2-5}$$

As shown in the discussion of the hydrogen problem, such an equation is separable into two equations, one depending on r and the other on θ and φ. The radial equation contains $U(r)$ which is, as yet, unknown except insofar as it must satisfy the general physical requirement $U(r) \to 0$ as $r \to \infty$.

The angular equation, however, is identical with (16.1-7) whose solutions are the spherical harmonics. Therefore the general solution to (19.2-5) consists of the one-electron functions

$$\varphi_{nlm_l}(r,\theta,\varphi) = R_{nl}(r)Y_{lm_l}(\theta,\varphi) = (1/r)P_{nl}(r)Y_{lm_l}(\theta,\varphi) \tag{19.2-6}$$

called *central field orbitals*. It is important not to confuse the radial function $P_{nl}(r)$ [or $R_{nl}(r)$] for an N-electron atom with the corresponding radial function (16.1-32) for the one-electron atom, despite the similarity in notation. In the former case $P_{nl}(r)$ depends on $U(r)$ and is likely to be a complicated function which may not be expressible in any simple analytical form.

When (19.2-6) is multiplied by a spin function $\xi(m_s)$ which is just α for $m_s = \frac{1}{2}$ and β for $m_s = -\frac{1}{2}$ we have a one-electron, central field, spin orbital

$$\psi_{nlm_lm_s}(\mathbf{r}, m_s) = \varphi_{nlm_l}(\mathbf{r})\xi(m_s) = (1/r)P_{nl}(r)Y_{lm_l}(\theta,\varphi)\xi(m_s) \tag{19.2-7}$$

characterized by the four quantum numbers n, l, m_l, and m_s where

$$n = 1, 2, \ldots, \qquad l = 0, 1, 2 \ldots n-1$$
$$m_l = l, l-1, \ldots -l, \qquad m_s = \pm\tfrac{1}{2}.$$

A general spin orbital need not be of the form (19.2-7)—any one-electron function that depends on the space and spin coordinates is a spin orbital. However, the central field spin orbital (19.2-7) is the most useful type and we shall have that in mind when we employ the term spin orbital without additional qualifications.

Finally, the spin orbitals may be organized into an antisymmetric N-electron wave function in order to satisfy the requirements of the Pauli principle. Such a wave function can be written in the form of one or more Slater determinants

$$\Psi(\lambda_1,\lambda_2,\ldots,\lambda_N) = \frac{1}{\sqrt{N!}}\begin{vmatrix} \psi_1(\lambda_1) & \psi_2(\lambda_1) & \cdots & \psi_N(\lambda_1) \\ \psi_1(\lambda_2) & \psi_2(\lambda_2) & \cdots & \psi_N(\lambda_2) \\ \vdots & & & \\ \psi_1(\lambda_N) & \psi_2(\lambda_N) & \cdots & \psi_N(\lambda_N) \end{vmatrix} \tag{19.2-8}$$

in which $\psi_i(\lambda_j)$ are spin orbitals. The subscript i identifies a particular choice of the four quantum numbers n, l, m_l, m_s, while λ_j represents the space and spin coordinates of the jth electron. Since the spin orbital $\psi_{nlm_lm_s}$ owes its parity entirely to the spherical harmonic Y_{lm_l} and the parity of the latter is determined by $(-1)^l$, the parity of a product function whether antisymmetrized or not is

$$(-1)^{l_1}(-1)^{l_2}\cdots(-1)^{l_N}. \tag{19.2-9}$$

Thus each Slater determinant has a definite parity.

In view of the fact that a determinant vanishes when two columns (or rows) are equal, the Slater determinants will vanish if two electrons have precisely the same values of the four quantum numbers n, l, m_l, m_s since these quantum numbers uniquely identify a spin orbital (apart from an arbitrary phase factor). Hence an alternative statement of the Pauli principle is that no two electrons can have the same values of n, l, m_l, and m_s. This imposes severe restrictions on the distribution of electrons in atomic orbitals and is responsible for the so-called "build-up" principle of atomic structure which states that in the ground state of an atom the electrons occupy those orbitals permitted by the Pauli principle and which result in the lowest energy. The distribution of electrons with respect to n and l is known as a *configuration.* Electrons in orbitals with the same values of n and l are said to be *equivalent.* The maximum number of equivalent electrons for a particular value of l is $2(2l + 1)$ which is just the product of the orbital and spin degeneracies. Such an assembly of equivalent electrons is known as a *closed shell.* Electrons outside of a closed shell in insufficient numbers to form another closed shell are known as *valence* electrons.

A few examples will help to fix these ideas. For $n = 1, l = 0, m_l = 0, m_s = \pm\frac{1}{2}$ there are just two possible electronic configurations: 1s which is H and $1s^2$ which is He. For $n = 2, l = 0, 1$; when $l = 0$ we have $1s^2 2s$(Li), $1s^2 2s^2$(Be). When $l = 1$, $m_l = 0, \pm 1$ and $m_s = \pm\frac{1}{2}$, six electrons can be accommodated with configurations $1s^2 2s^2 2p, 1s^2 2s^2 2p^2 \cdots 1s^2 2s^2 2p^6$ which correspond to the ground states of B, C, N, O, F, and Ne. This process may be continued as far as is necessary in order to accommodate all the electrons of a particular atom. However, the electrons are not compelled by the Pauli principle to occupy the orbitals in the order just described, and indeed, on energetic grounds, there are departures from such sequential occupation. The Pauli principle merely restricts more than one electron from occupying any particular spin orbital. Thus argon with $Z = 18$ has the closed shell configuration $1s^2 2s^2 2p^6 3s^2 3p^6$. In potassium ($Z = 19$) the first 18 electrons have the same configuration as that of argon but the last (19th) or valence electron must occupy an orbital outside of the closed shell which, one might suppose, would be the 3d orbital. In actuality, the valence electron is in a 4s orbital so that the electronic configuration of potassium is $1s^2 2s^2 2p^6 3s^2 3p^6 4s$.

If energy is imparted to an atom by whatever means, e.g., illumination with electromagnetic radiation, collisions, etc., one or more electrons will be excited to higher energy orbitals and the electronic configuration will correspondingly be altered, but in all cases the Pauli principle must be obeyed.

It is also common, particularly in x-ray spectroscopy to refer to electrons with principal quantum numbers $n = 1, 2, 3, \ldots$ as residing in the K, L, M, . . . shells, respectively.

In Section 16.1 we saw that in hydrogen the energy of an orbital depends only on the principal quantum number n and $E_{n+1} > E_n$. When the Coulomb potential of hydrogen is replaced by the more general central field $U(r)$, the energy of an orbital like (19.2-6) depends on both n and l. Generally speaking the energy increases with the sum $n + l$. Thus, as long as we confine ourselves to the central field approximation characterized by the Hamiltonian $\mathscr{H}_0'$ given by (19.2-2a), the energy of an atom is completely determined by the electronic configuration. However, this is no longer the case when $\mathscr{H}_1'$ [Eq. (19.2-2c)] is taken into account. When the latter is present, it is possible for electrons in a given configuration to have various energies depending on the mutual orientations of the orbital and spin angular momenta of the individual electrons.

The interaction $\mathscr{H}_1'$ [Eq. (19.2-2c)] produces a splitting in the energy of a configuration. Since all of the splitting is contained in

$$\sum_{i<j}^{N} e^2/r_{ij}$$

it is sufficient to calculate matrix elements of this operator alone. The energy of a configuration is then split into *terms* with which specific values of L and S are associated. If, subsequently, the spin-orbit interaction $\mathscr{H}_2$ [Eq. (19.1-4)] is applied, there is a further splitting into *levels* according to the allowed values of J. The level splitting is also known as the *fine structure* and the complete set of levels belonging to a given electronic configuration is a *multiplet*‡ (Fig. 19.1).

The spin-orbit interaction energy increases rapidly with atomic number (it is proportional to Z^4 in the one-electron case) while the interelectron repulsion energy varies much more slowly as a function of Z. Hence as we move up the periodic table, $\mathscr{H}_2$ assumes greater importance and eventually becomes the dominant perturbation relative to $\mathscr{H}_1$. Even in lighter elements, the electrostatic interaction is weakened if the atom is in a highly excited state since the electron involved in the excitation is generally situated far from the core which contains the nucleus and all the other electrons.

It is evident from (19.2-3) that $\mathscr{H}_0'$ is invariant with respect to an interchange of any two electrons by which we mean a simultaneous interchange of both space and spin coordinates. An equivalent statement is that $\mathscr{H}_0'$ is invariant under the permutation group of N particles S_N. Therefore the N-electron wave functions (of space and spin coordinates) which are solutions to the Schrödinger equation with the Hamiltonian (19.2-3) must be basis functions for the irreducible representations of S_N. The fact that electrons obey Fermi–Dirac statistics or, alternatively, that electrons satisfy the Pauli principle means that the only acceptable solutions are those that are

‡ Spectroscopists often refer to a multiplet as the set of *spectrum lines* arising from transitions between two sets of fine structure components whereas in our usage a multiplet refers to the energy levels themselves.

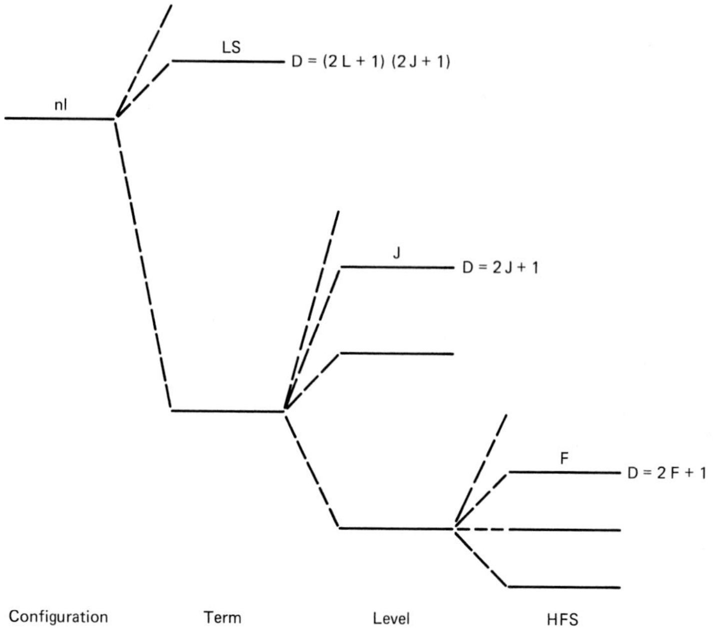

FIG. 19.1 Schematic arrangement of the energies of atomic states in *LS* coupling and their degeneracies *D*. The nonspherical part of the Coulomb repulsion splits a configuration into terms; spin-orbit coupling splits a term into levels; and the magnetic hyperfine interaction splits a level into hyperfine structure (HFS).

basis functions for the (one-dimensional) antisymmetric representation of S_N (see also Section 20.4). This feature is not altered by the inclusion of the perturbation terms.

19.3 Hartree–Fock Equations

We shall now consider the solution of the Schrödinger equation with the Hamiltonian

$$\mathscr{H} = \sum_{i}^{N} \mathscr{H}_0(i) + \sum_{i<j}^{N} e^2/r_{ij} \tag{19.3-1a}$$

$$\mathscr{H}_0(i) = (p_i^2/2m) - (Ze^2/r_i) \tag{19.3-1b}$$

from another point of view which ultimately will lead us to an evaluation of the radial wave functions in the central field orbitals (19.2-7). Hence, in effect, the computation will have determined the potentials $U(r_i)$.

Our main concern will be with the atomic ground state, that is, the state in which the atom spends most of its time. Such a state will be approximated by a single Slater determinant (19.2-8) which may be written in an abbreviated notation as

$$\Psi(\lambda_1, \lambda_2, \ldots, \lambda_N) = \frac{1}{\sqrt{N!}} \det\{\psi_1(\lambda_1)\psi_2(\lambda_2) \cdots \psi_N(\lambda_N)\}. \tag{19.3-2}$$

The $\psi_i(\lambda_j)$ are subject to the orthonormality condition

$$\langle \psi_i(\lambda_j) | \psi_k(\lambda_j) \rangle = \delta_{ik} \tag{19.3-3}$$

and are chosen to satisfy the variational principle (Section 14.1)

$$\delta \langle \Psi | \mathscr{H} | \Psi \rangle = 0 \tag{19.3-4}$$

where $\mathscr{H}$ is the Hamiltonian (19.3-1). At this stage the spin orbitals are quite general and are not necessarily of the central field type.

This is the *Hartree–Fock* approach in which it is assumed that the N-electron wave function Ψ, represented by a single Slater determinant and satisfying (19.3-4) together with the subsidiary condition (19.3-3), is an approximate solution to the N-electron Schrödinger equation (19.3-1).

By writing Ψ as a Slater determinant, the Pauli principle is automatically satisfied. However, confining Ψ to a *single* Slater determinant appears to be a drastic approximation since an arbitrary, normalizable, antisymmetric N-electron wave function can only be represented by an infinite sum of Slater determinants each constructed from a complete orthonormal set of one-electron functions and each multiplied by an appropriate coefficient. The Hartree–Fock method may therefore be regarded as a first step—and as it turns out a good first step—toward the construction of atomic wave functions.

In order to see the mechanics of the method it is sufficient to examine a two-electron system with a Hamiltonian

$$\mathscr{H} = \mathscr{H}_0(1) + \mathscr{H}_0(2) + (e^2/r_{12}) \tag{19.3-5}$$

where

$$\mathscr{H}_0(1) = (p_1{}^2/2m) - (Ze^2/r_1), \qquad \mathscr{H}_0(2) = (p_2{}^2/2m) - (Ze^2/r_2). \tag{19.3-6}$$

The Slater determinant for the ground state is

$$\Psi = \frac{1}{\sqrt{2}} \begin{vmatrix} u_a(\lambda_1) & u_b(\lambda_1) \\ u_a(\lambda_2) & u_b(\lambda_2) \end{vmatrix} \tag{19.3-7}$$

where $u_a(\lambda)$, $u_b(\lambda)$ are spin orbitals; λ_1 and λ_2 refer to the coordinates (space and spin) of electrons 1 and, 2 respectively. It will often be convenient to abbreviate functions like $u_a(\lambda_1)$ to $u_a(1)$, and if the reference to a particular electron is arbitrary, the spin orbital may be written without any argument. Thus the orthonormality condition may be written as

$$\langle u_a(\lambda_j)|u_b(\lambda_j)\rangle \equiv \langle u_a(j)|u_b(j)\rangle = \langle u_a|u_b\rangle = \delta_{ab} \tag{19.3-8}$$

which is equivalent to the three independent conditions

$$\langle u_a|u_a\rangle = 1, \qquad \langle u_b|u_b\rangle = 1, \tag{19.3-9}$$

$$\langle u_a|u_b\rangle = \langle u_b|u_a\rangle = 0. \tag{19.3-10}$$

With the aid of (11.1-16) and (11.1-19) the expectation value of the Hamiltonian is

$$\begin{aligned}\langle\Psi|\mathscr{H}|\Psi\rangle = {} & \langle u_a|\mathscr{H}_0|u_a\rangle + \langle u_b|\mathscr{H}_0|u_b\rangle + \langle u_a(1)u_b(2)|(e^2/r_{12})|u_a(1)u_b(2)\rangle \\ & - \langle u_a(1)u_b(2)|(e^2/r_{12})|u_b(1)u_a(2)\rangle.\end{aligned} \tag{19.3-11}$$

The implementation of the variational equation (19.3-4) requires that the auxiliary conditions (19.3-9) and (19.3-10) be incorporated. This is accomplished with the aid of Lagrange multipliers (also called variational parameters) $\lambda_{aa}, \lambda_{bb}, \lambda_{ab}, \lambda_{ba}$, with $\lambda_{ab}^* = \lambda_{ba}$. The variational equation now assumes the form

$$\begin{aligned}\delta\langle\Psi|\mathscr{H}|\Psi\rangle = \delta[& \langle u_a|\mathscr{H}_0|u_a\rangle + \langle u_b|\mathscr{H}_0|u_b\rangle + \langle u_a(1)u_b(2)|(e^2/r_{12})|u_a(1)u_b(2)\rangle \\ & - \langle u_a(1)u_b(2)|(e^2/r_{12})|u_b(1)u_a(2)\rangle - \lambda_{aa}\langle u_a(1)|u_a(1)\rangle \\ & - \lambda_{ab}\langle u_a(1)|u_b(1)\rangle - \lambda_{bb}\langle u_b(1)|u_b(1)\rangle - \lambda_{ba}\langle u_b(1)|u_a(1)\rangle] = 0\end{aligned} \tag{19.3-12}$$

and the two Euler equations (Section 14.1) are

$$\begin{aligned}&\mathscr{H}_0(1)u_a(1) + \left[\int u_b{}^*(2)(e^2/r_{12})u_b(2)\,d\tau_2\right]u_a(1) \\ &\quad - \left[\int u_b{}^*(2)(e^2/r_{12})u_a(2)\,d\tau_2\right]u_b(1) = \lambda_{aa}u_a(1) + \lambda_{ab}u_b(1)\end{aligned} \tag{19.3-13}$$

$$\begin{aligned}&\mathscr{H}_0(1)u_b(1) + \left[\int u_a{}^*(2)(e^2/r_{12})u_a(2)\,d\tau_2\right]u_b(1) \\ &\quad - \left[\int u_a{}^*(2)(e^2/r_{12})u_b(2)\,d\tau_2\right]u_a(1) = \lambda_{ba}u_a(1) + \lambda_{bb}u_b(1).\end{aligned} \tag{19.3-14}$$

The four Lagrange multipliers may be regarded as the elements of a matrix Λ and since $\lambda_{ab}^* = \lambda_{ba}$, the matrix is Hermitian. That being the case there must exist a unitary operator A that will diagonalize Λ by the transformation

$$A\Lambda A^{-1} = \varepsilon$$

where ε is a diagonal matrix with eigenvalues ε_a and ε_b. If then we transform to $\psi = Au$, the Lagrange multipliers λ_{ab} and λ_{ba} will be eliminated. This type of transformation does not alter the basic calculation because it is a characteristic of determinants, such as the Slater determinant $(N!)^{-1/2}\det\{\psi_1, \psi_2, \ldots, \psi_N\}$, that they are invariant under unitary transformations of the functions $\psi_1, \psi_2, \ldots, \psi_N$. The Hartree–Fock equations are, therefore, also invariant under a unitary transformation of the spin orbitals. We may, therefore suppose that such a procedure has been carried out from the outset, or, alternatively, that the coordinate system had been initially chosen (quite cleverly) so that the matrix of the Lagrange multipliers was diagonal. Indeed, had this been done, there would have been no need for the orthogonality conditions $\langle u_a|u_b\rangle = \langle u_b|u_a\rangle = 0$, which were responsible for the appearance of the off-diagonal multipliers in the first place. The conclusion is that the orthogonality conditions are not basic to the Hartree-Fock procedure and could have been omitted. The normalization conditions, on the other hand, must be employed.

There is an additional point to be made about the elimination of the off-diagonal multipliers. If the form of the spin orbitals is restricted in some manner, as for example by using central field spin orbitals of the type (19.2-7), and if the variation is applied only to the radial part of the spin orbital, there exist situations in which such additional restrictions may prevent the elimination of the off-diagonal multipliers [Slater (1960)]. Nevertheless, even if they cannot be eliminated, the off-diagonal terms are small and may often be neglected. We shall return to this matter in Section 19.5.

With these considerations in mind we may generalize (19.3-13) and (19.3-14) to an N-electron system. A typical equation for a single spin orbital will be of the form

$$\mathscr{H}_0(1)\psi_i(1) + \left[\sum_{j\neq i}\int \psi_j^*(2)(e^2/r_{12})\psi_j(2)\,d\tau_2\right]\psi_i(1)$$
$$- \sum_{j\neq i}\left[\int \psi_j^*(2)(e^2/r_{12})\psi_i(2)\,d\tau_2\right]\psi_j(1) = \varepsilon_i\psi_i(1) \qquad (19.3\text{-}15)$$

in which the summation over j extends over the N occupied spin orbitals. The restriction to $j \neq i$ is actually unnecessary since upon setting $j = i$ the first sum on the left will contain the term

$$\left[\int \psi_j^*(2)(e^2/r_{12})\psi_j(2)\,d\tau_2\right]\psi_j(1)$$

and precisely the same term will appear in the second sum but with a negative sign. Therefore it is often stated that a particle in a spin orbital ψ_i does not react with itself or that the self-energy of such a particle is zero. We then have for the Hartree–Fock equations

$$\mathscr{H}_0(1)\psi_i(1) + \left[\sum_j \int \psi_j^*(2)(e^2/r_{12})\psi_j(2)\,d\tau_2\right]\psi_i(1)$$
$$- \sum_j \left[\int \psi_j^*(2)(e^2/r_{12})\psi_i(2)\,d\tau_2\right]\psi_j(1) = \varepsilon_i\psi_i(1). \tag{19.3-16}$$

Coulomb and exchange operators $J_j(1)$ and $K_j(1)$ may be defined by analogy with (11.2-23) in which case

$$\left\{\mathscr{H}_0(1) + \sum_j [J_j(1) - K_j(1)]\right]\psi_i(1) = \varepsilon_i\psi_i(1), \tag{19.3-17}$$

$$J_j(1)\psi_i(1) = \left[\int \psi_j^*(2)(e^2/r_{12})\psi_j(2)\,d\tau_2\right]\psi_i(1),$$
$$K_j(1)\psi_i(1) = \left[\int \psi_j^*(2)(e^2/r_{12})\psi_i(2)\,d\tau_2\right]\psi_j(1). \tag{19.3-18}$$

Finally, we may define the *Fock* operator $F(1)$:

$$F(1) = \mathscr{H}_0(1) + V(1) \tag{19.3-19a}$$

with

$$V(1) = \sum_j [J_j(1) - K_j(1)] \tag{19.3-19b}$$

so that (19.3-17) assumes the form of an eigenvalue equation

$$F(1)\psi_i(1) = \varepsilon_i\psi_i(1). \tag{19.3-20}$$

It is worth noting that the Lagrange multiplier ε_i has, through the Hartree–Fock equations, acquired the significance of a one-electron energy eigenvalue.

The Hartree–Fock equations may also be expressed in terms of density matrices:

$$\mathscr{H}_0(1)\psi_i(1) + \left[\int (e^2/r_{12})\rho_1(2)\,d\tau_2\right]\psi_i(1) - \left[\int (e^2/r_{12})\rho_1(1;2)P_{12}\,d\tau_2\right]\psi_i(1)$$
$$= \varepsilon_i\psi_i(1) \tag{19.3-21}$$

in which

$$\rho_1(2) \equiv \rho_1(\lambda_2) \equiv \rho_1(\lambda_2;\lambda_2) = \sum_j \psi_j(2)\psi_j^*(2),$$
$$\rho_1(1;2) = \rho_1(\lambda_1;\lambda_2) = \sum_j \psi_j(1)\psi_j^*(2), \tag{19.3-22}$$
$$P_{12}\psi_i(1) = \psi_i(2).$$

Similarly, $V(1)$ (19.3-19b) may be written

$$V(1) \equiv \int (e^2/r_{12})(1 - P_{12})\rho_1(2;2')\,d\tau_2 \qquad (19.3\text{-}23)$$

since for multiplicative operators

$$\rho_1(2;2') = \rho_1(2;2) = \rho_1(2), \qquad P_{12}\rho_1(2;2') = \rho_1(1;2')P_{12} = \rho_1(1;2)P_{12}. \qquad (19.3\text{-}24)$$

$V(1)$ may be written as a sum of a Coulomb and an exchange potential:

$$V(1) = V_C(1) + V_{ex}(1) \qquad (19.3\text{-}25)$$

where

$$V_C(1) = \int (e^2/r_{12})\rho_1(2)\,d\tau_2 \qquad (19.3\text{-}26)$$

and

$$V_{ex}(1) = -\int (e^2/r_{12})P_{12}\rho_1(2;2')\,d\tau_2 = -\int (e^2/r_{12})\rho_1(1;2)P_{12}\,d\tau_2. \qquad (19.3\text{-}27)$$

The Hartree—Fock equations have been constructed on the basis of spin orbitals. For computational purposes the space and spin parts need to be separated. Using (11.2-24), the equation for a single spatial orbital is, from (19.3-17),

$$\mathscr{H}_0(1)\varphi_i(1) + \left[\sum_j \int \varphi_j^*(2)(e^2/r_{12})\varphi_j(2)\,d\mathbf{r}_2\right]\varphi_i(1)$$
$$- \sum_j \left[\delta(m_s^{\,i}, m_s^{\,j}) \int \varphi_j^*(2)(e^2/r_{12})\varphi_i(2)\,d\mathbf{r}_2\right]\varphi_j(1) = \varepsilon_i\varphi_i(1) \qquad (19.3\text{-}28)$$

in which the φs are functions of the space coordinates only. The Coulomb term

$$\sum_j J_j(1) = \sum_j \int \varphi_j^*(2)(e^2/r_{12})\varphi_j(2)\,d\mathbf{r}_2 \qquad (19.3\text{-}29)$$

is readily interpretable. The quantity $e^2\varphi_j^*(2)\varphi_j(2)$ is the charge density of electron 2 "smeared" out over the orbital φ_j. The integral is then the electrostatic repulsion potential due to electron 2 when the position of the latter is averaged over the orbital φ_j. The exchange term

$$\sum_j K_j(1) = \sum_j \delta(m_s^{\,i}, m_s^{\,j}) \int \varphi_j^*(2)(e^2/r_{12})\varphi_i(2)\,d\mathbf{r}_2 \qquad (19.3\text{-}30)$$

has no classical analog and it is therefore more difficult to endow it with a direct visual interpretation. Formally, it owes its presence to the antisymmetry of the total wave function Ψ [Eq. (19.3-2)] and requires the spin projections of electrons i and j to be parallel. If Ψ had been written as a simple product of one-electron orbitals (also known as a *Hartree* product) rather than as a Slater determinant, the exchange term would have been missing altogether. Further insight into the properties of the exchange term is obtainable by writing the Hartree–Fock equation (19.3-28) in terms of spinless density matrices (13.8-1) and (13.8-2):

$$\mathscr{H}_0(1)\varphi_i(1) + \left[\int (e^2/r_{12})P_1(2)\,d\mathbf{r}_2\right]\varphi_i(1) - \left[\delta(m_s{}^i, m_s{}^j)\int (e^2/r_{12})P_1(1;2)P_{12}\,d\mathbf{r}_2\right]\varphi_i(1) = \varepsilon_i\varphi_i(1) \quad (19.3\text{-}31)$$

where

$$P_1(2) \equiv P_1(\mathbf{r}_2) \equiv P_1(\mathbf{r}_2;\mathbf{r}_2)$$
$$P_1(1;2) = P_1(\mathbf{r}_1;\mathbf{r}_2), \qquad P_{12}\varphi_i(1) = \varphi_i(2). \quad (19.3\text{-}32)$$

The Coulomb and exchange potentials are

$$V_{\mathrm{C}} = \int (e^2/r_{12})P_1(2)\,d\mathbf{r}_2 \quad (19.3\text{-}33)$$

and

$$V_{\mathrm{ex}} = -\int (e^2/r_{12})P_1(1;2)P_{12}\,d\mathbf{r}_2 = -\int (e^2/r_{12})\rho_{\mathrm{ex}}(1;2)\,d\mathbf{r}_2 \quad (19.3\text{-}34)$$

in which $\rho_{\mathrm{ex}}(1;2)$ is known as an *exchange probability density*. More explicitly, it may be written as

$$\rho_{\mathrm{ex}}(1;2) = \sum_j \delta(m_s{}^i, m_s{}^j)\,\frac{\varphi_i{}^*(1)\varphi_j{}^*(2)\varphi_j(1)\varphi_i(2)}{\varphi_i{}^*(1)\varphi_i(1)}. \quad (19.3\text{-}35)$$

The quantity $-(e^2/r_{12})P_1(1;2)$ which appears in the integrand of (19.3-34) is often called the *nonlocal potential*. We note that

$$\int \rho_{\mathrm{ex}}(1;2)\,d\mathbf{r}_2 = \sum_j \delta(m_s{}^i, m_s{}^j)\,\frac{\varphi_j(1)}{\varphi_i(1)}\int \varphi_j{}^*(2)\varphi_i(2)\,d\mathbf{r}_2 = 1 \quad (19.3\text{-}36)$$

for orthogonal orbitals such as the Hartree–Fock orbitals. Thus the total exchange charge density corresponds to one electron but, because of the negative sign in front of the exchange term in (19.3-28), one speaks of an *exchange* (*or Fermi*) *hole* surrounding each electron. This may be visualized as a spherical hole surrounding a given electron from which electronic charge of the same spin as that of the given electron has been removed. The total charge that has been removed is equivalent to one electron. If we regard the Fermi hole as a uniformly charged sphere of density ρ and radius R, then

$$\rho_{\mathrm{ex}}(1;2)\,d\mathbf{r}_2 = \tfrac{4}{3}\pi R^3\rho = 1. \quad (19.3\text{-}37)$$

The potential energy at the center of such a sphere goes as $1/R$ which is then proportional to $\rho^{1/3}$. Clearly, this is the same phenomenon we encountered previously (Section 13.8) in connection with the Fermi correlation for electrons of the same spin and is therefore a manifestation of the Pauli principle.

What has emerged so far is that the variational method, based on the single Slater determinant (19.3-2) constructed from a set of orthonormal one-electron spin orbitals, when applied to the many-electron Schrödinger equation (19.3-1) results in a set of Hartree–Fock equations (19.3-16) which are coupled nonlinear integrodifferential equations for the one-electron spin orbitals and which consist entirely of one-electron operators. We may therefore picture a Hartree–Fock equation as the wave equation for a single electron moving in the field produced by the nucleus and the average field of the remaining electrons. This is precisely the objective of the central field approximation. The average field of the $(N-1)$ electrons is represented by the Coulomb and exchange operators which replace the operator e^2/r_{ij} in the exact Hamiltonian. This field consists of the field produced by electrons of opposite spin plus the field produced by electrons of the same spin but outside the Fermi hole. It is also necessary to bear in mind that since the Hartree–Fock method is based on the variational principle it only provides an upper bound to the energy of a state of given symmetry.

19.4 Properties of the Hartree–Fock Solutions

In the previous section we saw that the Hartree–Fock equations contained Lagrange multipliers which reflected the imposition of the normalization condition on the spin orbitals. However, the possibility of eliminating off-diagonal Lagrange multipliers by means of a unitary transformation was interpreted to mean that the spin orbitals need not necessarily be orthogonal. However, we shall now show that spin orbitals that satisfy the Hartree–Fock equations are indeed orthogonal.

Let $\psi_i(1)$ and $\psi_k(1)$ be two solutions of the Hartree–Fock equations:

$$\mathscr{H}_0(1)\psi_i(1) + \left[\sum_j \int \psi_j^*(2)(e^2/r_{12})\psi_j(2)\,d\tau_2\right]\psi_i(1) - \sum_j \left[\int \psi_j^*(2)(e^2/r_{12})\psi_i(2)\,d\tau_2\right]\psi_j(1) = \varepsilon_i\psi_i(1) \tag{19.4-1a}$$

and

$$\mathscr{H}_0(1)\psi_k(1) + \left[\sum_j \int \psi_j^*(2)(e^2/r_{12})\psi_j(2)\,d\tau_2\right]\psi_k(1) - \left[\sum_j \int \psi_j^*(2)(e^2/r_{12})\psi_k(2)\right]\psi_j(1) = \varepsilon_k\psi_k(1). \tag{19.4-1b}$$

If (19.4-1a) is multiplied by $\psi_k^*(1)$ and integrated, the result is

$$\langle\psi_k(1)|\mathscr{H}_0(1)|\psi_i(1)\rangle + \sum_j \langle\psi_k(1)\psi_j(2)|(e^2/r_{12})|\psi_i(1)\psi_j(2)\rangle$$
$$- \sum_j \langle\psi_k(1)\psi_j(2)|(e^2/r_{12})|\psi_j(1)\psi_i(2)\rangle = \varepsilon_i\langle\psi_k(1)|\psi_i(1)\rangle. \quad (19.4\text{-}2)$$

Similarly, on taking the complex conjugate of (19.4-1b), multiplying by $\psi_i(1)$, and integrating, the result is an expression which differs from (19.4-2) only in that ε_i is replaced by ε_k. Hence

$$\varepsilon_i\langle\psi_k(1)|\psi_i(1)\rangle = \varepsilon_k\langle\psi_k(1)|\psi_i(1)\rangle$$

or

$$\langle\psi_k|\psi_i\rangle = 0, \qquad \varepsilon_i \neq \varepsilon_k. \quad (19.4\text{-}3)$$

The total energy of the N-electron system is

$$E = \langle\Psi|\mathscr{H}|\Psi\rangle$$

which may be written, according to (11.2-30) and (11.2-31), as

$$E = \sum_i I_i + \sum_{i<j} [J(i,j) - K(i,j)] \quad (19.4\text{-}4)$$

where

$$I_i = \langle\psi_i|\mathscr{H}_0|\psi_i\rangle = \langle\varphi_i|\mathscr{H}_0|\varphi_i\rangle, \quad (19.4\text{-}5)$$
$$J(i,j) = \langle\psi_i(1)\psi_j(2)|(e^2/r_{12})|\psi_i(1)\psi_j(2)\rangle \quad (19.4\text{-}6a)$$
$$= \langle\varphi_i(1)\varphi_j(2)|(e^2/r_{12})|\varphi_i(1)\varphi_j(2)\rangle, \quad (19.4\text{-}6b)$$
$$K(i,j) = \langle\psi_i(1)\psi_j(2)|(e^2/r_{12})|\psi_j(1)\psi_i(2)\rangle \quad (19.4\text{-}7a)$$
$$= \delta(m_s{}^i, m_s{}^j)\langle\varphi_i(1)\varphi_j(2)|(e^2/r_{12})|\varphi_j(1)\varphi_i(2)\rangle. \quad (19.4\text{-}7b)$$

Also, when (19.4-1a) is multiplied by $\psi_i^*(1)$ and integrated we obtain an explicit expression for ε_i:

$$\varepsilon_i = I_i + \sum_j [J(i,j) - K(i,j)] \quad (19.4\text{-}8)$$

and on summing ε_i over all the electrons,

$$\sum_i \varepsilon_i = \sum_i I_i + \sum_{ij} [J(i,j) - K(i,j)] \quad (19.4\text{-}9a)$$
$$= \sum_i I_i + 2\sum_{i<j} [J(i,j) - K(i,j)]. \quad (19.4\text{-}9b)$$

Substituting in (19.4-4),

$$E = \sum_i \varepsilon_i - \sum_{i<j} [J(i,j) - K(i,j)] \quad (19.4\text{-}10)$$

which emphasizes the fact that the total energy is *not* simply the sum of the single electron energies ε_i.

A more specific interpretation of ε_i is achieved by considering the difference in energy between a system of $N+1$ electrons and a system of N electrons. Suppose we have a set of orbitals labeled $j = 1, 2, \ldots, N, i$. The energy, according to (19.4-4), is

$$E_{N+1} = \sum_{j=1}^{N} I_j + \sum_{j<k}^{N} [J(j,k) - K(j,k)] + I_i + \sum_{k}^{N} [J(i,k) - K(i,k)]$$

and

$$E_{N+1} - E_N = I_i + \sum_{k}^{N} [J(i,k) - K(i,k)] = \varepsilon_i. \tag{19.4-11}$$

Thus ε_i may be interpreted as the energy required to remove an electron from the ith orbital (*Koopmans' theorem*). However, this is certainly not rigorous inasmuch as it is assumed that the orbitals of the ion are the same as those of the neutral atom while the rapid readjustment in the orbitals that occurs after the removal of an electron is not taken into account.

A theorem pertaining to matrix elements between N-electron wave functions is now stated. Suppose

$$\Psi = \frac{1}{\sqrt{N!}} \det\{\psi_1, \psi_2, \ldots, \psi_k, \ldots, \psi_N\}, \tag{19.4-12a}$$

$$\Psi' = \frac{1}{\sqrt{N!}} \det\{\psi_1, \psi_2, \ldots, \psi_k', \ldots, \psi_N\} \tag{19.4-12b}$$

where Ψ' is the N-electron wave function associated with the excitation of an electron from the orbital ψ_k to ψ_k'. Using the theorems of Section 11.1,

$$\begin{aligned}
\langle \Psi' | \mathscr{H} | \Psi \rangle &= \langle \psi_k'(1) | \mathscr{H}_0(1) | \psi_k(1) \rangle + \sum_{t \neq k} \langle \psi_k'(1) \psi_t(2) | (e^2/r_{12}) | \psi_k(1) \psi_t(2) \rangle \\
&\quad - \sum_{t \neq k} \langle \psi_k'(1) \psi_t(2) | (e^2/r_{12}) | \psi_t(1) \psi_k(2) \rangle \\
&= \int \psi_k'^*(1) \Big\{ \mathscr{H}_0(1) + \sum_{i \neq k} \Big[\int \psi_t^*(2)(e^2/r_{12}) \psi_t(2)\, d\tau_2 \Big] \psi_k(1) \\
&\quad - \sum_{t \neq k} \Big[\int \psi_t^*(2)(e^2/r_{12}) \psi_k(2)\, d\tau_2 \Big] \psi_t(1) \Big\}\, d\tau_1.
\end{aligned} \tag{19.4-13}$$

If the orbitals satisfy the Hartree–Fock equations, we have, from (19.3-15),

$$\langle \Psi' | \mathscr{H} | \Psi \rangle = \int \psi_k'^*(1) \varepsilon_k \psi_k(1)\, d\tau_1 = 0 \tag{19.4-14}$$

since it has been shown that Hartree–Fock orbitals are orthogonal. This is the *Brillouin theorem* which states that matrix elements of the Hamiltonian (19.3-1) between a state represented by an N electron Slater determinant constructed of Hartree–Fock orbitals and a singly excited state vanish identically.

It is also of interest to investigate the form of the Hartree–Fock equations for a closed shell which is the simplest case that can be rigorously represented by a single Slater determinant. Assuming the closed shell contains $2N$ spin orbitals, there will be N spatial orbits, each one appearing twice in the Slater determinant—once with spin α and once with spin β. In terms of the spatial orbitals, the Hartree–Fock equations are

$$F(1)\varphi_i(1) = \varepsilon_i \varphi_i(1) \tag{19.4-15}$$

where

$$F(1) = \mathscr{H}_0(1) + \sum_k^N [2J_k(1) - K_k(1)], \tag{19.4-16}$$

$$J_k(1)\varphi_i(1) = \left[\int \varphi_k{}^*(2)(e^2/r_{12})\varphi_k(2)\,d\mathbf{r}_2\right]\varphi_i(1),$$

$$K_k(1)\varphi_i(1) = \left[\int \varphi_k{}^*(2)(e^2/r_{12})\varphi_i(2)\,d\mathbf{r}_2\right]\varphi_k(1). \tag{19.4-17}$$

The expressions for the energies are

$$\begin{aligned} E &= 2\sum_i^N I_i + \sum_{ik}^N [2J(i,k) - K(i,k)] \\ &= 2\sum_i^N \varepsilon_i - \sum_{ik}^N [2J(i,k) - K(i,k)] \end{aligned} \tag{19.4-18}$$

where the summations extend over the spatial orbitals and

$$\begin{aligned} I_i &= \langle \varphi_i | \mathscr{H}_0 | \varphi_i \rangle, \\ J(i,k) &= \langle \varphi_i(1)\varphi_k(2) | (e^2/r_{12}) | \varphi_i(1)\varphi_k(2) \rangle, \\ K(i,k) &= \langle \varphi_i(1)\varphi_k(2) | (e^2/r_{12}) | \varphi_k(1)\varphi_i(2) \rangle. \end{aligned} \tag{19.4-19}$$

An important property of closed shells is that the Hartree–Fock one-electron orbitals are, rigorously, solutions of a central field problem. In other words, each electron is a closed shell moves in a spherically symmetric potential. To prove this it will be assumed that the one-electron orbitals are of the form

$$\varphi(\mathbf{r}) = (1/r)P_{nl}(r)Y_{lm}(\theta, \varphi) \tag{19.4-20}$$

and that this assumption leads to spherically symmetric Hartree–Fock potentials so that the assumption is self-consistent.

Consider first the Coulomb term in (19.4-16). Since the summation is taken over all the spatial orbitals in the closed shell and n and l are constants for all such orbitals, the summation over k is equivalent to a summation over m. Therefore

$$\sum_{k=1}^{N} J_k(1) = \sum_{k=1}^{N} \int \varphi_k^*(2)(e^2/r_{12})\varphi_k(2)\, d\mathbf{r}_2$$
$$= \int P_{nl}^2(r_2)/r_{12} \sum_{m=-l}^{l} |Y_{lm}(\theta_2, \varphi_2)|^2\, dr_2\, d\Omega_2. \qquad (19.4\text{-}21)$$

This expression may be reduced by means of (1.2-22), (1.2-26), and (1.2-27):

$$\frac{1}{r_{12}} = 4\pi \sum_{L=0}^{\infty} \sum_{M=-L}^{L} \frac{1}{2L+1} \frac{r_<^{\,L}}{r_>^{L+1}} Y_{LM}^*(\theta_1, \varphi_1) Y_{LM}(\theta_2, \varphi_2), \qquad (19.4\text{-}22)$$

$$\sum_{m=-l}^{l} |Y_{lm}(\theta, \varphi)|^2 = (2l+1)/4\pi, \qquad (19.4\text{-}23)$$

$$\int Y_{LM}(\theta_2, \varphi_2)\, d\Omega_2 = \sqrt{4\pi}\, \delta(L, 0)\, \delta(M, 0). \qquad (19.4\text{-}24)$$

Substituting in (19.4-21),

$$\sum_{k=1}^{N} J_k(1) = \frac{(2l+1)e^2}{\sqrt{4\pi}} \int \frac{P_{nl}^2(r_2)}{r_>}\, dr_2 \qquad (19.4\text{-}25)$$

which is independent of angle and therefore spherically symmetric. For a point inside the closed shell $r_> = r_2$ and the potential is constant; outside of the closed shell $r_> = r_1$ and the potential is a function of r_1 only.

To demonstrate the spherical symmetry of the exchange term requires a bit more ingenuity. Consider the expression (de Shalit and Feshbach, 1974)

$$U_{\text{ex}}(\mathbf{r}_1, \mathbf{r}_1') = \sum_{kk'} \left[\int \varphi_k^*(\mathbf{r}_2)(e^2/r_{12})\varphi_{k'}(\mathbf{r}_2)\, d\mathbf{r}_2 \right] \varphi_{k'}^*(\mathbf{r}_1')\varphi_k(\mathbf{r}_1) \qquad (19.4\text{-}26)$$

and the integral

$$\int U_{\text{ex}}(\mathbf{r}_1, \mathbf{r}_1')\varphi_i(\mathbf{r}_1')\, d\mathbf{r}_1'$$
$$= \sum_{kk'} \left[\int \varphi_{k'}^*(\mathbf{r}_1')\varphi_i(\mathbf{r}_1')\, d\mathbf{r}_1' \int \varphi_k^*(\mathbf{r}_2)(e^2/r_{12})\varphi_{k'}(\mathbf{r}_2)\, d\mathbf{r}_2 \right] \varphi_k(\mathbf{r}_1). \qquad (19.4\text{-}27)$$

Since

$$\int \varphi_{k'}^*(\mathbf{r}_1')\varphi_i(\mathbf{r}_1')\, d\mathbf{r}_1' = \delta_{k',i}$$

eq. (19.4-27) simplifies to

$$\int U_{\text{ex}}(\mathbf{r}_1, \mathbf{r}_1')\varphi_i(\mathbf{r}_1')\,d\mathbf{r}_1' = \sum_k \left[\int \varphi_k{}^*(\mathbf{r}_2)(e^2/r_{12})\varphi_i(\mathbf{r}_2)\,d\mathbf{r}_2\right]\varphi_k(\mathbf{r}_1)$$

$$= \sum_k K_k(1)\varphi_i(1) \qquad (19.4\text{-}28)$$

which is just the exchange term in (19.4-17). It is therefore sufficient to investigate the spherical symmetry of $U_{\text{ex}}(\mathbf{r}_1, \mathbf{r}_1')$.

If the orbitals are written as in (19.4-20) and we use the expression (19.4-22) for $1/r_{12}$,

$$U_{\text{ex}}(\mathbf{r}_1, \mathbf{r}_1') = \sum_{kk'}\sum_{LM} \frac{4\pi e^2}{2L+1} G(n_k l_k, n_{k'} l_{k'}; r_1, r_1')$$

$$\times Y_{LM}^*(\theta_1, \varphi_1) Y_{l_{k'} m_{k'}}^*(\theta_1', \varphi_1') Y_{l_k m_k}(\theta_1, \varphi_1)$$

$$\times \int Y_{l_k m_k}^*(\theta_2, \varphi_2) Y_{LM}(\theta_2, \varphi_2) Y_{l_{k'} m_{k'}}(\theta_2, \varphi_2)\, d\Omega_2 \qquad (19.4\text{-}29)$$

where

$$G(n_k l_k, n_{k'} l_{k'}, r_1, r_1') = \frac{1}{r_1'} P_{n_{k'} l_{k'}}^*(r_1') \frac{1}{r_1} P_{n_k l_k}(r_1)$$

$$\times \int P_{n_k l_k}^*(r_2) \frac{r_<^L}{r_>^{L+1}} P_{n_{k'} l_{k'}}(r_2)\, dr_2. \qquad (19.4\text{-}30)$$

The product $Y_{LM}^*(\theta_1, \varphi_1) Y_{l_k m_k}(\theta_1, \varphi_1)$ may be written in the form of (1.2-35) while the integral of the product of the three spherical harmonics may be evaluated by means of (1.2-29) with the result that

$$U_{\text{ex}}(\mathbf{r}_1, \mathbf{r}_1')$$

$$= \sum_{kk'}\sum_{LM}\sum_{L'M'} \frac{4\pi e^2}{2L+1} G(n_k l_k, n_{k'} l_{k'}; r_1, r_1')$$

$$\times \frac{(2l_k+1)(2L+1)}{4\pi} \sqrt{(2L'+1)(2l_{k'}+1)} \begin{pmatrix} L & l_k & L' \\ 0 & 0 & 0 \end{pmatrix} \begin{pmatrix} L & l_k & l_{k'} \\ 0 & 0 & 0 \end{pmatrix}$$

$$\times \begin{pmatrix} L & l_k & L' \\ M & -m_k & M' \end{pmatrix} \begin{pmatrix} L & l_k & l_{k'} \\ M & -m_k & m_{k'} \end{pmatrix} Y_{l_{k'} m_{k'}}^*(\Omega_1') Y_{L'M'}(\Omega_1).$$

$$(19.4\text{-}31)$$

Further simplification is accomplished with the aid of the orthogonality property of the $3j$ symbols (1.5-50) and the addition theorem (1.2-21). The final result is

$$U_{ex}(\mathbf{r}_1, \mathbf{r}_1') = e^2 \sum_{kk'L} G(n_k l_k, n_{k'} l_{k'}; r_1, r_1')$$
$$\times \frac{(2l_k + 1)(2l_{k'} + 1)}{4\pi} \begin{pmatrix} L & l_k & l_{k'} \\ 0 & 0 & 0 \end{pmatrix}^2 P_{l_{k'}}(\cos\omega). \quad (19.4\text{-}32)$$

It is observed that $U_{ex}(\mathbf{r}_1, \mathbf{r}_1')$ depends on the magnitudes r_1 and r_1' and on the angle ω between the vectors $\mathbf{r}_1$ and $\mathbf{r}_1'$ but does not depend on the separate orientations of $\mathbf{r}_1$ and $\mathbf{r}_1'$. Hence $U_{ex}(\mathbf{r}_1, \mathbf{r}_1')$ is spherically symmetric.

Having shown that both the Coulomb and exchange potentials for a closed shell are spherically symmetric it follows that the total self-consistent field for a closed shell is spherically symmetric and the central field orbitals are exact solutions to the Hartree–Fock equations. It should be remarked that the proof hinged upon the fact that in a closed shell all the magnetic substates are populated thereby enabling us to carry out the summation over these substates which then resulted in a spherically symmetric field.

19.5 Computational Methods

Since the Hartree–Fock equations are coupled, nonlinear, integrodifferential equations there is no direct method for solving them. In fact, each Hartree–Fock equation requires prior knowledge of all the other orbitals. It is therefore necessary to resort to a *self-consistent* method which consists, in principle, of an initial more or less arbitrary selection of spin orbitals. These are used to calculate the Coulomb and exchange operators J_j and K_j and hence the Fock operator F. With this approximate Fock operator a new improved set of orbitals are calculated by solving the eigenvalue equation (19.3-20). The process is continued until the difference between two sets of orbitals in successive cycles of computation is sufficiently small to achieve the desired accuracy. This is the general method of solution and is known as the HF–SCF (Hartree–Fock–Self-Consistent Field) method.

In practice there are numerous variations but all of them involve iterative procedures on which certain criteria of self-consistency have been imposed. We shall now summarize some of these.

If it is assumed that the spin orbitals are of the central field variety (19.2-7), it becomes feasible to express the Hartree–Fock equations entirely in terms of the radial function $P_{nl}(r)$ since the angular parts of the orbitals are the spherical harmonics and their integrals are known. This is a great simplification inasmuch as the Hartree–Fock equations have been reduced from a dependence on three coordinates to a dependence on only one coordinate. It should be recognized, though, that this procedure implies that all the electrons in the same shell (same nl values) have the same radial function.

The constraint on the Hartree–Fock solutions to be in the form of central field orbitals requires attention to the possibility mentioned in Section 19.3 that the off-diagonal Lagrange multipliers cannot be eliminated by a unitary transformation. This matter is thoroughly discussed by Slater (1960) with the following conclusions.

For a closed shell the off-diagonal multipliers can be transformed away and the Hartree–Fock equations reduce to the radial equation (in Rydberg units)

$$\left[-\frac{d^2}{dr^2} - \varepsilon_i - \frac{2Z}{r} + \frac{l_i(l_i+1)}{r^2} + \frac{2}{r}\sum_j (4l_j+2)Y_0(j,j)\right]P_i(r)$$
$$-\frac{2}{r}\sum_j \sqrt{\frac{2l_j+1}{2l_i+1}}\sum_k c^k(i,j)Y_k(i,j)P_j(r) = 0 \qquad (19.5\text{-}1)$$

where the sum over j is a sum over shells,

$$Y_k(i,j) = \frac{1}{r^k}\int_0^r r_1{}^k P_i(r_1)P_j(r_1)\,dr_1 + r^{k+1}\int_r^\infty \frac{P_i(r_1)P_j(r_1)}{r_1^{k+1}}\,dr_1$$
$$P_i(r) \equiv P_{n_i l_i}(r), \qquad c^k(i,j) \equiv c^k(l_i\,0, l_j\,0). \qquad (19.5\text{-}2)$$

For nonclosed shells the effective potential is no longer spherically symmetric although the departure from nonsphericity is usually quite small. Since the spherical harmonics are no longer exact eigenfunctions it is generally not possible to remove the off-diagonal multipliers. In that case it is necessary to express the average energy of an electronic configuration in terms of integrals over the radial parts of the central field orbitals. A variational calculation is then carried out with respect to the radial function so as to minimize the energy. The result is the radial Hartree–Fock equation

$$\left[-\frac{d^2}{dr^2} - \varepsilon_i - \frac{2Z}{r} + \frac{l_i(l_i+1)}{r^2} + \frac{2}{r}\sum_j \omega_j Y_0(j,j) - \frac{2}{r}Y_0(i,i)\right.$$
$$\left. -\frac{2}{r}\frac{\omega_i - 1}{4l_i+1}\sum_{k=2,4,6,\ldots} c^k(i,i)Y_k(i,i)\right]P_i(r)$$
$$-\frac{2}{r}\sum_{j\neq i}\frac{\omega_j}{(4l_j+2)(4l_i+2)}\sum_k c^k(i,j)Y_k(i,j)P_j(r) = \sum_{j\neq i, l_j=l_i}\omega_j\varepsilon_{ij}P_j(r), \qquad (19.5\text{-}3)$$

which reduces to (19.5-1) for a closed shell. Here ε_i is the diagonal Lagrange multiplier which maintains normalization of the orbitals; similarly ε_{ij} maintains orthogonality of the orbitals with the same l. ω_i is the number of electrons in a particular shell (not necessarily a closed shell) with quantum numbers n_i, l_i; all ω_i electrons of one shell are assumed to have the same radial wavefunction and the $P_{nl}(r)$ satisfy the orthonormalization condition

$$\int_0^\infty P_{n_i l_i}(r) P_{n_j l_j}(r)\, dr = \begin{cases} 1 & \text{for} \quad i = j, \\ 0 & \text{for} \quad i \neq j, \quad l_i = l_j. \end{cases} \tag{19.5-4}$$

The computations begin with specifically chosen wave functions and values of ε_i. The Y_k terms are then calculated and the solution proceeds with the Y_k and P_j terms held constant. A new set of wave functions are then calculated and another iteration is begun until the criteria of self-consistency are satisfied. The computation is often simplified by first ignoring the off-diagonal terms since these are quite small. After the wave functions are obtained, using the diagonal multipliers only, the ε_{ij} are calculated and the energies are corrected. This method is known as the *restricted* Hartree–Fock method, owing to the assignment of the same radial function to all the orbitals in a given shell. In the unrestricted version, orbitals with different spins are permitted to have different radial functions.

Extensive numerical computations have been carried out by Mann (1967) and by Froese-Fischer (1972). From their tabulations one may obtain numerical values of the radial wavefunctions, the total energy of the atom at the center of gravity of a configuration ε_i, the integrals I, F^k, G^k, and certain radial expection values.

For many purposes analytic solutions of the Hartree–Fock equations are preferred over numerical solutions, especially if such solutions are to be further manipulated as, for example, in the evaluation of multicenter integrals which arise in molecular problems or if one wishes to improve upon the Hartree–Fock solutions by means of configuration interaction (Section 19.6). The Hartree–Fock orbitals are then expressed as linear combinations of a suitable basis set of functions. This is known as the *Roothaan* procedure.

A convenient basis set for this type of calculation consists of *Slater orbitals* whose radial part is

$$R_{nl}(r) = N r^{n-1} e^{-\xi r} \tag{19.5-5}$$

where

$$N = (2\xi)^{n+1/2}[(2n)!]^{-1/2} \tag{19.5-6}$$

is the normalization constant, n a positive integer, ξ the orbital exponent. The complete Slater orbitals are

$$\chi_{nlm}(\mathbf{r}) = R_{nl}(r) Y_{lm}(\theta, \varphi). \tag{19.5-7}$$

Slater orbitals have only an approximate resemblance to hydrogenic eigenfunctions; they have the proper asymptotic behavior but do not possess the necessary radial nodes. Nevertheless they serve as a convenient basis set.

The end result of a self-consistent procedure is a linear combination of the χ_{nlm} which minimizes the energy. As an example, for the ground state of boron whose electron configuration is $(1s)^2\,(2s)^2\,(2p)^1$ the spatial orbitals are (Clementi and Roetti, 1974):

$$\begin{aligned}\varphi(1s) &= 0.19030\chi_1 + 0.82091\chi_2 - 0.00364\chi_3 + 0.00251\chi_4\\ \varphi(2s) &= 0.00754\chi_1 - 0.25055\chi_2 + 0.87099\chi_3 + 0.18515\chi_4\\ \varphi(2p) &= 0.21526\chi_5 + 0.84052\chi_6\end{aligned} \tag{19.5-8}$$

where

$$\begin{aligned}\chi_1 &= N_1 r^0 \exp(-6.56657r)Y_{00}(\theta,\varphi), & \chi_2 &= N_2 r^0 \exp(-4.24927r)Y_{00}(\theta,\varphi),\\ \chi_3 &= N_3 r^1 \exp(-1.41314r)Y_{00}(\theta,\varphi), & \chi_4 &= N_4 r^1 \exp(-0.87564r)Y_{00}(\theta,\varphi),\\ \chi_5 &= N_5 r^1 \exp(-2.21734r)Y_{10}(\theta,\varphi), & \chi_6 &= N_6 r^1 \exp(-1.00551r)Y_{10}(\theta,\varphi),\end{aligned} \tag{19.5-9}$$

and the normalization constants are given by (19.5-6).

From the computational standpoint the exchange term is the one that is the most troublesome. In consequence, various simplifications have been employed among which is the Slater approximation based on the free-electron gas model. The Hartree–Fock–Slater equations based on this model are, in Rydberg units,

$$\left(-\nabla_1^{\,2} - \frac{2Z}{r_1}\right)\varphi_i(1) + \left[\sum_j 2\int \frac{\varphi_j^*(2)\varphi_j(2)}{r_{12}}\,d\tau_2\right]\varphi_i(1) - \left\{6\left[\frac{3}{8\pi}\sum_j \varphi_j^*(1)\varphi_j(1)\right]^{1/3}\right\}\varphi_i(1) = \varepsilon_i\varphi_i(1). \tag{19.5-10}$$

It is seen that the exchange integrals which appear in the Hartree–Fock equations (19.3-28) are no longer present. In their place we have terms which depend only on the local charge density. The origin of the $\frac{1}{3}$ power dependence is qualitatively similar to the $\frac{1}{3}$ power dependence of the potential on the charge density ρ in the case of the Fermi hole as a result of (19.3-37). Solutions to the Hartree–Fock–Slater equations have been obtained by Herman and Skillman (1963).

19.6 Correlation Error and Configuration Interaction

The Coulomb repulsion between electrons whose spins may be either parallel or antiparallel prevents two electrons from approaching arbitrarily close to one another. This means that the positions of all electrons, regardless of their spin, are correlated, or, in another terminology, all electrons are surrounded by a *Coulomb hole*. An exact N electron wave function would automatically give a correct description of the Coulomb hole. But a Hartree–Fock wave function is not exact; hence, the correlation is not properly accounted for. However, the error incurred for electrons with parallel spin is not the same as for electrons with opposite spin. The reason for this difference is associated with the antisymmetry property of the wave function, which we saw in Section 19.3, correlates the positions of electrons with parallel spins and prevents them from occupying the same spatial orbital, but has no effect on electrons with opposite spin. This feature was described by the *Fermi hole*, which is almost, though not entirely, equivalent to the Coulomb hole (for electrons with parallel spin). Thus, the major source of error in the Hartree–Fock method arises from a lack of proper correlation among electrons of opposite spin.

It is therefore expected that the Hartree–Fock energy E_{HF} will differ from the exact energy E_{ex} and the difference

$$E_{cor} = E_{ex} - E_{HF} \tag{19.6-1}$$

is known as the *correlation* energy. It is customary to regard E_{ex} as the exact *nonrelativistic* energy which is not quite the same as the experimentally measured energy since the latter contains contributions from relativistic[‡] and other terms not included in the nonrelativistic Hamiltonian. Since E_{HF} has been obtained by a variational method, it must lie above E_{ex} for the ground state so that the correlation energy is always negative. The error in the Hartree–Fock energy of an atom is about 1%. Although this may be an acceptable accuracy for the total energy, serious errors may nevertheless arise in various applications such as multiplet separations, hyperfine structure, molecular calculations, spectroscopy, etc., where small *differences* in energy are the important physical parameters. Estimates of correlation errors give −1.1 eV for He, −11 eV for Ne, and −21.5 eV for Ar.

As indicated previously (Section 19.3), to expand an N-electron antisymmetric wave function one needs a complete set of Slater determinants. The latter consists of all possible determinants containing N spin orbitals selected from a complete set of spin orbitals. Each Slater determinant constructed in this way corresponds to a single electron configuration; hence such an expansion of an antisymmetric wave function is also known as a superposition of configurations. If a single configuration is employed, i.e., the infinite series of Slater determinants is truncated after the first term and the spin orbitals are optimized with respect to the total energy, we have the

[‡] Relativistic effects which include the variation of mass with energy, spin-orbit and Fermi contact interactions are particularly important for inner shell electrons in heavy elements. In spectroscopic applications relativistic effects tend to be less important because, in the first place, only energy differences are relevant and, second, one is usually interested in transitions of outer shell electrons while those in inner shells remain relatively unchanged.

ordinary Hartree–Fock approximation. Clearly, improved results are to be expected by increasing the number of configurations in order to approach closer to the exact N-electron antisymmetric wave function. This method for obtaining improved eigenfunctions and eigenvalues for the N-electron problem is known as *configuration interaction.*

In this method N-electron wave functions are obtained by taking linear combinations of Slater determinants

$$\Psi(1,2,\ldots,N) = \sum_i c_i\psi_i \tag{19.6-2}$$

in which the c_i are coefficients and the ψ_i are Slater determinants which differ in the choice of spin orbitals which the electrons are assumed to occupy and hence correspond to different electronic configurations. The coefficients in (19.6-2) are again obtained by minimizing the energy. This leads to a system of linear equations

$$\sum_j (H_{ij} - E\,\delta_{ij})c_j = 0 \tag{19.6-3}$$

where

$$H_{ij} = \int \psi_i{}^* \mathscr{H} \psi_j \, d\tau \tag{19.6-4}$$

and a secular determinant

$$|H_{ij} - E\,\delta_{ij}| = 0$$

from which the required coefficients are obtained. Although conceptually this method should give an arbitrarily accurate N-electron wave function, the convergence is usually slow and the numerical work increases rapidly as the number of configurations is increased. For this reason one often includes only those configurations whose energies are close to one another and therefore interact more strongly than the configurations with widely separated energies.

CHAPTER 20

MULTIPLET WAVE FUNCTIONS

A spin orbital in the central field approximation is characterized by the quantum numbers n, l, s, m_l, and m_s although the inclusion of s is somewhat redundant since it always remains equal to $\frac{1}{2}$. Since the energy is independent of m_l and m_s, the degeneracy is $2(2l+1)$. Thus, even for a relatively small number of electrons the degeneracy can become quite high. For the pd configuration, for example, the degeneracy is 60. The implication for perturbation calculations is that it will be necessary to solve secular equations of high degree unless we take advantage of the inherent symmetry of the system to factor the secular determinant into a number of smaller ones. We shall refer to Section 19.1 for some of these symmetries with the objective of constructing antisymmetric wave functions in the $|LSM_LM_S\rangle$ basis as linear combinations of central field orbitals. The motivation for this stems from the fact that the operator $\sum_{i<j} e^2/r_{ij}$ is diagonal in this basis (see Sections 21.1 and 21.3).

20.1 Two-Electron Multiplets

We shall take

$$\mathscr{H} = \mathscr{H}_0(1) + \mathscr{H}_0(2) + (e^2/r_{12}) \tag{20.1-1}$$

with

$$\mathscr{H}_0 = (p^2/2m) - (Ze^2/r) \tag{20.1-2}$$

as the Hamiltonian for the two-electron atom. As shown in Section 19.1, for Hamiltonians of type (20.1-1),

$$[\mathscr{H}, \mathbf{L}] = 0, \qquad [\mathscr{H}, L^2] = 0, \tag{20.1-3}$$

$$[\mathscr{H}, \mathbf{S}] = 0, \qquad [\mathscr{H}, S^2] = 0, \tag{20.1-4}$$

where, for the two-electron atom

$$\mathbf{L} = \boldsymbol{l}_1 + \boldsymbol{l}_2, \tag{20.1-5}$$

$$\mathbf{S} = \mathbf{s}_1 + \mathbf{s}_2. \tag{20.1-6}$$

The rules for coupling angular momenta give

$$L = l_1 + l_2, \quad l_1 + l_2 - 1, \ldots, |l_1 - l_2| \tag{20.1-7}$$

$$M_L = m_{l_1} + m_{l_2} = L, \quad L - 1, \ldots, -L, \tag{20.1-8}$$

and since $s = \frac{1}{2}$ for each electron,

$$S = s_1 + s_2, \quad s_1 - s_2 = 1, \quad 0 \tag{20.1-9}$$

$$M_S = m_{s_1} + m_{s_2} = \begin{cases} 1, \quad 0, \quad -1 & \text{when} \quad S = 1, \\ 0 & \text{when} \quad S = 0. \end{cases} \tag{20.1-10}$$

In addition, it is necessary to satisfy the Pauli principle which requires the total wave function to be antisymmetric with respect to an interchange of all the coordinates (space and spin) of the two electrons.

With these considerations in mind, a sequence for constructing Russell–Saunders (LS) eigenfunctions of $\mathscr{H}$ on the basis of central field spin orbitals is given:

(a) Construct eigenfunctions of S^2 and S_z.
(b) Construct eigenfunctions of L^2 and L_z.
(c) Construct an antisymmetric wave function from the results of (a) and (b).

The spin part of the problem has already been solved in Section 1.7. The results are listed in Table 20.1; we note that the triplet states ($S = 1$) are symmetric and the singlet state ($S = 0$) is antisymmetric. We shall therefore

TABLE 20.1

Two-electron Spin Functions and their Symmetries[a]

Notation	Spin function	$2S + 1$	Symmetry	Young tableau
$\|1\ 1\rangle \equiv {}^3\xi_1$	$\alpha(1)\alpha(2)$	3	S	$\boxed{1}\boxed{2}$
$\|1\ 0\rangle \equiv {}^3\xi_0$	$\frac{1}{\sqrt{2}}[\alpha(1)\beta(2) + \beta(1)\alpha(2)]$	3	S	$\boxed{1}\boxed{2}$
$\|1\ -1\rangle \equiv {}^3\xi_{-1}$	$\beta(1)\beta(2)$	3	S	$\boxed{1}\boxed{2}$
$\|0\ 0\rangle \equiv {}^1\xi_0$	$\frac{1}{\sqrt{2}}[\alpha(1)\beta(2) - \beta(1)\alpha(2)]$	1	A	$\begin{array}{\|c\|}\hline 1 \\ \hline 2 \\ \hline\end{array}$

[a] The ket notation stands for $|SM\rangle$. Symmetry with respect to an interchange of electrons 1 and 2 is indicated by S (symmetric) and A (antisymmetric). $2S + 1$ is the multiplicity.

combine a symmetric (antisymmetric) spin function with an antisymmetric (symmetric) space function to form a totally antisymmetric two-electron wave function.

For eigenfunctions of L^2 and L_z the coupling scheme is (Section 1.5)

$$\begin{aligned}|LM_L\rangle &\equiv |l_1 l_2 LM_L\rangle \\ &= \sum_{m_{l_1} m_{l_2}} |l_1 l_2 m_{l_1} m_{l_2}\rangle\langle l_1 l_2 m_{l_1} m_{l_2}|l_1 l_2 LM_L\rangle \\ &= (-1)^{l_2 - l_1 - M_L} \sum_{m_{l_1} m_{l_2}} \sqrt{2L+1}\begin{pmatrix} l_1 & l_2 & L \\ m_{l_1} & m_{l_2} & -M_L \end{pmatrix} |l_1 l_2 LM_L\rangle.\end{aligned} \tag{20.1-11}$$

Since $|LM_L\rangle$ usually has no special symmetry, the required symmetrization or antisymmetrization may be achieved by means of the operators

$$\mathrm{S} = \frac{1}{\sqrt{2}}(1 + P_{12}), \tag{20.1-12}$$

$$\mathrm{A} = \frac{1}{\sqrt{2}}(1 - P_{12}), \tag{20.1-13}$$

where P_{12} interchanges the coordinates of electrons 1 and 2. The final antisymmetric wave function will be represented by $|LSM_L M_S\rangle$, and, in analogy with the one-electron case, states with $L = 0, 1, 2, 3, \ldots$ are called S, P, D, F, . . . states, respectively. In place of $|LSM_L M_S\rangle$ one may also use the notation $|^{2S+1}LM_L M_S\rangle$ where $2S + 1$ is the *multiplicity*: singlet ($S = 0$), doublet ($S = \frac{1}{2}$), triplet ($S = 1$), etc. If it is not necessary to distinguish among the various magnetic substates corresponding to different values of M_L and M_S, it is customary to write ^{2S+1}L, also known as a *term*. A parity superscript may also be included. Thus $^3\mathrm{P}^\circ$ means $L = 1$, $S = 1$, and odd parity. We shall now give several examples of the procedure outlined.

(1) $1\mathrm{s}^2$ $n_1 = n_2 = 1$, $l_1 = l_2 = 0$. The only possible value of L is $L = 0$. Therefore

$$|LM_L\rangle = |00\rangle = 1\mathrm{s}(1)\,1\mathrm{s}(2)$$

which is a symmetric function. Multiplying $|LM_L\rangle$ by the antisymmetric $^1\xi$ spin function we have

$$\begin{aligned}|LSM_L M_S\rangle &= |0000\rangle \equiv |^1\mathrm{S}00\rangle \equiv |^1\mathrm{S}M_L M_S\rangle \\ &= \frac{1}{\sqrt{2}} 1\mathrm{s}(1)\,1\mathrm{s}(2)[\alpha(1)\beta(2) - \beta(1)\alpha(2)] \\ &= \frac{1}{\sqrt{2}}\begin{vmatrix} 1\mathrm{s}(1)\alpha(1) & 1\mathrm{s}(1)\beta(1) \\ 1\mathrm{s}(2)\alpha(2) & 1\mathrm{s}(2)\beta(2) \end{vmatrix} \equiv |1\mathrm{s}^+\,1\mathrm{s}^-\rangle\end{aligned} \tag{20.1-14}$$

in which the $+$, $-$ notation refers to $\alpha(m_s = \frac{1}{2})$ and $\beta(m_s = -\frac{1}{2})$ spins, respectively. Note that for the closed shell configuration $1s^2$ there is but one Slater determinant and the only possible term is ^1S. The Pauli principle is satisfied because the two electrons have opposite values of m_s. On the other hand, a triplet state which would be in violation of the Pauli principle is impossible because the antisymmetric space function is identically zero.

(2) 1s2s $n_1 = 1, n_2 = 2, l_1 = l_2 = 0$. Here too, $L = 0$ is the only possible value of L and

$$|LM_L\rangle = 1s(1)\,2s(2).$$

But now it is possible to construct both symmetric (S) and antisymmetric (A) combinations:

$$|LM_L\rangle_S = |00\rangle_S = \psi_S = \frac{1}{\sqrt{2}}[1s(1)\,2s(2) + 2s(1)\,1s(2)],$$

$$|LM_L\rangle_A = |00\rangle_A = \psi_A = \frac{1}{\sqrt{2}}[1s(1)\,2s(2) - 2s(1)\,1s(2)]$$

so that the possible states are

$$|^1S00\rangle = \psi_S\,{}^1\xi = \frac{1}{\sqrt{2}}[1s(1)\,2s(2) + 2s(1)\,1s(2)]\frac{1}{\sqrt{2}}[\alpha(1)\beta(2) - \beta(1)\alpha(2)]$$

$$= \frac{1}{\sqrt{2}}\left\{\frac{1}{\sqrt{2}}\begin{vmatrix} 1s(1)\alpha(1) & 2s(1)\beta(1) \\ 1s(2)\alpha(2) & 2s(2)\beta(2) \end{vmatrix} - \frac{1}{\sqrt{2}}\begin{vmatrix} 1s(1)\beta(1) & 2s(1)\alpha(1) \\ 1s(2)\beta(2) & 2s(2)\alpha(2) \end{vmatrix}\right\}$$

$$= \frac{1}{\sqrt{2}}[|1s^+\,2s^-\rangle - |1s^-\,2s^+\rangle], \qquad (20.1\text{-}15a)$$

$$|^3S01\rangle = \psi_A\,{}^3\xi = \frac{1}{\sqrt{2}}\begin{vmatrix} 1s(1)\alpha(1) & 2s(1)\alpha(1) \\ 1s(2)\alpha(2) & 2s(2)\alpha(2) \end{vmatrix} = |1s^+\,2s^+\rangle$$

$$|^3S00\rangle = \frac{1}{\sqrt{2}}[|1s^+\,2s^-\rangle + |1s^-\,2s^+\rangle], \qquad |^3S0\,-1\rangle = |1s^-\,2s^-\rangle. \qquad (20.1\text{-}15b)$$

It is seen that the 1s2s configuration leads to ^1S and ^3S terms and that a single Slater determinant is generally not sufficient to represent a state.

(3) 1s2p $n_1 = 1,\ n_2 = 2,\ l_1 = 0,\ l_2 = 1$. In this case $L = 1$ and the possible terms are ^1P, ^3P.

$$|LM_L\rangle = |11\rangle = 1s(1)\,2p_{+1}(1),$$

$$|LM_L\rangle_A^S = |11\rangle_A^S = \frac{1}{\sqrt{2}}[1s(1)\,2p_{+1}(2) \pm 2p_{+1}(1)\,1s(2)],$$

$$|LM\rangle = |10\rangle = 1s(1)\,2p_0(2),$$

$$|LM\rangle_A^S = |10\rangle_A^S = \frac{1}{\sqrt{2}}[1s(1)\,2p_0(2) \pm 2p_0(1)\,1s(2)],$$

$$|LM_L\rangle = |1\ -1\rangle = 1s(1)\,2p_{-1}(2),$$

$$|LM_L\rangle_A^S = |1\ -1\rangle_A^S = \frac{1}{\sqrt{2}}[1s(1)\,2p_{-1}(2) \pm 2p_{-1}(1)\,1s(2)].$$

The states arising from the 1s2p configuration are listed in Table 20.2.

TABLE 20.2

Wave Functions for the 1s 2p *Configuration*[a]

$\|{}^1P\ 1\ 0\rangle = \frac{1}{\sqrt{2}}[\|1s^+\ 2p_{+1}^-\rangle - \|1s^-\ 2p_{+1}^+\rangle]$	$\|{}^3P\ 0\ 1\rangle = \|1s^+\ 2p_0{}^+\rangle$
$\|{}^1P\ 0\ 0\rangle = \frac{1}{\sqrt{2}}[\|1s^+\ 2p_0{}^-\rangle - \|1s^-\ 2p_0{}^+\rangle]$	$\|{}^3P\ 0\ 0\rangle = \frac{1}{\sqrt{2}}[\|1s^+\ 2p_0{}^-\rangle + \|1s^-\ 2p_0{}^+\rangle]$
$\|{}^1P\ -1\ 0\rangle = \frac{1}{\sqrt{2}}[\|1s^+\ 2p_{-1}^-\rangle - \|1s^-\ 2p_{-1}^+\rangle]$	$\|{}^3P\ 0\ -1\rangle = \|1s^-\ 2p_0{}^-\rangle$
$\|{}^3P\ 1\ 1\rangle = \|1s^+\ 2p_{+1}^+\rangle$	$\|{}^3P\ -1\ 1\rangle = \|1s^+\ 2p_{-1}^+\rangle$
$\|{}^3P\ 1\ 0\rangle = \frac{1}{\sqrt{2}}[\|1s^+\ 2p_{+1}^-\rangle + \|1s^-\ 2p_{+1}^+\rangle]$	$\|{}^3P\ -1\ 0\rangle = \frac{1}{\sqrt{2}}[\|1s^+\ 2p_{-1}^-\rangle + \|1s^-\ 2p_{-1}^+\rangle]$
$\|{}^3P\ 1\ -1\rangle = \|1s^-\ 2p_{+1}^-\rangle$	${}^3P\ -1\ -1\rangle = \|1s^-\ 2p_{-1}^-\rangle$

[a] Each ket on the right represents a Slater determinant. The representation $|LSM_LM_S\rangle$ is written $|{}^{2S+1}LM_LM_S\rangle$.

(4) 2p3p $n_1 = 2, n_2 = 3, l_1 = l_2 = 1$. The possible values of L are 0, 1, 2; the possible terms are therefore ${}^1S$, ${}^1P$, ${}^1D$, ${}^3S$, ${}^3P$, and ${}^3D$. To calculate $|LM_L\rangle$ we refer to a table of coupling coefficients. For example,

$$|LM_L\rangle = |00\rangle = \frac{1}{\sqrt{3}}[2p_{+1}(1)\,3p_{-1}(2) - 2p_0(1)\,3p_0(2) + 2p_{-1}(1)\,3p_{+1}(2)]$$

and the symmetric and antisymmetric combinations are

$$|00\rangle_A^S = \frac{1}{\sqrt{6}}[2p_{+1}(1)\,3p_{-1}(2) \pm 3p_{-1}(1)\,2p_{+1}(2) - 2p_0(1)\,3p_0(2)$$

$$\mp 3p_0(1)\,2p_0(2) + 2p_{-1}(1)\,3p_{+1}(2) \pm 3p_{+1}(1)\,2p_{-1}(2)].$$

On multiplication with the appropriate spin function we obtain all the states belonging to ^{1}S and ^{3}S. Thus

$$|^1\text{S}00\rangle = \frac{1}{\sqrt{6}}[|1^+\ -1^-\rangle - |1^-\ -1^+\rangle - |0^+\ 0^-\rangle + |0^-\ 0^+\rangle$$
$$+ |-1^+\ 1^-\rangle - |-1^-\ 1^+\rangle] \qquad (20.1\text{-}16)$$

in which the notation discloses the m_l and m_s values only, since all the other quantum numbers are constant. For example,

$$|1^+\ -1^-\rangle = \frac{1}{\sqrt{2}}\begin{vmatrix} 2\text{p}_{+1}(1)\alpha(1) & 3\text{p}_{-1}(1)\beta(1) \\ 2\text{p}_{+1}(2)\alpha(2) & 3\text{p}_{-1}(2)\beta(2) \end{vmatrix} \qquad (20.1\text{-}17)$$

but

$$|-1^-\ 1^+\rangle = \frac{1}{\sqrt{2}}\begin{vmatrix} 2\text{p}_{-1}(1)\beta(1) & 3\text{p}_{+1}(1)\alpha(1) \\ 2\text{p}_{-1}(2)\beta(2) & 3\text{p}_{+1}(2)\alpha(2) \end{vmatrix}. \qquad (20.1\text{-}18)$$

Two spin orbitals with the same values of n and l were said to be *equivalent*. Let us now examine the states arising from the configuration $(n\text{p})^2$. It is evident from (20.1-17) and (20.1-18), for example, that on setting $n_1 = n_2$,

$$|1^+\ -1^-\rangle = -|-1^-\ 1^+\rangle \qquad (n_1 = n_2)$$

and similarly

$$|1^-\ -1^+\rangle = -|-1^+\ 1^-\rangle, \qquad |0^+\ 0^-\rangle = -|0^-\ 0^+\rangle.$$

These relations may be substituted in (20.1-16). After readjusting the normalization constant we have, for two equivalent p electrons

$$|^1\text{S}00\rangle = \frac{1}{\sqrt{3}}[|1^+\ -1^-\rangle - |1^-\ -1^+\rangle - |0^+\ 0^-\rangle](\text{p}^2).$$

The net result is that instead of six terms (^{1}S ^{1}P ^{1}D ^{3}S ^{3}P ^{3}D) for the configuration $n\text{p}n'\text{p}$ there are only three terms—^{1}S ^{3}P ^{1}D—when the configuration is $(n\text{p})^2$. Here, too, we may recognize the intervention of the Pauli principle to limit the number of possible terms from a pair of equivalent electrons. The complete set of states for the p^2 configuration is shown in Table 20.3.

We note that for two equivalent electrons as, for example, in the p^2 configuration,

$$L + S = \text{even integer}. \qquad (20.1\text{-}19)$$

A general proof of this theorem may be constructed as follow: First let us recall one of the symmetry properties of the CG coefficients (1.5-46):

$$\langle j_1 j_2 m_1 m_2 | j_1 j_2 jm\rangle = (-1)^{j_1+j_2-j}\langle j_2 j_1 m_2 m_1 | j_2 j_1 jm\rangle \qquad (20.1\text{-}20)$$

TABLE 20.3

Wave Functions for the p² *Configuration*[a]

$\|{}^3\mathrm{P}\ 1\ 1\rangle = \|1^+\ 0^+\rangle$	$\|{}^3\mathrm{P}\ -1\ -1\rangle = \|0^-\ -1^-\rangle$
$\|{}^3\mathrm{P}\ 1\ 0\rangle = \frac{1}{\sqrt{2}}[\|1^+\ 0^-\rangle + \|1^-\ 0^+\rangle]$	$\|{}^1\mathrm{D}\ 2\ 0\rangle = \|1^+\ 1^-\rangle$
$\|{}^3\mathrm{P}\ 1\ -1\rangle = \|1^-\ 0^-\rangle$	$\|{}^1\mathrm{D}\ 1\ 0\rangle = \frac{1}{\sqrt{2}}[\|1^+\ 0^-\rangle - \|1^-\ 0^+\rangle]$
$\|{}^3\mathrm{P}\ 0\ 1\rangle = \|1^+\ -1^+\rangle$	$\|{}^1\mathrm{D}\ 0\ 0\rangle = \frac{1}{\sqrt{6}}[\|1^+\ -1^-\rangle - \|1^-\ -1^+\rangle + 2\|0^+\ 0^-\rangle]$
$\|{}^3\mathrm{P}\ 0\ 0\rangle = \frac{1}{\sqrt{2}}[\|1^+\ -1^-\rangle + \|1^-\ -1^+\rangle]$	$\|{}^1\mathrm{D}\ -1\ 0\rangle = \frac{1}{\sqrt{2}}[\|0^+\ -1^-\rangle - \|0^-\ -1^+\rangle]$
$\|{}^3\mathrm{P}\ 0\ -1\rangle = \|1^-\ -1^-\rangle$	$\|{}^1\mathrm{D}\ -2\ 0\rangle = \|-1^+\ -1^-\rangle$
$\|{}^3\mathrm{P}\ -1\ 1\rangle = \|0^+\ -1^+\rangle$	$\|{}^1\mathrm{S}\ 0\ 0\rangle = \frac{1}{\sqrt{3}}[\|1^+\ -1^-\rangle - \|1^-\ -1^+\rangle - \|0^+\ 0^-\rangle]$
$\|{}^3\mathrm{P}\ -1\ 0\rangle = \frac{1}{\sqrt{2}}[\|0^+\ -1^-\rangle + \|0^-\ -1^+\rangle]$	

[a] Each ket on the right represents a Slater determinant. The notation is the same as in Table 20.2.

from which we have

$$|j_1 j_2 jm\rangle = (-1)^{j_1+j_2-j}|j_2 j_1 jm\rangle. \tag{20.1-21}$$

Returning to (20.1-11) we shall make the notation more specific by including the reference to the individual electrons. Thus $|l_1(1)l_2(2)LM_L\rangle$ means that l_1 is associated with electron 1 and l_2 with electron 2. Suppose, for example, that we have the configuration 3d 4p and we are interested in the state with $L = 2$, $M_L = 1$. With the help of a table of coupling coefficients (Table 1.4):

$$\begin{aligned}|l_1(1)l_2(2)21\rangle = \sqrt{\tfrac{1}{3}}\,3\mathrm{d}_{+2}(1)\,4\mathrm{p}_{-1}(2) + \sqrt{\tfrac{1}{6}}\,3\mathrm{d}_{+1}(1)\,4\mathrm{p}_0(2)\\ - \sqrt{\tfrac{1}{2}}\,3\mathrm{d}_0(1)\,4\mathrm{p}_{+1}(2).\end{aligned} \tag{20.1-22}$$

To construct $|l_1(2)l_2(1)21\rangle$ we simply interchange electrons 1 and 2 in (20.1-22). The symmetric and antisymmetric combinations of $|LM_L\rangle$ may now be constructed with the operators (20.1-12) and (20.1-13):

$$|LM_L\rangle^{\mathrm{S}}_{\mathrm{A}} = \frac{1}{\sqrt{2}}[|l_1(1)l_2(2)LM_L\rangle \pm |l_1(2)l_2(1)LM_L\rangle]. \tag{20.1-23}$$

But, from (20.1-21),

$$|l_1(2)l_2(1)LM_L\rangle = (-1)^{l_1+l_2-L}|l_2(1)l_1(2)LM_L\rangle \tag{20.1-24}$$

so that

$$|LM_L\rangle^{\mathrm{S}}_{\mathrm{A}} = \frac{1}{\sqrt{2}}[|l_1(1)l_2(2)LM_L\rangle \pm (-1)^{l_1+l_2-L}|l_2(1)l_1(2)LM_L\rangle]. \tag{20.1-25}$$

In combination with the appropriate spin function we now have

$$|{}^1LM_LM_S\rangle = \frac{1}{\sqrt{2}}\left[|l_1(1)l_2(2)LM_L\rangle + (-1)^{l_1+l_2-L}|l_2(1)l_1(2)LM_L\rangle\right]{}^1\xi,$$

$$|{}^3LM_LM_S\rangle = \frac{1}{\sqrt{2}}\left[|l_1(1)l_2(2)LM_L\rangle - (-1)^{l_1+l_2-L}|l_2(1)l_1(2)LM_L\rangle\right]{}^3\xi.$$

(20.1-26)

These are general relations of which specific examples have been previously given. Now suppose that the electrons are in equivalent orbitals so that $l_1 = l_2 = l$. The singlet function ($S = 0$) will vanish unless L is even, whereas the triplet function ($S = 1$) will vanish unless L is odd. Hence for both singlets and triplets $L + S$ must be even. For equivalent electrons we have, in place of (20.1-26),

$$|LSM_LM_S\rangle = |l(1)l(2)LM_L\rangle \begin{cases} {}^1\xi, & L \text{ even}, \\ {}^3\xi, & L \text{ odd}. \end{cases} \tag{20.1-27}$$

As noted previously (Section 19.1) the Hamiltonian (20.1-1) is invariant under inversion. Therefore the eigenfunctions of the Hamiltonian have a definite parity. Since $|LSM_LM_S\rangle$ is constructed from products of central field spin orbitals, each of which contains a spherical harmonic, the parity of the state is determined by an expression of the type (19.2-8) which, for a two-electron system is $(-1)^{l_1}(-1)^{l_2}$. Hence the parity is odd or even depending on whether $l_1 + l_2$ is odd or even. One must be careful not to associate the parity with the value of L. Thus, for the configuration npn'p, $l_1 + l_2 = 2$ so that the parity of $|SLM_SM_L\rangle$ is always even despite the fact that the configuration gives rise to S ($L = 0$), P ($L = 1$), and D ($L = 2$) states. We also observe that for two equivalent electrons the parity is always even though L can be odd or even.

20.2 Terms from a Configuration of *n* Electrons

By following the ordinary rules for coupling angular momentum operators it is possible to proceed sequentially to identify all the multiplets arising from a given configuration. However, if the configuration contains equivalent electrons, the angular momentum rules will produce multiplets which violate the Pauli principle. Such multiplets must necessarily be eliminated. Several examples will illustrate these features.

First, suppose we have the configuration of nonequivalent electrons spd. There are various ways in which the angular momenta may be coupled. Thus

$$\begin{aligned} \mathrm{sp}({}^1\mathrm{P})\mathrm{d} &= {}^2\mathrm{P} \quad {}^2\mathrm{D} \quad {}^2\mathrm{F}, \\ \mathrm{sp}({}^3\mathrm{P})\mathrm{d} &= {}^2\mathrm{P} \quad {}^2\mathrm{D} \quad {}^2\mathrm{F} \quad {}^4\mathrm{P} \quad {}^4\mathrm{D} \quad {}^4\mathrm{F}. \end{aligned} \tag{20.2-1}$$

The same set of terms is obtained from the coupling sequence sd(^{1}D)p, sd(^{3}D)p or from pd(^{1}P)s, pd(^{3}P)s, pd(^{1}D)s, pd(^{3}D)s, pd(^{1}F)s, pd(^{3}F)s. To take another example, the terms from the configuration pp′p″ consisting of three nonequivalent p electrons are

$$\begin{aligned}
\mathrm{pp'}(^1\mathrm{S})\mathrm{p''} &= {}^2\mathrm{P},\\
\mathrm{pp'}(^1\mathrm{P})\mathrm{p''} &= {}^2\mathrm{S}\quad {}^2\mathrm{P}\quad {}^2\mathrm{D},\\
\mathrm{pp'}(^1\mathrm{D})\mathrm{p''} &= {}^2\mathrm{P}\quad {}^2\mathrm{D}\quad {}^2\mathrm{F},\\
\mathrm{pp'}(^3\mathrm{S})\mathrm{p''} &= {}^2\mathrm{P}\quad {}^4\mathrm{P},\\
\mathrm{pp'}(^3\mathrm{P})\mathrm{p''} &= {}^2\mathrm{S}\quad {}^2\mathrm{P}\quad {}^2\mathrm{D}\quad {}^4\mathrm{S}\quad {}^4\mathrm{P}\quad {}^4\mathrm{D},\\
\mathrm{pp'}(^3\mathrm{D})\mathrm{p''} &= {}^2\mathrm{P}\quad {}^2\mathrm{D}\quad {}^2\mathrm{F}\quad {}^4\mathrm{P}\quad {}^4\mathrm{D}\quad {}^4\mathrm{F},
\end{aligned} \tag{20.2-2}$$

and again the number of multiplets and their identity are the same regardless of the order in which the electrons are coupled. The method previously illustrated may be extended without difficulty to configurations consisting of any number of nonequivalent electrons.

For equivalent electrons we face a more difficult task because of the Pauli principle which forbids certain terms from appearing. For two electrons there is a simple rule, namely that $L + S$ is even, which enables us to identify the permissible terms immediately as was done in Section 20.1. But for more than two electrons there is no comparable rule. The general problem can be handled by group theoretical methods that will be described in Section 20.4. But it is also instructive to work through some simple configurations based on what is essentially a bookkeeping scheme to ensure that the orbitals are not populated in violation of the Pauli principle.

We shall use the configuration $n\mathrm{p}^3$, consisting of three equivalent p electrons, to illustrate the procedure. For this case

$$m_l = 1,\quad 0,\quad -1; \qquad m_s = \tfrac{1}{2},\quad -\tfrac{1}{2}, \tag{20.2-3}$$

and there are six possible combination of m_l and m_s, as shown in the accompanying tabulation. These represent six orbitals with a common value

m_l	m_s	Notation
1	$\frac{1}{2}$	(1^+)
0	$\frac{1}{2}$	(0^+)
-1	$\frac{1}{2}$	(-1^+)
1	$-\frac{1}{2}$	(1^-)
0	$-\frac{1}{2}$	(0^-)
-1	$-\frac{1}{2}$	(-1^-)

of n and $l\,(=1)$ but differing in m_l and m_s. If (1^+), (0^+), and (1^-) each contain an electron, we indicate such a distribution by $(1^+\ 0^+\ 1^-)$. Since the electrons are indistinguishable, they may be permuted without affecting the distribution; thus $(1^+\ 0^+\ 1^-)$, $(0^+\ 1^-\ 1^+)$, $(1^+\ 1^-\ 0^+)$, etc., correspond to the same distribution. Also since

$$M_L = \sum_i m_l; \qquad M_S = \sum_i m_{s_i} \tag{20.2-4}$$

a distribution like $(1^+\ 0^+\ 1^-)$ has associated with it the values $M_L = 2$, $M_S = \frac{1}{2}$.

The number of independent distributions of three (indistinguishable) electrons in six orbitals is

$$\frac{6 \times 5 \times 4}{3!} = 20.$$

These may be identified and grouped according to their respective values of M_L and M_S as is done in Table 20.4. In fact, the distributions per se are not needed—all that is needed is the number of distributions for each set of values (M_L, M_S). Moreover, the number of distributions is the same for M_L and $-M_L$ as well as for M_S and $-M_S$. The tabulation may therefore be simplified as shown in the upper part of Table 20.5. Starting with the highest value of M_L, it is apparent that the entry for $M_L = 2$, $M_S = \frac{1}{2}$ in Table 20.5—which corresponds to the four entries for $M_L = \pm 2$, $M_S = \pm\frac{1}{2}$ in Table 20.4—belongs to a ^{2}D term. But ^{2}D also includes $M_S = \frac{1}{2}$ for $M_L = 1, 0$ (corresponding to $M_S = \pm\frac{1}{2}$ for $M_L = \pm 1, 0$ in Table 20.4). One then removes the entries belonging to ^{2}D and the remaining tabulation is shown in the middle part of Table 20.5. Repetition of the process yields ^{2}P and finally ^{4}S.

TABLE 20.4

Orbital Distributions for p^3

M_S \ M_L	2	1	0	-1	-2
$\frac{3}{2}$			$(1^+\ 0^+\ -1^+)$		
$\frac{1}{2}$	$(1^+\ 0^+\ 1^-)$	$(1^+\ 0^+\ 0^-)$ $(1^+\ -1^+\ 1^-)$	$(1^+\ 0^+\ -1^-)$ $(1^+\ -1^+\ 0^-)$ $(0^+\ -1^+\ 1^-)$	$(1^+\ -1^+\ -1^-)$ $(0^+\ -1^+\ 0^-)$	$(0^+\ -1^+\ -1^-$
$-\frac{1}{2}$	$(1^+\ 1^-\ 0^-)$	$(1^+\ 1^-\ -1^-)$ $(0^+\ 1^-\ 0^-)$	$(1^+\ 0^-\ -1^-)$ $(0^+\ 1^-\ -1^-)$ $(-1^+\ 1^-\ 0^-)$	$(0^+\ 0^-\ -1^-)$ $(-1^+\ 1^-\ -1^-)$	$(-1^+\ 0^-\ -1^-$
$-\frac{3}{2}$			$(1^-\ 0^-\ -1^-)$		

TABLE 20.5

Derivation of Terms Arising from p^3

M_S \ M_L	2	1	0
$\frac{3}{2}$			1
$\frac{1}{2}$	1	2	3

$\downarrow$ ^2D $\downarrow$

M_S \ M_L	2	1	0
$\frac{3}{2}$			1
$\frac{1}{2}$		1	2

$\downarrow$ ^2P $\downarrow$

M_S \ M_L	2	1	0
$\frac{3}{2}$			1
$\frac{1}{2}$			1

$\downarrow$ ^4S $\downarrow$

Thus p^3 gives rise to ^2P, ^2D, and ^4S. Compared to the number of terms from the nonequivalent configuration $pp'p''$ (20.2-2) it is seen that the Pauli principle drastically reduces the number of terms.

It is also instructive to work out the terms from d^3. For this case

$$m_l = 2, \quad 1, \quad 0, \quad -1, \quad -2; \qquad m_s = \tfrac{1}{2}, \quad -\tfrac{1}{2}, \tag{20.2-5}$$

and the number of independent distributions is

$$\frac{10 \times 9 \times 8}{3!} = 120.$$

This would be rather tedious if it were necessary to identify all the distributions. However, the symmetry in the number of distributions for $\pm M_L$ and for $\pm M_S$. This reduces the number of distributions that one needs to calculate from 120 to 35. The resultant table is shown in Table 20.6. Repeating the same method, we find the terms belonging to d^3: ${}^2P$, ${}^2D$, ${}^2D$, ${}^2F$, ${}^2G$, ${}^2H$, ${}^4P$, ${}^4F$. Of particular interest is the appearance of two ${}^2D$ states.

TABLE 20.6

Derivation of Terms Arising from d^3

M_S \ M_L	5	4	3	2	1	0
$\frac{3}{2}$			1	1	2	2
$\frac{1}{2}$	1	2	4	6	8	8

↓ ${}^2H$

M_S \ M_L	5	4	3	2	1	0
$\frac{3}{2}$			1	1	2	2
$\frac{1}{2}$		1	3	5	7	7

↓ ${}^2G$

M_S \ M_L	5	4	3	2	1	0
$\frac{3}{2}$			1	1	2	2
$\frac{1}{2}$			2	4	6	6

↓ ${}^4F$

M_S \ M_L	5	4	3	2	1	0
$\frac{3}{2}$					1	1
$\frac{1}{2}$			1	3	5	5

↓ ${}^2F$

M_S \ M_L	5	4	3	2	1	0
$\frac{3}{2}$					1	1
$\frac{1}{2}$				2	4	4

↓ ${}^2D$

M_S \ M_L	5	4	3	2	1	0
$\frac{3}{2}$					1	1
$\frac{1}{2}$				1	3	3

↓ ${}^2D$

M_S \ M_L	5	4	3	2	1	0
$\frac{3}{2}$					1	1
$\frac{1}{2}$					2	2

↓ ${}^4P$

M_S \ M_L	5	4	3	2	1	0
$\frac{3}{2}$						
$\frac{1}{2}$					1	1

↓ ${}^2P$

A complete list of terms arising from configurations of equivalent s, p and d electrons is shown in Table 20.7. These features are to be noted:

(1) For a completely filled (closed) shell there is only one possible distribution of the electrons—namely, one electron in each orbital. In that case $M_L = 0$, $M_S = 0$ and the corresponding term is ${}^1S$.

TABLE 20.7

Terms Arising from l^n Configurations[a]

Configuration			Terms	Seniority number	$O^+(2l+1)$
s^0		s^2	^1S		
	s		^2S		
p^0		p^6	^1S		
p		p^5	$^2P^o$		
p^2		p^4	$^1S, {}^1D, {}^3P$		
	p^3		$^2P^o, {}^2D^o, {}^4S^o$		
d^0		d^{10}	^1S	0	(00)
d^1		d^9	^2D	1	(10)
d^2		d^8	^1S	0	(00)
			$^1D, {}^1G$	2	(20)
			$^3P, {}^3F$	2	(11)
d^3		d^7	^2D	1	(10)
			$^2P, {}^2D, {}^2F, {}^2G, {}^2H$	3	(21)
			$^4P, {}^4F$	3	(11)
d^4		d^6	^1S	0	(00)
			$^1D, {}^1G$	2	(20)
			$^3P, {}^3F$	2	(11)
			$^1S, {}^1D, {}^1F, {}^1G, {}^1I$	4	(22)
			$^3P, {}^3D, {}^3F, {}^3G, {}^3H$	4	(21)
			^5D	4	(10)
	d^5		^2D	1	(10)
			$^2P, {}^2D, {}^2F, {}^2G, {}^2H$	3	(21)
			$^4P, {}^4F$	3	(11)
			$^2S, {}^2D, {}^2F, {}^2G, {}^2I$	5	(22)
			$^4D, {}^4G$	5	(20)
			^6S	5	(00)

[a] The terms include the seniority number v and the classification with respect to $O^+(2l+1)$.

(2) The parity of a term is determined by

$$(-1)^{l_1}(-1)^{l_2}\cdots(-1)^{l_n},$$

that is, the parity is odd or even depending on whether $\sum l_i$ is odd or even.

(3) When the shell is half-full one of the possible electronic distributions consists of one electron in each spatial orbital and all the spin projections parallel. This results in the highest possible value of M_S which can only be

associated with the maximum value of S. With each spatial orbital singly occupied

$$M_L = \sum m_l = 0.$$

Since this is the only possible value of M_L, it follows that $L = 0$. This feature may be illustrated in the configuration p^3 for which the maximum spin is $S = \frac{3}{2}$. To realize such a state it must be possible to distribute the three electrons in the p orbitals in such a manner that $M_S = \sum m_{s_i} = \frac{3}{2}$. This can only be accomplished by putting one electron in each of the p orbitals, i.e., in the configuration $p_{+1}\alpha p_0\alpha p_{-1}\alpha$ which then results in $M_L = \sum m_{l_i} = 0$ and there is no other value of M_L consistent with $M_S = \frac{3}{2}$. Therefore $L = 0$ and the term of highest multiplicity in p^3 is ^4S. The general statement is that in a half-filled shell, the term of highest multiplicity is the spherically symmetric term ^{2l+2}S. This property acquires increased significance as a result of the fact that in *LS* coupling the term of highest multiplicity is usually the ground term (Hund's rule).

(4) Multiplets with the same values of L and S may appear more than once in the configurations d^n and f^n. We have already encountered one such case in d^3 in which ^2D occurs twice. To distinguish among such multiplets it is useful to introduce additional quantum numbers. For d^n configurations, one additional quantum number is sufficient; it is called the *seniority number v*. For f^n multiplets still other quantum numbers are required. We shall give an operational definition of the seniority number though its detailed origin requires consideration of the group $O^+(5)$ (Section 20.4).

For d^n each multiplet which occurs first for d^{n-2} is assigned a seniority number $n - 2$; each multiplet which occurs first for d^{n-4} is assigned a seniority number $n - 4$. All other multiplets have seniority numbers equal to n. These assignments are shown in Table 20.7. To illustrate, d^2 has the multiplets ^1S, ^3P, ^1D, ^3F, ^1G; the multiplet ^1S can be regarded as having occurred in the configuration d^0 which is equivalent to d^{10}, i.e., a closed shell. Therefore ^1S is assigned a seniority number $v = 0$ with the notation ^1_0S. The other multiplets from d^2 have not appeared in any lower configuration; their seniority numbers are all $v = 2$: ^3_2P, ^1_2D, ^3_2F, ^1_2G. If we now examine the multiplets from d^4 it is seen that we again have ^1S, ^3P, ^1D, ^3F, and ^1G plus others. The classification according to seniority number then becomes

$$^1_0S \quad ^3_2P\ ^1_2D\ ^3_2F\ ^1_2G \quad ^1_4S\ ^1_4D\ ^1_4F\ ^1_4G\ ^1_4I\ ^3_4P\ ^3_4D\ ^3_4F\ ^3_4G\ ^3_4H\ ^5_4D.$$

We note that the two ^1S multiplets are distinguished on the basis of the seniority number; similarly for ^1D, ^1G, ^3P, and ^3F. In the case of d^3 one of the ^2D multiplets has a seniority number $v = 1$ associated with the fact that ^2D occurs in d^1; the other ^2D in d^3 has $v = 3$. Hence we have ^2_1D and ^2_3D.

For f^n configurations the situation is substantially more complicated; a thorough discussion is to be found in the work of Judd (1963).

20.3 Construction of Multiplet Wave Functions

The construction of multiplet wave functions involves the coupling of angular momenta and antisymmetrization. The method was illustrated in Section 20.1 for two-electron multiplets. In principle, the same procedures can be applied to more complicated configurations but the amount of labor involved rapidly becomes prohibitive so that there is a strong motivation to seek more efficient methods. Before describing some of these it is instructive to give several examples of three-electron multiplet wave functions constructed by essentially the same methods as those employed in Section 20.1.

We consider the configuration p^2p' consisting of two equivalent p electrons and one nonequivalent p electron. It is assumed that the two-electron wave functions for p^2 have already been calculated (Table 20.3). The terms that may be constructed from p^2p' are

$$\begin{aligned} &p^2(^3P)p'\colon \quad {}^4D \quad {}^4P \quad {}^4S \quad {}^2D \quad {}^2P \quad {}^2S, \\ &p^2(^1D)p'\colon \quad {}^2F \quad {}^2D \quad {}^2P, \\ &p^2(^1S)p'\colon \quad {}^2P. \end{aligned} \tag{20.3-1}$$

If we regard the terms arising from p^2, namely ^3P, ^1D, and ^1S as *parents* of the terms arising from p^2p', it is seen that the parentage can be single or multiple. Thus ^4D, ^4P, ^4S, and ^2S have the single parent ^3P; similarly ^2F has the single parent ^1D. But ^2D has two parents, ^3P and ^1D, while ^2P has all three terms as parents.

Let us now calculate the antisymmetric wave function $|p^2(^3P)p'\ {}^4D2\tfrac{3}{2}\rangle$, namely ^4D with $M_L = 2$ and $M_S = \tfrac{3}{2}$. To construct this function it is necessary to couple the orbital angular momentum of ^3P ($L = 1$) with the orbital angular momentum of the p′ electron ($l' = 1$). Subsequently it will be necessary to couple the spin angular momenta of ^3P ($S = 1$) and the p′ electron ($s = \tfrac{1}{2}$) and, finally, to antisymmetrize.

The coupling of the orbital angular momenta gives

$$|p^2(^3P)p'\ {}^4D2M_S\rangle_{NA} = |^3P\ 1\ M_S\rangle p'_{+1} \tag{20.3-2}$$

where the subscript NA indicates that the wave function, at this stage, is *not antisymmetric* although the function $|^3P\ 1\ M_S\rangle$ from the p^2 configuration is assumed to have been antisymmetrized. When the spin coupling is included we have

$$|p^2(^3P)p'\ {}^4D2\tfrac{3}{2}\rangle_{NA} = |^3P11\rangle p'_{+1}\alpha. \tag{20.3-3}$$

From Table 20.3,

$$|^3\mathrm{P}11\rangle = |1^+\ 0^+\rangle = \frac{1}{\sqrt{2}}\begin{vmatrix} \mathrm{p}_{+1}(1)\alpha(1) & \mathrm{p}_0(1)\alpha(1) \\ \mathrm{p}_{+1}(2)\alpha(2) & \mathrm{p}_0(2)\alpha(2) \end{vmatrix}. \qquad (20.3\text{-}4)$$

Therefore

$$\begin{aligned} |\mathrm{p}^2(^3\mathrm{P})\mathrm{p}'\ ^4\mathrm{D}2\tfrac{3}{2}\rangle_{\mathrm{NA}} &= |^3\mathrm{P}11\rangle \mathrm{p}'_{+1}\alpha \\ &= \frac{1}{\sqrt{2}}[\mathrm{p}_{+1}(1)\alpha(1)\mathrm{p}_0(2)\alpha(2) \\ &\quad - \mathrm{p}_0(1)\alpha(1)\mathrm{p}_{+1}(2)\alpha(2)]\mathrm{p}'_{+1}(3)\alpha(3). \end{aligned} \qquad (20.3\text{-}5)$$

This function is antisymmetric with respect to interchanges of electrons 1 and 2 (the p^2 electrons), but is not antisymmetric with respect to interchanges of electrons 1 and 3 or 2 and 3. When (20.3-5) is completely antisymmetrized and properly normalized we obtain the final wave function:

$$\begin{aligned} \left|\mathrm{p}^2(^3\mathrm{P})\mathrm{p}'\ ^4\mathrm{D}2\,\frac{3}{2}\right\rangle &= \frac{1}{\sqrt{3!}}\begin{vmatrix} \mathrm{p}_{+1}(1)\alpha(1) & \mathrm{p}_0(1)\alpha(1) & \mathrm{p}'_{+1}(1)\alpha(1) \\ \mathrm{p}_{+1}(2)\alpha(2) & \mathrm{p}_0(2)\alpha(2) & \mathrm{p}'_{+1}(2)\alpha(2) \\ \mathrm{p}_{+1}(3)\alpha(3) & \mathrm{p}_0(3)\alpha(3) & \mathrm{p}'_{+1}(3)\alpha(3) \end{vmatrix} \\ &= |1^+\ 0^+\ 1^+\rangle \end{aligned} \qquad (20.3\text{-}6)$$

where 1, 0 and 1 in $|1^+\ 0^+\ 1^+\rangle$ refer to p_{+1}, p_0 and p'_{+1}, respectively.

In the example just given $^4\mathrm{D}$ had just one parent in p^2. Let us now carry out the calculation for a term that has more than one parent. We consider $|^2\mathrm{D}0\tfrac{1}{2}\rangle$ whose parentage is indicated by $|\mathrm{p}^2(^3\mathrm{P})\mathrm{p}'\ ^2\mathrm{D}0\tfrac{1}{2}\rangle$ and $|\mathrm{p}^2(^1\mathrm{D})\mathrm{p}'\ ^2\mathrm{D}0\tfrac{1}{2}\rangle$. For the first one we have, on coupling the orbital angular momenta,

$$\begin{aligned} |\mathrm{p}^2(^3\mathrm{P})\mathrm{p}'\ ^2\mathrm{D}0M_S\rangle_{\mathrm{NA}} &= \frac{1}{\sqrt{6}}|^3\mathrm{P}1M_S\rangle\mathrm{p}'_{-1} + \sqrt{\frac{2}{3}}|^3\mathrm{P}0M_S\rangle\mathrm{p}_0{}' \\ &\quad + \frac{1}{\sqrt{6}}|^3\mathrm{P}-1\,M_S\rangle\mathrm{p}'_{+1} \end{aligned}$$

and then the spin angular momenta

$$\begin{aligned} \left|\mathrm{p}^2(^3\mathrm{P})\mathrm{p}'\ ^2\mathrm{D}0\,\frac{1}{2}\right\rangle_{\mathrm{NA}} &= \sqrt{\frac{2}{3}}\left[\sqrt{\frac{1}{6}}|^3\mathrm{P}11\rangle\mathrm{p}'_{-1}\beta + \sqrt{\frac{2}{3}}|^3\mathrm{P}01\rangle\mathrm{p}_0{}'\beta \right. \\ &\quad \left. + \sqrt{\frac{1}{6}}|^3\mathrm{P}-1\ 1\rangle\mathrm{p}'_{+1}\beta\right] - \sqrt{\frac{1}{3}}\left[\sqrt{\frac{1}{6}}|^3\mathrm{P}10\rangle\mathrm{p}'_{-1}\alpha \right. \\ &\quad \left. + \sqrt{\frac{2}{3}}|^3\mathrm{P}00\rangle\mathrm{p}_0{}'\alpha + \sqrt{\frac{1}{6}}|^3\mathrm{P}-1\ 0\rangle\mathrm{p}'_{+1}\alpha\right]. \end{aligned} \qquad (20.3\text{-}7)$$

The ^{3}P functions, shown in Table 20.3, may then be substituted to give

$$|p^2(^3P)p'\ ^2D0\tfrac{1}{2}\rangle_{NA} = \tfrac{1}{6}[2|1^+\ 0^+\rangle p'_{-1}\beta - |1^+\ 0^-\rangle p'_{-1}\alpha - |1^-\ 0^+\rangle p'_{-1}\alpha$$
$$+ 4|1^+\ {-1}^+\rangle p_0'\beta - 2|1^+\ {-1}^-\rangle p_0'\alpha - 2|1^-\ {-1}^+\rangle p_0'\alpha$$
$$+ 2|0^+\ {-1}^+\rangle p'_{+1}\beta - |0^+\ {-1}^-\rangle p'_{+1}\alpha - |0^-\ {-1}^+\rangle p'_{+1}\alpha] \tag{20.3-8}$$

and, finally, on antisymmetrizing,

$$|p^2(^3P)p'\ ^2D0\tfrac{1}{2}\rangle = \tfrac{1}{6}[2|1^+\ 0^+\ {-1}^-\rangle - |1^+\ 0^-\ {-1}^+\rangle - |1^-\ 0^+\ {-1}^+\rangle$$
$$+ 4|1^+\ {-1}^+\ 0^-\rangle - 2|1^+\ {-1}^-\ 0^+\rangle - 2|1^-\ {-1}^+\ 0^+\rangle$$
$$+ 2|0^+\ {-1}^+\ 1^-\rangle - |0^+\ {-1}^-\ 1^+\rangle - |0^-\ {-1}^+\ 1^+\rangle]. \tag{20.3-9}$$

Following the same procedure it is found that

$$|p^2(^1D)p'\ ^2D0\tfrac{1}{2}\rangle_{NA} = \tfrac{1}{2}[|1^+\ 0^-\rangle p'_{-1}\alpha - |1^-\ 0^+\rangle p'_{-1}\alpha$$
$$- |0^+\ {-1}^-\rangle p_{+1}\alpha + |0^-\ {-1}^+\rangle p'_{+1}\alpha] \tag{20.3-10}$$

and after antisymmetrizing

$$|p^2(^1D)p'\ ^2D0\tfrac{1}{2}\rangle = \tfrac{1}{2}[|1^+\ 0^-\ {-1}^+\rangle - |1^-\ 0^+\ {-1}^+\rangle$$
$$- |0^+\ {-1}^-\ 1^+\rangle + |0^-\ {-1}^+\ 1^+\rangle]. \tag{20.3-11}$$

It is apparent that the two functions

$$|p^2(^3P)p'\ ^2D0\tfrac{1}{2}\rangle \equiv \psi_1 \quad \text{and} \quad |p^2(^1D)p'\ ^2D0\tfrac{1}{2}\rangle \equiv \psi_2$$

are not the same. Using ψ_1 and ψ_2 as a basis set, the Hamiltonian matrix will be a 2×2 matrix which, in general, will contain off-diagonal terms. The secular equation will be quadratic and we shall obtain two separate energy eigenvalues. If the interaction between the p′ electron and the p^2 configuration is weak, as would occur if n' the principal quantum number of the p′ electron is large compared with n the principal quantum number of the p^2 electrons, the off-diagonal terms in the Hamiltonian matrix will become small and the diagonal elements will approach the energies of the ^{3}P and ^{1}D terms of the p^2 configuration.

Let us now return to $|p^2(^3P)p'\ ^4D2\tfrac{3}{2}\rangle$ to investigate the consequences of setting p′ = p so that the configuration consists of three equivalent p electrons. Examining (20.3-6), it is seen that the effect of putting p′ = p is to cause the determinant to vanish. There is then no ^{4}D term from the p^3 configuration. By this route it may be verified that most, but not all, the terms in (20.3-1) disappear. Indeed we already know from the discussion Section 20.2 that the terms arising from p^3 are ^{2}P, ^{2}D, and ^{4}S. We may also examine (20.3-9) and

(20.3-11) under the assumption p′ = p. First we note the equalities

$$\begin{aligned} |1^+\,0^+\,-1^-\rangle &= -|1^+\,-1^-\,0^+\rangle = |0^+\,-1^-\,1^+\rangle,\\ |1^+\,0^-\,-1^+\rangle &= -|1^+\,-1^+\,0^-\rangle = |0^-\,-1^+\,1^+\rangle,\\ |1^-\,0^+\,-1^+\rangle &= -|1^-\,-1^+\,0^+\rangle = |0^+\,-1^+\,1^-\rangle. \end{aligned} \tag{20.3-12}$$

Then, after adjusting the normalization, it is found that

$$\begin{aligned} \left|p^2(^3P)p\ ^2D0\frac{1}{2}\right\rangle &= \frac{1}{\sqrt{6}}\left[|1^+\,0^+\,-1^-\rangle + |1^-\,0^+\,-1^+\rangle - 2|1^+\,0^-\,-1^+\rangle\right]\\ &= \left|p^2(^1D)p\ ^2D0\frac{1}{2}\right\rangle. \end{aligned} \tag{20.3-13}$$

The method we have followed appears to be extremely inefficient. Thus, to obtain the ^{2}D multiplet from p^3 we first calculated the two ^{2}D multiplets from p^{2}p′, each one containing numerous Slater determinants. Then on setting p′ = p, many of the Slater determinants disappeared and the two ^{2}D wave functions coalesced into a single wave function. All of this can be avoided by the use of *coefficients of fractional parentage* (*cfp*) as will now be explained.

Based on (20.3-1) the three terms ^{4}S, ^{2}D, and ^{2}P from p^3 have the following parentage in p^2: ^{4}S originates from ^{3}P; ^{2}D from ^{3}P and ^{1}D; and ^{2}P from ^{3}P, ^{1}D, and ^{1}S. Assuming one knows the wave functions for the terms in p^2, it is possible, by use of the cfp, to construct the wave functions for the terms in p^3. The pertinent theorem is now stated: If we have the antisymmetrized wave functions for $n-1$ equivalent electrons, the addition of one more equivalent electron results in wave functions given by

$$|l^n\alpha LS\rangle = \sum_{\alpha' L' S'} |l^{n-1}(\alpha' L' S')lLS\rangle\langle l^{n-1}(\alpha' L' S')lLS|\}l^n\alpha LS\rangle \tag{20.3-14}$$

in which $\langle l^{n-1}(\alpha' L' S')lLS|\}l^n\alpha LS\rangle$ is known as a coefficient of fractional parentage (cfp). It is important to note that the initial (parent) wave function $|l^{n-1}\alpha' L' S'\rangle$ and the final wave function $|l^n\alpha LS\rangle$ are both antisymmetric but the intermediate wave function $|l^{n-1}(\alpha' L' S')lLS\rangle$ is *not* antisymmetric.‡ Numerical values of cfp for p and d electrons are shown in Table 20.8; complete tables are given by Nielsen and Koster (1963).

We now illustrate the use of these coefficients by calculating $|p^3\ ^2D0\frac{1}{2}\rangle$. From Table 20.8,

$$\left|p^3\ ^2D0\frac{1}{2}\right\rangle = \frac{1}{\sqrt{2}}\left[\left|p^2(^3P)p\ ^2D0\frac{1}{2}\right\rangle_{NA} - \left|p^2(^1D)p\ ^2D0\frac{1}{2}\right\rangle_{NA}\right] \tag{20.3-15}$$

‡ We do not use the subscript NA as was done, for example, in (20.3-2) in order to conform to the standard notation for coefficients of fractional parentage.

in which the nonantisymmetrized functions

$$|p^2(^3P)p\ ^2D0\tfrac{1}{2}\rangle_{NA} \quad \text{and} \quad |p^2(^1D)p\ ^2D0\tfrac{1}{2}\rangle_{NA}$$

are given by (20.3-8) and (20.3-10) respectively, after setting $p' = p$. When these substitutions are made and the terms are grouped into 3×3 Slater determinants the result is precisely (20.3-13).

Some useful properties of the cfp are:

1. They are real, i.e.,

$$\langle l^{n-1}(\alpha' L' S')lLS|\}l^n\alpha LS\rangle = \langle l^n\alpha LS\{|l^{n-1}(\alpha' L' S')lLS\rangle. \quad (20.3\text{-}16)$$

2. They satisfy an orthogonality condition

$$\sum_{\alpha' L' S'} \langle l^{n-1}(\alpha' L' S')lLS|\}l^n\alpha LS\rangle\langle l^n\alpha LS\{|l^{n-1}(\alpha' L' S')lLS\rangle = 1. \quad (20.3\text{-}17)$$

Here the summation extends over all the terms $(\alpha' L' S')$ of the configuration l^{n-1}.

TABLE 20.8

Coefficients of Fractional Parentage[a,b,c]

p^3	N	p^2: ^1S	^3P	^1D
^4S	1	0	1	0
^2P	$18^{-1/2}$	2	-3	$-5^{1/2}$
^2D	$2^{-1/2}$	0	1	-1

d^3	N	d^2: ^1_0S	^3_2P	^1_2D	^3_2F	^1_2G
^2_3P	$30^{-1/2}$	0	$7^{1/2}$	$15^{1/2}$	$-8^{1/2}$	0
^4_3P	$15^{-1/2}$	0	$-8^{1/2}$	0	$-7^{1/2}$	0
^2_1D	$60^{-1/2}$	4	-3	$-5^{1/2}$	$-21^{1/2}$	-3
^2_3D	$140^{-1/2}$	0	-7	$45^{1/2}$	$21^{1/2}$	-5
^2_3F	$70^{-1/2}$	0	$28^{1/2}$	$-10^{1/2}$	$7^{1/2}$	-5
^4_3F	$5^{-1/2}$	0	-1	0	2	0
^2_3G	$42^{-1/2}$	0	0	$-10^{1/2}$	$21^{1/2}$	$11^{1/2}$
^2_3H	$2^{-1/2}$	0	0	0	-1	1

TABLE 20.8 (*continued*)

d^4	N	d^3							
		${}^{2}_{3}P$	${}^{4}_{3}P$	${}^{2}_{1}D$	${}^{2}_{3}D$	${}^{2}_{3}F$	${}^{4}_{3}F$	${}^{2}_{3}G$	${}^{2}_{3}H$
${}^{1}_{0}S$	1	0	0	1	0	0	0	0	0
${}^{1}_{4}S$	1	0	0	0	1	0	0	0	0
${}^{3}_{2}P$	$360^{-1/2}$	$-14^{1/2}$	β -8	$135^{1/2}$	$-35^{1/2}$	$-56^{1/2}$	$-56^{1/2}$	0	0
${}^{3}_{4}P$	$90^{-1/2}$	5	$-14^{1/2}$	0	$10^{1/2}$	-5	4	0	0
${}^{1}_{2}D$	$280^{-1/2}$	$-42^{1/2}$	0	$105^{1/2}$	$45^{1/2}$	$28^{1/2}$	0	$-60^{1/2}$	0
${}^{1}_{4}D$	$140^{-1/2}$	42	0	0	$20^{1/2}$	$63^{1/2}$	0	$15^{1/2}$	0
${}^{3}_{4}D$	$210^{-1/2}$	$-14^{1/2}$	7	0	$60^{1/2}$	$-21^{1/2}$	$-21^{1/2}$	$45^{1/2}$	0
${}^{5}_{4}D$	$10^{-1/2}$	0	$3^{1/2}$	0	0	0	7	0	0
${}^{1}_{4}F$	$560^{-1/2}$	$120^{1/2}$	0	0	$200^{1/2}$	$-105^{1/2}$	0	$-3^{1/2}$	$-132^{1/2}$
${}^{3}_{2}F$	$840^{-1/2}$	4	$-56^{1/2}$	$315^{1/2}$	$15^{1/2}$	$-14^{1/2}$	$224^{1/2}$	$90^{1/2}$	$110^{1/2}$
${}^{3}_{4}F$	$1680^{-1/2}$	$-200^{1/2}$	$-448^{1/2}$	0	$120^{1/2}$	$-175^{1/2}$	$-112^{1/2}$	$-405^{1/2}$	$220^{1/2}$
${}^{1}_{2}G$	$504^{-1/2}$	0	0	$189^{1/2}$	-5	$70^{1/2}$	0	$66^{1/2}$	$-154^{1/2}$
${}^{1}_{4}G$	$1008^{-1/2}$	0	0	0	$88^{1/2}$	$385^{1/2}$	0	$-507^{1/2}$	$-28^{1/2}$
${}^{3}_{4}G$	$1680^{-1/2}$	0	0	0	$200^{1/2}$	$315^{1/2}$	$-560^{1/2}$	$297^{1/2}$	$308^{1/2}$
${}^{3}_{4}H$	$60^{-1/2}$	0	0	0	0	$5^{1/2}$	$20^{1/2}$	-3	$26^{1/2}$
${}^{1}_{4}I$	$10^{-1/2}$	0	0	0	0	0	0	$3^{1/2}$	$7^{1/2}$

[a] N is a common normalization factor by which it is necessary to multiply each number in the corresponding row.

[b] Section p^2 reproduced from "Quantum Mechanics of Atomic Spectra and Atomic Structure", written by Masataka Mizushima, with permission of publishers. Addison-Wesley/W. A. Benjamin Inc., Advanced Book Program, Reading, Massachusetts.

[c] Sections d^3 and d^4 from (Sobelman, 1972).

TABLE 20.8 (*continued*)

d^5	N	d^4							
		${}^1_0S$	${}^1_4S$	${}^3_2P$	${}^3_4P$	${}^1_2D$	${}^1_4D$	${}^3_4D$	${}^5_4D$
${}^2_5S$	$5^{1/2}$	0	0	0	0	0	$-2^{1/2}$	$3^{1/2}$	0
${}^6_5S$	1	0	0	0	0	0	0	0	1
${}^2_3P$	$150^{-1/2}$	0	0	$14^{1/2}$	5	$30^{1/2}$	$15^{1/2}$	$10^{1/2}$	0
${}^4_3P$	$300^{-1/2}$	0	0	-8	$14^{1/2}$	0	0	$35^{1/2}$	$-75^{1/2}$
${}^2_1D$	$50^{-1/2}$	$6^{1/2}$	0	-3	0	$-5^{1/2}$	0	0	0
${}^2_3D$	$350^{-1/2}$	0	$-14^{1/2}$	-7	$-14^{1/2}$	$45^{1/2}$	$-10^{1/2}$	$60^{1/2}$	0
${}^2_5D$	$700^{-1/2}$	0	$-56^{1/2}$	0	$126^{1/2}$	0	$90^{1/2}$	$60^{1/2}$	0
${}^4_5D$	$700^{-1/2}$	0	0	0	$126^{1/2}$	0	0	$-135^{1/2}$	$-175^{1/2}$
${}^2_3F$	$2800^{-1/2}$	0	0	$448^{1/2}$	$-200^{1/2}$	$-160^{1/2}$	$180^{1/2}$	$120^{1/2}$	0
${}^2_5F$	$2800^{-1/2}$	0	0	0	$360^{1/2}$	0	-10	$600^{1/2}$	0
${}^4_3F$	$700^{-1/2}$	0	0	$-56^{1/2}$	-4	0	0	$-15^{1/2}$	$-175^{1/2}$
${}^2_3G$	$8400^{-1/2}$	0	0	0	0	$-800^{1/2}$	-10	$600^{1/2}$	0
${}^2_5G$	$18480^{-1/2}$	0	0	0	0	0	$1452^{1/2}$	$968^{1/2}$	0
${}^4_5G$	$420^{-1/2}$	0	0	0	0	0	0	5	$-105^{1/2}$
${}^2_3H$	$1100^{-1/2}$	0	0	0	0	0	0	0	0
${}^2_5I$	$550^{-1/2}$	0	0	0	0	0	0	0	0

TABLE 20.8 (*continued*)

d^5	N	d^4							
		${}^1_4F$	${}^3_2F$	${}^3_4F$	${}^1_2G$	${}^1_4G$	${}^3_4G$	${}^3_4H$	${}^1_4I$
${}^2_5S$	$5^{1/2}$	0	0	0	0	0	0	0	0
${}^6_5S$	1	0	0	0	0	0	0	0	0
${}^2_3P$	$150^{-1/2}$	$-15^{1/2}$	-4	-5	0	0	0	0	0
${}^4_3P$	$300^{-1/2}$	0	$-56^{1/2}$	$56^{1/2}$	0	0	0	0	0
${}^2_1D$	$50^{-1/2}$	0	$-21^{1/2}$	0	-3	0	0	0	0
${}^2_3D$	$350^{-1/2}$	$35^{1/2}$	$21^{1/2}$	$-21^{1/2}$	-5	$-11^{1/2}$	$45^{1/2}$	0	0
${}^2_5D$	$700^{-1/2}$	$35^{1/2}$	0	$189^{1/2}$	0	$99^{1/2}$	$45^{1/2}$	0	0
${}^4_5D$	$700^{-1/2}$	0	0	$-84^{1/2}$	0	0	$180^{1/2}$	0	0
${}^2_3F$	$2800^{-1/2}$	$105^{1/2}$	$112^{1/2}$	$-175^{1/2}$	-20	$275^{1/2}$	$-405^{1/2}$	$220^{1/2}$	0
${}^2_5F$	$2800^{-1/2}$	$-525^{1/2}$	0	$-315^{1/2}$	0	$495^{1/2}$	-3	$-396^{1/2}$	0
${}^4_3F$	$700^{-1/2}$	0	$224^{1/2}$	$14^{1/2}$	0	0	$-90^{1/2}$	$-110^{1/2}$	0
${}^2_3G$	$8400^{-1/2}$	$-7^{1/2}$	$1680^{1/2}$	$945^{1/2}$	$880^{1/2}$	$845^{1/2}$	$891^{1/2}$	$924^{1/2}$	$728^{1/2}$
${}^2_5G$	$18480^{-1/2}$	$2541^{1/2}$	0	$4235^{1/2}$	0	$-1215^{1/2}$	$-5577^{1/2}$	$-308^{1/2}$	$-2184^{1/2}$
${}^4_5G$	$420^{-1/2}$	0	0	$-70^{1/2}$	0	0	$-66^{1/2}$	$154^{1/2}$	0
${}^2_3H$	$1100^{-1/2}$	$33^{1/2}$	$-220^{1/2}$	$55^{1/2}$	$220^{1/2}$	$-5^{1/2}$	$-99^{1/2}$	$286^{1/2}$	$172^{1/2}$
${}^2_5I$	$550^{-1/2}$	0	0	0	0	$-45^{1/2}$	$99^{1/2}$	$231^{1/2}$	$-175^{1/2}$

3. The relation between the cfp for shells that are more than half-filled to the cfp for shells that are less than half-filled is

$$\langle l^{4l+1-n}(\alpha' L' S') lLS | \} l^{4l+2-n} \alpha LS \rangle$$
$$= (-1)^{L+L'+S+S'-l-(1/2)} \sqrt{\frac{(n+1)(2L'+1)(2S'+1)}{(4l+2-n)(2L+1)(2S+1)}}$$
$$\times \langle l^n(\alpha LS) lL'S' | \} l^{n+1} \alpha' L'S' \rangle. \tag{20.3-18}$$

In the example given previously we coupled a p′ electron to the configuration p^2. In principle we could have proceeded by coupling a p electron to the configuration pp′. But on physical grounds the latter would be less preferable simply because we expect the interaction p − p′ to be weaker than p − p. However, there are numerous situations where no *a priori* decision as to relative strengths of interaction can be made. The question may then be asked as to the consequences of following one or another coupling scheme. We shall take the configuration spd (20.2-1) as an example. Altogether there are nine terms: ^4F, ^4D, ^4P, $^2F(2)$, $^2D(2)$, $^2P(2)$. Regardless of the order in which the electrons are coupled we will obtain the same set of terms. Thus ^2F comes about from the coupling

$$\mathrm{dp}(^3\mathrm{F})\mathrm{s}\ ^2\mathrm{F}, \qquad \mathrm{dp}(^1\mathrm{F})\mathrm{s}\ ^2\mathrm{F} \tag{20.3-19a}$$

or

$$\mathrm{sp}(^3\mathrm{P})\mathrm{d}\ ^2\mathrm{F}, \qquad \mathrm{sp}(^1\mathrm{P})\mathrm{d}\ ^2\mathrm{F} \tag{20.3-19b}$$

or

$$\mathrm{sd}(^3\mathrm{D})\mathrm{p}\ ^2\mathrm{F}, \qquad \mathrm{sd}(^1\mathrm{D})\mathrm{p}\ ^2\mathrm{F}. \tag{20.3-19c}$$

It is possible to work out the combination of Slater determinants for these six functions and they will all be different. However, it is found that

$$\begin{aligned} |\mathrm{sd}(^3\mathrm{D})\mathrm{p}\ ^2\mathrm{F}\rangle &= -\tfrac{1}{2}|\mathrm{dp}(^3\mathrm{F})\mathrm{s}\ ^2\mathrm{F}\rangle + \tfrac{1}{2}\sqrt{3}|\mathrm{dp}(^1\mathrm{F})\mathrm{s}\ ^2\mathrm{F}\rangle, \\ |\mathrm{sd}(^1\mathrm{D})\mathrm{p}\ ^2\mathrm{F}\rangle &= -\tfrac{1}{2}|\sqrt{3}\mathrm{dp}(^3\mathrm{F})\mathrm{s}\ ^2\mathrm{F}\rangle - \tfrac{1}{2}|\mathrm{dp}(^1\mathrm{F})\mathrm{s}\ ^2\mathrm{F}\rangle, \end{aligned} \tag{20.3-20}$$

that is, the pair (20.3-19a) and the pair (20.3-19c) are connected by a unitary transformation with similar relations connecting other pairs. The Hamiltonian matrix can be constructed using any pair from among the three. Although the specific matrix elements will depend on which pair is chosen, the eigenvalues will be the same for all three as they must be when different sets of basis functions are related by a unitary transformation.

20.4 Symmetry Properties

When the spins of two electrons are coupled, the resulting spin functions are either symmetric ($S = 1$) or antisymmetric ($S = 0$) with respect to an interchange of the coordinates of the two electrons (Table 20.1). From the standpoint of the permutation group S_2 (Section 5.4), the symmetric functions are associated with the Young tableau

$$\boxed{1\,|\,2}$$

and the antisymmetric functions with the tableau

$$\begin{array}{|c|}\hline 1 \\ \hline 2 \\ \hline \end{array}\,.$$

To generalize this property we shall suppose that $\mathbf{J}_1$ and $\mathbf{J}_2$ are two operators that operate on electrons 1 and 2, respectively, and that

$$J_1^2\psi_m^j(1) = j(j+1)\psi_m^j(1), \qquad J_{1z}\psi_m^j(1) = m\psi_m^j(1), \tag{20.4-1a}$$

$$J_2^2\psi_m^j(2) = j(j+1)\psi_m^j(2), \qquad J_{2z}\psi_m^j(2) = m\psi_m^j(2). \tag{20.4-1b}$$

We shall now couple the angular momenta $\mathbf{J} = \mathbf{J}_1 + \mathbf{J}_2$ and construct the two-electron eigenfunctions as linear combinations of the ψ_m^j using appropriate coupling coefficients. The resulting functions, regardless of the value of j, will be either symmetric or antisymmetric as long as only two electrons are being coupled. In other words, the coupled eigenfunctions will be associated with either one of the two tableaux

$$\boxed{1\,|\,2} \qquad \begin{array}{|c|}\hline 1 \\ \hline 2 \\ \hline \end{array}\,.$$

This property has already been illustrated in the case of the spin functions where $j = s = \frac{1}{2}$. To give another example, let $j = 1$. In that case $J = 0, 1, 2$ and, with the aid of a table of coupling coefficients, all the eigenfunctions $|JM\rangle$ may be constructed. For example,

$$|JM\rangle = |00\rangle = \frac{1}{\sqrt{3}}[\psi_1^{\,1}(1)\psi_{-1}^{\,1}(2) - \psi_0^{\,1}(1)\psi_0^{\,1}(2) + \psi_{-1}^{\,1}(1)\psi_1^{\,1}(2)], \tag{20.4-2a}$$

$$|11\rangle = \frac{1}{\sqrt{2}}[\psi_1^{\,1}(1)\psi_0^{\,1}(2) - \psi_0^{\,1}(1)\psi_1^{\,1}(2)], \tag{20.4-2b}$$

$$|22\rangle = \psi_1^{\,1}(1)\psi_1^{\,1}(2). \tag{20.4-2c}$$

It is seen that $|00\rangle$ and $|2\ 2\rangle$ are symmetric while $|1\ 1\rangle$ is antisymmetric. All the eigenfunctions $|JM\rangle$ may be classified in this fashion with the result that those belonging to $J = 0$ and 2 are symmetric while those belonging to $J = 1$ are antisymmetric.

In summary, for two electrons, each with angular momentum j coupled to form a state with angular momentum J (which belongs to the irreducible

TABLE 20.9

Permutation Symmetries for Two Electrons

	j	J
$\square\square$	$\frac{1}{2}, \frac{3}{2}, \ldots$	$1, 3, 5, \ldots, 2j$
	$0, 1, \ldots$	$0, 2, 4, \ldots, 2j$
$\begin{smallmatrix}\square\\\square\end{smallmatrix}$	$\frac{1}{2}, \frac{3}{2}, \ldots$	$0, 2, 4, \ldots, 2j-1$
	$0, 1, \ldots$	$1, 3, 5, \ldots, 2j-1$

representation $D^{(J)}$ of $O^+(3)$), the correspondence between J and the permutation symmetry (Young tableaux) is shown in Table 20.9.

We call attention to the special case $|JM\rangle = |00\rangle$ in (20.4-2a) which, for all values of j, has the form

$$|JM\rangle = |00\rangle = \frac{1}{\sqrt{2j+1}} \sum_{m=-j}^{j} (-1)^{j-m} \psi_m{}^j(1)\psi^j_{-m}(2). \qquad (20.4\text{-}3)$$

This function is symmetric for integral values of j and antisymmetric for half-integral values.

The Pauli principle requires the total wave function to be antisymmetric. Therefore, the total wave function for two electrons is a product of a symmetric (antisymmetric) spin function and an antisymmetric (symmetric) function of the space coordinates. In terms of the Young tableaux this means that spin functions belonging to one tableau are to be multiplied by spatial functions that belong to the adjoint tableau (with rows and columns interchanged). For more than two electrons the total wave function must also be antisymmetric with respect to an interchange of any two sets of space and spin coordinates. However, the total wave function is generally not factorable into a product of two terms one of which depends only on the space coordinates and the other on the spin coordinates. Instead, as we shall see, there is a more complicated decomposition into functions of spatial and spin coordinates.

In Section 6.5 we saw that product functions of the type

$$\psi^j_{m_1}(1)\psi^j_{m_2}(2) \cdots \psi^j_{m_r}(r) \qquad (20.4\text{-}4)$$

were components of a Cartesian tensor of rank r with respect to $SU(2j+1)$. Such tensors are bases for representations of $SU(2j+1)$ but the representations are reducible. It must therefore be possible to find linear combinations of the product functions (20.4-4) that will transform irreducibly under the operations of $SU(2j+1)$. To construct such irreducible sets one may use the Young operator for each Young tableau. Since $O^+(3)$ is a subgroup of

$SU(2j+1)$, it is expected that functions that transform irreducibly under $O^+(3)$ will also belong to some irreducible representation of $SU(2j+1)$. What this means for a multielectron system is that when the one-electron functions $\psi_m{}^j$ are coupled to produce basis functions for $D^{(J)}$—the irreducible representations of $O^+(3)$—it is possible to subject these functions to the Young operator to produce basis functions for the irreducible representations of $SU(2j+1)$. Moreover, since rotations and permutations commute, the functions produced by the Young operator will still be basis functions for $D^{(J)}$. Therefore, with each tableau it is possible to associate certain values of J which arise from the coupling of the one-electron angular momenta.

For $r = 2$ (two particles) these features are already evident in Table 20.9. Let us now consider the case of three particles ($r = 3$) with $j = \frac{1}{2}$. The possible values of J are $\frac{1}{2}$ and $\frac{3}{2}$. Using the notation

$$|jm\rangle = |\tfrac{1}{2}\tfrac{1}{2}\rangle = \alpha, \qquad |jm\rangle = |\tfrac{1}{2}\, -\tfrac{1}{2}\rangle = \beta$$

and the abbreviation

$$(\alpha\beta\alpha) = \alpha(1)\beta(2)\alpha(3)$$

we have the following angular momentum eigenfunctions as basis functions for the irreducible representations of S_3, the permutation group for three particles:

$$\boxed{1\,|\,2\,|\,3} \qquad \begin{aligned} |JM\rangle = \left|\frac{3}{2}\frac{3}{2}\right\rangle &= (\alpha\alpha\alpha), \\ \left|\frac{3}{2}\frac{1}{2}\right\rangle &= \frac{1}{\sqrt{3}}[(\alpha\alpha\beta) + (\beta\alpha\alpha) + (\alpha\beta\alpha)], \\ \left|\frac{3}{2}\,-\frac{1}{2}\right\rangle &= \frac{1}{\sqrt{3}}[(\beta\beta\alpha) + (\alpha\beta\beta) + (\beta\alpha\beta)], \\ \left|\frac{3}{2}\,-\frac{3}{2}\right\rangle &= (\beta\beta\beta); \end{aligned} \tag{20.4-5}$$

$$\begin{array}{|c|c|}\hline 1 & 2 \\ \hline 3 & \\ \cline{1-1}\end{array} \qquad \begin{aligned} |JM\rangle = \left|\frac{1}{2}\frac{1}{2}\right\rangle &= \frac{1}{\sqrt{2}}[(\alpha\alpha\beta) - (\beta\alpha\alpha)], \\ \left|\frac{1}{2}\,-\frac{1}{2}\right\rangle &= \frac{1}{\sqrt{2}}[(\alpha\beta\beta) - (\beta\beta\alpha)]; \end{aligned} \tag{20.4-6}$$

$$\begin{array}{|c|c|}\hline 1 & 3 \\ \hline 2 & \\ \cline{1-1}\end{array} \qquad \begin{aligned} |JM\rangle = \left|\frac{1}{2}\frac{1}{2}\right\rangle &= \frac{1}{\sqrt{2}}[(\alpha\beta\alpha) - (\beta\alpha\alpha)], \\ \left|\frac{1}{2}\,-\frac{1}{2}\right\rangle &= \frac{1}{\sqrt{2}}[(\alpha\beta\beta) - (\beta\alpha\beta)]. \end{aligned} \tag{20.4-7}$$

Finally, for the tableau

$$\begin{array}{|c|}\hline 1 \\ \hline 2 \\ \hline 3 \\ \hline \end{array}$$

we require a basis function which is antisymmetric in all three indices. With only two spin functions, α and β, all such antisymmetric functions vanish identically.

For $j = \frac{1}{2}$, the situation may be summarized as follows: it is *not* possible to construct a wave function which is antisymmetric with respect to more than two particles. For electrons this means that spin functions are limited to basis functions for Young diagrams having no more than two rows:

$$\begin{array}{|c|c|c|c|c|c|c|c|c|c|}\hline & & & & & & & & & \\ \hline & & & & \\ \cline{1-5} \end{array} \; [\lambda_1, \lambda_2].$$

The spin S for the system is given by

$$S = \tfrac{1}{2}(\lambda_1 - \lambda_2) \tag{20.4-8}$$

and

$$\lambda_1 + \lambda_2 = r \tag{20.4-9}$$

where r is the number of electrons. In other words, a basis function for any of the standard tableaux associated with $[\lambda_1, \lambda_2]$ is an eigenfunction of S^2 with the eigenvalue $S(S + 1)$. The number of independent spin functions for a given S and M_S is equal to the number of standard tableaux associated with the Young diagram.

The application of these ideas to the construction of totally antisymmetric wave functions is based on the rule that spatial functions that belong to a specific Young tableau are to be multiplied by spin functions that belong to the adjoint tableau. The complete antisymmetric wave function consists of linear combinations of such products. For two electrons, we have seen that the rule is automatically satisfied; for three electrons, this rule may be illustrated by constructing the wave function $|p^3\,{}^2D2\tfrac{1}{2}\rangle$. This particular example was chosen for its simplicity since there is only one possible distribution of electrons resulting in a single Slater determinant:

$$|p^3\,{}^2D2\tfrac{1}{2}\rangle = |1^+\,0^+\,1^-\rangle. \tag{20.4-10}$$

To see how this can be written in terms of separate spatial and spin functions we first pick out the proper spin functions corresponding to $J = S = \frac{1}{2}$ and $M = M_S = \frac{1}{2}$. For three electrons we have, from (20.4-6) and (20.4-7),

$$\begin{aligned} &\begin{array}{|c|c|}\hline 1 & 2 \\ \hline 3 \\ \cline{1-1} \end{array} \quad \left|\frac{1}{2}\frac{1}{2}\right\rangle_{\mathrm{I}} = \frac{1}{\sqrt{2}}[(\alpha\alpha\beta) - (\beta\alpha\alpha)], \\ &\begin{array}{|c|c|}\hline 1 & 3 \\ \hline 2 \\ \cline{1-1} \end{array} \quad \left|\frac{1}{2}\frac{1}{2}\right\rangle_{\mathrm{II}} = \frac{1}{\sqrt{2}}[(\alpha\beta\alpha) - (\beta\alpha\alpha)]. \end{aligned} \tag{20.4-11}$$

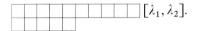

Next we construct spatial functions that are basis functions for the adjoint tableaux; in the present case, the latter are simply the same two tableaux as in (20.4-11). It may be readily verified that these functions are

$$\begin{array}{l|l} \begin{array}{|c|c|}\hline 1 & 2 \\ \hline 3 & \\ \cline{1-1}\end{array} & |2\,2\rangle_{\mathrm{I}} = \dfrac{1}{\sqrt{2}}[(1\;1\;0) - (0\;1\;1)], \\ \begin{array}{|c|c|}\hline 1 & 3 \\ \hline 2 & \\ \cline{1-1}\end{array} & |2\,2\rangle_{\mathrm{II}} = \dfrac{1}{\sqrt{2}}[(1\;0\;1) - (0\;1\;1)], \end{array} \qquad (20.4\text{-}12)$$

in which (110), for example, is an abbreviation for the product

$$p_{+1}(1)p_{+1}(2)p_0(3),$$

etc. Finally, let

$$\Psi = \sqrt{\tfrac{2}{3}}[|2\,2\rangle_{\mathrm{II}}|\tfrac{1}{2}\tfrac{1}{2}\rangle_{\mathrm{I}} - |2\,2\rangle_{\mathrm{I}}|\tfrac{1}{2}\tfrac{1}{2}\rangle_{\mathrm{II}}],$$

which, on substituting in (20.4-11) and (20.4-12) becomes

$$\begin{aligned} \Psi &= \sqrt{\tfrac{1}{6}}\{[(1\;0\;1) - (0\;1\;1)][\alpha\alpha\beta) - (\beta\alpha\alpha)] - [(1\;1\;0) - (0\;1\;1)][(\alpha\beta\alpha) - (\beta\alpha\alpha)]\} \\ &= \sqrt{\tfrac{1}{6}}\begin{vmatrix} p_{+1}(1)\alpha(1) & p_0(1)\alpha(1) & p_{+1}(1)\beta(1) \\ p_{+1}(2)\alpha(2) & p_0(2)\alpha(2) & p_{+1}(2)\beta(2) \\ p_{+1}(3)\alpha(3) & p_0(3)\alpha(3) & p_{+1}(3)\beta(3) \end{vmatrix} \\ &= |1^+\;0^+\;1^-\rangle \end{aligned} \qquad (20.4\text{-}13)$$

or

$$\begin{aligned} |p^3\,{}^2D2\tfrac{1}{2}\rangle &= |1^+\;0^+\;1^-\rangle = \Psi \\ &= \sqrt{\tfrac{2}{3}}[|2\,2\rangle_{\mathrm{II}}|\tfrac{1}{2}\tfrac{1}{2}\rangle_{\mathrm{I}} - |2\,2\rangle_{\mathrm{I}}|\tfrac{1}{2}\tfrac{1}{2}\rangle_{\mathrm{II}}]. \end{aligned} \qquad (20.4\text{-}14)$$

The consequences of this development for atomic wave functions may be summarized.

(1) For a given value of j, basis functions for Young diagrams with more than $2j + 1$ rows are identically zero. Thus, for $j = s = \frac{1}{2}$, it was shown that it is impossible to construct spin functions for diagrams with more than two rows.

(2) Because of the antisymmetry requirement on the total electronic wave function it is necessary to multiply spin functions belonging to a particular diagram with spatial functions belonging to adjoint diagrams. Since the former are limited to two rows, the latter are limited to two columns. In other words, diagrams for spatial functions are limited to at most, two columns and $2l + 1$ rows with $l = 0, 1, 2, \ldots$.

We shall now illustrate how this information, together with a group theoretical analysis, may be used to establish which terms, including their seniority assignments, arise from a configuration of equivalent electrons. In Section 6.5 it was shown that irreducible representations of $SU(2l+1)$ can be decomposed into irreducible representations of $O^+(2l+1)$ and, in turn, the latter can be decomposed into irreducible representations of $O^+(3)$. The results for p and d electrons are given in Table 6.1.

Consider the d^3 configuration. For a three-electron system, the spin functions, according to (20.4-5, 6, and 7) belong to the Young diagrams

[21] 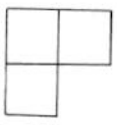 $S = \frac{1}{2}$

[2] 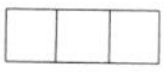 $S = \frac{3}{2}$.

Since the total wave function must be antisymmetric, the spatial functions must belong to the adjoint diagrams. The adjoint to [21] is [21] and the adjoint to [3] is $[1^3] \equiv [111]$. In other words, the only possible spatial functions are those that belong to the irreducible representations [21] or $[1^3]$ of $SU(2l+1)$. Those that belong to [21] will have $S = \frac{1}{2}$, and those that belong to $[1^3]$ will have $S = \frac{3}{2}$. Turning now to Table 6.1 for the decomposition $S(2l+1) \to O^+(2l+1) \to O^+(3)$, we find for d^3 ($l = 2, r = 3$)

$SU(2l+1)$	$O^+(2l+1)$	L	S	Term
[21]	(10)	2	$\frac{1}{2}$	^{2}D
	(21)	1, 2, 3, 4, 5	$\frac{1}{2}$	^{2}P, ^{2}D, ^{2}F, ^{2}G, ^{2}H
$[1^3]$	(11)	1, 3	$\frac{3}{2}$	^{4}P, ^{4}F

L is the label for the irreducible representations of $O^+(3)$. We note that the two ^{2}D terms are distinguishable on the basis of the irreducible representations of $O^+(2l+1)$, since one ^{2}D belongs to (10) and the other to (21). Hence, it is possible to devise a label which reflects this difference. Such a label is called the *seniority number*, v, defined as the smallest number of electrons for which a particular irreducible representation (μ_1, μ_2) of $O^+(2l+1)$ and the same value of S can occur. Consulting Table 6.1 again for $l = 2$, it is seen that (10) occurs for the configuration $d^1(S = \frac{1}{2})$; hence $v = 1$. The representation (21) appears for the first time in d^3; hence all the terms belonging to (21) will have $v = 3$. The representation (11) with $S = \frac{3}{2}$ occurs for the first time in d^3; hence, here too, $v = \frac{3}{2}$. These relations are summarized in Table 20.7.

20.5 *jj* Coupling

When angular momenta are combined as in (19.1-14), namely

$$\mathbf{j}_i = \boldsymbol{l}_i + \mathbf{s}_i, \qquad \mathbf{J} = \sum_i \mathbf{j}_i, \tag{20.5-1}$$

we have the *jj* coupling scheme. One electron states are designated by l_j as, for example, $s_{1/2}$, $p_{1/2}$, $p_{3/2}$, $d_{3/2}$, $d_{5/2}$, . . . , and each state has a degeneracy of $2j+1$. A two-electron state is symbolized by $(j_1 j_2)_J$ and the possible values of J range from $j_1 + j_2$ to $|j_1 - j_2|$. Thus, for the pd configuration, the possible *jj* coupling states are

$$(\tfrac{1}{2}\tfrac{3}{2})_{1,2}, \quad (\tfrac{1}{2}\tfrac{5}{2})_{2,3}, \quad (\tfrac{3}{2}\tfrac{3}{2})_{0,1,2,3}, \quad (\tfrac{3}{2}\tfrac{5}{2})_{1,2,3,4}.$$

For equivalent electrons the Pauli principle eliminates certain states which would otherwise be possible by angular momentum coupling. The rules for finding the states for two equivalent electrons are given:

(1) When $j_1 \neq j_2$, the value of J in $(j_1 j_2)_J$ is found by the rules for the coupling of angular momenta.

(2) When $j_1 = j_2$, the allowed values of J in $(j_1 j_2)_J$ are given by $J = 2j - 1$, $2j - 3, \ldots$.

Thus, for p^2, the possible states in *jj* coupling are

$$(\tfrac{1}{2}\tfrac{1}{2})_0, \quad (\tfrac{1}{2}\tfrac{3}{2})_{1,2}, \quad (\tfrac{3}{2}\tfrac{3}{2})_{0,2},$$

whereas in *LS* coupling the states are

$${}^1S_0, \quad {}^3P_{0,1,2}, \quad {}^1D_2.$$

The two sets of states may be uniquely correlated so that even in extreme cases of *jj* coupling it is still possible to employ an *LS* designation.

Since equivalent electrons reside in orbitals with the same values of n and l, the electrostatic interaction is expected to be large in which case *LS* coupling would be favored. Hence it is for nonequivalent electrons that *jj* coupling holds the greatest interest. There are many situations for which neither *LS* nor *jj* coupling are valid approximations; in such cases it is necessary to diagonalize the Coulomb repulsion and the spin-orbit interaction for each value of J. The coupling is then said to be *intermediate.*

CHAPTER 21

MATRIX ELEMENTS

In the lowest approximation an atom is described by a single Slater determinant $(N!)^{-1/2}\det\{\psi_1, \psi_2, \ldots, \psi_N\}$ in which the central-field spin orbitals $\psi = \psi_{nlm_lm_s}$ are of type (19.2-7). The energy of the atom, in this approximation, is therefore determined by the electronic configuration, that is, by the set of quantum numbers $nl, n'l', \ldots,$ and the number of electrons associated with each nl value.

The next higher approximation takes into account, by means of a perturbation treatment, the noncentral part of the electronic repulsion (19.2-2c). The energy of the atom will now depend on the specific details whereby electronic orbital and spin angular momenta $\mathbf{l}_i$ and $\mathbf{s}_i$ are coupled to form the total angular momentum. Thus several different energies may occur within a given configuration. In general, the differences in energy within a configuration are small in comparison with the total energy of the configuration. Nevertheless, they are of primary importance in numerous physical phenomena, e.g., spectroscopy, crystal field effects, magnetic resonance, and others.

This chapter will be devoted to an examination of the aforementioned perturbation as well as several others.

21.1 Electrostatic Matrix Elements—Two Electrons

Consider the Hamiltonian for a two-electron atom

$$\mathscr{H} = \mathscr{H}_0 + (e^2/r_{12}) \tag{21.1-1a}$$

with

$$\mathscr{H}_0 = \mathscr{H}_0(1) + \mathscr{H}_0(2), \tag{21.1-1b}$$

$$\mathscr{H}_0(i) = (p_i^2/2m) - (Ze^2/r_i). \tag{21.1-1c}$$

In order to use the central-field approximation, the Hamiltonian (21.1-1) is rewritten in the form (19.2-1):

$$\mathscr{H} = \mathscr{H}_0' + \mathscr{H}_1' \tag{21.1-2a}$$

with

$$\mathscr{H}_0' = \mathscr{H}_0(1) + \mathscr{H}_0(2) + \langle e^2/r_{12}\rangle_{\text{av}} \tag{21.1-2b}$$

$$\mathscr{H}_1' = (e^2/r_{12}) - \langle e^2/r_{12}\rangle_{\text{av}} \tag{21.1-2c}$$

where $\langle e^2/r_{12}\rangle_{\text{av}}$ is the spherical average of e^2/r_{12} and $\mathscr{H}_1'$ the perturbation that arises from the noncentral part of the electron repulsion term. Since e^2/r_{12} has been shown to commute with L^2, S^2, L_z, and S_z, the only nonzero matrix elements of e^2/r_{12} in the basis $|LSM_LM_S\rangle$ will be those that are diagonal. In the same basis, the contribution to the energy from $\langle e^2/r_{12}\rangle_{\text{av}}$ will be constant as long as the electronic configuration remains fixed. It is therefore sufficient to evaluate

$$E_C \equiv \langle LSM_LM_S|e^2/r_{12}|LSM_LM_S\rangle \tag{21.1-3}$$

in order to obtain the relative positions of the various terms.

Let us first employ a rather pedestrain method to illustrate the evaluation of (21.1-3). For the configuration 1s 2s we saw in (20.1-15) that

$$|{}^1\text{S}00\rangle = \frac{1}{\sqrt{2}}[|1\text{s}^+\,2\text{s}^-\rangle - |1\text{s}^-\,2\text{s}^+\rangle]. \tag{21.1-4}$$

The matrix element (21.1-3) then becomes

$$\begin{aligned}\langle {}^1\text{S}00|e^2/r_{12}|{}^1\text{S}00\rangle \\ &= \tfrac{1}{2}[\langle 1\text{s}^+\,2\text{s}^-|e^2/r_{12}|1\text{s}^+\,2\text{s}^-\rangle - \langle 1\text{s}^+\,2\text{s}^-|e^2/r_{12}|1\text{s}^-\,2\text{s}^+\rangle \\ &\quad - \langle 1\text{s}^-\,2\text{s}^+|e^2/r_{12}|1\text{s}^+\,2\text{s}^-\rangle + \langle 1\text{s}^-\,2\text{s}^+|e^2/r_{12}|1\text{s}^-\,2\text{s}^+\rangle].\end{aligned} \tag{21.1-5}$$

The four matrix elements in (21.1-5) may be evaluated by expanding each determinant; the result is

$$\langle {}^1\text{S}00|e^2/r_{12}|{}^1\text{S}00\rangle = J(1\text{s}, 2\text{s}) + K(1\text{s}, 2\text{s}) \tag{21.1-6}$$

where J and K are defined by (11.2-14) and (11.2-18). By the same procedure one finds

$$\begin{aligned}\langle {}^3\text{S}01|e^2/r_{12}|{}^3\text{S}01\rangle &= \langle {}^3\text{S}00|e^2/r_{12}|{}^3\text{S}00\rangle \\ &= \langle {}^3\text{S}\,0\,-1|e^2/r_{12}|{}^3\text{S}\,0\,-1\rangle = J(1\text{s}, 2\text{s}) - K(1\text{s}, 2\text{s}).\end{aligned} \tag{21.1-7}$$

We now have the Coulomb energies for the two terms ^{1}S and ^{3}S. For the latter the energy is independent of M_S as one might expect from the fact that the electrostatic energy for an isolated atom cannot depend on the orientation of the spin vector in space. Indeed we shall find that the matrix element (21.1-3) is independent of both M_L and M_S.

It is convenient to introduce the operator

$$\tfrac{1}{2}(1 + 4\mathbf{s}_1 \cdot \mathbf{s}_2) = \tfrac{1}{2}[1 + 2(S^2 - s_1{}^2 - s_2{}^2)].$$

Since the eigenvalues of $s_1{}^2$ and $s_2{}^2$ are $\frac{3}{4}$, we have

$$\tfrac{1}{2}(1 + 4\langle \mathbf{s}_1 \cdot \mathbf{s}_2 \rangle) = \begin{cases} -1 & \text{when} \quad S = 0, \\ +1 & \text{when} \quad S = 1. \end{cases} \tag{21.1-8}$$

so that (21.1-6) and (21.1-7) may be combined into

$$E_\mathrm{C} = J(1\mathrm{s}, 2\mathrm{s}) - \tfrac{1}{2}(1 + 4\langle s_1 \cdot s_2 \rangle)K(1\mathrm{s}, 2\mathrm{s}). \tag{21.1-9}$$

One might well inquire at this point why it is that singlet and triplet states have different energies, in view of the fact that the operator e^2/r_{12} is independent of spin. The answer resides in the spatial distributions of electrons in the singlet and triplet states. For the singlet, the spatial part of (21.1-4) is the symmetric function

$$\varphi_\mathrm{S} = \frac{1}{\sqrt{2}}[1\mathrm{s}(\mathbf{r}_1)\, 2\mathrm{s}(\mathbf{r}_2) + 2\mathrm{s}(\mathbf{r}_1)\, 1\mathrm{s}(\mathbf{r}_2)], \tag{21.1-10a}$$

while for the triplet state, the spatial part is the antisymmetric function

$$\varphi_\mathrm{A} = \frac{1}{\sqrt{2}}[1\mathrm{s}(\mathbf{r}_1)\, 2\mathrm{s}(\mathbf{r}_2) - 2\mathrm{s}(\mathbf{r}_1)\, 1\mathrm{s}(\mathbf{r}_2)]. \tag{21.1-10b}$$

Now let $\mathbf{r}_1 \to \mathbf{r}_2$. In that case $\varphi_\mathrm{A} \to 0$, which means that the probability of finding two electrons with parallel spin at the same point in space is zero. This feature, we have seen, is simply a manifestation of the Pauli principle or the antisymmetry requirement. On the other hand, the function φ_S associated with the singlet state does not vanish. Hence there is no restriction—on the basis of symmetry considerations—on the closeness of approach of two electrons with opposite spin. Since electrons with parallel spin avoid each other, whereas electrons of opposite spin do not, the Coulomb repulsion energy will be different in the two cases—higher for singlet states and lower for triplet states. Thus differences in the spin state can cause large differences in energy, but it must be recognized that such energy differences are electrostatic in origin and not magnetic. What has been said here is merely a rewording of the discussion on correlation effects in Section 13.8.

The J and K integrals in (21.1-9) may be expressed in terms of Slater–Condon parameters as in (11.2-17) and (11.2-21). With the aid of Tables 11.1 and 11.2,

$$J(1\mathrm{s}, 2\mathrm{s}) = F^0(1\mathrm{s}, 2\mathrm{s}), \qquad K(1\mathrm{s}, 2\mathrm{s}) = G^0(1\mathrm{s}, 2\mathrm{s}),$$

so that, for the 1s2s configuration, the Coulombic repulsion energies in the singlet and triplet states are

$$E_C(1s, 2s) = F^0(1s, 2s) - \tfrac{1}{2}(1 + 4\langle \mathbf{s}_1 \cdot \mathbf{s}_2 \rangle)G^0(1s, 2s) \qquad (21.1\text{-}11)$$

where F^0 and G^0 are radial integrals defined by (11.2-16) and (11.2-20) and calculated most often by Hartree–Fock methods. Finally, the average repulsion energy for the 1s2s configuration, weighted by the degeneracy of each term, is

$$\bar{E}_C = F^0(1s, 2s) - \tfrac{1}{2}G^0(1s, 2s). \qquad (21.1\text{-}12)$$

If both electrons occupy the 1s orbital, the Pauli principle eliminates the triplet state. Alternatively, we note that, according to (11.2-22), $F^k = G^k$ for equivalent electrons; hence the matrix element of e^2/r_{12} in the triplet state vanishes. In the singlet state we have

$$|{}^1S00\rangle = |1s^+ \; 1s^-\rangle$$

and

$$\langle {}^1S00|e^2/r_{12}|{}^1S00\rangle = \langle 1s^+ \; 1s^-|e^2/r_{12}|1s^+ \; 1s^-\rangle$$
$$= J(1s, 1s) = F^0(1s, 1s). \qquad (21.1\text{-}13)$$

For the 1s2p configuration the singlet and triplet energies may also be easily calculated; they are

$$\langle {}^1P|e^2/r_{12}|{}^1P\rangle = F^0(1s, 2p) + \tfrac{1}{3}G^1(1s, 2p), \qquad (21.1\text{-}14a)$$

$$\langle {}^3P|e^2/r_{12}|{}^3P\rangle = F^0(1s, 2p) - \tfrac{1}{3}G^1(1s, 2p), \qquad (21.1\text{-}14b)$$

or, combining the two expressions,

$$E_C(1s, 2p) = F^0(1s, 2p) - \tfrac{1}{6}(1 + 4\langle \mathbf{s}_1 \cdot \mathbf{s}_2 \rangle)G^1(1s, 2p). \qquad (21.1\text{-}14c)$$

The method just described for calculating electrostatic energies may, in principle, be used for any two-electron configuration but the labor involved soon becomes prohibitive. Clearly, it would be more efficient if it were possible to evaluate the matrix element (21.1-3) directly without expanding each $|LSM_LM_S\rangle$ into a sum of Slater determinants followed by an evaluation of matrix elements between pairs of determinants. A direct calculation of (21.1-3) is indeed possible.

We begin by substituting (20.1-26) into (21.1-3). For the singlet state of a configuration (n_1l_1, n_2l_2),

$$\begin{aligned}
&\langle {}^1LM_L|e^2/r_{12}|{}^1LM_L\rangle \\
&\quad = \tfrac{1}{2}[\langle l_1(1)l_2(2)LM_L|e^2/r_{12}|l_1(1)l_2(2)LM_L\rangle \\
&\qquad + \langle l_2(1)l_1(2)LM_L|e^2/r_{12}|l_2(1)l_1(2)LM_L\rangle \\
&\qquad + (-1)^{l_1+l_2-L}\langle l_1(1)l_2(2)LM_L|e^2/r_{12}|l_2(1)l_1(2)LM_L\rangle \\
&\qquad + (-1)^{l_1+l_2-L}\langle l_2(1)l_1(2)LM_L|e^2/r_{12}|l_1(1)l_2(2)LM_L\rangle]. \qquad (21.1\text{-}15)
\end{aligned}$$

From (1.2-24),

$$\frac{e^2}{r_{12}} \equiv \frac{e^2}{|\mathbf{r}_1 - \mathbf{r}_2|} = 4\pi e^2 \sum_{k=0}^{\infty} \frac{1}{2k+1} \frac{r_<^k}{r_>^{k+1}} \mathbf{Y}_1^{(k)} \cdot \mathbf{Y}_2^{(k)}. \tag{21.1-16}$$

Bearing in mind that a function like $|l_1(1)l_2(2)LM_L\rangle$ contains a radial factor of the form

$$(1/r_1 r_2)P_{n_1 l_1}(r_1)P_{n_2 l_2}(r_2),$$

the matrix elements in (21.1-15) may be written as

$$\left\langle l_1(1)l_2(2)LM_L \left| \frac{e^2}{r_{12}} \right| l_1(1)l_2(2)LM_L \right\rangle$$

$$= \sum_{k=0}^{\infty} \frac{4\pi}{2k+1} \langle l_1(1)l_2(2)LM_L | \mathbf{Y}_1^{(k)} \cdot \mathbf{Y}_2^{(k)} | l_1(1)l_2(2)LM_L \rangle$$

$$\times\, e^2 \int \frac{r_<^k}{r_>^{k+1}} P_{n_1 l_1}^2(r_1) P_{n_2 l_2}^2\, dr_1\, dr_2$$

$$= \sum_{k=0}^{\infty} \frac{4\pi}{2k+1} F^k(n_1 l_1, n_2 l_2) \langle l_1(1)l_2(2)LM_L | \mathbf{Y}_1^{(k)} \cdot \mathbf{Y}_2^{(k)} | l_1(1)l_2(2)LM_L \rangle$$

where we have used definition (11.2-16) for $F^k(n_1 l_1, n_2 l_2)$. In the same fashion

$$\left\langle l_2(1)l_1(2)LM_L \left| \frac{e^2}{r_{12}} \right| l_2(1)l_1(2)LM_L \right\rangle$$

$$= \sum_{k=0}^{\infty} \frac{4\pi}{2k+1} F^k(n_2 l_2, n_1 l_1) \langle l_2(1)l_1(2)LM_L | \mathbf{Y}_1^{(k)} \cdot \mathbf{Y}_2^{(k)} | l_2(1)l_1(2)LM_L \rangle,$$

$$\left\langle l_1(1)l_2(2)LM_L \left| \frac{e^2}{r_{12}} \right| l_2(1)l_1(2)LM_L \right\rangle$$

$$= \sum_{k=0}^{\infty} \frac{4\pi}{2k+1} \langle l_1(1)l_2(1)LM_L | \mathbf{Y}_1^{(k)} \cdot \mathbf{Y}_2^{(k)} | l_2(1)l_1(2)LM_L \rangle$$

$$\times\, e^2 \int \frac{r_<^k}{r_>^{k+1}} P_{n_1 l_1}(r_1) P_{n_2 l_2}(r_2) P_{n_2 l_2}(r_1) P_{n_1 l_1}(r_2)\, dr_1\, dr_2$$

$$= \sum_{k=0}^{\infty} \frac{4\pi}{2k+1} G^k(n_1 l_1, n_2 l_2) \langle l_1(1)l_2(2)LM_L | \mathbf{Y}_1^{(k)} \cdot \mathbf{Y}_2^{(k)} | l_2(1)l_1(2)LM_L \rangle$$

$$\left\langle l_2(1)l_1(2)LM_L \left| \frac{e^2}{r_{12}} \right| l_1(1)l_2(2)LM_L \right\rangle$$

$$= \sum_{k=0}^{\infty} \frac{4\pi}{2k+1} G^k(n_2 l_2, n_1 l_1) \langle l_2(1)l_1(2) | \mathbf{Y}_1^{(k)} \cdot \mathbf{Y}_2^{(k)} | l_1(1)l_2(2)LM_L \rangle$$

where G^k is defined by (11.2-20). From the definitions of F^k and G^k,

$$F^k(n_1l_1, n_2l_2) = F^k(n_2l_2, n_1l_1) \equiv F^k,$$
$$G^k(n_1l_1, n_2l_2) = G^k(n_2l_2, n_1l_1) \equiv G^k.$$

At this stage (21.1-15) takes the form

$$\left\langle {}^1LM_L \left| \frac{e^2}{r_{12}} \right| {}^1LM_L \right\rangle$$
$$= \frac{1}{2}\sum_{k=0}^{\infty} \frac{4\pi}{2k+1} [F^k\{\langle l_1(1)l_2(2)LM_L|\mathbf{Y}_1^{(k)} \cdot \mathbf{Y}_2^{(k)}|l_1(1)l_2(2)LM_L\rangle$$
$$+ \langle l_2(1)l_1(2)LM_L|\mathbf{Y}_1^{(k)} \cdot \mathbf{Y}_2^{(k)}|l_2(1)l_1(2)LM_L\}$$
$$+ (-1)^{l_1+l_2-L}G^k\{\langle l_1(1)l_2(2)LM_L|\mathbf{Y}_1^{(k)} \cdot \mathbf{Y}_2^{(k)}|l_2(1)l_1(2)LM_L\rangle$$
$$+ \langle l_2(1)l_1(2)LM_L|\mathbf{Y}_1^{(k)} \cdot \mathbf{Y}_2^{(k)}|l_1(1)l_2(2)LM_L\rangle\}]. \quad (21.1\text{-}17)$$

Our next task is to evaluate the angular matrix elements. This may be accomplished by means of (6.3-37):

$$\langle j_1j_2jm|\mathbf{T}^{(k)} \cdot \mathbf{U}^{(k)}|j_1'j_2'j'm'\rangle$$
$$= (-1)^{j+j_2+j_1'}\delta_{jj'}\delta_{mm'}\begin{Bmatrix} j_1' & j_2' & j \\ j_2 & j_1 & k \end{Bmatrix}\langle j_1\| \mathbf{T}^{(k)}\| j_1'\rangle\langle j_2\| \mathbf{U}^{(k)}\| j_2'\rangle. \quad (21.1\text{-}18)$$

With a change of notation

$$j_1 = l_1, \quad j_2 = l_2, \qquad j_1' = l_1', \quad j_2' = l_2',$$
$$j = j' = L, \qquad m = m' = M_L,$$
$$\mathbf{T}^{(k)} = \mathbf{Y}_1^{(k)}, \qquad \mathbf{U}^{(k)} = \mathbf{Y}_2^{(k)}$$

we now have

$$\langle l_1l_2LM_L|\mathbf{Y}_1^{(k)} \cdot \mathbf{Y}_2^{(k)}|l_1'l_2'LM_L\rangle$$
$$= (-1)^{L+l_2+l_1'}\langle l_1\| \mathbf{Y}_1^{(k)}\| l_1'\rangle\langle l_2\| \mathbf{Y}_2^{(k)}\| l_2'\rangle\begin{Bmatrix} l_1' & l_2' & L \\ l_2 & l_1 & k \end{Bmatrix}. \quad (21.1\text{-}19)$$

With this formula one may now evaluate the four matrix elements in (21.1-17):

$$\langle l_1(1)l_2(2)LM_L|\mathbf{Y}_1^{(k)} \cdot \mathbf{Y}_2^{(k)}|l_1(1)l_2(2)LM_L\rangle$$
$$= (-1)^{L+l_2+l_1}\langle l_1(1)\| \mathbf{Y}_1^{(k)}\| l_1(1)\rangle\langle l_2(2)\| \mathbf{Y}_2^{(k)}\| l_2(2)\rangle\begin{Bmatrix} l_1 & l_2 & L \\ l_2 & l_1 & k \end{Bmatrix}, \quad (21.1\text{-}20)$$

$$\langle l_2(1)l_1(2)LM_L|\mathbf{Y}_1^{(k)}\cdot\mathbf{Y}_2^{(k)}|l_2(1)l_1(2)LM_L\rangle$$
$$=(-1)^{L+l_2+l_1}\langle l_2(1)|\,|\mathbf{Y}_1^{(k)}|\,|l_2(1)\rangle\langle l_1(2)|\,|\mathbf{Y}_2^{(k)}|\,|l_1(2)\rangle\begin{Bmatrix} l_2 & l_1 & L\\ l_1 & l_2 & k\end{Bmatrix}. \tag{21.1-21}$$

These two expressions may be combined by means of the Wigner–Eckart theorem,

$$\langle l'm'|Y_{kq}|lm\rangle=(-1)^{l'-m'}\begin{pmatrix} l' & k & l\\ -m' & q & m\end{pmatrix}\langle l'|\,|\mathbf{Y}^{(k)}|\,|l\rangle \tag{21.1-22}$$

from which it follows that

$$\langle l_1(1)|\,|\mathbf{Y}_1^{(k)}|\,|l_1(1)\rangle=\langle l_1(2)|\,|\mathbf{Y}_2^{(k)}|\,|l_1(2)\rangle\equiv\langle l_1|\,|\mathbf{Y}^{(k)}|\,|l_1\rangle,$$
$$\langle l_2(1)|\,|\mathbf{Y}_1^{(k)}|\,|l_2(1)\rangle=\langle l_2(2)|\,|\mathbf{Y}_2^{(k)}|\,|l_2(2)\rangle\equiv\langle l_2|\,|\mathbf{Y}^{(k)}|\,|l_2\rangle.$$

Also, from the symmetry properties of the $6j$ coefficients

$$\begin{Bmatrix} l_1 & l_2 & L\\ l_2 & l_1 & k\end{Bmatrix}=\begin{Bmatrix} l_2 & l_1 & L\\ l_1 & l_2 & k\end{Bmatrix}.$$

Therefore the sum of (21.1-20) and (21.1-21) is

$$\langle l_1(1)l_2(2)LM_L|\mathbf{Y}_1^{(k)}\cdot\mathbf{Y}_2^{(k)}|l_1(1)l_2(2)LM_L\rangle$$
$$+\langle l_2(1)l_1(2)LM_L|\mathbf{Y}_1^{(k)}\cdot\mathbf{Y}_2^{(k)}|l_2(1)l_1(2)LM_L\rangle$$
$$=2(-1)^{L+l_2+l_1}\langle l_1|\,|\mathbf{Y}^{(k)}|\,|l_1\rangle\langle l_2|\,|\mathbf{Y}^{(k)}|\,|l_2\rangle\begin{Bmatrix} l_1 & l_2 & L\\ l_2 & l_1 & k\end{Bmatrix}. \tag{21.1-23}$$

For the last two matrix elements in (21.1-17) we have

$$(-1)^{l_1+l_2-L}\langle l_1(1)l_2(2)LM_L|\mathbf{Y}_1^{(k)}\cdot\mathbf{Y}_2^{(k)}|l_2(1)l_1(2)LM_L\rangle$$
$$=(-1)^{l_1+l_2-L}(-1)^{L+2l_2}\langle l_1(1)|\,|\mathbf{Y}_1^{(k)}|\,|l_2(1)\rangle\langle l_2(2)|\,|\mathbf{Y}_2^{(k)}|\,|l_1(2)\rangle\begin{Bmatrix} l_2 & l_1 & L\\ l_2 & l_1 & k\end{Bmatrix},$$

$$(-1)^{l_1+l_2-L}\langle l_2(1)l_1(2)LM_L|\mathbf{Y}_1^{(k)}\cdot\mathbf{Y}_2^{(k)}|l_1(1)l_2(2)LM_L\rangle$$
$$=(-1)^{l_1+l_2-L}(-1)^{L+2l_1}\langle l_2(1)|\,|\mathbf{Y}_1^{(k)}|\,|l_1(1)\rangle\langle l_1(2)|\,|\mathbf{Y}_2^{(k)}|\,|l_2(2)\rangle\begin{Bmatrix} l_1 & l_2 & L\\ l_1 & l_2 & k\end{Bmatrix}.$$

These two may also be combined. On the basis of (21.1-22),

$$\langle l_1(1)|\,|\mathbf{Y}_1^{(k)}|\,|l_2(1)\rangle=\langle l_1(2)|\,|\mathbf{Y}_2^{(k)}|\,|l_2(2)\rangle=\langle l_1|\,|\mathbf{Y}^{(k)}|\,|l_2\rangle,$$
$$\langle l_2(1)|\,|\mathbf{Y}_1^{(k)}|\,|l_1(1)\rangle=\langle l_2(2)|\,|\mathbf{Y}_2^{(k)}|\,|l_1(2)\rangle=\langle l_2|\,|\mathbf{Y}^{(k)}|\,|l_1\rangle.$$

But, from (6.3-23),

$$\langle l_1|\,|\mathbf{Y}^{(k)}|\,|l_2\rangle=(-1)^{l_2-l_1}\langle l_2|\,|\mathbf{Y}^{(k)}|\,|l_1\rangle.$$

Therefore

$$
\begin{aligned}
&(-1)^{l_1+l_2-L}\langle l_1(1)l_2(2)LM_L|\mathbf{Y}_1^{(k)}\cdot\mathbf{Y}_2^{(k)}|l_2(1)l_1(2)LM_L\rangle \\
&\quad+\langle l_2(1)l_1(2)LM_L|\mathbf{Y}_1^{(k)}\cdot\mathbf{Y}_2^{(k)}|l_1(1)l_2(2)LM_L\rangle \\
&\quad=2\langle l_1|\,|\mathbf{Y}^{(k)}|\,|l_2\rangle^2\begin{Bmatrix} l_1 & l_2 & L\\ l_1 & l_2 & k\end{Bmatrix}. \qquad (21.1\text{-}24)
\end{aligned}
$$

Substituting (21.1-23) and (21.1-24) in (21.1-17) we obtain

$$
\begin{aligned}
&\left\langle {}^1LM_L\left|\frac{e^2}{r_{12}}\right|{}^1LM_L\right\rangle \\
&\quad=\sum_{k=0}^{\infty}\frac{4\pi}{2k+1}\Bigg[(-1)^{l_1+l_2+L}\langle l_1|\,|\mathbf{Y}^{(k)}|\,|l_1\rangle\langle l_2|\,|\mathbf{Y}^{(k)}|\,|l_2\rangle \\
&\quad\times\begin{Bmatrix} l_1 & l_2 & L\\ l_2 & l_1 & k\end{Bmatrix}F^k+\langle l_1|\,|\mathbf{Y}^{(k)}|\,|l_2\rangle^2\begin{Bmatrix} l_1 & l_2 & L\\ l_1 & l_2 & k\end{Bmatrix}G^k\Bigg]. \qquad (21.1\text{-}25)
\end{aligned}
$$

The reduced matrix elements are given by (6.3-22). The final expression for the electronic repulsion energy may now be written as

$$
\begin{aligned}
E_{\mathrm{C}}&=\left\langle l_1l_2LS\left|\frac{e^2}{r_{12}}\right|l_1l_2LS\right\rangle \\
&=\sum_k(f_kF^k\pm g_kG^k) \\
&=\sum_{k=0}^{\infty}(2l_1+1)(2l_2+1)\Bigg[(-1)^L\begin{pmatrix} l_1 & k & l_1\\ 0 & 0 & 0\end{pmatrix} \\
&\quad\times\begin{pmatrix} l_2 & k & l_2\\ 0 & 0 & 0\end{pmatrix}\begin{Bmatrix} l_1 & l_2 & L\\ l_2 & l_1 & k\end{Bmatrix}F^k\pm\begin{pmatrix} l_1 & k & l_2\\ 0 & 0 & 0\end{pmatrix}^2\begin{Bmatrix} l_1 & l_2 & L\\ l_1 & l_2 & k\end{Bmatrix}G^k\Bigg] \qquad (21.1\text{-}26a)
\end{aligned}
$$

in which the plus sign applies to singlet states and the minus sign to triplets. An alternative form of (21.1-26a) is

$$
\begin{aligned}
E_{\mathrm{C}}&=\left\langle l_1l_2LS\left|\frac{e^2}{r_{12}}\right|l_1l_2LS\right\rangle \\
&=\sum_k(f_kF^k\pm g_kG^k) \\
&=\sum_{k=0}^{\infty}(2l_1+1)(2l_2+1)\Bigg[(-1)^L\begin{pmatrix} l_1 & k & l_1\\ 0 & 0 & 0\end{pmatrix}\begin{pmatrix} l_2 & k & l_2\\ 0 & 0 & 0\end{pmatrix} \\
&\quad\times\begin{Bmatrix} l_1 & l_2 & L\\ l_2 & l_1 & k\end{Bmatrix}F^k-\tfrac{1}{2}(1+4\langle\mathbf{s}_1\cdot\mathbf{s}_2\rangle)\begin{pmatrix} l_1 & k & l_2\\ 0 & 0 & 0\end{pmatrix}^2\begin{Bmatrix} l_1 & l_2 & L\\ l_1 & l_2 & k\end{Bmatrix}G^k\Bigg]. \qquad (21.1\text{-}26b)
\end{aligned}
$$

We note that E_C is independent of M_L and M_S, and, therefore, of $M\,(= M_L + M_S)$. To the extent that the spin-orbit interaction can be ignored, E_c is also independent of J.

An important special case of (21.1-26) arises when the two electrons are in equivalent orbitals. In that event the wave functions (20.1-27) are

$$|LSM_LM_S\rangle = |l(1)l(2)LM_L\rangle \begin{cases} {}^1\xi, & L \text{ even}, \\ {}^3\xi, & L \text{ odd}, \end{cases}$$

and

$$\begin{aligned} E_C &= \left\langle l^2LS\left|\frac{e^2}{r_{12}}\right|l^2LS\right\rangle \\ &= \left\langle l(1)l(2)LM_L\left|\frac{e^2}{r_{12}}\right|l(1)l(2)LM_L\right\rangle \\ &= \sum_{k=0}^{\infty}\frac{4\pi}{2k+1}F^k(nl,nl)\langle l(1)l(2)LM_L|\mathbf{Y}_1^{(k)}\cdot\mathbf{Y}_2^{(k)}|l(1)l(2)LM_L\rangle. \end{aligned} \quad (21.1\text{-}27)$$

But, from (21.1-19),

$$\begin{aligned} &\langle l(1)l(2)LM_L|\mathbf{Y}_1^{(k)}\cdot\mathbf{Y}_2^{(k)}|l(1)l(2)LM_L\rangle \\ &\quad = (-1)^L\langle l(1)\|\,\mathbf{Y}_1^{(k)}\|\,l(1)\rangle\langle l(2)\|\,\mathbf{Y}_2^{(k)}\|\,l(2)\rangle\begin{Bmatrix} l & l & L \\ l & l & k \end{Bmatrix} \\ &\quad = (-1)^L\langle l\|\,\mathbf{Y}^{(k)}\|\,l\rangle^2\begin{Bmatrix} l & l & L \\ l & l & k \end{Bmatrix}. \end{aligned} \quad (21.1\text{-}28)$$

Substituting (21.1-28) in (21.1-27) and replacing the reduced matrix elements by (6.3-23), the Coulombic energy for two equivalent electrons is

$$\begin{aligned} E_C &= \left\langle l^2LS\left|\frac{e^2}{r_{12}}\right|l^2LS\right\rangle = \sum_k f_kF^k \\ &= \sum_{k=0}^{\infty}(2l+1)^2F^k(-1)^L\begin{pmatrix} l & k & l \\ 0 & 0 & 0 \end{pmatrix}^2\begin{Bmatrix} l & l & L \\ l & l & k \end{Bmatrix} \end{aligned} \quad (21.1\text{-}29)$$

An odd value of L corresponds to a triplet state and an even value to a singlet.

Let us now illustrate the use of (21.1-26) and (21.1-29) in several special cases. Tables 1.6 and 1.8 provide numerical values of the $3j$ and $6j$ symbols. We also recall that

$$\begin{pmatrix} j_1 & j_2 & j_3 \\ 0 & 0 & 0 \end{pmatrix} = 0 \qquad \text{unless} \qquad \begin{cases} \triangle(j_1j_2j_3) \\ j_1 + j_2 + j_3 \text{ is an even integer.} \end{cases}$$

For the 1s2p configuration, $l_1 = 0, l_2 = 1$ and

$$\begin{pmatrix} l_1 & k & l_1 \\ 0 & 0 & 0 \end{pmatrix} = \begin{pmatrix} 0 & k & 0 \\ 0 & 0 & 0 \end{pmatrix} = \begin{cases} 1 & \text{when} \quad k = 0, \\ 0 & \text{when} \quad k \neq 0, \end{cases}$$

$$\begin{pmatrix} l_1 & k & l_2 \\ 0 & 0 & 0 \end{pmatrix} = \begin{pmatrix} 0 & k & 1 \\ 0 & 0 & 0 \end{pmatrix} = \begin{cases} -\sqrt{\frac{1}{3}} & \text{when} \quad k = 1, \\ 0 & \text{when} \quad k \neq 1, \end{cases}$$

$$\begin{Bmatrix} 0 & 1 & 1 \\ 1 & 0 & 0 \end{Bmatrix} = \sqrt{\tfrac{1}{3}}, \qquad \begin{Bmatrix} 0 & 1 & 1 \\ 0 & 1 & 1 \end{Bmatrix} = \tfrac{1}{3}.$$

Substituting these values into (21.1-26) we obtain (21.1-14).

Next, consider the $(np)^2$ configuration. The applicable equation is (21.1-29). The possible values of k that result in a nonvanishing $3j$ symbol are 0 and 2. With the help of Tables 1.6, 1.8, 11.1, and 11.3 it is found that

$$\begin{aligned} \left\langle {}^1\mathrm{S} \left| \frac{e^2}{r_{12}} \right| {}^1\mathrm{S} \right\rangle &= F^0 + \frac{2}{5} F^2 = F_0 + 10F_2, \\ \left\langle {}^3\mathrm{P} \left| \frac{e^2}{r_{12}} \right| {}^3\mathrm{P} \right\rangle &= F^0 - \frac{1}{5} F^2 = F_0 - 5F_2, \\ \left\langle {}^1\mathrm{D} \left| \frac{e^2}{r_{12}} \right| {}^1\mathrm{D} \right\rangle &= F^0 + \frac{1}{25} F^2 = F_0 + F_2. \end{aligned} \tag{21.1-30}$$

The average repulsion energy of the configuration np^2 weighted according to the degeneracy $(2L + 1)(2S + 1)$ of each term is

$$\bar{E}_\mathrm{C} = \left\langle \frac{e^2}{r_{12}} \right\rangle_{(np)^2} = F^0 - \frac{2}{25} F^2 = F_0 - 2F_2. \tag{21.1-31}$$

For $(nd)^2$ the energies are

$$\begin{aligned} \left\langle {}^3\mathrm{F} \left| \frac{e^2}{r_{12}} \right| {}^3\mathrm{F} \right\rangle &= F^0 - \frac{8}{49} F^2 - \frac{1}{49} F^4 = F_0 - 8F_2 - 9F_4 = A - 8B, \\ \left\langle {}^3\mathrm{P} \left| \frac{e^2}{r_{12}} \right| {}^3\mathrm{P} \right\rangle &= F^0 + \frac{7}{49} F^2 - \frac{4}{21} F^4 = F_0 + 7F_2 - 84F_4 = A + 7B, \\ \left\langle {}^1\mathrm{G} \left| \frac{e^2}{r_{12}} \right| {}^1\mathrm{G} \right\rangle &= F^0 + \frac{4}{49} F^2 + \frac{1}{441} F^4 = F_0 + 4F_2 + F_4 = A + 4B + 2C, \\ \left\langle {}^1\mathrm{D} \left| \frac{e^2}{r_{12}} \right| {}^1\mathrm{D} \right\rangle &= F^0 - \frac{3}{49} F^2 + \frac{4}{49} F^4 = F_0 - 3F_2 + 36F_4 = A - 3B + 2C, \\ \left\langle {}^1\mathrm{S} \left| \frac{e^2}{r_{12}} \right| {}^1\mathrm{S} \right\rangle &= F^0 + \frac{14}{49} F^2 + \frac{6}{21} F^4 = F_0 + 14F_2 + 126F_4 = A + 14B + 7C, \end{aligned} \tag{21.1-32}$$

with an average energy for the $(nd)^2$ configuration of

$$\bar{E}_C = \left\langle \frac{e^2}{r_{12}} \right\rangle_{(nd)^2} = F^0 - \frac{2}{63} F^2 - \frac{2}{63} F^4$$

$$= F_0 - \frac{14}{9} F_2 - 14F_4 = A - \frac{14}{9} B + \frac{7}{9} C. \quad (21.1\text{-}33)$$

As a final example, we use (21.1-26) to calculate the repulsion energies of the $(npn'p)$ configuration.

$$\begin{aligned}
\left\langle {}^1\mathrm{S} \left| \frac{e^2}{r_{12}} \right| {}^1\mathrm{S} \right\rangle &= F^0 + \frac{2}{5} F^2 + G^0 + \frac{2}{5} G^2, \\
\left\langle {}^3\mathrm{S} \left| \frac{e^2}{r_{12}} \right| {}^3\mathrm{S} \right\rangle &= F^0 + \frac{2}{5} F^2 - G^0 - \frac{2}{5} G^2, \\
\left\langle {}^1\mathrm{P} \left| \frac{e^2}{r_{12}} \right| {}^1\mathrm{P} \right\rangle &= F^0 - \frac{1}{5} F^2 - G^0 + \frac{1}{5} G^2, \\
\left\langle {}^3\mathrm{P} \left| \frac{e^2}{r_{12}} \right| {}^3\mathrm{P} \right\rangle &= F^0 - \frac{1}{5} F^2 + G^0 - \frac{1}{5} G^2, \\
\left\langle {}^1\mathrm{D} \left| \frac{e^2}{r_{12}} \right| {}^1\mathrm{D} \right\rangle &= F^0 + \frac{1}{25} F^2 + G^0 + \frac{1}{25} G^2, \\
\left\langle {}^3\mathrm{D} \left| \frac{e^2}{r_{12}} \right| {}^3\mathrm{D} \right\rangle &= F^0 + \frac{1}{25} F^2 - G^0 - \frac{1}{25} G^2,
\end{aligned} \quad (21.1\text{-}34)$$

$$\bar{E}_C = \left\langle \frac{e^2}{r_{12}} \right\rangle_{npn'p} = F^0 - \frac{1}{6} G^0 - \frac{1}{15} G^2. \quad (21.1\text{-}35)$$

The average energies in these examples were computed from the term energies. If all we wanted were the average energies for a two-electron configuration, it would be possible to bypass the calculation of the individual term energies. By calculating averages of $[J(i, j) - K(i, j)]$, Slater (1960) obtains expressions for the weighted (according to degeneracy) average repulsion energy for a pair of electrons. For equivalent electrons,

$$\bar{E}_C = \left\langle \frac{e^2}{r_{12}} \right\rangle = F^0(l, l) - \frac{1}{4l + 1} [c^2(l0, l0)F^2(l, l) + c^4(l0, l0)F^4(l, l) + \cdots]. \quad (21.1\text{-}36)$$

For nonequivalent electrons,

$$\bar{E}_C = \left\langle \frac{e^2}{r_{12}} \right\rangle = F^0(l_1, l_2) - \frac{1}{2} \frac{1}{\sqrt{(2l_1 + 1)(2l_2 + 1)}} \sum_k c^k(l_1 0, l_2 0) G^k(l_1, l_2) \quad (21.1\text{-}37)$$

in which the coefficients c^k are the integrals over spherical harmonics that are tabulated in Table 11.1.

In summary, the effect of the Coulomb repulsion is to shift the multiplet energy upwards as well as to split the multiplet into terms of different energy. We notice that for the configuration $(np)^2$ the ^{3}P term has the lowest energy among the possible terms ^{1}S, ^{3}P, and ^{1}D. Also, for $(nd)^2$ the term of lowest energy among ^{1}S, ^{3}P, ^{1}D, ^{3}F, and ^{1}G is ^{3}F. These examples are illustrative of *Hund's rules* which state that for equivalent electrons the term with the lowest energy is the one that, first, has the highest value of S (highest multiplicity) and, second, has the highest value of L.

As far as the spins are concerned, Hund's rules for a two-electron system merely state what has already been shown, namely, that triplet states lie lower than singlets and the basic reason for this is that on account of the Pauli principle, the electron distribution in a triplet state results in a lower Coulomb repulsion energy as compared with the singlet state. The second part of Hund's rules stem from the fact that electrons in states with high orbital angular momentum tend to be further apart relative to states with lower angular momentum. Hence the Coulomb repulsion is reduced as L is increased.

21.2 Some *n*-Electron Matrix Elements

Let F be a sum of one-electron operators

$$F = \sum_i^n f_i \tag{21.2-1}$$

and let it be required to evaluate the matrix element

$$\langle l^n\alpha LSM_LM_S|F|l^n\alpha' L'S'M_L'M_S'\rangle \tag{21.2-2}$$

in a configuration of n equivalent electrons. Since each operator f_i acts on one of the equivalent electrons and the contribution to the matrix element from each electron is the same we have

$$\langle l^n\alpha LSM_LM_S|F|l^n\alpha' L'S'M_L'M_S'\rangle = n\langle l^n\alpha LSM_LM_S|f_n|l^n\alpha' L'S'M_L'M_S'\rangle \tag{21.2-3}$$

in which f_n is the operator acting on the nth electron, although we could have used any of the other one-electron operators.

The wave functions may now be expressed in terms of fractional parentage coefficients as in (20.3-14):

$$|l^n\alpha LS\rangle = \sum_{\alpha_1L_1S_1} |l^{n-1}(\alpha_1L_1S_1)lLS\rangle\langle l^{n-1}(\alpha_1L_1S_1)lLS|l^n\alpha LS\rangle. \tag{21.2-4a}$$

In a more compact notation (Sobelman, 1972)

$$|l^n\alpha LS\rangle = \sum_{\alpha_1 L_1 S_1} G^{\alpha LS}_{\alpha_1 L_1 S_1}|l^{n-1}(\alpha_1 L_1 S_1)lLS\rangle \tag{21.2-4b}$$

and the matrix element (21.2-3) becomes

$$\begin{aligned}\langle l^n\alpha LSM_LM_S|F|l^n\alpha' L'S'M_L'M_S'\rangle \\ &= n\sum_{\alpha_1 L_1 S_1} G^{\alpha LS}_{\alpha_1 L_1 S_1} G^{\alpha' L'S'}_{\alpha_1 L_1 S_1} \\ &\quad\times \langle l^{n-1}(\alpha_1 L_1 S_1)l_n LSM_LM_S|f_n|l^{n-1}(\alpha_1 L_1 S_1)l_n L'S'M_L'M_S'\rangle. \end{aligned}\tag{21.2-5}$$

The operator f_n acting on the nth electron now occupies a special position because all the rest of the $n-1$ electrons have already been coupled to form the state with quantum numbers α_1, L_1, and S_1.

To proceed further it is necessary to specify the tensor character of the operators so that we may use the apparatus of the Wigner–Eckart formalism. Therefore let the operators in (21.2-1) be written as

$$\mathbf{F}^{(k)} = \sum_i \mathbf{f}_i^{(k)} \tag{21.2-6}$$

where $\mathbf{f}_i^{(k)}$ is an irreducible tensor operator of rank k operating on the coordinates of the ith electron. If the Wigner–Eckart theorem is applied to both sides of (21.2-5), the symmetry factors will cancel leaving a relation between the reduced matrix elements identical in form to (21.2-5):

$$\begin{aligned}\langle l^n\alpha LS\|\mathbf{F}^{(k)}\|l^n\alpha' L'S'\rangle \\ &= n\sum_{\alpha_1 L_1 S_1} G^{\alpha LS}_{\alpha_1 L_1 S_1} G^{\alpha' L'S'}_{\alpha_1 L_1 S_1} \\ &\quad\times \langle l^{n-1}(\alpha_1 L_1 S_1)l_n LS\|\mathbf{f}_n^{(k)}\|l^{n-1}(\alpha_1 L_1 S_1)l_n L'S'\rangle. \end{aligned}\tag{21.2-7}$$

But the reduced matrix element of $\mathbf{f}_n^{(k)}$ may be expressed, as in (6.3-39), as

$$\begin{aligned}\langle l^{n-1}(\alpha_1 L_1 S_1)l_n LS\|\mathbf{f}_n^{(k)}\|l^{n-1}(\alpha_1 L_1 S_1)l_n L'S'\rangle \\ &= (-1)^{L_1+l+L+k}\sqrt{(2L+1)(2L'+1)}\begin{Bmatrix} L & k & L' \\ l & L_1 & l \end{Bmatrix}\langle l\|\mathbf{f}_n^{(k)}\|l\rangle\,\delta_{SS'} \end{aligned}\tag{21.2-8}$$

which, when substituted in (21.2-7), yields

$$\begin{aligned}\langle l^n\alpha LS\|\mathbf{F}^{(k)}\|l^n\alpha' L'S'\rangle \\ &= n\sum_{\alpha_1 L_1 S_1} G^{\alpha LS}_{\alpha_1 L_1 S_1} G^{\alpha' L'S'}_{\alpha_1 L_1 S_1}\,\delta_{SS'} \\ &\quad\times (-1)^{L_1+l+L+k}\sqrt{(2L+1)(2L'+1)}\begin{Bmatrix} L & k & L' \\ l & L_1 & l \end{Bmatrix}\langle l\|\mathbf{f}_n^{(k)}\|l\rangle. \end{aligned}\tag{21.2-9}$$

We may now use the Wigner–Eckart theorem (6.3-1) to obtain

$$\langle l^n\alpha LSM_LM_S|F_q^{(k)}|l^n\alpha' L'S'M_L'M_S'\rangle$$

$$= (-1)^{L-M_L}\begin{pmatrix} L & k & L' \\ -M_L & q & M_L' \end{pmatrix}$$

$$\times \langle l^n\alpha LS\| \mathbf{F}^{(k)}\| l^n\alpha' L'S'\rangle\, \delta_{SS'}\,\delta_{M_SM_S'} \tag{21.2-10}$$

$$= (-1)^{L_1+l+M_L+k} n\sqrt{(2L+1)(2L'+1)}\,\delta_{SS'}\,\delta_{M_SM_S'}$$

$$\times \sum_{\alpha_1L_1S_1} G^{\alpha LS}_{\alpha_1L_1S_1} G^{\alpha' L'S'}_{\alpha_1L_1S_1}\begin{pmatrix} L & k & L' \\ -M_L & q & M_L' \end{pmatrix}\begin{Bmatrix} L & k & L' \\ l & L_1 & l \end{Bmatrix}\langle l\| \mathbf{f}_n^{(k)}\| l\rangle. \tag{21.2-11}$$

A special case of $\mathbf{F}^{(k)}$ is the operator

$$\mathbf{U}^{(k)} = \sum_i \mathbf{u}_i^{(k)} \tag{21.2-12}$$

where each $\mathbf{u}^{(k)}$ is defined by the general relation

$$\langle \alpha j\| \mathbf{u}^{(k)}\| \alpha' j'\rangle = \delta_{\alpha\alpha'}\,\delta_{jj'}. \tag{21.2-13}$$

$\mathbf{u}^{(k)}$ is known as a *unit tensor operator* of rank k; thus $\mathbf{U}^{(k)}$ is a *sum* of unit tensor operators. The reduced matrix element of $\mathbf{U}^{(k)}$, based on (21.2-9), is

$$\langle l^n\alpha LS\| \mathbf{U}^{(k)}\| l^n\alpha' L'S\rangle$$

$$= n\sum_{\alpha_1L_1S_1} G^{\alpha LS}_{\alpha_1L_1S_1} G^{\alpha' L'S}_{\alpha_1L_1S_1}(-1)^{L_1+L+k+l}$$

$$\times \sqrt{(2L+1)(2L'+1)}\begin{Bmatrix} L & k & L' \\ l & L_1 & l \end{Bmatrix} \tag{21.2-14a}$$

$$= (-1)^{L-L'}\langle l^n\alpha' L'S\| \mathbf{U}^{(k)}\| l^n\alpha LS\rangle. \tag{21.2-14b}$$

The case $k = 0$ is particularly important. The triangle conditions on the $6j$ symbol require $L' = L$, and from Table 1.7 it is found that

$$\begin{Bmatrix} L & 0 & L \\ l & L_1 & l \end{Bmatrix} = (-1)^{L_1+l+L}\frac{1}{\sqrt{(2l+1)(2L+1)}}. \tag{21.2-15}$$

Further, assuming $\alpha = \alpha'$,

$$\sum_{\alpha_1L_1S_1} |G^{\alpha\,L\,S}_{\alpha_1L_1S_1}|^2 = 1. \tag{21.2-16}$$

Therefore (21.2-9) reduces to

$$\langle l^n\alpha LS\| \mathbf{F}^{(0)}\| l^n\alpha LS\rangle = n\sqrt{(2L+1)/(2l+1)}\langle l\| \mathbf{f}_n^{(0)}\| l\rangle. \tag{21.2-17}$$

If we examine (21.2-10) for a closed shell, we find that since $L = L' = 0$ all matrix elements of one-electron tensor operators of rank k will vanish except those with $k = 0$. This means that only scalar operators have matrix elements in closed shell configurations. Alternatively, we may ignore closed shell configurations when calculating matrix elements of one-electron nonscalar irreducible tensor operators.

21.3 Electrostatic Matrix Elements—*n* Electrons

In Section 21.1 we calculated matrix elements of e^2/r_{12} for two electrons. Our present objective is to compute the electrostatic interaction within a configuration of n equivalent electrons; hence the main problem is the evaluation of the matrix elements

$$\left\langle l^n \alpha LS \left| \sum_{i<j} \frac{e^2}{r_{ij}} \right| l^n \alpha LS \right\rangle \tag{21.3-1}$$

in which α represents the radial part of the term. Reference to M_L and M_S has been omitted since, as shown in the two-electron case and as will also be shown in the present case, the electrostatic interaction does not depend on them.

As in (1.2-24) we write

$$\sum_{i<j} \frac{e^2}{r_{ij}} = e^2 \sum_{k=0}^{\infty} \frac{4\pi}{2k+1} \frac{r_<^k}{r_>^{k+1}} \sum_{i<j} \mathbf{Y}_i^{(k)} \cdot \mathbf{Y}_j^{(k)}, \tag{21.3-2}$$

and upon substitution in (21.3-1) the matrix element becomes

$$4\pi \sum_{k=0}^{\infty} F^k(nl, nl) \frac{1}{2k+1} \left\langle l^n LS \left| \sum_{i<j} \mathbf{Y}_i^{(k)} \cdot \mathbf{Y}_j^{(k)} \right| l^n LS \right\rangle \tag{21.3-3}$$

in which the radial function $F^k(nl, nl)$ is defined by (11.2-16).

The matrix element (21.3-3) is best handled by means of the unit tensor operator $\mathbf{u}^{(k)}$ as defined by (21.2-13) which in the present context takes the form

$$\langle l \| \mathbf{u}^{(k)} \| l' \rangle = \delta_{ll'}. \tag{21.3-4}$$

To see how the unit tensor operator enters into the problem, we first refer to (21.1-20) which, for two equivalent electrons, is

$$\langle l(1)l(2)LM_L | \mathbf{Y}_1^{(k)} \cdot \mathbf{Y}_2^{(k)} | l(1)l(2)LM_L \rangle$$

$$= (-1)^L \langle l(1) \| \mathbf{Y}_1^{(k)} \| l(1) \rangle \langle l(2) \| \mathbf{Y}_2^{(k)} \| l(2) \rangle \begin{Bmatrix} l & l & L \\ l & l & k \end{Bmatrix}. \tag{21.3-5}$$

Similarly

$$\langle l(1)l(2)LM_L|\mathbf{u}_1^{(k)}\cdot\mathbf{u}_2^{(k)}|l(1)l(2)LM_L\rangle = (-1)^L\begin{Bmatrix} l & l & L \\ l & l & k \end{Bmatrix} \tag{21.3-6}$$

in which property (21.3-4) has been used. Combining (21.3-5) and (21.3-6) we obtain

$$\left\langle l^nLS\left|\sum_{i<j}\mathbf{Y}_i^{(k)}\cdot\mathbf{Y}_j^{(k)}\right|l^nlS\right\rangle$$
$$= \sum_k \langle l\|\mathbf{Y}^{(k)}\|l\rangle^2\left\langle l^nLS\left|\sum_{i<j}\mathbf{u}_i^{(k)}\cdot\mathbf{u}_j^{(k)}\right|l^nLS\right\rangle. \tag{21.3-7}$$

But

$$\sum_{i<j}\mathbf{u}_i^{(k)}\cdot\mathbf{u}_j^{(k)} = \tfrac{1}{2}\sum_{i\neq j}\mathbf{u}_i^{(k)}\cdot\mathbf{u}_j^{(k)} = \tfrac{1}{2}\left[\sum_i\mathbf{u}_i^{(k)}\cdot\sum_j\mathbf{u}_j^{(k)} - \sum_i\mathbf{u}_i^{(k)}\cdot\mathbf{u}_i^{(k)}\right] \tag{21.3-8}$$

and if we let

$$\mathbf{U}^{(k)} = \sum_i \mathbf{u}_i^{(k)} \tag{21.3-9}$$

as in (21.2-12), then

$$\sum_{i<j}\mathbf{u}_i^{(k)}\cdot\mathbf{u}_j^{(k)} = \tfrac{1}{2}\mathbf{U}^{(k)}\cdot\mathbf{U}^{(k)} - \tfrac{1}{2}\sum_i\mathbf{u}_i^{(k)}\cdot\mathbf{u}_i^{(k)}. \tag{21.3-10}$$

At this stage, after combining (21.3-3), (21.3-7), and (21.3-10), we have

$$\left\langle l^n\alpha LS\left|\sum_{i<j}\frac{e^2}{r_{ij}}\right|l^n\alpha LS\right\rangle$$
$$= 4\pi\sum_{k=0}\left\{F^k(nl,nl)\frac{1}{2k+1}\langle l\|\mathbf{Y}^{(k)}\|l\rangle^2\left[\langle l^n\alpha LS|\tfrac{1}{2}\mathbf{U}^{(k)}\cdot\mathbf{U}^{(k)}|l^n\alpha LS\rangle\right.\right.$$
$$\left.\left. - \left\langle l^n\alpha LS\left|\frac{1}{2}\sum_i\mathbf{u}^{(k)}\cdot\mathbf{u}^{(k)}\right|l^n\alpha LS\right\rangle\right]\right\}. \tag{21.3-11}$$

The first matrix element in the square brackets can be evaluated by means of (6.3-34) which gives

$$\langle l^n\alpha LS|\mathbf{U}^{(k)}\cdot\mathbf{U}^{(k)}|l^n\alpha LS\rangle$$
$$= \frac{1}{2L+1}\sum_{\alpha'L'}(-1)^{L-L'}\langle l^n\alpha LS\|\mathbf{U}^{(k)}\|l^n\alpha'L'S\rangle$$
$$\times\langle l^n\alpha'L'S\|\mathbf{U}^{(k)}\|l^n\alpha LS\rangle. \tag{21.3-12}$$

The reduced matrix elements have already been evaluated in (21.2-14).

It remains to calculate matrix elements of

$$\mathbf{T}^{(0)} \equiv \sum_i \mathbf{u}_i^{(k)} \cdot \mathbf{u}_i^{(k)} \equiv \sum_i \mathbf{t}_i^{(0)}. \tag{21.3-13}$$

From the Wigner–Eckart theorem

$$\begin{aligned} &\langle l^n\alpha LSM_LM_S|T_0^{(0)}|l^n\alpha LSM_LM_S\rangle \\ &= (-1)^{L-M_L}\begin{pmatrix} L & 0 & L \\ -M_L & 0 & M_L \end{pmatrix}\langle l^n\alpha LS|\,|T^{(0)}|\,|l^n\alpha LS\rangle \\ &= \frac{1}{\sqrt{2L+1}}\langle l^n\alpha LS|\,|T^{(0)}|\,|l^n\alpha LS\rangle \end{aligned} \tag{21.3-14}$$

in which we have replaced the $3j$ symbol by its value from Table 1.5. But, according to (21.2-17),

$$\langle l^n\alpha LS|\,|\mathbf{T}^{(0)}|\,|l^n\alpha LS\rangle = n\sqrt{(2L+1)/(2l+1)}\langle l|\,|\mathbf{t}^{(0)}|\,|l\rangle, \tag{21.3-15}$$

and applying the Wigner–Eckart theorem again to $t_0^{(0)}$,

$$\langle lm|t_0^{(0)}|lm\rangle = \langle l|\,|\mathbf{t}^{(0)}|\,|l\rangle/\sqrt{2l+1}. \tag{21.3-16}$$

However, using (6.3-34) and (21.3-4),

$$\begin{aligned} \langle lm|t_0^{(0)}|lm\rangle &= \langle lm|\mathbf{u}_i^{(k)} \cdot \mathbf{u}_i^{(k)}|lm\rangle \\ &= \frac{1}{2l+1}\sum_{l'}(-1)^{l-l'}\langle l|\,|\mathbf{u}^{(k)}|l'\rangle\langle l'|\,|\mathbf{u}^{(k)}|\,|l\rangle \\ &= \frac{1}{2l+1} \end{aligned} \tag{21.3-17}$$

Therefore

$$\langle l|\,|\mathbf{t}^{(0)}|\,|l\rangle = \sqrt{2l+1}\langle lm|t^{(0)}|lm\rangle = \frac{1}{\sqrt{2l+1}}, \tag{21.3-18}$$

which, when substituted in (21.3-15) gives

$$\langle l^n\alpha LS|\,|\mathbf{T}^{(0)}|\,|l^n\alpha LS\rangle = n\sqrt{(2L+1)/(2l+1)}. \tag{21.3-19}$$

Equation (21.3-14) now becomes

$$\begin{aligned} &\langle l^n\alpha LSM_LM_S|T_0^{(0)}|l^n\alpha LSM_LM_S\rangle \\ &= \left\langle l^n\alpha LSM_LM_S\left|\sum_i \mathbf{u}_i^{(k)} \cdot \mathbf{u}_i^{(k)}\right| l^n\alpha LSM_LM_S\right\rangle = n/(2l+1). \end{aligned} \tag{21.3-20}$$

From (6.3-22),

$$\langle l \| \mathbf{Y}^{(k)} \| l \rangle = (-1)^l (2l+1)\sqrt{(2k+1)/4\pi}\begin{pmatrix} l & k & l \\ 0 & 0 & 0 \end{pmatrix}. \tag{21.3-21}$$

We may now combine (21.3-12), (21.3-13), (21.3-20), and (21.3-21) to achieve the final result

$$\left\langle l^n \alpha LS \left| \sum_{i<j} \frac{e^2}{r_{ij}} \right| l^n \alpha LS \right\rangle = \sum_k f_k F^k \tag{21.3-22}$$

where

$$\begin{aligned} f_k = {} & \frac{1}{2}(2l+1)^2 \begin{pmatrix} l & k & l \\ 0 & 0 & 0 \end{pmatrix}^2 \\ & \times \left\{ \frac{1}{2L+1} \sum_{\alpha' L'} |\langle l^n \alpha LS \| \mathbf{U}^{(k)} \| l^n \alpha' L' S \rangle|^2 - \frac{n}{2l+1} \right\} \end{aligned} \tag{21.3-23}$$

and the reduced matrix element is given by (21.2-14).

For f_0 we note that according to (21.2-17) and (21.3-4),

$$\begin{aligned} \langle l^n \alpha LS \| \mathbf{U}^{(0)} \| l^n \alpha LS \rangle &= n\sqrt{(2L+1)/(2l+1)} \langle l \| \mathbf{u}^{(0)} \| l \rangle \\ &= n\sqrt{(2L+1)/(2l+1)} \end{aligned} \tag{21.3-24}$$

and there are no off-diagonal matrix elements. Also, according to Table 1.5,

$$\begin{pmatrix} l & 0 & l \\ 0 & 0 & 0 \end{pmatrix}^2 = 1/(2l+1). \tag{21.3-25}$$

Inserting (21.3-24) and (21.3-25) into (21.3-23),

$$f_0 = n(n-1)/2. \tag{21.3-26}$$

To illustrate the use of these formulas in a simple case consider the configuration p^3 which contains the multiplets ^4S, ^2P, and ^2D. For the $3j$ symbol

$$\begin{pmatrix} 1 & k & 1 \\ 0 & 0 & 0 \end{pmatrix}$$

the possible values of k are limited by the triangle condition $\triangle(1k1)$ and the requirement that $2+k$ be even. Hence $k = 0, 2$. For $k = 0$ we obtain f_0 immediately from (21.3-26), i.e., $f_0 = 3$; the same applies to all three terms.

For f_2 it is necessary to evaluate the reduced matrix elements of $\mathbf{U}^{(2)}$. Although we have an appropriate expression in (21.2-14) it is best handled in practice by means of tables such as those by Nielsen and Koster (1963). For the chosen example the only nonvanishing contribution comes from

$$|\langle p^3\ {}^2D\| \mathbf{U}^{(2)} \| p^3\ {}^2P\rangle|^2 = 3.$$

Since

$$\begin{pmatrix} 1 & 2 & 1 \\ 0 & 0 & 0 \end{pmatrix}^2 = \frac{2}{15},$$

we have, for the electrostatic repulsion energies,

$$E_C(^4S) = 3F^0 - \tfrac{3}{5}F^2, \quad E_C(^2P) = 3F^0, \quad E_C(^2D) = 3F^0 - \tfrac{6}{25}F^2 \quad (21.3\text{-}27)$$

Table 21.1 gives electrostatic energies for equivalent electrons; more extensive tables are given by Slater (1960) and by Nielson and Koster (1963). The reader should be cautioned, however, that the tabulated energy of a

TABLE 21.1

Electrostatic Matrix Elements

p^2	F^0	F^2
^3P ^3P	1	$-\frac{1}{5}$
^1S ^1S	1	$\frac{2}{5}$
^1D ^1D	1	$\frac{1}{25}$

p^3	F^0	F^2
^4S ^4S	3	$-\frac{3}{5}$
^2P ^2P	3	0
^2D ^2D	3	$-\frac{6}{25}$

TABLE 21.1 *(continued)*

d^2	F^0	F^2	F^4
${}^3_2\mathrm{P}$ ${}^3_2\mathrm{P}$	1	$\frac{1}{7}$	$-\frac{4}{21}$
${}^3_2\mathrm{F}$ ${}^3_2\mathrm{F}$	1	$-\frac{8}{49}$	$-\frac{1}{49}$
${}^1_0\mathrm{S}$ ${}^1_0\mathrm{S}$	1	$\frac{2}{7}$	$\frac{2}{7}$
${}^1_2\mathrm{D}$ ${}^1_2\mathrm{D}$	1	$-\frac{3}{49}$	$\frac{4}{49}$
${}^1_2\mathrm{G}$ ${}^1_2\mathrm{G}$	1	$\frac{4}{49}$	$\frac{1}{441}$

d^4	F^0	F^2	F^4
${}^5_4\mathrm{D}$ ${}^5_4\mathrm{D}$	6	$-\frac{3}{7}$	$-\frac{3}{7}$
${}^3_2\mathrm{P}$ ${}^3_2\mathrm{P}$	6	$-\frac{1}{7}$	$-\frac{2}{63}$
${}^3_2\mathrm{P}$ ${}^3_4\mathrm{P}$	0	$\frac{4\sqrt{14}}{49}$	$-\frac{20\sqrt{14}}{441}$
${}^3_4\mathrm{P}$ ${}^3_4\mathrm{P}$	6	$-\frac{3}{49}$	$-\frac{139}{441}$
${}^3_4\mathrm{D}$ ${}^3_4\mathrm{D}$	6	$-\frac{5}{49}$	$-\frac{43}{147}$
${}^3_2\mathrm{F}$ ${}^3_2\mathrm{F}$	6	$-\frac{2}{49}$	$-\frac{13}{147}$
${}^3_2\mathrm{F}$ ${}^3_4\mathrm{F}$	0	$\frac{12}{49}$	$-\frac{20}{147}$
${}^3_4\mathrm{F}$ ${}^3_4\mathrm{F}$	6	$-\frac{8}{49}$	$-\frac{38}{147}$
${}^3_4\mathrm{G}$ ${}^3_4\mathrm{G}$	6	$-\frac{12}{49}$	$-\frac{94}{441}$
${}^3_4\mathrm{H}$ ${}^3_4\mathrm{H}$	6	$-\frac{17}{49}$	$-\frac{23}{147}$
${}^1_0\mathrm{S}$ ${}^1_0\mathrm{S}$	6	$\frac{2}{7}$	$\frac{2}{7}$

d^4	F^0	F^2	F^4
${}^1_0\mathrm{S}$ ${}^1_4\mathrm{S}$	0	$\frac{6\sqrt{21}}{49}$	$-\frac{10\sqrt{21}}{147}$
${}^1_4\mathrm{S}$ ${}^1_4\mathrm{S}$	6	$\frac{6}{49}$	$-\frac{38}{147}$
${}^1_2\mathrm{D}$ ${}^1_2\mathrm{D}$	6	$\frac{15}{49}$	$-\frac{6}{49}$
${}^1_2\mathrm{D}$ ${}^1_4\mathrm{D}$	0	$\frac{12\sqrt{2}}{49}$	$-\frac{20\sqrt{2}}{147}$
${}^1_4\mathrm{D}$ ${}^1_4\mathrm{D}$	6	$\frac{3}{49}$	$-\frac{11}{49}$
${}^1_4\mathrm{F}$ ${}^1_4\mathrm{F}$	6	0	$-\frac{4}{21}$
${}^1_2\mathrm{G}$ ${}^1_2\mathrm{G}$	6	$-\frac{6}{49}$	$\frac{17}{147}$
${}^1_2\mathrm{G}$ ${}^1_4\mathrm{G}$	0	$\frac{4\sqrt{11}}{49}$	$-\frac{20\sqrt{11}}{441}$
${}^1_4\mathrm{G}$ ${}^1_4\mathrm{G}$	6	$-\frac{4}{49}$	$-\frac{64}{441}$
${}^1_4\mathrm{I}$ ${}^1_4\mathrm{I}$	6	$-\frac{15}{49}$	$-\frac{1}{49}$

TABLE 21.1 *(continued)*

d^3	F^0	F^2	F^4
${}^4_3P$ ${}^4_3P$	3	0	$-\frac{1}{3}$
${}^4_3F$ ${}^4_3F$	3	$-\frac{15}{49}$	$-\frac{8}{49}$
${}^2_3P$ ${}^2_3P$	3	$-\frac{6}{49}$	$-\frac{4}{147}$
${}^2_1D$ ${}^2_1D$	3	$\frac{1}{7}$	$\frac{1}{7}$
${}^2_1D$ ${}^2_3D$	0	$\frac{3\sqrt{21}}{49}$	$-\frac{5\sqrt{21}}{147}$

d^3	F^0	F^2	F^4
${}^2_3D$ ${}^2_3D$	3	$\frac{3}{49}$	$-\frac{19}{147}$
${}^2_3F$ ${}^2_3F$	3	$\frac{9}{49}$	$-\frac{29}{147}$
${}^2_3G$ ${}^2_3G$	3	$-\frac{11}{49}$	$\frac{13}{441}$
${}^2_3H$ ${}^2_3H$	3	$-\frac{6}{49}$	$-\frac{4}{147}$

d^5	F^0	F^2	F^4
${}_5S$ ${}^6_5S$	10	$-\frac{5}{7}$	$-\frac{5}{7}$
P ${}^4_3P$	10	$-\frac{4}{7}$	$-\frac{5}{21}$
D ${}^4_5D$	10	$-\frac{18}{49}$	$-\frac{25}{49}$
F ${}^4_3F$	10	$-\frac{13}{49}$	$-\frac{20}{49}$
G ${}^4_5G$	10	$-\frac{25}{49}$	$-\frac{190}{441}$
S ${}^2_5S$	10	$-\frac{3}{49}$	$-\frac{65}{147}$
P ${}^2_3P$	10	$\frac{20}{49}$	$-\frac{80}{147}$
D ${}^2_1D$	10	0	0
D ${}^2_3D$	0	0	0
D ${}^2_5D$	0	$-\frac{6\sqrt{14}}{49}$	$\frac{10\sqrt{14}}{147}$

d^5	F^0	F^2	F^4
${}^2_3D$ ${}^2_3D$	10	$-\frac{4}{49}$	$-\frac{40}{147}$
${}^2_3D$ ${}^2_5D$	0	0	0
${}^2_5D$ ${}^2_5D$	10	$-\frac{6}{49}$	$-\frac{20}{49}$
${}^2_3F$ ${}^2_3F$	10	$-\frac{25}{49}$	$-\frac{5}{147}$
${}^2_3F$ ${}^2_5F$	0	0	0
${}^2_5F$ ${}^2_5F$	10	$-\frac{9}{49}$	$-\frac{55}{147}$
${}^2_3G$ ${}^2_3G$	10	$\frac{3}{49}$	$-\frac{155}{441}$
${}^2_3G$ ${}^2_5G$	0	0	0
${}^2_5G$ ${}^2_5G$	10	$-\frac{13}{49}$	$-\frac{145}{441}$
${}^2_3H$ ${}^2_3H$	10	$-\frac{22}{49}$	$-\frac{10}{147}$
${}^2_5I$ ${}^2_5I$	10	$-\frac{24}{49}$	$-\frac{10}{49}$

given term may vary depending on the reference energy with respect to which the term energy is given. Thus, Slater (1960) refers term energies to the average energy of a configuration. Special formulas for electronic configurations other than l^n are to be found in the work of Slater (1960) and Sobelman (1972).

The average energy of a configuration may be computed on the basis of (19.4-4):

$$E_{\text{av}} = \sum I(nl) + \sum_{\text{pairs}} \langle e^2/r_{12} \rangle.$$

The first sum ranges over all the electrons; the second sum over all pairs of electrons with $\langle e^2/r_{12} \rangle$ given by (21.1-36) for a pair of equivalent electrons or by (21.1-37) for nonequivalent electrons. To give an example we compute the average energy of boron in the ground state with the configuration $1s^2\,2s^2\,2p$. For this case

$$\sum I(nl) = 2I(1\text{s}) + 2I(2\text{s}) + I(2\text{p}),$$

$$\sum_{\text{pairs}} \text{Interactions} = E_{\text{C}}(1\text{s}, 1\text{s}) + E_{\text{C}}(2\text{s}, 2\text{s}) + 4E_{\text{C}}(1\text{s}, 2\text{s}) + 2E_{\text{C}}(1\text{s}, 2\text{p}) + 2E_{\text{C}}(2\text{s}, 2\text{p}),$$

$$E_{\text{C}}(1\text{s}, 1\text{s}) = F^0(1\text{s}, 1\text{s}), \qquad E_{\text{C}}(2\text{s}, 2\text{s}) = F^0(2\text{s}, 2\text{s}),$$

$$4E_{\text{C}}(1\text{s}, 2\text{s}) = 4[F^0(1\text{s}, 2\text{s}) - \tfrac{1}{2}G^0(1\text{s}, 2\text{s})],$$

$$2E_{\text{C}}(1\text{s}, 2\text{p}) = 2[F^0(1\text{s}, 2\text{p}) - \tfrac{1}{6}G^1(1\text{s}, 2\text{p})],$$

$$2E_{\text{C}}(2\text{s}, 2\text{p}) = 2[F^0(2\text{s}, 2\text{p}) - \tfrac{1}{6}G^1(2\text{s}, 2\text{p})].$$

Summing all the contributions

$$E_{\text{av}} = 2I(1\text{s}) + 2I(2\text{s}) + I(2\text{p}) + F^0(1\text{s}, 1\text{s}) + F^0(2\text{s}, 2\text{s}) + 4F^0(1\text{s}, 2\text{s}) + 2F^0(1\text{s}, 2\text{p}) + 2F^0(2\text{s}, 2\text{p}) - 2G^0(1\text{s}, 2\text{s}) - \tfrac{1}{3}G^1(1\text{s}, 2\text{p}) - \tfrac{1}{3}G^1(2\text{s}, 2\text{p}). \tag{21.3-28}$$

which is also obtainable from (19.4-4). The Hartree–Fock tables give (in Rydbergs)

$$F^0(1\text{s}, 1\text{s}) = 5.784091, \qquad G^0(1\text{s}, 2\text{s}) = 0.0771609, \qquad I(1\text{s}) = -24.87798,$$

$$F^0(1\text{s}, 2\text{s}) = 1.297264, \qquad G^1(1\text{s}, 2\text{p}) = 0.0863852, \qquad I(2\text{s}) = -5.210903,$$

$$F^0(1\text{s}, 2\text{p}) = 1.200249, \qquad G^1(2\text{s}, 2\text{p}) = 0.5465473, \qquad I(2\text{p}) = -4.558612.$$

$$F^0(2\text{s}, 2\text{s}) = 0.9205248,$$

$$F^0(2\text{s}, 2\text{p}) = 0.8746893,$$

The total energy is -49.05812 Ry.

For a closed shell there is just one term ^{1}S. Therefore

$$E(^1\mathrm{S}) = E_{\mathrm{av}} = (4l+2)I(nl) + \tfrac{1}{2}(4l+2)(4l+1)\langle e^2/r_{12}\rangle \quad (21.3\text{-}29)$$

in which the coefficient in front of the electron repulsion term is the number of pairs in a configuration of $4l + 2$ electrons. Referring to (21.1-36) the closed shell energy E_{cs} is

$$\begin{aligned} E_{\mathrm{cs}} &= (4l+2)I(nl) + \tfrac{1}{2}(4l+2)(4l+1)F^0 \\ &\quad - (2l+1)[c^2(l0,l0)F^2(l,l) + c^4(l0,l0)F^4(l,l) + \cdots]. \quad (21.3\text{-}30) \end{aligned}$$

In calculating the relative spacing of energy levels the closed shell contribution can be disregarded. Spacings between levels in the configuration $l^{2(2l+1)}l'^x$ are the same as those in the configuration l'^x.

When a configuration contains both equivalent and nonequivalent electrons it is necessary first of all to find the possible terms for the group of equivalent electrons and then, using the rules for the addition of angular momentum, couple to this group as a whole the remaining electrons of the configuration. Thus to obtain the terms from p^{2}d we have, first for p^2, the terms ^{1}S, ^{1}D, ^{3}P; when the d electron is coupled to these terms one obtains ^{1}S + d → ^{2}D; ^{1}D + d → ^{2}G, ^{2}F, ^{2}D, ^{2}P, ^{2}S; ^{3}P + d → ^{2}F, ^{2}D, ^{2}P, ^{4}F, ^{4}D, ^{4}P. If the configuration contains two groups of equivalent electrons, it is necessary first to find the terms of each group separately and then to find the terms of the overall configuration by the rule for addition of angular momentum.

Hund's rules which were stated in Section 21.1 apply to the configuration l^n provided LS coupling is valid.

21.4 Spin-Orbit Interaction

The spin-orbit interaction for an N-electron system

$$\mathscr{H}_{\mathrm{SO}} = \sum_i^N \xi(r_i)\mathbf{l}_i \cdot \mathbf{s}_i \quad (21.4\text{-}1)$$

is merely an extension of (16.3-3), the spin-orbit interaction for a single electron. Assuming LS coupling which implies that the electronic repulsion energy is much greater than the spin-orbit energy, it will be supposed that we have the functions $|LSM_LM_S\rangle$, that is, the wave functions for the various terms whose energies differ from one another as a consequence of the Coulomb repulsion among the electrons. As in the one-electron case (Section 16.3), it will be convenient to construct the operator

$$\mathbf{J} = \mathbf{L} + \mathbf{S}$$

to permit a change of representation from $|LSM_LM_S\rangle$ to $|LSJM_J\rangle$ with $J = L+S, L+S-1, \ldots, |L-S|$. Thus for ^{3}P, $L = 1$ and $S = 1$; hence the possible values of J are 2, 1, 0. This is designated by writing the value of J as a subscript: ${}^3P_2$, ${}^3P_1$, ${}^3P_0$, or, in general, ${}^{2S+1}L_J$. If it is necessary to designate M_J, we might write, for example, for ${}^3P_2$ with $M_J = 0$,

$$|LSJM_J\rangle = |1120\rangle = |{}^3P_2 0\rangle.$$

Having constructed the wave functions $|LSJM_J\rangle$ using the standard rules for the coupling of angular momenta we now wish to obtain the energies associated with the spin-orbit interaction. This entails evaluation of matrix elements of the operator (21.4-1). For this purpose we invoke (6.3-37):

$$\langle j_1 j_2 jm|\mathbf{T}^{(k)} \cdot \mathbf{U}^{(k)}|j_1' j_2' j' m'\rangle$$
$$= (-1)^{j+j_2+j_1'} \delta_{jj'} \delta_{mm'} \begin{Bmatrix} j_1' & j_2' & j \\ j_2 & j_1 & k \end{Bmatrix} \langle j_1 \| \mathbf{T}^{k} \| j_1' \rangle \langle j_2 \| \mathbf{U}^{(k)} \| j_2' \rangle \tag{21.4-2}$$

which, with a change in notation,

$$\begin{aligned} j_1 &= L, & j_1' &= L', \\ j_2 &= S, & j_2' &= S', \\ j &= J, & j' &= J', \\ m &= M_J, & m' &= M_J', \\ \mathbf{T}^{(k)} &= \mathbf{l}_i, & \mathbf{U}^{(k)} &= \mathbf{s}_i, \end{aligned}$$

and $k = 1$, becomes

$$\langle LSJM_J|\mathbf{l}_i \cdot \mathbf{s}_i|L'S'J'M_J'\rangle$$
$$= (-1)^{J+S+L'} \delta(J, J')\delta(M_J, M_J') \begin{Bmatrix} L' & S' & J \\ S & L & 1 \end{Bmatrix} \langle L \| \mathbf{l}_i \| L' \rangle \langle S \| \mathbf{s}_i \| S' \rangle. \tag{21.4-3}$$

The $6j$ symbol vanishes unless the triangle conditions $\triangle(L'L1)$ and $\triangle(S'S1)$ are satisfied. Hence the matrix element of $\mathbf{l}_i \cdot \mathbf{s}_i$ vanishes unless

$$\begin{aligned} \Delta S &= 0, \pm 1, & S' + S &\geqslant 1, \\ \Delta L &= 0, \pm 1, & L' + L &\geqslant 1, \end{aligned} \tag{21.4-4}$$

and

$$\Delta J = 0, \qquad \Delta M_J = 0. \tag{21.4-5}$$

It is important to note that the spin-orbit interaction has nonvanishing matrix elements between states which differ in spin by one unit. Thus, for example, spin-orbit interaction can produce an admixture of singlet and triplet states. The diagonal matrix elements of (21.4-3) are of special importance since they represent the spin-orbit interaction within a particular term which in most cases is well separated from other terms. With $L' = L$ and $S' = S$ the $6j$ symbol has the form (Table 1.7).

$$\begin{Bmatrix} L & S & J \\ S & L & 1 \end{Bmatrix} = (-1)^{L+S+J} \frac{J(J+1) - L(L+1) - S(S+1)}{\sqrt{L(2L+1)(2L+2)S(2S+1)(2S+2)}}. \qquad (21.4\text{-}6)$$

We may therefore write

$$\begin{aligned} &\langle \alpha LSJM_J | \mathscr{H}_{\mathrm{SO}} | \alpha LSJM_J \rangle \\ &= \left\langle \alpha LSJM_J \middle| \sum_i \xi(r_i) \mathbf{l}_i \cdot \mathbf{s}_i \middle| \alpha LSJM_J \right\rangle \\ &= \tfrac{1}{2} A(\alpha LS)[J(J+1) - L(L+1) - S(S+1)] = E(J). \end{aligned} \qquad (21.4\text{-}7)$$

Here α represents the dependence of the wave function on the radial coordinate and $A(\alpha LS)$ is a proportionality factor which depends on α, L, and S but not on J. The reduced matrix elements, phase factors and the dependence on $\xi(r_i)$ are all contained in $A(\alpha LS)$. The virtue in writing the spin-orbit matrix element in this fashion stems from the fact that

$$\begin{aligned} \langle \alpha LSJM_J | \mathbf{L} \cdot \mathbf{S} | \alpha LSJM_J \rangle &= \tfrac{1}{2} \langle \alpha LSJM_J | J^2 - L^2 - S^2 | \alpha LSJM_J \rangle \\ &= \tfrac{1}{2}[J(J+1) - L(L+1) - S(S+1)], \end{aligned} \qquad (21.4\text{-}8)$$

which then permits the identification

$$\langle \alpha LSJM_J | \mathscr{H}_{\mathrm{SO}} | \alpha LSJM_J \rangle = A(\alpha LS) \langle \alpha LSJM_J | \mathbf{L} \cdot \mathbf{S} | \alpha LSJM_J \rangle. \qquad (21.4\text{-}9)$$

Thus within the manifold of states comprising a given LS term the matrix element of $\mathscr{H}_{\mathrm{SO}}$ is proportional to the matrix element of $\mathbf{L} \cdot \mathbf{S}$.

Equation (21.4-7) also contains the Landé interval rule which states that

$$\Delta E(J) \equiv E(J) - E(J-1) = A(\alpha LS) J. \qquad (21.4\text{-}10)$$

Multiplets which obey this rule are said to be *normal.*

Let us now give a more explicit interpretation of the constant $A(\alpha LS)$ for a configuration of equivalent electrons. For this purpose we note that

$$\begin{aligned} &\langle LSM_L M_S | \mathbf{L} \cdot \mathbf{S} | LSM_L M_s \rangle \\ &= \langle LSM_L M_S | -L_{+1} S_{-1} + L_0 S_0 - L_{-1} S_{+1} | LSM_L M_S \rangle \\ &= M_L M_S. \end{aligned} \qquad (21.4\text{-}11a)$$

Also in view of the fact that

$$M_L = \sum_i m_l^{\,i}, \qquad M_S = \sum_i m_s^{\,i}, \qquad (21.4\text{-}11b)$$

the matrix element of $\mathscr{H}_{\text{SO}}$ becomes

$$\begin{aligned}\langle \alpha LSM_LM_S|\mathscr{H}_{\text{SO}}|\alpha LSM_LM_S\rangle &= \left\langle \alpha LSM_LM_S \left| \sum_i \xi(r_i)\mathbf{l}_i \cdot \mathbf{s}_i \right| \alpha LSM_LM_S \right\rangle \\ &= \sum_i \langle LSm_l{}^i m_s{}^i|\mathbf{l}_i \cdot \mathbf{s}_i|LSm_l{}^i m_s{}^i\rangle\langle\xi(r)\rangle_{nl} \\ &= \sum_i \langle LSm_l{}^i m_s{}^i| - l^i_{+1}s^i_{-1} + l_0{}^i s_0{}^i \\ &\qquad - l^i_{-1}s^i_{+1}|LSm_l{}^i m_s{}^i\rangle\langle\xi(r)\rangle_{nl} \\ &= \langle\xi(r)\rangle_{nl} \sum_i m_l{}^i m_s{}^i. \end{aligned} \tag{21.4-12}$$

Here $\langle\xi(r)\rangle_{nl}$ is the expectation value of $\xi(r)$ over the radial part of the wave function which, for equivalent electrons, is the same for all electrons. But since (21.4-9) must also hold in the $|LSM_LM_L\rangle$ representation we have

$$\langle\xi(r)\rangle_{nl} \sum_i m_l{}^i m_s{}^i = A(\alpha LS)M_LM_S$$

or

$$A(\alpha LS) = \langle\xi(r)\rangle_{nl} \sum_i m_l{}^i m_s{}^i/M_LM_S. \tag{21.4-13}$$

In this way we find, for example, that

$$A(\mathrm{p}^2, {}^3\mathrm{P}) = \tfrac{1}{2}\langle\xi(r)\rangle_{\mathrm{p}}, \qquad A(\mathrm{d}^2, {}^3\mathrm{F}) = \tfrac{1}{2}\langle\xi(r)\rangle_{\mathrm{d}}.$$

Since $A(\alpha LS)$ is independent of M_L and M_S, these may be chosen arbitrarily, the most convenient choice being the maximum values of M_L and M_S. For a closed shell, there is, of course no splitting. Multiplet splittings for the configurations p^2 and d^2 are shown in Figs. 21.1 and 21.2.

Returning to (21.4-3) it is possible to evaluate the reduced matrix elements by means of (6.3-39) and (6.3-41). The method will be illustrated by applying it to a pair of equivalent electrons. In this case

$$\langle L\| \mathbf{l}_i \|L'\rangle \equiv \langle l^2L\| \mathbf{l} \|l^2L'\rangle$$

which, according to (6.3-39), is

$$\begin{aligned}\langle l^2L\| \mathbf{l} \|l^2L'\rangle &= (-1)^{2l+L'+1}\sqrt{(2L+1)(2L'+1)} \\ &\quad \times \begin{Bmatrix} L & 1 & L' \\ l & l & l \end{Bmatrix} \langle l\| \mathbf{l} \|l\rangle. \end{aligned} \tag{21.4-14}$$

But, from (6.3-25),

$$\langle l\| \mathbf{l} \|l\rangle = \sqrt{(2l+1)(l+1)l} \tag{21.4-15}$$

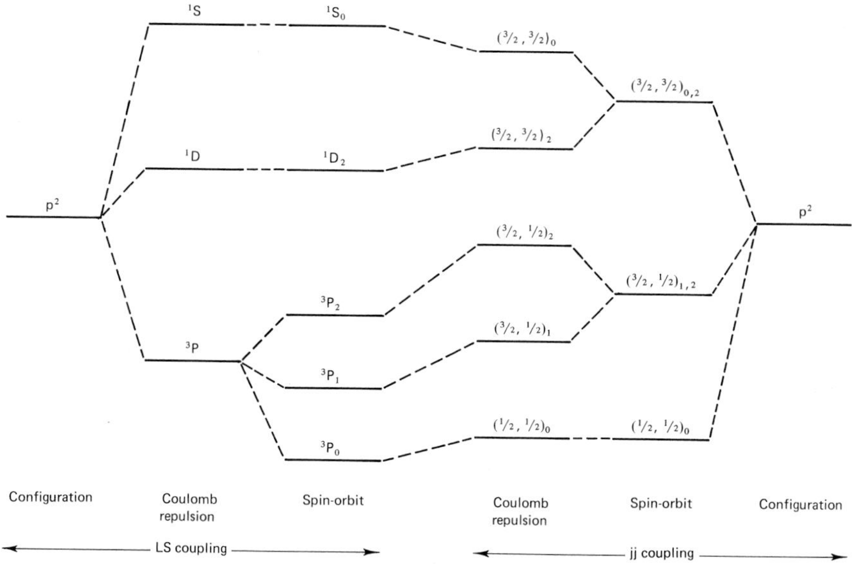

FIG. 21.1 Correlation between *LS* and *jj* coupling schemes.

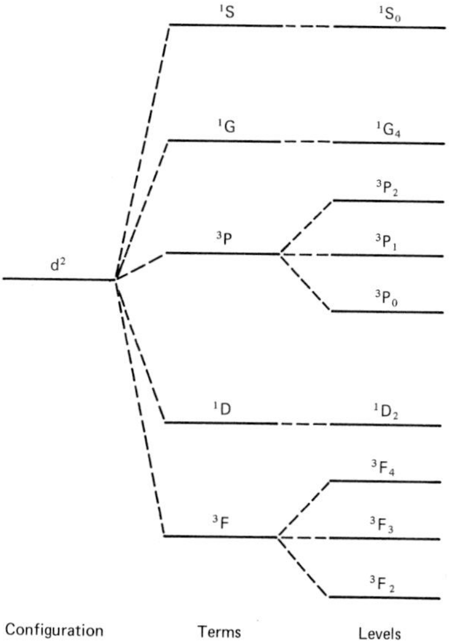

FIG. 21.2 Splitting of d^2 configuration by the nonspherical portion of the Coulomb repulsion (terms) and the spin-orbit interaction (levels).

and for the second matrix element we have similarly from (6.3-41)

$$\begin{aligned}\langle S\|\mathbf{s}_i\|S'\rangle &= \langle s^2S\|\mathbf{s}\|s^2S'\rangle \\ &= (-1)^{2s+S+1}\sqrt{(2S+1)(2S'+1)}\begin{Bmatrix} S & 1 & S' \\ s & s & s \end{Bmatrix}\langle s\|\mathbf{s}\|s\rangle \end{aligned} \tag{21.4-16}$$

with

$$\langle s\|\mathbf{s}\|s\rangle = \sqrt{(2s+1)(s+1)s}. \tag{21.4-17}$$

These expressions may be substituted into (21.4-3) remembering that $s = \frac{1}{2}$. The result is

$$\begin{aligned}&\langle l^2LSJM_J|\mathbf{l}\cdot\mathbf{s}|l^2L'S'JM_J\rangle \\ &\quad = (-1)^{J+S'+S+1}\sqrt{\tfrac{3}{2}} \\ &\quad\quad \times \sqrt{(2l+1)(l+1)l}\sqrt{(2L+1)(2L'+1)}\sqrt{(2S+1)(2S'+1)} \\ &\quad\quad \times \begin{Bmatrix} L' & S' & J \\ S & L & 1 \end{Bmatrix}\begin{Bmatrix} L & 1 & L' \\ l & l & l \end{Bmatrix}\begin{Bmatrix} S & 1 & S' \\ \frac{1}{2} & \frac{1}{2} & \frac{1}{2} \end{Bmatrix}. \end{aligned} \tag{21.4-18}$$

Finally we sum over the two electrons and include the integral over the radial part of the wave function to obtain the final result

$$\begin{aligned}&\langle l^2LSJM_J|\mathscr{H}_{\mathrm{SO}}|l^2L'S'JM_J\rangle \\ &\quad = \sqrt{6}(-1)^{J+S'+S+1} \\ &\quad\quad \times \langle\xi(r)\rangle_{nl}\sqrt{(2l+1)(l+1)l(2L+1)(2L'+1)(2S+1)(2S'+1)} \\ &\quad\quad \times \begin{Bmatrix} L' & S' & J \\ S & L & 1 \end{Bmatrix}\begin{Bmatrix} L & 1 & L' \\ l & l & l \end{Bmatrix}\begin{Bmatrix} S & 1 & S' \\ \frac{1}{2} & \frac{1}{2} & \frac{1}{2} \end{Bmatrix}. \end{aligned} \tag{21.4-19}$$

The phase factor may be written in various ways in view of the fact that both $L + S$ and $L' + S'$ are even.

The generalization to n equivalent electrons proceeds along analogous lines. For the diagonal elements, based on (21.4-3), the one-electron matrix element is

$$\langle\alpha lsjm|\xi(r)\mathbf{l}\cdot\mathbf{s}|\alpha lsjm\rangle = (-1)^{j+l+s}\langle\xi(r)\rangle_{nl}\begin{Bmatrix} l & s & j \\ s & l & 1 \end{Bmatrix}\langle l\|\mathbf{l}\|l\rangle\langle s\|\mathbf{s}\|s\rangle. \tag{21.4-20}$$

It is convenient to introduce the operator $\mathbf{v}^{(11)}$ defined by

$$\langle ls\|\mathbf{v}^{(11)}\|ls\rangle \equiv \langle l\|\mathbf{u}^{(1)}\|l\rangle\langle s\|\mathbf{s}\|s\rangle \tag{21.4-21}$$

in which the unit operator $\mathbf{u}^{(1)}$ has been defined as in (21.2-13). But, from (21.4-15),

$$\frac{\langle l\|\mathbf{l}\|l\rangle}{\sqrt{(2l+1)(l+1)l}} = 1 = \langle l\|\mathbf{u}^{(1)}\|l\rangle. \tag{21.4-22}$$

Hence (21.4-20) may be written in terms of $\mathbf{v}^{(11)}$:

$$\langle \alpha lsjm|\xi(r)\mathbf{l}\cdot\mathbf{s}|\alpha lsjm\rangle$$

$$= (-1)^{j+l+s}\langle\xi(r)\rangle_{nl}\begin{Bmatrix} l & s & j \\ s & l & 1 \end{Bmatrix}\langle ls\| \mathbf{v}^{(11)}\| ls\rangle\sqrt{(2l+1)(l+1)l}. \tag{21.4-23}$$

For n equivalent electrons the matrix element of the spin-orbit interaction is expressible in terms of $\mathbf{V}^{(11)}$ where

$$\mathbf{V}^{(11)} = \sum_i \mathbf{v}_i^{(11)} \tag{21.4-24}$$

which is a special case of the general tensor operator $\mathbf{R}^{(kl)}$ defined by (6.3-42). The result is (Sobelman, 1972)

$$\left\langle l^n\alpha LSJM_J\left|\sum_i \xi(r_i)\mathbf{l}_i\cdot\mathbf{s}_i\right| l^n\alpha' L'S'J'M_J{}'\right\rangle$$

$$= (-1)^{L+S'+J}\langle\xi(r)\rangle_{nl}\sqrt{(2l+1)(l+1)l}\;\delta_{JJ'}\,\delta_{M_JM_J'}$$

$$\times\begin{Bmatrix} L & S & J \\ S' & L' & 1 \end{Bmatrix}\langle l^n\alpha LS\|\mathbf{V}^{(11)}\|l^n\alpha' L'S'\rangle. \tag{21.4-25}$$

Tables 21.2 and 21.3 give reduced matrix elements of $\mathbf{V}^{(11)}$; more extensive tabulations are to be found in the books by Slater (1960), Sobelman (1972), and Nielson and Koster (1963).

TABLE 21.2

Matrices of $\mathbf{V}^{(11)}$ *for* p *Electrons*

$\langle p^2LS\|\mathbf{V}^{(11)}\|p^2L'S'\rangle$

p^2	^1S	^3P	^1D
^1S	0	1	0
^3P	1	$(\frac{3}{2})^{1/2}$	$-\frac{1}{2}(5)^{1/2}$
^1D	0	$-\frac{1}{2}(5)^{1/2}$	0

$\langle p^3LS\|\mathbf{V}^{(11)}\|p^3L'S'\rangle$

p^3	^4S	^2P	^2D
^4S	0	$(2)^{1/2}$	0
^2P	$(2)^{1/2}$	0	$(\frac{5}{2})^{1/2}$
^2D	0	$-(\frac{5}{2})^{1/2}$	0

Equation (21.4-25) includes the diagonal matrix elements, that is, the spin-orbit interaction within a given term with fixed values of L and S. Specifically, the diagonal elements are

$$\left\langle l^n\alpha LSJM_J \left| \sum_i \xi(r_i)\mathbf{l}_i \cdot \mathbf{s}_i \right| l^n\alpha LSJM_J \right\rangle$$

$$= (-1)^{L+S+J}\langle\xi(r)\rangle_{nl}\sqrt{(2l+1)(l+1)l}\begin{Bmatrix} L & S & J \\ S & L & 1 \end{Bmatrix}$$

$$\times \langle l^n\alpha LS\| \mathbf{V}^{(11)} \| l^n\alpha LS\rangle \qquad (21.4\text{-}26a)$$

$$= \tfrac{1}{2}A(l^n\alpha LS)[J(J+1) - L(L+1) - S(S+1)] \qquad (21.4\text{-}26b)$$

in which, using (21.4-6),

$$A(l^n\alpha LS) = \langle\xi(r)\rangle_{nl}\sqrt{\frac{(2l+1)(l+1)l}{L(2L+1)(L+1)S(2S+1)(S+1)}}$$

$$\times \langle l^n\alpha LS\| \mathbf{V}^{(11)} \| l^n\alpha LS\rangle. \qquad (21.4\text{-}27)$$

The Landé interval rule (21.4-10) is seen to be also contained in (21.4-26b). Since the present development is based on LS coupling it should be kept in mind that the Landé interval rule holds only to the degree that the LS coupling condition is satisfied.

TABLE 21.3

Matrices of $\mathbf{V}^{(11)}$ *for* d *Electrons*

$\langle d^2LS\|\mathbf{V}^{(11)}\|d^2L'S'\rangle$					
d^2	^1S	^3P	^1D	^3F	^1G
^1S	0	$\left(\frac{3}{5}\right)^{1/2}$	0	0	0
^3P	$\left(\frac{3}{5}\right)^{1/2}$	$\left(\frac{3}{10}\right)^{1/2}$	$-\left(\frac{21}{20}\right)^{1/2}$	0	0
^1D	0	$-\left(\frac{21}{20}\right)^{1/2}$	0	$\left(\frac{6}{5}\right)^{1/2}$	0
^3F	0	0	$\left(\frac{6}{5}\right)^{1/2}$	$\left(\frac{21}{5}\right)^{1/2}$	$-\left(\frac{9}{10}\right)^{1/2}$
^1G	0	0	0	$-\left(\frac{9}{10}\right)^{1/2}$	0

TABLE 21.3 *(continued)*

d^3	${}^2_3P$	${}^4_3P$	${}^2_1D$	${}^2_3D$	${}^2_3F$	${}^4_3F$	${}^2_3G$	${}^2_3H$
	$\langle d^3 vLS \Vert \sqrt{30}\mathbf{V}^{(11)} \Vert d^3 v'L's' \rangle$							
${}^2_3P$	2	$-2\,(14)^{1/2}$	$-\frac{1}{2}(42)^{1/2}$	$\frac{9}{2}(2)^{1/2}$	0	0	0	0
${}^4_3P$	$2\,(14)^{1/2}$	$(10)^{1/2}$	$-4\,(3)^{1/2}$	0	0	0	0	0
${}^2_1D$	$\frac{1}{2}(42)^{1/2}$	$-4\,(3)^{1/2}$	$\frac{3}{2}(5)^{1/2}$	$-\frac{1}{2}(105)^{1/2}$	$(42)^{1/2}$	$-(42)^{1/2}$	0	0
${}^2_3D$	$-\frac{9}{2}(2)^{1/2}$	0	$-\frac{1}{2}(105)^{1/2}$	$-\frac{1}{2}(5)^{1/2}$	$(2)^{1/2}$	$5\,(2)^{1/2}$	0	0
${}^2_3F$	0	0	$-(42)^{1/2}$	$-(2)^{1/2}$	$-\frac{1}{2}(14)^{1/2}$	$-(14)^{1/2}$	$-\frac{3}{2}(10)^{1/2}$	0
${}^4_3F$	0	0	$-(42)^{1/2}$	$5\,(2)^{1/2}$	$(14)^{1/2}$	$2\,(35)^{1/2}$	$-3\,(10)^{1/2}$	0
${}^2_3G$	0	0	0	0	$\frac{3}{2}(10)^{1/2}$	$-3\,(10)^{1/2}$	$\frac{9}{10}(30)^{1/2}$	$\frac{6}{5}(55)^{1/2}$
${}^2_3H$	0	0	0	0	0	0	$-\frac{6}{5}(55)^{1/2}$	$\frac{3}{5}(55)^{1/2}$

TABLE 21.3 (continued)

$\langle d^4 v L S \| \sqrt{30}$

d^4	${}^1_0S$	${}^1_4S$	${}^3_2P$	${}^3_4P$	${}^1_2D$	${}^1_4D$	${}^3_4D$	${}^5_4D$
${}^1_0S$	0	0	$3\,(3)^{1/2}$	0	0	0	0	0
${}^1_4S$	0	0	$-(7)^{1/2}$	$2\,(2)^{1/2}$	0	0	0	0
${}^3_2P$	$3\,(3)^{1/2}$	$-(7)^{1/2}$	1	$-2\,(14)^{1/2}$	$-\frac{1}{2}(14)^{1/2}$	$2\,(7)^{1/2}$	0	$-4\,(5)^{1/2}$
${}^3_4P$	0	$2\,(2)^{1/2}$	$-2\,(14)^{1/2}$	2	2	$\frac{1}{2}(2)^{1/2}$	$\frac{9}{2}(2)^{1/2}$	$\frac{1}{2}(70)^{1/2}$
${}^1_2D$	0	0	$-\frac{1}{2}(14)^{1/2}$	2	0	0	$2\,(10)^{1/2}$	0
${}^1_4D$	0	0	$2\,(7)^{1/2}$	$\frac{1}{2}(2)^{1/2}$	0	0	$-(5)^{1/2}$	0
${}^3_4D$	0	0	0	$-\frac{9}{2}(2)^{1/2}$	$-2\,(10)^{1/2}$	$(5)^{1/2}$	$-\frac{5}{2}(5)^{1/2}$	$\frac{5}{2}(7)^{1/2}$
${}^5_4D$	0	0	$-4\,(5)^{1/2}$	$\frac{1}{2}(70)^{1/2}$	0	0	$-\frac{5}{2}(7)^{1/2}$	$\frac{15}{2}$
${}^1_4F$	0	0	0	0	0	0	$-2\,(5)^{1/2}$	0
${}^3_2F$	0	0	0	0	2	$(2)^{1/2}$	$-5\,(2)^{1/2}$	$-(70)^{1/2}$
${}^3_4F$	0	0	0	0	4	$-4\,(2)^{1/2}$	$-(2)^{1/2}$	$(70)^{1/2}$
${}^1_2G$	0	0	0	0	0	0	0	0
${}^1_4G$	0	0	0	0	0	0	0	0
${}^3_4G$	0	0	0	0	0	0	0	0
${}^3_4H$	0	0	0	0	0	0	0	0
${}^1_4I$	0	0	0	0	0	0	0	0

$^{11)}\|d^4 v' L' S'\rangle$

${}^1_4F$	${}^3_2F$	${}^3_4F$	${}^1_2G$	${}^1_4G$	${}^3_4G$	${}^3_4H$	${}^1_4I$
0	0	0	0	0	0	0	0
0	0	0	0	0	0	0	0
0	0	0	0	0	0	0	0
0	0	0	0	0	0	0	0
0	2	4	0	0	0	0	0
0	$(2)^{1/2}$	$-4\,(2)^{1/2}$	0	0	0	0	0
$-2\,(5)^{1/2}$	$5\,(2)^{1/2}$	$(2)^{1/2}$	0	0	0	0	0
0	$-(70)^{1/2}$	$(70)^{1/2}$	0	0	0	0	0
0	$(35)^{1/2}$	$\frac{1}{2}(35)^{1/2}$	0	0	$-\frac{9}{2}$	0	0
$-(35)^{1/2}$	$(14)^{1/2}$	$-(14)^{1/2}$	$-(3)^{1/2}$	$(33)^{1/2}$	$3\,(10)^{1/2}$	0	0
$\frac{1}{2}(35)^{1/2}$	$-(14)^{1/2}$	$-\frac{1}{2}(14)^{1/2}$	$5\,(3)^{1/2}$	$-\frac{1}{2}(33)^{1/2}$	$-\frac{3}{2}(10)^{1/2}$	0	0
0	$-(3)^{1/2}$	$5\,(3)^{1/2}$	0	0	3	$-(66)^{1/2}$	0
0	$(33)^{1/2}$	$-\frac{1}{2}(33)^{1/2}$	0	0	$\frac{3}{2}(11)^{1/2}$	$2\,(6)^{1/2}$	0
$-\frac{9}{2}$	$-3\,(10)^{1/2}$	$\frac{3}{2}(10)^{1/2}$	-3	$-\frac{3}{2}(11)^{1/2}$	$\frac{9}{10}(30)^{1/2}$	$\frac{6}{5}(55)^{1/2}$	0
0	0	0	$-(66)^{1/2}$	$2\,(6)^{1/2}$	$-\frac{6}{2}(55)^{1/2}$	$\frac{3}{5}(55)^{1/2}$	$-\frac{3}{2}(26)$
0	0	0	0	0	0	$-\frac{3}{5}(26)^{1/2}$	0

TABLE 21.3 (continued)

$\langle d^5 vLS \| \sqrt{30}\mathbf{V}^{(11}$

d^5	${}^{2}_{5}S$	${}^{6}_{5}S$	${}^{2}_{3}P$	${}^{4}_{3}P$	${}^{2}_{1}D$	${}^{2}_{3}D$	${}^{2}_{5}D$	${}^{4}_{5}D$
${}^{2}_{5}S$	0	0	-4	$(14)^{1/2}$	0	0	0	0
${}^{6}_{5}S$	0	0	0	$3\,(10)^{1/2}$	0	0	0	0
${}^{2}_{3}P$	4	0	0	0	$-(14)^{1/2}$	0	1	$2\,(2)^{1/2}$
${}^{4}_{3}P$	$(14)^{1/2}$	$3\,(10)^{1/2}$	0	0	-8	0	$-2\,(14)^{1/2}$	$-(70)^{1/2}$
${}^{2}_{1}D$	0	0	$(14)^{1/2}$	-8	0	$-(35)^{1/2}$	0	0
${}^{2}_{3}D$	0	0	0	0	$-(35)^{1/2}$	0	$(10)^{1/2}$	$-4\,(5)^{1/2}$
${}^{2}_{5}D$	0	0	-1	$-2\,(14)^{1/2}$	0	$(10)^{1/2}$	0	0
${}^{4}_{5}D$	0	0	$2\,(2)^{1/2}$	$(70)^{1/2}$	0	$4\,(5)^{1/2}$	0	0
${}^{2}_{3}F$	0	0	0	0	$-2\,(14)^{1/2}$	0	-8	$4\,(2)^{1/2}$
${}^{2}_{5}F$	0	0	0	0	0	$2\,(10)^{1/2}$	0	0
${}^{4}_{3}F$	0	0	0	0	$-2\,(14)^{1/2}$	0	-2	$4\,(5)^{1/2}$
${}^{2}_{3}G$	0	0	0	0	0	0	0	0
${}^{2}_{5}G$	0	0	0	0	0	0	0	0
${}^{4}_{5}G$	0	0	0	0	0	0	0	0
${}^{2}_{3}H$	0	0	0	0	0	0	0	0
${}^{2}_{5}I$	0	0	0	0	0	0	0	0

21.5 Conjugate Configurations

A configuration arising from a completely filled shell with n electrons missing, i.e., n holes, is said to be conjugate to the configuration in which the shell contains n electrons. Thus the configurations l^n and $l^{2(2l+1)-n}$ are conjugate configurations as, for example, p and p^5 or d^2 and d^8, etc. In other words we may speak of configurations of electrons or holes in which the holes are the number of electrons missing from a closed shell. We list a number of properties of conjugate configurations.

(1) The terms arising from a configuration l^n are the same as those arising from l^{4l+2-n}. For example, p and p^5 both produce ${}^2P$; p^2 and p^4 both produce ${}^1S$, ${}^1D$, ${}^3P$; the terms from d^3 and d^7 are the same, etc.

$d^5 v' L' S' \rangle$

${}^2_3F$	${}^2_5F$	${}^4_5F$	${}^2_3G$	${}^2_5G$	${}^4_5G$	${}^2_3H$	${}^2_5I$
0	0	0	0	0	0	0	0
0	0	0	0	0	0	0	0
0	0	0	0	0	0	0	0
0	0	0	0	0	0	0	0
$2\,(14)^{1/2}$	0	$-2\,(14)^{1/2}$	0	0	0	0	0
0	$-2\,(10)^{1/2}$	0	0	0	0	0	0
8	0	-2	0	0	0	0	0
$4\,(2)^{1/2}$	0	$-4\,(5)^{1/2}$	0	0	0	0	0
0	$-(70)^{1/2}$	0	0	$-\frac{1}{2}(66)^{1/2}$	$5\,(6)^{1/2}$	0	0
$-\frac{1}{2}(70)^{1/2}$	0	$-(70)^{1/2}$	$\frac{9}{2}(2)^{1/2}$	0	0	0	0
0	$(70)^{1/2}$	0	0	$-(66)^{1/2}$	$(15)^{1/2}$	0	0
0	$-\frac{3}{2}(2)^{1/2}$	0	0	$-\frac{3}{2}(22)^{1/2}$	$-3\,(2)^{1/2}$	0	0
$\frac{1}{2}(66)^{1/2}$	0	$-(66)^{1/2}$	$-\frac{3}{2}(22)^{1/2}$	0	0	$-4\,(3)^{1/2}$	0
$5\,(6)^{1/2}$	0	$2\,(15)^{1/2}$	$3\,(2)^{1/2}$	0	0	$-2\,(33)^{1/2}$	0
0	0	0	0	$4\,(3)^{1/2}$	$-2\,(33)^{1/2}$	0	$-3\,(13)^{1/2}$
0	0	0	0	0	0	$3\,(13)^{1/2}$	0

(2) The electrostatic energy matrices for conjugate configurations are identical except for the addition of a constant diagonal term. Hence the level separations for n holes will be the same as for n electrons. The constant term is due to the overall number of electrons and is different in the two cases; this is evident from (21.3-23).

(3) Within a given LS term the reduced matrix elements of $\mathbf{V}^{(11)}$ for conjugate configurations are related by

$$\langle l^n \alpha LS \| \mathbf{V}^{(11)} \| l^n \alpha LS \rangle = -\langle l^{4l+2-n} \alpha LS \| \mathbf{V}^{(11)} \| l^{4l+2-n} \alpha LS \rangle. \quad (21.5\text{-}1)$$

For a configuration l^n, the (diagonal) reduced matrix element of $\mathbf{V}^{(11)}$ is positive for $n < 2l + 1$, negative for $n > 2l + 1$, and zero for $n = 2l + 1$. It is customary to define a *regular* multiplet as one in which the energy increases with increasing values of J and an *inverted* multiplet as one in which the energy increases with decreasing values of J. Since the Landé rule is based on (21.4-26b) and $A(l^n\alpha LS)$ is proportional to $\langle l^n\alpha LS\| \mathbf{V}^{(11)} \| l^n\alpha LS\rangle$, we have the correspondence:

Multiplet	$A(l^n\alpha LS)$	Shell
Regular	>0	less than half-full, $n < 2l + 1$
Inverted	<0	more than half-full, $n > 2l + 1$
No splitting	0	half-full, $n = 2l + 1$

The absence of any multiplet splitting in a shell that is half-full is correct to first order in the spin-orbit interaction, that is, within an LS term. In higher order, when spin-orbit interaction *between* terms is taken into account, a half-filled shell will generally exhibit some multiplet splitting.

21.6 Other Interactions

In an external magnetic field the interaction Hamiltonian for an N-electron atom is of the same form as (17.1-25):

$$\mathscr{H}_{\mathrm{m}} = -\boldsymbol{\mu}_J \cdot \mathbf{B} \tag{21.6-1}$$

where

$$\boldsymbol{\mu}_J = -\beta g_J \mathbf{J} \tag{21.6-2}$$

and

$$g_J = 1 + \frac{J(J+1) + S(S+1) - L(L+1)}{2J(J+1)} \tag{21.6-3}$$

is the Landé g factor. The effect of the field is to remove the M_J degeneracy so that a level with a given J value splits into $2J + 1$ components. When the magnetic interaction is small compared to the spin-orbit interaction ("weak" fields), the splitting

$$\begin{aligned} E_{\mathrm{m}} &= \langle \alpha JM | \mathscr{H}_{\mathrm{m}} | \alpha' J' M' \rangle \\ &= \beta g_J M_J B\, \delta_{\alpha\alpha'}\, \delta_{JJ'}\, \delta_{MM'} \end{aligned} \tag{21.6-4}$$

is linear in B and symmetrical about the unsplit level. In "strong" fields ($E_{\mathrm{m}} \gg A\mathbf{L} \cdot \mathbf{S}$) the splitting is

$$E_{\mathrm{m}} = \beta B(M_L + 2M_S). \tag{21.6-5}$$

This is the Paschen–Back region. The splitting is also linear in B but degeneracies may occur because it is possible for more than one combination of

M_L and M_S to give the same value of $M_L + 2M_S$. At intermediate fields both spin-orbit and magnetic field interactions must be taken into account at the same time resulting in a nonlinear dependence of the splitting on the magnetic field.

For an N-electron system in an external electric field the Hamiltonian, by extension of (17.2-1), is

$$\mathscr{H}_s = eE \sum_i^N z_i = -e D_z \tag{21.6-6}$$

in which $D_z = -e \sum z_i$ and it is assumed that the z axis of the coordinate system has been aligned with the electric field.

The operator D_z has matrix elements only between states of opposite parity. Hence it has no diagonal elements, and there is no first-order Stark splitting if states with opposite parity are well separated in energy. It is then necessary to go at least to second order in perturbation theory which gives

$$\Delta E(\alpha JM) = E^2 \sum_{\alpha' J'} \frac{|\langle \alpha JM | D_z | \alpha' J' M' \rangle|^2}{E(\alpha J) - E(\alpha' J')} \tag{21.6-7}$$

in which the sum is taken over all states whose parity is opposite to $|\alpha JM\rangle$. Thus, for all atoms except hydrogen which has an accidental degeneracy (Section 17.2), the Stark splitting produced by a relatively weak electric field is proportional to the square of the electric field intensity. The matrix element in (21.6-7) may be evaluated by the Wigner–Eckart theorem. Since D_z is a component of an irreducible tensor of rank one (a vector), the appropriate form of the Wigner–Eckart theorem is

$$\langle \alpha JM | T_0^{(1)} | \alpha' J' M' \rangle = (-1)^{J-M} \begin{pmatrix} J & 1 & J' \\ -M & 0 & M' \end{pmatrix} \langle \alpha J \| \mathbf{T}^k \| \alpha' J' \rangle \tag{21.6-8}$$

which leads immediately to the requirement

$$\Delta J = 0, \pm 1; \qquad \Delta M = 0.$$

The $3j$ symbol may be given an explicit form with the help of Table 1.5. It is then found that

$$\Delta E \text{ is proportional to } \begin{cases} J^2 - M^2 & \text{when} \quad J' = J - 1, \\ M^2 & \text{when} \quad J' = J, \\ (J+1)^2 - M^2 & \text{when} \quad J' = J + 1, \end{cases} \tag{21.6-9}$$

plus a constant that depends on J but not on M. It is seen that the dependence of ΔE on M is quadratic; hence the application of an electric field splits J

into components

$$|M| = J, J-1, J-2, \ldots \begin{cases} 0 & \text{if } J \text{ is an integer} \\ \frac{1}{2} & \text{if } J \text{ is a half-integer} \end{cases}$$

Since all levels except $M = 0$ are twofold degenerate with respect to the sign of M there is no splitting when $J = \frac{1}{2}$ and, of course, none when $J = 0$, although all levels are shifted.

For the magnetic hyperfine interaction we may extend (18.1-59) to an N-electron atom:

$$\mathscr{H}_{\mathrm{h}} = 2\beta\gamma\hbar \sum_{i}^{N} \left[\frac{(\mathbf{l}_i - \mathbf{s}_i) \cdot \mathbf{I}}{r_i^{\,3}} + \frac{3(\mathbf{r}_i \cdot \mathbf{s}_i)(\mathbf{r}_i \cdot \mathbf{I})}{r_i^{\,5}} + \frac{8\pi}{3} \delta(\mathbf{r}_i)\mathbf{s}_i \cdot \mathbf{I} \right]. \tag{21.6-10}$$

where r_i is the distance of the ith electron from the nucleus and the summation extends over all electrons; $\mathbf{l}_i$ and $\mathbf{s}_i$ are the orbital and spin angular momentum operators for the ith electron, respectively. If $\mathscr{H}_{\mathrm{d}}$ represents the dipole–dipole part of $\mathscr{H}_{\mathrm{h}}$, it is possible to express the matrix elements of $\mathscr{H}_{\mathrm{h}}$ in the $|JIFM_F\rangle$ representation in a manner very similar to that employed in the one-electron case (Section 18.2). Thus

$$\langle \alpha JIF | \mathscr{H}_{\mathrm{d}} | \alpha JIF \rangle = 2\beta\gamma\hbar \frac{\mu_I}{I} A'(J) \frac{K}{2} \left\langle \frac{1}{r^3} \right\rangle \tag{21.6-11}$$

where

$$K = F(F+1) - I(I+1) - J(J+1)$$

and $A'(J)$ is a constant of proportionality. It is common to define the magnetic hyperfine structure constant a_{d} as

$$a_{\mathrm{d}} = 2\beta\gamma\hbar \frac{\mu_I}{I} A'(J) \left\langle \frac{1}{r^3} \right\rangle \tag{21.6-12}$$

so that

$$\langle \alpha JIF | \mathscr{H}_{\mathrm{d}} | \alpha JIF \rangle = a_{\mathrm{d}} \frac{K}{2}. \tag{21.6-13}$$

For the contact term only s electrons contribute. However, the contribution from an electron with α spin is positive while that from an electron with β spin is negative. Hence a closed shell configuration $(\mathrm{ns})^2$ does not contribute to the contact interaction. Nevertheless there is an important indirect effect known as *core polarization* which gives rise to a large contribution from the Fermi contact term even when the s shells are filled. This comes about as a result of exchange interactions between outer electrons—most often d electrons—and s electrons and depends on whether the spin of an s electron is parallel or antiparallel to the spin of the electronic state associated with the d^n configuration. The effect is to alter slightly the radial charge distribution of s electrons with α spin compared to that of s electrons with β spin. This unbalance in the radial charge distribution of s electrons with opposite spin is sufficient to produce an extremely high magnetic field ($3{-}5 \times 10^5$ G) at the nucleus in a direction parallel to the net 3d spin.

PART V

Electromagnetic Interactions

CHAPTER 22

INTERACTION BETWEEN ATOMS AND RADIATION

22.1 Hamiltonian of the Radiation Field

The transition from classical to quantum mechanics is facilitated if the classical system is first described by a Hamiltonian function. This holds both for particles and for fields. We shall be interested, at first, in a pure radiation field in which **E** and **B** are perpendicular to one another and both are perpendicular to the direction of propagation. Subsequently, the interaction of the field with a physical system will be considered.

The electromagnetic field is describable by a vector potential **A**; a complete description requires the specification of the three components of **A** at each point in space and at each instant of time. Such a description leads to problems of normalization which can be avoided by assuming that the radiation field is enclosed in a cavity. For simplicity the cavity is assumed to be a cube of side L with perfectly conducting walls. Provided L is large compared with the dimensions of any physical system with which the radiation field may interact, physical results will be independent of the size and shape of the cavity.

It is convenient to impose boundary conditions that are periodic at each face of the cube. In the x direction, for example, it will be required that all plane waves satisfy

$$e^{ik_x x} = e^{ik_x(x+L)} \tag{22.1-1}$$

as a result of which

$$k_x = (2\pi/L)N_x \qquad (N_x = 0, \pm 1, \pm 2, \ldots). \tag{22.1-2}$$

Similarly, for propagation in the y and z directions:

$$k_y = (2\pi/L)N_y \qquad (N_y = 0, \pm 1, \pm 2, \ldots). \tag{22.1-3}$$

$$k_z = (2\pi/L)N_z \qquad (N_z = 0, \pm 1, \pm 2, \ldots). \tag{22.1-4}$$

These components define the propagation vector or wave vector

$$\mathbf{k} = (2\pi/L) \qquad (N_x\hat{\mathbf{i}} + N_y\hat{\mathbf{j}} + N_z\hat{\mathbf{k}}) \tag{22.1-5}$$

which may be normalized to give the unit vector

$$\hat{\mathbf{k}} = \mathbf{k}/k \qquad (k = |\mathbf{k}|) \tag{22.1-6}$$

in the direction of propagation. For a plane wave

$$k = \omega_{\mathbf{k}}/c. \tag{22.1-7}$$

The imposition of periodic boundary conditions gives rise to a discrete set of modes in the cavity. Although the number of modes is infinite, it is a denumerable infinity which is mathematically simpler than the continuous infinity of modes existing in free space.

Each set of integers (N_x, N_y, N_z) defines a mode in the cavity (apart from the polarization). The number of modes ΔN contained in an interval specified by ΔN_x, ΔN_y, and ΔN_z is simply the product

$$\Delta N = \Delta N_x \, \Delta N_y \, \Delta N_z = (L/2\pi)^3 \, \Delta k_x \, \Delta k_y \, \Delta k_z. \tag{22.1-8}$$

For a box whose dimensions are large compared to the wavelength of the radiation we may regard the modes as forming a quasi-continuous distribution in which case (22.1-8) may be replaced by

$$\begin{aligned} dN &= (L/2\pi)^3 \, dk_x \, dk_y \, dk_z \\ &= (L/2\pi)^3 k^2 \sin\theta \, dk \, d\theta \, d\varphi \\ &= (L/2\pi)^3 k^2 \, dk \, d\Omega = (V/(2\pi)^3) k^2 \, dk \, d\Omega. \end{aligned} \tag{22.1-9}$$

This gives the number of propagation modes contained in an interval dk and in the direction defined by the element of solid angle $d\Omega$.

We now express the vector potential $\mathbf{A}(\mathbf{r}, t)$ as a linear superposition of plane waves

$$\mathbf{A}(\mathbf{r}, t) = \sum_{\mathbf{k}\lambda} \hat{\mathbf{e}}_{\mathbf{k}\lambda} \{A_{\mathbf{k}\lambda} \exp[i(\mathbf{k} \cdot \mathbf{r} - \omega_{\mathbf{k}} t)] + A^*_{\mathbf{k}\lambda} \exp[-i(\mathbf{k} \cdot \mathbf{r} - \omega_{\mathbf{k}} t)]\}, \tag{22.1-10}$$

where $\hat{\mathbf{e}}_{\mathbf{k}\lambda}$ is a real unit vector which indicates the linear polarization; $\hat{\mathbf{e}}_{\mathbf{k}\lambda}$ depends on the propagation direction $\mathbf{k}$ and has two independent components $\hat{\mathbf{e}}_{\mathbf{k}1}$ and $\hat{\mathbf{e}}_{\mathbf{k}2}$ which satisfy

$$\hat{\mathbf{e}}_{\mathbf{k}\lambda} \cdot \hat{\mathbf{e}}_{\mathbf{k}\lambda'} = \delta_{\lambda\lambda'} \qquad (\lambda, \lambda' = 1, 2). \tag{22.1-11}$$

$A_{\mathbf{k}\lambda}$ is a constant amplitude for the mode $\mathbf{k}\lambda$. The Coulomb gauge

$$\nabla \cdot \mathbf{A} = 0 \tag{22.1-12}$$

ensures the transversality of the electromagnetic fields so that

$$\hat{\mathbf{e}}_{\mathbf{k}\lambda} \cdot \hat{\mathbf{k}} = 0. \tag{22.1-13}$$

Thus $\hat{\mathbf{e}}_{\mathbf{k}1}$, $\hat{\mathbf{e}}_{\mathbf{k}2}$, and $\hat{\mathbf{k}}$ form a right-handed set of mutually orthogonal unit vectors. The vector potential in (22.1-10) then consists of plane waves each one labeled by the propagation vector **k** and the real polarization vector $\hat{\mathbf{e}}_{\mathbf{k}\lambda}$. In place of the single vector index **k** one may also use three scalar indices based on (22.1-5). Since the two sets of terms in (22.1-10) are complex conjugates, $\mathbf{A}(\mathbf{r}, t)$ is real.

We may, if we so desire, replace the linear polarization vectors $\hat{\mathbf{e}}_{\mathbf{k}1}$ and $\hat{\mathbf{e}}_{\mathbf{k}2}$ by unit vectors which indicate circular polarizations. This is accomplished by defining, as in Section 6.1,

$$\hat{\mathbf{e}}_{\mathbf{k}\,+1} = -\frac{1}{\sqrt{2}}(\hat{\mathbf{e}}_{\mathbf{k}1} + i\hat{\mathbf{e}}_{\mathbf{k}2}), \qquad \hat{\mathbf{e}}_{\mathbf{k}\,-1} = \frac{1}{\sqrt{2}}(\hat{\mathbf{e}}_{\mathbf{k}1} - i\hat{\mathbf{e}}_{\mathbf{k}2}). \quad (22.1\text{-}14)$$

These vectors satisfy (7.2-10) which, in the present context, is written

$$\hat{\mathbf{e}}_{\mathbf{k}q}^* \times \hat{\mathbf{e}}_{\mathbf{k}q'} = iq\hat{\mathbf{k}}\,\delta_{qq'} \qquad (q, q' = \pm 1). \quad (22.1\text{-}15)$$

With $q = +1$ the cross product gives a vector parallel to the direction of propagation whereas with $q = -1$ the cross product is antiparallel. For this reason one refers to $\hat{\mathbf{e}}_{\mathbf{k}+1}$ and $\hat{\mathbf{e}}_{\mathbf{k}-1}$ as positive and negative *helicity* (unit) vectors. Also, as shown in Section 7.2, $\hat{\mathbf{e}}_{\mathbf{k}+1}$ and $\hat{\mathbf{e}}_{\mathbf{k}-1}$ represent left and right circular polarization, respectively, according to standard optical convention.

In Section 7.1 we saw that the rotational properties of a vector field are such that one may associate an intrinsic spin $S = 1$ with the field. It was shown that the unit vectors $\hat{\mathbf{e}}_{+1}$, $\hat{\mathbf{e}}_0$ and $\hat{\mathbf{e}}_{-1}$ were simultaneous eigenvectors of S^2 and S_z and that the eigenvalues of S_z were 1, 0, -1. Since the electromagnetic field is a vector field we expect these spin properties to be applicable. Indeed, the helicity vectors $\hat{\mathbf{e}}_{\mathbf{k}\,+1}$ and $\hat{\mathbf{e}}_{\mathbf{k}\,-1}$ are two of the eigenvectors in question for the mode **k** and they define two of the three possible spin states. But what about the third spin state? It is nonexistent because of the transversability condition (22.1-13) which rules out a polarization parallel to the direction of propagation. Hence a radiation field in which **E** and **B** are transverse to the direction of propagation is characterized by two spin states despite the fact that $S = 1$.

The electric and magnetic fields are obtained directly from the vector potential. For the electric field

$$\begin{aligned}\mathbf{E}(\mathbf{r}, t) &= -\frac{1}{c}\frac{\partial}{\partial t}\mathbf{A}(\mathbf{r}, t)\\ &= \frac{i}{c}\sum_{\mathbf{k}\lambda} \omega_{\mathbf{k}}\hat{\mathbf{e}}_{\mathbf{k}\lambda}\{A_{\mathbf{k}\lambda}\exp[i(\mathbf{k}\cdot\mathbf{r} - \omega_{\mathbf{k}}t)] - A_{\mathbf{k}\lambda}^*\exp[-i(\mathbf{k}\cdot\mathbf{r} - \omega_{\mathbf{k}}t)]\}.\end{aligned} \quad (22.1\text{-}16)$$

For the magnetic field it is necessary to evaluate $\nabla \times \mathbf{A}$. Noting that

$$\nabla \times (\varphi\mathbf{A}) = \nabla\varphi \times \mathbf{A} + \varphi\nabla \times \mathbf{A}$$

we have

$$\nabla \times A_{\mathbf{k}\lambda} e^{i\mathbf{k}\cdot\mathbf{r}} \hat{\mathbf{e}}_{\mathbf{k}\lambda} = \nabla A_{\mathbf{k}\lambda} e^{i\mathbf{k}\cdot\mathbf{r}} \times \hat{\mathbf{e}}_{\mathbf{k}\lambda} + A_{\mathbf{k}\lambda} e^{i\mathbf{k}\cdot\mathbf{r}} \nabla \times \hat{\mathbf{e}}_{\mathbf{k}\lambda}.$$

Since both the amplitude $A_{\mathbf{k}\lambda}$ and the polarization vector $\hat{\mathbf{e}}_{\mathbf{k}\lambda}$ are constant, the second term vanishes and

$$\nabla \times A_{\mathbf{k}\lambda} e^{i\mathbf{k}\cdot\mathbf{r}} \hat{\mathbf{e}}_{\mathbf{k}\lambda} = A_{\mathbf{k}\lambda} \nabla e^{i\mathbf{k}\cdot\mathbf{r}} \times \hat{\mathbf{e}}_{\mathbf{k}\lambda} = i\frac{\omega_{\mathbf{k}}}{c} A_{\mathbf{k}\lambda} e^{i\mathbf{k}\cdot\mathbf{r}} \hat{\mathbf{k}} \times \hat{\mathbf{e}}_{\mathbf{k}\lambda}. \qquad (22.1\text{-}17)$$

Hence the magnetic field is

$$\mathbf{B}(\mathbf{r},t) = \nabla \times \mathbf{A}(\mathbf{r},t)$$

$$= \frac{i}{c} \sum_{\mathbf{k}\lambda} \omega_{\mathbf{k}} (\hat{\mathbf{k}} \times \hat{\mathbf{e}}_{\mathbf{k}\lambda}) \{A_{\mathbf{k}\lambda} \exp[i(\mathbf{k}\cdot\mathbf{r} - \omega_{\mathbf{k}} t)] - A^*_{\mathbf{k}\lambda} \exp[-i(\mathbf{k}\cdot\mathbf{r} - \omega_{\mathbf{k}} t)]\}. \qquad (22.1\text{-}18)$$

To obtain the Hamiltonian for the electromagnetic field in the cavity it is necessary to express the field energy W in terms of canonical variables. From electromagnetic theory

$$W = (1/8\pi) \int_V (\mathbf{E}\cdot\mathbf{E} + \mathbf{B}\cdot\mathbf{B})\, dV \qquad (22.1\text{-}19)$$

where V is the volume of the cavity and the fields $\mathbf{E}$ and $\mathbf{B}$ are given by (22.1-16) and (22.1-18). As a result of the boundary conditions of the type (22.1-1),

$$\int_V e^{\pm i\mathbf{k}\cdot\mathbf{r}}\, dV = \begin{cases} 0 & \text{for} \quad \mathbf{k} \neq 0, \\ V & \text{for} \quad \mathbf{k} = 0. \end{cases} \qquad (22.1\text{-}20)$$

Therefore

$$\int_V e^{\pm i(\mathbf{k}+\mathbf{k}')\cdot\mathbf{r}}\, dV = \delta_{\mathbf{k}',-\mathbf{k}} V, \qquad (22.1\text{-}21a)$$

$$\int_V e^{\pm i(\mathbf{k}-\mathbf{k}')}\, dV = \delta_{\mathbf{k}'\mathbf{k}} V. \qquad (22.1\text{-}21b)$$

If we write

$$A_{\mathbf{k}\lambda}(t) = A_{\mathbf{k}\lambda} e^{-i\omega_{\mathbf{k}} t} \qquad (22.1\text{-}22)$$

in (22.1-16) and use the orthogonality condition on the polarization vectors (22.1-11) and the relation $\omega_{\mathbf{k}} = \omega_{-\mathbf{k}}$,

$$\int_V \mathbf{E}\cdot\mathbf{E}\, dV = \frac{2V}{c^2} \sum_{\mathbf{k}\lambda} \omega_{\mathbf{k}}^2 A_{\mathbf{k}\lambda}(t) A^*_{\mathbf{k}\lambda}(t)$$

$$- \frac{V}{c^2} \sum_{\mathbf{k}\lambda\lambda'} \omega_{\mathbf{k}}^2 (\hat{\mathbf{e}}_{\mathbf{k}} \cdot \hat{\mathbf{e}}_{-\mathbf{k}\lambda}) [A_{\mathbf{k}\lambda}(t) A_{-\mathbf{k}\lambda'}(t) + A^*_{\mathbf{k}\lambda}(t) A^*_{-\mathbf{k}\lambda'}(t)]. \qquad (22.1\text{-}23)$$

To compute the field energy associated with $\mathbf{B}$ we employ the vector identity

$$(\mathbf{A} \times \mathbf{B}) \cdot (\mathbf{C} \times \mathbf{D}) = (\mathbf{A} \cdot \mathbf{C})(\mathbf{B} \cdot \mathbf{D}) - (\mathbf{A} \cdot \mathbf{D})(\mathbf{B} \cdot \mathbf{C}) \qquad (22.1\text{-}24a)$$

to show that

$$(\hat{\mathbf{k}} \times \hat{\mathbf{e}}_{\mathbf{k}\lambda}) \cdot (\hat{\mathbf{k}} \times \hat{\mathbf{e}}_{\mathbf{k}\lambda'}) = \delta_{\lambda\lambda'} \qquad (22.1\text{-}24b)$$

in which the orthogonality property (22.1-13) has been used. Also

$$(\hat{\mathbf{k}} \times \hat{\mathbf{e}}_{\mathbf{k}\lambda}) \cdot (-\hat{\mathbf{k}} \times \hat{\mathbf{e}}_{-\mathbf{k}\lambda'}) = -\hat{\mathbf{e}}_{\mathbf{k}\lambda} \cdot \hat{\mathbf{e}}_{-\mathbf{k}\lambda'}. \qquad (22.1\text{-}25)$$

Therefore in computing the integral of $\mathbf{B} \cdot \mathbf{B}$ using (22.1-18) one finds two terms identical to the two terms on the right side of (22.1-23) except for the sign of the second term. The field energy then becomes

$$W = \frac{V}{2\pi c^2} \sum_{\mathbf{k}\lambda} \omega_{\mathbf{k}}{}^2 A_{\mathbf{k}\lambda}(t) A_{\mathbf{k}\lambda}^*(t). \qquad (22.1\text{-}26)$$

Actually, W is independent of time because of the exponential time dependence of $A_{\mathbf{k}\lambda}(t)$ as given by (22.1-22). Hence

$$W = \frac{V}{2\pi c^2} \sum_{\mathbf{k}\lambda} \omega_{\mathbf{k}}{}^2 A_{\mathbf{k}\lambda} A_{\mathbf{k}\lambda}^*. \qquad (22.1\text{-}27)$$

It is observed that the total energy is merely the sum of the energies in the individual modes as a consequence of the orthogonality conditions (22.1-21); furthermore the energy is shared equally by the electric and magnetic fields.

We now introduce a new set of variables

$$A_{\mathbf{k}\lambda} = \frac{c}{\omega_{\mathbf{k}}} \sqrt{\frac{\pi}{V}} (\omega_{\mathbf{k}} Q_{\mathbf{k}\lambda} + iP_{\mathbf{k}\lambda}), \qquad A_{\mathbf{k}\lambda}^* = \frac{c}{\omega_{\mathbf{k}}} \sqrt{\frac{\pi}{V}} (\omega_{\mathbf{k}} Q_{\mathbf{k}\lambda} - iP_{\mathbf{k}\lambda}) \qquad (22.1\text{-}28)$$

which, when substituted into (22.1-27), gives

$$W = \tfrac{1}{2} \sum_{\mathbf{k}\lambda} (P_{\mathbf{k}\lambda}^2 + \omega_{\mathbf{k}}{}^2 Q_{\mathbf{k}\lambda}^2). \qquad (22.1\text{-}29)$$

It will now be shown that $P_{\mathbf{k}\lambda}$ and $Q_{\mathbf{k}\lambda}$ satisfy the Hamilton equations; this will identify these quantities as canonical variables and therefore W in the form (22.1-29) will be interpretable as the Hamiltonian of the electromagnetic field. Upon inverting (22.1-28),

$$Q_{\mathbf{k}\lambda} = \sqrt{\frac{V}{\pi}} \frac{1}{2c} (A_{\mathbf{k}\lambda} + A_{\mathbf{k}\lambda}^*), \qquad P_{\mathbf{k}\lambda} = -i \sqrt{\frac{V}{\pi}} \frac{\omega_{\mathbf{k}}}{2c} (A_{\mathbf{k}\lambda} - A_{\mathbf{k}\lambda}^*). \qquad (22.1\text{-}30)$$

When the time dependence (22.1-22) is included,

$$\dot{Q}_{\mathbf{k}\lambda} = P_{\mathbf{k}\lambda}, \qquad \dot{P}_{\mathbf{k}\lambda} = -\omega_{\mathbf{k}}{}^2 Q_{\mathbf{k}\lambda}. \qquad (22.1\text{-}31)$$

Using (22.1-31) we obtain Hamilton equations:

$$\partial W/\partial Q_{\mathbf{k}\lambda} = \omega^2 Q_{\mathbf{k}\lambda} = -\dot{P}_{\mathbf{k}\lambda}, \qquad \partial W/\partial P_{\mathbf{k}\lambda} = P_{\mathbf{k}\lambda} = \dot{Q}_{\mathbf{k}\lambda}. \quad (22.1\text{-}32)$$

Hence $P_{\mathbf{k}\lambda}$ and $Q_{\mathbf{k}\lambda}$ are canonical variables so that W in (22.1-29) is indeed the Hamiltonian $\mathscr{H}$. It may also be verified that the Maxwell equations

$$\nabla \times \mathbf{E} = -\frac{1}{c}\frac{\partial \mathbf{B}}{\partial t}, \qquad \nabla \times \mathbf{B} = \frac{1}{c}\frac{\partial \mathbf{E}}{\partial t}$$

reduce to the Hamilton equations (22.1-32).

Finally it is noted that the Hamiltonian (22.1-29) is precisely of the same form as that for an assembly of simple harmonic oscillators whose Hamiltonians are given by (10.2-1). Each mode of the radiation field is therefore formally equivalent to a single harmonic oscillator.

22.2 Quantization of the Radiation Field

Having identified the Hamiltonian of each mode of a radiation field with that of a harmonic oscillator we may proceed with the quantization exactly as in the case of the harmonic oscillator (Section 10.2). The transition to quantum mechanics is accomplished by interpreting $Q_{\mathbf{k}\lambda}$ and $P_{\mathbf{k}\lambda}$ as operators that satisfy the commutation relations

$$[Q_{\mathbf{k}\lambda}, P_{\mathbf{k}'\lambda'}] = i\hbar\, \delta_{\mathbf{k}\mathbf{k}'}\, \delta_{\lambda\lambda'}, \qquad [Q_{\mathbf{k}\lambda}, Q_{\mathbf{k}'\lambda'}] = [P_{\mathbf{k}\lambda}, P_{\mathbf{k}'\lambda'}] = 0. \quad (22.2\text{-}1)$$

By analogy with (10.2-3) we define

$$a_{\mathbf{k}\lambda} = \frac{1}{\sqrt{2\hbar\omega_{\mathbf{k}}}}(\omega_{\mathbf{k}} Q_{\mathbf{k}\lambda} + iP_{\mathbf{k}\lambda}), \qquad a_{\mathbf{k}\lambda}^{\dagger} = \frac{1}{\sqrt{2\hbar\omega_{\mathbf{k}}}}(\omega_{\mathbf{k}} Q_{\mathbf{k}\lambda} - iP_{\mathbf{k}\lambda}) \quad (22.2\text{-}2)$$

which then must obey the commutation rules

$$[a_{\mathbf{k}\lambda}, a_{\mathbf{k}'\lambda'}^{\dagger}] = \delta_{\mathbf{k}\mathbf{k}'}\delta_{\lambda\lambda'}, \qquad [a_{\mathbf{k}\lambda}, a_{\mathbf{k}'\lambda'}] = [a_{\mathbf{k}\lambda}^{\dagger}, a_{\mathbf{k}'\lambda'}^{\dagger}] = 0. \quad (22.2\text{-}3)$$

From the discussion in Section 12.1 it is recognized that these relations are characteristic of boson operators.

In terms of these operators the Hamiltonian (22.1-29) becomes

$$\mathscr{H} = \sum_{\mathbf{k}\lambda} \mathscr{H}_{\mathbf{k}\lambda} = \sum_{\mathbf{k}\lambda} \hbar\omega_{\mathbf{k}}(a_{\mathbf{k}\lambda}^{\dagger} a_{\mathbf{k}\lambda} + \tfrac{1}{2}) = \sum_{\mathbf{k}\lambda} \hbar\omega_{\mathbf{k}}(N_{\mathbf{k}\lambda} + \tfrac{1}{2}) \quad (22.2\text{-}4)$$

in which $N_{\mathbf{k}\lambda}$ is known as the *number operator* for the mode $\mathbf{k}$ and polarization λ and is given by

$$N_{\mathbf{k}\lambda} = a_{\mathbf{k}\lambda}^{\dagger} a_{\mathbf{k}\lambda}. \quad (22.2\text{-}5)$$

The eigenvalues of $N_{\mathbf{k}\lambda}$ are $n_{\mathbf{k}\lambda} = 0, 1, 2, \ldots$, that is,

$$N_{\mathbf{k}\lambda}|n_{\mathbf{k}\lambda}\rangle = n_{\mathbf{k}\lambda}|n_{\mathbf{k}\lambda}\rangle \quad (22.2\text{-}6)$$

where $|n_{\mathbf{k}\lambda}\rangle$ represents an eigenstate of $N_{\mathbf{k}\lambda}$. We see then, from (22.2-4), that the possible energies of the system which are the eigenvalues of $\mathscr{H}$ are

$$E = \sum_{\mathbf{k}\lambda} E_{\mathbf{k}\lambda} = \sum_{\mathbf{k}\lambda} \hbar\omega_{\mathbf{k}}(n_{\mathbf{k}\lambda} + \tfrac{1}{2}). \tag{22.2-7}$$

Furthermore, the harmonic oscillator analogy permits us to write

$$a_{\mathbf{k}\lambda}|n_{\mathbf{k}\lambda}\rangle = \sqrt{n_{\mathbf{k}\lambda}}\,|n_{\mathbf{k}\lambda} - 1\rangle, \qquad a^{\dagger}_{\mathbf{k}\lambda}|n_{\mathbf{k}\lambda}\rangle = \sqrt{n_{\mathbf{k}\lambda} + 1}\,|n_{\mathbf{k}\lambda} + 1\rangle, \qquad a_{\mathbf{k}\lambda}|0\rangle = 0, \tag{22.2-8}$$

consistent with (10.2-18), (10.2-19), and (10.2-20).

The transition to quantum mechanics by means of the preceding formalism lends itself to the following interpretation: $a_{\mathbf{k}\lambda}$ and $a^{\dagger}_{\mathbf{k}\lambda}$ are annihilation and creation operators, respectively, for a photon with propagation vector $\mathbf{k}$, polarization vector $\hat{\mathbf{e}}_{\mathbf{k}\lambda}$, frequency $\omega_{\mathbf{k}}$, momentum $\hbar\mathbf{k}$, and energy $\hbar\omega_{\mathbf{k}}$. The quantities $n_{\mathbf{k}\lambda}$, also known as *occupation numbers*, are eigenvalues of the number operator $N_{\mathbf{k}\lambda}$ and give the number of photons in the mode $\mathbf{k}$ and polarization λ. The corresponding oscillator will have a set of levels with an energy difference $\hbar\omega_{\mathbf{k}}$ between adjacent levels (Fig. 22.1).

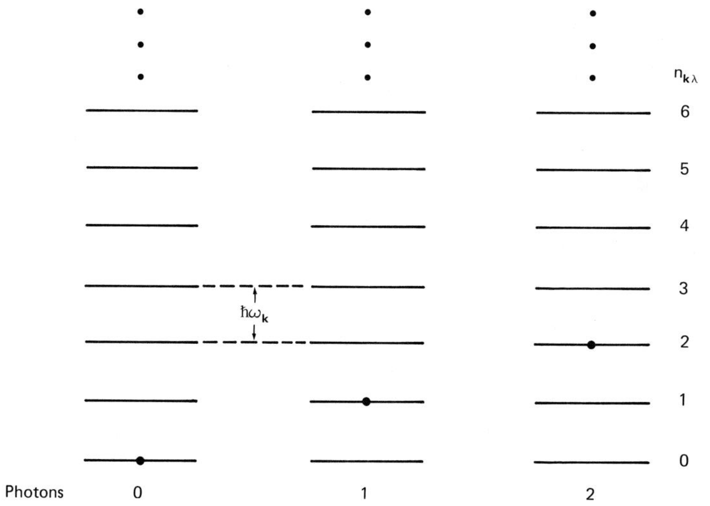

FIG. 22.1 Photon description of a radiation field.

A complete description of the radiation field consists of an enumeration of the occupation numbers $n_{\mathbf{k}\lambda}$. Since each mode is independent the product of all the eigenstates, $|n_{\mathbf{k}\lambda}\rangle$ is an eigenstate of the total field. A many-photon state is therefore described by

$$|n_{\mathbf{k}_1\lambda_1}\rangle|n_{\mathbf{k}_2\lambda_2}\rangle\cdots|n_{\mathbf{k}_i\lambda_i}\rangle\cdots\equiv|n_{\mathbf{k}_1\lambda_1},n_{\mathbf{k}_2\lambda_2},\ldots,n_{\mathbf{k}_i\lambda_i},\ldots\rangle. \quad (22.2\text{-}9)$$

The notation is becoming quite cumbersome; we shall therefore make the replacement

$$n_{\mathbf{k}_i\lambda_i}\equiv n_i \quad (22.2\text{-}10)$$

so that the many-photon state is written as

$$|n_1,n_2,\ldots,n_i,\ldots\rangle. \quad (22.2\text{-}11a)$$

With the orthogonality requirement

$$\langle n_1',n_2',\ldots,n_i',\ldots|n_1,n_2,\ldots,n_i,\ldots\rangle=\delta_{n_1'n_1}\delta_{n_2'n_2}\cdots\delta_{n_i'n_i}\cdots \quad (22.2\text{-}11b)$$

For operators we shall also replace the indices $\mathbf{k}_i\lambda_i$ by the single index i. Thus, the many-photon analog of (22.2-8) is

$$\begin{aligned}
a_i|n_1,n_2,\ldots,n_i,\ldots\rangle&=\sqrt{n_i}|n_1,n_2,\ldots,n_i-1,\ldots\rangle,\\
a_i^\dagger|n_1,n_2,\ldots,n_i,\ldots\rangle&=\sqrt{n_i+1}|n_1,n_2,\ldots,n_i+1,\ldots\rangle, \quad (22.2\text{-}12)\\
a_i|n_1,n_2,\ldots,0_i,\ldots\rangle&=0,
\end{aligned}$$

and the analog of (22.2-6) is

$$N_i|n_1,n_2,\ldots,n_i,\ldots\rangle=n_i|n_1,n_2,\ldots,n_i,\ldots\rangle, \quad (22.2\text{-}13)$$

with

$$E=\sum_i\hbar\omega_i(n_i+\tfrac{1}{2}). \quad (22.2\text{-}14)$$

When $n_{\mathbf{k}\lambda}=0$, there are no photons in the mode $\mathbf{k}\lambda$; nevertheless the energy in the mode, according to (22.2-7), is $\frac{1}{2}\hbar\omega_{\mathbf{k}}$. This value, which is characteristic of the harmonic oscillator, is known as the *zero point energy*. For the radiation field as a whole with an infinite number of modes, the zero point energy becomes infinite. This is one of the peculiarities of the quantum-mechanical description of the radiation field. A formal explanation is based on the noncommutativity of the number operator $N_{\mathbf{k}\lambda}$ with the annihilation and creation operators $a_{\mathbf{k}\lambda}$ and $a_{\mathbf{k}\lambda}^\dagger$ as a result of which $\sum N_{\mathbf{k}\lambda}$ does not commute with the fields $\mathbf{E}$ and $\mathbf{B}$. Since observables associated with noncommuting operators are subject to the uncertainty principle, an increase in precision in the number of photons means an increase in the uncertainty of the fields. When no photons are present, the fluctuations in the field strengths are responsible for the infinite zero point energy. Fortunately, a consistent description of physical processes in practically all cases is obtained by simply ignoring the infinite zero point energy of the radiation field.

The rotational properties of the radiation field are carried along in the process of quantization. In the context of the photon picture we say that the photon has an intrinsic spin $S = 1$. However, the transversality condition limits the values of M_S to ± 1 and does not permit a state with $M_S = 0$. That is to say, if the photon is regarded as a particle, its spin vector can be oriented either parallel or antiparallel to its momentum vector $\hbar\mathbf{k}$. In the former case the photon is said to have positive helicity and in the latter, negative. This result can be shown to be associated with the fact that the photon has zero mass, or alternatively with the fact that photons travel at the velocity of light.

The transition from classical to quantum mechanics embodied in (22.2-1) implies that the classical vector potential $\mathbf{A}(\mathbf{r}, t)$ in (22.1-10) becomes a quantum-mechanical operator. This transition may be accomplished by means of (22.1-30) which relate $A_{\mathbf{k}\lambda}$ and $A^*_{\mathbf{k}\lambda}$ to $Q_{\mathbf{k}\lambda}$ and $P_{\mathbf{k}\lambda}$; the latter in turn are related to the annihilation and creation operators $a_{\mathbf{k}\lambda}$ and $a^\dagger_{\mathbf{k}\lambda}$ via (22.2-2). Hence the required replacement is

$$A_{\mathbf{k}\lambda} \to \sqrt{(2\pi\hbar c^2/V\omega_{\mathbf{k}})}a_{\mathbf{k}\lambda}, \qquad A^*_{\mathbf{k}\lambda} \to \sqrt{(2\pi\hbar c^2/V\omega_{\mathbf{k}})}a^\dagger_{\mathbf{k}\lambda}. \tag{22.2-15}$$

We may now write the quantum-mechanical expression for the vector potential by substituting (22.2-15) into (22.1-10):

$$\mathbf{A}_{\mathrm{H}}(\mathbf{r}, t) = \sum_{\mathbf{k}\lambda} \sqrt{(2\pi\hbar c^2/V\omega_{\mathbf{k}})}\hat{\mathbf{e}}_{\mathbf{k}\lambda}\{a_{\mathbf{k}\lambda} \exp[i(\mathbf{k}\cdot\mathbf{r} - \omega_{\mathbf{k}}t)] + a^\dagger_{\mathbf{k}\lambda} \exp[-i(\mathbf{k}\cdot\mathbf{r} - \omega_{\mathbf{k}}t)]\}. \tag{22.2-16}$$

The subscript H in $\mathbf{A}_{\mathrm{H}}(\mathbf{r}, t)$ is intended to indicate that the operator is in the Heisenberg representation because of its time dependence. In fact, on referring to the harmonic oscillator, we find that according to (10.3-8) the operators $ae^{-i\omega t}$ and $a^\dagger e^{+i\omega t}$ are in the Heisenberg representation while a and $a^\dagger$ are in the Schrödinger representation. Consequently, the vector potential in the Schrödinger representation is

$$\mathbf{A}(\mathbf{r}) = \sum_{\mathbf{k}\lambda} \sqrt{(2\pi\hbar c^2/V\omega_{\mathbf{k}})}\hat{\mathbf{e}}_{\mathbf{k}\lambda}[a_{\mathbf{k}\lambda}e^{i\mathbf{k}\cdot\mathbf{r}} + a^\dagger_{\mathbf{k}\lambda}e^{-i\mathbf{k}\cdot\mathbf{r}}]. \tag{22.2-17}$$

We note that $\mathbf{A}(\mathbf{r})$ is Hermitian although the annihilation and creation operators themselves are not.

The electric and magnetic field operators may also be expressed in terms of $a_{\mathbf{k}\lambda}$ and $a^\dagger_{\mathbf{k}\lambda}$. Using (22.1-16) and (22.2-15),

$$\mathbf{E}(\mathbf{r}) = i \sum_{\mathbf{k}\lambda} \sqrt{(2\pi\hbar\omega_{\mathbf{k}}/V)}\hat{\mathbf{e}}_{\mathbf{k}\lambda}[a_{\mathbf{k}\lambda}e^{i\mathbf{k}\cdot\mathbf{r}} - a^\dagger_{\mathbf{k}\lambda}e^{-i\mathbf{k}\cdot\mathbf{r}}]. \tag{22.2-18}$$

Similarly, with (22.1-18) and (22.2-15) the magnetic field operator is

$$\mathbf{B}(\mathbf{r}) = i \sum_{\mathbf{k}\lambda} \sqrt{(2\pi\hbar\omega_{\mathbf{k}}/V)}(\hat{\mathbf{k}} \times \hat{\mathbf{e}}_{\mathbf{k}\lambda})[a_{\mathbf{k}\lambda}e^{i\mathbf{k}\cdot\mathbf{r}} - a_{\mathbf{k}\lambda}^{\dagger}e^{-i\mathbf{k}\cdot\mathbf{r}}]. \quad (22.2\text{-}19)$$

22.3 Interaction Hamiltonian and Matrix Elements

The total Hamiltonian for a radiation field interacting with an atomic system may be written as a sum of three terms

$$\mathscr{H} = \mathscr{H}_{\text{rad}} + \mathscr{H}_{\text{atom}} + \mathscr{H}_{\text{int}}. \quad (22.3\text{-}1)$$

Here

$$\mathscr{H}_{\text{rad}} = \sum_{\mathbf{k}\lambda} \hbar\omega_{\mathbf{k}}(a_{\mathbf{k}\lambda}^{\dagger}a_{\mathbf{k}\lambda} + \tfrac{1}{2}) \quad (22.3\text{-}2)$$

is the Hamiltonian for the free field. The atomic Hamiltonian is

$$\mathscr{H}_{\text{atom}} = \sum_i (p_i^{\,2}/2m) + V \quad (22.3\text{-}3)$$

where V contains whatever terms are necessary to define the atomic state, e.g., Coulomb interaction with the nucleus, Coulomb repulsion among electrons, spin-orbit interaction, external fields, etc. For $\mathscr{H}_{\text{int}}$ we refer to the Schrödinger equation (15.2-32) and pick out all the terms which contain the vector potential. These are

$$\frac{e}{2mc}\,\mathbf{p}\cdot\mathbf{A} + \frac{e}{2mc}\,\mathbf{A}\cdot\mathbf{p} + \frac{e^2}{2mc^2}\,A^2 + \frac{e\hbar}{2mc}\,\boldsymbol{\sigma}\cdot\nabla\times\mathbf{A}.$$

However, the Coulomb gauge $\nabla\cdot\mathbf{A} = 0$ for the radiation field also leads to

$$\mathbf{p}\cdot\mathbf{A} = \mathbf{A}\cdot\mathbf{p};$$

hence the interaction Hamiltonian is

$$\mathscr{H}_{\text{int}} = \frac{e}{mc}\,\mathbf{p}\cdot\mathbf{A} + \frac{e^2}{2mc^2}\,A^2 + \frac{e\hbar}{2mc}\,\boldsymbol{\sigma}\cdot\nabla\times\mathbf{A}. \quad (22.3\text{-}4)$$

The total Hamiltonian (22.3-1) may now be written as

$$\mathscr{H} = \mathscr{H}_0 + \mathscr{H}_{\text{int}} \quad (22.3\text{-}5)$$

in which $\mathscr{H}_0$ contains (22.3-2) and (22.3-3) and $\mathscr{H}_{\text{int}}$ is given by (22.3-4). It will now be assumed that $\mathscr{H}_{\text{int}}$ may be regarded as a perturbation on $\mathscr{H}_0$ although, as in previous cases, the justification is not apparent until the calculations have been completed. We shall concentrate on

$$\mathscr{H}_1 = (e/mc)\mathbf{p}\cdot\mathbf{A} \quad (22.3\text{-}6)$$

which is most often the dominant term in electromagnetic interactions. With the vector potential in the Schrödinger representation (22.2-17),

$$\mathscr{H}_1 = \sum_{\mathbf{k}\lambda} (e/m)\sqrt{2\pi\hbar/\omega_\mathbf{k} V}(\mathbf{e}_{\mathbf{k}\lambda} \cdot \mathbf{p})[a_{\mathbf{k}\lambda}e^{i\mathbf{k}\cdot\mathbf{r}} + a_{\mathbf{k}\lambda}^\dagger e^{-i\mathbf{k}\cdot\mathbf{r}}]. \tag{22.3-7}$$

$$= \mathscr{H}_1^{(-)} + \mathscr{H}_1^{(+)} \tag{22.3-8}$$

where

$$\begin{aligned} \mathscr{H}_1^{(-)} &= \sum_{\mathbf{k}\lambda} (e/m)\sqrt{2\pi\hbar/\omega_\mathbf{k} V}(\hat{\mathbf{e}}_{\mathbf{k}\lambda} \cdot \mathbf{p})a_{\mathbf{k}\lambda}e^{i\mathbf{k}\cdot\mathbf{r}} \\ \mathscr{H}_1^{(+)} &= \sum_{\mathbf{k}\lambda} (e/m)\sqrt{2\pi\hbar/\omega_\mathbf{k} V}(\hat{\mathbf{e}}_{\mathbf{k}\lambda} \cdot \mathbf{p})a_{\mathbf{k}\lambda}^\dagger e^{-i\mathbf{k}\cdot\mathbf{r}}. \end{aligned} \tag{22.3-9}$$

In zero order, that is in the absence of an interaction, the wave function of the total system will be a product of an atomic wave function and a wave function describing the radiation field:

$$\psi_a|n_1, n_2, \ldots, n_i, \ldots\rangle \equiv |\psi_a; n_1, n_2, \ldots, n_i, \ldots\rangle. \tag{22.3-10}$$

Here ψ_a is an eigenfunction of the atomic Hamiltonian (22.3-3); $|n_1, n_2, \ldots, n_i, \ldots\rangle$, with $n_i \equiv n_{\mathbf{k}_i\lambda_i}$ is the wave function describing the radiation field in the occupation number (or Fock space) representation.

Let us now investigate the matrix elements

$$\langle\psi_b; n_1', n_2', \ldots, n_i', \ldots|\mathscr{H}_1^{(-)}|\psi_a; n_1, n_2, \ldots, n_i, \ldots\rangle \tag{22.3-11a}$$

$$\langle\psi_b; n_1', n_2', \ldots, n_i', \ldots|\mathscr{H}_1^{(+)}|\psi_a; n_1, n_2, \ldots, n_i, \ldots\rangle. \tag{22.3-11b}$$

In view of (22.2-12) and the orthogonality relation (22.2-11b), a nonvanishing result for the matrix element (22.3-11a) will be obtained only when there is at least one mode such that $n_i' = n_i - 1$. Similarly, (22.3-11b) will be nonvanishing only when there is at least one mode for which $n_i' = n_i + 1$. For a single mode characterized by $\mathbf{k}\lambda$ or what amounts to the same thing, for photons of momentum $\hbar\mathbf{k}$ and polarization λ, the notation may be simplified to

$$\begin{aligned} \langle b; n_{\mathbf{k}\lambda} - 1|\mathscr{H}_1^{(-)}|a; n_{\mathbf{k}\lambda}\rangle &= \frac{e}{m}\sqrt{\frac{2\pi\hbar}{\omega_\mathbf{k} V}}\langle b; n_{\mathbf{k}\lambda} - 1|(\hat{\mathbf{e}}_{\mathbf{k}\lambda} \cdot \mathbf{p})a_{\mathbf{k}\lambda}e^{i\mathbf{k}\cdot\mathbf{r}}|a; n_{\mathbf{k}\lambda}\rangle \\ &= \frac{e}{m}\sqrt{\frac{2\pi\hbar n_{\mathbf{k}\lambda}}{\omega_\mathbf{k} V}}\langle b|(\hat{\mathbf{e}}_{\mathbf{k}\lambda} \cdot \mathbf{p})e^{i\mathbf{k}\cdot\mathbf{r}}|a\rangle \end{aligned} \tag{22.3-12}$$

in which $|a\rangle \equiv |\psi_a\rangle$ represents the initial atomic state and $|b\rangle \equiv |\psi_b\rangle$ represents the final atomic state. In the same fashion we have

$$\langle b; n_{\mathbf{k}\lambda} + 1|\mathscr{H}_1^{(+)}|a; n_{\mathbf{k}\lambda}\rangle = \frac{e}{m}\sqrt{\frac{2\pi\hbar(n_{\mathbf{k}\lambda} + 1)}{\omega_\mathbf{k} V}}\langle b|(\hat{\mathbf{e}}_{\mathbf{k}\lambda} \cdot \mathbf{p})e^{-i\mathbf{k}\cdot\mathbf{r}}|a\rangle. \tag{22.3-13}$$

Matrix elements of the type

$$\langle b|(\hat{e}_{\mathbf{k}\lambda} \cdot \mathbf{p})e^{i\mathbf{k}\cdot\mathbf{r}}|a\rangle \equiv \hat{\mathbf{e}}_{\mathbf{k}\lambda} \cdot \langle b|\mathbf{p}e^{i\mathbf{k}\cdot\mathbf{r}}|a\rangle \tag{22.3-14}$$

may be simplified if $\mathbf{k} \cdot \mathbf{r} \ll 1$ so that $e^{i\mathbf{k}\cdot\mathbf{r}} \approx 1$. Using the atomic Hamiltonian (22.3-3) and the commutation law (1.1-8c)

$$[r_i, p^2] = 2i\hbar p_i \qquad (r_i = x, y, z)$$

we have

$$[\mathbf{r}, \mathscr{H}_{\text{atom}}] = (i\hbar/m)\mathbf{p}. \tag{22.3-15a}$$

Therefore

$$\begin{aligned}\langle b|\mathbf{p}|a\rangle &= (m/i\hbar)\langle b|[\mathbf{r}, \mathscr{H}_{\text{atom}}]|a\rangle \\ &= (im/\hbar)(E_b - E_a)\langle b|\mathbf{r}|a\rangle \\ &= im\omega_{ba}\langle b|\mathbf{r}|a\rangle = im\omega_{\mathbf{k}}\langle b|\mathbf{r}|a\rangle.\end{aligned} \tag{22.3-15b}$$

For an atom, $\mathbf{r}$ is the position vector of an electron referred, most conveniently, to the nucleus; E_a and E_b are the eigenvalues of the atomic Hamiltonian corresponding to the eigenstates $|a\rangle$ and $|b\rangle$, respectively; and $\omega_{ba} = \omega_{\mathbf{k}}$ by energy conservation.

Now, in Section 7.4 it was demonstrated that the approximation $e^{ikr} \approx 1$ in the vector potential corresponded to the approximation in which all the terms in the multipole expansion of the field were neglected except for the electric dipole (E1) term. Therefore within the electric dipole approximation

$$\hat{\mathbf{e}}_{\mathbf{k}\lambda} \cdot \langle b|\mathbf{p}e^{i\mathbf{k}\cdot\mathbf{r}}|a\rangle \quad \hat{\mathbf{e}}_{\mathbf{k}\lambda} \cdot \langle b|\mathbf{p}|a\rangle = im\omega_{\mathbf{k}}\hat{\mathbf{e}}_{\mathbf{k}\lambda} \cdot \langle b|\mathbf{r}|a\rangle, \tag{22.3-16}$$

and

$$\langle b; n_{\mathbf{k}\lambda} - 1|\mathscr{H}_1^{(-)}|a; n_{\mathbf{k}\lambda}\rangle_{\text{E1}} = ie\sqrt{\frac{2\pi\hbar\omega_{\mathbf{k}}n_{\mathbf{k}\lambda}}{V}}\,\hat{\mathbf{e}}_{\mathbf{k}\lambda} \cdot \langle b|\mathbf{r}|a\rangle \tag{22.3-17a}$$

$$\langle b; n_{\mathbf{k}\lambda} + 1|\mathscr{H}_1^{(+)}|a; n_{\mathbf{k}\lambda}\rangle_{\text{E1}} = ie\sqrt{\frac{2\pi\hbar\omega_{\mathbf{k}}(n_{\mathbf{k}\lambda} + 1)}{V}}\,\hat{\mathbf{e}}_{\mathbf{k}\lambda} \cdot \langle b|\mathbf{r}|a\rangle. \tag{22.3-17b}$$

The next higher approximation is

$$e^{i\mathbf{k}\cdot\mathbf{r}} \approx 1 + i\mathbf{k} \cdot \mathbf{r}. \tag{22.3-18}$$

To analyze the matrix element containing the operator $\mathbf{k} \cdot \mathbf{r}$ $(= \mathbf{r} \cdot \mathbf{k})$ it is convenient to write

$$\begin{aligned}i(\hat{\mathbf{e}}_{\mathbf{k}\lambda} \cdot \mathbf{p})(\mathbf{k} \cdot \mathbf{r}) &\equiv i\hat{\mathbf{e}}_{\mathbf{k}\lambda} \cdot \mathbf{pr} \cdot \mathbf{k} \\ &= \tfrac{1}{2}i\hat{\mathbf{e}}_{\mathbf{k}\lambda} \cdot (\mathbf{pr} - \mathbf{rp}) \cdot \mathbf{k} + \tfrac{1}{2}i\hat{\mathbf{e}}_{\mathbf{k}\lambda} \cdot (\mathbf{pr} + \mathbf{rp}) \cdot \mathbf{k} \\ &\equiv O_{\text{A}} + O_{\text{S}}\end{aligned} \tag{22.3-19}$$

where O_A stands for the antisymmetric combination and O_S for the symmetric combination. We consider the two terms separately. For O_A, using the vector identity (22.1-24a),

$$\begin{aligned}
O_A &= \tfrac{1}{2} i \hat{\mathbf{e}}_{\mathbf{k}\lambda} \cdot (\mathbf{pr} - \mathbf{rp}) \cdot \mathbf{k} \\
&= \tfrac{1}{2} i (\hat{\mathbf{e}}_{\mathbf{k}\lambda} \cdot \mathbf{p})(\mathbf{r} \cdot \mathbf{k}) - (\hat{\mathbf{e}}_{\mathbf{k}\lambda} \cdot \mathbf{r})(\mathbf{p} \cdot \mathbf{k}) \\
&= \tfrac{1}{2} i (\mathbf{k} \times \hat{\mathbf{e}}_{\mathbf{k}\lambda}) \cdot (\mathbf{r} \times \mathbf{p}) \\
&= \tfrac{1}{2} i\hbar (\mathbf{k} \times \hat{\mathbf{e}}_{\mathbf{k}\lambda}) \cdot \mathbf{L} \\
&= (i\hbar\omega_{\mathbf{k}}/2c)(\hat{\mathbf{k}} \times \hat{\mathbf{e}}_{\mathbf{k}\lambda}) \cdot \mathbf{L} \\
&= i\beta(m\omega_{\mathbf{k}}/e)(\hat{\mathbf{k}} \times \hat{\mathbf{e}}_{\mathbf{k}\lambda}) \cdot \mathbf{L} \\
&= -i(m\omega_{\mathbf{k}}/e)(\hat{\mathbf{k}} \times \hat{\mathbf{e}}_{\mathbf{k}\lambda}) \cdot \boldsymbol{\mu}_L
\end{aligned} \tag{22.3-20}$$

in which β is the Bohr magneton (17.1-7) and $\boldsymbol{\mu}_L$ the magnetic moment operator associated with the orbital angular momentum $\mathbf{L}$. The vector $\hat{\mathbf{k}} \times \hat{\mathbf{e}}_{\mathbf{k}\lambda}$ lies in the direction of the magnetic field as is evident from (22.2-19). The connection with the magnetic field can be made more explicit by writing the complete interaction associated with O_A. From (22.3-9),

$$\begin{aligned}
\mathscr{H}_1^{(-)} = (e/m)\sqrt{(2\pi\hbar/\omega_{\mathbf{k}} V)}\, a_{\mathbf{k}\lambda} O_A &= -i\sqrt{(2\pi\hbar\omega_{\mathbf{k}}/V)}\, a_{\mathbf{k}\lambda}(\hat{\mathbf{k}} \times \hat{\mathbf{e}}_{\mathbf{k}\lambda}) \cdot \boldsymbol{\mu}_L \\
&= -\mathbf{B}_{\mathbf{k}\lambda}^{(-)} \cdot \boldsymbol{\mu}_L
\end{aligned} \tag{22.3-21}$$

where $\mathbf{B}_{\mathbf{k}\lambda}^{(-)}$ is the term containing $a_{\mathbf{k}\lambda}$ in (22.2-19).

To determine the multipole character of the interaction $-\mathbf{B}_{\mathbf{k}\lambda}^{(-)} \cdot \boldsymbol{\mu}_L$ we may refer to the development in Section 7.4. Taking the propagation vector $\hat{\mathbf{k}}$ in the z direction and the polarization vector $\hat{\mathbf{e}}_{\mathbf{k}\lambda}$ in the x direction, $\hat{\mathbf{k}} \times \hat{\mathbf{e}}_{\mathbf{k}\lambda}$ is in the y direction and

$$O_A = \tfrac{1}{2} ik(\hat{\mathbf{k}} \times \hat{\mathbf{e}}_{\mathbf{k}\lambda}) \cdot (\mathbf{r} \times \mathbf{p}) = \tfrac{1}{2} ik(\mathbf{r} \times \mathbf{p})_y = \tfrac{1}{2} ik(zp_x - xp_z). \tag{22.3-22}$$

But this is precisely the result obtained using (7.4-31), namely

$$\begin{aligned}
(\mathbf{A}_x^{(2)})_{M1} \cdot \mathbf{p} &= \tfrac{1}{2} ik(z\hat{\mathbf{e}}_x - x\hat{\mathbf{e}}_z) \cdot (p_x\hat{\mathbf{e}}_x + p_y\hat{\mathbf{e}}_y + p_z\hat{\mathbf{e}}_z) \\
&= \tfrac{1}{2} ik(zp_x - xp_z).
\end{aligned} \tag{22.3-23}$$

For the M1 component of $\mathbf{A}_y^{(2)}$ we have

$$(\mathbf{A}_y^{(2)})_{M1} \cdot \mathbf{p} = \tfrac{1}{2} ik(zp_y - yp_z) \tag{22.3-24}$$

which corresponds to the polarization vector $\hat{\mathbf{e}}_{\mathbf{k}\lambda}$ pointing in the y direction since then $\hat{\mathbf{k}} \times \hat{\mathbf{e}}_{\mathbf{k}\lambda}$ is in the negative x direction giving

$$O_A = \tfrac{1}{2} ik(\hat{\mathbf{k}} \times \hat{\mathbf{e}}_{\mathbf{k}\lambda}) \cdot (\mathbf{r} \times \mathbf{p}) = -\tfrac{1}{2} ik(\mathbf{r} \times \mathbf{p})_x = \tfrac{1}{2} ik(zp_y - yp_z). \tag{22.3-25}$$

The result of this analysis is that O_A or $-\mathbf{B}_{\mathbf{k}\lambda}^{(-)} \cdot \boldsymbol{\mu}_L$ or any of the equivalent forms of these operators are magnetic dipole (M1) operators.

There is really no need to limit the magnetic moment operators to those which are associated with the orbital angular momentum. We may as well include the spin which will now take into amount the term proportional to $\boldsymbol{\sigma} \cdot \nabla \times \mathbf{A}$ in (22.3-4). This amounts to adding $2\mathbf{S}$ to $\mathbf{L}$ in (22.3-20). Hence the matrix elements for magnetic dipole (M1) transitions may now be written as

$$\begin{aligned}\langle b; n_{\mathbf{k}\lambda} - 1|\mathscr{H}_1^{(-)}|a; n_{\mathbf{k}\lambda}\rangle_{\mathrm{M1}} &= \langle b; n_{\mathbf{k}\lambda} - 1|-\boldsymbol{\mu} \cdot \mathbf{B}_{\mathbf{k}\lambda}^{(-)}|a, n_{\mathbf{k}\lambda}\rangle \\ &= -i\sqrt{\frac{2\pi\hbar\omega_{\mathbf{k}} n_{\mathbf{k}\lambda}}{V}}\,(\hat{\mathbf{k}} \times \hat{\mathbf{e}}_{\mathbf{k}\lambda}) \cdot \langle b|\boldsymbol{\mu}|a\rangle \\ &= i\sqrt{\frac{2\pi\hbar\omega_{\mathbf{k}} n_{\mathbf{k}\lambda}}{V}}\,(\hat{\mathbf{k}} \times \hat{\mathbf{e}}_{\mathbf{k}\lambda}) \cdot \langle b|\beta(\mathbf{L} + 2\mathbf{S})|a\rangle,\end{aligned} \tag{22.3-26}$$

and

$$\begin{aligned}\langle b; n_{\mathbf{k}\lambda} + 1|\mathscr{H}_1^{(+)}|a; n_{\mathbf{k}\lambda}\rangle_{\mathrm{M1}} &= \langle b; n_{\mathbf{k}\lambda} + 1|-\boldsymbol{\mu} \cdot \mathbf{B}_{\mathbf{k}\lambda}^{(+)}|a, n_{\mathbf{k}\lambda}\rangle \\ &= i\sqrt{\frac{2\pi\hbar\omega_{\mathbf{k}}(n_{\mathbf{k}\lambda} + 1)}{V}}\,(\hat{\mathbf{k}} \times \hat{\mathbf{e}}_{\mathbf{k}\lambda}) \cdot \langle b|\boldsymbol{\mu}|a\rangle \\ &= -i\sqrt{\frac{2\pi\hbar\omega_{\mathbf{k}}(n_{\mathbf{k}\lambda} + 1)}{V}}\,(\hat{\mathbf{k}} \times \hat{\mathbf{e}}_{\mathbf{k}\lambda}) \cdot \langle b|\beta(\mathbf{L} + 2\mathbf{S})|a\rangle\end{aligned} \tag{22.3-27}$$

in which $\boldsymbol{\mu}$ is the total magnetic moment operator associated with both orbital and spin angular momenta.

For the symmetric operator in (22.3-19),

$$O_{\mathrm{S}} = \tfrac{1}{2} i \hat{\mathbf{e}}_{\mathbf{k}\lambda} \cdot (\mathbf{pr} + \mathbf{rp}) \cdot \mathbf{k}. \tag{22.3-28}$$

The replacement of $\mathbf{p}$ according to (22.3-15a) gives

$$\begin{aligned}O_{\mathrm{S}} &= -(m/2\hbar)\hat{\mathbf{e}}_{\mathbf{k}\lambda} \cdot \{[\mathscr{H}_{\mathrm{atom}}, \mathbf{r}]\mathbf{r} + \mathbf{r}[\mathscr{H}_{\mathrm{atom}}, \mathbf{r}]\} \cdot \mathbf{k} \\ &= (m/2\hbar)\hat{\mathbf{e}}_{\mathbf{k}\lambda} \cdot [\mathbf{rr}, \mathscr{H}_{\mathrm{atom}}] \cdot \mathbf{k}\end{aligned}$$

and

$$\begin{aligned}\langle b|O_{\mathrm{S}}|a\rangle &= -(m/2\hbar)(E_b - E_a)\hat{\mathbf{e}}_{\mathbf{k}\lambda} \cdot \langle b|\mathbf{rr}|a\rangle \cdot \mathbf{k} \\ &= -(m\omega_{ba}/2)\hat{e}_{\mathbf{k}\lambda} \cdot \langle b|\mathbf{rr}|a\rangle \cdot \mathbf{k}.\end{aligned}$$

The operator $\mathbf{rr}$ is a symmetric Cartesian tensor of rank 2. Recalling the discussion in Section 6.4, such tensors are not irreducible with respect to orthogonal transformations. This is somewhat inconvenient because we should like to take advantage of the Wigner–Eckart theorem but the latter requires an operator that transforms as an irreducible tensor. It is observed, however, that

$$\hat{\mathbf{e}}_{\mathbf{k}\lambda} \cdot \langle b|r^2\,\delta_{ij}|a\rangle \cdot \mathbf{k} = \langle b|r^2\,\delta_{ij}|a\rangle \hat{\mathbf{e}}_{\mathbf{k}\lambda} \cdot \mathbf{k} = 0$$

as a consequence of the transversality condition. Therefore **rr** may be replaced by a new operator

$$Q = \mathbf{rr} - \tfrac{1}{3}r^2\,\delta_{ij} \tag{22.3-29}$$

which, as shown in Section 6.4, is an irreducible tensor of rank 2. We now have

$$\begin{aligned} \langle b|O_{\mathrm{S}}|a\rangle &= -(m\omega_{ba}/2)\hat{\mathbf{e}}_{\mathbf{k}\lambda} \cdot \langle b|Q|a\rangle \cdot \mathbf{k} \\ &= -(m\omega_{\mathbf{k}}/2)\hat{\mathbf{e}}_{\mathbf{k}\lambda} \cdot \langle b|Q|a\rangle \cdot \mathbf{k} \\ &= -(m\omega_{\mathbf{k}}{}^2/2c)\hat{\mathbf{e}}_{\mathbf{k}\lambda} \cdot \langle b|Q|a\rangle \cdot \hat{\mathbf{k}} \end{aligned} \tag{22.3-30}$$

where again, because of energy conservation, we have set $\omega_{ba} = \omega_{\mathbf{k}}$.

The multipole character of O_{S} (22.3-28) may be investigated in the same fashion as that of the operator O_{A}. If **k** is in the z direction and $\hat{\mathbf{e}}_{\mathbf{k}\lambda}$ in the x direction,

$$O_{\mathrm{S}} = \tfrac{1}{2}ik(zp_x + xp_z), \tag{22.3-31}$$

which, according to (7.4-33), identifies the operator as being electric quadrapole (E2). A similar identification is obtained by taking the polarization in the y direction.

The matrix elements for electric quadrapole (E2) transitions may now be written as

$$\langle b; n_{\mathbf{k}\lambda} - 1|\mathscr{H}_1^{(-)}|a; n_{\mathbf{k}\lambda}\rangle_{\mathrm{E2}} = -\frac{e}{2c}\sqrt{\frac{2\pi\hbar\omega_{\mathbf{k}}{}^3 n_{\mathbf{k}\lambda}}{V}}\,\hat{\mathbf{e}}_{\mathbf{k}\lambda} \cdot \langle b|Q|a\rangle \cdot \hat{\mathbf{k}}, \tag{22.3-32a}$$

$$\langle b; n_{\mathbf{k}\lambda} + 1|\mathscr{H}_1^{(+)}|a; n_{\mathbf{k}\lambda}\rangle_{\mathrm{E2}} = \frac{e}{2c}\sqrt{\frac{2\pi\hbar\omega_{\mathbf{k}}{}^3 (n_{\mathbf{k}\lambda} + 1)}{V_{\mathrm{L}}}}\,\hat{\mathbf{e}}_{\mathbf{k}\lambda} \cdot \langle b|Q|a\rangle \cdot \hat{\mathbf{k}}. \tag{22.3-32b}$$

The dipole matrix elements given by (22.3-17) deserves additional comment. Consider the operator

$$O = \exp[-(ie/\hbar c)\mathbf{A}(t) \cdot \mathbf{r}] \tag{22.3-33}$$

in which **A** is the vector potential, assumed to be a function of time but independent of position within a region of atomic dimensions. Such a situation will occur when the wavelength of the radiation is long compared with atomic dimensions. Since the operator in the exponent is Hermitian, O

itself is unitary. As shown in Section 9.1, if ψ satisfies

$$i\hbar\,\partial\psi/\partial t = \mathscr{H}\psi,$$

then $\psi' = O^\dagger\psi$ satisfies

$$i\hbar\,\partial\psi'/\partial t = \mathscr{H}'\psi'$$

where

$$\mathscr{H}' = O^\dagger\,\mathscr{H}O - i\hbar O^\dagger\,\partial O/\partial t \tag{22.3-34}$$

as in (9.1-21). Let us then transform the Hamiltonian

$$\frac{1}{2m}\left(\mathbf{p} + \frac{e}{c}\mathbf{A}\right)^2$$

using operator (22.3-33). We have

$$\mathbf{p}O = -(e/c)\mathbf{A}O + O\mathbf{p}$$

or

$$[\mathbf{p} + (e/c)\mathbf{A}]O = O\mathbf{p}$$

and

$$[\mathbf{p} + (e/c)\mathbf{A}]^2 O = [\mathbf{p} + (e/c)\mathbf{A}]O\mathbf{p} = Op^2.$$

Therefore

$$O^\dagger\,[\mathbf{p} + (e/c)\mathbf{A}]^2 O = O^\dagger\,Op^2 = p^2.$$

Also,

$$\frac{\partial O}{\partial t} = \frac{-ie}{\hbar c}\mathbf{r}\cdot\frac{\partial \mathbf{A}}{\partial t}O$$

so that

$$-i\hbar O^\dagger\,\frac{\partial O}{\partial t} = \frac{-e}{c}\mathbf{r}\cdot\frac{\partial \mathbf{A}}{\partial t} = e\mathbf{r}\cdot\mathbf{E}.$$

Thus

$$\mathscr{H}' = \frac{1}{2m}O^\dagger\left(\mathbf{p} + \frac{e}{c}\mathbf{A}\right)^2 O - i\hbar O^\dagger\,\frac{\partial O}{\partial t} = \frac{p^2}{2m} + e\mathbf{r}\cdot\mathbf{E}. \tag{22.3-35}$$

It is seen that the effect of the canonical transformation (22.3-34) has been to transform

$$\frac{1}{2m}\left(\mathbf{p} + \frac{e}{c}\mathbf{A}\right)^2 = \frac{p^2}{2m} + \frac{e}{mc}\mathbf{p}\cdot\mathbf{A} + \frac{2}{2mc^2}A^2$$

into

$$\frac{p^2}{2m} + e\mathbf{r} \cdot \mathbf{E}.$$

We may therefore regard the interaction Hamiltonian

$$\mathscr{H}'_{\text{int}} = e\mathbf{r} \cdot E \equiv -\mathbf{d} \cdot \mathbf{E} \tag{22.3-36}$$

as equivalent to $(e/mc)\mathbf{p} \cdot \mathbf{A} + (e^2/2mc^2)A^2$. In (22.3-36), $\mathbf{d} = -e\mathbf{r}$ is the electric dipole moment operator.

Two points should be kept in mind concerning (22.3-36). One is that classically $-\mathbf{d} \cdot \mathbf{E}$ is the interaction energy for an electric dipole in an electric field. Quantum mechanically, the operator $-\mathbf{d} \cdot \mathbf{E}$ contains contributions from both the linear and quadratic terms in the vector potential. The second point is that (22.3-36) is valid only when the vector potential is constant over the region of interest which, it was shown previously, corresponds to the electric dipole (E1) approximation.

Using (22.3-36) and (22.2-18) with $kr \ll 1$ we have

$$\begin{aligned}\langle b; n_{\mathbf{k}\lambda} - 1 | -\mathbf{d} \cdot \mathbf{E} | a; n_{\mathbf{k}\lambda} \rangle &= -i \sqrt{\frac{2\pi\hbar\omega_{\mathbf{k}} n_{\mathbf{k}\lambda}}{V}} \, \hat{\mathbf{e}}_{\mathbf{k}\lambda} \cdot \langle b | \mathbf{d} | a \rangle \\ &= ie \sqrt{\frac{2\pi\hbar\omega_{\mathbf{k}} n_{\mathbf{k}\lambda}}{V}} \, \hat{\mathbf{e}}_{\mathbf{k}\lambda} \cdot \langle b | \mathbf{r} | a \rangle, \end{aligned} \tag{22.3-37a}$$

$$\begin{aligned}\langle b; n_{\mathbf{k}\lambda} + 1 | -\mathbf{d} \cdot \mathbf{E} | a; n_{\mathbf{k}\lambda} \rangle &= -i \sqrt{\frac{2\pi\hbar\omega_{\mathbf{k}} (n_{\mathbf{k}\lambda} + 1)}{V}} \, \hat{\mathbf{e}}_{\mathbf{k}\lambda} \cdot \langle b | \mathbf{d} | a \rangle \\ &= ie \sqrt{\frac{2\pi\hbar\omega_{\mathbf{k}} (n_{\mathbf{k}\lambda} + 1)}{V}} \, \hat{\mathbf{e}}_{\mathbf{k}\lambda} \cdot \langle b | \mathbf{r} | a \rangle. \end{aligned} \tag{22.3-37b}$$

The identity of these expressions with those in (22.3-17) might at first seem surprising, since the latter were derived on the basis of the interaction Hamiltonian (22.3-6), which contains the $\mathbf{p} \cdot \mathbf{A}$ term but not the A^2 term, yet we have just seen that $-\mathbf{d} \cdot \mathbf{E}$ contains both terms. However, as will be shown in Section 24.1, the A^2 term does not contribute to one-photon processes, which is all we are concerned with at present.

The discussion up to this point involved the interaction of a radiation field with a single electron. We shall now summarize the important expressions and put them in a form applicable to a multielectron system. However, we shall continue to use a single mode characterized by $\mathbf{k}\lambda$. For a system containing N electrons let

$$\mathbf{R} = \sum_{j=1}^{N} \mathbf{r}_j \tag{22.3-38}$$

where $\mathbf{r}_j$ is the position vector of the jth electron. For an atom the natural reference is the nucleus; more generally, as in molecules, $\mathbf{r}_j$ is referred to the center of gravity of the positive charge. The interaction matrix elements are then the following:

Electric Dipole (E1):

$$\langle b; n_{\mathbf{k}\lambda} - 1|\mathscr{H}_1^{(-)}|a; n_{\mathbf{k}\lambda}\rangle_{\mathrm{E1}} = ie\sqrt{\frac{2\pi\hbar\omega_{\mathbf{k}}n_{\mathbf{k}\lambda}}{V}}\,\hat{\mathbf{e}}_{\mathbf{k}\lambda}\cdot\langle b|\mathbf{R}|a\rangle, \tag{22.3-39a}$$

$$\langle b; n_{\mathbf{k}\lambda} + 1|\mathscr{H}_1^{(+)}|a; n_{\mathbf{k}\lambda}\rangle_{\mathrm{E1}} = ie\sqrt{\frac{2\pi\hbar\omega_{\mathbf{k}}(n_{\mathbf{k}\lambda}+1)}{V}}\,\hat{\mathbf{e}}_{\mathbf{k}\lambda}\cdot\langle b|\mathbf{R}|a\rangle. \tag{22.3-39b}$$

Magnetic Dipole (M1):

$$\langle b; n_{\mathbf{k}\lambda} - 1|\mathscr{H}_1^{(-)}|a; n_{\mathbf{k}\lambda}\rangle_{\mathrm{M1}} = -i\sqrt{\frac{2\pi\hbar\omega_{\mathbf{k}}n_{\mathbf{k}\lambda}}{V}}\,(\hat{\mathbf{k}}\times\hat{\mathbf{e}}_{\mathbf{k}\lambda})\cdot\left\langle b\left|\sum_j \boldsymbol{\mu}_j\right|a\right\rangle, \tag{22.3-40a}$$

$$\langle b; n_{\mathbf{k}\lambda} + 1|\mathscr{H}_1^{(+)}|a; n_{\mathbf{k}\lambda}\rangle_{\mathrm{M1}} = i\sqrt{\frac{2\pi\hbar\omega_{\mathbf{k}}(n_{\mathbf{k}\lambda}+1)}{V}}\,(\hat{\mathbf{k}}\times\hat{\mathbf{e}}_{\mathbf{k}\lambda})\cdot\left\langle b\left|\sum_j \boldsymbol{\mu}_j\right|a\right\rangle. \tag{22.3-40b}$$

Electric Quadrupole (E2):

$$\langle b; n_{\mathbf{k}\lambda} - 1|\mathscr{H}_1^{(-)}|a; n_{\mathbf{k}\lambda}\rangle_{\mathrm{E2}} = \frac{-e}{2c}\sqrt{\frac{2\pi\hbar\omega_{\mathbf{k}}{}^3 n_{\mathbf{k}\lambda}}{V}}\,\hat{\mathbf{e}}_{\mathbf{k}\lambda}\cdot\left\langle b\left|\sum_j Q_j\right|a\right\rangle\cdot\hat{\mathbf{k}}, \tag{22.3-41a}$$

$$\langle b; n_{\mathbf{k}\lambda} + 1|\mathscr{H}_1^{(+)}|a; n_{\mathbf{k}\lambda}\rangle_{\mathrm{E2}} = \frac{e}{2c}\sqrt{\frac{2\pi\hbar\omega_{\mathbf{k}}{}^3(n_{\mathbf{k}\lambda}+1)}{V}}\,\hat{\mathbf{e}}_{\mathbf{k}\lambda}\cdot\left\langle b\left|\sum_j Q_j\right|a\right\rangle\cdot\hat{\mathbf{k}} \tag{22.3-41b}$$

22.4 Selection Rules and Angular Distributions

We shall write

$$|a\rangle = |J_a M_a\rangle, \qquad |b\rangle = |J_b M_b\rangle, \tag{22.4-1}$$

$$g_a = 2J_a + 1, \qquad g_b = 2J_b + 1 \tag{22.4-2}$$

where J_a, M_a and J_b, M_b are the angular momentum quantum numbers and g_a, g_b the degeneracies associated with the states $|a\rangle$ and $|b\rangle$ that participate in a radiative transition. Since operators for the various multipole transitions are irreducible tensors, the selection rules are obtained directly from the Wigner–Eckart theorem. For E1 and M1 transitions the operators are irreducible tensors of rank 1 (vectors). Hence

$$\text{E1 and M1:} \quad \Delta J = 0, \pm 1, \qquad \Delta M = 0, \pm 1, \qquad J_a + J_b \geqslant 1. \tag{22.4-3}$$

For E2, the operator is an irreducible tensor of rank 2 so that

$$\text{E2:} \quad \Delta J = 0, \pm 1, \pm 2, \qquad \Delta M = 0, \pm 1, \pm 2, \qquad J_a + J_b \geqslant 2. \tag{22.4-4}$$

This is as much as we can obtain from the Wigner–Eckart theorem. Referring to Table 7.2 we find that the electromagnetic field associated with an E1 multipole has odd parity ($\pi_{\text{em}} = -1$), while the fields associated with M1 and E2 multipoles have even parity ($\pi_{\text{em}} = +1$). The same is true of the quantum-mechanical operators that govern these multipole transitions as is evident from an inspection of the operators in (22.3-39)–(22.3-41). This is not surprising in view of the retention of symmetry properties in the transition from classical to quantum mechanics. Thus $\pi_{\text{E1}} = -1$ and $\pi_{\text{M1}} = \pi_{\text{E2}} = +1$. The conservation of parity then requires

$$\pi_a \pi_b \pi_{\text{op}} = +1 \tag{22.4-5}$$

where π_a and π_b are the parities of the initial and final states, respectively, and π_{op} is the parity of the operator for the particular multipole transition. If the product of the parities in (22.4-5) is -1, the integrand of the transition matrix element is odd and the integral vanishes. Alternatively, the matrix element theorem of Section 3.7 leads to the same conclusion. In either case we have the parity selection rules

$$\begin{aligned} \text{E1:} \quad & \pi_a = -\pi_b \qquad (\Delta \textstyle\sum_i l_i = \pm 1), \\ \text{M1 and E2:} \quad & \pi_a = \pi_b. \end{aligned} \tag{22.4-6}$$

The selection rules that have been developed to this point follow from general symmetry considerations and are independent of the scheme whereby the angular momenta are coupled. When the latter is taken into account, additionai restrictions emerge. For LS coupling, which is the most important case, the orbital and spin angular momenta are independent, and since the E1 and E2 operators are independent of spin we must have $\Delta S = 0$. Thus

$$\text{E1:} \quad \Delta S = 0, \qquad \Delta L = 0, \pm 1, \qquad L_a + L_b \geqslant 1, \qquad \Delta M_L = 0, \pm 1. \tag{22.4-7}$$

For a single electron $\Delta l = \pm 1$ only, $\quad \Delta m_l = 0, \pm 1,$

$$\text{E2:} \quad \Delta S = 0, \qquad \Delta L = 0, \pm 1, \pm 2, \qquad L_a + L_b \geqslant 2, \qquad \Delta M_L = 0, \pm 1, \pm 2 \tag{22.4-8}$$

For a single electron $\Delta l = 0, \pm 2$ only,
$l_a + l_b \geqslant 1, \quad \Delta m_l = 0, \pm 1, \pm 2$

In exact LS coupling M1 transitions cannot occur. The reason for this may be seen as follows: The operator for an M1 transition is $\mathbf{L} + 2\mathbf{S}$. Consider the matrix element

$$\langle L'S'M_L'M_S'|L + 2S|LSM_LM_S\rangle.$$

The components of $\mathbf{L} + 2\mathbf{S}$ operating on $|LSM_LM_S\rangle$ will have no effect on the values of L and S but may alter the values of M_L and M_S. Since the wave functions are orthogonal, the matrix element will vanish unless $L' = L$ and $S' = S$. But then the energies are identical, since in LS coupling the states are degenerate with respect to M_L and M_S and the photon energy in the transition is zero. If there is spin-orbit interaction, the coupling of $\mathbf{L}$ and $\mathbf{S}$ results in states whose energies depend on J, thereby departing from exact LS coupling. Then

$$\langle LSJ'M_J'|\mathbf{L} + 2\mathbf{S}|LSJM_J\rangle$$

can have off diagonal elements with $\Delta J = \pm 1$ since the operator is a vector operator.

A summary of the selection rules for E1, M1 and E2 transitions is given in Table 22.1.

TABLE 22.1

Selection Rules for Radiative Transitions from an Initial State ψ_a to a Final State ψ_b‡

	Electric dipole E1	Magnetic dipole M1	Electric quadrupole E2
ΔJ	$0, \pm 1$ $J_a + J_b \geqslant 1$	$0, \pm 1$ $J_a + J_b \geqslant 1$	$0, \pm 1, \pm 2$ $J_a + J_b \geqslant 1$
ΔM	$0, \pm 1$	$0, \pm 1$	$0, \pm 1, \pm 2$
Parity	$\pi_a = -\pi_b$	$\pi_a = \pi_b$	$\pi_a = \pi_b$
Δl	± 1	0 $(\Delta n = 0)$	$0, \pm 2$ $l_a = 0 \rightarrow l_b = 0$ $(\Delta m_l = 0, \pm 1, \pm 2)$
ΔS	0 $(\Delta M_S = 0)$	0	0 $(\Delta M_S = 0)$
ΔL	$0, \pm 1$ $L_a + L_b \geqslant 2$ $(\Delta M_L = 0, \pm 1)$	0	$0, \pm 1, \pm 2$ $L_a + L_b \geqslant 2$

‡ The rules for ΔJ, ΔM, and the parity are rigorous; Δl refers to one-electron transitions; the rules for ΔS and ΔL hold only for LS coupling. For exact LS coupling there are no magnetic dipole transitions.

The angular distribution associated with a transition from $|J_a M_a\rangle$ to $|J_b M_b\rangle$ may also be inferred from the Wigner–Eckart theorem. Consider again an E1 transition which, as we saw before, has the dual interpretation namely the operator in the matrix element is an electric dipole operator or the multipolarity of the radiation associated with the transition is electric dipole. Referring to Table 7.2 we find that $\mathbf{B}_E$ is proportional to $\mathbf{Y}_{11m}$. The angular distribution is then given by $\mathbf{Y}_{11m} \cdot \mathbf{Y}_{11m}$ which has already been calculated in (7.2-13) and (7.2-14). Since

$$M_a = M_b + m \tag{22.4-9}$$

as required by the Wigner–Eckart theorem or

$$\Delta M = m,$$

we have the angular distributions

$$\begin{aligned} \Delta M = 0: \quad & \mathbf{Y}_{110} \cdot \mathbf{Y}_{110} = \frac{3}{8\pi} \sin^2 \theta, \\ \Delta \mathrm{M} = \pm 1: \quad & \mathbf{Y}_{11\pm1} \cdot \mathbf{Y}_{11\pm1} = \frac{3}{16\pi} (1 + \cos^2 \theta). \end{aligned} \tag{22.4-10}$$

If the three transitions occur with equal probability the angular distribution is given by the sum

$$\mathbf{Y}_{110} \cdot \mathbf{Y}_{110} + 2\mathbf{Y}_{11\pm1} \cdot \mathbf{Y}_{11\pm1} = 3/4\pi \tag{22.4-11}$$

which is totally independent of angle. This simply means that the system has no preferred direction as indeed there is none in an atom.

For an M1 transition the situation is precisely the same except for the fact that it is $\mathbf{E}_M$ which is proportional to $f_1(kr)\mathbf{Y}_{11m}$ so that the **B** and **E** fields are interchanged but the intensity distributions are the same. Thus E1 and M1 transitions are indistinguishable on the basis of the angular distributions alone.

For E2 transitions $\mathbf{B}_E$ is proportional to $\mathbf{Y}_{22m}$. Hence, from the selection rules (22.4-4), we have

$$\begin{aligned} \Delta M = 0: \quad & \mathbf{Y}_{220} \cdot \mathbf{Y}_{220} = \frac{15}{8\pi} \sin^2 \theta \cos^2 \theta, \\ \Delta M = \pm 1: \quad & \mathbf{Y}_{22\pm1} \cdot \mathbf{Y}_{22\pm1} = \frac{5}{16\pi} (1 - 3\cos^2 \theta + 4\cos^4 \theta), \\ \Delta M = \pm 2: \quad & \mathbf{Y}_{22\pm2} \cdot \mathbf{Y}_{22\pm2} = \frac{5}{16\pi} (1 - \cos^4 \theta) \end{aligned} \tag{22.4-12}$$

and

$$\mathbf{Y}_{2\,2\,0} \cdot \mathbf{Y}_{2\,2\,0} + 2\mathbf{Y}_{2\,2\,\pm1} \cdot \mathbf{Y}_{2\,2\,\pm1} + 2\mathbf{Y}_{2\,2\,\pm2} \cdot \mathbf{Y}_{2\,2\,\pm2} = 5/4\pi \qquad (22.4\text{-}13)$$

We note again, as in Section 7.2, that if $\theta = 0$, the only nonvanishing transition for any multipole is the one with $\Delta M = \pm 1$—a result that goes back to the transversality of the electromagnetic field.

CHAPTER 23

ABSORPTION AND EMISSION

23.1 Transition Probabilities

In the description of absorption and emission processes in atoms it is often permissible to replace the totality of atomic states by just two discrete states between which the transition occurs (Fig. 23.1). Such a simplification is possible if the energy separation between two states of an atom correspond to the photon energy $\hbar\omega$ whereas all other levels are spaced so that there are no energy differences close to $\hbar\omega$. Assuming this to be the case we consider

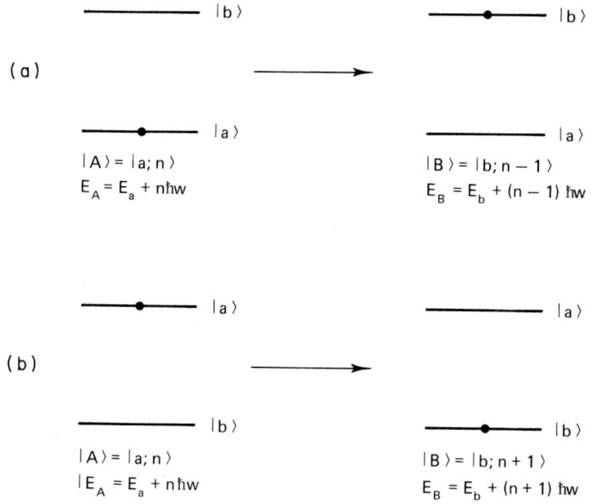

FIG. 23.1 Nomenclature for (a) absorption and (b) emission of photons. Zero point energy has been omitted.

an atom in the state $|a\rangle$ interacting with a radiation field described by $|n_{\mathbf{k}\lambda}\rangle$. The initial state of the entire system, i.e., atom plus field, is $|A\rangle = |a; n_{\mathbf{k}\lambda}\rangle$. If absorption takes place, the atom makes a transition to the state $|b\rangle$ and there is one photon fewer in the field. Hence the final state of the system is $|B\rangle = |b; n_{\mathbf{k}\lambda} - 1\rangle$. In this notation reference to other atomic states or other photons is suppressed since they are not involved in the physical process. Thus

$$|A\rangle = |a, n_{\mathbf{k}\lambda}\rangle, \qquad E_A = E_a + (n_{\mathbf{k}\lambda} + \tfrac{1}{2})\hbar\omega_{\mathbf{k}},$$
$$|B\rangle = |b, n_{\mathbf{k}\lambda} - 1\rangle, \qquad E_B = E_b + (n_{\mathbf{k}\lambda} - \tfrac{1}{2})\hbar\omega_{\mathbf{k}}, \tag{23.1-1}$$

$$E_B - E_A = E_b - E_a - \hbar\omega_{\mathbf{k}} = \hbar(\omega_{ba} - \omega_{\mathbf{k}}), \tag{23.1-2}$$

in which E_a and E_b are the energies of the initial and final atomic states, respectively, and

$$\hbar\omega_{ba} = E_b - E_a. \tag{23.1-3}$$

We shall describe the absorption of a photon by means of the time-dependent perturbation theory developed in Section 14.4 where it was shown that the probability per unit time for a transition from an arbitrary state $|m\rangle$ to a state $|l\rangle$, to first order, is given by

$$W_{lm} = (2\pi/\hbar)|\langle l|V|m\rangle|^2\,\delta(E_l - E_m) \tag{23.1-4}$$

in which V is the perturbing potential in the Schrödinger representation. For photon absorption (and emission) in atoms the most prevalent situation is the one in which $kr \ll 1$ where $2\pi/k$ is the wavelength and r an atomic dimension. Thus, for a wavelength of 1000 Å and $r = 1$ Å, $kr = 2\pi \times 10^{-3}$. In that case the vector potential is essentially constant over the region of the atom and electric dipole matrix elements are applicable. From (22.3-39a) we therefore have, for absorption,

$$W_{\text{abs}} = (4\pi^2 e^2 \omega_{\mathbf{k}} n_{\mathbf{k}\lambda}/V)|\hat{\mathbf{e}}_{\mathbf{k}\lambda} \cdot \langle b|\mathbf{R}|a\rangle|^2\,\delta(E_B - E_A) \tag{23.1-5}$$

in which the δ function

$$\delta(E_B - E_A) = \delta(E_b - E_a - \hbar\omega_{\mathbf{k}}) = (1/\hbar)\,\delta(\omega_{ba} - \omega_{\mathbf{k}}) \tag{23.1-6}$$

ensures the conservation of energy for the system as a whole (atom plus radiation field).

The assumption of a radiation field with a single frequency $\omega_{\mathbf{k}}$ or single mode $\mathbf{k}\lambda$ is clearly unrealistic since it is impossible to achieve infinitely sharp frequencies. Physically it is more meaningful to assume a distribution of modes within a narrow range of frequencies. For a radiation field in a cubical enclosure, according to (22.1-9), the number of modes per unit energy interval is

$$\frac{1}{\hbar}\frac{dN}{d\omega_{\mathbf{k}}} = \frac{V}{(2\pi)^3\hbar c^3}\,\omega_{\mathbf{k}}^2\,d\Omega. \tag{23.1-7}$$

We shall therefore replace the infinitely sharp density distribution $\delta(E_b - E_a - \hbar\omega_{\mathbf{k}})$ in (23.1-5) by (23.1-7) to obtain

$$dW_{\text{abs}} = \frac{\alpha\omega_{\mathbf{k}}{}^3 n_{\mathbf{k}\lambda}}{2\pi c^2} |\hat{\mathbf{e}}_{\mathbf{k}\lambda} \cdot \langle b|\mathbf{R}|a\rangle|^2 \, d\Omega. \tag{23.1-8}$$

Here $\alpha = e^2/\hbar c$ is the fine structure constant and quantities such as $\omega_{\mathbf{k}}$ and $n_{\mathbf{k}\lambda}$ must be understood as averages within the interval.

For emission, the calculation goes through in exactly the same way but with the matrix element (22.3-39b). The final result, analogous to that in (23.1-8), is

$$dW_{\text{em}} = \frac{\alpha\omega_{\mathbf{k}}{}^3(n_{\mathbf{k}\lambda} + 1)}{2\pi c^2} |\hat{\mathbf{e}}_{\mathbf{k}\lambda} \cdot \langle b|\mathbf{R}|a\rangle|^2 \, d\Omega. \tag{23.1-9}$$

The two equations (23.1-8) and (23.1-9) are the probabilities per unit time for absorption and emission, respectively, of a photon with propagation vector $\mathbf{k}$, polarization λ, and frequency $\omega_{\mathbf{k}}$ contained within an element of solid angle $d\Omega$. We may now sum over the two independent polarizations and integrate over the solid angle. Referring to Fig. 23.2

$$\begin{aligned} \hat{e}_{\mathbf{k}1} \cdot \langle b|\mathbf{R}|a\rangle &= |\langle b|\mathbf{R}|a\rangle| \sin\theta_k \cos\varphi_k, \\ \hat{e}_{\mathbf{k}2} \cdot \langle b|\mathbf{R}|a\rangle &= |\langle b|\mathbf{R}|a\rangle| \sin\theta_k \sin\varphi_k, \end{aligned} \tag{23.1-10}$$

$$\sum_{\lambda=1}^{2} |\hat{e}_{\mathbf{k}\lambda} \cdot \langle b|\mathbf{R}|a\rangle|^2 = |\langle b|\mathbf{R}|a\rangle|^2 \sin^2\theta_k,$$

$$\int \sin^2\theta_k \, d\Omega = 8\pi/3. \tag{23.1-11}$$

Therefore the total probability per unit time for absorption of a photon regardless of direction of propagation or polarization is

$$W_{\text{abs}} = (4\alpha\omega^3 n/3c^2)|\langle b|\mathbf{R}|a\rangle|^2 \tag{23.1-12}$$

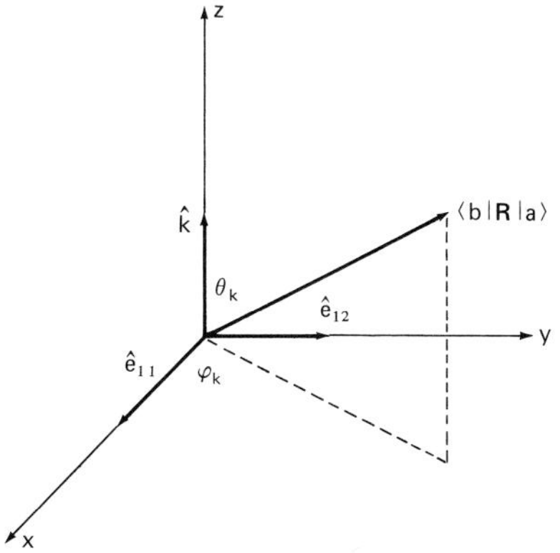

FIG. 23.2 Coordinate system for Eq. (23.1-10).

in which the subscripts referring to the propagation vector and the polarization are no longer needed. The corresponding expression for emission is

$$W_{\mathrm{em}} = (4\alpha\omega^3(n+1)/3c^2)|\langle b|\mathbf{R}|a\rangle|^2. \tag{23.1-13}$$

The close connection between absorption and emission is apparent from a comparison of the expressions for W_{abs} and W_{em}. Nevertheless there is an important distinction inasmuch as the absorption probability is proportional to n while the emission probability is proportional to $n+1$ which then permits W_{em} to be written as a sum of two terms:

$$W_{\mathrm{em}} = W_{\mathrm{i}} + W_{\mathrm{s}} \tag{23.1-14}$$

where

$$W_{\mathrm{i}} = (4\alpha\omega^3 n/3c^2)|\langle b|\mathbf{R}|a\rangle|^2, \tag{23.1-15}$$

$$W_{\mathrm{s}} = (4\alpha\omega^3/3c^2)|\langle b|\mathbf{R}|a\rangle|^2. \tag{23.1-16}$$

W_{i} is identical with W_{abs} (23.1-12), and is known as the *induced* (or stimulated) emission probability per unit time since it is proportional to n, the number of photons of frequency ω. On the other hand W_{s} is independent of n; it is therefore a *spontaneous* emission probability per unit time. Thus, if $n = 0$, there can be no absorption or induced emission, but since W_{s} is nonzero, spontaneous emission can occur.

If $W(ba)$ represents the transition probability per unit time from the state $|\alpha_a J_a M_a\rangle$ with degeneracy g_a to the state $|\alpha_b J_b M_b\rangle$ with degeneracy g_b where α_a and α_b represent the quantum numbers required to complete the specification of the states $|a\rangle$ and $|b\rangle$, respectively, we shall have, in place of (23.1-12) and (23.1-13):

$$W_{\mathrm{abs}}(ba) = \frac{1}{g_a} \sum_{M_a M_b} \frac{4\alpha\omega^3 n}{3c^2} |\langle \alpha_b J_b M_b|\mathbf{R}|\alpha_a J_a M_a\rangle|^2, \tag{23.1-17}$$

$$W_{\mathrm{em}}(ba) = \frac{1}{g_a} \sum_{M_a M_b} \frac{4\alpha\omega^3(n+1)}{3c^2} |\langle \alpha_b J_b M_b|\mathbf{R}|\alpha_a J_a M_a\rangle|^2. \tag{23.1-18}$$

In these expressions the probabilities for all possible transitions $|\alpha_a J_a M_a\rangle \to |\alpha_b J_b M_b\rangle$ have been added, but since the probability of an electron being in any one of the initial states $|\alpha_a J_a M_a\rangle$ is $1/g_a$, the sum of the probabilities must be multiplied by this factor. W_{i} and W_{s} are correspondingly modified. Note that

$$W(ab) = (g_a/g_b)W(ba) \tag{23.1-19}$$

where $W(ba)$ stands for either $W_{\mathrm{abs}}(ba)$ or $W_{\mathrm{em}}(ba)$ and the same for $W(ab)$.

It is often more convenient to work with quantities that do not contain the degeneracy factors g_a, g_b and are therefore symmetric in the initial and final states. Such a quantity is the *line strength S* defined by

$$S(ab) = S(ba) = e^2 \sum_{M_a M_b} |\langle \alpha_b J_b M_b | \mathbf{R} | \alpha_a J_a M_a \rangle|^2. \qquad (23.1\text{-}20)$$

In terms of the line strength

$$\begin{aligned} W_{\text{abs}}(ba) &= (4\omega^3 n/3\hbar c^3 g_a) S(ab) = W_{\text{i}}(ba), \\ W_{\text{abs}}(ab) &= (4\omega^3 n/3\hbar c^3 g_b) S(ab) = W_{\text{i}}(ab), \end{aligned} \qquad (23.1\text{-}21)$$

$$W_{\text{s}}(ba) = (4\omega^3/3\hbar c^3 g_a) S(ab), \qquad W_{\text{s}}(ab) = (4\omega^3/3\hbar c^3 g_b) S(ab). \qquad (23.1\text{-}22)$$

Several properties of the line strength are of interest. In terms of individual components

$$S(ab) = e^2 \sum_{M_a M_b} \{ |\langle \alpha_b J_b M_b | R_x | \alpha_a J_a M_a \rangle|^2 + |\langle \alpha_b J_b M_b | R_y | \alpha_a J_a M_a \rangle|^2 + |\langle \alpha_b J_b M_b | R_z | \alpha_a J_a M_a \rangle|^2 \}.$$

Applying (6.3-28) with $k = 1$, the three terms in the sum are each equal to

$$\tfrac{1}{3} |\langle \alpha_b J_b \| \mathbf{R} \| \alpha_a J_a \rangle|^2$$

so that

$$S(ab) = S(ba) = e^2 |\langle \alpha_b J_b \| \mathbf{R} \| \alpha_a J_a \rangle|^2. \qquad (23.1\text{-}23)$$

In LS coupling

$$\langle \alpha_b J_b \| \mathbf{R} \| \alpha_a J_a \rangle = \langle \alpha_b L_b S_b J_b \| \mathbf{R} \| \alpha_a L_a S_a J_a \rangle.$$

Since **R** does not contain spin operators we may use (6.3-39) to write

$$\begin{aligned} &\langle \alpha_b L_b S_b J_b \| \mathbf{R} \| \alpha_a L_a S_a J_a \rangle \\ &= (-1)^{L_b + S_b + J_a + 1} \delta(S_a, S_b) \\ &\quad \times \sqrt{(2J_b + 1)(2J_a + 1)} \begin{Bmatrix} J_b & 1 & J_a \\ L_a & S_b & L_b \end{Bmatrix} \langle \alpha_b L_b \| \mathbf{R} \| \alpha_a L_a \rangle. \end{aligned} \qquad (23.1\text{-}24)$$

The Kronecker δ and the $6j$ symbol contain the selection rules

$$\Delta S = 0, \qquad \Delta L = 0, \pm 1, \qquad L_a + L_b \geqslant 1, \qquad \Delta J = 0, \pm 1, \qquad J_a + J_b \geqslant 1$$

which have already been encountered in Section 22.4. Therefore the line strength in LS coupling becomes

$$\begin{aligned} S(ab) &= S(ba) \\ &= e^2 |\langle \alpha_b J_b \| \mathbf{R} \| \alpha_a J_a \rangle|^2 \\ &= e^2 (2J_b + 1)(2J_a + 1) \begin{Bmatrix} J_b & 1 & J_a \\ L_a & S & L_b \end{Bmatrix}^2 |\langle \alpha_b L_b \| \mathbf{R} \| \alpha_a L_a \rangle|^2 \end{aligned} \qquad (23.1\text{-}25)$$

with $S = S_a = S_b$.

23.2 Einstein Coefficients and Planck's Law

The Einstein A coefficient in dipole approximation is defined by

$$A = W_s = (4\alpha\omega^3/3c^2)|\langle b|\mathbf{R}|a\rangle|^2 \tag{23.2-1}$$

where W_s is the spontaneous emission probability per unit time (23.1-16). One may also define the *spontaneous lifetime* τ of a state $|a\rangle$ by

$$\tau = 1/A. \tag{23.2-2}$$

The Einstein B coefficient is related to the absorption (or induced emission) probability per unit time by the relation

$$W_{\text{abs}} = W_{\text{i}} = BU_\omega \tag{23.2-3}$$

where $U_\omega\, d\omega$ is the energy per unit volume for photons in the interval ω to $\omega + d\omega$. This quantity consists of the product of (a) the number of modes in the interval $d\omega\, d\Omega$ (23.1-7), (b) the number of independent polarizations (2), (c) the average number of photons (n) in $d\omega$, (d) the average energy of a photon ($\hbar\omega$); this product is then integrated over $d\Omega$ and divided by the volume V. Thus

$$U_\omega\, d\omega = \frac{V\omega^2\, d\omega}{(2\pi c)^3}\, 2n\hbar\omega\, \frac{1}{V}\int d\Omega \tag{23.2-4}$$

or

$$U_\omega = n\hbar\omega^3/\pi^2 c^3 \tag{23.2-5}$$

and upon replacing W_{abs} in (23.2-3) by (23.1-12),

$$B = (4\pi^2\alpha c/3\hbar)|\langle b|\mathbf{R}|a\rangle|^2 \tag{23.2-6}$$

$$= (\pi^2 c^3/\hbar\omega^3)A. \tag{23.2-7}$$

When we take cognizance of possible degeneracies the Einstein A coefficient is written as

$$\begin{aligned} A(ba) &= W_s(ba) \\ &= (4\alpha\omega^3/3c^2 g_a) \sum_{M_a M_b} |\langle \alpha_b J_b M_b|\mathbf{R}|\alpha_a J_a M_a\rangle|^2 \\ &= (4\omega^3/3\hbar c^3 g_a)S(ab), \end{aligned} \tag{23.2-8a}$$

$$A(ab) = (4\omega^3/3\hbar c^3/g_b)S(ab), \tag{23.2-8b}$$

and the Einstein B coefficient is

$$B(ba) = (\pi^2 c^3/\hbar\omega^3)A(ba) = (4\pi^2/3\hbar g_a)S(ab), \tag{23.2-9a}$$

$$B(ab) = (4\pi^2/3\hbar^2 g_b)S(ab). \tag{23.2-9b}$$

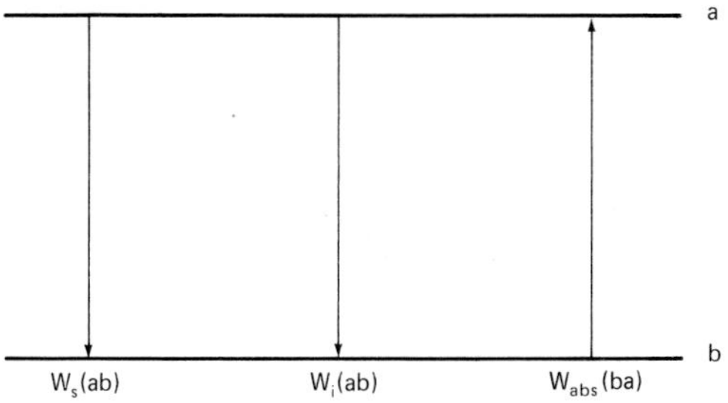

FIG. 23.3 Radiative transitions between two levels. W_s, W_i, and W_{abs} are probabilities for spontaneous emission, induced emission, and absorption, respectively.

Let us now suppose that the populations (atoms per cm^3) in $|a\rangle$ and $|b\rangle$ are N_a and N_b, respectively. When the system is in equilibrium, the number of transitions per unit time $|a\rangle \to |b\rangle$ will be equal to the number of transitions per unit time $|b\rangle \to |a\rangle$ or

$$N_a W_{abs}(ba) = N_b[W_i(ab) + W_s(ab)] \tag{23.2-10}$$

where a refers to the lower state and b to the upper state (Fig. 23.3). In terms of the Einstein coefficients

$$N_a B(ba) U_\omega = N_b[B(ab) U_\omega + A(ab)]. \tag{23.2-11}$$

But

$$B(ab) = \frac{\pi^2 c^3}{\hbar\omega^3} A(ab), \qquad B(ba) = \frac{\pi^2 c^3}{\hbar\omega^3} \frac{g_b}{g_a} A(ab) \tag{23.2-12}$$

from (23.2-9) and (23.1-21). Also, the equilibrium population satisfies the Boltzmann distribution

$$N_a/N_b = (g_a/g_b) e^{\hbar\omega/kT} \tag{23.2-13}$$

where $\hbar\omega$ is the separation in energy between the states $|a\rangle$ and $|b\rangle$, i.e., $E_b - E_a = \hbar\omega$. Inserting these relations into (23.2-11) we obtain

$$U_\omega^{eq} = \frac{\hbar\omega^3}{\pi^2 c^3} \frac{1}{e^{\hbar\omega/kT} - 1}. \tag{23.2-14}$$

This is known as *Planck's distribution law*, where $U_\omega^{eq}\, d\omega$ is the energy per unit volume, under conditions of equilibrium, for photons in the interval ω to $\omega + d\omega$. Note that Planck's law is independent of degeneracies.

The reader is cautioned not to confuse U_ω with U_ν where $\omega = 2\pi\nu$. Since

$$U_\omega\, d\omega = U_\nu\, d\nu,$$

the relation between U_ω and U_ν is

$$U_\omega = U_\nu/2\pi \tag{23.2-15}$$

and, in place of (23.2-7), the Einstein coefficient B_ν referred to U_ν is related to the A coefficient by

$$B_\nu = (c^3/8\pi h\nu^3)A \tag{23.2-16}$$

which differs from (23.2-7) by a factor of 2π. Therefore

$$W_{\text{abs}} = W_{\text{i}} = B_\nu U_\nu = B_\omega U_\omega. \tag{23.2-17}$$

It is pertinent to remark that in semiclassical theory the spontaneous part of the emission probability emerges only after one attempts to reconcile Planck's radiation law, taken to be established experimentally, with thermodynamic requirements. On the other hand, the quantum-mechanical treatment of the radiation field gives the spontaneous emission automatically.

23.3 Oscillator Strengths and Sum Rules

For two states $|a\rangle$ and $|b\rangle$ with $E_b > E_a$ (Fig. 23.3), the oscillator strength $f(ba)$ for absorption $|a\rangle \to |b\rangle$ is defined by

$$f(ba) = (m\hbar\omega_{ba}/2\pi^2 e^2)B(ba) \tag{23.3-1}$$

where

$$\hbar\omega_{ba} = E_b - E_a. \tag{23.3-2}$$

In the dipole approximation, from (23.2-9) and (23.1-20),

$$\begin{aligned} f(ba) &= (2m\omega_{ba}/3\hbar e^2 g_a)S(ab) \\ &= (2m\omega_{ba}/3\hbar g_a) \sum_{M_a M_b} |\langle \alpha_b J_b M_b|\mathbf{R}|\alpha_a J_a M_a\rangle|^2. \end{aligned} \tag{23.3-3}$$

The oscillator strength $f(ab)$ for induced emission $|b\rangle \to |a\rangle$ is also defined as in (23.3-1), i.e.,

$$f(ab) = (m\hbar\omega_{ab}/2\pi^2 e^2)B(ab). \tag{23.3-4}$$

But since

$$g_a B(ba) = g_b B(ab)$$

and

$$\omega_{ab} = -\omega_{ba},$$

we have

$$f(ab) = -(g_a/g_b)f(ba). \tag{23.3-5}$$

By convention, the *absorption* oscillator strength is taken to be *positive* so that the *emission* oscillator for strength is *negative*.

It is of interest to remark that the oscillator strength, as the name suggests, has its origin in the classical theory where an atom emitting or absorbing radiation is modeled by a set of oscillators interacting with the classical fields. The quantum-mechanical oscillator strength for an atomic transition, as previously defined, corresponds to the number of classical oscillators that would emit or absorb the same amount of radiation.

Oscillator strengths satisfy a sum rule which will now be derived. In view of the commutation relations (1.1-8) among components of $\mathbf{r}$ and $\mathbf{p}$,

$$\mathbf{r}\cdot\mathbf{p} - \mathbf{p}\cdot\mathbf{r} = 3i\hbar. \tag{23.3-6}$$

and

$$\sum_{j=1}^{Z} [\mathbf{r}_j\cdot\mathbf{p}_j - \mathbf{p}_j\cdot\mathbf{r}_j] = 3i\hbar Z \tag{23.3-7}$$

where Z is the total number of electrons. Since

$$\langle b|\mathbf{p}|a\rangle = im\omega_{ba}\langle b|\mathbf{r}|a\rangle \tag{23.3-8}$$

from (22.3-15b), we have

$$\begin{aligned}\langle a|\mathbf{r}\cdot\mathbf{p}|a\rangle - \langle a|\mathbf{p}\cdot\mathbf{r}|a\rangle &= \sum_b \langle a|\mathbf{r}|b\rangle\cdot\langle b|\mathbf{p}|a\rangle - \sum_b \langle a|\mathbf{p}|b\rangle\cdot\langle b|\mathbf{r}|a\rangle \\ &= \sum_b [im\omega_{ab}\langle a|\mathbf{r}|b\rangle\cdot\langle b|\mathbf{r}|a\rangle - im\omega_{ab}\langle a|\mathbf{r}|b\rangle\cdot\langle b|\mathbf{r}|a\rangle] \\ &= 2im\sum_b \omega_{ba}|\langle b|\mathbf{r}|a\rangle|^2, \end{aligned}\tag{23.3-9}$$

and

$$\sum_j \langle a|[\mathbf{r}_j\cdot\mathbf{p}_j - \mathbf{p}_j\cdot\mathbf{r}_j]|a\rangle = 2im\sum_b \omega_{ba}\left|\left\langle b\left|\sum_j \mathbf{r}_j\right| a\right\rangle\right|^2 = 3i\hbar Z,$$

or

$$\sum_b \omega_{ba}\left|\left\langle b\left|\sum_j \mathbf{r}_j\right| a\right\rangle\right|^2 = \sum_b \omega_{ba}|\langle b|\mathbf{R}|a\rangle|^2 = \frac{3}{2}\frac{\hbar}{m}Z. \tag{23.3-10}$$

When the atomic states are represented as in (22.4-1) and (22.4-2), the previous equation is replaced by

$$\sum_b \omega_{ba}|\langle b|\mathbf{R}|a\rangle|^2 = \sum_b \omega_{ba}\sum_{M_b} |\langle \alpha_b J_b M_b|\mathbf{R}|\alpha_a J_a M_a\rangle|^2. \tag{23.3-11}$$

However,

$$\sum_{M_b} |\langle \alpha_b J_b M_b|\mathbf{R}|\alpha_a J_a M_a\rangle|^2 = (1/g_a)\sum_{M_a M_b} |\langle \alpha_b J_b M_b|\mathbf{R}|\alpha_a J_a M_a\rangle|^2 \tag{23.3-12a}$$

$$= 3\hbar f(ba)/2m\omega_{ba} \tag{23.3-12b}$$

from (23.3-3). Hence, from (23.3-10),

$$\sum_b f(ba) = Z. \tag{23.3-13}$$

This is the Reiche–Thomas–Kuhn sum rule which states that the sum of all the oscillator strengths from a given level a is equal to the number of electrons Z.

Another sum rule emerges from the expression for the line strength in LS coupling (23.1-25). Writing

$$S(ba) = S(ab) = S(L_bS_bJ_b\, L_aS_aJ_a)$$

and summing over J_a while keeping J_b fixed, we have

$$\sum_{J_a} S(L_bS_bJ_b\, L_aS_aJ_a)$$
$$= e^2 \sum_{J_a} (2J_b + 1)(2J_a + 1)\begin{Bmatrix} J_b & 1 & J_a \\ L_a & S & L_b \end{Bmatrix}^2 |\langle \alpha_b L_b \| \mathbf{R} \| \alpha_a L_a \rangle|^2.$$

The sum may be evaluated with the aid of the orthogonality property of the $6j$ symbols (1.6-16):

$$\sum_{J_a} (2J_a + 1)\begin{Bmatrix} J_b & 1 & J_a \\ L_a & S & L_b \end{Bmatrix}^2 = \frac{1}{2L_b + 1},$$

which then gives

$$\sum_{J_a} S(L_bS_bJ_b\, L_aS_aJ_a) = e^2 \frac{2J_b + 1}{2L_b + 1} |\langle L_b \| \mathbf{R} \| L_a \rangle|^2. \tag{23.3-14}$$

Thus the sum of the line strengths for the transitions that end on the same value of J_b is proportional to $2J_b + 1$. From the symmetry of the expression for the line strength with respect to initial and final states, it is seen that the sum of the line strengths for the transitions that begin on the same value of J_a is proportional to $2J_a + 1$. In terms of intensities these rules state that

(a) The sum of the intensities of all components of a transition $LSJ \to LSJ'$ having a given *initial* value J is proportional to $2J + 1$.

(b) The sum of the intensities of all components of a transition $LSJ \to LSJ'$ having a given *final* value J' is proportional to $2J' + 1$.

Since

$$\sum_J (2J + 1) = (2L + 1)(2S + 1)$$

we may sum (23.3-14) over J_b to obtain

$$\sum_{J_aJ_b} S(L_bS_bJ_b\, L_aS_aJ_a) = e^2(2S + 1)|\langle L_b \| \mathbf{R} \| L_a \rangle|^2 \tag{23.3-15}$$

which provides an expression for the line strength summed over all allowed transitions $J_a \to J_b$.

However, departures from these rules occur when

(1) conditions in the emitting or absorbing medium are not isotropic as, for example, when large d.c. fields are present;

(2) the levels J' in (a) or the levels J in (b) have an appreciable separation in energy compared to kT, in which case the intensities are further modulated by the Boltzmann distribution; and

(3) there is a significant departure from LS coupling.

23.4 Numerical Computations

The essential part in the calculation of transition probabilities is the calculation of the line strengths defined in (23.1-20). Once these have been determined other quantities such as the A coefficient (23.2-8), B coefficient (23.2-9), oscillator strength (23.3-3), and others follow immediately. Numerical work is simplified by the use of atomic units in terms of which the unit of line strength is (Wiese *et al.*, 1966, 1969)

$$a_0{}^2 e^2 = 6.459 \times 10^{-36} \quad \text{cm}^2\ \text{esu}^2. \tag{23.4-1}$$

With S in atomic units and λ (wavelength) in angstroms,

$$A(ab) = 2.026 \times 10^{18}\, S(ab)/g_b \lambda^3 \quad \text{sec}^{-1}, \tag{23.4-2}$$

$$f(ba) = 303.7\, S(ab)/g_a \lambda. \tag{23.4-3}$$

For the purpose of illustration let us examine the 1s–2p transition in hydrogen. To calculate the line strength the matrix elements in (23.1-20) are required. Since only one electron is involved, $\mathbf{R} = \mathbf{r}$ and its components in the spherical basis are

$$\begin{aligned} r_{\pm 1} &= \mp \frac{1}{\sqrt{2}}(x \pm iy) = \sqrt{\frac{4\pi}{3}}\, r Y_{1\pm 1}, \\ r_0 &= z = \sqrt{\frac{4\pi}{3}}\, r Y_{10}. \end{aligned} \tag{23.4-4}$$

Either the Gaunt formula (1.2-29) or its tabulation (Table 11.1) may now be used to evaluate the matrix elements of the spherical harmonics. The results are

$$\begin{aligned} \frac{4\pi}{3} |\langle Y_{11} | Y_{1+1} | Y_{00} \rangle|^2 &= \frac{4\pi}{3} |\langle Y_{10} | Y_{10} | Y_{00} \rangle|^2 \\ &= \frac{4\pi}{3} \langle Y_{1-1} | Y_{1-1} | Y_{00} \rangle|^2 = \frac{1}{3} \end{aligned} \tag{23.4-5}$$

and all other matrix elements are zero. For the radial integral

$$|\langle 2\mathrm{p}|r|1\mathrm{s}\rangle|^2 = \int P_{2\mathrm{p}}(r)P_{1\mathrm{s}}(r)r\,dr = 1.666 \tag{23.4-6}$$

which may be obtained by direct integration with hydrogen radial functions or, more conveniently, from a tabulation by Bethe and Salpeter (1957). Combining (23.4-5) and (23.4-6),

$$\begin{aligned}|\langle 2\mathrm{p}_{+1}|r_{+1}|1\mathrm{s}\rangle|^2 &= |\langle 2\mathrm{p}_0|r_0|1\mathrm{s}\rangle|^2 = |\langle 2\mathrm{p}_{-1}|r_{-1}|1\mathrm{s}\rangle|^2\\ &= \tfrac{5}{9} \quad \text{a.u.}\end{aligned} \tag{23.4-7}$$

In rectangular coordinates

$$x = -\frac{1}{\sqrt{2}}(r_{+1} - r_{-1}), \qquad y = \frac{i}{\sqrt{2}}(r_{+1} + r_{-1}), \qquad z = r_0$$

and

$$\begin{aligned}|\langle 2\mathrm{p}_{+1}|x|1\mathrm{s}\rangle|^2 &= |\langle 2\mathrm{p}_{-1}|x|1\mathrm{s}\rangle|^2\\ &= |\langle 2\mathrm{p}_{+1}|y|1\mathrm{s}\rangle|^2 = |\langle 2\mathrm{p}_{-1}|y|1\mathrm{s}\rangle|^2 = \tfrac{5}{18} \quad \text{a.u.}\\ |\langle 2\mathrm{p}_0|z|1\mathrm{s}\rangle|^2 &= \tfrac{5}{9} \quad \text{a.u.}\end{aligned} \tag{23.4-8}$$

The line strength is now obtained by adding the matrix elements in (23.4-7) or in (23.4-8). In either case

$$S(1\mathrm{s}, 2\mathrm{p}) = e^2 \sum |\langle 2\mathrm{p}|\mathbf{r}|1\mathrm{s}\rangle|^2 = \tfrac{5}{3} \quad \text{a.u.} \tag{23.4-9}$$

For the 1s–2p transition in hydrogen

$$\lambda = 1216.68 \quad \text{Å}, \qquad g_{2\mathrm{p}} = 3.$$

The Einstein A coefficient from (23.2-8) is

$$A(1\mathrm{s}, 2\mathrm{p}) = 6.25 \times 10^8 \quad \mathrm{sec}^{-1} \tag{23.4-10}$$

which then gives for the spontaneous radiative lifetime of the 2p state

$$\tau = 1/A = 0.16 \times 10^{-8} \quad \mathrm{sec}. \tag{23.4-11}$$

The oscillator strength for absorption from 1s to 2p is obtained from (23.3-3) with $g_{1\mathrm{s}} = 1$,

$$f(2\mathrm{p}, 1\mathrm{s}) = 0.416 \tag{23.4-12}$$

and the oscillator strength for emission is given by (23.3-5):

$$f(1\mathrm{s}, 2\mathrm{p}) = -(g_{1\mathrm{s}}/g_{2\mathrm{p}})f(2\mathrm{p}, 1\mathrm{s}) = -0.139. \tag{23.4-13}$$

The results just obtained hold for light that is unpolarized. If the light is polarized linearly, we shall get contributions to the line strength only from

x, y, or z matrix elements in which case S is reduced by a factor of 3. The same reduction occurs if the light is circularly polarized.

If the computation is performed in the $|lsjm\rangle$ representation, the same result is obtained, but in addition it is found that the line strength for $^2S_{1/2} - {}^2P_{3/2}$ is twice that of $^2S_{1/2} - {}^2P_{1/2}$ because of the fourfold degeneracy of $^2P_{3/2}$ compared with the twofold degeneracy of $^2P_{1/2}$. The sum of the two line strengths is, of course, the same as before.

23.5 Line Broadening

The theory of radiative processes developed up to this point is not entirely self-consistent. The reason is that the atomic states $|a\rangle$ and $|b\rangle$ between which transitions are assumed to occur are solutions of a time-independent Schrödinger equation and hence are stationary states. But it was also found that there exists a spontaneous emission process so that all states except for the lowest one must eventually decay and therefore cannot be stationary. This feature must now be incorporated into the formalism.

One approach is to use time-dependent perturbation theory in the form developed in Section 14.3. Thus, from (14.3-6),

$$i\hbar\, \partial c_l(t)/\partial t = \sum_k \langle \varphi_l|V|\varphi_k\rangle c_k(t) \exp[i(E_l - E_k)t/\hbar]. \qquad (23.5\text{-}1)$$

We shall first modify this expression, as was done previously, to the extent of reducing the sum to a single term so that only two states, φ_k (initial) and φ_l (final), are involved in the transition. It will also be assumed that the initial state decays at a rate given by

$$c_k(t) = e^{-\Gamma t/2\hbar} = e^{-t/2\tau} \qquad (23.5\text{-}2)$$

where

$$\tau = \hbar/\Gamma = A^{-1} \qquad (23.5\text{-}3)$$

is the mean life (or lifetime) of the state. With these assumptions

$$c_l(t) = -(i/\hbar)\langle \varphi_l|V|\varphi_k\rangle \int_0^t \exp[i(E_l - E_k)t_1/\hbar] \exp[-\Gamma t_1/2\hbar]\, dt_1$$

and since

$$\int_0^t e^{(\alpha+i\beta)t_1}\, dt_1 = (e^{(\alpha+i\beta)t} - 1)/(\alpha + i\beta),$$

$$\lim_{t\to\infty} |(e^{(\alpha+i\beta)t} - 1)/(\alpha + i\beta)|^2 = 1/(\alpha^2 + \beta^2)$$

we have

$$|c_l(\infty)|^2 = |\langle \varphi_l|V|\varphi_k\rangle|^2/[(E_l - E_k)^2 + \tfrac{1}{4}\Gamma^2]. \qquad (23.5\text{-}4)$$

Using the Fermi golden rule (23.1-4) with the δ function replaced by the density of states dN/dE, an equivalent form of (23.5-4) is

$$|c_l(\infty)|^2 \frac{dN}{dE} = \frac{\hbar}{2\pi} \frac{W_{lk}}{(E_l - E_k)^2 + \frac{1}{4}\Gamma^2}. \tag{23.5-5}$$

So far the development has been quite general. We shall now adapt it to the radiation process–specifically, to spontaneous emission. The general transition $|k\rangle$ to $|l\rangle$, governed by the probability per unit time W_{lk}, will be replaced by the transition $|A\rangle$ to $|B\rangle$ governed by the spontaneous emission probability per unit time W_s. For the initial state $|A\rangle$ we may take a state of zero photons and for $|B\rangle$ a state with one photon. Thus

$$|A\rangle = |a;0\rangle, \qquad |B\rangle = |b;1\rangle \tag{23.5-6}$$

where $|a\rangle$ and $|b\rangle$ are the initial and final atomic states and

$$E_B - E_A = E_b - E_a + \hbar\omega.$$

From (23.2-1) and (23.5-3),

$$W_s = A = \Gamma/\hbar.$$

Equation (23.5-5) now translates into

$$|c_B(\infty)|^2 \frac{dN}{dE} = \frac{1}{2\pi} \frac{\Gamma}{(E_a - E_b - \hbar\omega)^2 + \frac{1}{4}\Gamma^2} \tag{23.5-7a}$$

or

$$|c_B(\infty)|^2 \frac{dN}{d\omega} = \frac{\hbar}{2\pi} \frac{\Gamma}{(E_a - E_b - \hbar\omega)^2 + \frac{1}{4}\Gamma^2} \equiv L(\omega). \tag{23.5-7b}$$

Since $|c_B(\infty)|^2$ is the probability of a transition $|A\rangle \to |B\rangle$ in a time which is long compared with $\tau = \hbar/\Gamma$, the product $|c_B(\infty)|^2\, dN/d\omega$ has the significance of the number of modes per unit (angular) frequency interval that are excited by spontaneous emission irrespective of polarization or direction of propagation. The function $L(\omega)$ is known as a *Lorentzian line shape function*; it is a normalized function since

$$\int L(\omega)\, d\omega = 1. \tag{23.5-8}$$

The net result of this analysis is that the excited atomic state is broadened to a width $\Gamma = \hbar/\tau$ as a result of the spontaneous emission process, resulting in a Lorentzian line shape in the emission spectrum whose full width at half-maximum is Γ.

A similar situation must exist in absorption; that is, the absorption line is broadened due to the finite lifetime of the excited state. Although this conclusion may be reached on the basis of the intrinsic symmetry between absorption and emission, it may also be shown explicitly by returning to (23.5-1). Using the two-state model again, we assume that the final state φ_l decays according to

$$c_l(t) = e^{-\Gamma t/2\hbar} = e^{-t/2\tau}. \tag{23.5-9}$$

If the depletion of the initial state φ_k by absorption is small, as is usually the case, it may be assumed that

$$c_k(t) = c_k(0) = 1. \tag{23.5-10}$$

With these assumptions, (23.5-1) is modified to

$$\frac{\partial c_l(t)}{\partial t} = \frac{1}{i\hbar}\langle\varphi_l|V|\varphi_k\rangle \exp\frac{i(E_l - E_k)t}{\hbar} - \frac{\Gamma}{2\hbar}c_l(t), \tag{23.5-11}$$

in which the first term on the right represents a growth in φ_l due to absorption from φ_k while the second term represents the decay in φ_l due to its radiative lifetime.

The general solution to the differential equation

$$\dot{y} + ay = be^{i\alpha t}$$

is

$$y = Ae^{-at} + \frac{be^{-\alpha t}}{a + i\alpha}.$$

Therefore (23.5-11) has solutions of the form

$$c_l(t) = Ae^{-\Gamma t/2\hbar} + \frac{1}{i\hbar}\frac{\langle\varphi_l|V|\varphi_k\rangle \exp[i(E_l - E_k)t/\hbar]}{(\Gamma/2\hbar) + i(E_l - E_k)/\hbar},$$

and with the initial condition $c_l(0) = 0$, it is found that for times which are long compared to $\hbar/\Gamma$,

$$|c_l(\infty)|^2 = |\langle\varphi_l|V|\varphi_k\rangle|^2/[(E_l - E_k)^2 + \tfrac{1}{4}\Gamma^2]$$

which is identical with (23.5-4). Hence the absorption line shape is also a Lorentzian.

What has been described so far is broadening due to the radiation process itself and the line width associated with it is sometimes called the "natural" line width. There are a number of other processes that contribute to the broadening of a spectral line; we mention several of the most commonly observed effects.

Collisions between atoms may have the effect of interrupting the wave train of the radiation emitted by an atom. This is a classical viewpoint. Quantum mechanically one would say that an excited atom cannot only relax to the ground state by spontaneous emission but may also release its excitation energy to another atom in the course of an impact. The lifetime is therefore shortened and may be completely dominated by the time between collisions. Since the collision frequency increases with pressure this is known as *pressure* or *collision broadening*. Analysis shows that collision broadening produces a Lorentzian profile, the same as the natural profile but of increased width.

The random motions of the atoms in a gas give rise to the Doppler effect which will also broaden a line. The profile due to Doppler broadening is Gaussian as a result of the Maxwellian distribution of velocities.

Doppler and collision broadening are independent. When both effects are present the resulting profile is a convolution of the two and is known as a *Voigt profile*. The basic integral is of the form

$$V(x) = \int_{-\infty}^{\infty} L(y)G(x-y)\,dy \tag{23.5-12}$$

where L and G are the Lorentzian and Gaussian functions, respectively.

23.6 Cross Sections

The absorption cross section per atom σ_A is defined by

$$\sigma_A = P_\omega / I_\omega \tag{23.6-1}$$

where $P_\omega\,d\omega$ is the energy absorbed by one atom per unit time in the interval $d\omega$ and $I_\omega\,d\omega$ the incident energy per unit time in the interval $d\omega$ crossing unit area. Therefore σ_A has dimensions of an area. Since $U_\omega\,d\omega$ (23.2-4) is the energy per unit volume in $d\omega$,

$$I_\omega = cU_\omega. \tag{23.6-2}$$

For the total absorbed energy per unit time, integrated over all frequencies we have

$$\begin{aligned} P = \int P_\omega\,d\omega &= \hbar\omega W_{\text{abs}} = \hbar\omega B(ba)U_\omega \\ &= (\hbar\omega/c)B(ba)I_\omega = (\hbar\omega/c)(g_b/g_a)A(ab)I_\omega \\ &= (\pi^2c^2/\hbar\omega^2)(g_b/g_a)\Gamma I_\omega \end{aligned} \tag{23.6-3}$$

in which (23.2-12) and (23.5-3) have been used and the states a and b are as shown in Fig. 23.3.

If the frequency dependence of P_ω is contained in a normalized line shape function $g(\omega)$ (which need not be a Lorentzian) we may write

$$P_\omega = (\pi^2c^2/\hbar\omega^2)(g_b/g_a)\Gamma I_\omega g(\omega). \tag{23.6-4}$$

with the understanding that ω and I_ω represent average values within the bandwidth of $g(\omega)$. The absorption cross section then becomes

$$\sigma_A = (\pi^2c^2/\hbar\omega^2)(g_b/g_a)\Gamma g(\omega). \tag{23.6-5}$$

For induced emission the cross section is very similar except for the appearance of $B(ab)$ in place of $B(ba)$ in the relations leading to (23.6-3). This removes the degeneracy factors from (23.6-5) giving

$$\sigma_{\mathrm{i}} = (\pi^2 c^2/\hbar\omega^2)\Gamma g(\omega), \tag{23.6-6}$$

and

$$\sigma_{\mathrm{i}} = (g_a/g_b)\sigma_{\mathrm{A}}. \tag{23.6-7}$$

With (23.6-5 and 23.6-6) it is possible to derive the attenuation coefficient, μ, for a beam traversing a medium in which absorption and induced emission takes place. Let N_a and N_b be the populations in states $|a\rangle$ and $|b\rangle$ in a sample of thickness dx in the direction of the beam. Let dI_ω be the loss in I_ω as the beam traverses the medium. We then have

$$-dI_\omega = (N_a\sigma_A - N_b\sigma_i)I_\omega\, dx \tag{23.6-8}$$

The first term on the right represents a loss in energy due to absorption and the second term is a gain in energy due to induced emission. Using (23.6-7),

$$-dI_\omega/I_\omega = (N_a - N_b g_a/g_b)\sigma_{\mathrm{A}}\, dx = \mu\, dx \tag{23.6-9}$$

or

$$\mu = (N_a - N_b g_a/g_b)\sigma_{\mathrm{A}} \tag{23.6-10}$$

Under conditions of thermal equilibrium, the populations satisfy the Boltzmann distribution (23.2-13), which guarantees that $N_a > N_b g_a/g_b$. A beam traversing such a medium is, therefore, attenuated. Frequently, the effect of induced emission may be neglected altogether, in which case the integration of (23.6-9) gives (after dropping subscripts),

$$I = I_0 e^{-\mu x} = I_0 e^{-\sigma N x} \tag{23.6-11}$$

in which I_0 is the incident intensity, I the transmitted intensity, μ the absorption coefficient (cm^{-1}), x the thickness of the absorber (cm) in the direction of the beam, σ the absorption cross section (cm^2), and N the number of absorbers per unit volume (cm^{-3}). It is also possible to create conditions under which $N_a < N_b g_a/g_b$. In that case, the population is said to be inverted and the induced emission exceeds the absorption. The transmitted intensity is then greater than the incident intensity. This is the situation in laser action.

23.7 Photoelectric Effect

The effect whereby the absorption of a photon by an atom results in the removal of an electron from a bound state, leaving the atom in an ionized condition, is known as the *photoelectric effect*. We shall calculate the cross section for this process under certain simplifying assumptions.

The initial state $|i\rangle$ is assumed to be a 1s state

$$|i\rangle = \frac{Z^{3/2}}{\sqrt{\pi a^3}} e^{-Zr/a} \tag{23.7-1}$$

where a is the Bohr radius, and the final state $|f\rangle$ that of a free electron

$$|f\rangle = \frac{1}{\sqrt{V}} e^{i\mathbf{q}\cdot\mathbf{r}} \tag{23.7-2}$$

normalized in the volume V. Then $\hbar\mathbf{q}$ is the momentum of the electron, that is,

$$\mathbf{p}|f\rangle = \mathbf{p}\frac{1}{\sqrt{V}} e^{i\mathbf{q}\cdot\mathbf{r}} = \hbar\mathbf{q}\frac{1}{\sqrt{V}} e^{i\mathbf{q}\cdot\mathbf{r}}$$
$$= \hbar\mathbf{q}|f\rangle. \tag{23.7-3}$$

Hence $|f\rangle$ is a momentum eigenfunction with eigenvalue $\hbar\mathbf{q}$.

For photon absorption it is necessary to evaluate the matrix element

$$\langle f|e^{i\mathbf{k}\cdot\mathbf{r}}\hat{\mathbf{e}}_{\mathbf{k}\lambda}\cdot\mathbf{p}|i\rangle \tag{23.7-4}$$

which now becomes

$$= \left(\frac{Z}{a}\right)^{3/2} \frac{1}{\sqrt{\pi V}} \hat{\mathbf{e}}_{\mathbf{k}\lambda}\cdot\langle e^{i\mathbf{q}\cdot\mathbf{r}}|\mathbf{p}e^{i\mathbf{k}\cdot\mathbf{r}}|e^{-Zr/a}\rangle$$
$$= \left(\frac{Z}{a}\right)^{3/2} \frac{\hbar}{\sqrt{\pi V}} \hat{\mathbf{e}}_{\mathbf{k}\lambda}\cdot\mathbf{q}\langle e^{i\mathbf{q}\cdot\mathbf{r}}|e^{i\mathbf{k}\cdot\mathbf{r}}|e^{-Zr/a}\rangle.$$

Defining

$$\mathbf{K} = \mathbf{k} - \mathbf{q}, \tag{23.7-5}$$

we have

$$\langle e^{i\mathbf{q}\cdot\mathbf{r}}|e^{i\mathbf{k}\cdot\mathbf{r}}|e^{-Zr/a}\rangle = \int e^{i\mathbf{K}\cdot\mathbf{r}}e^{-Zr/a}\,d\mathbf{r}$$
$$= 2\pi\int_0^\infty dr\,r^2 e^{-Zr/a}\int_0^\pi e^{i\mathbf{K}\cdot\mathbf{r}}\sin\theta'\,d\theta' \tag{23.7-6}$$

in spherical coordinates. If the coordinate system is oriented so that the z axis is along the direction of $\mathbf{K}$,

$$\int_0^\pi e^{i\mathbf{K}\cdot\mathbf{r}}\sin\theta'\,d\theta' = \frac{2}{Kr}\sin Kr$$

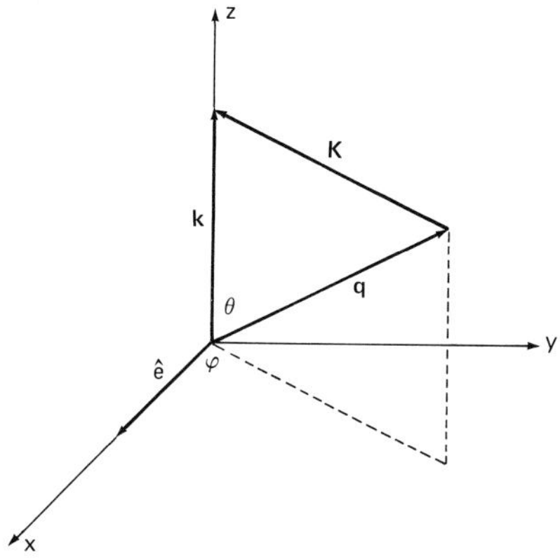

FIG. 23.4 Coordinate system for Eq. (23.7-9).

and

$$\langle e^{i\mathbf{q}\cdot\mathbf{r}}|e^{i\mathbf{k}\cdot\mathbf{r}}|e^{-Zr/a}\rangle = \frac{4\pi}{K}\int_0^\infty dr\, e^{-Zr/a} r \sin Kr$$

$$= \frac{8\pi Z a^3}{(Z^2 + a^2K^2)^2}. \tag{23.7-7}$$

Thus

$$\langle \mathrm{f}|e^{i\mathbf{k}\cdot\mathbf{r}}\hat{\mathbf{e}}_{\mathbf{k}\lambda}\cdot\mathbf{p}|\mathrm{i}\rangle = \frac{8\sqrt{\pi}(aZ)^{3/2}Z\hbar}{\sqrt{V}(Z^2 + a^2K^2)^2}(\hat{\mathbf{e}}_{\mathbf{k}\lambda}\cdot\mathbf{q}). \tag{23.7-8}$$

The relationship among the various vectors is shown in Fig. 23.4; it is noted that

$$\hat{\mathbf{e}}_{\mathbf{k}\lambda}\cdot\mathbf{q} = q\sin\theta\cos\varphi, \tag{23.7-9a}$$

$$K^2 = (\mathbf{k} - \mathbf{q})^2 = k^2 + q^2 - 2kq\cos\theta. \tag{23.7-9b}$$

It will now be assumed that the photon energy is large compared with the binding energy of the electron in the atom but small compared to the rest mass of the electron. The first assumption ensures that the final state is that of a free electron and the second avoids the necessity of a relativistic treatment. Fortunately both conditions are satisfied when

$$ka \approx Z \tag{23.7-10}$$

where k is the photon wave number, a the Bohr radius, and Z the atomic number of the atom. To see the justification of this criterion we first calculate the photon energy corresponding to (23.7-10):

$$E = \hbar ck = \hbar cZ/a = (Z^2me^4/2\hbar^2)(2\hbar c/Ze^2). \tag{23.7-11}$$

The first factor on the right is of the order of the binding energy I and the second factor is $2 \times 137/Z$; hence E is at least several times larger than I. At the same time (23.7-11) may be rearranged to give

$$E = (Ze^2/\hbar c)mc^2 = (Z/137)mc^2 \tag{23.7-12}$$

so that $E \ll mc^2$ provided Z is not too high.

With the assumption that $E \gg I$, the photon energy becomes comparable to the kinetic energy of the electron

$$\hbar ck = mv^2/2$$

from which it follows that

$$k/q = v/2c = Z/qa$$

or

$$qa \gg Z \tag{23.7-13}$$

and

$$\begin{aligned} Z^2 + a^2K^2 &= Z^2 + a^2(k^2 + q^2 - 2qk\cos\theta) \\ &\approx q^2a^2(1 - \beta\cos\theta) \qquad (\beta = v/c). \end{aligned} \tag{23.7-14}$$

Therefore the matrix element (23.7-4) is

$$\langle \mathrm{f}|e^{i\mathbf{k}\cdot\mathbf{r}}\hat{\mathbf{e}}_{\mathbf{k}\lambda}\cdot\mathbf{p}|\mathrm{i}\rangle = \frac{8\sqrt{\pi}(aZ)^{3/2}Z\hbar q\sin\theta\cos\varphi}{\sqrt{V}[q^2a^2(1-\beta\cos\theta)]^2} \tag{23.7-15}$$

and the complete interaction matrix element is

$$\frac{e}{m}\sqrt{\frac{2\pi\hbar n}{\omega V}}\langle \mathrm{f}|e^{i\mathbf{k}\cdot\mathbf{r}}\hat{\mathbf{e}}_{\mathbf{k}\lambda}\cdot\mathbf{p}|\mathrm{i}\rangle. \tag{23.7-16}$$

On inserting these expressions into the Fermi golden rule with the density of states

$$\frac{dN}{dE} = \frac{V}{(2\pi)^3}\frac{mq}{\hbar^2}\,d\Omega, \tag{23.7-17}$$

the probability per unit time for photoelectric absorption is

$$W = \frac{n}{V}\frac{e^2}{\hbar c}\frac{32Z^5\hbar\sin^2\theta\cos^2\varphi}{q^5a^5km(1-\beta\cos\theta)^4}\,d\Omega \tag{23.7-18}$$

and the differential cross section is

$$\frac{d\sigma}{d\Omega} = \frac{WV}{\hbar c} = \frac{e^2}{\hbar c}\frac{32Z^5\hbar\sin^2\theta\cos^2\varphi}{kcq^5a^5m(1-\beta\cos\theta)^4}. \tag{23.7-19}$$

When $\theta = \pi/2$, $\varphi = 0$ or π, the cross section is a maximum, that is, when the momentum of the electron is parallel to the polarization of the photon. The angular factor in the denominator causes the electron emission to have a slight forward tilt which increases as v/c increases. Ignoring the factor $(1 - \beta \cos \theta)^4$, the total cross section is

$$\sigma = \int \frac{d\sigma}{d\Omega} \, d\Omega = \frac{128\pi}{3} \frac{\alpha Z^5 \hbar}{\omega m (qa)^5}. \tag{23.7-20}$$

23.8 Survey of Atomic Spectra

Various compilations of atomic spectra, such as those published by the National Bureau of Standards (Moore, 1971; Wiese *et al.*, 1966, 1969), serve as references for the reader who requires specific information. Here we must, of necessity, confine our remarks to a few general features:

(1) In most cases the spectra arise from single-electron excitations. For spectra in or near the visible region a transition of one of the outermost electrons is responsible for the spectrum; in the x-ray region an inner shell electron is involved.

(2) Most of the atomic spectra are of the electric dipole type with transition probabilities in the range of 10^7 to 10^9 sec^{-1}. Magnetic dipole and electric quadrupole transitions are less probable by a factor of 10^5 but may nevertheless be observed in special situations.

(3) Very few spectra conform to pure *LS* or *jj* coupling. In passing from light to heavy atoms there is a continuous transition from predominantly *LS* to predominantly *jj* coupling.

(4) As the departure from *LS* coupling increases, intercombination lines ($\Delta S \neq 0$) become more probable. Such lines are weak in the spectra of the light atoms and become fairly strong in heavy atoms.

The classification of spectra is facilitated by separating the atoms into groups based on the number and type of electrons that are excitable. The best-understood spectrum is, of course, that of hydrogen whose main features are contained in the expression

$$\frac{1}{\lambda} = R_{\mathrm{H}} \left[\frac{1}{{n_1}^2} - \frac{1}{{n_2}^2} \right]$$

where

$$R_{\mathrm{H}} = \frac{R_\infty}{1 + (m/M)} = \frac{\mu e^4}{2\hbar^2} = 109{,}678.758 \quad \mathrm{cm}^{-1} = 13.595 \quad \mathrm{eV},$$

$$R_\infty = \frac{me^4}{2\hbar^2} = 109{,}737.311 \quad \mathrm{cm}^{-1} = 13.605 \quad \mathrm{eV},$$

and m is the electron mass, M the proton mass, $\mu = mM/(m + M)$, R_H, which is based on the reduced mass of the electron–proton system, is the ionization potential of hydrogen, and R_∞ the Rydberg constant. The major series are those associated with $n_1 = 1, \ldots, 5$ (Table 23.1 and Fig. 23.5); another classification which includes reference to angular momentum quantum numbers is shown in Table 23.2.

The alkali elements (Li, Na, K, Rb, Cs, Fr) have a single electron outside of a filled s or p shell; the ground state is $^2S_{1/2}$. Because of the spherical symmetry of the filled shells, the outer electron moves in a central field as in hydrogen. Therefore the spectra of these elements resemble the spectrum of hydrogen and consist of analogous series. In the ground state, the outer s electron is more loosely bound than any of the other electrons in the filled shells and is also more loosely bound than the s electron in hydrogen. Hence the ionization potentials of these elements are lower; however, the doublet splitting due to spin-orbit interaction is much larger than in hydrogen.

Cu, Ag, and Au in their ground states have an s electron outside of a filled d shell. However, the binding energies of the s and d electrons are comparable so that the spectra of these elements are interpretable on the basis of either s or d excitation.

Elements with a single p electron outside of closed shells are B, Al, Ga, In, and Tl. In boron, for example, the ground state configuration is $1s^2 2s^2 2p$; the spectrum therefore resembles that of an alkali atom but with the lowest ^2S missing. In the same category are the elements F, Cl, Br, I, and At whose ground state configurations are np^5 which gives rise to the single term ^2P as in the case of np but the multiplet structure is inverted because the shell is more than half-full.

TABLE 23.1

Main Series in the Hydrogen Spectra

Series	n_1	n_2	Longest wavelength (A^0)
Lyman	1	$2, 3, \ldots$	1215.68
Balmer	2	$3, 4, \ldots$	6562.79
Ritz-Paschen	3	$4, 5, \ldots$	18,751
Brackett	4	$5, 6, \ldots$	40,510
Pfund	5	$6, 7, \ldots$	74,560

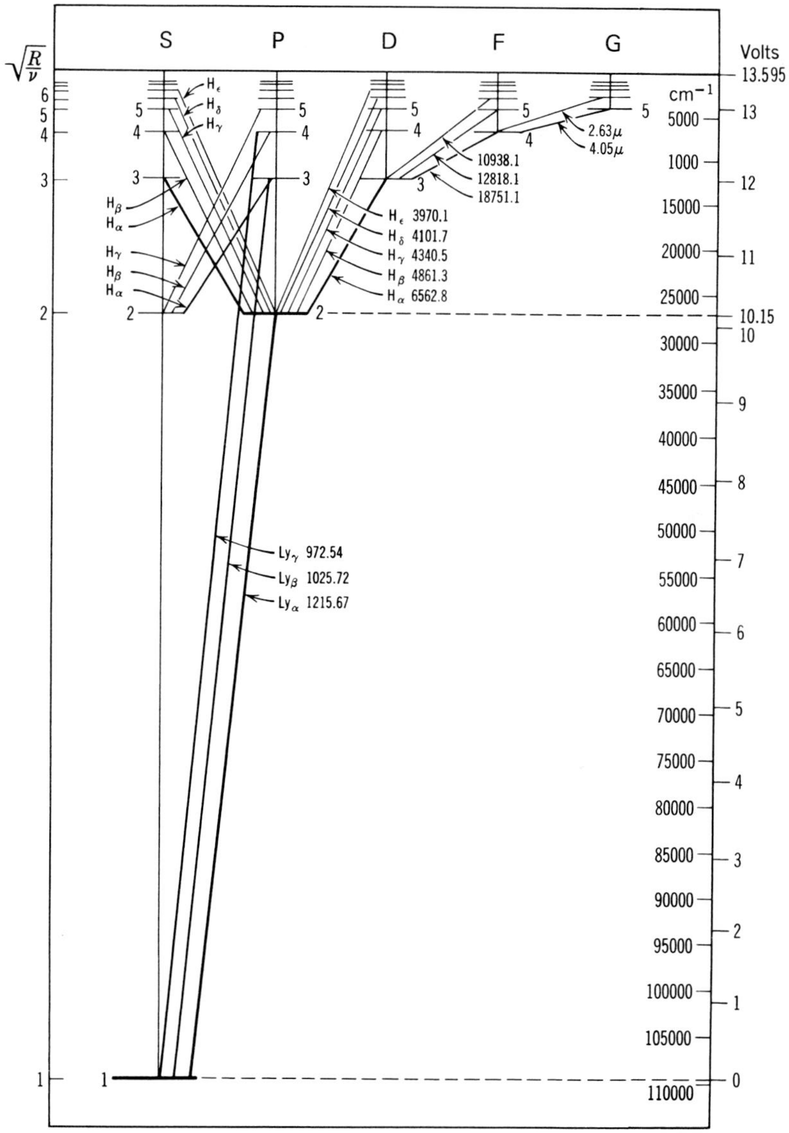

FIG. 23.5 Energy levels and main transitions in hydrogen (Shore and Menzel, 1968).

Helium, the lightest element with a closed shell configuration in the ground state can produce two-term systems (Fig. 23.6)—singlets (parahelium) and triplets (orthohelium); however, both singlet and triplet S states must remain single. In the ground state the configuration is $1s^2$; hence the only

TABLE 23.2

Alternate Classification of the Hydrogen Spectrum

Series	Transition	n	Number of lines
Principal	$1\ ^2S_{1/2}$–$n\ ^2P_{1/2,3/2}$	2, 3, . . .	2
Sharp	$2\ ^2P_{1/2,3/2}$–$n\ ^2S_{1/2}$	3, 4, . . .	2
Diffuse	$2\ ^2P_{1/2}$–$n\ ^2D_{3/2}$ $2\ ^2P_{3/2}$–$n\ ^2D_{3/2,5/2}$	3, 4, . . .	3
Fundamental	$3\ ^2D_{3/2}$–$n\ ^2F_{5/2}$ $3\ ^2D_{5/2}$–$n\ ^2F_{5/2,7/2}$	4, 5, . . .	3

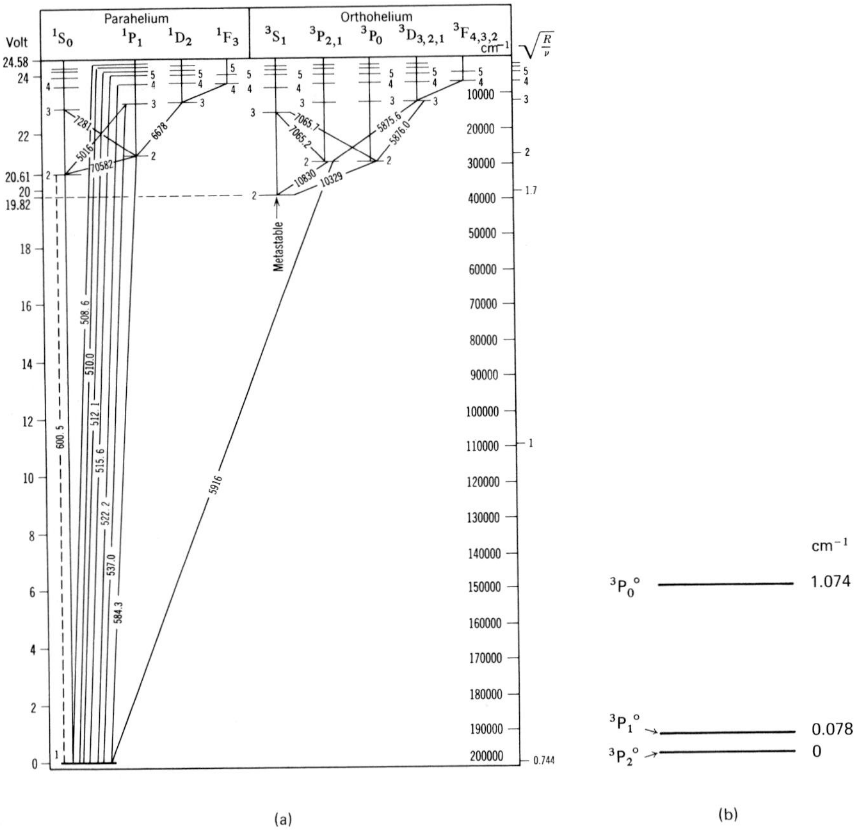

FIG. 23.6 Energy levels and main transitions in helium (a) and details of the lowest ^3P state (b). (Shore and Menzel, 1968.)

possible term consistent with the Pauli principle is $^1S_0$. The lowest triplet state $^3S_1$ arises from the configuration 1s2s and each triplet term is lower than the singlet term with the same values of n. The two-term systems do not mix since transitions between singlets and triplets are forbidden in the LS coupling approximation, and as a result there are two separate spectral systems, one associated with singlet states and the other with triplet states. The lowest triplet state 1s2s $^3S_1$ is metastable since there is no allowed transition to a lower state. There is a curious anomaly in the 2 ^3P term which shows a reversal in the order of the levels, i.e., the levels increase in energy in the order 2 $^3P_2$, 2 $^3P_1$, 2 $^3P_0$. This is surprising since LS coupling predicts the opposite order. An explanation of this phenomenon has been given on the basis of spin-spin interactions and relativistic corrections.

The alkaline earth elements, Be, Mg, Ca, Sr, Ba, Ra, Zn, Cd, and Hg, contain two electrons (ns^2) outside of closed shells. Hence there are singlet and triplet terms as in helium with the triplet terms lying lower than the corresponding singlets. However, the selection rule $\Delta S = 0$ is not as strict as in helium so the spectra contain intercombination lines which increase in strength as one progresses from the lighter to the heavier elements due to increasing departure from LS coupling. Also, the ionization potentials are smaller than in helium. These elements also provide examples in which it is necessary to consider configuration interaction. Thus, in calcium, 3d5s ^3D lies in the same energy region as 4s8d ^3D and 4s9d ^3D and the states are all of the same type, that is, in regard to L, S, and the parity; hence each of these states are properly regarded as a mixture of configurations.

For the elements C, Si, Ge, Sn, and Pb the ground state configuration is np^2 and the ground term is $^3P_0$ as required by Hund's rule for a shell that is less than half-full. In configurations of the type $np(n+1)s$ like 2p3s in C, 3p4s in Si, etc., there is a gradual transition from LS coupling in the light elements to jj coupling in the heavier elements. O, S, Se, Te, Po have the configuration np^4 which produces the same terms as those arising from np^2 but with an inverted multiplet structure. A half-full shell np^3 is found in the ground states of N, P, As, Sb, and Bi. The lowest term is ^4S while the other terms from np^3—^2P, ^2D—are metastable because of the selection rule $\Delta S = 0$, but some intercombinations are observed in the heavier elements. The closed shell configuration np^6 occurs in Ne, Ar, Kr, Xe, and Rn; in each case the lowest term is $^1S_0$.

There are several groups of elements in which the occupation of s orbitals competes with the occupation of d orbitals. As a result, these elements have one or two s electrons outside of *partially* filled d shells. The groups are classified according to their unfilled d shells: the 3d (iron) group (Sc, Ti, V, Cr, Mn, Fe, Co, Ni), the 4d (palladium) group (Y, Zr, Nb, Mo, Tc, Ru, Rb, Pd), and the 5d (platinum) group (Lu, Hf, Ta, W, Re, Os, Ir, Pt). In view of the large number of terms that are produced by several d electrons, these elements have rich spectra in the visible and ultraviolet. Among these elements are those with important magnetic properties which also are attributable to the unfilled d shells.

Finally we have the elements whose properties are mainly determined by their f electrons. These are the lanthanides (4f) group (Ce, Pr, Nd, Pn, Sm, Eu, Gd, Tb, Dy, Ho, Er, Tm, Yb) and the actinides (5f) group (Ac, Th, Pa, U, Np, Pu, Am, Cm, Bk, Cf). Here too there are numerous terms arising from f electron configurations and consequently many lines in the spectra. The magnetic properties of the lanthanides are also determined by the partially filled 4f shell which is shielded by the outer 5s and 5p shells.

Spectra which originate from transitions by inner shell (core) electrons are known as x-ray spectra. In order to produce such spectra it is first necessary to remove (ionize) an electron from an inner shell thus creating a vacancy or hole. This may be accomplished by electron bombardment, x-ray irradiation, or by other means, and the deeper the shell, the more energy is required and the higher is the excitation of the ionized atom. Upon reverting to its ground state the atom may undergo a transition whereby the vacancy is filled by an electron from an outer shell with the release of an x-ray photon. These events are shown schematically in Fig. 23.7.

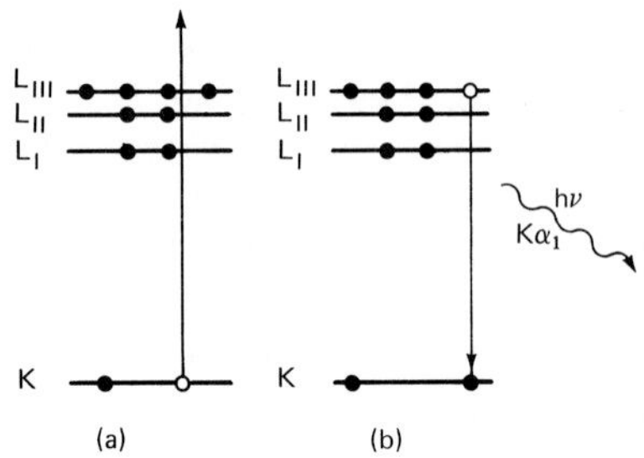

FIG. 23.7 X-ray absorption and fluorescence. (a) Excitation; (b) deexcitation by fluorescence.

As shown previously, a vacancy in a closed shell produces the same term as a single electron but with an inverted multiplet structure. X-ray states are therefore characterized by the missing electron and the selection rules are those pertaining to a single electron, namely, $\Delta l = \pm 1$, $\Delta j = 0, \pm 1$. A diagram of x-ray terms and possible transitions is shown in Fig. 23.8. Thus, for the K level, $n = 1, l = 0, j = \frac{1}{2}$; for the L level

$$n = 2, \quad l = \begin{cases} 0, & j = \frac{1}{2} \quad (\mathrm{L_I}) \\ 1, & j = \frac{1}{2}, \frac{3}{2} \quad (\mathrm{L_{II}}, \mathrm{L_{III}}) \end{cases}$$

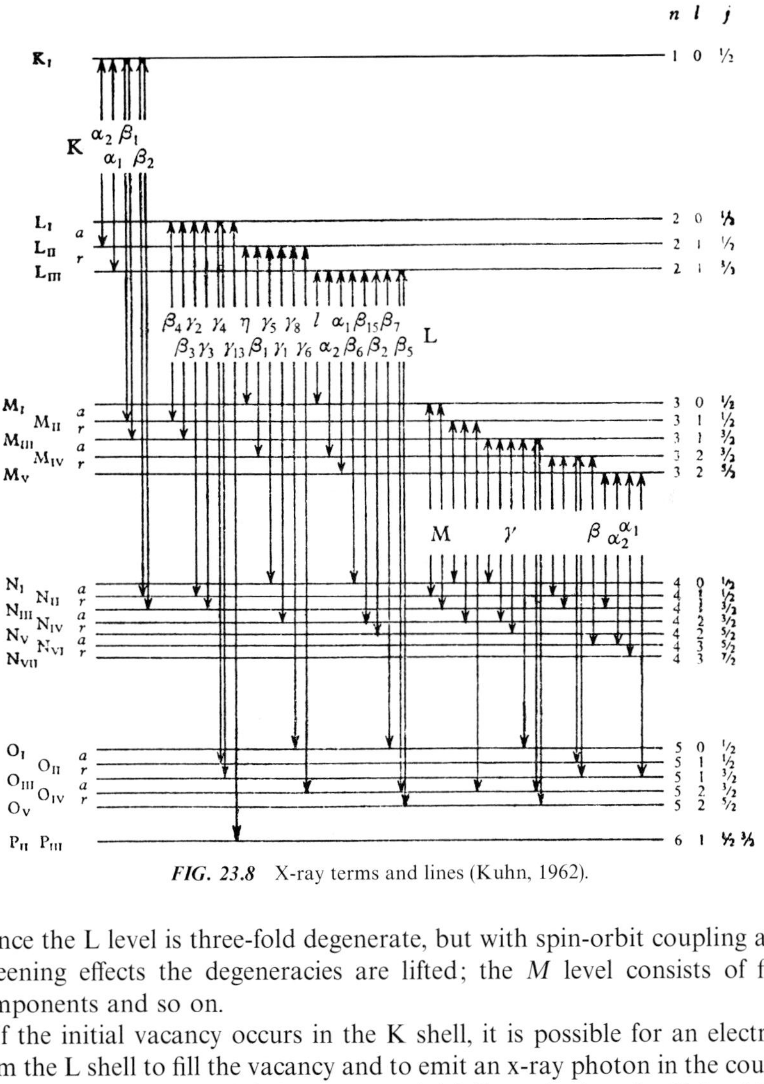

FIG. 23.8 X-ray terms and lines (Kuhn, 1962).

Hence the L level is three-fold degenerate, but with spin-orbit coupling and screening effects the degeneracies are lifted; the *M* level consists of five components and so on.

If the initial vacancy occurs in the K shell, it is possible for an electron from the L shell to fill the vacancy and to emit an x-ray photon in the course of the transition. More briefly, an atom initially excited to the K level has the possibility of making a radiative transition to the L level. The selection rules permit two transitions of this type which give rise to emission lines labeled K_{α_1} and K_{α_2}. If the vacancy in the K shell is filled by an electron from the M shell, the resulting lines are known as K_{β_1} and K_{β_2}. Now after the emission of a K_α photon the vacancy has moved to the L shell (or the atom is now in the L level) and again we may have a series of transitions analogous to those which filled the K shell. The corresponding lines are shown in Fig. 23.8. In short, the emission of the K series is accompanied by the L, M, N . . . series; the emission of the L series is accompanied by M, N, . . . series, and so on.

The general form of an x-ray absorption spectrum is shown in Fig. 23.9. It may be understood as follows: as the energy of the incident photons is increased from zero, first the outer shell electrons are excited and as the energy reaches a few electron volts some of the electrons will be ionized. This is the onset of the photoelectric effect. As the photon energy is increased further the electrons in inner shells begin to absorb energy and when the photon energy just reaches the binding energy of a particular shell there is a sharp rise in absorption. Such discontinuities in the absorption spectrum are called *absorption edges*. Beyond a particular edge the absorption declines in a manner governed by the photoelectric cross section until the next edge is reached where again there is a sudden rise in absorption. The highest energy absorption edge is the K edge and corresponds to an energy just sufficient to remove an electron from the K shell but not to impart any kinetic energy to it. From the previous discussion it is seen that each absorption edge is accompanied by emission lines also known as the *fluorescence spectrum*.

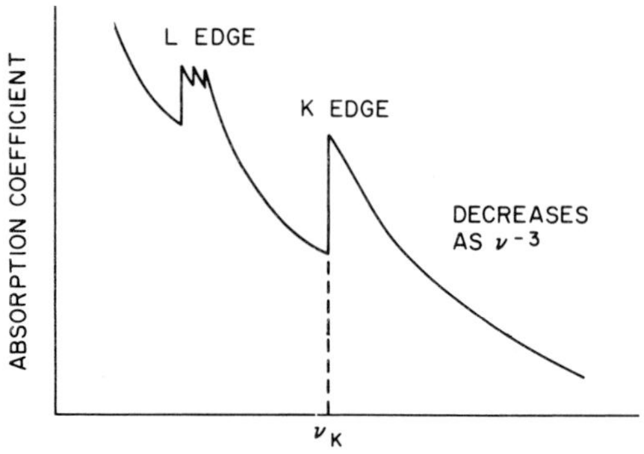

FIG. 23.9 X-ray absorption spectrum.

Consider again the creation of a vacancy in an inner shell of an atom, say, the K shell. When the atom relaxes by the emission of photons we have the situation just described. But this is not the only mode of relaxation. The excitation energy of the atom can be imparted to the ejection of an electron from one of the outer shells without any photons being produced in the process. Thus, for example, the K shell vacancy may be filled by an electron initially in the L_I shell and the excess energy is used to eject an electron from L_{III} (Fig. 23.10). Such radiationless deexcitation of an excited atom by emission of electrons rather than x-rays is known as the *Auger* effect. It must

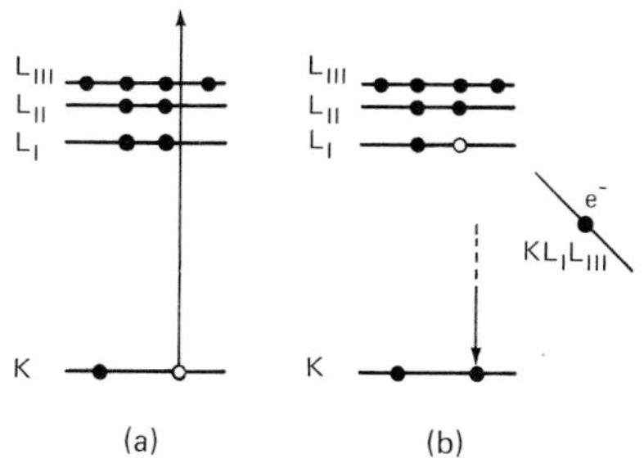

FIG. 23.10 Auger effect. (a) Excitation; (b) by Auger effect.

be recognized that an Auger process leaves the atom in a doubly ionized state which can further relax by photon emission so that an Auger electron is usually accompanied by x-rays.

The effect of an external **B** or **E** field is to remove degeneracies of the atomic states in whole or in part. A magnetic field that produces splittings that are small compared to the spin-orbit interaction energies—a so-called weak field—will split each J level into $2J + 1$ components whose splitting is given by (21.6-4). Each line in the spectrum will then have a number of components (Zeeman components) consistent with the selection rules

$$\Delta J = 0, \quad \pm 1; \qquad \Delta M = 0(\pi), \quad \pm 1(\sigma).$$

The spectrum viewed parallel to the field contains only those σ components that are circularly polarized. In a direction perpendicular to the field both π and σ components are seen and they are plane polarized—the π components are polarized parallel to the field and the σ components, perpendicular to the field. At an arbitrary angle to the field one sees elliptically polarized components. These results are summarized in Table 23.3. An example of an energy level diagram and the possible Zeeman components is shown in Fig. 23.11. At much stronger fields **L** and **S** are no longer coupled

TABLE 23.3

Polarization of the Zeeman Components

Direction of magnetic field	Direction of observation	Polarization	
		σ ($\Delta M = \pm 1$)	π ($\Delta m = 0$)
z	z	Circular	Not observed
z	x	Plane (y)	Plane (z)
z	y	Plane (x)	Plane (z)

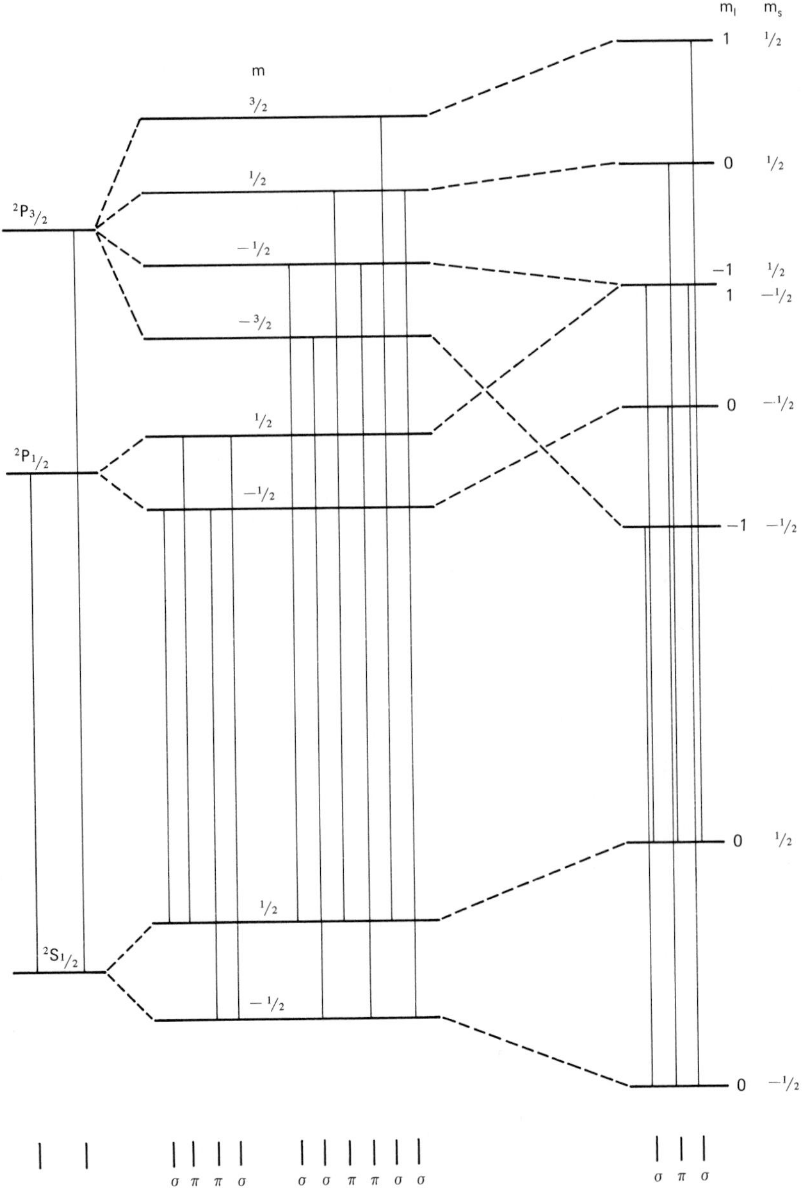

FIG. 23.11 Zeeman and Paschen–Back effects in the ^2S–^2P transitions.

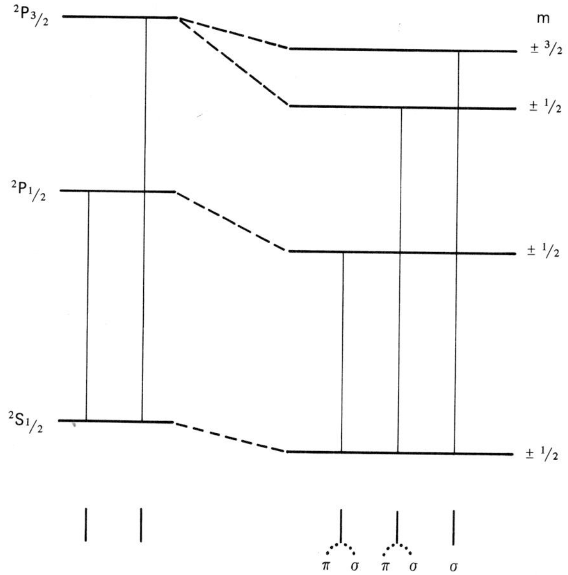

FIG. 23.12 Stark effect in the ^2S–^2P transitions. Stark lines are shifted to longer wavelengths relative to the field-free transitions.

and the pertinent selection rules are

$$\Delta M_L = 0(\pi), \quad \pm 1(\sigma), \qquad \Delta M_S = 0.$$

This is the region of the Paschen–Back effect; an example is shown in Fig. 23.11.

In an **E** field, all atoms except hydrogen show a splitting (Stark effect) which is proportional to E^2. The selection rules are the same as in the Zeeman case; hence Table 23.3 is applicable here too. However, in contrast to the Zeeman patterns the Stark patterns are not symmetric about the original line (Fig. 23.12). If the field is sufficiently strong to cause splittings that are large compared to the fine structure (separation between terms), the Stark effect passes from a quadratic to a linear dependence on the field.

CHAPTER 24

HIGHER ORDER ELECTROMAGNETIC INTERACTIONS

24.1 The Kramers–Heisenberg Formula

The processes of absorption and emission described in the previous chapter are one-photon processes since in every transition the field either loses or gains a photon while the atom undergoes a corresponding change of state to keep the total energy (atom plus field) constant. There are other important interactions between an atom and a radiation field which involve a change of more than one photon. In a general scattering event, for example, the field loses an incident photon characterized by $\mathbf{k}\lambda$ and gains another photon—the scattered one—characterized by $\mathbf{k}'\lambda'$. With laser sources, numerous multiphoton processes have been observed although the present discussion will be restricted to several aspects of two-photon interactions.

In the absence of an external magnetic field the Hamiltonian describing the interaction of an electron with a radiation field is

$$\begin{gathered}\mathscr{H}_{\text{int}} = \mathscr{H}_1 + \mathscr{H}_2, \\ \mathscr{H}_1 = (e/mc)\mathbf{A}\cdot\mathbf{p}, \qquad \mathscr{H}_2 = (e^2/2mc^2)A^2\end{gathered} \tag{24.1-1}$$

where $\mathbf{A}$ is given by (22.2-17). As in (22.3-8) let

$$\mathscr{H}_1 = \mathscr{H}_1^{(-)} + \mathscr{H}_1^{(+)}$$

with $\mathscr{H}_1^{(-)}$ and $\mathscr{H}_1^{(+)}$ given by (22.3-9). For $\mathscr{H}_2$ we have

$$\begin{aligned}\mathscr{H}_2 = {} & \frac{\pi e^2\hbar}{mV}\sum_{\mathbf{k}\lambda}\sum_{\mathbf{k}'\lambda'}\frac{1}{\sqrt{\omega_{\mathbf{k}}\omega_{\mathbf{k}'}}}(\hat{\mathbf{e}}_{\mathbf{k}\lambda}\cdot\hat{\mathbf{e}}_{\mathbf{k}'\lambda'}) \\ & \times(a_{\mathbf{k}\lambda}e^{i\mathbf{k}\cdot\mathbf{r}} + a_{\mathbf{k}\lambda}^\dagger e^{-i\mathbf{k}\cdot\mathbf{r}})(a_{\mathbf{k}'\lambda'}e^{i\mathbf{k}'\cdot\mathbf{r}} + a_{\mathbf{k}'\lambda'}^\dagger e^{-i\mathbf{k}'\cdot\mathbf{r}})\end{aligned} \tag{24.1-2}$$

$$= \frac{\pi e^2 \hbar}{mV} \sum_{\mathbf{k}\lambda} \sum_{\mathbf{k}'\lambda'} \frac{1}{\sqrt{\omega_{\mathbf{k}}\omega_{\mathbf{k}'}}} (\hat{\mathbf{e}}_{\mathbf{k}\lambda} \cdot \hat{\mathbf{e}}_{\mathbf{k}'\lambda'})$$

$$\times \{a^\dagger_{\mathbf{k}\lambda} a^\dagger_{\mathbf{k}'\lambda'} \exp[-i(\mathbf{k} + \mathbf{k}') \cdot \mathbf{r}] + a_{\mathbf{k}\lambda} a_{\mathbf{k}'\lambda'} \exp[i(\mathbf{k} + \mathbf{k}') \cdot \mathbf{r}]$$

$$+ a^\dagger_{\mathbf{k}\lambda} a_{\mathbf{k}'\lambda'} \exp[-i(\mathbf{k} - \mathbf{k}') \cdot \mathbf{r}] + a_{\mathbf{k}\lambda} a^\dagger_{\mathbf{k}'\lambda'} \exp[i(\mathbf{k} - \mathbf{k}') \cdot \mathbf{r}]\}$$

$$\equiv \mathscr{H}^{(+)}_{\mathbf{k}\lambda}\mathscr{H}^{(+)}_{\mathbf{k}'\lambda'} + \mathscr{H}^{(-)}_{\mathbf{k}\lambda}\mathscr{H}^{(-)}_{\mathbf{k}'\lambda'} + \mathscr{H}^{(+)}_{\mathbf{k}\lambda}\mathscr{H}^{(-)}_{\mathbf{k}'\lambda'} + \mathscr{H}^{(-)}_{\mathbf{k}\lambda}\mathscr{H}^{(+)}_{\mathbf{k}'\lambda'} \tag{24.1-3}$$

in which $\mathscr{H}^{(+)}_{\mathbf{k}\lambda}\mathscr{H}^{(+)}_{\mathbf{k}'\lambda'}$ refers to the term containing $a^\dagger_{\mathbf{k}\lambda} a^\dagger_{\mathbf{k}'\lambda'}$; $\mathscr{H}^{(-)}_{\mathbf{k}\lambda}\mathscr{H}^{(-)}_{\mathbf{k}'\lambda'}$ to the term with $a_{\mathbf{k}\lambda} a_{\mathbf{k}'\lambda'}$, etc. Matrix elements of $a_{\mathbf{k}\lambda}$ and $a^\dagger_{\mathbf{k}\lambda}$ may be obtained from (22.2-8):

$$\langle n_{\mathbf{k}\lambda} + 1 | a^\dagger_{\mathbf{k}\lambda} | n_{\mathbf{k}\lambda} \rangle = \sqrt{n_{\mathbf{k}\lambda} + 1}, \qquad \langle n_{\mathbf{k}\lambda} - 1 | a_{\mathbf{k}\lambda} | n_{\mathbf{k}\lambda} \rangle = \sqrt{n_{\mathbf{k}\lambda}} \tag{24.1-4}$$

which lead directly to

$$\begin{aligned}
\langle n_{\mathbf{k}\lambda} + 1, n_{\mathbf{k}'\lambda'} + 1 | a^\dagger_{\mathbf{k}\lambda} a^\dagger_{\mathbf{k}'\lambda'} | n_{\mathbf{k}\lambda}, n_{\mathbf{k}'\lambda'} \rangle &= \sqrt{(n_{\mathbf{k}\lambda} + 1)(n_{\mathbf{k}'\lambda'} + 1)}, \\
\langle n_{\mathbf{k}\lambda} + 1, n_{\mathbf{k}'\lambda'} - 1 | a^\dagger_{\mathbf{k}\lambda} a_{\mathbf{k}'\lambda'} | n_{\mathbf{k}\lambda}, n_{\mathbf{k}'\lambda'} \rangle &= \sqrt{(n_{\mathbf{k}\lambda} + 1) n_{\mathbf{k}'\lambda'}}, \\
\langle n_{\mathbf{k}\lambda} - 1, n_{\mathbf{k}'\lambda'} + 1 | a_{\mathbf{k}\lambda} a^\dagger_{\mathbf{k}'\lambda'} | n_{\mathbf{k}\lambda}, n_{\mathbf{k}'\lambda'} \rangle &= \sqrt{n_{\mathbf{k}\lambda}(n_{\mathbf{k}'\lambda'} + 1)}, \\
\langle n_{\mathbf{k}\lambda} - 1, n_{\mathbf{k}'\lambda'} - 1 | a_{\mathbf{k}\lambda} a_{\mathbf{k}'\lambda'} | n_{\mathbf{k}\lambda}, n_{\mathbf{k}'\lambda'} \rangle &= \sqrt{n_{\mathbf{k}\lambda} n_{\mathbf{k}'\lambda'}}.
\end{aligned} \tag{24.1-5}$$

Since the photon modes are independent we also have

$$[a_{\mathbf{k}\lambda}, a_{\mathbf{k}'\lambda'}] = [a_{\mathbf{k}\lambda}, a^\dagger_{\mathbf{k}'\lambda'}] = [a^\dagger_{\mathbf{k}\lambda}, a^\dagger_{\mathbf{k}'\lambda'}] = 0.$$

The only nonvanishing matrix elements of $a^\dagger_{\mathbf{k}\lambda}$ and $a_{\mathbf{k}\lambda}$ are those which involve a change of one photon. For all bilinear combinations of $a^\dagger_{\mathbf{k}\lambda}$ and $a_{\mathbf{k}\lambda}$ the change in the number of photons must be zero or two.

We consider a two-level system (Fig. 24.1). Initially the atom is in the state $|\mathrm{i}\rangle$ and two photon modes with occupation numbers $n_{\mathbf{k}\lambda}$ and $n_{\mathbf{k}'\lambda'}$ are present. The total system is in the state

$$|\mathrm{I}\rangle = |\mathrm{i}; n_{\mathbf{k}\lambda}, n_{\mathbf{k}'\lambda'}\rangle \tag{24.1-6}$$

with energy (not including the zero point energy)

$$E_{\mathrm{I}} = E_{\mathrm{i}} + n_{\mathbf{k}\lambda}\hbar\omega_{\mathbf{k}} + n_{\mathbf{k}'\lambda'}\hbar\omega_{\mathbf{k}'}. \tag{24.1-7}$$

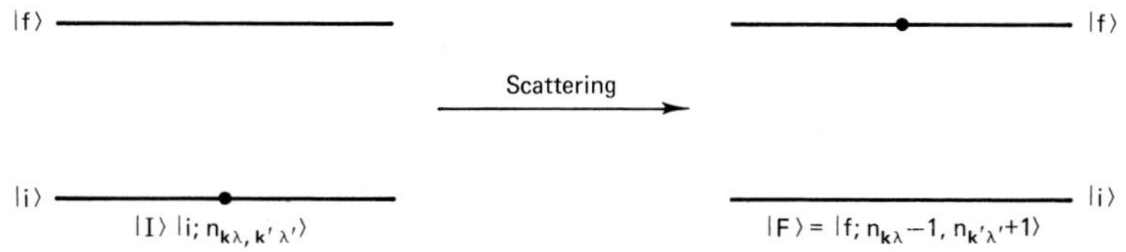

FIG. 24.1 Nomenclature for photon scattering. $\mathbf{k}\lambda$ and $\mathbf{k}'\lambda'$ refer to the incident and scattered photons, respectively.

After scattering has occurred, $n_{\mathbf{k}'\lambda'}$ is increased by one photon and $n_{\mathbf{k}\lambda}$ is decreased by one photon so that the total system is now in the state

$$|\mathrm{F}\rangle = |\mathrm{f}; n_{\mathbf{k}\lambda} - 1, n_{\mathbf{k}'\lambda'} + 1\rangle \tag{24.1-8}$$

with energy

$$E_\mathrm{F} = E_\mathrm{f} + (n_{\mathbf{k}\lambda} - 1)\hbar\omega_\mathbf{k} + (n_{\mathbf{k}'\lambda'} + 1)\hbar\omega_{\mathbf{k}'}. \tag{24.1-9}$$

The energy E_i relative to E_f is arbitrary and the scattered photon may have more or less energy compared with the incident photon. This represents the general case of Raman scattering and includes elastic scattering as a special case.

From (24.1-5) it follows that

$$\begin{aligned}
\langle \mathrm{F}|\mathscr{H}^{(+)}_{\mathbf{k}\lambda}\mathscr{H}^{(+)}_{\mathbf{k}'\lambda'}|\mathrm{I}\rangle &= \langle \mathrm{F}|\mathscr{H}^{(-)}_{\mathbf{k}\lambda}\mathscr{H}^{(-)}_{\mathbf{k}'\lambda'}|\mathrm{I}\rangle = 0,\\
\langle \mathrm{F}|\mathscr{H}^{(+)}_{\mathbf{k}\lambda}\mathscr{H}^{(-)}_{\mathbf{k}'\lambda'}|\mathrm{I}\rangle &= \langle \mathrm{F}|\mathscr{H}^{(-)}_{\mathbf{k}\lambda}\mathscr{H}^{(+)}_{\mathbf{k}'\lambda'}|\mathrm{I}\rangle\\
&= \frac{\pi e^2\hbar}{mV}\sqrt{\frac{n_{\mathbf{k}\lambda}(n_{\mathbf{k}'\lambda'}+1)}{\omega_\mathbf{k}\omega_{\mathbf{k}'}}}(\hat{\mathbf{e}}_{\mathbf{k}\lambda}\cdot\hat{\mathbf{e}}_{\mathbf{k}'\lambda'})\langle \mathrm{f}|e^{i(\mathbf{k}-\mathbf{k}')\cdot\mathbf{r}}|\mathrm{i}\rangle.
\end{aligned}$$

Therefore

$$\langle \mathrm{F}|\mathscr{H}_2|\mathrm{I}\rangle = \frac{2\pi e^2\hbar}{mV}\sqrt{\frac{n_{\mathbf{k}\lambda}(n_{\mathbf{k}'\lambda'}+1)}{\omega_\mathbf{k}\omega_{\mathbf{k}'}}}(\hat{\mathbf{e}}_{\mathbf{k}\lambda}\cdot\hat{\mathbf{e}}_{\mathbf{k}'\lambda'})\langle \mathrm{f}|e^{i(\mathbf{k}-\mathbf{k}')\cdot\mathbf{r}}|\mathrm{i}\rangle. \tag{24.1-10}$$

The scattering process is regarded as a two-photon process in the sense that an incident photon of specified energy, polarization, and propagation direction symbolized by the indices $\mathbf{k}\lambda$ is scattered so that the emerging photon may have a different energy, polarization, and direction of propagation symbolized by the indices $\mathbf{k}'\lambda'$. Hence a $\mathbf{k}\lambda$ photon has been lost and a $\mathbf{k}'\lambda'$ photon has been gained. Matrix elements of $\mathscr{H}_1$ will be of type (24.1-4) in which there is a change of one photon. Therefore $\mathscr{H}_1$ cannot contribute to scattering in the first order of perturbation theory but may contribute in a higher order. The matrix elements of $\mathscr{H}_2$, on the other hand, are nonvanishing when there is a change of two photons; first order contributions from $\mathscr{H}_2$ are therefore expected.

The contribution of $\mathscr{H}_1$ in second order may be obtained from (14.4-7):

$$W_{km} = (2\pi/\hbar)|\langle k|R|m\rangle|^2\,\delta(E_k - E_m) \tag{24.1-11}$$

with $\langle k|R|m\rangle$ to second order given by

$$\langle k|R|m\rangle = \langle k|V|m\rangle + \sum_l \frac{\langle k|V|l\rangle\langle l|V|m\rangle}{E_m - E_l}. \tag{24.1-12}$$

Now let

$$|m\rangle = |\mathrm{I}\rangle, \qquad |k\rangle = |\mathrm{F}\rangle$$

where $|\mathrm{I}\rangle$ and $|\mathrm{F}\rangle$ are given by (24.1-6) and (24.1-8). For the intermediate

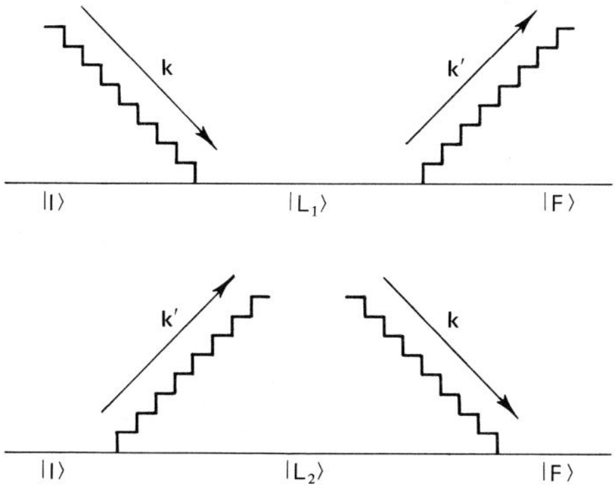

FIG. 24.2 Contributions from $\mathbf{p}\cdot\mathbf{A}$ term in second order to photon scattering [see Eqs. (24.1-15) and (24.1-16)].

states $|l\rangle$ there are two possibilities

$$|l\rangle = \begin{cases} |L_1\rangle = |l_1; n_{\mathbf{k}\lambda} - 1, n_{\mathbf{k}'\lambda'}\rangle, \\ |L_2\rangle = |l_2; n_{\mathbf{k}\lambda}, n_{\mathbf{k}'\lambda'} + 1\rangle. \end{cases} \tag{24.1-13}$$

Hence there are two pathways from $|\mathrm{I}\rangle$ to $|\mathrm{F}\rangle$:

$$\begin{aligned} &\text{path 1:} \quad |\mathrm{I}\rangle \to |L_1\rangle \to |\mathrm{F}\rangle, \\ &\text{path 2:} \quad |\mathrm{I}\rangle \to |L_2\rangle \to |\mathrm{F}\rangle. \end{aligned} \tag{24.1-14}$$

In the first path the transition $|\mathrm{I}\rangle \to |L_1\rangle$ involves the loss (absorption) of a $\mathbf{k}\lambda$ photon accompanied by an atomic transition $|\mathrm{i}\rangle \to |l_1\rangle$; the transition $|L_1\rangle \to |\mathrm{F}\rangle$ involves the gain (emission) of a $\mathbf{k}'\lambda'$ photon with an atomic transition $|l_1\rangle \to |\mathrm{f}\rangle$. In the second path there is an emission of a $\mathbf{k}'\lambda'$ photon with an atomic transition $|\mathrm{i}\rangle \to |l_2\rangle$ in the step $|\mathrm{I}\rangle \to |L_2\rangle$ and an absorption of a $\mathbf{k}\lambda$ photon with an atomic transition $|l_2\rangle \to |\mathrm{f}\rangle$ in the step $|L_2\rangle \to |\mathrm{F}\rangle$. It is important to note that in each path the intermediate state differs from the initial and final states by just one photon. These steps may be symbolized by diagrams as in Fig. 24.2.

In Section 22.3 we saw that $\mathscr{H}_1^{(-)}$ was associated with the absorption of a single photon and $\mathscr{H}_1^{(+)}$ with the emission of a single photon. Therefore the matrix elements in the two pathways are the following.

For $|\mathrm{I}\rangle \to |L_1\rangle \to |\mathrm{F}\rangle$:

$$\begin{aligned} \langle \mathrm{F}|\mathscr{H}_1^{(+)}|L_1\rangle\langle L_1|\mathscr{H}_1^{(-)}|\mathrm{I}\rangle &= \langle \mathrm{f}; n_{\mathbf{k}\lambda} - 1, n_{\mathbf{k}'\lambda'} + 1|\mathscr{H}_1^{(+)}|l_1; n_{\mathbf{k}\lambda} - 1, n_{\mathbf{k}'\lambda'}\rangle \\ &\quad \times \langle l_1; n_{\mathbf{k}\lambda} - 1, n_{\mathbf{k}'\lambda'}|\mathscr{H}_1^{(+)}|\mathrm{i}; n_{\mathbf{k}\lambda}, n_{\mathbf{k}'\lambda'}\rangle \\ &= \frac{2\pi e^2\hbar}{m^2 V}\sqrt{\frac{n_{\mathbf{k}\lambda}(n_{\mathbf{k}'\lambda'} + 1)}{\omega_{\mathbf{k}}\omega_{\mathbf{k}'}}}\langle \mathrm{f}|e^{-i\mathbf{k}'\cdot\mathbf{r}}(\hat{\mathbf{e}}_{\mathbf{k}'\lambda'}\cdot\mathbf{p})|l_1\rangle \\ &\quad \times \langle l_1|e^{i\mathbf{k}\cdot\mathbf{r}}(\hat{\mathbf{e}}_{\mathbf{k}\lambda}\cdot\mathbf{p})|\mathrm{i}\rangle. \end{aligned} \tag{24.1-15}$$

For $|\mathrm{I}\rangle \to |L_2\rangle \to |\mathrm{F}\rangle$:

$$\begin{aligned}\langle \mathrm{F}|\mathscr{H}_1^{(-)}|L_2\rangle\langle L_2|\mathscr{H}_1^{(+)}|\mathrm{I}\rangle &= \langle \mathrm{f}; n_{\mathbf{k}\lambda} - 1, n_{\mathbf{k}'\lambda'} + 1|\mathscr{H}_1^{(-)}|l_2; n_{\mathbf{k}\lambda}, n_{\mathbf{k}'\lambda'} + 1\rangle \\ &\quad \times \langle l_2; n_{\mathbf{k}\lambda}, n_{\mathbf{k}'\lambda'} + 1|\mathscr{H}_1^{(+)}|\mathrm{i}; n_{\mathbf{k}\lambda}, n_{\mathbf{k}'\lambda'}\rangle \\ &= \frac{2\pi e^2\hbar}{m^2 V}\sqrt{\frac{n_{\mathbf{k}\lambda}(n_{\mathbf{k}'\lambda'}+1)}{\omega_{\mathbf{k}}\omega_{\mathbf{k}'}}}\langle \mathrm{f}|e^{i\mathbf{k}\cdot\mathbf{r}}(\hat{\mathbf{e}}_{\mathbf{k}\lambda}\cdot\mathbf{p})|l_2\rangle \\ &\quad \times \langle l_2|e^{-i\mathbf{k}\cdot\mathbf{r}}(\hat{\mathbf{e}}_{\mathbf{k}'\lambda'}\cdot\mathbf{p})|\mathrm{i}\rangle. \end{aligned} \tag{24.1-16}$$

Both pathways contribute to the second-order term in (24.1-12). On setting

$$E_{\mathrm{I}} - E_{L_1} = E_{\mathrm{i}} - E_{l_1} + \hbar\omega_{\mathbf{k}}, \qquad E_{\mathrm{I}} - E_{L_2} = E_{\mathrm{i}} - E_{l_2} - \hbar\omega_{\mathbf{k}'}, \tag{24.1-17}$$

the second-order term becomes

$$\begin{aligned}&\frac{2\pi e^2\hbar}{m^2 V}\sqrt{\frac{n_{\mathbf{k}\lambda}(n_{\mathbf{k}'\lambda'}+1)}{\omega_{\mathbf{k}}\omega_{\mathbf{k}'}}} \\ &\times \sum_l \left[\frac{\langle \mathrm{f}|e^{-i\mathbf{k}'\cdot\mathbf{r}}(\hat{\mathbf{e}}_{\mathbf{k}'\lambda'}\ \mathbf{p})|l_1\rangle\langle l_1|e^{i\mathbf{k}\cdot\mathbf{r}}(\hat{\mathbf{e}}_{\mathbf{k}\lambda}\cdot\mathbf{p})|\mathrm{i}\rangle}{E_{\mathrm{i}} - E_l + \hbar\omega_{\mathbf{k}}}\right. \\ &\left. + \frac{\langle \mathrm{f}|e^{i\mathbf{k}\cdot\mathbf{r}}(\hat{\mathbf{e}}_{\mathbf{k}\lambda}\cdot\mathbf{p})|l_2\rangle\langle l_2|e^{-i\mathbf{k}'\cdot\mathbf{r}}(\hat{\mathbf{e}}_{\mathbf{k}'\lambda'}\cdot\mathbf{p})|\mathrm{i}\rangle}{E_{\mathrm{i}} - E_l - \hbar\omega_{\mathbf{k}'}}\right]\end{aligned} \tag{24.1-18}$$

in which the general summation index l indicates the sum over all intermediate states accessible to one-photon transitions.

The first-order contribution due to the Hamiltonian $\mathscr{H}_2$ was given by (24.1-10) and is to be added to (24.1-18) to obtain the total matrix element $\langle \mathrm{F}|\mathscr{H}_{\mathrm{int}}|\mathrm{I}\rangle$ for the scattering process. We shall also adopt the dipole approximation whereby all exponentials are set equal to unity. Thus

$$\begin{aligned}\langle \mathrm{F}|\mathscr{H}_{\mathrm{int}}|\mathrm{I}\rangle = \frac{2\pi e^2\hbar}{mV}\sqrt{\frac{n_{\mathbf{k}\lambda}(n_{\mathbf{k}'\lambda'}+1)}{\omega_{\mathbf{k}}\omega_{\mathbf{k}'}}}\Bigg\{&(\hat{e}_{\mathbf{k}\lambda}\cdot\hat{e}_{\mathbf{k}'\lambda'})\delta_{\mathrm{fi}} \\ &+ \frac{1}{m}\sum_l\left[\frac{\langle \mathrm{f}|\hat{\mathbf{e}}_{\mathbf{k}'\lambda'}\cdot\mathbf{p}|l\rangle\langle l|\hat{\mathbf{e}}_{\mathbf{k}\lambda}\cdot\mathbf{p}|\mathrm{i}\rangle}{E_{\mathrm{i}} - E_l + \hbar\omega_{\mathbf{k}}}\right. \\ &\left. + \frac{\langle \mathrm{f}|\hat{\mathbf{e}}_{\mathbf{k}\lambda}\cdot\mathbf{p}|l\rangle\langle l|\hat{\mathbf{e}}_{\mathbf{k}'\lambda'}\cdot\mathbf{p}|\mathrm{i}\rangle}{E_{\mathrm{i}} - E_l - \hbar\omega_{\mathbf{k}'}}\right]\Bigg\}.\end{aligned} \tag{24.1-19}$$

This matrix element may now be inserted into (24.1-11) with the δ function replaced by a density of final states (23.1-7). The transition probability per unit time that an incident $\mathbf{k}\lambda$ photon has been scattered into the element of solid angle $d\Omega$ as a $\mathbf{k}'\lambda'$ photon then becomes

$$W = \left(\frac{e^2}{mc^2}\right)^2 \frac{c}{V}\frac{\omega_{\mathbf{k}'}}{\omega_{\mathbf{k}}}\, d\Omega\, n_{\mathbf{k}\lambda}(n_{\mathbf{k}'\lambda'} + 1)\Bigg|(\hat{\mathbf{e}}_{\mathbf{k}\lambda}\cdot\hat{\mathbf{e}}_{\mathbf{k}'\lambda'})\,\delta_{\mathrm{fi}} + \frac{1}{m}\sum_l \left[\frac{\langle \mathrm{f}|\hat{\mathbf{e}}_{\mathbf{k}'\lambda'}\cdot\mathbf{p}|l\rangle\langle l|\hat{\mathbf{e}}_{\mathbf{k}\lambda}\cdot\mathbf{p}|\mathrm{i}\rangle}{E_\mathrm{i} - E_l + \hbar\omega_{\mathbf{k}}} + \frac{\langle \mathrm{f}|\hat{\mathbf{e}}_{\mathbf{k}\lambda}\cdot\mathbf{p}|l\rangle\langle l|\hat{\mathbf{e}}_{\mathbf{k}'\lambda'}\cdot\mathbf{p}|\mathrm{i}\rangle}{E_\mathrm{i} - E_l - \hbar\omega_{\mathbf{k}'}}\right]\Bigg|^2. \tag{24.1-20}$$

Equation (24.1-20) is known as the *Kramers–Heisenberg* dispersion formula; it may also be expressed as a differential scattering cross section,

$$\frac{d\sigma}{d\Omega} = \frac{\text{number of } \mathbf{k}'\lambda' \text{ photons scattered/sec sr}}{\text{number of incident } \mathbf{k}\lambda \text{ photons/sec cm}^2} = \frac{W/d\Omega}{n_{\mathbf{k}}c/V}$$

$$= r_0{}^2\frac{\omega_{\mathbf{k}'}}{\omega_{\mathbf{k}}}(n_{\mathbf{k}'\lambda'} + 1)\Bigg|(\hat{\mathbf{e}}_{\mathbf{k}\lambda}\cdot\hat{\mathbf{e}}_{\mathbf{k}'\lambda'})\,\delta_{\mathrm{fi}} + \frac{1}{m}\sum_l\left[\frac{(\hat{\mathbf{e}}_{\mathbf{k}'\lambda'}\cdot\mathbf{p}_{\mathrm{f}l})(\hat{\mathbf{e}}_{\mathbf{k}\lambda}\cdot\mathbf{p}_{l\mathrm{i}})}{E_\mathrm{i} - E_l + \hbar\omega_{\mathbf{k}}} + \frac{(\hat{\mathbf{e}}_{\mathbf{k}\lambda}\cdot\mathbf{p}_{\mathrm{f}l})(\hat{\mathbf{e}}_{\mathbf{k}'\lambda'}\cdot\mathbf{p}_{l\mathrm{i}})}{E_\mathrm{i} - E_l - \hbar\omega_{\mathbf{k}'}}\right]\Bigg|^2 \tag{24.1-21}$$

where

$$r_0 = e^2/mc^2 = \text{classical radius of the electron},$$
$$\mathbf{p}_{\mathrm{f}l} = \langle \mathrm{f}|\mathbf{p}|l\rangle, \qquad \mathbf{p}_{l\mathrm{i}} = \langle l|\mathbf{p}|\mathrm{i}\rangle.$$

Formula (24.1-21) for the differential cross section contains the factor $(n_{\mathbf{k}'\lambda'} + 1)$; hence the cross section is a sum of two terms one of which is proportional to $n_{\mathbf{k}'\lambda'}$, the number of *scattered* photons. This term corresponds to stimulated Raman scattering; it contributes negligibly at ordinary intensities, but produces important effects at high intensities.

The differential cross section (24.1-21) becomes infinite when one of the denominators vanishes. This is a consequence of the assumption of infinitely sharp atomic states. Near resonance it is therefore necessary to include a damping factor Γ as was done in Section 23.5. The differential cross section, including damping, is

$$\frac{d\sigma}{d\Omega} = r_0{}^2\frac{\omega_{\mathbf{k}'}}{\omega_{\mathbf{k}}}(n_{\mathbf{k}'\lambda'} + 1)\Bigg|(\hat{\mathbf{e}}_{\mathbf{k}\lambda}\cdot\hat{\mathbf{e}}_{\mathbf{k}'\lambda'})\,\delta_{\mathrm{fi}} + \frac{1}{m}\sum_l\left[\frac{(\hat{\mathbf{e}}_{\mathbf{k}'\lambda'}\cdot\mathbf{p}_{\mathrm{f}l})(\hat{\mathbf{e}}_{\mathbf{k}\lambda}\cdot\mathbf{p}_{l\mathrm{i}})}{E_\mathrm{i} - E_l + \hbar\omega_{\mathbf{k}} + \frac{1}{2}i\Gamma} + \frac{(\hat{\mathbf{e}}_{\mathbf{k}\lambda}\cdot\mathbf{p}_{\mathrm{f}l})(\hat{\mathbf{e}}_{\mathbf{k}'\lambda'}\cdot\mathbf{p}_{l\mathrm{i}})}{E_\mathrm{i} - E_l - \hbar\omega_{\mathbf{k}'} + \frac{1}{2}i\Gamma}\right]\Bigg|^2. \tag{24.1-22}$$

The intermediate states $|l\rangle$ which appear in the Kramers–Heisenberg dispersion formula, as well as the initial state $|\mathrm{i}\rangle$ and the final state $|\mathrm{f}\rangle$ are all eigenstates of the atom, i.e., solutions of the atomic Schrödinger equation. By virtue of the existence of matrix elements between $|\mathrm{i}\rangle$ and $|l\rangle$ and $|l\rangle$ and $|\mathrm{f}\rangle$, the two states $|\mathrm{i}\rangle$ and $|\mathrm{f}\rangle$ can interact via two-photon processes. However, the transitions $|\mathrm{i}\rangle \to |l\rangle$ and $|l\rangle \to |\mathrm{f}\rangle$ are not observable, and it is because of this feature that the states $|l\rangle$ are sometimes called "virtual". We prefer to avoid such terminology because it may give the impression of states whose existence is nebulous and whose properties are not defined which, of course, is not the case at all.

24.2 Scattering—Special Cases

A number of applications of the differential cross section formulas (24.1-21 or 24.1-22), not including the stimulated portion, will now be given.

If $\omega_{\mathbf{k}} = \omega_{\mathbf{k}'}$ in (24.1-21) and $\hbar\omega \gg |E_\mathrm{i} - E_l|$, the sum over intermediate states may be neglected and the differential cross section is

$$\begin{aligned} d\sigma/d\Omega &= r_0{}^2\, \delta_{\mathrm{fi}}(\hat{e}_{\mathbf{k}\lambda} \cdot \hat{e}_{\mathbf{k}'\lambda'}) \\ &= r_0{}^2 \cos^2\theta\, \delta_{\mathrm{fi}} \end{aligned} \tag{24.2-1}$$

where θ is the angle between the two polarizations. Since the incident and scattered photons are of the same energy (elastic scattering), the scattering atom remains in the same state and $\delta_{\mathrm{fi}} = 1$. The total cross section then becomes

$$\sigma = \int (d\sigma/d\Omega)\, d\Omega = 8\pi/3 r_0{}^2. \tag{24.2-2}$$

This is known as the *Thomson cross section*; it is due entirely to the A^2 term in the interaction Hamiltonian. Since the atomic states do not appear in the cross section, Thomson scattering may be regarded as the scattering of photons from free electrons.

Formula (24.2-2) is the total cross section when the incident light is unpolarized. If this is not the case, a more detailed geometrical analysis is required (Louisell, 1973).

The Kramers–Heisenberg formula (24.1-21) may be converted to another form which is more adaptable for discussions of Raman and Rayleigh scattering (Dirac, 1958). The notation will first be simplified by writing

$$\begin{aligned} \hat{\mathbf{e}}_{\mathbf{k}\lambda} &= \hat{\mathbf{e}} & \hat{\mathbf{e}}_{\mathbf{k}'\lambda'} &= \hat{\mathbf{e}}' \\ n_{\mathbf{k}\lambda} &= n & n_{\mathbf{k}'\lambda'} &= n' \\ \omega_{\mathbf{k}} &= \omega & \omega_{\mathbf{k}'} &= \omega' \end{aligned} \tag{24.2-3}$$

Now consider the commutator

$$[\hat{\mathbf{e}}' \cdot \mathbf{r}, \hat{\mathbf{e}} \cdot \mathbf{p}].$$

Since

$$[r_j, p_k] = i\hbar\, \delta_{jk}, \tag{24.2-4}$$

the expansion of (24.2-4) into its components results in

$$[\hat{\mathbf{e}}' \cdot \mathbf{r}, \hat{\mathbf{e}} \cdot \mathbf{p}] = i\hbar \hat{\mathbf{e}}' \cdot \hat{\mathbf{e}} \tag{24.2-5}$$

from which one obtains

$$\langle \mathrm{f}|[\hat{\mathbf{e}}' \cdot \mathbf{r}, \hat{\mathbf{e}} \cdot \mathbf{p}]|\mathrm{i}\rangle = i\hbar\langle \mathrm{f}|\hat{\mathbf{e}}' \cdot \hat{\mathbf{e}}|\mathrm{i}\rangle = i\hbar\hat{\mathbf{e}} \cdot \hat{\mathbf{e}}\,\delta_{\mathrm{fi}}. \tag{24.2-6}$$

But

$$\langle \mathrm{f}|[\hat{\mathbf{e}}' \cdot \mathbf{r}, \hat{\mathbf{e}} \cdot \mathbf{p}]|\mathrm{i}\rangle = \sum_l [\langle \mathrm{f}|\hat{\mathbf{e}}' \cdot \mathbf{r}|l\rangle\langle l|\hat{\mathbf{e}} \cdot \mathbf{p}|\mathrm{i}\rangle - \langle \mathrm{f}|\hat{\mathbf{e}} \cdot \mathbf{p}|l\rangle\langle l|\hat{\mathbf{e}}' \cdot \mathbf{r}|\mathrm{i}\rangle] \tag{24.2-7}$$

and making the replacement

$$\langle \mathrm{f}|\hat{\mathbf{e}} \cdot \mathbf{p}|l\rangle = \frac{im}{\hbar}(E_\mathrm{f} - E_l)\langle \mathrm{f}|\hat{\mathbf{e}} \cdot \mathbf{r}|l\rangle \tag{24.2-8}$$

with a corresponding expression for $\langle l|\hat{\mathbf{e}} \cdot \mathbf{p}|\mathrm{i}\rangle$, we have

$$\begin{aligned}\sum_l [&(E_l - E_\mathrm{i})\langle \mathrm{f}|\hat{\mathbf{e}}' \cdot \mathbf{r}|l\rangle\langle l|\hat{\mathbf{e}} \cdot \mathbf{r}|\mathrm{i}\rangle \\ &- (E_\mathrm{f} - E_l)\langle \mathrm{f}|\hat{\mathbf{e}} \cdot \mathbf{r}|l\rangle\langle l|\hat{\mathbf{e}}' \cdot \mathbf{r}|\mathrm{i}\rangle] = (\hbar^2/m^2)\hat{\mathbf{e}}' \cdot \hat{\mathbf{e}}\,\delta_{\mathrm{fi}}\end{aligned} \tag{24.2-9}$$

Also,

$$\begin{aligned}\sum_l &[\langle \mathrm{f}|\hat{\mathbf{e}}' \cdot \mathbf{r}|l\rangle\langle l|\hat{\mathbf{e}} \cdot \mathbf{r}|\mathrm{i}\rangle - \langle \mathrm{f}|\hat{\mathbf{e}} \cdot \mathbf{r}|l\rangle\langle l|\hat{\mathbf{e}}' \cdot \mathbf{r}|\mathrm{i}\rangle] \\ &= \langle \mathrm{f}|\hat{\mathbf{e}}' \cdot \mathbf{r}\hat{\mathbf{e}} \cdot \mathbf{r}|\mathrm{i}\rangle - \langle \mathrm{f}|\hat{\mathbf{e}} \cdot \mathbf{r}\hat{\mathbf{e}}' \cdot \mathbf{r}|\mathrm{i}\rangle \\ &= \langle \mathrm{f}|[\hat{\mathbf{e}}' \cdot \mathbf{r}, \hat{\mathbf{e}} \cdot \mathbf{r}]|\mathrm{i}\rangle = 0\end{aligned} \tag{24.2-10}$$

since the components of **r** commute with one another. When (24.2-10) is multiplied by $\hbar\omega'$ and added to (24.2-9) the result is

$$\begin{aligned}\frac{m^2}{\hbar^2}[&(E_l - E_\mathrm{i} + \hbar\omega')\langle \mathrm{f}|\hat{\mathbf{e}}' \cdot \mathbf{r}|l\rangle\langle l|\hat{\mathbf{e}} \cdot \mathbf{r}|\mathrm{i}\rangle \\ &- (E_\mathrm{f} - E_l + \hbar\omega')\langle \mathrm{f}|\hat{\mathbf{e}} \cdot \mathbf{r}|l\rangle\langle l|\hat{\mathbf{e}}' \cdot \mathbf{r}|\mathrm{i}\rangle] = \hat{\mathbf{e}}' \cdot \hat{\mathbf{e}}\delta_{\mathrm{fi}}\end{aligned} \tag{24.2-11}$$

We now replace the matrix elements in (24.1-21) by expressions of the type (24.2-8) and $\hat{\mathbf{e}}' \cdot \hat{\mathbf{e}}\, \delta_{\mathrm{fi}}$ by (24.2-11). The result is

$$\begin{aligned} \frac{d\sigma}{d\Omega} &= r_0^2 \frac{\omega'}{\omega}(n'+1)\frac{m^2}{\hbar^4} \\ &\times \Bigg|\sum_l \Bigg\{ \langle \mathrm{f}|\hat{\mathbf{e}}' \cdot \mathbf{r}|l\rangle\langle l|\hat{\mathbf{e}} \cdot \mathbf{r}|\mathrm{i}\rangle \\ &\times \frac{(E_l - E_\mathrm{i} + \hbar\omega')(E_\mathrm{i} - E_l + \hbar\omega) - (E_\mathrm{f} - E_l)(E_l - E_\mathrm{i})}{E_\mathrm{i} - E_l + \hbar\omega} \\ &- \langle \mathrm{f}|\hat{\mathbf{e}} \cdot \mathbf{r}|l\rangle\langle l|\hat{\mathbf{e}}' \cdot \mathbf{r}|\mathrm{i}\rangle \\ &\times \frac{(E_\mathrm{f} - E_l + \hbar\omega')(E_\mathrm{i} - E_l - \hbar\omega') - (E_\mathrm{f} - E_l)(E_l - E_\mathrm{i})}{E_\mathrm{i} - E_l - \hbar\omega'} \end{aligned} \qquad (24.2\text{-}12)$$

But

$$E_\mathrm{f} - E_\mathrm{i} = \hbar\omega - \hbar\omega'$$

which then gives

$$\begin{aligned} \frac{d\sigma}{d\Omega} &= r_0^2 \omega\,\omega'^3 (n'+1) m^2 \\ &\times \left|\sum_l \left\{ \frac{\langle \mathrm{f}|\hat{\mathbf{e}}' \cdot \mathbf{r}|l\rangle\langle l|\hat{\mathbf{e}} \cdot \mathbf{r}|\mathrm{i}\rangle}{E_\mathrm{i} - E_l + \hbar\omega} + \frac{\langle \mathrm{f}|\hat{\mathbf{e}} \cdot \mathbf{r}|l\rangle\langle l|\hat{\mathbf{e}}' \cdot \mathbf{r}|\mathrm{i}\rangle}{E_\mathrm{i} - E_l - \hbar\omega'} \right\}\right|^2 \end{aligned} \qquad (24.2\text{-}13)$$

This is the general form for Raman scattering. The energy of the scattered radiation ($\hbar\omega'$) can be lower or higher than the energy of the incident radiation ($\hbar\omega$). The process is, therefore, inelastic and involves energetic changes in internal degrees of freedom. *Stokes* and *anti-Stokes* shifts are terms used to describe scattering in which $\hbar\omega' < \hbar\omega$ and $\hbar\omega' < \hbar\omega$, respectively. Raman scattering from molecular vibrational modes is discussed in Section 27.4.

Rayleigh scattering occurs when $\omega' = \omega$ and $\hbar\omega \ll |E_\mathrm{i} - E_l|$. Inserting these conditions into (24.2-13), with $|\mathrm{i}\rangle = |\mathrm{f}\rangle$,

$$\frac{d\sigma}{d\Omega} = r_0^2 m^2 \omega^4 \left|\sum_l \left[\frac{\langle \mathrm{i}|\hat{\mathbf{e}}' \cdot \mathbf{r}|l\rangle\langle l|\hat{\mathbf{e}} \cdot \mathbf{r}|\mathrm{i}\rangle}{E_\mathrm{i} - E_l} + \frac{\langle \mathrm{i}|\hat{\mathbf{e}} \cdot \mathbf{r}|l\rangle\langle l|\hat{\mathbf{e}}' \cdot \mathbf{r}|\mathrm{i}\rangle}{E_\mathrm{i} - E_l} \right]\right|^2. \qquad (24.2\text{-}14)$$

which yields the familiar result that at long wavelengths the scattering cross section varies inversely as the fourth power of the wavelength.

24.3 Diagrams

Starting with the Fermi golden rule (24.1-11) in which the general form of the matrix element is

$$\langle k|R|m\rangle \equiv R_{km} = R_{km}^{(1)} + R_{km}^{(2)} + \cdots \qquad (24.3\text{-}1)$$

where

$$R_{km}^{(1)} = \langle k|V|m\rangle,$$

$$R_{km}^{(2)} = \sum_l \frac{\langle k|V|l\rangle\langle l|V|m\rangle}{E_m - E_l},$$

$$R_{km}^{(3)} = \sum_l \sum_{l'} \frac{\langle k|V|l\rangle\langle l|V|l'\rangle\langle l'|V|m\rangle}{(E_m - E_{l'})(E_m - E_l)}, \qquad (24.3\text{-}2)$$

it is possible to symbolize these expressions by means of diagrams which are helpful in writing the pertinent matrix elements associated with a particular process. We use the simplified notation (24.2-3).

Single photon absorption.

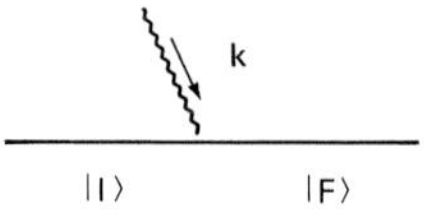

$$|\mathrm{I}\rangle = |\mathrm{i}; n\rangle$$

$$|\mathrm{F}\rangle = |\mathrm{f}; n-1\rangle$$

$$R_{\mathrm{fi}}^{(1)} = \langle \mathrm{F}|\mathscr{H}_1|\mathrm{I}\rangle = \langle \mathrm{F}|\mathscr{H}_1^{(-)}|\mathrm{I}\rangle \qquad (24.3\text{-}3)$$

$$= \frac{e}{m}\sqrt{\frac{2\pi\hbar n}{\omega V}}\,\langle \mathrm{f}|e^{i\mathbf{k}\cdot\mathbf{r}}(\hat{e}\cdot\mathbf{p})|\mathrm{i}\rangle.$$

Single photon emission.

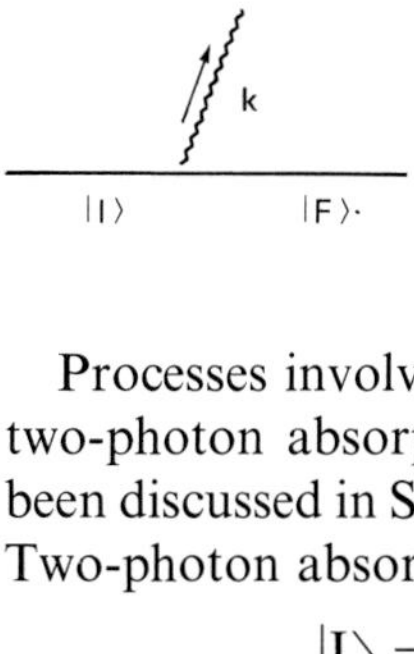

$$|\mathrm{I}\rangle = |\mathrm{i}; n\rangle$$

$$|\mathrm{F}\rangle = |\mathrm{f}; n+1\rangle$$

$$R_{\mathrm{fi}}^{(1)} = \langle \mathrm{F}|\mathscr{H}_1|\mathrm{I}\rangle = \langle \mathrm{F}|\mathscr{H}_1^{(+)}|\mathrm{I}\rangle \qquad (24.3\text{-}4)$$

$$= \frac{e}{m}\sqrt{\frac{2\pi\hbar(n+1)}{\omega V}}\,\langle \mathrm{f}|e^{-i\mathbf{k}\cdot\mathbf{r}}(\hat{\mathbf{e}}\cdot\mathbf{p})|\mathrm{i}\rangle.$$

Processes involving two photons (second order processes) are scattering, two-photon absorption and two-photon emission. Scattering has already been discussed in Section 24.1; the pertinent diagrams are shown in Fig. 24.2.

Two-photon absorption.

$$|\mathrm{I}\rangle = |\mathrm{i}; n, n'\rangle, \qquad |L_1\rangle = |l_1; n-1, n'\rangle,$$

$$|\mathrm{F}\rangle = |\mathrm{f}; n-1, n'-1\rangle, \qquad |L_2\rangle = |l_2; n, n'-1\rangle,$$

$$R_{\mathrm{fi}}^{(2)} = \sum_{L_1} \frac{\langle \mathrm{F}|\mathscr{H}_1^{(-)}|L_1\rangle\langle L_1|\mathscr{H}_1^{(-)}|\mathrm{I}\rangle}{E_\mathrm{I} - E_{L_1}} + \sum_{L_2} \frac{\langle \mathrm{F}|\mathscr{H}_1^{(-)}|L_2\rangle\langle L_2|\mathscr{H}_1^{(-)}|\mathrm{I}\rangle}{E_\mathrm{I} - E_{L_2}}$$

$$\equiv M_3 + M_4, \qquad (24.3\text{-}5)$$

$$M_3 = \frac{e^2}{m^2}\frac{2\pi\hbar}{V}\sqrt{\frac{nn'}{\omega\omega'}}\sum_{l_1} \frac{\langle \mathrm{f}|e^{i\mathbf{k}'\cdot\mathbf{r}}(\hat{\mathbf{e}}'\cdot\mathbf{p})|l_1\rangle\langle l_1|e^{i\mathbf{k}\cdot\mathbf{r}}(\hat{\mathbf{e}}\cdot\mathbf{p})|\mathrm{i}\rangle}{E_\mathrm{i} - E_{l_1} + \hbar\omega}, \qquad (24.3\text{-}6)$$

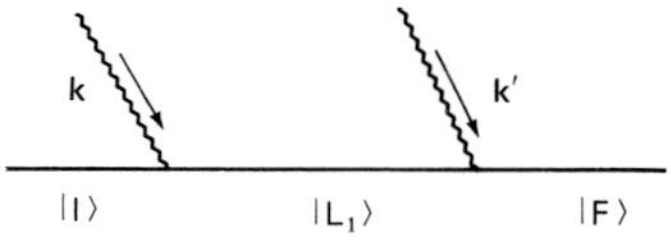

$$M_4 = \frac{e^2}{m^2}\frac{2\pi\hbar}{V}\sqrt{\frac{nn'}{\omega\omega'}}\sum_{l_2}\frac{\langle \mathrm{f}|e^{i\mathbf{k}\cdot\mathbf{r}}(\hat{\mathbf{e}}\cdot\mathbf{p})|l_2\rangle\langle l_2|e^{i\mathbf{k}'\cdot\mathbf{r}}(\hat{\mathbf{e}}'\cdot\mathbf{p})|\mathrm{i}\rangle}{E_\mathrm{i} - E_{l_2} + \hbar\omega'}, \tag{24.3-7}$$

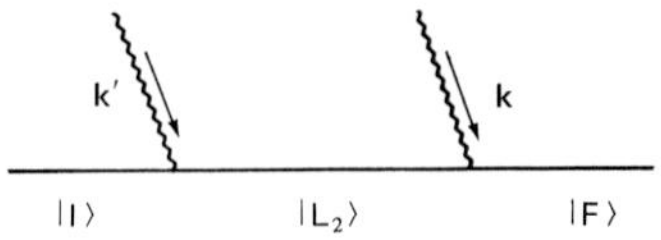

Two-photon emission.

$$|\mathrm{I}\rangle = |\mathrm{i}; n, n'\rangle, \qquad |L_1\rangle = |l_1; n+1, n'\rangle,$$
$$|\mathrm{F}\rangle = |\mathrm{f}; n+1, n'+1\rangle, \qquad |L_2\rangle = |l_2; n, n'+1\rangle,$$

$$R_{\mathrm{fi}}^{(2)} = \sum_{L_1}\frac{\langle \mathrm{F}|\mathscr{H}_1^{(+)}|L_1\rangle\langle L_1|\mathscr{H}_1^{(+)}|\mathrm{I}\rangle}{E_\mathrm{I} - E_{L_1}} + \sum_{L_2}\frac{\langle \mathrm{F}|\mathscr{H}_1^{(+)}|L_2\rangle\langle L_2|\mathscr{H}_1^{(+)}|\mathrm{I}\rangle}{E_\mathrm{I} - E_{L_2}}$$
$$\equiv M_5 + M_6, \tag{24.3-8}$$

$$M_5 = \frac{e^2}{m^2}\frac{2\pi\hbar}{V}\sqrt{\frac{(n+1)(n'+1)}{\omega\omega'}}\sum_{l_1}\frac{\langle \mathrm{f}|e^{-i\mathbf{k}'\cdot\mathbf{r}}(\hat{\mathbf{e}}'\cdot\mathbf{p})|l_1\rangle\langle l_1|e^{-i\mathbf{k}\cdot\mathbf{r}}(\hat{\mathbf{e}}\cdot\mathbf{p})|\mathrm{i}\rangle}{E_\mathrm{i} - E_{l_1} - \hbar\omega}, \tag{24.3-9}$$

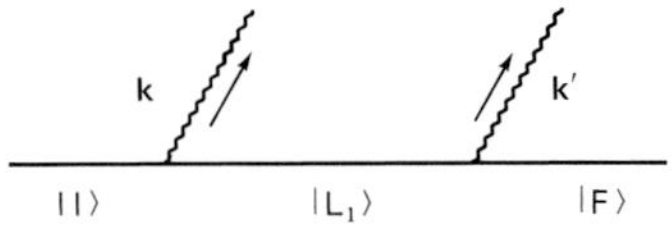

$$M_6 = \frac{e^2}{m^2}\frac{2\pi\hbar}{V}\sqrt{\frac{(n+1)(n'+1)}{\omega\omega'}}\sum_{l_2}\frac{\langle \mathrm{f}|e^{-i\mathbf{k}\cdot\mathbf{r}}(\hat{\mathbf{e}}\cdot\mathbf{p})|l_2\rangle\langle l_2|e^{-i\mathbf{k}'\cdot\mathbf{r}}(\hat{\mathbf{e}}'\cdot\mathbf{p})|\mathrm{i}\rangle}{E_\mathrm{i} - E_{l_2} - \hbar\omega'} \tag{24.3-10}$$

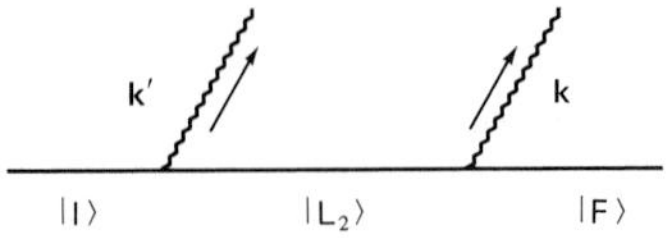

All the matrix elements considered thus far were of the operator $\mathscr{H}_1$ in (24.1-1); for $\mathscr{H}_2$ (24.1-2) and (24.1-3) we have the following:

Scattering

$$|\mathrm{I}\rangle = |\mathrm{i}; n, n'\rangle, \qquad |\mathrm{F}\rangle = |\mathrm{f}; n-1, n'+1\rangle,$$

$$R_{\mathrm{fi}}^{(1)} = \langle \mathrm{F}|\mathscr{H}_2|\mathrm{I}\rangle = 2\langle \mathrm{F}|\mathscr{H}_{\mathbf{k}\lambda}^{(+)}\mathscr{H}_{\mathbf{k}'\lambda'}^{(-)}|\mathrm{I}\rangle$$
$$= 2\langle \mathrm{F}|\mathscr{H}_{\mathbf{k}\lambda}^{(-)}\mathscr{H}_{\mathbf{k}'\lambda'}^{(+)}|\mathrm{I}\rangle$$
$$= \frac{e^2}{m}\frac{2\pi\hbar}{V}\sqrt{\frac{n(n'+1)}{\omega\omega'}}(\hat{\mathbf{e}}\cdot\hat{\mathbf{e}}')\langle \mathrm{f}|\exp[-i(\mathbf{k}-\mathbf{k}')\cdot\mathbf{r}]|\mathrm{i}\rangle. \tag{24.3-11}$$

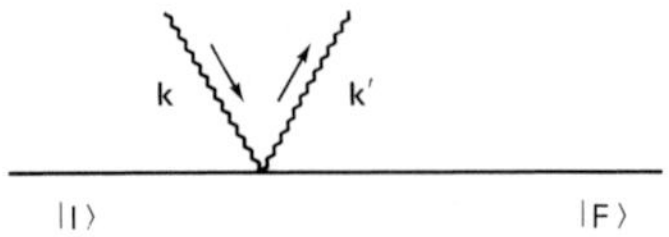

Two-photon absorption.

$$|\mathrm{I}\rangle = |\mathrm{i}; n, n'\rangle, \qquad |\mathrm{F}\rangle = |\mathrm{f}; n-1, n'-1\rangle,$$

$$R_{\mathrm{fi}}^{(1)} = \langle \mathrm{F}|\mathscr{H}_2|\mathrm{I}\rangle = \langle \mathrm{F}|\mathscr{H}_{\mathbf{k}\lambda}^{(-)}\mathscr{H}_{\mathbf{k}'\lambda'}^{(-)}|\mathrm{I}\rangle$$
$$= \frac{e^2}{2m}\frac{2\pi\hbar}{V}\sqrt{\frac{nn'}{\omega\omega'}}(\hat{\mathbf{e}}\cdot\hat{\mathbf{e}}')\langle \mathrm{f}|\exp[i(\mathbf{k}+\mathbf{k}')\cdot\mathbf{r}]|\mathrm{i}\rangle. \tag{24.3-12}$$

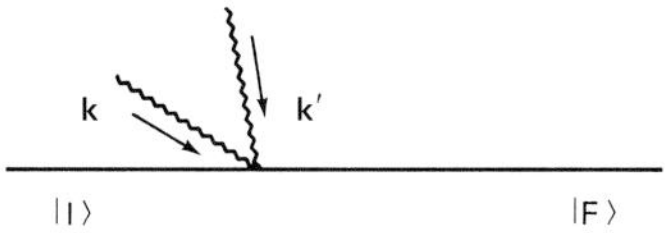

Two-photon emission.

$$|\mathrm{I}\rangle = |\mathrm{i}; n, n'\rangle, \qquad |\mathrm{F}\rangle = |\mathrm{f}; n+1, n'+1\rangle$$

$$R_{km}^{(1)} = \langle \mathrm{F}|\mathscr{H}_2|\mathrm{I}\rangle = \langle \mathrm{F}|\mathscr{H}_{\mathbf{k}\lambda}^{(+)}\mathscr{H}_{\mathbf{k}'\lambda'}^{(+)}|\mathrm{I}\rangle$$
$$= \frac{e^2}{2m}\frac{2\pi\hbar}{V}\sqrt{\frac{(n+1)(n'+1)}{\omega\omega'}}(\hat{\mathbf{e}}\cdot\hat{\mathbf{e}}')\langle \mathrm{f}|\exp[-i(\mathbf{k}+\mathbf{k}')\cdot\mathbf{r}]|\mathrm{i}\rangle. \tag{24.3-13}$$

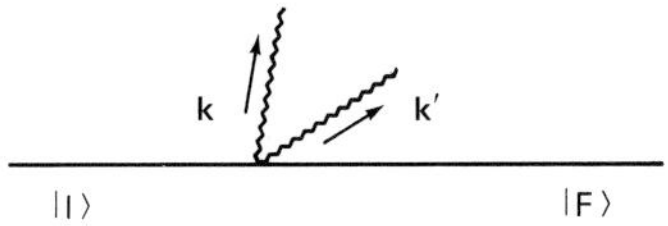

Note that in dipole approximation there is no contribution from the A^2 term to two-photon absorption or emission; as shown previously the contribution to scattering occurs only in the elastic case.

24.4 Optical Susceptibility and Nonlinear Effects

For a system of N identical particles (e.g., atoms, molecules, etc.) let the Hamiltonian be

$$\mathscr{H} = \mathscr{H}_0 + \mathscr{H}' \tag{24.4-1}$$

with

$$\mathscr{H}' = -\mathbf{E}(\mathbf{r}, t)\cdot\mathbf{D} \tag{24.4-2}$$

where $\mathbf{D}$ is the dipole moment operator $e\mathbf{R}$ and $\mathscr{H}' = 0$ at $t = t_0$. In terms of the individual particles

$$\mathscr{H}_0 = \sum_n \mathscr{H}_n^{(0)}, \qquad \mathbf{D} = \sum_n \mathbf{D}_n \tag{24.4-3}$$

with the assumption that operators associated with different particles commute. The macroscopic polarization is defined by

$$\mathbf{P} = \langle\mathbf{D}\rangle = \mathrm{Tr}\,\rho\mathbf{D}. \tag{24.4-4}$$

In order to calculate $\mathbf{P}$ it is necessary to calculate the density operator ρ in the presence of the field $\mathbf{E}$. For this purpose we refer to (13.5-9) which gives the expansion of the density operator in the interaction representation:

$$\rho_{\mathrm{I}}(t) = \rho_{\mathrm{I}}^{(0)}(t) + \rho_{\mathrm{I}}^{(1)}(t) + \rho_{\mathrm{I}}^{(2)}(t) + \cdots \tag{24.4-5}$$

where

$$\begin{aligned}
\rho_{\mathrm{I}}^{(0)}(t) &= \rho_{\mathrm{I}}(t_0),\\
\rho_{\mathrm{I}}^{(1)}(t) &= (i/\hbar)\int_{t_0}^{t} dt_1\, [\rho_{\mathrm{I}}(t_0), \mathscr{H}_{\mathrm{I}}'(t_1)]\\
\rho_{\mathrm{I}}^{(2)}(t) &= (i/\hbar)^2\int_{t_0}^{t} dt_1 \int_{t_0}^{t_1} dt_2\, [[\rho_{\mathrm{I}}(t_0), \mathscr{H}_{\mathrm{I}}'(t_2)], \mathscr{H}_{\mathrm{I}}'(t_1)], \quad \text{etc.,}
\end{aligned} \tag{24.4-6}$$

and

$$\begin{aligned}
\mathscr{H}_{\mathrm{I}}' &= e^{i\mathscr{H}_0 t/\hbar}\mathscr{H}' e^{-i\mathscr{H}_0 t/\hbar}\\
&= -\mathbf{E}(\mathbf{r}, t)\cdot e^{i\mathscr{H}_0 t/\hbar}\mathbf{D}e^{-i\mathscr{H}_0 t/} = -\mathbf{E}(\mathbf{r}, t)\cdot \mathbf{D}_{\mathrm{I}}(t).
\end{aligned} \tag{24.4-7}$$

If the system is in thermal equilibrium at the time the perturbation is applied,

$$\rho_{\mathrm{I}}^{(0)}(t) = \rho_{\mathrm{I}}(t_0) = \rho_{\mathrm{T}} \tag{24.4-8}$$

where, according to (13.4-7),

$$\rho_{\mathrm{T}} = e^{-\mathscr{H}_0/kT}/\mathrm{Tr}\, e^{-\mathscr{H}_0/kT} \equiv (1/Z)e^{-\mathscr{H}_0/kT}. \tag{24.4-9}$$

One may now expand **P** to various orders:

$$\mathbf{P} = \mathbf{P}^{(0)} + \mathbf{P}^{(1)} + \mathbf{P}^{(2)} + \cdots \tag{24.4-10}$$

with the leading term

$$\mathbf{P}^{(0)} = \mathrm{Tr}\,\rho_{\mathrm{T}}\mathbf{D}. \tag{24.4-11}$$

The next term is

$$\mathbf{P}^{(1)}(t) = \mathrm{Tr}\,\rho^{(1)}\mathbf{D} = \mathrm{Tr}[e^{-i\mathscr{H}_0 t/\hbar}\rho_{\mathrm{I}}^{(1)}e^{i\mathscr{H}_0 t/\hbar}\mathbf{D}] \tag{24.4-12}$$

or, in component form,

$$P_\mu^{(1)}(t) = \mathrm{Tr}[e^{-i\mathscr{H}_0 t/\hbar}\rho_{\mathrm{I}}^{(1)}e^{i\mathscr{H}_0 t/\hbar} D_\mu] \tag{24.4-13}$$

$$= \mathrm{Tr}\,\frac{i}{\hbar}\, e^{-i\mathscr{H}_0 t/\hbar}\int_{t_0}^{t} dt_1\, \mathbf{E}(\mathbf{r}, t_1)\cdot[\mathbf{D}_{\mathrm{I}}(t_1), \rho_{\mathrm{T}}]e^{i\mathscr{H}_0 t/\hbar}D_\mu. \tag{24.4-14}$$

Expanding the scalar product,

$$\begin{aligned}
P_\mu^{(1)}(t) &= \frac{i}{\hbar}\,\mathrm{Tr}\, e^{-i\mathscr{H}_0 t/\hbar}\left\{\int_{t_0}^{t} dt_1 \sum_\alpha E_\alpha(\mathbf{r}, t_1)[D_{\mathrm{I}\alpha}(t_1), \rho_{\mathrm{T}}]\right\} e^{i\mathscr{H}_0 t/\hbar} D_\mu\\
&= \frac{i}{\hbar}\int_{t_0}^{t} dt_1 \sum_\alpha E_\alpha(\mathbf{r}, t_1)\,\mathrm{Tr}\{e^{-i\mathscr{H}_0 t/\hbar}[D_{\mathrm{I}\alpha}(t_1), \rho_{\mathrm{T}}]e^{i\mathscr{H}_0 t/\hbar} D_\mu\}.
\end{aligned} \tag{24.4-15}$$

The expression inside the trace may be simplified somewhat since

$$e^{-i\mathscr{H}_0 t/\hbar}[D_{I\alpha}(t_1), \rho_T]e^{i\mathscr{H}_0 t/\hbar} = e^{i\mathscr{H}_0(t_1-t)}[D_\alpha, \rho_T]e^{-i\mathscr{H}_0(t_1-t)}$$
$$= [D_{I\alpha}(t_1 - t), \rho_T] = [D_{I\alpha}(t'), \rho_T] \quad (24.4\text{-}16)$$

where

$$t' = t_1 - t. \quad (24.4\text{-}17)$$

Also, since the trace of a product of operators is invariant under a cyclic permutation of their order,

$$\mathrm{Tr}\{[D_{I\alpha}(t'), \rho_T]\, D_\mu\} = \mathrm{Tr}\{\rho_T[D_\mu, D_{I\alpha}(t')]\}. \quad (24.4\text{-}18)$$

The electric field $E_\alpha(\mathbf{r}, t_1)$ may be expressed in terms of its Fourier components. Suppressing the dependence of the field on the spatial coordinates,

$$E_\alpha(t_1) = \int_{-\infty}^{\infty} E_\alpha(\omega)e^{-i\omega t_1}\, d\omega$$
$$= \int_{-\infty}^{\infty} d\omega\, E_\alpha(\omega)e^{-i\omega t}e^{-i\omega(t_1-t)}$$
$$= \int_{-\infty}^{\infty} d\omega\, E_\alpha(\omega)e^{-i\omega t}e^{-i\omega t'}. \quad (24.4\text{-}19)$$

Inserting (24.4-18) and (24.4-19) into (24.4-15) with $t_0 = -\infty$ and $t = 0$,

$$P_\mu^{(1)}(t) = \int_{-\infty}^{\infty} d\omega \sum_\alpha \left[(i/\hbar) \int_{-\infty}^{0} dt'\, \mathrm{Tr}\{\rho_T[D_\mu, D_{I\alpha}(t')]\}e^{-i\omega t'} \right] E_\alpha(\omega)e^{-i\omega t}. \quad (24.4\text{-}20)$$

This expression may be compared with (14.6-29) written in component form;

$$P_\mu^{(1)}(t) = \int_{-\infty}^{\infty} \sum_\alpha \chi_{\mu\alpha}^{(1)}(\omega)E_\alpha(\omega)\, d\omega\, e^{-i\omega t} \quad (24.4\text{-}21)$$

which then permits us to identify the quantity in the square brackets in (24.4-20) as a component of the second-rank susceptibility tensor, i.e.,

$$\chi_{\mu\alpha}^{(1)} = (i/\hbar) \int_{-\infty}^{0} dt'\, \mathrm{Tr}\{\rho_T[D_\mu, D_{I\alpha}(t')]\}e^{-i\omega t'}. \quad (24.4\text{-}22)$$

Let us now choose a basis set which consists of the eigenfunctions of $\mathscr{H}_0$, i.e., let

$$\mathscr{H}_0|k\rangle = E_k|k\rangle. \quad (24.4\text{-}23)$$

In this basis set

$$\mathrm{Tr}\{\rho_T[D_\mu, D_{I\alpha}(t')]\} = \sum_k \langle k|\rho_T[D_\mu, D_{I\alpha}(t')]|k\rangle$$
$$= \sum_k \sum_l \langle k|\rho_T|l\rangle\langle l|[D_\mu, D_{I\alpha}(t')]|k\rangle. \quad (24.4\text{-}24)$$

But

$$\begin{aligned}\langle k|\rho_{\mathrm{T}}|l\rangle &= (1/Z)\langle k|e^{-i\mathscr{H}_0/kT}|l\rangle \\ &= (1/Z)\langle k|e^{-i\mathscr{H}_0/kT}|k\rangle\,\delta_{kl} = \langle k|\rho_{\mathrm{T}}|k\rangle\,\delta_{kl}, \qquad (24.4\text{-}25)\end{aligned}$$

$$\begin{aligned}\langle k|D_\mu D_{\mathrm{I}\alpha}(t')|k\rangle &= \sum_l \langle k|D_\mu|l\rangle\langle l|D_{\mathrm{I}\alpha}(t')|k\rangle \\ &= \sum_l \langle k|D_\mu|l\rangle\langle l|e^{i\mathscr{H}_0 t'/\hbar} D_\alpha e^{-i\mathscr{H}_0 t'/\hbar}|k\rangle \\ &= \sum_l \langle k|D_\mu|l\rangle\langle l|D_\alpha|k\rangle \exp[i(E_l - E_k)t'/\hbar]. \qquad (24.4\text{-}26)\end{aligned}$$

When (24.4-25) and (24.4-26) are substituted in (24.4-24) we obtain

$$\begin{aligned}\mathrm{Tr}\{\rho_{\mathrm{T}}[D_\mu, D_{\mathrm{I}\alpha}(t')]\} &= \sum_k \sum_l \langle k|\rho_{\mathrm{T}}|k\rangle\,\delta_{kl}\langle l|[D_\mu, D_{\mathrm{I}\alpha}(t')]|k\rangle \\ &= \sum_k \langle k|\rho_{\mathrm{T}}|k\rangle\langle k|[D_\mu, D_{\mathrm{I}\alpha}(t')]|k\rangle \\ &= \sum_k \sum_l \langle k|\rho_{\mathrm{T}}|k\rangle[\langle k|D_\mu|l\rangle\langle l|D_\alpha|k\rangle \exp[i(E_l - E_k)t'/\hbar] \\ &\quad - \langle k|D_\alpha|l\rangle\langle l|D_\mu|k\rangle \exp[i(E_k - E_l)t'/\hbar]]. \qquad (24.4\text{-}27)\end{aligned}$$

When this expression is substituted in (24.4-22), the integration may be carried out keeping in mind the adiabatic approximation which implies that the perturbation vanishes at $t' = -\infty$. Writing

$$E_l - E_k = \hbar\omega_{lk}, \qquad E_k - E_l = \hbar\omega_{kl},$$

$$\chi^{(1)}_{\mu\alpha}(\omega) = \frac{1}{\hbar}\sum_{kl} \langle k|\rho_{\mathrm{T}}|k\rangle \left[\frac{\langle k|D_\mu|l\rangle\langle l|D_\alpha|k\rangle}{\omega_{lk} - \omega} + \frac{\langle k|D_\alpha|l\rangle\langle l|D_\mu|k\rangle}{\omega_{lk} + \omega}\right], \qquad (24.4\text{-}28)$$

which is another form of the Kramers–Heisenberg dispersion formula (24.1-21).

The next term in the expansion of (24.4-10) is $\mathbf{P}^{(2)}$ whose general form is

$$\mathbf{P}^{(2)}(\omega_1 + \omega_2) = \chi^{(2)}(\omega_1, \omega_2)\mathbf{E}(\omega_1)\mathbf{E}(\omega_2). \qquad (24.4\text{-}29)$$

The susceptibility $\chi^{(2)}(\omega_1, \omega_2)$ is a tensor of rank 3 and is associated with processes involving the mixing of two light waves such as parametric amplification and second harmonic generation.

The symmetry of the medium imposes important restrictions on the susceptibility functions. Thus, for a medium with a center of symmetry $\chi^{(2)} = 0$. More generally $\chi^{(2)}$ must be invariant under any operation of the symmetry group of the medium. $\chi^{(3)}$ is the first nonlinear term for atoms and molecules that have centers of inversion. $\chi^{(3)}$ is a tensor of rank 4 and encompasses numerous interactions which may occur among three distinct frequency components (Bloembergen, 1965; Butcher, 1965; Loudon, 1973).

PART VI

Molecules

CHAPTER 25

GENERAL PROPERTIES OF MOLECULES

In passing from atoms to molecules we ascend to a new, higher level of organization in the structure of matter. Although atomic properties play an essential role in the description of molecular properties, the latter are not merely sums over atomic properties. Indeed, the very fact that molecules exist means that the atoms interact with one another in such a way as to produce a stable, organized system with its own distinctive characteristics. Thus, a molecule is not only quantitatively but *qualitatively* different from its constituent atoms.

The complexities of molecular structure require a variety of approximations whose general objective is to isolate different features of the problem and to treat them separately. We examine some of these methods in the present chapter; subsequent chapters will deal with representative molecular types.

25.1 Born–Oppenheimer Approximation

For a system of electrons and nuclei interacting entirely through Coulombic interactions, the Hamiltonian is

$$\mathscr{H} = \sum_{\alpha} \frac{P_{\alpha}^{2}}{2M_{\alpha}} + \sum_{i} \frac{p_{i}^{2}}{2m} + V(Q_{\alpha}, q_{i}) \qquad (25.1\text{-}1)$$

in which $P_{\alpha}^{2}/2M_{\alpha}$ is the kinetic energy operator for a nucleus of mass M_{α}, $p_{i}^{2}/2m$ the kinetic energy operator for the ith electron of mass m, Q_{α} a set of nuclear coordinates, q_i a set of electronic coordinates, and

$$V(Q_{\alpha}, q_{i}) = \sum_{i<j} \frac{e^{2}}{r_{ij}} + \sum_{\alpha<\beta} \frac{Z_{\alpha}Z_{\beta}e^{2}}{r_{\alpha\beta}} - \sum_{i\alpha} \frac{Z_{\alpha}e^{2}}{r_{i\alpha}}. \qquad (25.1\text{-}2)$$

$V(Q_\alpha, q_i)$ is the potential energy of the entire system of electrons and nuclei; it consists of all the Coulomb interactions between all pairs of charges. The first term in (25.1-2) is the sum of the Coulomb interactions between all pairs of electrons, the second is the same for all pairs of nuclei with atomic numbers Z_α, Z_β, and the third term is the sum of the Coulomb interactions between nuclei and electrons.

It is essentially hopeless to attempt a frontal attack on the solution of the Schrödinger equation with the Hamiltonian (25.1-1) even for simple molecules. The essence of the Born–Oppenheimer approximation is to decompose the problem into separate nuclear and electronic motions based on the large disparity between the masses of nuclei and the mass of the electron. In seeking an approximate solution to

$$\mathscr{H}\Psi(Q_\alpha, q_i) = E\Psi(Q_\alpha, q_i), \tag{25.1-3}$$

the Born–Oppenheimer procedure consists of five steps:

(1) Assume the nuclei are clamped in fixed positions described by the coordinates Q_α'. This will eliminate the nuclear kinetic energy term in the Hamiltonian (25.1-1).

(2) Solve the Schrödinger equation under assumption (1):

$$\left[\sum_i \frac{p_i^2}{2m} + V(Q_\alpha', q_i)\right]\psi(Q_\alpha', q_i) = E'(Q_\alpha')\psi(Q_\alpha', q_i). \tag{25.1-4}$$

The wave function $\psi(Q_\alpha', q_i)$ is an electronic wave function and $E'(Q_\alpha')$ is the corresponding electronic energy eigenvalue when the nuclei are clamped in fixed positions at $Q_\alpha = Q_\alpha'$. When the nuclear positions are assigned a different set of values Q_α'', the potential energy (25.1-2) will be altered to $V(Q_\alpha'', q_i)$ and another set of electronic energy eigenvalues and eigenfunctions will be obtained. Therefore, it is possible, at least in principle to repeat the computation for a sufficiently large series of fixed nuclear positions to define the dependence of the wave functions and the energies on both Q_α and q_i.

(3) Assume

$$\Psi(Q_\alpha, q_i) = \psi(Q_\alpha, q_i)v(Q_\alpha). \tag{25.1-5}$$

At this stage $v(Q_\alpha)$, known as a vibrational function because of its dependence solely on the nuclear coordinates, is unknown.

(4) Substitute (25.1-5) into (25.1-3):

$$\begin{aligned}\mathscr{H}\psi(Q_\alpha, q_i)v(Q_\alpha) &= E\psi(Q_\alpha, q_i)v(Q_\alpha)\\ &= \left[\sum_\alpha \frac{P_\alpha^2}{2M_\alpha} + \sum_i \frac{p_i^2}{2m} + V(Q_\alpha, q_i)\right]\psi(Q_\alpha, q_i)v(Q_\alpha). \end{aligned} \tag{25.1-6}$$

(5) Assume

$$\begin{aligned}\frac{P_\alpha^2}{2M_\alpha}\psi(Q_\alpha,q_i)v(Q_\alpha) &= -\frac{\hbar^2}{2M_\alpha}\nabla_\alpha^2\psi(Q_\alpha,q_i)v(Q_\alpha)\\ &= -\frac{\hbar^2}{2M_\alpha}[\psi(Q_\alpha,q_i)\nabla_\alpha^2 v(Q_\alpha) + \nabla_\alpha\psi(Q_\alpha,q_i)\cdot\nabla_\alpha v(Q_\alpha)\\ &\quad + v(Q_\alpha)\nabla_\alpha^2\psi(Q_\alpha,q_i)]\\ &\approx -\frac{\hbar^2}{2M_\alpha}\psi(Q_\alpha,q_i)\nabla_\alpha^2 v(Q_\alpha)\\ &= \psi(Q_\alpha,q_i)\frac{P_\alpha^2}{2M_\alpha}v(Q_\alpha).\end{aligned} \tag{25.1-7}$$

This is the central approximation in the Born–Oppenheimer approach. Its justification stems from the fact that $m/M_\alpha \ll 1$ which, from a classical view point, implies that nuclear velocities are small in comparison with electronic velocities. As a consequence, the electronic wave function $\psi(Q_\alpha, q_i)$ is relatively insensitive to changes in the nuclear positions and momenta, and is, therefore, capable of adjusting itself quasi-statically to the nuclear motion. This is also known as the *adiabatic approximation*. Substituting (25.1-7) into (25.1-6), we have

$$\psi(Q_\alpha,q_i)\frac{P_\alpha^2}{2M_\alpha}v(Q_\alpha) + \left[\sum_i \frac{p_i^2}{2m} + V(Q_\alpha,q_i)\right]\psi(Q_\alpha,q_i)v(Q_\alpha) = E\psi(Q_\alpha,q_i)v(Q_\alpha). \tag{25.1-8}$$

But in view of (25.1-4) the second term on the left is $E'\psi(Q_\alpha,q_i)v(Q_\alpha)$. Therefore

$$\left[\sum_\alpha \frac{P_\alpha^2}{2M_\alpha} + E'(Q_\alpha)\right]v(Q_\alpha) = Ev(Q_\alpha). \tag{25.1-9}$$

Equation (25.1-9) is an important result; it states that the wave function for the nuclear motion is calculated in an effective potential $E'(Q_\alpha)$ which is obtained from the energy eigenvalues of a particular electronic state as a function of nuclear position.

The form of the total wave function $\Psi(Q_\alpha, q_i)$ (25.1-5) as a product of $\psi(Q_\alpha, q_i)$ and $v(Q_\alpha)$ indicates that electronic and nuclear motions may be treated separately. Since the electronic motion can be regarded as occurring in a potential field created by stationary nuclei, the electronic wave function may be evaluated with the nuclei in their equilibrium positions and the small displacements of the nuclei about their equilibrium positions will have little effect on the electronic wave function. The energy eigenvalues of a particular electronic state, as a function of nuclear position, provide the effective potential for the nuclear motion which is then solved as a separate problem.

The terms that have been omitted in the Born–Oppenheimer approximation may not be negligible if the zero-order electronic eigenfunctions are degenerate or nearly so. In that case the neglected terms behave as perturbations which may remove the degeneracy. The applicability of the Born–Oppenheimer approximation must then be carefully investigated.

We return to (25.1-4), the Schrödinger equation for the electronic wave function in the field of the fixed nuclei whose coordinates are collectively represented by Q_α:

$$\mathscr{H}_{\mathrm{e}}\psi(Q_\alpha, q_i) = E'(Q_\alpha)\psi(Q_\alpha, q_i)$$

with

$$\mathscr{H}_{\mathrm{e}} = \sum_i \frac{p_i^2}{2m} + V(Q_\alpha, q_i).$$

Multiplying on the left by $\psi^*(Q_\alpha, q_i)$ and integrating over the electronic coordinates,

$$E'(Q_\alpha) = \int \psi^*(Q_\alpha, q_i)\mathscr{H}_{\mathrm{e}}\psi(Q_\alpha, q_i)\,d\tau_q.$$

Now consider

$$\begin{aligned}\frac{\partial E'(Q_\alpha)}{\partial Q_\alpha} &= \frac{\partial}{\partial Q_\alpha}\int \psi^*(Q_\alpha, q_i)\mathscr{H}_{\mathrm{e}}\psi(Q_\alpha, q_i)d\tau_q \\ &= \int \frac{\partial \psi^*(Q_\alpha, q_i)}{\partial Q_\alpha}\mathscr{H}_{\mathrm{e}}\psi(Q_\alpha, q_i)d\tau_q + \int \psi^*(Q_\alpha, q_i)\frac{\partial \mathscr{H}_{\mathrm{e}}}{\partial Q_\alpha}\psi(Q_\alpha, q_i)d\tau_q \\ &\quad + \int \psi^*(Q_\alpha, q_i)\mathscr{H}_{\mathrm{e}}\frac{\partial \psi(Q_\alpha, q_i)}{\partial Q_\alpha}d\tau_q.\end{aligned}$$

Since $\mathscr{H}_{\mathrm{e}}$ is Hermitian, the third integral may be written as

$$\begin{aligned}\int \psi^*(Q_\alpha, q_i)\mathscr{H}_{\mathrm{e}}\frac{\partial \psi(Q_\alpha, q_i)}{\partial Q_\alpha}d\tau_q &= \int \frac{\partial \psi(Q_\alpha, q_i)}{\partial Q_\alpha}\mathscr{H}_{\mathrm{e}}\psi^*(Q_\alpha, q_i)\,d\tau_q \\ &= \int \frac{\partial \psi(Q_\alpha, q_i)}{\partial Q_\alpha}E'\psi^*(Q_\alpha, q_i)\,d\tau_q\end{aligned}$$

as a result of the Schrödinger equation $\mathscr{H}_e\psi = E'\psi$. Combining the first and third integrals we have

$$\int \frac{\partial\psi^*}{\partial Q_\alpha}\mathscr{H}_e\psi\,d\tau_q + \int \psi^*\mathscr{H}_e\frac{\partial\psi}{\partial Q_\alpha}\,d\tau_q = E'\int\left(\frac{\partial\psi^*}{\partial Q_\alpha}\psi + \psi^*\frac{\partial\psi}{\partial Q_\alpha}\right)d\tau_q$$

$$= E'\int\frac{\partial}{\partial Q_\alpha}(\psi^*\psi)\,d\tau_q$$

$$= E'\frac{\partial}{\partial Q_\alpha}\int \psi^*\psi\,d\tau_q = 0.$$

We are left with the result that

$$\frac{\partial E'(Q_\alpha)}{\partial Q_\alpha} = \int \psi^*(Q_\alpha, q_i)\frac{\partial \mathscr{H}_e}{\partial Q_\alpha}\psi(Q_\alpha, q_i)\,d\tau_q$$

$$= \int \psi^*(Q_\alpha, q_i)\frac{\partial V(Q_\alpha, q_i)}{\partial Q_\alpha}\psi(Q_\alpha, q_i)\,d\tau_q$$

since the kinetic energy operator does not depend on Q_α.

This is known as the *Hellmann–Feynman theorem* (Hellmann, 1937; Feynman, 1938). Its significance is that within the Born–Oppenheimer approximation the forces acting on nuclei are derivable from the classical potential $V(Q_\alpha, q_i)$ (25.1-2) and the electronic wave function. Thus, if a force component F_α is written as

$$F_\alpha = -\partial E'(Q_\alpha)/\partial Q_\alpha,$$

it is seen that F_α is obtained from the classical expression for

$$-\partial V(Q_\alpha, q_i)/\partial Q_\alpha$$

averaged over the electronic wave function.

25.2 Molecular Orbitals and the Self-Consistent Field Method

Adopting the Born–Oppenheimer approach we face the problem of solving for the electronic and nuclear wave functions. In this section and the next we consider some general aspects of the electronic problem.

In the atomic case, multielectron wave functions are constructed from orbitals which are populated by electrons in accordance with the Pauli principle. The three-dimensional rotation group, particularly through its formulation in terms of angular momentum operators, plays a significant role in this theory. For molecules it is possible to adopt an analogous approach in which the multielectron molecular wave functions are constructed from a set of *molecular* orbitals which have the following characteristics:

(1) A molecular orbital is an eigenfunction of a one-electron Hamiltonian, that is, a Hamiltonian consisting of one-electron operators; hence a molecular orbital is a wave function that depends on the coordinates of a single electron.

(2) A molecular orbital can extend over any number of atoms in a molecule. Therefore an electron in a molecular orbital is not localized but may have a nonvanishing probability density in various parts of the molecule.

(3) No more than two electrons can occupy a molecular orbital; two electrons in a molecular orbital must have opposite spin.

(4) Molecular orbitals reflect the basic symmetry of the molecule. Each orbital belongs to an irreducible representation of the molecular symmetry group which is the group of transformations under which the molecular Hamiltonian is invariant.

(5) A molecular orbital multiplied by a spin function (α or β) is a molecular spin orbital.

The close analogy between atomic and molecular orbitals is evident from these properties.

The basic approach to the calculation of molecular orbitals consists of efforts to solve the Hartree–Fock equations which are formally identical for atoms and molecules. We have seen that the problem is already quite complicated for atoms; in molecules, the absence of three-dimensional rotational symmetry and the occurrence of multicenter integrals increase the complexity to an extent that direct, iterative, solutions of the Hartree–Fock equations are, for all practical purposes, impossible. Instead, one postulates at the outset that molecular orbitals are to be expressed as linear combinations of atomic orbitals (LCAO). This is known as the *Roothaan method* and it will be shown that such a starting point simplifies the problem enormously.

We shall confine the discussion to closed shells. Such systems are the easiest ones to handle and, fortuitously, many molecules in their ground state have closed shell configurations. Since all the electrons in a closed shell have their spins paired ($S = 0$), the total antisymmetric wave function can be represented by a single Slater determinant

$$\Psi = \frac{1}{\sqrt{(2N)!}} \begin{vmatrix} \psi_1(1) & \bar{\psi}_1(1) & \cdots & \psi_N(1) & \bar{\psi}_N(1) \\ \psi_1(2) & \bar{\psi}_1(2) & \cdots & \psi_N(2) & \bar{\psi}_N(2) \\ \vdots & \vdots & & \vdots & \vdots \\ \psi_1(2N) & \bar{\psi}_1(2N) & \cdots & \psi_N(2N) & \bar{\psi}_N(2N) \end{vmatrix} \tag{25.2-1}$$

where the orthonormal molecular spin orbitals $\psi_j(\lambda)$ and $\bar{\psi}_j(\lambda)$ refer to the same spatial orbital but with $m_s = \frac{1}{2}$ in the former and $m_s = -\frac{1}{2}$ in the latter. Thus the $2N$ electrons are distributed in $2N$ spin orbitals or N spatial orbitals. As in the atomic case, the objective is to find the orbitals that will minimize the energy

$$E = \langle \Psi | \mathscr{H} | \Psi \rangle \tag{25.2-2}$$

where $\mathscr{H}$ is the complete electronic Hamiltonian of the molecule:

$$\mathscr{H} = \sum_i \frac{p_i^2}{2m} + \sum_{i<j} \frac{e^2}{r_{ij}} - \sum_{i,\alpha} \frac{Z_\alpha e^2}{r_{i\alpha}} + \sum_{\alpha<\beta} \frac{Z_\alpha Z_\beta e^2}{r_{\alpha\beta}}. \tag{25.2-3}$$

The indices i, j refer to electrons; α, β to nuclei; Z is a nuclear charge and r an interparticle distance. $\mathscr{H}$ contains all Coulombic terms under the assumption that the nuclear skeleton is clamped in its equilibrium configuration.

The molecular orbitals that satisfy the variational principle are obtained from solutions of the Hartree–Fock equations; for a closed shell (Section 19.4) they are of the form

$$F(1)\psi_i(1) = \varepsilon_i \psi_i(1) \tag{25.2-4a}$$

where

$$F(1) = \mathscr{H}_0(1) + \sum_k [2J_k(1) - K_k(1)], \tag{25.2-4b}$$

$$\mathscr{H}_0(1) = \frac{p_1^{\ 2}}{2m} - \sum_\alpha \frac{Z_\alpha e^2}{r_{1\alpha}}. \tag{25.2-4c}$$

J_k and K_k are the Coulomb and exchange operators, respectively, which depend on the molecular orbitals themselves and are defined as in (11.2-23):

$$\begin{aligned} J_k(1)\psi_i(1) &= \left[\int \psi_k{}^*(2) \frac{e^2}{r_{12}} \psi_k(2)\, d\tau_2\right] \psi_i(1), \\ K_k(1)\psi_i(1) &= \left[\int \psi_k{}^*(2) \frac{e^2}{r_{12}} \psi_i(2)\, d\tau_2\right] \psi_k(1). \end{aligned} \tag{25.2-5a}$$

The total energy is

$$E = 2\sum_i I_i + \sum_{ik} [2J(i,k) - K(i,k)]. \tag{25.2-5b}$$

We now introduce the constraint, due to Roothaan (1951), that the molecular orbitals are to be expressed as linear combinations of atomic orbitals:

$$\psi_i = \sum_\mu c_{\mu i} \varphi_\mu, \tag{25.2-6a}$$

or, in matrix form,

$$\Psi = \Phi C \tag{25.2-6b}$$

where Ψ and Φ are row matrices and C is a square matrix.

The φ_μ are the atomic orbitals which, in the aggregate, constitute the basis set. It is important to recognize that the φ_μ need not be true atomic orbitals in the sense that they must be eigenfunctions of an atomic Hamiltonian. They are "atomic" only to the extent that they are one-electron functions centered on the various nuclei and chosen largely for their ability to provide a flexible starting point for the calculation. The φ_μ are not necessarily orthogonal. Hence we define

$$S_{\mu\nu} = \langle \varphi_\mu | \varphi_\nu \rangle = \int \varphi_\mu{}^*(1)\varphi_\nu(1)\, d\tau_1 \tag{25.2-7}$$

as the *overlap integral* for the orbitals φ_μ and φ_ν. The molecular orbitals, on the other hand, are required to satisfy the orthogonality condition

$$\langle \psi_i | \psi_j \rangle = \delta_{ij} \tag{25.2-8}$$

so that in the LCAO approximation,

$$\sum_{\mu\nu} c_{\mu i}^* c_{\nu j} \langle \varphi_\mu | \varphi_\nu \rangle = \sum_{\mu\nu} c_{\mu i}^* c_{\nu j} S_{\mu\nu} = \delta_{ij}, \tag{25.2-9}$$

which, in matrix form, is

$$C^\dagger SC = 1. \tag{25.2-10}$$

The charge density at the point $\mathbf{r}$ is defined by

$$\rho(\mathbf{r}) = 2 \sum_i^{\text{occ}} \psi_i{}^*(\mathbf{r})\psi_i(\mathbf{r}), \tag{25.2-11}$$

the factor of 2 appearing because of the double occupancy of the molecular orbitals by two electrons with opposite spin. In terms of atomic orbitals,

$$\begin{aligned} \rho(\mathbf{r}) &= 2 \sum_i \sum_{\mu\nu} c_{\mu i}^* c_{\nu i} \varphi_\mu{}^*(\mathbf{r})\varphi_\nu(\mathbf{r}) \\ &= \sum_{\mu\nu} P_{\mu\nu} \varphi_\mu{}^*(\mathbf{r})\varphi_\nu(\mathbf{r}) \end{aligned} \tag{25.2-12}$$

where

$$P_{\mu\nu} = 2 \sum_i^{\text{occ}} c_{\mu i}^* c_{\nu i} \tag{25.2-13}$$

is the *density matrix*.

If N is the number of occupied (spatial) orbitals, the closed shell will contain $2N$ electrons; therefore the integral of the charge density over all space must be equal to $2N$:

$$\int \rho(\mathbf{r})\,d\mathbf{r} = \sum_{\mu\nu} P_{\mu\nu} \int \varphi_\mu{}^*(\mathbf{r})\varphi_\nu(\mathbf{r})\,d\mathbf{r}$$

$$= \sum_{\mu\nu} P_{\mu\nu} S_{\mu\nu} = 2N. \qquad (25.2\text{-}14)$$

With these relations

$$\mathscr{H}_0(1)\psi_i(1) = \sum_\nu \mathscr{H}_0(1)c_{\nu i}\varphi_\nu(1) \qquad (25.2\text{-}15)$$

$$J_k(1)\psi_i(1) = \sum_{\lambda\sigma\nu} c_{\lambda k}^* c_{\sigma k} c_{\nu i}\left[\int \varphi_\lambda{}^*(2)\frac{e^2}{r_{12}}\varphi_\sigma(2)\,d\tau_2\right]\varphi_\nu(1), \qquad (25.2\text{-}16)$$

$$K_k(1)\psi_i(1) = \sum_{\lambda\sigma\nu} c_{\lambda k}^* c_{\sigma k} c_{\nu i}\left[\int \varphi_\lambda{}^*(2)\frac{e^2}{r_{12}}\varphi_\nu(2)\,d\tau_2\right]\varphi_\sigma(1), \qquad (25.2\text{-}17)$$

$$F(1)\psi_i(1) = \sum_\nu \mathscr{H}_0(1)c_{\nu i}\varphi_\nu(1)$$

$$+ \sum_{k\lambda\sigma\nu} c_{\lambda k}^* c_{\sigma k} c_{\nu i}\left\{2\left[\int \varphi_\lambda{}^*(2)\frac{e^2}{r_{12}}\varphi_\sigma(2)\,d\tau_2\right]\varphi_\nu(1)\right.$$

$$\left. - \left[\int \varphi_\lambda{}^*(2)\frac{e^2}{r_{12}}\varphi_\nu(2)\,d\tau_2\right]\varphi_\sigma(1)\right\}$$

$$= \sum_\nu \varepsilon_i c_{\nu i}\varphi_\nu(1). \qquad (25.2\text{-}18)$$

Equation (25.2-18) is the Hartree–Fock equation in LCAO form. At this stage it is useful to introduce the notation

$$\langle \varphi_\mu(1)\varphi_\lambda(2)|e^2/r_{12}|\varphi_\nu(1)\varphi_\sigma(2)\rangle \equiv \langle \mu\nu|\lambda\sigma\rangle \qquad (25.2\text{-}19)$$

in which the convention is to put functions referring to the same electron in the same half-bracket and to put the complex conjugates on the left side of each half-bracket. Definition (25.2-19) implies

$$\langle \mu\nu|\lambda\sigma\rangle = \langle \lambda\sigma|\mu\nu\rangle, \qquad \langle \mu\nu|\lambda\sigma\rangle = \langle \nu\mu|\sigma\lambda\rangle^*. \qquad (25.2\text{-}20)$$

Upon multiplying both sides of (25.2-18) by $\varphi_\mu{}^*(1)$ and integrating over the coordinates of electron (1) we obtain

$$\sum_\nu c_{\nu i}H_{\mu\nu} + \sum_{k\lambda\sigma\nu} c_{\lambda k}^* c_{\sigma k} c_{\nu i}[2\langle \mu\nu|\lambda\sigma\rangle - \langle \mu\sigma|\lambda\nu\rangle] = \sum_\nu \varepsilon_i c_{\nu i} S_{\mu\nu} \qquad (25.2\text{-}21)$$

in which

$$H_{\mu\nu} = \langle \varphi_\mu(1)|\mathscr{H}_0(1)|\varphi_\nu(1)\rangle,$$

or, inserting the density matrix,

$$\sum_{\nu} c_{\nu i} H_{\mu\nu} + \sum_{\lambda\sigma\nu} P_{\lambda\sigma}[\langle\mu\nu|\lambda\sigma\rangle - \tfrac{1}{2}\langle\mu\sigma|\lambda\nu\rangle] = \sum_{\nu} \varepsilon_i c_{\nu i} S_{\mu\nu}. \quad (25.2\text{-}22)$$

Finally, by writing

$$F_{\mu\nu} = H_{\mu\nu} + \sum_{\lambda\sigma} P_{\lambda\sigma}[\langle\mu\nu|\lambda\sigma\rangle - \tfrac{1}{2}\langle\mu\sigma|\lambda\nu\rangle] \quad (25.2\text{-}23)$$

the Hartree–Fock equations assume the compact form

$$\sum_{\nu} (F_{\mu\nu} - \varepsilon_i S_{\mu\nu}) c_{\nu i} = 0 \quad (25.2\text{-}24)$$

or, in matrix notation,

$$(F - \varepsilon S)C = 0 \quad (25.2\text{-}25)$$

where ε is a diagonal matrix whose diagonal elements ε_i are the orbital energies. The set of homogeneous equations (25.2-24) or (25.2-25) will have nontrivial solutions only when

$$\det|F_{\mu\nu} - \varepsilon_i S_{\mu\nu}| = 0. \quad (25.2\text{-}26)$$

It is now apparent that the LCAO approximation has converted the original partial differential equations (Hartree–Fock) into algebraic equations (Roothaan). This is a significant simplification without which it is virtually impossible to calculate credible molecular orbitals for any but the simplest molecules. The Roothaan equations (also known as LCAO–SCF equations) are the basic equations of molecular orbital theory.

The total energy of a closed shell was shown to be

$$E = 2\sum_{i} I_i + \sum_{ik} [2J(i,k) - K(i,k)]. \quad (25.2\text{-}27)$$

In the LCAO approximation we have

$$I_i = \langle\psi_i(1)|\mathscr{H}_0(1)|\psi_i(1)\rangle = \sum_{\mu\nu} c_{\mu i}^* c_{\nu i} H_{\mu\nu}, \quad (25.2\text{-}28)$$

$$\begin{aligned} J(i,k) &= \langle\psi_i(1)\psi_k(2)|e^2/r_{12}|\psi_i(1)\psi_k(2)\rangle \\ &= \sum_{\mu\lambda\nu\sigma} c_{\mu i}^* c_{\lambda k}^* c_{\nu i} c_{\sigma k}\langle\mu\nu|\lambda\sigma\rangle, \end{aligned} \quad (25.2\text{-}29)$$

$$\begin{aligned} K(i,k) &= \langle\psi_i(1)\psi_k(2)|e^2/r_{12}|\psi_k(1)\psi_i(2)\rangle \\ &= \sum_{\mu\lambda\nu\sigma} c_{\mu i}^* c_{\lambda k}^* c_{\nu i} c_{\sigma k}\langle\mu\sigma|\lambda\nu\rangle. \end{aligned} \quad (25.2\text{-}30)$$

Substituting in (25.2-27) the expression for the energy becomes

$$\begin{aligned} E &= \sum_{\mu\nu} P_{\mu\nu} H_{\mu\nu} + \tfrac{1}{2} \sum_{\mu\nu\lambda\sigma} P_{\mu\nu} P_{\lambda\sigma}[\langle\mu\nu|\lambda\sigma\rangle - \tfrac{1}{2}\langle\mu\sigma|\lambda\nu\rangle] \\ &= \tfrac{1}{2}\sum_{\mu\nu} P_{\mu\nu}(H_{\mu\nu} + F_{\mu\nu}). \end{aligned} \quad (25.2\text{-}31)$$

The preceding development has been based on the Hamiltonian (25.2-3) which, apart from kinetic energy operators, consists entirely of Coulombic interactions. In the case of atoms it is also necessary to consider spin-orbit interactions whose importance relative to Coulombic interactions depends on the atomic number of the atom. It is recalled that the general form of the spin-orbit interaction (Section 21.4) is

$$\mathscr{H}_{\mathrm{so}} = \sum_i \xi(r_i)\mathbf{l}_i \cdot \mathbf{s}_i. \tag{25.2-32}$$

Implicit in this expression is the assumption of spherical symmetry of the nuclear field—a condition which ceases to exist in a molecule. Nevertheless, the spin-orbit interaction is quite insensitive to the molecular environment so that, to a good approximation, (25.2-32) is also valid for a molecule. The operator $\mathbf{l}_i$ can no longer be regarded as an angular momentum operator when the field departs from spherical symmetry although it still retains the definition

$$\mathbf{l}_i = -i\hbar\mathbf{r}_i \times \nabla_i.$$

Hence spin-orbit effects in molecules are treated in a manner that closely resembles the atomic case.

25.3 Computational Methods

The Roothaan equations into which the Hartree–Fock equations have evolved as a result of the LCAO approximation still require an iterative procedure because the matrix elements $F_{\mu\nu}$ depend on unknown coefficients (or density matrices). A typical calculation involves the following steps:

(1) Specify a set of atomic orbitals (the basis set). Among the popular choices are the Slater orbitals (19.5-5) and Gaussian orbitals

$$\chi_{nlm}(\mathbf{r}) = Nr^{n-1}e^{-\alpha r^2}Y_{lm}(\theta, \varphi) \tag{25.3-1}$$

which are usually converted into real form by appropriate linear combinations (see, for example, Table 1.2).

(2) Compute the overlap integrals

$$S_{\mu\nu} = \int \varphi_\mu(1)\varphi_\nu(1)\,d\tau_1. \tag{25.3-2}$$

(3) Compute the core Hamiltonian matrix

$$H_{\mu\nu} = \int \varphi_\mu(1)\mathscr{H}_0(1)\varphi_\nu(1)\,d\tau_1. \tag{25.3-3}$$

(4) Compute the two-electron integrals

$$\langle\mu\nu|\lambda\sigma\rangle = \int \varphi_\mu(1)\varphi_\nu(1)(e^2/r_{12})\varphi_\lambda(2)\varphi_\sigma(2)\,d\tau_1\,d\tau_2. \tag{25.3-4}$$

(5) Compute the eigenvectors of $H_{\mu\nu}$. This gives a starting set of LCAO coefficients (and the initial set of molecular orbitals).

(6) Assign electrons in pairs to the lowest molecular orbitals until all the electrons have been assigned.

(7) Compute the density matrix

$$P_{\mu\nu} = 2 \sum_{i}^{\text{occ}} c_{\mu i}^{*} c_{\nu i}. \tag{25.3-5}$$

(8) Compute the total electronic energy

$$E = \sum_{\mu\nu} P_{\mu\nu} H_{\mu\nu} + \tfrac{1}{2} \sum_{\mu\nu\lambda\sigma} P_{\mu\nu} P_{\lambda\sigma} [\langle \mu\nu | \lambda\sigma \rangle - \tfrac{1}{2} \langle \mu\lambda | \nu\sigma \rangle]. \tag{25.3-6}$$

(9) Compute the Fock Hamiltonian

$$F_{\mu\nu} = H_{\mu\nu} + \sum_{\lambda\sigma} P_{\lambda\sigma} [\langle \mu\nu | \lambda\sigma \rangle - \tfrac{1}{2} \langle \mu\lambda | \nu\sigma \rangle]. \tag{25.3-7}$$

(10) Compute the eigenvectors of $F_{\mu\nu}$ to obtain a second set of LCAO coefficients.

(11) Continue until the total energy E remains constant to the required accuracy.

Examples of this procedure and some of its variants are given by Pople and Beveridge (1970).

Despite the simplification in the Hartree–Fock formalism resulting from the introduction of the LCAO approximation, the most serious limitation in molecular-orbital calculations is the large number of integrals that must be evaluated. For a comparatively simple molecule like CO_2 there are about 10^5 integrals that must be computed in order to achieve moderately accurate molecular orbitals. Many integrals are of the two-electron, multi-center type which consume most of the computer time. In consequence, numerous variations in molecular orbital theory have come into existence with the objective of reducing the number of integrals without, hopefully, suffering excessive loss in accuracy. In this section we shall summarize some of these approaches and indicate their general area of applicability.

Practically all molecular orbital methods use certain approximations to a greater or lesser extent. They are:

(1) Neglect of differential overlap (NDO). This consists of setting

$$\varphi_\mu(1)\varphi_\nu(1) = 0 \qquad (\mu \neq \nu) \tag{25.3-8}$$

in the electron repulsion integrals, and has the effect of eliminating all three- and four-center integrals.

(2) Neglect of all but valence electrons. Atomic orbitals, chosen exclusively from the valence shell of each atom, constitute the basis set for the

LCAO expansion. Inner shell electrons, because they are more tightly bound, do not participate to any significant extent in the formation of molecular orbitals. Empirically, it is known that chemical effects are largely due to the valence electrons.

(3) Except in the very simplest cases, some use is made of experimental information such as atomic ionization potentials and electron affinities. These are used to assign numerical values to a number of theoretical parameters.

A popular version of molecular-orbital theory is known as the CNDO method (complete neglect of differential overlap) in which condition (25.3-8) is used for all two-electron interaction integrals. In this scheme, using the convention that φ_μ is on atom A and φ_ν on atom B,

$$F_{\mu\mu} = U_{\mu\mu} + \tfrac{1}{2}P_{\mu\mu}\gamma_{\mu\mu} + \sum_{\nu \neq \mu} (P_{\nu\nu} - Z_\nu)\gamma_{\mu\nu}, \tag{25.3-9}$$

$$F_{\mu\nu} = H_{\mu\nu} - \tfrac{1}{2}P_{\mu\nu}\gamma_{\mu\nu} \qquad (\mu \neq \nu) \tag{25.3-10}$$

In these expressions $U_{\mu\mu}$ is a core integral which is approximated by

$$U_{\mu\mu} = -I_\mu$$

where I_μ is the atomic ionization potential for the orbital φ_μ. The condition of zero differential overlap is employed so that

$$\langle \mu\lambda | \nu\sigma \rangle = \langle \mu\mu | \nu\nu \rangle \, \delta_{\mu\lambda} \, \delta_{\nu\sigma}. \tag{25.3-11}$$

$\gamma_{\mu\nu}$ is the two-electron repulsion integral

$$\gamma_{\mu\nu} = \langle \mu\mu | \nu\nu \rangle = \langle \varphi_\mu(1)\varphi_\nu(2) | (e^2/r_{12}) | \varphi_\mu(1)\varphi_\nu(2) \rangle. \tag{25.3-12}$$

When $\mu = \nu$, the integrals are approximated by

$$\gamma_{\mu\mu} = I_\mu - A_\mu \tag{25.3-13}$$

where A_μ is the electron affinity in the orbital φ_μ. $H_{\mu\nu}$ is a core resonance integral which is often approximated by setting

$H_{\mu\nu} = 0$ for μ and ν not nearest neighbors,

$H_{\mu\nu} =$ empirical constant when μ and ν are bonded nearest neighbors.

Z_ν is the core charge of atom B and is equal to the nuclear charge minus the number of inner shell electrons.

The CNDO method has been applied mainly to small molecules composed of light atoms. The programs employed by Pople and Beveridge (1970), for example, are limited to molecules containing up to 35 atoms or 80 basis functions, whichever is smaller. Such limits are, however, rapidly being expanded by advances in computer technology.

For larger molecules, the approximations are, of necessity, more crude. In this category we mention the Extended Hückel Method (EHM) in which all, or almost all, integrals are replaced by empirical parameters thereby eliminating the self-consistent, iterative procedure. If the secular equation (25.2-26) is written as

$$\det\left|H_{\mu\nu} - \varepsilon_i S_{\mu\nu}\right| = 0, \tag{25.3-14}$$

the diagonal elements are assigned empirical values depending on the orbitals and the type of molecule. The off-diagonal elements are approximated by

$$H_{\mu\nu} = 0.5K(H_{\mu\mu} + H_{\nu\nu})S_{\mu\nu} \tag{25.3-15}$$

in which K is an empirical parameter and $S_{\mu\nu}$ the overlap integral for atomic orbitals φ_μ and φ_ν centered on the appropriate atoms. Having made the numerical assignments, solutions to (25.3-14) yield the eigenvalues ε_i and hence the coefficients $c_{\mu i}$ in the molecular orbitals (25.2-6).

Several other approximate methods, applicable to special types of molecules, are described in succeeding chapters.

Although the atomic orbitals in the LCAO expansion were not assumed to be orthogonal, it is possible, in principle, to orthogonalize them. Thus let

$$\langle\varphi_\mu|\varphi_\nu\rangle = S_{\mu\nu}, \tag{25.3-16}$$

$$l_i = \sum_\mu \varphi_\mu A_{\mu i}, \qquad l_j = \sum_\nu \varphi_\nu A_{\nu j}, \tag{25.3-17}$$

$$\langle l_i|l_j\rangle = \delta_{ij}, \tag{25.3-18}$$

that is, we seek the transformation matrix A which, when applied to the orbitals $\varphi_\mu, \varphi_\nu, \ldots$, will produce a new set of orbitals $l_i, l_j, \ldots$ which satisfy the orthogonality condition. Now

$$\langle l_i|l_j\rangle = \sum_{\mu\nu} \langle\varphi_\mu A_{\mu i}|\varphi_\nu A_{\nu j}\rangle = \sum_{\mu\nu} A^*_{i\mu}\langle\varphi_\mu|\varphi_\nu\rangle A_{\nu j} = \sum_{\mu\nu} A^*_{i\mu} S_{\mu\nu} A_{\nu j}.$$

In matrix notation, the orthogonality condition requires that

$$\langle l_i|l_j\rangle = A^\dagger S A = I \tag{25.3-19}$$

where I is the unit matrix. If it is assumed that A is Hermitian ($A^\dagger = A$), (25.3-19) may be symbolically written as

$$A = S^{-1/2}. \tag{25.3-20}$$

If S were diagonal, the components A_{ii} would simply be $1/\sqrt{S_{ii}}$, but S is generally not diagonal; hence it is necessary to transform it to S' where

$$S' = U^{-1}SU$$

is a diagonal matrix. This is always possible because S is Hermitian (or real symmetric). One now evaluates $(S')^{-1/2}$ and transforms back to give

$$S^{-1/2} = U(S')^{-1/2}U^{-1}$$

which then defines the matrix A and transformation (25.3-17). Orbitals such as $l_i, l_j, \ldots$ are known as *Löwdin* orbitals.

Finally, it is necessary to remark that the limitations inherent in the Hartree–Fock method, described in Section 19.6 for the atomic case, apply as well to molecules. In the former case relativistic corrections and configuration interaction result in improved wave functions. The same is true for molecules although the feasibility of applying such corrections is more restricted.

CHAPTER 26

ELECTRONIC STATES OF MOLECULES

The extremely wide range of molecular sizes and shapes has led to the development of numerous special techniques adapted to specific types of molecules, but in one form or another most methods employ the molecular-orbital approach with the LCAO approximation and hence may be regarded as variants of the Roothaan procedure for solving the Hartree–Fock equations. In this section we shall examine some of these methods beginning with the simplest molecule H_2^+, which consists of two protons and a single electron, and progressing along the scale of structural complexity to the conjugated molecules of organic chemistry. The importance of symmetry properties is a constantly recurring theme throughout this discussion.

26.1 Hydrogen Molecule Ion (H_2^+)

Assuming the two protons have been clamped in a fixed position (Fig. 26.1) as required by the Born–Oppenheimer approximation, the Hamiltonian for the system in Rydberg units is

$$\mathcal{H} = -\nabla^2 - \frac{2}{r_a} - \frac{2}{r_b} + \frac{2}{R}. \tag{26.1-1}$$

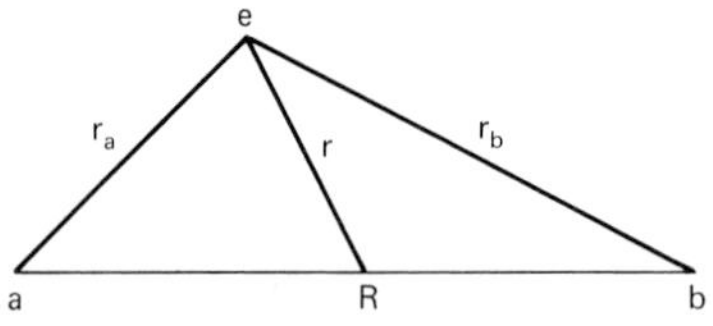

FIG. 26.1 Nomenclature for H_2^+. The two protons are at *a* and *b*; the electron is at *e*.

This Hamiltonian is unique because it is possible to obtain exact solutions to the corresponding Schrödinger equation by adopting spheroidal coordinates (Slater, 1963):

$$\lambda = (r_a + r_b)/R, \qquad \mu = (r_a - r_b)/R. \tag{26.1-2}$$

However, this approach does not permit extension to more complicated molecules. We shall therefore not pursue it; instead we go directly to the LCAO method in which it is assumed that ψ, the eigenfunction of $\mathscr{H}$ can be written

$$\psi = c_a\varphi_a + c_b\varphi_b \tag{26.1-3}$$

in which φ_a and φ_b are atomic orbitals of the same type centered on atoms a and b; c_a and c_b are constants. The criterion for choosing the constants is that E, the total energy, defined by

$$E = \langle\psi|\mathscr{H}|\psi\rangle/\langle\psi|\psi\rangle \tag{26.1-4}$$

shall be a minimum when $\mathscr{H}$ is given by (26.1-1).

It will be assumed henceforth in molecular calculations that atomic orbitals such as φ_a and φ_b are real and are normalized. Hence the substitution of (26.1-3) into (26.1-4) gives

$$E(c_a^2 + c_b^2 + 2c_ac_bS) = c_a^2H_{aa} + c_b^2H_{bb} + 2c_ac_bH_{ab} \tag{26.1-5}$$

where S is the overlap integral

$$S = \langle\varphi_a|\varphi_b\rangle$$

with values lying between zero and one and

$$\begin{aligned} H_{aa} &= \langle\varphi_a|\mathscr{H}|\varphi_a\rangle, \qquad H_{bb} = \langle\varphi_b|\mathscr{H}|\varphi_b\rangle, \\ H_{ab} &= H_{ba} = \langle\varphi_a|\mathscr{H}|\varphi_b\rangle = \langle\varphi_b|\mathscr{H}|\varphi_a\rangle. \end{aligned} \tag{26.1-6}$$

If E is to be a minimum with respect to variations in c_a and c_b,

$$\partial E/\partial c_a = \partial E/\partial c_b = 0. \tag{26.1-7}$$

Differentiating both sides of (26.1-5) with respect to c_a and then with respect to c_b, one obtains the two homogeneous equations

$$c_a(H_{aa} - E) + c_b(H_{ab} - ES) = 0, \qquad c_a(H_{ab} - ES) + c_b(H_{bb} - E) = 0 \tag{26.1-8}$$

in which conditions (26.1-7) have been used.

Solutions to (26.1-8) exist if

$$\begin{vmatrix} H_{aa} - E & H_{ab} - ES \\ H_{ba} - ES & H_{bb} - E \end{vmatrix} = 0. \tag{26.1-9}$$

When the two nuclei are alike, as in the case of H_2^+, $H_{aa} = H_{bb}$; hence the energy eigenvalues, obtained as solutions to (26.1-9), are

$$E_1 = (H_{aa} + H_{ab})/(1 + S), \qquad E_2 = (H_{aa} - H_{ab})/(1 - S) \quad (26.1\text{-}10)$$

and the corresponding normalized eigenfunctions are

$$\psi_1 = (\varphi_a + \varphi_b)/\sqrt{2(1 + S)}, \qquad \psi_2 = (\varphi_a - \varphi_b)/\sqrt{2(1 - S)}. \quad (26.1\text{-}11)$$

Thus we have found two molecular orbitals of form (26.1-3). To give more substance to this development let us take φ_a and φ_b to be 1s atomic orbitals of hydrogen:

$$\varphi_a = 1s_a = \frac{1}{\sqrt{\pi}} e^{-r_a}, \qquad \varphi_b = 1s_b = \frac{1}{\sqrt{\pi}} e^{-r_b}. \quad (26.1\text{-}12)$$

The energies may now be explicitly evaluated [see, for example, Kauzmann (1957) or Slater (1963)]. These are shown in Fig. 26.2b as a function of the internuclear distance R. Since $E_1 \leqslant E_2$ at all internuclear distances, H_{ab} must be negative. Also, E_1 has a minimum at a certain value of R which may be called R_{eq}; this must then correspond to a stable configuration of H_2^+ with an equilibrium separation between the two nuclei equal to R_{eq}. Thus

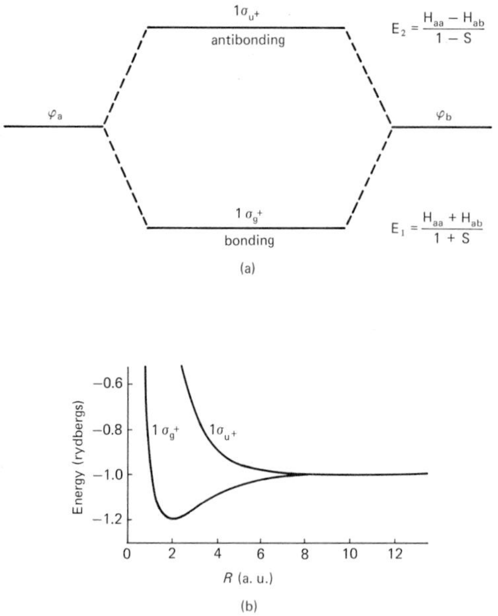

FIG. 26.2 (a) The two lowest molecular orbitals of H_2^+ and (b) their energies as a function of internuclear distance.

ψ_1 is the molecular orbital that corresponds to the lowest (or ground) state of H_2^+.

ψ_2 may be regarded as an excited state but since E_2 does not have a minimum at any finite value of R, the two nuclei simply move apart and the molecule dissociates. For this reason ψ_1 is known as a *bonding* (molecular) orbital while ψ_2 is an *antibonding* (molecular) orbital. Antibonding orbitals are often designated by an asterisk so that in place of ψ_1 and ψ_2 one might write ψ and ψ^* (not to be confused with the complex conjugate) for the bonding and antibonding orbitals, respectively. The qualitative features we have just described are quite general and are independent of the choice of atomic orbitals. An energy level diagram is shown in Fig. 26.2a.

Referring to (26.1-10), it is observed that the upward displacement of the antibonding orbital E_2 is greater than the downward displacement of the bonding orbital. For computational purposes, the overlap integral S in (26.1-10) and (26.1-11) may, at times, be set equal to zero. This simplifies the normalization of the molecular orbitals, as well as the integrals that depend on them, and also causes the energies of the bonding and antibonding orbitals to be shifted symmetrically. However, the qualitative features of the bonding and antibonding orbitals are not altered by the purely formal device of setting $S = 0$.

It is of interest to compute $|\psi_1|^2$ and $|\psi_2|^2$ which are proportional to the charge distributions in the two molecular orbitals. From (26.1-11)

$$|\psi_1|^2 = \frac{1}{1+S}\left[\frac{1}{2}(\varphi_a^{\,2} + \varphi_b^{\,2})\right] + \frac{S}{1+S}\frac{\varphi_a\varphi_b}{S} \tag{26.1-13a}$$

$$|\psi_2|^2 = \frac{1}{1-S}\left[\frac{1}{2}(\varphi_a^{\,2} + \varphi_b^{\,2})\right] - \frac{S}{1-S}\frac{\varphi_a\varphi_b}{S}. \tag{26.1-13b}$$

These functions are plotted in Fig. 26.3. The important feature to note is that for the bonding orbital the electronic charge tends to concentrate in the region between the nuclei, whereas for the antibonding orbital the region of

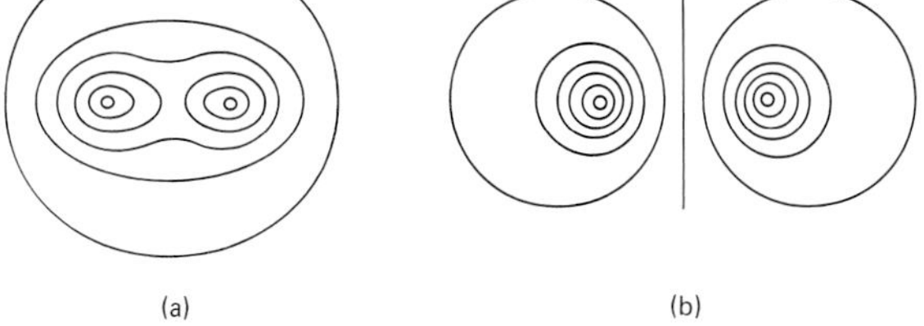

FIG. 26.3 Charge distribution in (a) bonding and (b) antibonding molecular orbitals in H_2^+. (Holden 1971.)

high charge concentration lies outside of the two nuclei. If $\varphi_a\varphi_b$ were zero everywhere, as would be the case if the nuclei were sufficiently far apart, we would have, in place of (26.1-13),

$$|\psi_1|^2 = |\psi_2|^2 = \tfrac{1}{2}(\varphi_a{}^2 + \varphi_b{}^2). \tag{26.1-14}$$

This simply represents one half of an electron in each of the two atomic orbitals φ_a and φ_b. Comparing this expression with (26.1-13a) one arrives at the interpretation that, in the bonding molecular orbital (ψ_1), the fraction of an electron that has been moved from each atomic orbital into the overlap region is $S/2(1+S)$. Twice this quantity, or $S/(1+S)$, has been given the name *overlap population* (Mulliken, 1955). In the same fashion $-S/(1-S)$ is the overlap population in the antibonding molecular orbital (ψ_2) and represents a shift of electronic charge *from* the overlap region toward the two centers.

The $H_2{}^+$ molecule in the bonding orbital owes its stability to the overlap population. When there is a greater density of electrons between the two nuclei as compared with the density of the two underformed atoms, the negative electronic charge can overcome the nuclear repulsion and give rise to a net attraction. This is possible provided R, the separation between the two nuclei, is not too small. Should this occur, the internuclear repulsion would become too large and some electronic charge would actually leave the region between the nuclei and would contribute to the internuclear repulsion. Hence, there is a value of R at which the system is in equilibrium. The antibonding molecular orbital is characterized by a depletion of electronic charge in the region between the two nuclei with an accompanying pile-up of charge in the regions to the right and left of the molecule, which then enhances the nuclear repulsion and leads to dissociation.

The $H_2{}^+$ bond described above is the prototype of a *covalent* bond found in numerous molecules and solids. Although the qualitative picture has been couched in classical language, it is also valid from a strictly quantum mechanical viewpoint. The basis for this is the Hellmann–Feynman theorem (Section 25.1), which states that within the Born–Oppenheimer approximation the force acting on a nucleus α is given by $-\partial V(Q_\alpha, q_i)/\partial Q_\alpha$ averaged over the electronic wave function where $V(Q_\alpha, q_i)$ is the classical potential energy as a function of nuclear coordinates Q_α and electronic coordinates q_i.

For $H_2{}^+$ there exists an exact solution which may be used as a standard of reference to judge the accuracy of any particular approximate solution. For other molecules comparison with experimental data is often possible. However, there is an internal check based on the quantum mechanical *virial theorem*,

$$\langle T\rangle = \tfrac{1}{2}t\langle V\rangle \tag{26.1-15}$$

where $\langle T\rangle$ and $\langle V\rangle$ are the expectation values of the kinetic and potential energies, respectively, and the potential energy is assumed to be a homogeneous function of degree t in the coordinates. To prove this theorem let the Schrödinger equation for a system of N particles be

$$\sum_{i=1}^{3N} -\frac{\hbar^2}{2m_i}\frac{\partial^2\psi}{\partial x_i^2} + (V - E)\psi = 0. \tag{26.1-16}$$

When (26.1-16) is differentiated with respect to x_j and multiplied on the left by $x_j\psi^*$, one obtains

$$\sum_{i=1}^{3N} -\frac{\hbar^2}{2m_i} x_j\psi^* \frac{\partial^3\psi}{\partial x_j \partial x_i^2} + x_j\psi^* \frac{\partial V}{\partial x_j}\psi + x_j\psi^*(V - E)\frac{\partial\psi}{\partial x_j} = 0. \tag{26.1-17}$$

The complex conjugate of (26.1-16) may be substituted in (26.1-17) and the result summed over j:

$$\sum_i -\frac{\hbar^2}{2m_i}\sum_j x_j\left(\psi^* \frac{\partial^3\psi}{\partial x_j \partial x_i^2} - \frac{\partial^2\psi^*}{\partial x_i^2}\frac{\partial\psi}{\partial x_j}\right) + \psi^* \sum_j x_j\psi^* \frac{\partial V}{\partial x_j}\psi = 0.$$

Integration of this equation yields (Slater, 1963)

$$\sum_i -\frac{\hbar^2}{2m_i}\int \psi^* \frac{\partial^2\psi}{\partial x_i^2}\psi\, dx_i = \frac{1}{2}\int \psi^* \sum_j\left(x_j\frac{\partial V}{\partial x_j}\right)\psi\, dx_j. \tag{26.1-18}$$

Now, according to Euler's theorem, if $V(x_1, x_2, \ldots, x_n)$ is a homogeneous function of degree t and has continuous first derivatives,

$$\sum_j x_j(\partial V/\partial x_j) = tV. \tag{26.1-19}$$

Therefore the right side of (26.1-18) is just $\frac{1}{2}t\langle V\rangle$ and the left side is $\langle T\rangle$. For Coulomb interactions

$$\langle T\rangle = -\tfrac{1}{2}\langle V\rangle. \tag{26.1-20}$$

We may now apply this criterion to the bonding orbital ψ_1 in (26.1-11) with

$$\langle T\rangle = \langle\psi_1|-\nabla^2|\psi_1\rangle, \qquad \langle V\rangle = \left\langle\psi_1\left|\frac{2}{R} - \frac{2}{r_a} - \frac{2}{r_b}\right|\psi_1\right\rangle. \tag{26.1-21}$$

At the calculated equilibrium distance it is found that $\langle T\rangle = 10.37\,\text{eV}$ and $-\frac{1}{2}\langle V\rangle = 12.84\,\text{eV}$. Evidently ψ_1 does not satisfy the virial theorem very well. However, the discrepancy suggests that an improved molecular orbital could be constructed by incorporating an additional parameter whose value would be adjusted to comply with the virial theorem. Let such a parameter be Z and let

$$\psi_1(Z) = \frac{Z^{3/2}}{\sqrt{(1+S)}}\frac{1}{\sqrt{\pi}}\left[e^{-Zr_a} + e^{-Zr_b}\right] \tag{26.1-22}$$

as if the molecular orbital had been constructed from

$$\varphi_a(Z) = \frac{Z^{3/2}}{\sqrt{\pi}} e^{-Zr_a}, \qquad \varphi_b(Z) = \frac{Z^{3/2}}{\sqrt{\pi}} e^{-Zr_b} \tag{26.1-23}$$

in which $\varphi_a(Z)$ and $\varphi_b(Z)$ are 1s atomic orbitals for a hydrogen-like atom whose nucleus carries a charge Ze. On evaluating $T(Z)$ where

$$\langle T(Z)\rangle = \langle \psi_1(z)| - \nabla^2 |\psi_1(Z)\rangle \tag{26.1-24}$$

and comparing with $\langle T\rangle$ in (26.1-21), it is seen that

$$\langle T(Z)\rangle = Z^2\langle T\rangle \tag{26.1-25}$$

which is to be expected since the kinetic energy is a homogeneous quadratic function of the velocities. Similarly

$$\langle V(Z)\rangle = Z\langle V\rangle. \tag{26.1-26}$$

The virial theorem will now be satisfied if

$$Z^2\langle T\rangle = -\tfrac{1}{2}Z\langle V\rangle.$$

Hence the free parameter Z should be assigned the value

$$Z = -\langle V\rangle/2\langle T\rangle. \tag{26.1-27}$$

With the numerical values for $\langle T\rangle$ and $-\frac{1}{2}\langle V\rangle$ as just given, $Z = 1.23$. This procedure for modifying a molecular orbital so as to satisfy the virial theorem is known as *scaling*.

The energy associated with the scaled orbital $\psi_1(Z)$ is

$$E(Z) = \langle T(Z)\rangle + \langle V(Z)\rangle = Z^2\langle T\rangle + Z\langle V\rangle. \tag{26.1-28}$$

Let us now minimize $E(Z)$ with respect to variations in Z; setting

$$\partial E(Z)/\partial Z = 0$$

gives precisely the same value of Z as was found in (26.1-27). Hence the scaled wave function $\psi_1(Z)$ with $Z = 1.23$ will have a lower energy than ψ_1 (with $Z = 1$). By the variational theorem, $\psi_1(Z)$ must be closer to the true wave function. A comparison of numerical values is shown in Table 26.1 in which R_{eq} is the separation between the two protons at equilibrium, i.e., the separation at minimum energy, E_{eq} the energy required to dissociate $H_2{}^+$ according to

$$H_2{}^+ \rightarrow H^+ + H^+ + e,$$

and E_D is the dissociation energy, namely the energy required to dissociate $H_2{}^+$ according to

$$H \rightarrow H^+ + H.$$

TABLE 26.1

Parameters of H_2^+

Parameter	R_{eq} (Å)	E_{eq} (eV)	E_D (eV)
$Z = 1$ (no scaling)	1.32	−15.37	−1.76
$Z = 1.23$ (scaling)	1.06	−15.85	−2.25
Experimental	1.06	−16.39	−2.79

26.2 Symmetry Considerations—H_2^+

The Hamiltonian of the hydrogen molecule ion is invariant under the transformations of the symmetry group $D_{\infty h}$. The molecular orbitals must therefore transform according to the irreducible representations of this group. We shall now examine the symmetry properties of several types of molecular orbitals constructed within the LCAO approximation and having the general form

$$\psi = \varphi_a + \varphi_b.$$

For the present purpose the normalizing factors are unimportant and will be omitted.

As a simple example let φ_a and φ_b be s type orbitals centered on atoms a and b, respectively (Fig. 26.4a), i.e.,

$$\varphi_a = e^{-r_a}, \qquad \varphi_b = e^{-r_b}, \tag{26.2-1}$$

and let

$$\psi_1 = e^{-r_a} + e^{-r_b}, \qquad \psi_2 = e^{-r_a} - e^{-r_b}. \tag{26.2-2}$$

With the center of symmetry located midway between the two nuclei and lying on the internuclear axis it is readily verified that the effect of the symmetry operations on ψ_1 and ψ_2 are:

$$\begin{aligned}
&P_E\psi_1 = \psi_1, && P_E\psi_2 = \psi_2,\\
&P_{C_\varphi}\psi_1 = \psi_1, && P_{C_\varphi}\psi_2 = \psi_2,\\
&P_{C_2}\psi_1 = \psi_1, && P_{C_2}\psi_2 = -\psi_2,\\
&P_i\psi_1 = \psi_1, && P_i\psi_2 = -\psi_2,\\
&P_{iC_\varphi}\psi_1 = P_iP_{C_\varphi}\psi_1 = \psi_1, && P_{iC_\varphi}\psi_2 = P_iP_{C_\varphi}\psi_2 = -\psi_2,\\
&P_{iC_2}\psi_1 = P_iP_{C_2} = \psi_1, && P_{iC_2}\psi_2 = P_iP_{C_2}\psi_2 = \psi_2.
\end{aligned}$$

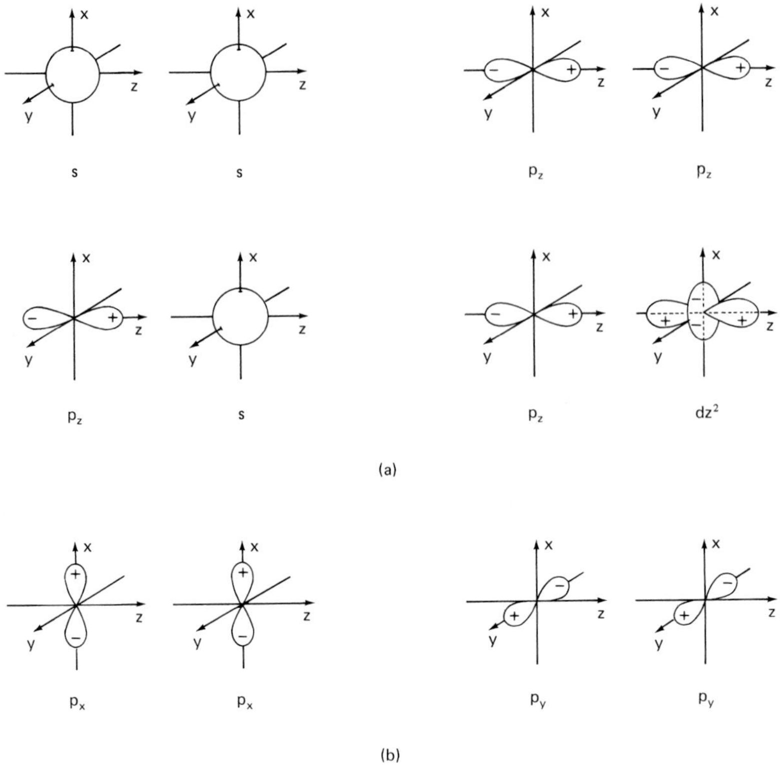

FIG. 26.4 Molecular orbitals: (a) σ type; (b) π type.

From the character table for $D_{\infty h}$ it is clear that ψ_1 is a basis function for the irreducible representation Σ_g^+. We therefore label ψ_1 as a σ_g^+ molecular orbital adhering to the convention of representing one-electron wave functions, such as molecular orbitals, by lower case letters. In similar fashion it is found that ψ_2 is a σ_u^+ molecular orbital.

Now let

$$\psi_3 = z_a + z_b, \qquad \psi_4 = z_a - z_b. \tag{26.2-3}$$

The atomic orbitals are of the p_z type as illustrated in Fig. 1.1. The transformation properties of ψ_3 and ψ_4 are established in much the same way, namely ψ_3 transforms as σ_u^+ and ψ_4 as σ_g^+. However, it is important to pay attention to the relative orientations of the two coordinate systems centered on atoms a and b. Our convention is shown in Fig. 26.4. If the two z axes point at each other, the transformation properties of ψ_3 and ψ_4 are interchanged. A somewhat more complicated situation arises when molecular orbitals are constructed from p_x and p_y atomic orbitals as shown in Fig. 26.4b:

$$\psi_5 = x_a + x_b, \qquad \psi_6 = y_a + y_b. \tag{26.2-4}$$

The coordinate transformation C_φ is a rotation of coordinates about the z axis through an angle φ. Therefore

$$C_\varphi = \begin{pmatrix} \cos\varphi & \sin\varphi \\ -\sin\varphi & \cos\varphi \end{pmatrix},$$

$$P_{C_\varphi} f(\mathbf{x}) = f(C_\varphi^{-1}\mathbf{x}) = f(x\cos\varphi - y\sin\varphi,\ x\sin\varphi + y\cos\varphi,\ z),$$

and

$$P_{C_\varphi}\psi_5 = \psi_5\cos\varphi - \psi_6\sin\varphi, \qquad P_{C_\varphi}\psi_6 = \psi_5\sin\varphi + \psi_6\cos\varphi \tag{26.2-5}$$

or

$$P_{C_\varphi}(\psi_5, \psi_6) = (\psi_5, \psi_6)\begin{pmatrix} \cos\varphi & \sin\varphi \\ -\sin\varphi & \cos\varphi \end{pmatrix}. \tag{26.2-6}$$

The coordinate transformation C_2 is a rotation of coordinates about any axis perpendicular to z. Let us choose the x axis, remembering that the center of symmetry is at the midpoint between the two nuclei. Then

$$\begin{aligned} P_{C_2} f(\mathbf{r}) &= f(C_2^{-1}\mathbf{r}) = f(x, -y, -z) \\ P_{C_2}(\psi_5, \psi_6) &= (\psi_5, \psi_6)\begin{pmatrix} 1 & 0 \\ 0 & -1 \end{pmatrix}. \end{aligned} \tag{26.2-7}$$

The other symmetry operators are handled in the same fashion and it is ultimately found that the characters of the representation generated by ψ_5 and ψ_6 are

$$\begin{aligned} \chi(E) &= 2, & \chi(i) &= -2, \\ \chi(C_\varphi) &= 2\cos\varphi, & \chi(iC_\varphi) &= -2\cos\varphi, \\ \chi(C_2) &= 0, & \chi(iC_2) &= 0. \end{aligned} \tag{26.2-8}$$

By comparing these characters with the character table for $D_{\infty h}$ it may be verified that (ψ_5, ψ_6) belong to the two-dimensional representation π_u; the two eigenfunctions ψ_5 and ψ_6 are therefore degenerate.

Other LCAO functions may be analyzed in similar fashion; some of the more common ones are listed in Table 26.2. LCAO molecular orbitals that belong to a σ irreducible representation are called σ (molecular) orbitals or are said to form σ bonds. Similarly, molecular orbitals that transform according to a π irreducible representation are π (molecular) orbitals and are said to form π bonds. One also encounters, although less often, δ bonds.

TABLE 26.2

Classification of Diatomic LCAO Molecular Orbitals in Terms of the Irreducible Representations of $D_{\infty h}$[a]

LCAO molecular orbital	Irreducible representation of $D_{\infty h}$
$s_a + s_b$	σ_g^+
$s_a - s_b$	σ_u^+
$p_{za} + p_{zb}$	σ_u^+
$p_{za} - p_{zb}$	σ_g^+
$p_{xa} + p_{xb}$, $p_{ya} + p_{yb}$	π_u
$p_{xa} - p_{xb}$, $p_{ya} - p_{yb}$	π_g
$s + p_z$	σ_g^+
$p_z + d_z$	σ_g^+

[a] The orientation of the coordinate system is shown below the table (see also Fig. 26.4).

Since the field due to several nuclei is not spherically symmetric, there is no law for the conservation of total orbital angular momentum. An equivalent statement is that **L** the orbital angular momentum operator does not commute with the Hamiltonian. In diatomic molecules, however, the system is symmetric with respect to rotation about the internuclear axis, that is, the Hamiltonian is invariant under the operation C_φ. Therefore the component of the orbital angular moment on the axis L_z commutes with the Hamiltonian

$$[L_z, \mathscr{H}] = [-i\,\partial/\partial\varphi, \mathscr{H}] = 0, \tag{26.2-9}$$

and one may classify the electronic states of diatomic molecules according to the eigenvalues of L_z. Classically, this means that the projection of the electronic (orbital) angular momentum on the internuclear axis is a constant of the motion. But, in contrast to atomic systems, L_x and L_y do not commute with the Hamiltonian because $\mathscr{H}$ is not invariant under all rotations about the x and y axes, but only for rotations of 180°.

It is quite easy to verify (26.2-9) explicitly for the Hamiltonian (26.1-1) for H_2^+. Thus, referring to Fig. 26.5,

$$r_b^2 = r_a^2 + R^2 - 2r_a R\cos\theta. \tag{26.2-10}$$

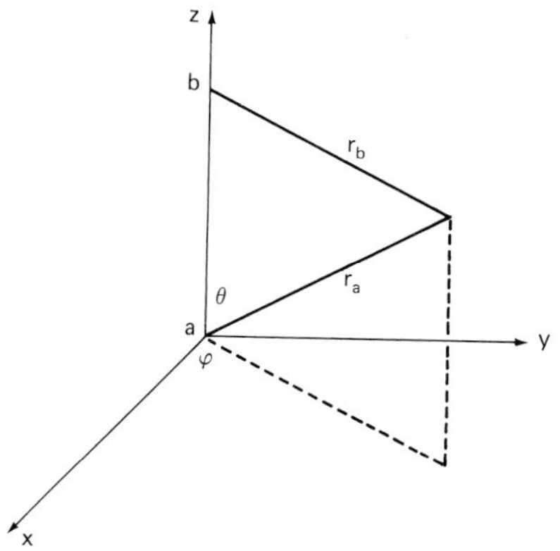

FIG. 26.5 Coordinate system for H_2^+.

If R lies along the z axis, then r_a, r_b, and R are independent of the azimuth angle φ; also, since ∇^2 commutes with L_z, $\mathscr{H}$ in (26.1-1) commutes with L_z.

In view of (26.2-9) we expect eigenfunctions of $\mathscr{H}$ to be, simultaneously, eigenfunctions of L_z. Thus

$$L_z\psi_1 = L_z\psi_2 = 0$$

since ψ_1 and ψ_2 (26.2-2) are independent of φ. Similarly

$$L_z\psi_3 = L_z\psi_4 = 0$$

since $z_a = r\cos\theta_a$ and $z_b = r\cos\theta_b$ so that ψ_3 and ψ_4 are also independent of φ. Indeed for all σ orbitals,

$$L_z\sigma = 0, \tag{26.2-11}$$

which implies that electrons in σ orbitals have no angular momentum about the z axis. The electronic charge cloud is cylindrical and there is no net velocity of electrons about the internuclear axis, hence no magnetic field along the axis.

Let us now examine ψ_5 and ψ_6 (26.2-4) which, we saw previously transform according to π_u. It is possible to construct a new set of molecular orbitals

$$\pi_{u\,+1} = -\sqrt{\frac{3}{8\pi}}(\psi_5 + i\psi_6) = -\sqrt{\frac{3}{8\pi}}[(x_a + iy_a) + (x_b + iy_b)]$$

$$= Y_{11}^a + Y_{11}^b \tag{26.2-12a}$$

$$\pi_{u\,-1} = \sqrt{\frac{3}{8\pi}}(\psi_5 - i\psi_6) = \sqrt{\frac{3}{8\pi}}[(x_a - iy_a) + (x_b - iy_b)]$$

$$= Y_{1\,-1}^a + Y_{1\,-1}^b. \tag{26.2-12b}$$

Operating with L_z we have

$$L_z\pi_{u+1} = L_{za}Y^a_{11} + L_{zb}Y^a_{11} = -\pi_{u-1}. \tag{26.2-13a}$$

$$L_z\pi_{u-1} = L_{za}Y^a_{1-1} + L_{zb}Y^b_{1-1} = -\pi_{u-1}. \tag{26.2-13b}$$

Thus

$$L_z\pi = \pm\pi \tag{26.2-14}$$

which means that π orbitals have one unit of projected angular momentum. The two-fold degeneracy of the π orbitals is apparent once more. In the same way it may be shown that

$$L_z\delta = \pm 2\delta. \tag{26.2-15}$$

It is also clear from the foregoing that the φ dependence of the molecular orbitals is given by $e^{im_l\varphi}$ in which the requirement of single-valuedness forces m_l to be a positive or negative integer. Thus

$$L_z e^{im_l\varphi} = m_l e^{im}{}_{l\varphi}$$

with

$$m_l = 0, \quad \pm 1, \quad \pm 2, \ldots .$$

If we define

$$\Lambda = |m_l| \tag{26.2-16}$$

as the absolute value of the projected orbital angular momentum along the axis of the molecule, the molecular-orbital notation for molecules that belong to $D_{\infty h}$, which includes all homonuclear molecules, may be summarized as:

Λ	0	1	2	$\cdots$
m_l	0	± 1	± 2	$\cdots$
Notation	$\sigma^{\pm}_{g,u}$	$\pi_{g,u}$	$\delta_{g,u}$	$\cdots$

The indices g (gerade) and u (ungerade) refer to even and odd states with respect to inversion (i) at the center of symmetry, midway between the two nuclei. Thus, if the electron's coordinates x, y, and z with respect to the midpoint are replaced by $-x$, $-y$, and $-z$, the electronic wave function will either remain unchanged (g) or undergo a change in sign (u). The + and − refer to even and odd states with respect to reflection in any plane containing the internuclear axis (σ_v). Note that the characters of σ_v are zero for all representations other than Σ; hence there is no $\pm$ classification for $\pi, \delta, \ldots$ orbitals. States with nonzero values of Λ are doubly degenerate; the two states with the same energy differ only in the direction of the projection of the orbital angular momentum on the internuclear axis. All this is reminiscent of an atom in an **E** field where the energy depends on $|M_L|$ rather than on M_L.

Finally, we note that the Hamiltonian is independent of spin; hence it commutes with S^2 and S_z.

There is another way of treating the electronic structure of H_2^+ called the *united atom treatment*. In this method the states of H_2^+ are represented by atomic orbitals centered at the middle of the bond with the objective of constructing orbitals having a single center rather than two. In place of the Hamiltonian (26.1-1) one may write

$$\mathscr{H} = \mathscr{H}_0 + \mathscr{H}' \tag{26.2-17}$$

where

$$\mathscr{H}_0 = -\nabla^2 - \frac{2Z}{r}, \qquad \mathscr{H}' = \frac{2Z}{r} - \frac{2}{r_a} - \frac{2}{r_b}, \tag{26.1-18}$$

and r is the distance from the electron to the midpoint between the two protons (Fig. 26.1). $\mathscr{H}_0$ is regarded as the zero-order Hamiltonian whose eigenfunctions are hydrogen-like atomic wave functions with $Z = 2$; $\mathscr{H}'$ is treated as a perturbation.

Each molecular orbital of H_2^+, based on two centers (also known as the separated atom treatment) can be correlated with an atomic orbital in the united atom treatment because both types of orbitals must be basis functions for irreducible representations of $D_{\infty h}$. Such a correlation diagram is shown in Fig. 26.6; it is noted that the energies of orbitals of the same symmetry do not cross in going from the separated to the united atom, as to be expected from general quantum-mechanical arguments.

26.3 Hydrogen Molecule

The addition of another electron to H_2^+ to form H_2, as in the analogous atomic case, requires consideration of effects due to electron repulsion and the Pauli principle. The theoretical development of Section 25.2 is therefore applicable. In the ground state the molecule will be represented by a single Slater determinant

$$\Psi_0 = \frac{1}{\sqrt{2}} \begin{vmatrix} \psi_1(1)\alpha(1) & \psi_1(1)\beta(1) \\ \psi_1(2)\alpha(2) & \psi_1(2)\beta(2) \end{vmatrix} \tag{26.3-1}$$

in which ψ_1 is a molecular orbital chosen to minimize the energy

$$E = \langle \Psi_0 | \mathscr{H} | \Psi_0 \rangle \tag{26.3-2}$$

and the Hamiltonian (in Rydberg units) with the nuclei a and b in fixed positions (Fig. 26.7) is given by

$$\mathscr{H} = -\nabla_1{}^2 - \nabla_2{}^2 - \frac{2}{r_{a1}} - \frac{2}{r_{b1}} - \frac{2}{r_{a2}} - \frac{2}{r_{b2}} + \frac{2}{r_{12}} + \frac{2}{R}. \tag{26.3-3}$$

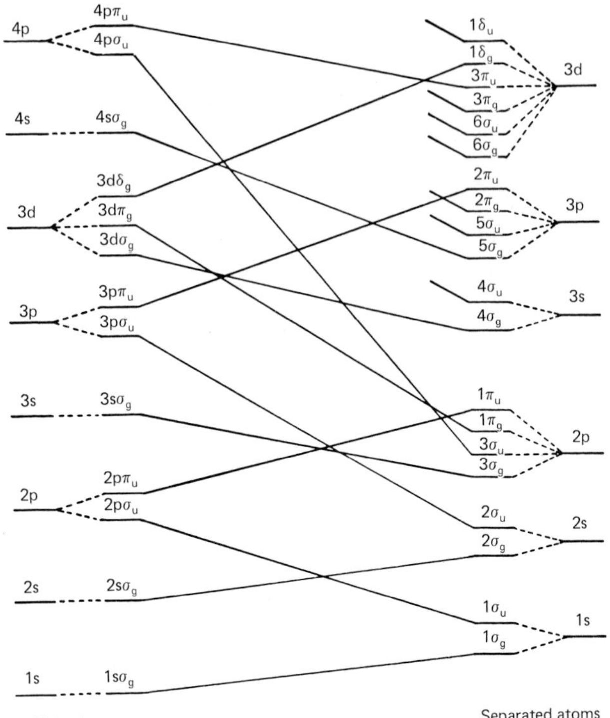

FIG. 26.6 Correlation diagrams for H_2^+ (Daudel *et al.*, 1959.)

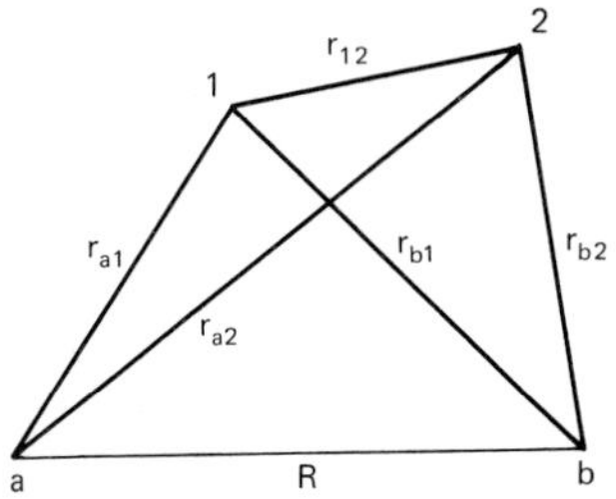

FIG. 26.7 Coordinates for H_2.

The Roothaan procedure is to represent molecular orbitals in the LCAO approximation by

$$\psi = c_a\varphi_a + c_b\varphi_b \tag{26.3-4}$$

and then to perform the necessary self-consistent calculations to obtain the energies of the molecular orbitals and their LCAO coefficients. Moreover, since H_2 is a homonuclear molecule, it belongs to the group $D_{\infty h}$; hence the molecular orbitals of H_2 belong to the irreducible representations of $D_{\infty h}$ and the total wave function for the ground state Ψ_0 or any excited state must also belong to an irreducible representation of $D_{\infty h}$.

We shall not carry out the complete self-consistent field calculation; the interested reader is referred to the literature (e.g., Slater, 1963). Qualitatively it is clear that the molecular orbitals of lowest energy in H_2 will be those arising from atomic orbitals which closely resemble the 1s orbitals of (atomic) hydrogen. Thus, if φ_a and φ_b are 1s orbitals centered on nuclei a and b, respectively, we expect to find two molecular orbitals, ψ_1 with symmetry σ_g^+ and ψ_2 with symmetry σ_u^+, corresponding to symmetric and antisymmetric combinations (Table 26.2) of the 1s orbitals. As in H_2^+ (Fig. 26.2) the σ_g^+ orbital has the lower energy and has a minimum at a particular internuclear separation while σ_u^+ has a higher energy and no minimum. The electronic configuration in the ground state of H_2 would then be $(1\sigma_g^+)^2$ in which a numerical index has been included in the molecular orbital designation to distinguish among various possible molecular orbitals with the same symmetry, e.g., $1\sigma_g^+$ for the lowest one; $2\sigma_g^+, 3\sigma_g^+, \ldots$ for those at higher energy.

The molecular wave function Ψ_0 (26.3-1) may now be written as

$$\Psi_0 = \frac{1}{\sqrt{2}}\begin{vmatrix} 1\sigma_g^+(1)\alpha(1) & 1\sigma_g^+(1)\beta(1) \\ 1\sigma_g^+(2)\alpha(2) & 1\sigma_g^+(2)\beta(2) \end{vmatrix}$$
$$\equiv |1\sigma_g^+\alpha\ 1\sigma_g^+\beta\rangle. \tag{26.3-5}$$

Since both electrons are in σ_g^+ orbitals, the symmetry of Ψ_0 is Σ_g^+ and since the two electrons have opposite spin, the state is a singlet ($S = 0$). Hence the complete designation for the lowest molecular state of H_2 is

$$(1\sigma_g^+)^2\ {}^1\Sigma_g^+ = |1\sigma_g^+\alpha\ 1\sigma_g^+\beta\rangle. \tag{26.3-6}$$

By analogy with the atomic case, upper case letters are used for total molecular wave functions to distinguish them from molecular orbitals which, we noted, are represented by lower case letters.

If the Hamiltonian (26.3-3) is written as

$$\mathscr{H} = \mathscr{H}_0(1) + \mathscr{H}_0(2) + \frac{2}{r_{12}} + \frac{2}{R} \tag{26.3-7}$$

with

$$\mathscr{H}_0(1) = -\nabla_1^2 - \frac{2}{r_{a1}} - \frac{2}{r_{b1}}, \qquad \mathscr{H}_0(2) = -\nabla_2^2 - \frac{2}{r_{a2}} - \frac{2}{r_{b2}}, \tag{26.3-8}$$

we may define

$$I(1\sigma_g^+) = \langle 1\sigma_g^+(1)|\mathscr{H}_0(1)|1\sigma_g^+(1)\rangle = \langle 1\sigma_g^+(2)|\mathscr{H}_0(2)|1\sigma_g^+(2)\rangle \tag{26.3-9}$$

$$J(1\sigma_g^+, 1\sigma_g^+) = \langle 1\sigma_g^+(1)1\sigma_g^+(2)|2/r_{12}|1\sigma_g^+(1)1\sigma_g^+(2)\rangle. \tag{26.3-10}$$

The total electronic energy (26.3-2) of the molecule in the ground state will have the form of (25.2-27), namely,

$$E_0 = E[(1\sigma_g^+)^2\ {}^1\Sigma_g^+] = 2I(1\sigma_g^+) + J(1\sigma_g^+, 1\sigma_g^+). \tag{26.3-11}$$

Excited states arise when one or both electrons are placed in ψ_2. Suppose that ψ_1 and ψ_2 each contain one electron; it is then possible for both singlet and triplet states to occur. Since ψ_2 has the symmetry σ_u^+, both singlet and triplet states will have the symmetry Σ; also, since the product of two + states is + and the product of a g and a u state is u, the singlet and triplet states will be ${}^1\Sigma_u^+$ and ${}^3\Sigma_u^+$, respectively.

These states may be expressed in terms of Slater determinants. As in the case of the two-electron atom, the spin function in the singlet state is antisymmetric and in the triplet, symmetric. We therefore have for the total wave function, as a consequence of the antisymmetry requirement

$$\begin{aligned}(1\sigma_g^+\ 1\sigma_u^+)\,{}^1\Sigma_u^+ &= \frac{1}{\sqrt{2}}[1\sigma_g^+(1)1\sigma_u^+(2) + 1\sigma_u^+(1)1\sigma_g^+(2)] \\ &\quad\times \frac{1}{\sqrt{2}}[\alpha(1)\beta(2) - \beta(1)\alpha(2)] \\ &\equiv \frac{1}{\sqrt{2}}[|1\sigma_g^+\alpha 1\sigma_u^+\beta\rangle - |1\sigma_g^+\beta 1\sigma_u^+\alpha\rangle]\end{aligned} \tag{26.3-12a}$$

$$\begin{aligned}&(1\sigma_g^+\ 1\sigma_u^+)\,{}^3\Sigma_u^+ \\ &= [1\sigma_g^+(1)1\sigma_u^+(2) - 1\sigma_u^+(1)1\sigma_g^+(2)]\begin{cases}\alpha(1)\alpha(2), \\ \dfrac{1}{\sqrt{2}}[\alpha(1)\beta(2) + \beta(1)\alpha(2)], \\ \beta(1)\beta(2),\end{cases} \\ &\equiv \begin{cases}|1\sigma_g^+\alpha 1\sigma_u^+\alpha\rangle, \\ \dfrac{1}{\sqrt{2}}[|1\sigma_g^+\alpha 1\sigma_u^+\beta\rangle + |1\sigma_g^+\beta 1\sigma_u^+\alpha\rangle], \\ |1\sigma_g^+\beta 1\sigma_u^+\beta\rangle.\end{cases}\end{aligned} \tag{26.3-12b}$$

If both electrons are excited into the $\psi_2(1\sigma_u^+)$ molecular orbital, we again have a singlet state ${}^1\Sigma_g^+$ (since u × u = g) where

$$(1\sigma_u)^2\ {}^1\Sigma_g^+ = |1\sigma_u^+\alpha 1\sigma_u^+\beta\rangle. \tag{26.3-13}$$

The electronic energies of these states will contain integrals of the form (26.3-9) and (26.3-10) and may also include exchange integrals. Writing

$$I(1\sigma_u^+) = \langle 1\sigma_u^+(1)|\mathscr{H}_0(1)|1\sigma_u^+(1)\rangle = \langle 1\sigma_u^+(2)|\mathscr{H}_0(2)|1\sigma_u^+(2)\rangle, \tag{26.3-14}$$

$$J(1\sigma_g^+, 1\sigma_u^+) = \langle 1\sigma_g^+(1)1\sigma_u^+(2)|2/r_{12}|1\sigma_g^+(1)1\sigma_u^+(2)\rangle, \tag{26.3-15}$$

$$J(1\sigma_u^+, 1\sigma_u^+) = \langle 1\sigma_u^+(1)1\sigma_u^+(2)|2/r_{12}|1\sigma_u^+(1)1\sigma_u^+(2)\rangle, \tag{26.3-16}$$

$$K(1\sigma_g^+, 1\sigma_u^+) = \langle 1\sigma_g^+(1)1\sigma_u^+(2)|2/r_{12}|1\sigma_u^+(1)1\sigma_g^+(2)\rangle, \tag{26.3-17}$$

the energies are

$$\begin{aligned} E[(1\sigma_g^+\ 1\sigma_u^+)^1\Sigma_u^+] &= I(1\sigma_g^+) + I(1\sigma_u^+) + J(1\sigma_g^+, 1\sigma_u^+) + K(1\sigma_g^+, 1\sigma_u^+) \\ E[(1\sigma_g^+\ 1\sigma_u^+)^3\Sigma_u^+] &= I(1\sigma_g^+) + I(1\sigma_u^+) + J(1\sigma_g^+, 1\sigma_u^+) - K(1\sigma_g^+, 1\sigma_u^+) \\ E[(1\sigma_u^+)^2\ {}^1\Sigma_g^+] &= 2I(1\sigma_u^+) + J(1\sigma_u^+, 1\sigma_u^+). \end{aligned} \tag{26.3-18}$$

To these energies it is necessary to add the nuclear electrostatic energy $2/R$ in order to obtain the total energy of the system, assuming the nuclei are stationary. To elevate H_2 from $(1\sigma_g^+)^2\ {}^1\Sigma_g^+$ to $(1\sigma_g^+\ 1\sigma_u^+)^1\Sigma_u^+$ requires an energy of about 96000 cm^{-1}. Other numerical values are given by Slater (1963).

The states of H_2 that have been described thus far have all been based on 1s atomic orbitals. Clearly, the formation of molecular orbitals from atomic orbitals of higher energy (e.g., 2s, 2p, etc.) will result in states of higher energy for the molecule. Figure 26.8 shows the general features of the energy levels for molecular orbitals constructed from several kinds of atomic orbitals.

An alternative treatment of the hydrogen molecule (as well as other molecules) is based on the *valence bond* method in which the ground state wave function is written as

$${}^1\Sigma_g^+ = N[1s_a(1)1s_b(2) + 1s_b(1)1s_a(2)][\alpha(1)\beta(2) - \beta(1)\alpha(2)] \quad \text{(VB)} \tag{26.3-19}$$

with N a normalizing factor. This function may be compared with Ψ_0 in (26.3-1) which, in expanded form, is

$$\begin{aligned} {}^1\Sigma_g^+ = N[&1s_a(1)1s_a(2) + 1s_a(1)1s_b(2) + 1s_b(1)1s_a(2) + 1s_b(1)1s_b(2)] \\ &\times [\alpha(1)\beta(2) - \beta(1)\alpha(2)] \quad \text{(MO)}. \end{aligned}$$

It is seen that the valence bond function, compared with the molecular orbital omits terms containing the products $1s_a(1)1s_a(2)$ and $1s_b(1)1s_b(2)$

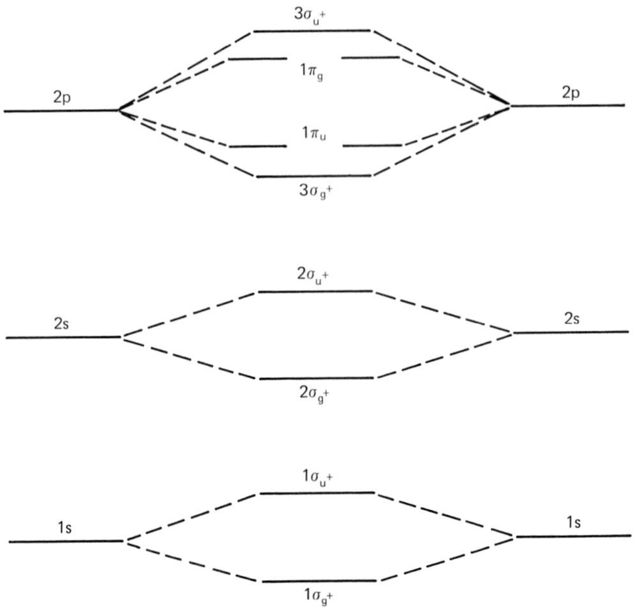

FIG. 26.8 Energy level diagram for homonuclear diatomic molecules.

which allow both electrons to be located on the same atom. Historically, the valence bond method provided a good description of H_2 but as a general method it appears to be less preferable, from a computational standpoint, than the molecular-orbital method.

26.4 Diatomic and Linear Molecules

The formalism described in the last section for H_2 is readily extended to other homonuclear diatomic molecules. In particular, the energy level diagram (Fig. 26.8) serves as a model for molecules such as Li_2, N_2, O_2, etc. The ground state configuration is achieved by assigning electrons to the lowest set of molecular orbitals in conformity with the Pauli principle until all the electrons have been assigned. Since σ orbitals are nondegenerate, while π (and other) orbitals are doubly degenerate, the maximum number of electrons in a σ orbital is two and in a π orbital, four. A closed shell configuration is one in which a certain number of molecular orbitals are fully occupied. Such configurations do not contribute to orbital or spin angular momentum. Therefore, as in the atomic case, the only possible molecular state (or term) is ${}^1\Sigma_g{}^+$. Thus Li_2 and N_2 with configurations $(1\sigma_g{}^+)^2(1\sigma_u{}^+)^2(2\sigma_g{}^+)^2$ and $(1\sigma_g{}^+)^2(1\sigma_u{}^+)^2(2\sigma_g{}^+)^2(2\sigma_u{}^+)^2(3\sigma_g{}^+)^2(1\pi_u)^4$, respectively, both have ${}^1\Sigma_g^+$ ground states (Table 26.4). He_2 would have the configuration $(1\sigma_g^+)^2(1\sigma_u^+)^2$. Since the number of electrons in the antibonding orbital $(1\sigma_u^+)$ is equal to the number of electrons in the bonding orbital $(1\sigma_g^+)$, there is no net bonding and He_2 is therefore unstable. On the other hand, He_2^+ with a configuration $(1\sigma_g)^2 1\sigma_u$ is stable because the number of bonding electrons exceeds the number of antibonding electrons, and it is the former that stabilize a molecule.

TABLE 26.3(a)

Molecular States Obtainable from a π^2 Configuration

	M_{l_1}	M_{l_2}	M_L	M_{s_1}	M_{s_2}	M_S
$\pi_{g+1}\alpha\pi_{g+1}\beta$	1	1	2	$\frac{1}{2}$	$-\frac{1}{2}$	0
$\pi_{g+1}\alpha\pi_{g-1}\alpha$	1	-1	0	$\frac{1}{2}$	$\frac{1}{2}$	1
$\pi_{g+1}\alpha\pi_{g-1}\beta$	1	-1	0	$\frac{1}{2}$	$-\frac{1}{2}$	0
$\pi_{g+1}\beta\pi_{g-1}\alpha$	1	-1	0	$-\frac{1}{2}$	$\frac{1}{2}$	0
$\pi_{g+1}\beta\pi_{g-1}\beta$	1	-1	0	$-\frac{1}{2}$	$-\frac{1}{2}$	-1
$\pi_{g-1}\alpha\pi_{g-1}\beta$	-1	-1	-2	$\frac{1}{2}$	$-\frac{1}{2}$	0

TABLE 26.3(b)

M_L	M_S	Molecular state
2	0	${}^1\Delta_g$
-2	0	
0	1	${}^3\Sigma_g$
0	0	
0	-1	
0	0	${}^1\Sigma_g$

In O_2 the situation is more complicated. With two more electrons than N_2, the configuration may be written as $N_2(1\pi_g)^2$. The $1\pi_g$ orbital has a maximum capacity of four electrons but in O_2 this orbital contains only two electrons. To determine the possible states that can arise from a π_g^2 configuration we note that

$$M_L = \sum_i m_{l_i}, \qquad M_S = \sum_i m_{s_i}. \tag{26.4-1}$$

As in the one-electron case (Section 26.2), $\Lambda = |M_L|$ is the component of the resultant electronic orbital angular momentum parallel to the internuclear axis. States with $\Lambda = 0, 1, 2, \ldots$ are designated by $\Sigma, \Pi, \Delta, \ldots$ in complete conformity with the one-electron case. If the two components of the doubly degenerate π_g-orbital are written π_{g+1} and π_{g-1} by analogy with (26.2-12),

where

$$\pi_{g\,+1} = Y_{1\,1}^{a} - Y_{1\,1}^{b}$$
$$\pi_{g\,-1} = Y_{1\,-1}^{a} - Y_{1\,-1}^{b} \tag{26.4-2}$$

the possible Slater determinants for two electrons in π_g and the corresponding angular momentum components are those shown in Table 26.3a. The entries may be regrouped as in Table 26.3b to conform to the terms $^1\Delta_g$, $^3\Sigma_g$, and $^1\Sigma_g$. It is also necessary to determine whether the Σ states are $+$ or $-$ (symmetry under reflection in a vertical plane). For this purpose we consider $|\pi_{g\,+1}\alpha\pi_{g\,-1}\alpha\rangle$ which describes the state $^3\Sigma_g$ with $M_S = 1$. By definition

$$|\pi_{g\,+1}\alpha\pi_{g\,-1}\alpha\rangle = \frac{1}{\sqrt{2}}\begin{vmatrix} \pi_{g\,+1}(1)\alpha(1) & \pi_{g\,-1}(1)\alpha(1) \\ \pi_{g\,+1}(2)\alpha(2) & \pi_{g\,-1}(2)\alpha(2) \end{vmatrix}$$
$$= \frac{1}{\sqrt{2}}\left[\pi_{g\,+1}(1)\pi_{g\,-1}(2) - \pi_{g\,-1}(1)\pi_{g\,+1}(2)\right]\alpha(1)\alpha(2).$$

Now $\pi_{g\,\pm 1}$ depend on φ as $e^{\pm i\varphi}$. Under reflection in a plane containing the z axis (vertical plane) $e^{i\varphi}$ goes into $e^{-i\varphi}$ and vice versa. Therefore $|\pi_{g\,+1}\alpha\pi_{g\,-1}\alpha\rangle$ changes sign and the triplet state is $^3\Sigma_g^-$. The previous argument can be put in more formal terms. If the internuclear axis is along z and we consider reflections in the yz plane,

$$P_{\sigma_v}f(x, y, z) = f(-x, y, z)$$

and

$$P_{\sigma_v}Y_{11} = Y_{1\,-1}, \qquad P_{\sigma_v}Y_{1\,-1} = Y_{11}. \tag{26.4-3}$$

In view of (26.4-2) we have

$$P_{\sigma_v}\pi_{g\,+1} = \pi_{g\,-1}, \qquad P_{\sigma_v}\pi_{g\,-1} = \pi_{g\,+1} \tag{26.4-4}$$

and

$$P_{\sigma_v}|\pi_{g\,+1}\alpha\pi_{g\,-1}\beta\rangle = -|\pi_{g\,+1}\alpha\pi_{g\,-1}\alpha\rangle. \tag{26.4-5}$$

Exactly the same result is obtained if the reflection is considered to take place in the xz plane, and by a similar argument we arrive at the conclusion that the singlet state is $^1\Sigma_g^+$.

The three molecular states $^1\Sigma_g^+$, $^1\Delta_g$, and $^3\Sigma_g^-$ would be degenerate were it not for the coulombic repulsion among the electrons. When these energies are calculated, the results, shown in Fig. 26.9, indicate that the ground state of O_2 is $^3\Sigma_g^-$ in agreement with Hund's rule. Indeed whenever a degenerate molecular orbital ($\pi, \delta, \ldots$) is occupied by two electrons, Hund's rule will

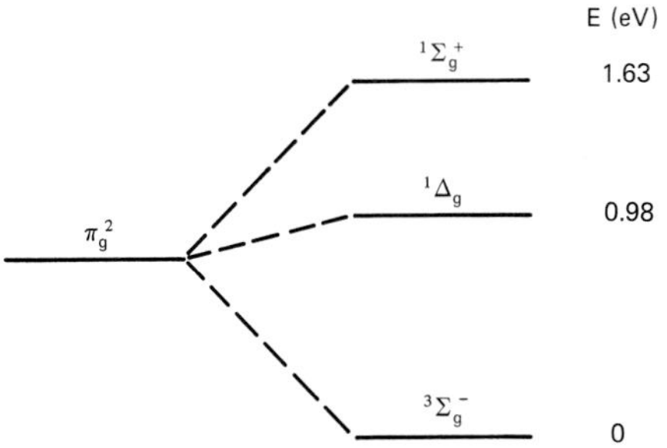

FIG. 26.9 The lowest energy levels in O_2.

operate to make the ground state a triplet. An immediate consequence is that such molecules, including O_2, are endowed with paramagnetic properties.

Homonuclear diatomic molecules, we have seen, belong to the group $D_{\infty h}$. When the two nuclei are different (heteronuclear), the system no longer has inversion symmetry and the appropriate symmetry group is $C_{\infty v}$ (Table 5.2). Electronic states of heteronuclear diatomic molecules therefore lack the g, u classification; otherwise the classification is the same as in the homonuclear case. Thus, molecular orbitals are classified as $\sigma^{\pm}$ $(m_l = 0)$, π $(m_l = \pm 1)$, δ $(m_l = \pm 2)$, etc. and molecular states by corresponding upper case letters, i.e., $\Sigma^{\pm}$ $(M_L = 0)$, $\Pi(M_L = \pm 1)$, $\Delta(M_L = \pm 2)$, ... The multiplicity $2S + 1$ is indicated as previously by a superscript on the left. Σ states are nondegenerate; Π, Δ states are doubly degenerate.

A molecular-orbital calculation carried out along the same lines as in Section 26.1 would result in a set of homogeneous equations as in (26.1-8) and a secular equation as in (26.1-9) but with the important difference that in the heteronuclear case $H_{aa} \neq H_{bb}$. The energies are then given by

$$E = \tfrac{1}{2}\{H_{aa} + H_{bb} \pm \sqrt{(H_{aa} - H_{bb})^2 + 4(H_{ab} - ES)^2}\}. \qquad (26.4\text{-}6)$$

If $|H_{aa} - H_{bb}| \gg 2|(H_{ab} - ES)|$, the two solutions are $E_1 \approx H_{aa}$ and $E_2 \approx H_{bb}$ and the corresponding molecular orbitals are $\psi_1 \approx \varphi_a$ and $\psi_2 \approx \varphi_b$. In such cases the molecular orbitals practically coincide with one or the other atomic orbitals and are therefore highly localized on the individual atoms. This implies that, in effect, an electron has been transferred from one atom to the other. The molecule is then said to be *ionically* bonded. When $|H_{aa} - H_{bb}| \approx H_{ab}$, the molecular orbital is partly delocalized but may favor one of the two atoms. This contrasts with the homonuclear case where $H_{aa} = H_{bb}$ as a result of which the molecular orbital is completely delocalized. In the latter case the molecule is said to be *covalently* bonded. In general, a bond is partially ionic and partially covalent. A typical energy level diagram for the molecular orbitals in a heteronuclear molecule is shown in Fig. 26.10.

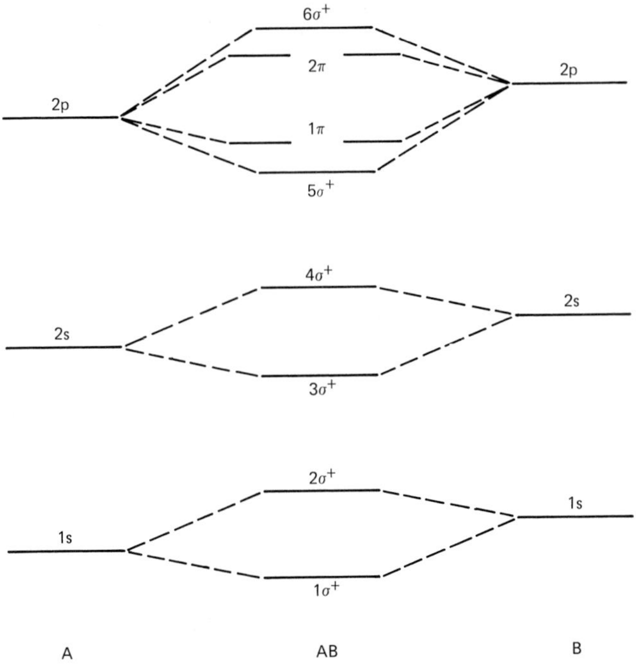

FIG. 26.10 Energy level diagram for a heteronuclear diatomic molecule AB. Interaction between the $4\sigma^+$ and $5\sigma^+$ levels may push $5\sigma^+$ above 1π.

Suppose $n = 0, 1, 2$ is the number of electrons in a nondegenerate molecular orbital ψ. In the LCAO approximation

$$\psi = c_a\varphi_a + c_b\varphi_b \tag{26.4-7}$$

and

$$\int n|\psi|^2\,d\tau = n(c_a{}^2 + 2c_ac_bS_{ab} + c_b{}^2) = n \tag{26.4-8}$$

where $S_{ab} = \langle\varphi_a|\varphi_b\rangle$ is the overlap integral for the atomic orbitals φ_a and φ_b. The quantities $nc_a{}^2$ and $nc_b{}^2$ are known as *net atomic populations* on atoms a and b, respectively and $2nc_ac_bS_{ab}$ is the *overlap population* between atoms a and b (Mulliken, 1955). We have already encountered an example of these quantities in the case of $H_2{}^+$ (Section 26.1). Overlap populations can be positive or negative. If they are positive, they contribute to the formation of a bonding orbital; if negative, to an antibonding orbital. The *partial gross population* on atom a due to the atomic orbital φ_a in the molecular orbital ψ is obtained by adding half of the overlap population to the net atomic population on atom a thus giving $nc_a{}^2 + nc_ac_bS_{ab}$. The other half of the overlap population is assigned to atom b which then acquires a partial gross population equal to $nc_b^2 + nc_a c_b S_{ab}$.

At this stage it would appear to be desirable to summarize the fundamental ideas pertaining to the electronic structure of diatomic molecules and to establish contact with corresponding chemical concepts although occasional allusions to the latter have already been made.

A bonding-type molecular orbital is characterized by a charge distribution concentrated in the region between the two nuclei and close to the internuclear axis. Two electrons with opposite spin occupying such an orbital are said to form a single bond. In the LCAO approximation, the strength of the bond increases the greater the overlap between the atomic orbitals. Molecular orbitals transform according to the irreducible representations of the symmetry groups $D_{\infty h}$ for homonuclear diatomic molecules and $C_{\infty v}$ for heteronuclear diatomic molecules. The most important symmetry types are σ and π and bonds associated with σ and π molecular orbitals are known as σ and π bonds, respectively. σ orbitals have cylindrical symmetry about the internuclear axis, i.e., they are eigenfunctions of L_z with an eigenvalue equal to zero. Hence the electronic charge cloud in a σ bond possesses cylindrical symmetry but has no angular momentum along the internuclear axis. π orbitals are antisymmetric in some vertical plane (containing the internuclear axis), i.e., with respect to such a plane, π orbitals have positive and negative lobes and are eigenfunctions of L_z with eigenvalues ± 1. In a π bond, therefore, the charge cloud has a plane of nodes (a plane in which the charge density is zero) and with respect to this plane the charge cloud is symmetric and has one unit of angular momentum along the axis, either in the positive or negative direction.

A double bond consists of a σ and a π bond; a triple bond contains two π bonds in perpendicular orientation and a σ bond. When all the electrons of a molecule are assigned to molecular orbitals in compliance with the Pauli principle, we have the electronic configuration of the molecule. One or more antisymmetric molecular wave functions (or terms) can be constructed from a given configuration. As an empirical rule, it is found that in most cases chemically stable diatomic molecules have ground states which are ${}^1\Sigma_g^+$ (homonuclear) or ${}^1\Sigma^+$ (heteronuclear). Some notable exceptions are O_2 (${}^3\Sigma_g^-$) and NO (${}^2\Pi$).

The *formal charge* on an atom is the net charge that would result if all the electrons that are shared in bonds, i.e., the total overlap population associated with all the occupied molecular orbitals were divided equally between the atoms. In a covalent compound, the *oxidation number* of each atom is the charge of the atom when each shared electron pair is assigned into the more electronegative of the two bonding atoms. An electron pair shared by two atoms of the same element is divided between them.

The carbon dioxide molecule is linear in the ground state (O=C=O) and belongs to the symmetry group $D_{\infty h}$. Both carbon and oxygen have electrons residing in 1s, 2s, and 2p atomic orbitals. To construct approximate molecular wavefunctions it is sufficient to consider just the 2s and 2p orbitals since the 1s electrons are the most tightly bound to their respective atoms and therefore do not contribute significantly to the bonding. To construct molecular orbitals we refer to Fig. 26.11. The carbon atomic orbitals classified

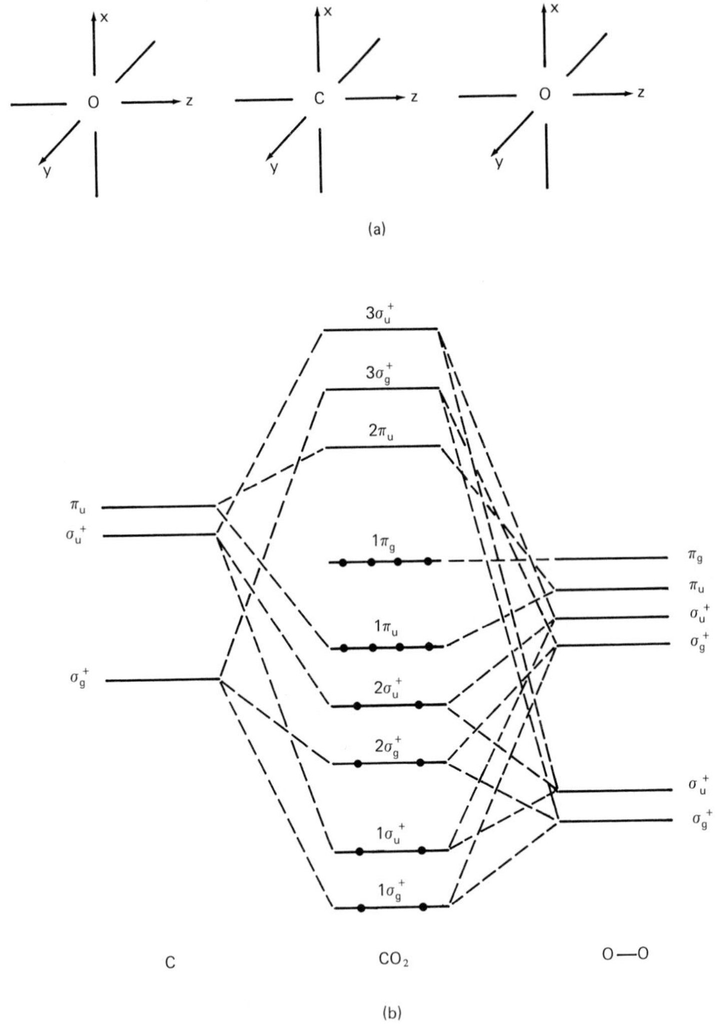

FIG. 26.11 Molecular orbitals in CO_2. (a) Coordinate system. (b) Energy level diagram with the distribution of the 16 valence electrons in the ground state. The symmetry is $D_{\infty h}$.

according to $D_{\infty h}$ are

$$2s(C)\sigma_g^+, \quad 2p_z(C)\sigma_u^+,$$

$$\left.\begin{matrix} 2p_x(C) \\ 2p_y(C) \end{matrix}\right\} \pi_u. \tag{26.4-9}$$

The oxygen atomic orbitals can be grouped in pairs so as to transform according to $D_{\infty h}$. Thus, if the two oxygen atoms are identified as O_a and O_b, then, from Table 26.2 we have

$$\begin{aligned} &[2s(O_a) + 2s(O_b)]\sigma_g^+, \qquad [2s(O_a) - 2s(O_b)]\sigma_u^+, \\ &[2p_z(O_a) + 2p_z(O_b)]\sigma_u^+, \qquad [2p_z(O_a) - 2p_z(O_b)]\sigma_g^+, \\ &\left.\begin{matrix} [2p_x(O_a) + 2p_x(O_b)] \\ [2p_y(O_a) + 2p_y(O_b)] \end{matrix}\right\} \pi_u, \\ &\left.\begin{matrix} [2p_x(O_a) - 2p_x(O_b)] \\ [2p_y(O_a) - 2p_y(O_b)] \end{matrix}\right\} \pi_g. \end{aligned} \tag{26.4-10}$$

The molecular orbitals of CO_2 then consist of linear combinations of (26.4-9) and (26.4-10). An energy level diagram is shown in Fig. 26.11 from which it is seen that the ground state configuration of CO_2 is

$$(1\sigma_g^+)^2(1\sigma_u^+)^2(2\sigma_g^+)^2(2\sigma_u^+)^2(1\pi_u)^4(1\pi_g^4).$$

TABLE 26.4

Ground State Electronic Configurations and Molecular States for some Linear Molecules[a]

Molecule	Configuration	Molecular state
H_2	$(1\sigma_g^+)^2$	$^1\Sigma_g^+$
Li_2	$(1\sigma_g^+)^2(1\sigma_u^+)^2(2\sigma_g^+)^2$	$^1\Sigma_g^+$
N_2	$(1\sigma_g^+)^2(1\sigma_u^+)^2(2\sigma_g^+)^2(2\sigma_u^+)^2(1\pi_u)^4(3\sigma_g^+)^2$	$^1\Sigma_g^+$
O_2	$(1\sigma_g^+)^2(1\sigma_u^+)^2(2\sigma_g^+)^2(2\sigma_u^+)^2(1\pi_u)^4(3\sigma_g^+)^2(1\pi_g)^2$	$^3\Sigma_g^-$
LiH	$(1\sigma^+)^2(2\sigma^+)^2$	$^1\Sigma^+$
LiF	$(1\sigma^+)^2(2\sigma^+)^2(3\sigma^+)^2(4\sigma^+)^2(1\pi)^4$	$^1\Sigma^+$
CO_2	$(1\sigma_g^+)^2(1\sigma_u^+)^2(2\sigma_g^+)^2(2\sigma_u^+)^2(1\pi_u)^4(1\pi_g)^4$	$^1\Sigma^+$

[a] In CO_2 the configuration is based on the 16 valence electrons.

26.5 Hybrid Orbitals

In compounds like methane (CH_4) the angle between each pair of bonds is the tetrahedral angle 109° 28′ (Fig. 26.12b) and all four bonds are chemically indistinguishable. Such a structure belongs to the symmetry group T_d

(Table 5.2) since under the operations of this group each bond is transformed into another (or itself). Bonds which are related in this fashion under a symmetry operation are said to be equivalent. Now the atomic configuration of carbon is $(1s)^2(2s)^2(2p)^2$—a configuration from which it is certainly not apparent how one can construct a set of four molecular orbitals consisting of carbon and hydrogen orbitals which will be consistent with the chemical structure.

The problem can be solved by constructing four new atomic orbitals called *hybrids* consisting of linear combinations of the 2s and 2p carbon orbitals such that the four hybrids form a basis for a representation of T_d. Omitting reference to the principal quantum number, the four hybrid orbitals are

$$\begin{aligned} \varphi_1 &= \tfrac{1}{2}(\mathrm{s} + \mathrm{p}_x + \mathrm{p}_y + \mathrm{p}_z), \\ \varphi_2 &= \tfrac{1}{2}(\mathrm{s} + \mathrm{p}_x - \mathrm{p}_y - \mathrm{p}_z), \\ \varphi_3 &= \tfrac{1}{2}(\mathrm{s} - \mathrm{p}_x + \mathrm{p}_y - \mathrm{p}_z), \\ \varphi_4 &= \tfrac{1}{2}(\mathrm{s} - \mathrm{p}_x - \mathrm{p}_y + \mathrm{p}_z). \end{aligned} \quad (T_d). \qquad (26.5\text{-}1)$$

These are known as sp^3 hybrids and are shown in Fig. 26.12b; their spatial distribution is such that the directions of maximum density point from the center of the tetrahedron to its corners. It is important to observe that the linear combinations in (26.5-1) are *not* molecular orbitals but *atomic* orbitals because the s and p orbitals belong to the *same* atom. Orbitals such as $\varphi_1, \ldots, \varphi_4$ are siad to be *symmetry adapted* because they belong to a representation of the symmetry group of the molecule. But the representation need not be irreducible. Thus the highest dimension of an irreducible representation of T_d is 3 so that $\varphi_1, \ldots, \varphi_4$ must belong to a reducible representation of T_d.

One may now use $\varphi_1, \ldots, \varphi_4$ as the atomic orbitals of carbon which, when combined with the 1s orbitals of the four hydrogens, leads to the construction

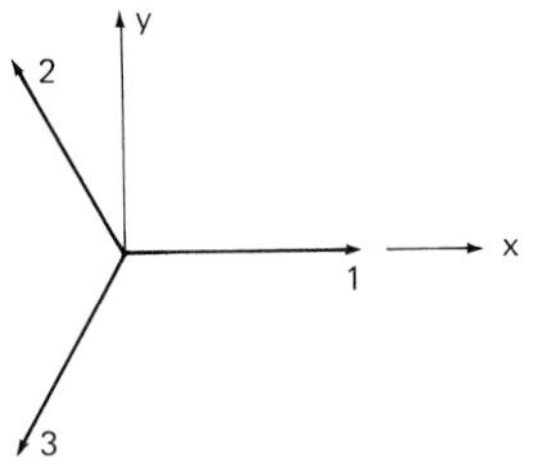

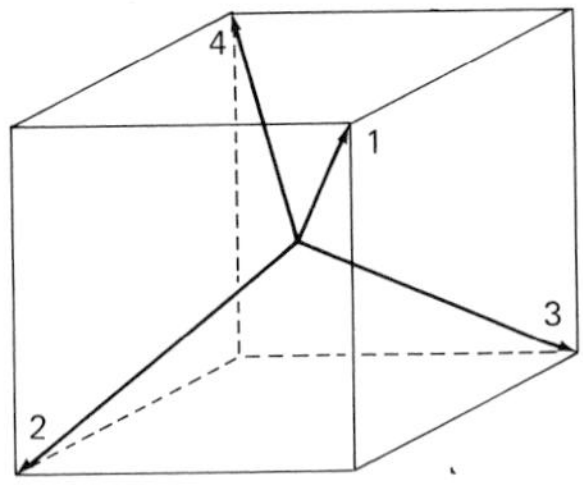

FIG. 26.12 Principal directions of (a) sp^2 and (b) sp^3 hybrid orbitals.

of a set of σ molecular orbitals in which the regions of maximum overlap conform closely to the known chemical structure of CH_4. We have omitted the 1s carbon orbitals since their participation in the bonding is minimal. Hybrid atomic orbitals thus have the virtue that they provide additional flexibility in the construction of molecular orbitals so that a close correspondence between the orbital and chemical pictures can be maintained.

In addition to sp^3 hybrids there are other combinations which are required for various types of molecules. For example, it is possible to combine

$$\mathrm{s},\quad \mathrm{d}_{xy},\quad \mathrm{d}_{xz},\quad \mathrm{d}_{yz}\qquad (\mathrm{sd}^3) \tag{26.5-2}$$

to form hybrids which also have T_d symmetry. Some other commonly occurring hybrids are:

sp^2 *hybrids* (D_{3h})

$$\begin{aligned}
\varphi_1 &= \frac{1}{\sqrt{3}}\,\mathrm{s} + \sqrt{\frac{2}{3}}\,\mathrm{p}_x,\\
\varphi_2 &= \frac{1}{\sqrt{3}}\,\mathrm{s} - \frac{1}{\sqrt{6}}\,\mathrm{p}_x + \frac{1}{\sqrt{2}}\,\mathrm{p}_y,\\
\varphi_3 &= \frac{1}{\sqrt{3}}\,\mathrm{s} - \frac{1}{\sqrt{6}}\,\mathrm{p}_x - \frac{1}{\sqrt{2}}\,\mathrm{p}_y.
\end{aligned} \tag{26.5-3}$$

The directions of maximum density lie in one plane and the angle between adjacent directions is 120° (Fig. 26.12a). Other combinations with D_{3h} symmetry can be constructed from the orbitals

$$\begin{aligned}
&\mathrm{s},\quad \mathrm{d}_{xy},\quad \mathrm{d}_{x^2-y^2} && (\mathrm{sd}^2),\\
&\mathrm{d}_{z^2},\quad \mathrm{p}_x,\quad \mathrm{p}_y && (\mathrm{dp}^2),\\
&\mathrm{d}_{z^2},\quad \mathrm{d}_{xy},\quad \mathrm{d}_{x^2-y^2} && (\mathrm{d}^3).
\end{aligned} \tag{26.5-4}$$

sp hybrids

$$\varphi_1 = \alpha\mathrm{s} + \sqrt{1-\alpha^2}\,\mathrm{p}_z,\qquad \varphi_2 = \sqrt{1-\alpha^2}\,\mathrm{s} - \alpha\mathrm{p}_z \tag{26.5-5}$$

with $0 \leqslant \alpha^2 \leqslant 1$. The directions of maximum density lie along the z axis. Hybrids with D_{4h} symmetry are obtainable from

$$\begin{aligned}
&\mathrm{s},\quad \mathrm{d}_{x^2-y^2},\quad \mathrm{p}_x,\quad \mathrm{p}_y && (\mathrm{sdp}^2),\\
&\mathrm{d}_{z^2},\quad \mathrm{d}_{x^2-y^2},\quad \mathrm{p}_x,\quad \mathrm{p}_y && (\mathrm{d}^2\mathrm{p}^2),
\end{aligned} \tag{26.5-6}$$

and those with O_h symmetry can be constructed from

$$\mathrm{s},\quad \mathrm{d}_{z^2},\quad \mathrm{d}_{x^2-y^2},\quad \mathrm{p}_x,\quad \mathrm{p}_z\qquad (\mathrm{sd}^2\mathrm{p}^3). \tag{26.5-7}$$

It should be remarked that it costs a certain amount of energy to produce hybrid orbitals. In the case of carbon the energy of the four valence electrons ($n = 2$) is $2E(2s) + 2E(2p)$. If we form trigonal (D_{3h}) hybrids of the sp^2 type and put three electrons into these orbitals and one electron into the p_z orbital, the total energy will be $E(2s) + 3E(2p)$. It is therefore necessary to supply an amount of energy equal to $E(2p) - E(2s)$ to form the hybrid, i.e. to change the configuration from $(1s)^2(2s)^2(2p)^2$ to $(1s)^2(2p)^1(sp^2)^3$. However, this expenditure of energy is more than regained in the formation of the bonds.

As an initial example of the application of hybrid orbitals we consider the water molecule which belongs to the symmetry group C_{2v} (Table 5.2). Molecular orbitals of H_2O must therefore belong to irreducible representations of C_{2v} which consist of the four one-dimensional representations A_1, A_2, B_1 and B_2. In the LCAO approximation the molecular orbitals will be constructed from 2s and 2p orbitals belonging to oxygen (again omitting the 1s orbitals) and the 1s orbitals from the two hydrogens. Their classification with respect to C_{2v} are:

$$\begin{aligned} A_1&:\quad 2s,\quad 2p_z,\quad 1s_1 + 1s_2,\\ B_1&:\quad 2p_x + 2p_y,\quad 1s_1 - 1s_2,\\ B_2&:\quad 2p_x - 2p_y. \end{aligned} \tag{26.5-8}$$

These are the only symmetry-adapted atomic orbitals which can be constructed from the original set. A molecular orbital belonging to A_1 must then have the general form

$$a_1 = c_1(2s) + c_2(2p_z) + c_3(1s_1 + 1s_2) \tag{26.5-9a}$$

in which the cs are constants. Similarly for B_1 and B_2 we have

$$b_1 = d_1(2p_x + 2p_y) + d_2(1s_1 - 1s_2), \tag{26.5-9b}$$

$$b_2 = 2p_x - 2p_y. \tag{26.5-9c}$$

When the variational principle is applied to these orbitals to obtain the coefficients, the a_1 molecular orbital will result in a third-degree secular equation with three roots and three sets of coefficients. Hence there will be three orbitals of this type $1a_1, 2a_1$, and $3a_1$. Similarly there will be two molecular orbitals of the b_1 type and one b_2. These are shown in Fig. 26.13 from which we conclude that the ground state configuration in H_2O is

$$(1a_1)^2(1b_1)^2(2a_1)^2(b_2)^2$$

which is a closed shell configuration leading to the molecular state $^1A_1$.

An interesting feature of this analysis is that the two electrons in the b_2 orbital are *nonbonding* since there is no combination of 1s orbitals on the two

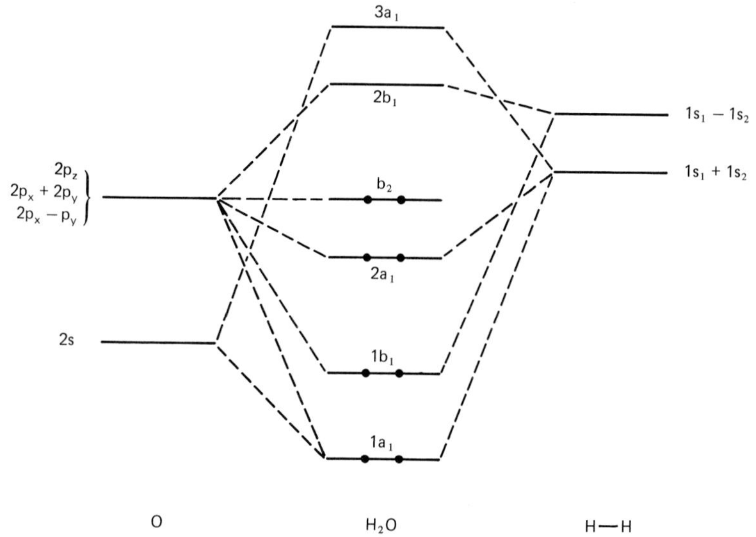

FIG. 26.13 Molecular orbitals in H_2O showing the distribution of the eight valence electrons in the ground state. The symmetry is C_{2v}.

hydrogens which belong to the B_2 irreducible representation of C_{2v}. These two electrons remain in p orbitals on the oxygen atom and do not contribute to the bonding or antibonding. In an alternative terminology an electron pair of this type is known as a *lone pair.*

It is also possible to manipulate the orbitals a_1 and b_1 to show the emergence of sp^3 hybrids on the oxygen atom. Thus, let

$$c_1 = c_2 = d_1 = \alpha/2, \qquad c_3 = d_2 = \beta/2.$$

Then

$$a_1 + b_1 = \alpha[\tfrac{1}{2}(2s + 2p_x + 2p_y + 2p_z)] + \beta(1s_1),$$
$$a_1 - b_1 = \alpha[\tfrac{1}{2}(2s - 2p_x - 2p_y + 2p_z)] + \beta(1s_1).$$

These combinations lend themselves to the interpretation that the water molecule contains two σ bonds each consisting of a tetrahedral hybrid orbital on oxygen and a 1s orbital on hydrogen. The two σ bonds would be separated by the tetrahedral angle which, to a first approximation is indeed quite close to the actual HOH angle.

Formaldehyde (H_2CO) is another molecule with C_{2v} symmetry; it is planar in the ground state and the angles between adjacent bonds on the carbon are approximately 120° (Fig. 26.14). The chemical picture is that

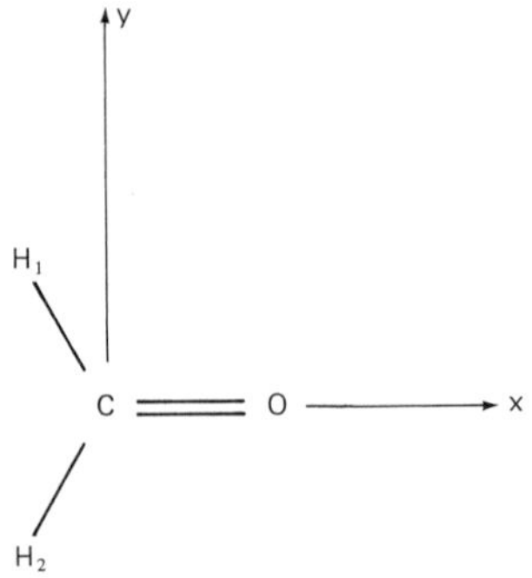

FIG. 26.14 Coordinate system for H_2CO.

there are three σ bonds (C—H_1, C—H_2, C—O) and one π bond (C—O) thus forming a double bond between carbon and oxygen. The corresponding orbital picture is formulated in terms of an sp^2 hybrid on carbon and an sp hybrid on oxygen. On this basis the atomic configurations are

$$\begin{aligned} H_1:&\quad 1s, \qquad H_2:\ 1s, \\ C:&\quad (1s)^2(2p_z)^1(sp^2)^3, \\ O:&\quad (1s)^2(2p_z)^1(2p_y)^2(sp_x)^3. \end{aligned}$$

From these orbitals one constructs three σ molecular orbitals $C(sp^2)$—$H_1(1s)$, $C(sp^2)$—$H_2(1s)$, and $C(sp^2)$—$O(sp_x)$ with two electrons in each orbital and one π molecular orbital $C(2p_z)$—$O(2p_z)$ containing two electrons. The rest of the orbitals, C(1s), O(1s) $O(2p_y)$ and the second $O(sp_x)$, each containing two electrons are nonbonding.

Other molecules are discussed by Ballhausen and Gray (1965).

26.6 The π-Electron Approximation

Groups of atoms interconnected by alternating single and double bonds are said to be *conjugated* as, for example, the carbon atoms in butadiene (Fig. 26.15a) and benzene (Fig. 26.16a). Since each double bond consists of a σ and a π bond, the electrons in such molecules can be divided into two groups—those in σ orbitals and those in π orbitals. One may then write the Hamiltonian for the system as a sum of three terms

$$\mathscr{H} = \mathscr{H}_\sigma + \mathscr{H}_\pi + \mathscr{H}_{\sigma\pi} \tag{26.6-1}$$

where the separate terms refer to σ and π electrons and the Coulombic interaction between the two groups. The π-electron approximation focuses on the π-electron system and regards the nuclei and the σ electrons, rather

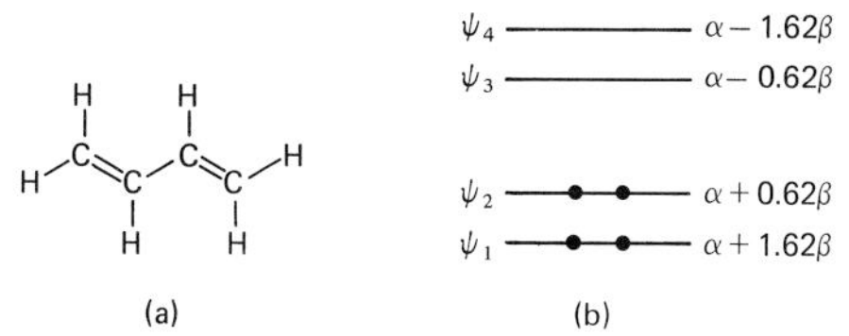

FIG. 26.15 Hückel molecular orbitals in butadiene. (a) Structure and bonding. (b) Energy levels and the distribution of π electrons in the ground state. ψ_1 and ψ_2 are bonding orbitals; ψ_3 and ψ_4 are antibonding.

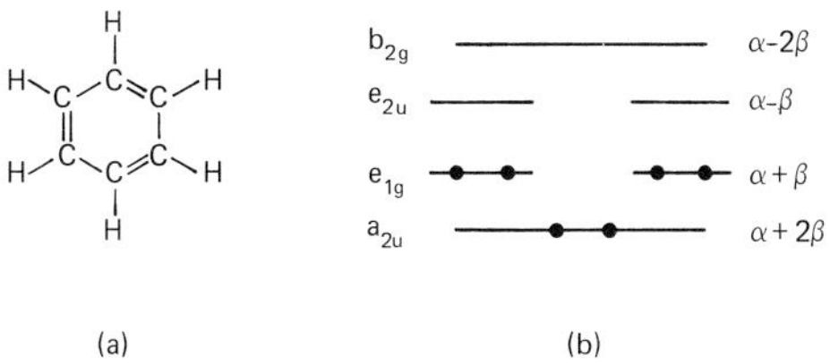

FIG. 26.16 Hückel molecular orbitals in benzene. (a) Structure and bonding. (b) Energy levels and the distribution of π electrons in the ground state. The symmetry is D_{6h}; a_{2u} and e_{1g} are bonding orbitals; e_{2u} and b_{2g} are antibonding.

than just the nuclei, as providing a framework in which the π electrons execute their motion. Experimentally, there is a variety of chemical and spectroscopic evidence to indicate that the main properties of conjugated molecules are attributable largely to the π electrons.

The simplest form of the π-electron approximation is known as the *Hückel method* which we now demonstrate using butadiene as an example. The molecule is planar and the angle between adjacent bonds on each carbon is 120°. It is therefore natural to employ trigonal (sp^2) hybrid atomic orbitals in describing the atomic configuration of carbon: $(1s)^2(2p_z)^1(sp^2)^3$. Each carbon—carbon σ bond may then be represented by the overlap of two sp^2 hybrids and a carbon—hydrogen σ bond by an sp^2 hybrid from carbon overlapping the 1s orbital from hydrogen. Three of the four carbon valence electrons are therefore tied up in σ bonds and the remaining carbon electron is in a $2p_z$ orbital which is oriented in a direction perpendicular to the plane of the molecule. The four $2p_z$ orbitals, one from each carbon, may be combined into proper linear combinations to form molecular orbitals and the four electrons that occupy these orbitals constitute the π-electron system.

This example also illustrates the physical basis for the σ–π separability. The $2p_z$ orbitals are antisymmetric with respect to the molecular plane since they have positive lobes on one side of the plane and negative on the other but the sp^2 orbitals which are linear combinations of 2s, $2p_x$, and $2p_y$ orbitals are symmetric with respect to the molecular plane. Since the Hamiltonian is invariant under reflection in the molecular plane, matrix elements of the type $\langle 2p_z|\mathscr{H}|sp^2\rangle$ must vanish, and unless one goes to higher orders in perturbation theory there will be no mixing between σ and π electrons.

The 30 electrons in butadiene have now been replaced, in the π-electron approximation, by four electrons. The Hückel version of the approximation now proceeds:

1. Let $\varphi_1, \varphi_2, \varphi_3$, and φ_4 be atomic orbitals associated with each of the carbon cores, respectively. One may think of the φs as $2p_z$ orbitals centered on the carbon atoms but for the purpose of the Hückel approximation a detailed specification of the φs is unnecessary apart from the requirement that

$$S_{ij} \equiv \langle\varphi_i|\varphi_j\rangle = \delta_{ij}. \tag{26.6-2}$$

We note that the overlap integrals have been set equal to zero; therefore an alternative viewpoint is to regard the φs as orthogonalized Löwdin orbitals.

2. Construct molecular orbitals using the LCAO approximation

$$\psi = c_1\varphi_1 + c_2\varphi_2 + c_3\varphi_3 + c_4\varphi_4 \tag{26.6-3}$$

with the coefficients c_i chosen so as to minimize the energy E given by

$$E = \langle\psi|\mathscr{H}|\psi\rangle/\langle\psi|\psi\rangle. \tag{26.6-4}$$

The conditions for the minimum are

$$\partial E/\partial c_1 = \partial E/\partial c_2 = \partial E/\partial c_3 = \partial E/\partial c_4 = 0. \tag{26.6-5}$$

3. All diagonal matrix elements are set equal to a common value α, i.e.,

$$\alpha = \langle\varphi_i|\mathscr{H}|\varphi_i\rangle, \tag{26.6-6}$$

and all integrals of the form $\langle\varphi_i|\mathscr{H}|\varphi_j\rangle$ are zero if atom i and j are not connected by a bond. When there is a bond between atoms i and j, all such integrals are set equal to a common value β. Thus

$$\langle\varphi_i|\mathscr{H}|\varphi_j\rangle = \begin{cases} \beta & \text{if i and j are bonded neighbors,} \\ 0 & \text{otherwise.} \end{cases} \tag{26.6-7}$$

The values of α and β are to be chosen empirically; hence there is no need to display the explicit form of the Hamiltonian.

When (26.6-3) is substituted into (26.6-4) with conditions (26.6-6) and (26.6-7), the four equations (26.6-5) become

$$\begin{aligned} &c_1(\alpha - E) + c_2\beta = 0, \qquad c_1\beta + c_2(\alpha - E) + c_3\beta = 0, \\ &c_2\beta + c_3(\alpha - E) + c_4\beta = 0, \qquad c_3\beta + c_4(\alpha - E) = 0 \end{aligned} \tag{26.6-8}$$

and the corresponding secular equation is

$$\begin{vmatrix} \alpha - E & \beta & 0 & 0 \\ \beta & \alpha - E & \beta & 0 \\ 0 & \beta & \alpha - E & \beta \\ 0 & 0 & \beta & \alpha - E \end{vmatrix} = 0. \tag{26.6-9}$$

The secular determinant could actually be constructed by inspection since all the diagonal elements are $\alpha - E$ and the off-diagonal elements are zero except in positions that correspond to the bonding between two atoms and in that case the matrix element is β. The normalized molecular orbitals and their energies are then found to be

$$\begin{aligned} \psi_1 &= 0.37\varphi_1 + 0.60\varphi_2 + 0.60\varphi_3 + 0.37\varphi_4, \\ \psi_2 &= 0.60\varphi_1 + 0.37\varphi_2 - 0.37\varphi_3 - 0.60\varphi_4, \\ \psi_3 &= 0.60\varphi_1 - 0.37\varphi_2 - 0.37\varphi_3 + 0.60\varphi_4, \\ \psi_4 &= 0.37\varphi_1 - 0.60\varphi_2 + 0.60\varphi_3 - 0.37\varphi_4; \end{aligned} \tag{26.6-10}$$

$$E_1 = \alpha + 1.62\beta, \qquad E_2 = \alpha + 0.62\beta, \qquad E_3 = \alpha - 0.62\beta, \qquad E_4 = \alpha - 1.62\beta. \tag{26.6-11}$$

β is taken to be negative as may be verified by a more refined calculation; the energy level diagram therefore has the appearance shown in Fig. 26.15b. A typical (empirical) value of β for conjugated carbon systems is -2.4 eV. The value of α is less important since the level spacing is determined entirely by β. An empirical value of $\alpha = -6.9$ eV is often used‡ (Avery, 1972).

The four π electrons may now be distributed among the molecular orbitals. In the ground state of the molecule the lowest two molecular orbitals will each contain two electrons with opposite spin resulting in a total π-electron energy

$$E_\pi = 2(\alpha + 1.62\beta) + 2(\alpha + 0.62\beta) = 4\alpha + 4.4\beta.$$

The bond order matrix for the π system is defined in the same way as the density matrix (25.2-13), i.e.,

$$P_{\mu\nu} = \sum_i^{\text{occ}} n_i c_{\mu i}^* c_{\nu i} \tag{26.6-12}$$

where μ, ν are labels for the atomic orbitals and i, j are labels for the molecular orbitals; n_i is the number of electrons in the ith molecular orbital and may assume values of 0, 1, or 2. For $\mu \neq \nu$, the bond order matrix element provides a measure of the strength of the bond between atoms μ and ν; for $\mu = \nu$ the matrix element is interpreted as the total electron density

‡ Experimental estimates of α and β vary over a wide range.

TABLE 26.5

Bond and Charge Order Matrix for Butadiene[a]

	1	2	3	4
1	1	0.89	0	0
2	0.89	1	0.45	0
3	0	0.45	1	0.89
4	0	0	0.89	1

[a] $C_1 \overset{0.89}{=\!=} C_2 \overset{0.45}{-\!-} C_3 \overset{0.89}{=\!=} C_4$

on the μth atom

$$P_{\mu\mu} = \rho_\mu = \sum_i^{\text{occ}} n_i c_{\mu i}^* c_{\mu i}. \tag{26.6-13}$$

In butadiene the bond order matrix may be obtained from (26.6-10); it is shown in Table 26.5.

The relations in (26.6-11) are all of the form

$$E_j = \alpha + m_j \beta \tag{26.6-14}$$

where m_j are constants. This permits us to make the association

$$\begin{aligned} m_j &> 0, \quad &&\text{bonding orbital,} \\ m_j &= 0, \quad &&\text{nonbonding orbital,} \\ m_j &< 0, \quad &&\text{antibonding orbital.} \end{aligned} \tag{26.6-15}$$

It is also customary to refer to two electrons in a nonbonding orbital as a *lone pair*. In the ground state of butadiene all four π electrons are bonding.

The structure of the molecular orbitals (26.6-10) indicates that the four π electrons are delocalized, that is, there is a nonvanishing probability for an electron to be found in the vicinity of any one of the carbons. Such electrons are often said to be *mobile* and, in major aspects, resemble the delocalization of conduction electrons in a metal.

The Hückel method appears to be a rather drastic approximation insofar as it neglects (a) the σ electrons, (b) the σ-π interactions, (c) the overlap integral, and (d) the Coulomb repulsion among π electrons. Despite these shortcomings the method is nevertheless useful for qualitative purposes and enables one to systematize a large number of organic molecules. In some cases it has even yielded reasonable quantitative information.

As another example, we consider the benzene molecule (Fig. 26.16a) which is also a planar structure. The carbon electronic configuration, as in the previous case is $(1s)^2(2p_z)^1(sp^2)^3$ with the $2p_z$ orbital perpendicular to the molecular plane. The π-electron system therefore consists of six electrons, one from each carbon. The secular determinant may now be written by inspection:

$$\begin{vmatrix} \alpha - E & \beta & 0 & 0 & 0 & \beta \\ \beta & \alpha - E & \beta & 0 & 0 & 0 \\ 0 & \beta & \alpha - E & \beta & 0 & 0 \\ 0 & 0 & \beta & \alpha - E & \beta & 0 \\ 0 & 0 & 0 & \beta & \alpha - E & \beta \\ \beta & 0 & 0 & 0 & \beta & \alpha - E \end{vmatrix} = 0 \qquad (26.6\text{-}16)$$

which has solutions

$$E_1 = \alpha + 2\beta, \qquad E_2 = E_3 = \alpha + \beta, \qquad E_4 = E_5 = \alpha - \beta, \qquad E_6 = \alpha - 2\beta. \qquad (26.6\text{-}17)$$

It is seen (Fig. 26.16b) that in this case there are two two-fold degenerate states and in the ground state of benzene the six π electrons are distributed as shown in the diagram. The total π-electron energy is

$$E_\pi = 6\alpha + 8\beta \qquad (26.6\text{-}18)$$

and the molecular orbitals are

$$\begin{aligned} \psi_1 &= \frac{1}{\sqrt{6}}[\varphi_1 + \varphi_2 + \varphi_3 + \varphi_4 + \varphi_5 + \varphi_6], \\ \psi_2 &= \frac{1}{\sqrt{3}}[\tfrac{1}{2}\varphi_1 - \tfrac{1}{2}\varphi_2 - \varphi_3 - \tfrac{1}{2}\varphi_4 + \tfrac{1}{2}\varphi_5 + \varphi_6], \\ \psi_3 &= \frac{1}{2}[\varphi_1 + \varphi_2 - \varphi_4 - \varphi_5], \\ \psi_4 &= \frac{1}{\sqrt{3}}[\tfrac{1}{2}\varphi_1 + \tfrac{1}{2}\varphi_2 - \varphi_3 + \tfrac{1}{2}\varphi_4 + \tfrac{1}{2}\varphi_5 - \varphi_6], \\ \psi_5 &= \frac{1}{\sqrt{2}}[\varphi_1 - \varphi_2 + \varphi_4 - \varphi_5], \\ \psi_6 &= \frac{1}{\sqrt{6}}[\varphi_1 - \varphi_2 + \varphi_3 - \varphi_4 + \varphi_5 - \varphi_6]. \end{aligned} \qquad (26.6\text{-}19)$$

Again the π electrons move around the hexagon in much the same way as conduction electrons.

There is an alternative way of labeling the molecular orbitals and their energies. Let

$$E_j = \alpha + 2\beta \cos(2\pi j/6) \qquad (j = 0, \quad \pm 1, \quad \pm 2, \quad +3) \tag{26.6-20}$$

$$\psi_j = (1/\sqrt{6}) \sum_{r=1}^{6} \exp[i2\pi jr/6]\varphi_r. \tag{26.6-21}$$

Comparing (26.6-20) with (26.6-17) we find

$$\begin{aligned} E_0 &= E_1 = \alpha + 2\beta, \\ E_{+1} &= E_{-1} = E_2 = E_3 = \alpha + \beta, \\ E_{+2} &= E_{-2} = E_4 = E_5 = \alpha - \beta, \\ E_{+3} &= E_6. \end{aligned} \tag{26.6-22}$$

The correspondence between (26.6-19) and (26.6-21) is given by

$$\begin{aligned} &\psi_0 = \psi_1, && \psi_1 = \psi_0, \\ &\psi_{+1} = \frac{1}{\sqrt{2}}(\psi_2 + i\psi_3), && \psi_2 = \frac{1}{\sqrt{2}}(\psi_{+1} + \psi_{-1}), \\ &\psi_{-1} = \frac{1}{\sqrt{2}}(\psi_2 - i\psi_3), && \psi_3 = \frac{1}{i\sqrt{2}}(\psi_{+1} - \psi_{-1}), \\ &\psi_{+2} = -\frac{1}{\sqrt{2}}(\psi_4 - i\psi_5), && \psi_4 = -\frac{1}{\sqrt{2}}(\psi_{+2} + \psi_{-2}), \\ &\psi_{-2} = -\frac{1}{\sqrt{2}}(\psi_4 + i\psi_5), && \psi_5 = \frac{1}{i\sqrt{2}}(\psi_{+2} - \psi_{-2}), \\ &\psi_{+3} = \psi_6, && \psi_6 = \psi_{+3}. \end{aligned} \tag{26.6-23}$$

The molecular orbitals written in the form of (26.6-21) are seen to be identical with the Bloch orbitals in solid state theory. The energies (26.6-17) or (26.6-20) constitute an elementary band in which the Fermi energy, equal to α, lies midway between the (filled) bonding and (empty) antibonding orbitals.

The extension of (26.6-20) and (26.6-21) to other cyclic polyenes is, quite obviously,

$$\begin{aligned} E_j &= \alpha + 2\beta \cos(2\pi j/N), \\ \psi_j &= 1/\sqrt{N} \sum_{r=1}^{N} \exp[i2\pi jr/N]\varphi_r. \end{aligned} \tag{26.6-24}$$

The benzene molecule belongs to the symmetry group D_{6h}. When the symmetry operators are applied to the $2p_z$ orbitals centered on the various carbons, it may be verified that the π molecular orbitals of benzene belong to certain irreducible representations of D_{6h}:

$$\begin{aligned} &a_{2u}: && \psi_1\ (\psi_0), \\ &e_{1g}: && \psi_2, \quad \psi_3\ (\psi_{+1}, \psi_{-1}), \\ &e_{2u}: && \psi_4, \quad \psi_5\ (\psi_2, \psi_{-2}), \\ &b_{2g}: && \psi_6\ (\psi_{+3}). \end{aligned} \qquad (26.6\text{-}25)$$

In the ground state of the molecule the configuration is $(a_{2u})^2(e_{1g})^4$; since this is a closed shell, the molecular state is $^1A_{1g}$. If one electron is excited to an e_{2u} orbital, the electronic configuration becomes $(a_{2u})^2(e_{1g})^3(e_{2u})^1$. This is no longer a closed shell and we may expect several terms to arise from this configuration. To identify such terms it is necessary to examine the product representation

$$a_{2u} \times a_{2u} \times e_{1g} \times e_{1g} \times e_{1g} \times e_{2u}.$$

The product representation $a_{2u} \times a_{2u}$ leads to A_{1g} and will therefore have no effect on the final terms. The terms arising from $e_{1g} \times e_{1g} \times e_{1g}$ may be identified most easily on the basis of the hole-particle formalism namely, that the states arising from a configuration $(\Gamma)^l$ are the same as those obtained from $(\Gamma)^{n-l}$ where n is the number of electrons in the closed shell. For e_{1g}, which is two-fold degenerate, $n = 4$. Therefore, for purposes of symmetry considerations, $(e_{1g})^3$ may be simply replaced by e_{1g}. We then examine the product representation $e_{1g} \times e_{2u}$ which in turn is reducible into irreducible representations of D_{6h}:

$$e_{1g} \times e_{2u} = b_{1u} + b_{2u} + e_{1u}. \qquad (26.6\text{-}26)$$

We therefore conclude that the configuration $(a_{2u})^2(e_{1g})^3e_{2u}$ produces the three molecular states (terms) B_{1u}, B_{2u}, and E_{1u} which are degenerate provided we ignore the electronic repulsions. For each term there can be a singlet and a triplet state.

The discussion in this section has thus far been based on the simple Hückel model; a somewhat more elaborate version is known as the extended Hückel method which was mentioned in Section 25.3 and which can also be applied to a π-electron system. In both versions one begins with the LCAO approximation

$$\psi_i = \sum_{j=1}^{N} c_{ij}\varphi_j \qquad (26.6\text{-}27)$$

which leads to the set of homogeneous equations

$$\sum_j (H_{ij} - ES_{ij})c_{ij} = 0 \qquad (26.6\text{-}28)$$

where

$$H_{ij} = \langle \varphi_i | \mathscr{H} | \varphi_j \rangle, \tag{26.6-29}$$

$$S_{ij} = \langle \varphi_i | \varphi_j \rangle. \tag{26.6-30}$$

The Hamiltonian is not specified but assumed to be a one-electron Hamiltonian, i.e., without Coulombic interactions among the π electrons. At this point the simple and extended methods begin to diverge; we shall refer to the former as H and the latter, EH.

$$S_{ij}\begin{cases}\delta_{ij} & \text{(H)},\\ \text{calculated from an assumed atomic basis set} & \text{(EH)},\end{cases}$$

$$H_{ii}\begin{cases}\alpha, & \text{an empirical parameter (H)},\\ \alpha_i, & \text{valence state ionization energy (EH)},\end{cases}$$

$$H_{ij}\begin{cases}\beta \ \text{ if } i \text{ and } j \text{ are bonded neighbors, zero otherwise} & \text{(H)},\\ 0.5K(H_{ii} + H_{jj})S_{ij},\ K \text{ is an empirical parameter} & \text{(EH)}.\end{cases}$$

The most glaring omission in the Hückel method, whether in its simple or extended version, is that of electron repulsion. The *Pariser–Parr–Pople* (PPP) method is designed to take the latter into account while maintaining the π-electron approximation. The method will be illustrated by a calculation on ethylene

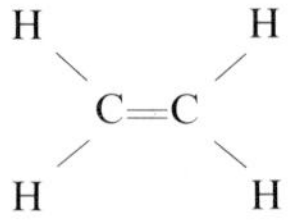

which contains just two π electrons.

Starting with the simple Hückel calculations the two π molecular orbitals and their respective energies are

$$\begin{aligned} \psi_1 &= \frac{1}{\sqrt{2}}(\varphi_a + \varphi_b); \qquad E_1 = \alpha + \beta, \\ \psi_2 &= \frac{1}{\sqrt{2}}(\varphi_a - \varphi_b); \qquad E_2 = \alpha - \beta, \end{aligned} \tag{26.6-31}$$

where φ_a and φ_b are two atomic orbitals of the same type but centered on carbon atoms a and b, respectively. In the ground state both electrons are in ψ_1 and the antisymmetrized molecular state is

$${}^1\Psi_0 = |\psi_1 \alpha \psi_1 \beta\rangle. \tag{26.6-32}$$

It is now assumed that the Hamiltonian of the π-electron system has the form

$$\mathscr{H} = \mathscr{H}_0(1) + \mathscr{H}_0(2) + (e^2/r_{12}). \qquad (26.6\text{-}33)$$

There is no intention of specifying $\mathscr{H}_0$ in detail, although it may be visualized as the kinetic and potential energy of a π electron in the field produced by the nuclei and the σ electrons.

By analogy with (26.3-9)–(23.3-11) the energy of the two-electron π system in the ground state is

$$E_0 = 2I_1 + J(1,1) \qquad (26.6\text{-}34)$$

where

$$I_1 = \langle\psi_1(1)|\mathscr{H}_0(1)|\psi_1(1)\rangle = \langle\psi_1(2)|\mathscr{H}_0(2)|\psi_1(2)\rangle, \qquad (26.6\text{-}35)$$

$$J(1,1) = \langle\psi_1(1)\psi_1(2)|e^2/r_{12}|\psi_1(1)\psi_1(2)\rangle. \qquad (26.6\text{-}36)$$

In terms of the atomic orbitals φ_a and φ_b,

$$I_1 = \langle\varphi_a(1)|\mathscr{H}_0(1)|\varphi_a(1)\rangle + \langle\varphi_a(1)|\mathscr{H}_0(1)|\varphi_b(1)\rangle, \qquad (26.6\text{-}37)$$

$$J(1,1) = \tfrac{1}{2}\langle aa|aa\rangle + \tfrac{1}{2}\langle aa|bb\rangle + 2\langle aa|ab\rangle + \langle ab|ab\rangle \qquad (26.6\text{-}38)$$

with

$$\langle ab|cd\rangle \equiv \int \varphi_a(1)\varphi_b(1)(e^2/r_{12})\varphi_c(2)\varphi_d(2)\,d\tau_1\,d\tau_2. \qquad (26.6\text{-}39)$$

Since $\mathscr{H}_0$ has not been given an explicit form, the integrals in (26.6-37) cannot be evaluated; instead they are assigned empirical values:

$$\begin{aligned}\langle\varphi_a(1)|\mathscr{H}_0(1)|\varphi_a(1)\rangle &= \langle\varphi_b(1)|\mathscr{H}_0(1)|\varphi_b(1)\rangle = \alpha,\\ \langle\varphi_a(1)|\mathscr{H}_0(1)|\varphi_b(1)\rangle &= \beta.\end{aligned} \qquad (26.6\text{-}40)$$

The Coulomb integral $J(1,1)$ is simplified by the assumption of zero differential overlap:

$$\langle ab|cd\rangle = \delta_{ab}\,\delta_{cd}\langle aa|cc\rangle. \qquad (26.6\text{-}41)$$

With these assumptions, the ground state of ethylene, in the π-electron approximation but with π-electron repulsion taken into account has an energy,

$$E_0 = 2(\alpha + \beta) + \tfrac{1}{2}[\langle aa|aa\rangle + \langle aa|bb\rangle]. \qquad (26.6\text{-}42)$$

In some cases it is feasible to carry out a Hartree–Fock SCF calculation for the π-electron system. The method parallels that described in Section 25.2 and involves an iterative solution to the Roothaan equations (Daudel *et al.*, 1959; Richards and Horsley, 1970).

CHAPTER 27

MOLECULAR SPECTRA

Thus far our attention has been focused on the electronic states of molecules under the assumption that the nuclei are maintained in fixed positions. We must now relax this assumption and take the nuclear motion into account. The general picture that emerges is that each electronic state has its own pattern of vibrational and rotational levels. In orders of magnitude, electronic states are separated by 10^4 to 10^5 cm^{-1} (visible, ultraviolet), vibrational states by 10^2 to 10^3 cm^{-1} (infrared), and rotational states by 0.1 to 1.0 cm^{-1} (microwave). Molecular absorption spectra therefore span the spectral region from the ultraviolet to the microwave and appear in the form of quasi-continuous bands in contrast to atomic spectra which are characterized by discrete lines.

27.1 Vibrations and Rotations of Diatomic Molecules

Neglecting electronic motion, the kinetic energy of a diatomic molecule is

$$T = (p_1{}^2/2m_1) + (p_2{}^2/2m_2) \tag{27.1-1}$$

where $\mathbf{p}_1$ and $\mathbf{p}_2$ are the momenta and m_1 and m_2 the masses of the two nuclei. Part of this kinetic energy is a result of the motion of the molecule as a whole, that is, the center of mass motion, and the rest of the kinetic energy is associated with the relative motion of the two nuclei, namely rotation and vibration. The center of mass motion is not relevant as far as intrinsic molecular properties are concerned and may be removed from consideration by writing (27.1-1) in the form of

$$T = (p^2/2\mu) + (P^2/2M) \tag{27.1-2}$$

in which

$$\mathbf{p} = [1/(m_1 + m_2)][m_1\mathbf{p}_2 - m_2\mathbf{p}_1], \tag{27.1-3a}$$

$$\mu = m_1 m_2/(m_1 + m_2) = \text{reduced mass}, \tag{27.1-3b}$$

$$\mathbf{P} = \mathbf{p}_1 + \mathbf{p}_2, \tag{27.1-3c}$$

$$M = m_1 + m_2. \tag{27.1-3d}$$

The term $P^2/2M$ is clearly the kinetic energy of the center of mass, thus leaving $p^2/2\mu$ for the kinetic energy of relative motion.

The potential energy V of the vibrating–rotating system is a function of $r = r_2 - r_1$, the internuclear distance. If r_e, the equilibrium distance, is defined as the value of r at which $V(r)$ is a minimum, the expansion of $V(r)$ about r_e gives

$$V(r) = V(r_e) + (dV/dr)_{r=r_e}(r - r_e) + \tfrac{1}{2}(d^2V/dr^2)_{r=r_e}(r - r_e)^2 + \cdots .$$

Since the first derivative is zero by definition, the potential energy up to quadratic terms is

$$V(r) = -D_e + \tfrac{1}{2}k(r - r_e)^2 \tag{27.1-4}$$

with

$$V(r_e) \equiv -D_e, \qquad (d^2V/dr^2)_{r=re} \equiv k.$$

It is not to be expected that such a potential, derived without reference to the electronic motion as, for example, by means of the Born–Oppenheimer approximation, can be valid over all values of r. Furthermore, a harmonic oscillator potential would make the molecule infinitely stable which, of course, cannot be true since all molecules will dissociate if sufficient energy is imparted to them. Nevertheless, in the region of the minimum, (27.1-4) serves as a useful first approximation which may later be improved by the addition of higher order terms.

With the Hamiltonian

$$\mathscr{H} = (p^2/2\mu) + V(r) \tag{27.1-5}$$

the Schrödinger equation is

$$[-(\hbar^2/2\mu)V^2 + V(r)]\psi = E\psi, \tag{27.1-6}$$

or in spherical coordinates

$$r^2\left(\frac{\partial^2}{\partial r^2} + \frac{2}{r}\frac{\partial}{\partial r}\right)\psi + \frac{2\mu r^2}{\hbar^2}[E - V(r)]\psi = L^2\psi. \tag{27.1-7}$$

Adopting the notation

$$L^2\psi = J(J+1)\psi, \tag{27.1-8}$$

we have, for the radial part of (27.1-7),

$$\frac{d^2P}{dr^2} + \frac{2\mu}{\hbar^2}[E - V(r)]P(r) - \frac{J(J+1)}{r^2}P(r) = 0, \qquad (27.1\text{-}9)$$

where

$$\psi = \frac{1}{r}P(r)Y_{JM}(\theta, \varphi). \qquad (27.1\text{-}10)$$

$$J = 0, 1, 2, \ldots \qquad M = J, J-1, \ldots, \quad -J.$$

To eliminate the rotational motion we set $J = 0$. Then with $V(r)$ as in (27.1-4),

$$\frac{d^2P(r)}{dr^2} + \frac{2\mu}{\hbar^2}[E - V(r)]P(r) = 0 \qquad (27.1\text{-}11)$$

is the Schrödinger equation of a simple harmonic oscillator whose energies are given by

$$E_v = \hbar\omega_0(v + \tfrac{1}{2}) - D_e \qquad (\omega_0 = \sqrt{k/\mu}, \quad v = 0, 1, 2, \ldots) \quad (27.1\text{-}12)$$

and the difference in energy between adjacent levels is

$$\omega_e = \frac{\hbar\omega_0(v + 1 + \frac{1}{2})}{2\pi\hbar c} - \frac{\hbar\omega_0(v + \frac{1}{2})}{2\pi\hbar c} = \frac{\omega_0}{2\pi c} \quad (\text{cm}^{-1}). \qquad (27.1\text{-}13)$$

In H_2 and HCl the values of $\omega_e(\text{cm}^{-1})$, the force constant k (dyn/cm) and the period of oscillation T(sec) are shown:

	ω_e	k	T
H_2	4403.2	5.756×10^5	7.57×10^{-15}
HCl	2885.9	4.806×10^5	1.15×10^{-14}

The vibrational levels are occupied in accordance with the Boltzmann distribution

$$N_v \propto e^{-E_v/kT} \qquad (27.1\text{-}14)$$

where N_v is the number of molecules in the state v, E_v is the energy of the vibrational state; k the Boltzmann constant and T the absolute temperature. The population decreases exponentially as v is increased. At room temperature practically all molecules are in their lowest vibrational state $v = 0$.

More realistic potentials have been devised empirically (Levine, 1975). Among these is the Morse potential

$$V = D[1 - e^{-\beta(r - r_e)}]^2 \qquad (27.1\text{-}15)$$

which is shown in Fig. 27.1. D is the depth of the potential energy minimum below the asymptote and represents the bond dissociation energy. This is the energy required to separate two atoms from their equilibrium distance r_e to infinite separation. The constant β is obtained by solving the Schrödinger equation with the potential (27.1-15). The energies in reciprocal centimeters are given by

$$G(v) = \omega_e(v + \tfrac{1}{2}) - \omega_e\chi_e(v + \tfrac{1}{2})^2 \tag{27.1-16a}$$

with

$$\omega_e = \beta\sqrt{\frac{D\hbar}{\pi c\mu}} \tag{27.1-16b}$$

$$\chi_e = \frac{\hbar\beta^2}{4\pi c\mu}. \tag{27.1-16c}$$

$\omega_e\chi_e$ is known as the *anharmonicity constant.*

Evidently the harmonic oscillator (parabolic) potential is a good approximation to the Morse potential when v is small. But at large v, the potential becomes anharmonic and the energy levels are no longer equidistant but rather crowd together with increasing v.

We now turn to some aspects of the rotational motion under the assumption that r is constant and hence $V(r)$ is constant and may be taken to be zero. In that case (27.1-7) becomes

$$(2I/\hbar^2)E\psi = L^2\psi = J(J+1)\psi \tag{27.1-17a}$$

or

$$E = (\hbar^2/2I)J(J+1) \qquad (J = 0, 1, 2, \ldots) \tag{27.1-17b}$$

with

$$I = \mu r^2 \tag{27.1-18}$$

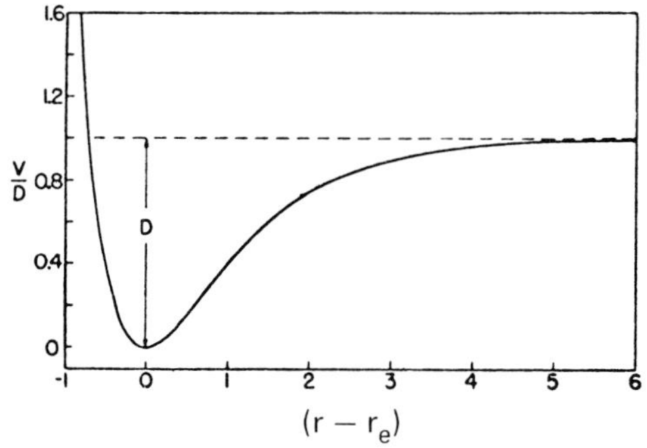

FIG. 27.1 Morse potential with $\beta = 1$. (From Kauzmann, 1957.)

as the moment of inertia of the diatomic molecule. It is customary to write the energy as

$$F(J) = B_v J(J + 1) \tag{27.1-19}$$

where the rotational constant (in reciprocal centimeters)

$$B_v = \hbar/4\pi I c \tag{27.1-20}$$

may depend on the vibrational quantum number since the moment of inertia is expected to depend on the vibrational energy. From (27.1-19) the separation between two adjacent rotational levels (Fig. 27.2) is

$$F(J) - F(J - 1) = 2B_v J. \tag{27.1-21}$$

Thus, rotational levels open-up with increasing energy, whereas vibrational levels, according to (27.1-16) tend to converge. In CO, for example, $B_v = 1.92\ \text{cm}^{-1}$ (v = 0) which puts the spacing between rotational levels in the microwave regions.

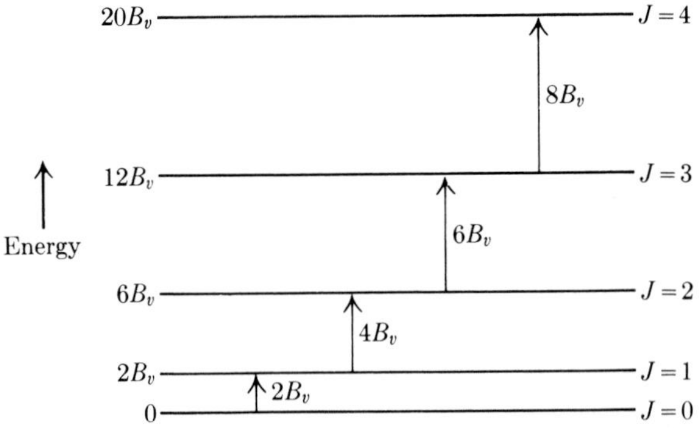

FIG. 27.2 Rotational levels in a diatomic molecule (From Dunford: "Elements of Diatomic Molecular Spectra," 1968, Addison-Wesley, Reading, Massachusetts).

Since each rotational level has a degeneracy of $2J + 1$, the occupation of rotational levels is proportional to the product of $2J + 1$ and the Boltzmann factor:

$$N \propto (2J + 1)e^{-F(J)/kT}. \tag{27.1-22}$$

This is shown in Fig. 27.3; the presence of a maximum in the distribution should be noted. At room temperature the molecules are distributed over a considerable number of rotational states.

The energy for combined vibration and rotation is obtained from the sum of (27.1-16) and (27.1-19):

$$G(v) + F(J) = \omega_e(v + \tfrac{1}{2}) - \omega_e\chi_e(v + \tfrac{1}{2})^2 + B_v J(J + 1). \tag{27.1-23}$$

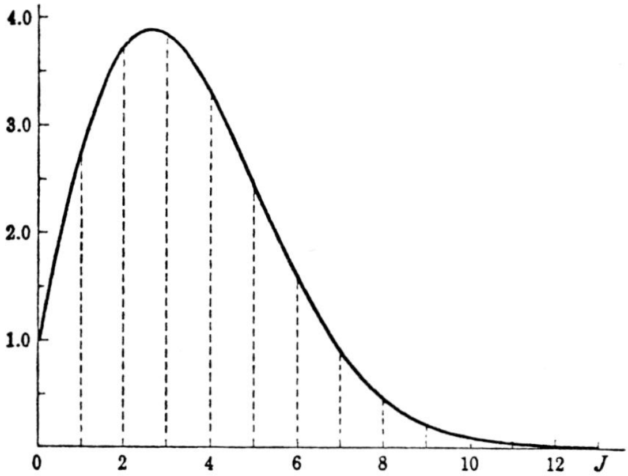

FIG. 27.3 Thermal distribution of diatomic molecules in rotational levels. (From Herzberg, 1950.)

When high precision is required, this expression may not be sufficiently accurate. In that event it is necessary to add higher order terms as well as the coupling between vibrational and rotational motions. A typical vibrational–rotational energy level diagram is shown in Fig. 27.4.

The simple rotor with a single axis of rotation is the simplest model for a rotating diatomic molecule. More realistically, electronic rotations should be included in which case one must consider more general rotations of a rigid body with three principal moments of inertia I_A, I_B, and I_C. When $I_A = I_B = I_C$, the rigid body is called a *spherical top* and when $I_A \neq I_B \neq I_C$, it is an *asymmetric top*. A diatomic molecule lies between these two extremes.

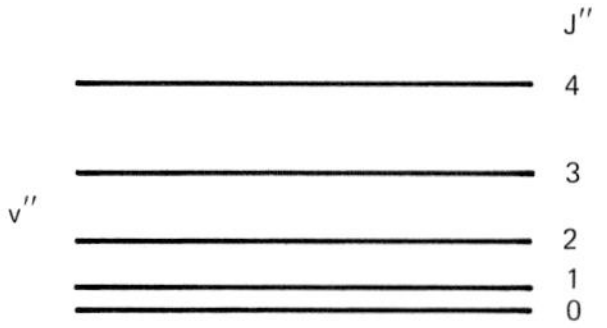

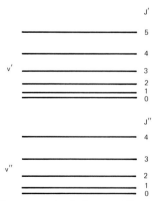

FIG. 27.4 Vibration–rotation energy level diagram. The rotational levels $J' = 0, 1, 2, \ldots$ are associated with the vibrational state v' and the levels $J'' = 0, 1, 2, \ldots$ with v''. High rotational levels belonging to v'' may overlap rotational levels in v'.

We take the A axis along the internuclear direction with the B and C axes along perpendicular directions. In the rigid, or simple, rotor $I_B = I_C$, $I_A = 0$; hence there is just one moment of inertia which we have previously called I. If $I_B = I_C$ but $I_A \neq 0$, the system is called a *symmetric top*. In this model of the diatomic molecule, I_A is the moment of inertia of the electrons about the internuclear axis and $I_B = I_C$ is the moment of inertia of the nuclei about axes perpendicular to the internuclear axis. It should be noted that although $I_A \ll I_B = I_C$, the angular momentum of the electrons (about the internuclear axis) may be comparable to the angular momentum of the nuclei (about perpendicular axes).

The total energy of rotation of a rigid body is

$$E = \tfrac{1}{2}[I_A\omega_A^2 + I_B\omega_B^2 + I_C\omega_C^2].$$

If P_A, P_B, and P_C are the angular momenta‡ about the A, B and C principal axes respectively, the Hamiltonian may be written

$$\mathscr{H} = \frac{P_A^2}{2I_A} + \frac{P_B^2}{2I_B} + \frac{P_C^2}{2I_C}. \tag{27.1-24}$$

This also becomes the quantum mechanical Hamiltonian when P_A, P_B, and P_C are reinterpreted as appropriate operators. Thus, for a symmetric top ($I_A \neq I_B = I_C$) the Hamiltonian is

$$\mathscr{H} = P_A^2\left[\frac{1}{2I_A} - \frac{1}{2I_B}\right] + \frac{P^2}{2I_B} \tag{27.1-25}$$

where

$$P^2 = P_A^2 + P_B^2 + P_C^2. \tag{27.1-26}$$

The eigenvalues of P^2 are $J(J+1)\hbar^2$. If a space-fixed coordinate system (x, y, z) is oriented so that the z and A axes coincide, then $P_A = P_z$ and the eigenvalues of P_A^2 are $\Lambda^2\hbar^2$ where $\Lambda \equiv |M_L| = 0, 1, 2, \ldots$ as in Section 26.4. The energy may then be written

$$E = \left[\frac{1}{2I_A} - \frac{1}{2I_B}\right]\hbar^2\Lambda^2 + \frac{\hbar^2}{2I_B} J(J+1) \tag{27.1-27a}$$

or, in reciprocal centimeters

$$F(J) = BJ(J+1) + (A - B)\Lambda^2 \tag{27.1-27b}$$

with

$$A = \hbar/4\pi c I_A, \qquad B = \hbar/4\pi c I_B. \tag{27.1-28}$$

‡ We use **P** to represent an angular momentum rather than **L** as in the case of the simple rotor, in order to conform to standard molecular nomenclature.

The rotational levels of the symmetric top according to (27.1-27) are the same as those of the simple rotor except that there is a shift in magnitude of $(A - B)\Lambda^2$. If $\Lambda = 0$, the energy levels are just those of the simple rotor. The electronic and nuclear angular momenta are related as shown in Fig. 27.5. The component of the total angular momentum (**J**) perpendicular to the internuclear axis is designated by **N** and represents, essentially, the rotation of the two nuclei. For a given value of Λ, the possible values of the quantum number J are

$$J = \Lambda, \Lambda + 1, \Lambda + 2, \ldots . \qquad (27.1\text{-}29)$$

Thus, since J cannot be smaller than Λ, the rotational levels with $J < \Lambda$ are missing. A further point is that states with $\Lambda > 0$ are two-fold degenerate because of the two senses of rotation of the electrons about the internuclear axis associated with the two values of M_L. Hence for each value of J (for a given **N**) there are two directions of Λ. Quite often there is sufficient coupling between the electronic motion and the nuclear rotation to split the two-fold degeneracy; this is known as Λ-*doubling*.

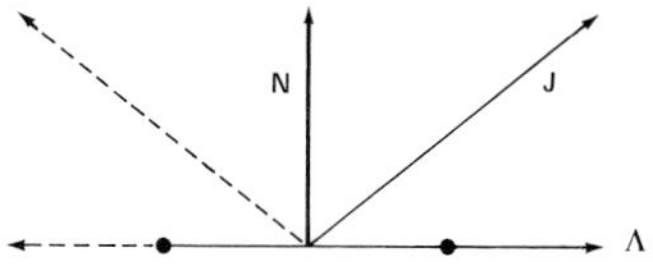

FIG. 27.5 Angular momenta of a symmetric top molecule ($S = 0$, $\Lambda \neq 0$).

The two models that have been discussed so far are the rigid rotor and the symmetric top. In both models there is no reference to the spin of the electrons. Hence the models are applicable to molecules with $S = 0$. We now need to consider the case $S \neq 0$.

If S is the total spin quantum number of the electrons and M_s is the projection on the internuclear axis, the possible values of M_s are $S, S - 1, \ldots, -S$. We denote this by writing

$$\Sigma = S, S - 1, \ldots, -S. \qquad (27.1\text{-}30)$$

The quantum number Σ should not be confused with the Σ state ($\Lambda = 0$). When $\Lambda \neq 0$ there is an internal magnetic field in the direction of the internuclear axis resulting from the orbital motion of the electrons. The internuclear axis may then serve as an axis of quantization upon which the spin may be projected to give the components Σ. But when $\Lambda = 0$, there is no net orbital motion, i.e., no net rotation of the electron cloud about the internuclear axis, hence no magnetic field to define a direction in space. In that event Σ is not defined. However, the total spin (and the multiplicity) retains its usual meaning. We also note that, in contrast to Λ, Σ can be positive or negative.

The quantum number for the total electronic angular momentum about the internuclear axis is obtained by adding Λ and Σ. It is denoted by Ω where

$$\Omega = |\Lambda + \Sigma| = |\Lambda + S|, \quad |\Lambda + S - 1|, \ldots, |\Lambda - S| = 0, \tfrac{1}{2}, 1, \ldots . \qquad (27.1\text{-}31)$$

If $\Lambda > 0$, there are $2S + 1$ values of Ω (for a given value of Λ) corresponding to the $2S + 1$ values of Σ. The values of Ω can be integral or half-integral depending on whether the number of electrons is even or odd. In the usual designation, the value of Ω is written as a subscript and the multiplicity $(2S + 1)$ as a left superscript. Thus, for $\Lambda = 2$ and $S = 1$, we have $\Omega = 1, 2, 3$ and the molecular states are written as

$$^3\Delta_1, \quad ^3\Delta_2, \quad ^3\Delta_3.$$

The different angular momenta in the molecule that are associated with nuclear motion, orbital electronic motion and electronic spin form a resultant which is always designated by $\mathbf{J}$.

The interactions among the various angular momenta give rise to several limiting cases. The two most important ones are known as Hund's case (a) and (b). In case (a) the electronic spin is strongly coupled to the internuclear axis by virtue of the interaction between the spin magnetic moment and the axial magnetic field produced by the orbital motion of the electrons. This case is therefore very similar to the symmetric top except for the replacement of Λ by Ω. Thus, in place of (27.1-29) we now have, for a given value of Ω,

$$J = \Omega, \Omega + 1, \Omega + 2, \ldots . \qquad (27.1\text{-}32)$$

Different values of Ω correspond to different energies and the missing levels are those for which $J < \Omega$. Since Ω has integral values when the number of electrons is even and half-integral values when the number of electrons is odd, the same is true for J. The relations among the various angular momentum vectors are shown in Fig. 27.6a. $\mathbf{\Omega}$ and $\mathbf{N}$ form the resultant $\mathbf{J}$ which remains

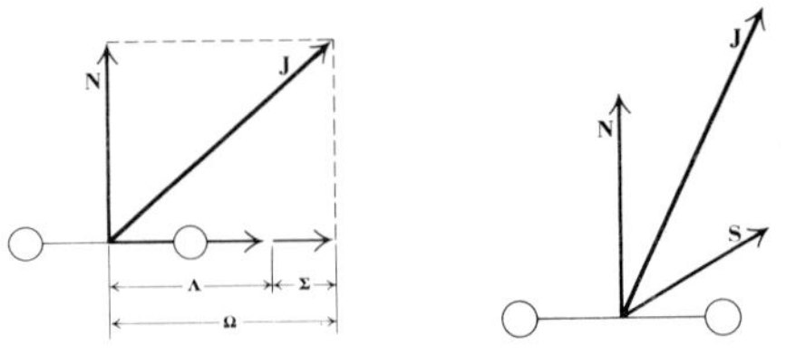

FIG. 27.6 Angular momenta for (a) Hund's case a ($S \neq 0$, $\Lambda \neq 0$) and (b) Hund's case b ($S \neq 0$, $\Lambda = 0$). (From Dunford: "Elements of Diatomic Molecular Spectra," 1968, Addison-Wesley, Reading, Massachusetts.)

constant in magnitude and direction. As an example, for the $^2\Pi$ state in NO, $S = \frac{1}{2}, \Lambda = 1, \Sigma = \pm\frac{1}{2}, \Omega = \frac{1}{2}, \frac{3}{2}$ and the two components of $^2\Pi$ are $^2\Pi_{1/2}$ and $^2\Pi_{3/2}$. In $^2\Pi_{1/2}$ the rotational levels are $J = \frac{1}{2}, \frac{3}{2}, \frac{5}{2}, \ldots$ and in $^2\Pi_{3/2}$ they are $J = \frac{3}{2}, \frac{5}{2}, \ldots$.

In Hund's case (b) the electronic spin is not coupled to the internuclear axis, i.e., there is no spin-axis interaction. Although there are situations in which case (b) applies to states with $\Lambda \neq 0$ if the spin-axis interaction is sufficiently weak, the most common application is to Σ states ($\Lambda = 0$) illustrated in Fig. 27-6b. The electron spin **S** couples directly to the angular momentum **N** of the nuclear rotation to form the total angular momentum **J**. The allowed values of **J** are

$$J = N + S, N + S - 1, \ldots, |N - S| \tag{27.1-33}$$

with $N = 0, 1, 2, \ldots$ and $S = 0, \frac{1}{2}, 1, \frac{3}{2}, \ldots$. As an example, the ground state of O_2 is $^3\Sigma_g^-$. Thus $\Lambda = 0, S = 1$ and the possible values of J are given by

$$\begin{array}{ll} N = 0, & J = 1 \\ N = 1, & J = 2, 1, 0 \\ N = 2, & J = 3, 2, 1 \\ & \vdots \end{array}$$

For $^1\Sigma$ states there is clearly no distinction between N and J. A summary of angular momentum relations and notations is shown in Table 27.1.

27.2 Transitions in Diatomic Molecules

A vibrating diatomic molecule possessing an electric dipole moment whose magnitude is a function of internuclear distance behaves like an oscillating dipole and possesses a vibrational absorption spectrum. Heteronuclear molecules such as CO, NO, HCl fall in this category whereas homonuclear molecules (e.g., H_2, N_2, O_2) which have no dipole moments do not have vibrational absorption spectra. Similar considerations apply to rotations since only in a molecule with a permanent electric moment can there be periodic variations of charge distribution along some axis fixed in space when the molecule is rotating. The same results will also emerge from the quantum mechanical selection rules. It should nevertheless be remarked that other mechanisms in homonuclear diatomic molecules (e.g., collisions at high pressure, magnetic dipole transitions in triplet states) may give rise to absorption of infrared or microwave radiation.

In the harmonic oscillator model for a vibrating diatomic molecule the selection rule for electric dipole transitions may be obtained with the help of

TABLE 27.1

Summary of Relationships among Angular Momentum Quantum Numbers for a Diatomic Molecule

Angular momentum	Symbol	Possible values	
Projection of electronic orbital angular momentum on internuclear axis	M_L $\Lambda = \|M_L\|$	$0, \pm 1, \pm 2, \ldots$ $0, 1, 2, \ldots$ $\Sigma, \Pi, \Delta, \ldots$	
Projection of electronic spin angular momentum on internuclear axis	S Σ	$0, \frac{1}{2}, 1, \ldots$ $S, S-1, \ldots, -S = 0, \pm\frac{1}{2}, \pm 1, \ldots (\Lambda \neq 0)$ Σ is undefined when $\Lambda = 0$	
Total electronic angular momentum about the internuclear axis	Ω	$\|\Lambda + \Sigma\| = 0, \frac{1}{2}, 1, \ldots$ Ω is undefined when $\Lambda = 0, S \neq 0$	
Angular momentum of nuclear rotation	N	$0, 1, 2, \ldots$	
Total angular momentum including total electronic angular momentum and the angular momentum of nuclear rotation	J	$\Omega, \Omega + 1, \ldots$ $N + S, N + S - 1, \ldots, \|N - S\|$	Hund's case (a) Hund's case (b)

the identity (10.1-10) satisfied by harmonic oscillator eigenfunctions ψ_v:

$$r\psi_v = \sqrt{\frac{v+1}{2}}\,\psi_{v+1} + \sqrt{\frac{v}{2}}\,\psi_{v-1}.$$

Computing the matrix element with another state $\psi_{v'}$ we have

$$\langle \psi_{v'}|r|\psi_v\rangle = \sqrt{\frac{v+1}{2}}\,\delta_{v',v+1} + \sqrt{\frac{v}{2}}\,\delta_{v',v-1}$$

or

$$\Delta v = \pm 1 \tag{27.2-1}$$

The spectrum then consists of only one line since the levels of the harmonic oscillator have a constant spacing. More realistically, the vibrations of a diatomic molecule are anharmonic and the selection rule (27.2-1) is replaced by

$$\Delta v = \pm 1, \pm 2, \pm 3, \ldots . \tag{27.2-2}$$

which then gives rise to numerous lines in the vibrational spectrum. The most intense transitions are those with $\Delta v = \pm 1$, and as Δv increases, the intensity drops rapidly. At room temperature most molecules occupy the lowest vibrational state $v = 0$ so the most intense absorption band is the $0 \to 1$ (zero to one) transition which is called *fundamental*. Bands corresponding to the transitions $0 \to 2$, $0 \to 3$, etc., are called *overtones*.

For rotational transitions we note that the eigenfunctions of L^2 are the spherical harmonics, hence the selection rule for dipole radiation is

$$\Delta J = \pm 1 \tag{27.2-3}$$

with $\Delta J = 0$ forbidden because of parity restrictions. From (27.1-21) it is seen that for the simple rotor the spectrum consists of evenly spaced lines with a separation of $2B_v$.

Transitions from rotational levels of one vibrational state to the rotational levels of another vibrational state with no change in the electronic state are said to produce a *vibration–rotation* spectrum. Referring to Fig. 27.4, let the upper set of states be labeled by the quantum numbers (v', J') and the lower set by (v'', J''). Transitions between the two level systems are governed by the selection rule $\Delta J = \pm 1$ which results in an absorption spectrum consisting of two branches:

$$\begin{aligned} &\text{R branch,} \quad \Delta J = J' - J'' = +1, \\ &\text{P branch,} \quad \Delta J = J' - J'' = -1. \end{aligned} \tag{27.2-4}$$

If it is assumed, as a first approximation, that $B_{v'} = B_{v''} = B$, then the energy of a rotational level $(v'J')$ can be written as

$$F(J') + G(v') = BJ'(J' + 1) + G(v') \tag{27.2-5}$$

and the energy of a rotational level $(v''J'')$ will be

$$F(J'') + G(v'') = BJ''(J'' + 1) + G(v''). \tag{27.2-6}$$

If we let ν_0 be the difference in the vibrational energies, i.e.,

$$\nu_0 = G(v') - G(v''), \tag{27.2-7}$$

then the transition energy of a line belonging to the R branch will be

$$\begin{aligned} \nu_R \equiv R(J'') &= \nu_0 + F(J') - F(J'') \\ &= \nu_0 + BJ'(J' + 1) - BJ''(J'' + 1) \\ &= \nu_0 + B(J'' + 1)(J'' + 2) - BJ''(J'' + 1) \\ &= \nu_0 + 2BJ'' + 2B \qquad (J'' = 0, 1, 2, \ldots), \end{aligned} \tag{27.2-8}$$

and a line belonging to the P branch will have an energy

$$\nu_P \equiv P(J'') = \nu_0 - 2BJ'' \qquad (J'' = 1, 2, 3, \ldots). \tag{27.2-9}$$

The two branches may be represented by

$$\nu = \nu_0 + 2Bm \tag{27.2-10}$$

where

$$\begin{aligned} m &= 1, 2, 3, \ldots \qquad &\text{(R branch)}, \\ m &= -1, -2, -3, \ldots \qquad &\text{(P branch)}. \end{aligned} \tag{27.2-11}$$

The value $m = 0$ corresponds to the forbidden transition $\Delta J = 0$ and is known as the *band gap* (or band origin, or zero line).

Were we to take into account the fact that $B_{v'}$ and $B_{v''}$ are not quite equal, expressions (27.2-8) and (27.2-9) would be replaced by

$$\begin{aligned} \nu_R \equiv R(J'') = \nu_0 + 2B_{v'} + (3B_{v'} - B_{v''})J'' + (B_{v'} - B_{v''})J''^2 \\ (J'' = 0, 1, 2, \ldots), \end{aligned} \tag{27.2-12}$$

$$\begin{aligned} \nu_P \equiv P(J'') = \nu_0 - (B_{v'} + B_{v''})J'' + (B_{v'} - B_{v''})J''^2 \\ (J'' = 1, 2, 3, \ldots). \end{aligned} \tag{27.2-13}$$

These reduce to (27.2-8) and (27.2-9) when $B_{v'} = B_{v''}$. The analog to (27.2-10) when $B_{v'} \neq B_{v''}$ is

$$\nu = \nu_0 + (B_{v'} + B_{v''})m + (B_{v'} - B_{v''})m^2 \tag{27.2-14}$$

with m as in (27.2-11). Whereas ν is linear in m in (27.2-10), it is quadratic in (27.2-14) and results in the appearance of a *vertex* (or band head) associated with the value of m such that $d\nu/dm = 0$. This value is given by

$$m = -(B_{v'} + B_{v''})/2(B_{v'} - B_{v''}). \tag{27.2-15}$$

If $B_{v'} - B_{v''}$ is negative, as is generally the case, the vertex occurs at a positive value of m. This means that in the R branch, ν goes through a maximum as m is varied (Fig. 27.7). The curve representing the dependence of ν_R on m is known as a *Fortrat* parabola.

There are several important differences between the spectra of simple rotors as compared with symmetric tops. For selection rules we have

$$\text{Simple rotor:}\quad \Delta J = \pm 1, \tag{27.2-16}$$

$$\text{Symmetric top:}\quad \Delta J = \begin{cases} 0, \pm 1 & \text{if}\quad \Lambda \neq 0, \\ \pm 1 & \text{if}\quad \Lambda = 0. \end{cases} \tag{27.2-17}$$

Thus, when $\Lambda = 0$, the selection rule of the symmetric top matches that of the rigid rotor, but when $\Lambda \neq 0$, transitions are allowed when $\Delta J = 0$. This

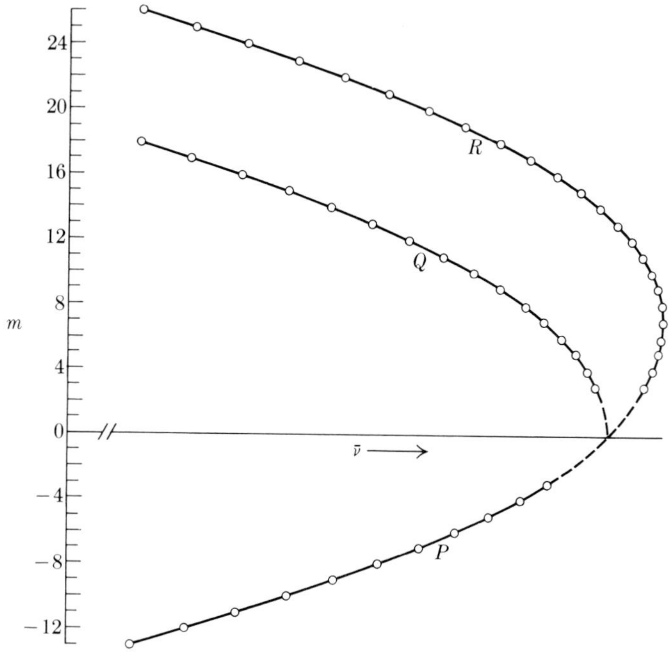

FIG. 27.7 Fortrat parabola showing the R, P, and Q branches. The circles represent transitions (From Dunford: "Elements of Diatomic Molecular Spectra," 1968, Addison-Wesley, Reading, Massachusetts).

leads to a new branch in each band, called the Q branch (Fig. 27.7). For reference we list all three branches of the symmetric top

$$\nu_R = \nu_0 + (B_{v''} - B_{v'})\Lambda^2 + 2B_{v'} + (3B_{v'} - B_{v''})J + (B_{v'} - B_{v''})J^2$$
$$(J = J'' = 0, 1, 2, \ldots), \qquad (27.2\text{-}18)$$

$$\nu_P = \nu_0 + (B_{v''} - B_{v'})\Lambda^2 - (B_{v'} + B_{v''})J + (B_{v'} - B_{v''})J^2$$
$$(J = J'' = 1, 2, 3, \ldots) \qquad (27.2\text{-}19)$$

$$\nu_Q = \nu_0 + (B_{v''} - B_{v'})\Lambda^2 + (B_{v'} - B_{v''})J + (B_{v'} - B_{v''})J^2$$
$$(J = J'' = 0, 1, 2, \ldots). \qquad (27.2\text{-}20)$$

In this example it was assumed that $\Delta\Lambda = 0$ which is usually, although not always, the case.

Each electronic state of a diatomic molecule has its own complement of vibrational and rotational levels. A general transition between two electronic states A and B (Fig. 27.8) therefore usually involves simultaneous changes in vibrational and rotational states. The interplay of the three kinds of transitions results in complex spectra governed by various types of selection rules.

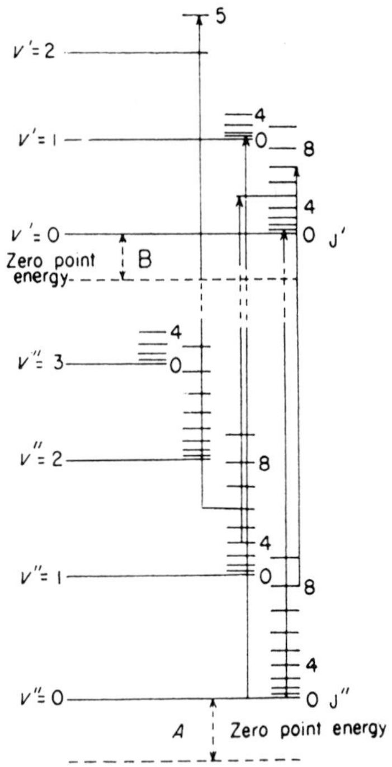

FIG. 27.8 A general transition in a diatomic molecule including changes in electronic, vibrational, and rotational states.

To simplify the discussion we shall, at first, ignore the rotational aspect and concentrate on the changes in electronic and vibrational states.

In considering transitions between two electronic states there is an important principle known as the *Franck–Condon principle.* In essence, it states that the radiative transitions from vibrational levels of one electronic state to vibrational levels of another electronic state take place in a time much shorter than that of a single vibration. Hence the nuclei, because of their large masses, have very nearly the same positions and velocities before and after an electronic transition which takes about 10^{-15} sec. These are known as vertical transitions and are illustrated in Fig. 27.9. The figure also shows that the transition from the lower electronic state occurs from points on the potential curve. This is to be expected classically since such points correspond to extremes of the oscillatory motion where the velocity is zero. Since the molecule spends more time in the vicinity of the extremes, the probability of a transition from these positions is increased. It will be shown that the same conclusion is obtained quantum mechanically.

Assume that the wave function of a vibrating molecule, without rotation, is written in the Born–Oppenheimer approximation as

$$\psi = \psi_e \psi_v$$

where ψ_e is a function of electronic coordinates with the nuclei in their equilibrium positions and ψ_v is a vibrational eigenfunction. It is also assumed that

$$\langle \psi_{ef} | \psi_{ei} \rangle = 0 \tag{27.2-21}$$

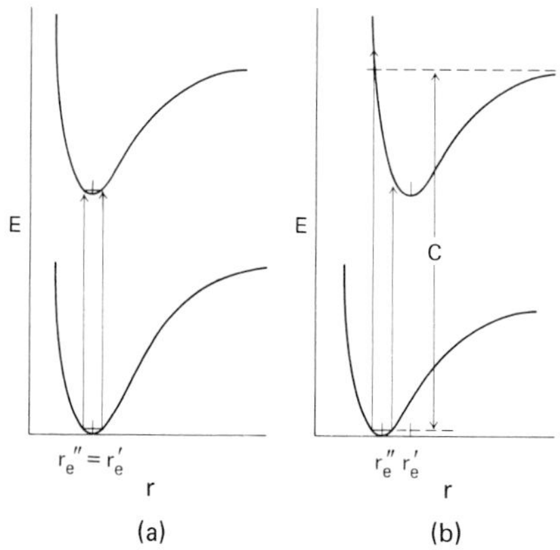

FIG. 27.9 Vertical transitions between two electronic states: in (a) $r_e'' = r_e'$ and in (b) $r_e'' < r_e'$. The transition C represents a dissociation. (From Dunford: "Elements of Diatomic Molecular Spectra," 1968, Addison-Wesley, Reading, Massachusetts.)

where i and f refer to initial and final states. The matrix element for the transition is

$$\langle\psi_f|\mathbf{R}|\psi_i\rangle = \langle\psi_{ef}\psi_{vf}|\mathbf{R}_e + \mathbf{R}_v|\psi_{ei}\psi_{vi}\rangle, \tag{27.2-22}$$

in which $\mathbf{R}_e$ is the transition moment operator for transitions between electronic states and $\mathbf{R}_v$ is the same for vibrational states. Expanding (27.2-22):

$$\langle\psi_f|\mathbf{R}|\psi_i\rangle = \langle\psi_{vf}|\psi_{vi}\rangle\langle\psi_{ef}|\mathbf{R}_e|\psi_{ei}\rangle + \langle\psi_{ef}|\psi_{ei}\rangle\langle\psi_{vf}|\mathbf{R}_v|\psi_{vi}\rangle.$$

In view of the orthogonality of the electronic wave functions, the second term will vanish leaving

$$\langle\psi_f|\mathbf{R}|\psi_i\rangle = \langle\psi_{vf}|\psi_{vi}\rangle\langle\psi_{ef}|\mathbf{R}_e|\psi_{ei}\rangle. \tag{27.2-23}$$

The two vibrational functions ψ_{vi} and ψ_{vf} belong to two different electronic states with different potential functions. There is, therefore, no requirement for ψ_{vi} and ψ_{vf} to be orthogonal. The integral $\langle\psi_{vf}|\psi_{vi}\rangle$, known as the *overlap* integral or the *Franck–Condon* integral, multiplies the electronic transition matrix element (27.2-23) and, therefore, exerts an important influence on the intensity of a transition. We note further that there are no restrictions on Δv for a vibrational-electronic transition.

Vibrational eigenfunctions oscillate much less rapidly near the end points of the motion than in the region in-between. Since rapid oscillations of a wave function correspond to high momenta, we have here the quantum-mechanical counterpart of the classical description previously given. Furthermore, the contribution to the vibrational overlap integral will be small from regions where the wave function oscillates rapidly with changes in sign. Near the end points, the overlap integral is greater, hence the probability of a transition is increased. Thus, either classically or quantum mechanically the most probable transitions are those that occur when the nuclei scarcely move.

The relative positions of the potential curves involved in the transition have an important bearing on the absorption spectrum. In Fig. 27.9(a) the equilibrium separation between the nuclei is about the same in the ground and excited electronic states. At room temperature, with most molecules in the lowest vibrational state ($v = 0$) of the ground electronic state, the most intense absorption band, according to the Franck–Condon principle, would be $0 \to 0$. If the equilibrium separation between the two nuclei is greater in the excited electronic state than in the ground state, a vertical transition will leave the molecule in a vibrationally excited state or may even lead to dissociation (Fig. 27.9(b)).

According to the Born–Oppenheimer approximation, the total wave function of any molecule, including diatomic molecules, may be written as

$$\psi_t = \psi_e\psi_v\psi_r \tag{27.2-24}$$

where ψ_e, ψ_v, and ψ_r are electronic, vibrational, and rotational eigenfunctions, respectively. It is important to bear in mind that while ψ_e specifies an electronic state, we need the *product* $\psi_e\psi_v$ in order to specify a *vibrational level* since such levels are associated with some particular electronic state. In the same fashion a *rotational level* is fully described only when the *product* $\psi_e\psi_v\psi_r = \psi_t$ is given. We shall refer to ψ_v and ψ_r, by themselves, as vibrational and rotational eigenfunctions, respectively.

The Hamiltonian of an assembly of nuclei and electrons, as for any system of particles is invariant under an inversion, that is, under a simultaneous change in sign of the coordinates of *all* the particles (nuclei and electrons). However such invariance implies that wave functions describing the entire assembly of particles must have either even or odd parity. In the former case the wave functions remain unchanged by a reversal of sign of all the coordinates, and in the latter, the wave functions change their sign. This theorem must apply to the total molecular wave function which includes the motion of all the electrons and nuclei. Hence each rotational level ψ_t (27.2-24) must have even or odd parity which, according to standard molecular terminology, is designated by $+$ and $-$, respectively.

It should be remarked that the parity property is independent of the Born–Oppenheimer approximation and holds regardless of the form chosen for the total wave function of the system. However, form (27.2-24) is a very convenient one for sorting out the contributions from the various types of motions within the molecule.

The vibrational eigenfunction ψ_v depends only on the magnitude of the internuclear distance; hence it remains unchanged by reflection in the origin. Therefore, as far as symmetry properties are concerned, we may examine

$$\psi = \psi_e\psi_r \tag{27.2-25}$$

in place of (27.2-24). We shall be interested in certain special cases that are encountered in diatomic molecules and it will be well to bear in mind that the *electronic eigenfunction* ψ_e *is described with respect to a set of coordinates rigidly attached to the nuclei.* If the nuclei rotate (with respect to a set of axes fixed in space), ψ_e remains unaltered. Finally, we note that there are situations in which nuclear spin must be taken into account, but this feature will be omitted for the time being.

1. *Simple Rotor, $S = 0$, $\Lambda = 0$, Heteronuclear.* The rotational eigenfunctions are simply the spherical harmonics Y_{JM} whose parity is $(-1)^J$. This means that under a reflection in the origin Y_{JM} changes sign for odd J and remains unchanged for even J. The electronic eigenfunction ψ_e is $^1\Sigma^+$ or $^1\Sigma^-$. We now wish to ascertain what happens to ψ_e under a reflection in the origin, remembering that all particles—electrons and nuclei—are being reflected.

Such an operation may be regarded as a product of two operations: a rotation of the molecule through 180° about an axis perpendicular to the internuclear axis followed by a reflection in a plane perpendicular to the rotational axis and containing the internuclear axis. As previously stated, the rotation of the nuclei has no effect on ψ_e since the latter depends only on the coordinates of the electrons relative to the nuclei (also on the internuclear distance which does not change during a rotation). Under the reflection $^1\Sigma^+$ remains unchanged and $^1\Sigma^-$ changes sign. Hence, under an inversion (of all particles), $^1\Sigma^+$ is unchanged while $^1\Sigma^-$ changes sign. We then have the following scheme (Fig. 27.10a):

ψ_e	J	Parity of ψ
$^1\Sigma^+$	Even	+
$^1\Sigma^+$	Odd	−
$^1\Sigma^-$	Even	−
$^1\Sigma^-$	Odd	+

2. *Symmetric Top,* $S = 0, \Lambda > 0$, *Heteronuclear.* The permissible values of J are given by (27.1-29) and each rotational level is doubly degenerate on account of the two possible directions of the angular momentum about the axis of the molecule. The two rotational levels of equal energy associated with a value of J transform into each other under an inversion of all the particles because the effect of such an inversion is merely to reverse the direction of the angular momentum about the internuclear axis (see Fig. 27.5). It is therefore possible to construct both + and − states from the two degenerate rotational levels and, hence, for every value of J there is one + and one − rotational level (Fig. 27.10b).

3. *Hund's Case* (*a*), $S \neq 0, \Lambda > 0$, *Heteronuclear.* This case is similar to the symmetric top but with Λ replaced by Ω as defined by (27.1-31) and the permissible values of J given by (27.1-32). Each rotational level is doubly degenerate with one + and one − component (Fig. 27.10c).

4. *Hund's Case* (*b*), $S \neq 0, \Lambda = 0$, *Heteronuclear.* The allowed values of J are given by (27.1-33). Each rotational level has a degeneracy of $2S + 1$ (except when $N = 0$). The analysis proceeds in a manner similar to that of the rigid rotor except that ψ_r is determined by N rather than J. Examples are shown in Fig. 27.10d.

We come now to homonuclear molecules. As in the heteronuclear case, the complete Hamiltonian of the system, which includes the motions of all the electrons and the nuclei, is invariant under a reflection in the origin of

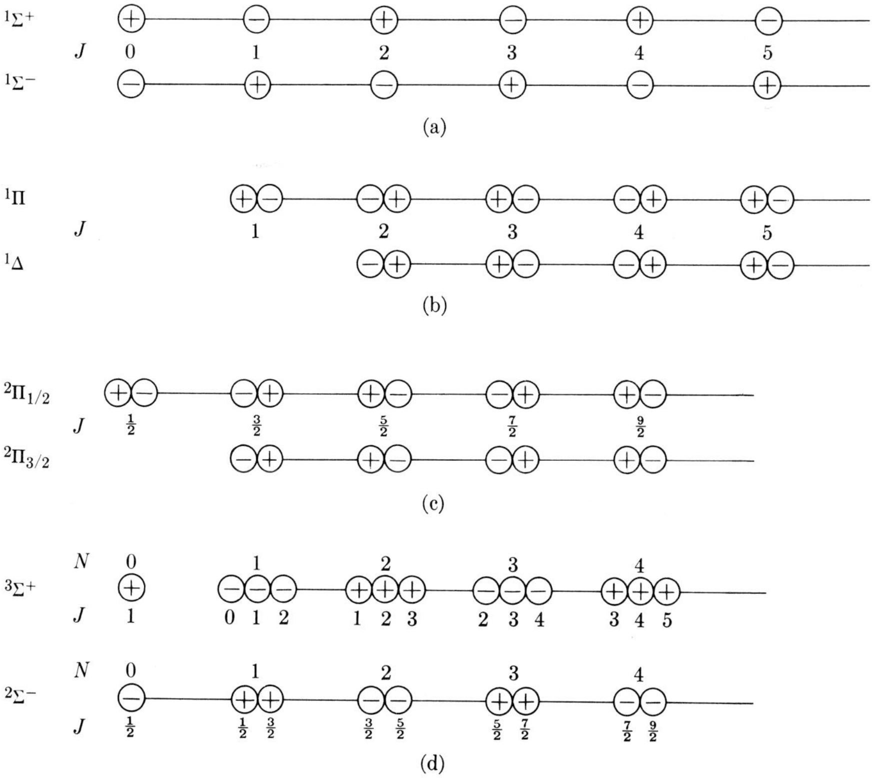

FIG. 27.10 Symmetry properties of rotational levels in a heteronuclear diatomic molecule. (a) Rigid rotor; (b) symmetric top; (c) Hund's case a; and (d) Hund's case b. (From Dunford: "Elements of Molecular Spectra," 1968, Addison-Wesley, Reading, Massachusetts.)

all the particles. As we have seen, this invariance implies that each rotational level ψ_t $(= \psi_e \psi_v \psi_r)$ is either positive or negative with respect to an inversion. But, in addition, for homonuclear molecules, the Hamiltonian is invariant with respect to an interchange of the nuclei. This additional invariance implies that ψ_t either remains unchanged or changes sign under an interchange of the nuclei. The terminology that is used to describe this feature is:

ψ_t is symmetric (s) if it remains unchanged;
ψ_t is antisymmetric (a) if it changes sign.

Thus, for homonuclear molecules, each rotational level carries two symmetry labels: plus or minus with respect to a reflection in the origin of all the particles, symmetric or antisymmetric with respect to an interchange of nuclei.

Also, as before, the vibrational eigenfunction remains unchanged so that the symmetry of ψ_t $(= \psi_e\psi_v\psi_r)$ is the same as that of ψ $(= \psi_e\psi_r)$. We now need to examine the effect on ψ_e and ψ_r when the nuclei are interchanged. To accomplish this we define three operations: R_1 is the reflection of all particles (electrons and nuclei) in the origin, R_2 the reflection of the electrons alone in the origin, and $R_3 = R_1R_2$. Evidently R_3 corresponds to the interchange of the two nuclei. From the discussion of the heteronuclear molecules we already know the behavior of ψ_e and ψ_r under R_1. Also, ψ_r remains unchanged under R_2 since only the electrons are involved. Beyond this point we need to consider a number of special cases.

5. *Rigid Rotor, $S = 0$, $\Lambda = 0$, Homonuclear.* Σ^+ and Σ^- are positive and negative, respectively, under R_1; Σ_g and Σ_u are positive and negative, respectively, under R_2. Hence under R_3 $(= R_1R_2)$, $\Sigma_g{}^+$ and $\Sigma_u{}^-$ are both positive while $\Sigma_g{}^-$ and $\Sigma_u{}^+$ are both negative. These properties are combined with the already established properties of the rotational eigenfunctions ψ_r to give the $(+, -)$ and (s, a) classification of ψ. This is shown in Table 27.2 and in Fig. 27.11.

TABLE 27.2

Symmetries of Σ States in Homonuclear Diatomic Molecules[a]

	Behavior of ψ_e under operations					Behavior of ψ_t under operations	
ψ_e	R_1	R_2	R_3	J	Behavior of ψ_r under R_1 or R_3	R_1	R_3
$^1\Sigma_g{}^+$	+	+	+	Even	+	+	s
				Odd	−	−	a
$^1\Sigma_u{}^+$	+	−	−	Even	+	+	a
				Odd	−	−	s
$^1\Sigma_g{}^-$	−	+	−	Even	+	−	a
				Odd	−	+	s
$^1\Sigma_u{}^-$	−	−	+	Even	+	−	s
				Odd	−	+	a

[a] R_1 is a reflection of all particles (electrons and nuclei) in the origin. R_1 can be represented by $R_1{}^aR_1{}^b$ where $R_1{}^a$ is a rotation of the molecule through 180° about an axis perpendicular to the internuclear axis and $R_1{}^b$ a reflection in a plane containing the internuclear axis and perpendicular to the rotational axis. R_2 is a reflection of the electrons alone in the origin and $R_3 = R_1R_2$ represents an interchange of nuclei.

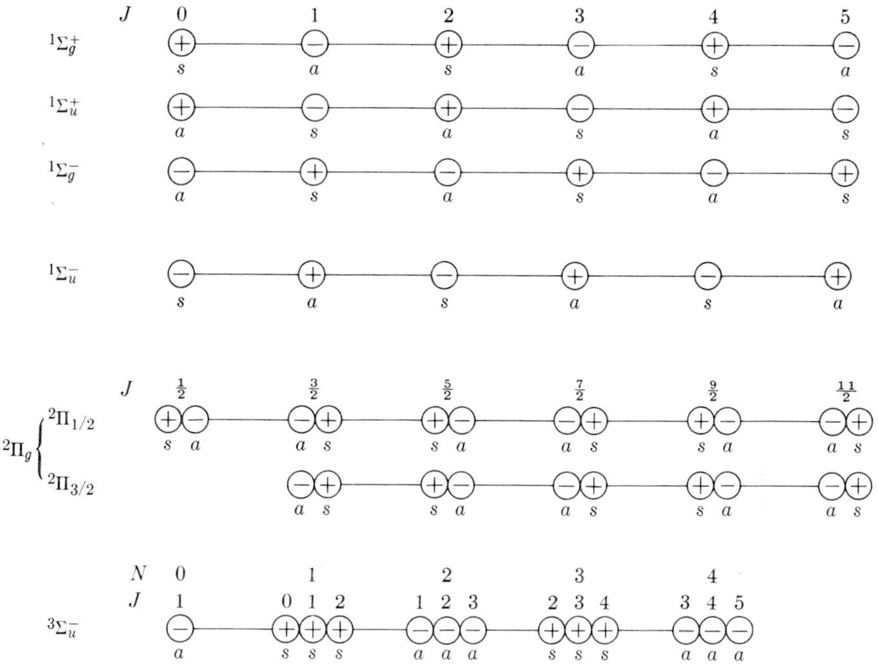

FIG. 27.11 Symmetry properties of rotational levels in a homonuclear diatomic molecule. (From Dunford: "Elements of Molecular Spectra," 1968, Addison-Wesley, Reading, Massachusetts.)

For $S \neq 0$ and $\Lambda = 0$, the behavior is entirely analogous to that of the rigid rotor. When $\Lambda > 0$, positive levels are symmetric, and negative levels are antisymmetric when the electronic eigenfunction ψ_e is positive (under R_2) and conversely when ψ_e is negative (Fig. 27.11). The reader is referred to the work of Herzberg (1950) for more extensive discussions of these and other matters.

The selection rules for radiative transitions in diatomic molecules are based on the symmetry properties of the total wave function $\psi_t = \psi_e \psi_v \psi_r$, ignoring, for the time being, effects that are attributable to the nuclear spin. For *all diatomic molecules* the *dipole selection rule* is

$$+ \leftrightarrow - \tag{27.2-26}$$

where $+, -$ refers to the behavior of ψ_t under a reflection in the origin of all the electrons and the nuclei. For *homonuclear* molecules we have both

$$+ \leftrightarrow - \tag{27.2-27a}$$

and

$$s \leftrightarrow s, \qquad a \leftrightarrow a \tag{27.2-27b}$$

where s and a refer to the behavior of ψ_t under an interchange of the two nuclei. The basis for the selection rules (27.2-27b) is contained in the fact that the dipole moment **d** remains unchanged if the two identical nuclei are interchanged. Therefore the integrand of a matrix element such as $\langle\psi_s|\mathbf{d}|\psi_a\rangle$ where ψ_s refers to a symmetric and ψ_a to an antisymmetric state changes sign under an interchange of nuclei. Since the integral cannot depend on the labeling of the nuclei, it must vanish. The group theoretical counterpart to this argument states that since **d** is invariant under inversion of the two nuclei, the matrix element will vanish unless the wave functions belong to the same irreducible representation of the inversion group. We shall now consider several types of transitions:

1. *Rotational Transitions.* These are radiative transitions within a particular electronic and vibrational state, that is, with ψ_e and ψ_v remaining fixed during the transition

(a) *Heteronuclear*:

$$\begin{gathered} + \leftrightarrow - \\ \Delta J = \pm 1. \end{gathered} \tag{27.2-28}$$

(b) *Homonuclear.* It is necessary to satisfy $+ \leftrightarrow -$, $s \leftrightarrow s$, $a \leftrightarrow a$, $\Delta J = \pm 1$. A glance at Fig. 27.11 reveals that it is impossible to satisfy these conditions simultaneously. Hence there are no pure rotation spectra in homonuclear diatomic molecules as we have previously concluded on the basis of classical considerations.

2. *Vibration–Rotation Spectra.* The electronic state does not change; the transition involves changes in the vibrational and rotational states and the variation in intensity of the lines within a band as a function of J is essentially determined by the thermal distribution of rotational levels (Fig. 27.3).

(a) *Heteronuclear.*

$$\begin{array}{ll} & + \leftrightarrow - \\ \Delta J = \pm 1 & (\Lambda = 0), \\ \Delta J = 0, \pm 1 & (\Lambda \neq 0), \\ \Delta v = \pm 1 & (\text{harmonic oscillator model}), \\ \Delta v = \pm 1, \pm 2, \ldots & (\text{anharmonic oscillator}). \end{array} \tag{27.2-29}$$

(b) *Homonuclear.* The selection rules $s \leftrightarrow s$, $a \leftrightarrow a$, $+ \leftrightarrow -$ cannot be satisfied simultaneously. Hence there is no vibration–rotation spectrum in a homonuclear diatomic molecule as in the previous case.

3. *Electronic Spectra.* The selection rules are

$$\begin{aligned} &\Delta S = 0, \\ &\Delta\Lambda = 0, \pm 1, \\ &+ \leftrightarrow - \qquad (\psi_t) \\ &\Sigma^+ \leftrightarrow \Sigma^+, \qquad \Sigma^- \leftrightarrow \Sigma^- \\ &g \leftrightarrow u, \qquad s \leftrightarrow s, \qquad a \leftrightarrow a \qquad \text{(homonuclear)} \\ &\Delta J = \begin{cases} 0, \pm 1 & (\Lambda \neq 0), \qquad J = 0 \not\leftrightarrow J = 0, \\ \pm 1 & (\Lambda = 0). \end{cases} \end{aligned} \tag{27.2-30}$$

A few examples will help to clarify the application of these rules. ${}^1\Sigma \leftrightarrow {}^1\Sigma$: The allowed transitions of this type are

$$ {}^1\Sigma^+ \leftrightarrow {}^1\Sigma^+, \qquad {}^1\Sigma^- \leftrightarrow {}^1\Sigma^- \qquad \text{(heteronuclear)}, \tag{27.2-31a}$$

$$ {}^1\Sigma_g^+ \leftrightarrow {}^1\Sigma_u^+, \qquad {}^1\Sigma_g^- \leftrightarrow {}^1\Sigma_u^- \qquad \text{(homonuclear)}, \tag{27.2-31b}$$

$$\Delta J = \pm 1 \qquad (\Lambda = 0). \tag{27.2-31c}$$

With these selection rules we automatically satisfy $+ \leftrightarrow -$ for ψ_t in both the hetero- and homonuclear cases as well as $s \leftrightarrow s$, $a \leftrightarrow a$ in the homonuclear case (Figs. 27.10 and 27.11).

There is another feature that should be taken into account in the discussion of homonuclear molecules; this is the effect of nuclear spin. It has been stated that each rotational level $\psi_t = \psi_e \psi_v \psi_r$ is either symmetric or antisymmetric with respect to an interchange of nuclei. ψ_t is, of course, a function of coordinates. Let us suppose that ψ_t is multiplied by a nuclear spin function ψ_n to form the function $\psi_{tn} = \psi_t \psi_n$. This construction is analogous to the construction of the total wave function of a two-electron atom which consists of a spatial wave function multiplied by a spin function. Now in the atomic case it was necessary for the total wave function to be antisymmetric with respect to an interchange of two electrons because electrons obey Fermi statistics. But nuclei can obey either Fermi or Bose statistics, the determining criterion being whether the spin I of the nucleus is half-integral or integral. In the former case, Fermi statistics apply and in the latter, Bose; it is further recalled that the total wave function for fermions is antisymmetric and for bosons, symmetric. Thus, for nuclei with $I = \frac{1}{2}, \frac{3}{2}, \ldots$ (fermions), an interchange of the two nuclei changes the sign of the molecular wave function, but for nuclei with $I = 0, 1, 2, \ldots$ (bosons), there is no change in sign. We then have the scheme:

I	Statistics	ψ_{tn}
Half-integral	Fermi	Antisymmetric
Integral	Bose	Symmetric

Let us now see how this goes in a few special cases. Suppose $I = 0$ (as in O_2); the nuclei satisfy Bose statistics. Hence $\psi_{tn} = \psi_t$ is symmetric. This means that the antisymmetric levels are not populated. Suppose $I = \frac{1}{2}$ (as in H_2). Such nuclei satisfy Fermi statistics requiring that ψ_{tn} be antisymmetric. When two identical nuclei each with spin $I = \frac{1}{2}$ are coupled, the total spin $T = 0, 1$. The spin function $\psi_n{}^0$ corresponding to $T = 0$ is antisymmetric with respect to an interchange of the two nuclei while the spin function $\psi_n{}^1$ corresponding to $T = 1$ is symmetric. We then have the combinations:

ψ_{tn}	T	ψ_n	ψ_t	$2T+1$	Statistical weight of the rotational levels
a	0	a	s	1	$1(2J+1)$
a	1	s	a	3	$3(2J+1)$

Thus the antisymmetric rotational levels (ψ_t) will have a statistical weight of $3(2J + 1)$ whereas the symmetric rotational levels will have a statistical weight of $(2J + 1)$.

As a more complex example, let $I = 1$ (as in D_2, N_2). Now we have again Bose statistics and ψ_{tn} is symmetric. The coupling of the two spins gives $T = 0, 1, 2$. Either on the basis of a general theorem or by actually constructing the spin functions by means of a table of coupling coefficients it may be verified that $\psi_n{}^0$ is symmetric, $\psi_n{}^1$ is antisymmetric, and $\psi_n{}^2$ is symmetric. The possible combinations are:

ψ_{tn}	T	ψ_n	ψ_t	$2T+1$	Statistical weight of the rotational levels
s	0	s	s	1	$1(2J+1)$
s	1	a	a	3	$3(2J+1)$
s	2	s	s	5	$5(2J+1)$

In this case the symmetric rotational levels ψ_t will have a statistical weight of $6(2J + 1)$ compared to a statistical weight of $3(2J + 1)$ for the antisymmetric rotational levels.

It is seen, then, that the symmetry requirements imposed by the two identical nuclei have an effect on the statistical weights of the symmetric and antisymmetric rotational levels. These differences manifest themselves in the intensities of the observed spectra as will now be illustrated. To be specific, let the molecule be H_2 in which $I = \frac{1}{2}$ and let the transition be ${}^1\Sigma_g{}^+ \leftrightarrow {}^1\Sigma_u{}^+$. From Table 27.2 we learn that for ${}^1\Sigma_g{}^+$ the wave function ψ_t is symmetric for even values of J and antisymmetric for odd values; for ${}^1\Sigma_u{}^+$ even (odd) values of J are associated with an antisymmetric (symmetric) ψ_t. We then have the accompanying tabulation.

ψ_e	ψ_{tn}	T	ψ_n	ψ_t	J	$2T+1$
${}^1\Sigma_g{}^+$	a	0	a	s	Even	1
${}^1\Sigma_g{}^+$	a	1	s	a	Odd	3
${}^1\Sigma_u{}^+$	a	0	a	s	Odd	1
${}^1\Sigma_u{}^+$	a	1	s	a	Even	3

Now, in the transition ${}^1\Sigma_g{}^+ \leftrightarrow {}^1\Sigma_u{}^+$ the selection rules s$\leftrightarrow$s, a$\leftrightarrow$a must be satisfied. The ratio of the intensities of these two types of transitions will be in the ratio of the statistical weights (or the values of $2T+1$). Thus

$$\frac{\mathrm{s} \leftrightarrow \mathrm{s}}{\mathrm{a} \leftrightarrow \mathrm{a}} = \frac{{}^1\Sigma_g{}^+ \text{ (even } J) \leftrightarrow {}^1\Sigma_u{}^+ \text{ (odd } J)}{{}^1\Sigma_g{}^+ \text{ (odd } J) \leftrightarrow {}^1\Sigma_u{}^+ \text{ (even } J)} = \frac{1}{3} \tag{27.2-32}$$

and the spectrum will show an intensity alternation of 1:3. Note further that there are actually two types of H_2 molecules—those with $T = 0$ (antiparallel spin) occupy rotational levels with $J = 0, 2, 4, \ldots$ and those with $T = 1$ (parallel spin) occupy levels with $J = 1, 3, 5, \ldots$. The former type is known as *parahydrogen* and the latter *orthohydrogen.* In view of the selection rules s$\leftrightarrow$s and a$\leftrightarrow$a there are no radiative transitions between the two kinds of molecules, nor will these selection rules be violated during collisions, so that, under normal circumstances, H_2 is a mixture of ortho and parahydrogen in the ratio of 3:1. For other molecules, like D_2 in which $T = 0, 1, 2$, the ortho modification refers to the symmetric nuclear spin functions, that is, $T = 0, 2$ in D_2. Since the ground electronic state in D_2 is ${}^1\Sigma_g{}^+$ and the nuclei are bosons, D_2 molecules with $T = 0$ and 2 occupy even J rotational levels, and those with $T = 1$ are in odd J levels.

A general expression for the intensity ratio of strong to weak transitions is $(I+1)/I$ where I is the spin of the nucleus. In O_2, with $I = 0$ the ratio is infinite which means that every other line in the spectrum is missing.

27.3 Vibration of Polyatomic Molecules

A molecule consisting of N atoms requires $3N$ coordinates in order to specify the position of each atom. But the translational motion of the center of mass and the rotational motion of the molecule as a whole account for 6 degrees of freedom. Therefore a system of N particles has $3N$-6 vibrational degrees of freedom except for linear molecules which have $3N$-5 degrees of freedom since the nuclei do not rotate about the molecular axis.

To describe the motion of the nuclei it is convenient to imagine that a rectangular coordinate system has been affixed to the equilibrium position of each nucleus. Let ξ_i be a displacement along one of the axes associated with the coordinate system of the ith nucleus. The total energy, up to quadratic terms in the displacement, is

$$E = \tfrac{1}{2} \sum_{i}^{3N} m_i \dot{\xi}_i^2 + \tfrac{1}{2} \sum_{ij}^{3N} (\partial^2 V / \partial \xi_i \partial \xi_j)_0 \, \xi_i \xi_j = T + U \qquad (27.3\text{-}1)$$

in which the second derivatives are evaluated at the equilibrium. Now let

$$q_i = \sqrt{m_i} \xi_i. \qquad (27.3\text{-}2)$$

This eliminates the masses from the energy expression and we have

$$E = \tfrac{1}{2} \sum_{i} \dot{q}_i^2 + \tfrac{1}{2} \sum_{ij} (\partial^2 V / \partial q_i \partial q_j)_0 \, q_i q_j \qquad (27.3\text{-}3)$$

or, setting

$$(\partial^2 V / \partial q_i \partial q_j)_0 \equiv V_{ij}, \qquad (27.3\text{-}4)$$

the energy becomes

$$E = \tfrac{1}{2} \sum_{i} \dot{q}_i^2 + \tfrac{1}{2} \sum_{ij} V_{ij} q_i q_j. \qquad (27.3\text{-}5)$$

Note that, according to (27.3-4), $V_{ij} = V_{ji}$ so that the V_{ij} are components of a symmetric tensor of rank 2. Alternatively we may write the energy in matrix form

$$E = \tfrac{1}{2} \tilde{\dot{q}} \dot{q} + \tfrac{1}{2} \tilde{q} V q \qquad (27.3\text{-}6)$$

in which

$$q = \begin{pmatrix} q_1 \\ q_2 \\ \vdots \end{pmatrix}, \qquad \tilde{q} = (q_1 q_2 \cdots), \qquad (27.3\text{-}7)$$

$$V = \begin{pmatrix} V_{11} & V_{12} & \cdots \\ V_{21} & & \cdots \\ & & \cdots \\ \vdots & \vdots & \cdots \end{pmatrix}. \qquad (27.3\text{-}8)$$

Since V is symmetric it is possible to find an orthogonal matrix A which will diagonalize V,

$$A^{-1} V A = \Omega \qquad (27.3\text{-}9)$$

where Ω is a diagonal matrix whose nonvanishing components are the eigenvalues of V. Let these eigenvalues be ω_i^2. Now define new coordinates

Q_i which are related to the q_i by the matrix equation

$$q = AQ. \tag{27.3-10}$$

Substituting (27.3-10) into (27.3-6) we have

$$\begin{aligned} E &= \tfrac{1}{2}(\widetilde{A\dot{Q}})(A\dot{Q}) + \tfrac{1}{2}(\widetilde{AQ})V(AQ) \\ &= \tfrac{1}{2}\tilde{\dot{Q}}\tilde{A}A\dot{Q} + \tfrac{1}{2}\tilde{Q}\tilde{A}VAQ = \tfrac{1}{2}\tilde{\dot{Q}}\dot{Q} + \tfrac{1}{2}\tilde{Q}\Omega Q \end{aligned} \tag{27.3-11}$$

in which we have used (27.3-9) and the orthogonality property of A. But

$$\tilde{\dot{Q}}\dot{Q} = \sum_i \dot{Q}_i^{\,2}, \qquad \tilde{Q}\Omega Q = \sum_i \omega_i^{\,2} Q_i^{\,2}. \tag{27.3-12}$$

Therefore the energy can now be written as a sum of squares:

$$E = \tfrac{1}{2}\sum_i (\dot{Q}_i^{\,2} + \omega_i^{\,2} Q_i^{\,2}). \tag{27.3-13}$$

The Q_i are known as *normal coordinates.* Further, (27.3-13) may be regarded as the Hamiltonian of the system because the canonical momentum which is defined as

$$P_i = \partial T/\partial Q_i \tag{27.3-14}$$

is, from (27.3-13), equal to $\dot{Q}_i$. Therefore the kinetic energy is just

$$T = \tfrac{1}{2}\sum_i P_i^{\,2} \tag{27.3-15}$$

and

$$\mathscr{H} = \tfrac{1}{2}\sum_i^{3N} (P_i^{\,2} + \omega_i^{\,2} Q_i^{\,2}). \tag{27.3-16}$$

The equations of motion are now readily obtained from Hamilton's equations (or the Lagrange equations):

$$\dot{P}_i = -\partial\mathscr{H}/\partial Q_i = -\omega_i^{\,2} Q_i = \ddot{Q}_i \tag{27.3-17}$$

or

$$\ddot{Q}_i + \omega_i^{\,2} Q_i = 0. \tag{27.3-18}$$

Thus in terms of the normal coordinates we have a set of simple harmonic oscillator equations. The advantage of the normal coordinate (or mode) description resides in the fact that the complex motion of a vibrating molecule can be described as a linear superposition of normal modes with each mode having a fixed frequency.

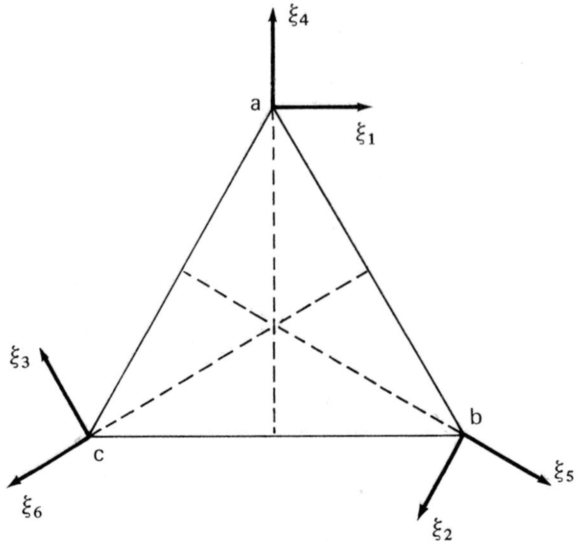

FIG. 27.12 Coordinate system for the vibrations of an A_3 (equilateral triangle) molecule.

To make matters more specific we will consider the vibrations of a triatomic molecule in the shape of an equilateral triangle with three equal masses (Fig. 27.12). It is assumed that the vibrations are confined to the plane of the triangle. At each nucleus we construct a two-dimensional orthogonal coordinate system. Although the orientations of the individual coordinate systems are arbitrary there are, nevertheless, certain orientations which result in simpler expressions for the pertinent physical observables. The kinetic energy is

$$T = \tfrac{1}{2}m \sum_{i=1}^{6} \dot{\xi}_i^2 \tag{27.3-19}$$

and the potential energy is

$$\begin{aligned} U = \tfrac{1}{2}k\{[\tfrac{1}{2}(\xi_2 - \xi_1) + \tfrac{1}{2}\sqrt{3}(\xi_4 + \xi_5)]^2 + [\tfrac{1}{2}(\xi_3 - \xi_2) + \tfrac{1}{2}\sqrt{3}(\xi_5 + \xi_6)]^2 \\ + [\tfrac{1}{2}(\xi_1 - \xi_3) + \tfrac{1}{2}\sqrt{3}(\xi_4 + \xi_6)]^2\} \end{aligned} \tag{27.3-20}$$

with

$$k = (\partial^2 V/\partial\xi_i\,\partial\xi_j)_0$$

where, for example, $\frac{1}{2}(\xi_2 - \xi_1)$ represents the relative displacement of atoms a and b along the line joining them when the displacement of a from equilibrium is ξ_1 and that of b is ξ_2. Introducing the mass-weighted coordinates

(27.3-2), the kinetic energy is

$$T = \tfrac{1}{2} \sum_i \dot{q}_i \tag{27.3-21}$$

and the potential energy is

$$U = \tfrac{1}{2}(k/m)\{[\tfrac{1}{2}(q_2 - q_1) + \tfrac{1}{2}\sqrt{3}(q_4 + q_5)]^2 + [\tfrac{1}{2}(q_3 - q_2) + \tfrac{1}{2}\sqrt{3}(q_5 + q_6)]^2 + [\tfrac{1}{2}(q_1 - q_3) + \tfrac{1}{2}\sqrt{3}(q_4 + q_6)]^2\}. \tag{27.3-22}$$

The matrix V, from (27.3-5) and (27.3-22), is

$$V = \frac{k}{m}\begin{pmatrix} 1/2 & -1/4 & -1/4 & 0 & -\sqrt{3}/4 & \sqrt{3}/4 \\ -1/4 & 1/2 & -1/4 & \sqrt{3}/4 & 0 & -\sqrt{3}/4 \\ -1/4 & -1/4 & 1/2 & -\sqrt{3}/4 & \sqrt{3}/4 & 0 \\ 0 & \sqrt{3}/4 & -\sqrt{3}/4 & 3/2 & 3/4 & 3/4 \\ -\sqrt{3}/4 & 0 & \sqrt{3}/4 & 3/4 & 3/2 & 3/4 \\ \sqrt{3}/4 & -\sqrt{3}/4 & 0 & 3/4 & 3/4 & 3/2 \end{pmatrix}. \tag{27.3-23}$$

The eigenvalues of V, which we call ω_i^2, are

$$\omega_i^2 = 3\frac{k}{m},\quad \frac{3}{2}\frac{k}{m},\quad \frac{3}{2}\frac{k}{m},\quad 0,\quad 0,\quad 0. \tag{27.3-24}$$

These are the frequencies of the normal modes. Having found the eigenvalues we may then compute the normalized eigenvectors; the latter define the orthogonal transformation A in (27.3-10) from which we subsequently obtain the normal coordinates Q_i. We shall not carry out this computation which is straightforward but tedious. Instead we shall examine the problem from the standpoint of group theory.

The fundamental postulate is that the Hamiltonian must be invariant under all symmetry operations applicable to the molecule in its equilibrium configuration. In particular, the potential energy must be an invariant, i.e., U must belong to the totally symmetric (unit) representation of the symmetry group or

$$P_R U = U. \tag{27.3-25}$$

From (27.3-13),

$$U = \tfrac{1}{2} \sum_i \omega_i^2 Q_i^2. \tag{27.3-26}$$

Therefore the invariance of U imposes certain symmetry requirements on the normal coordinates Q_i. If Q_i is nondegenerate, i.e., there is only one Q_i

for the frequency ω_i, then

$$P_R Q_i = \pm Q_i \tag{27.3-27}$$

because then $P_R Q_i^2 = P_R Q_i P_R Q_i = Q_i^2$. Relation (27.3-27) is interpretable as meaning that Q_i is a basis function for a one-dimensional representation of the symmetry group. If Q_i is degenerate, there are several normal modes with the same frequency. If, for example, the degeneracy is twofold, there will be a frequency ω_k for which there are two normal modes say, Q_1 and Q_2. The invariance of U then implies the invariance of $(Q_1^2 + Q_2^2)\omega_k^2$. Assume now that the effect of a symmetry operator on Q_1 is to transform it into a linear combination of Q_1 and Q_2 and that a similar effect is produced on Q_2. Thus

$$P_R Q_1 = Q_1 a_{11} + Q_2 a_{21}, \qquad P_R Q_2 = Q_1 a_{12} + Q_2 a_{22} \tag{27.3-28}$$

or

$$P_R(Q_1, Q_2) = (Q_1 Q_2)\begin{pmatrix} a_{11} & a_{12} \\ a_{21} & a_{22} \end{pmatrix}, \tag{27.3-29}$$

or

$$P_R \tilde{Q} \equiv \tilde{Q} A.$$

This simply means that (Q_1, Q_2) is a basis for an (irreducible) representation of the symmetry group. Such representations can be assumed to be unitary and in the present example they are orthogonal; hence

$$A\tilde{A} = \tilde{A}A = 1 \tag{27.3-30}$$

and

$$P_R \tilde{Q} Q = Q A \tilde{A} Q = \tilde{Q} Q. \tag{27.3-31}$$

Thus

$$P_R(Q_1^2 + Q_2^2) = Q_1^2 + Q_2^2. \tag{27.3-32}$$

We see then that if (27.3-28) is valid, the potential energy associated with a degenerate mode such as $\frac{1}{2}(Q_1^2 + Q_2^2)\omega_k^2$ remains invariant under a symmetry operation. The argument is readily extended to n-fold degeneracies. Additionally, it is seen that Q_i^2 is an invariant whether or not Q_i is degenerate. The general conclusion, then, is that the normal modes are basis functions for irreducible representations of the symmetry group of the molecule and that the degeneracy of the mode is equal to the dimensionality of the irreducible representation. This also implies that the normal coordinates are orthogonal to one another. Alternatively it may be said that to each normal mode frequency there corresponds some irreducible representation of the symmetry group of the molecule.

We now return to the specific example of the triatomic molecule which belongs to the symmetry group D_3. Under the various symmetry operators, the coordinates q_i (which are proportional to ξ_i) transform as

$$\begin{aligned}
P_E(q_1, \ldots, q_6) &= (q_1, \ldots, q_6), \\
P_A(q_1, \ldots, q_6) &= (-q_1, -q_3, -q_2, q_4, q_6, q_5), \\
P_B(q_1, \ldots, q_6) &= (-q_3, -q_2, -q_1, q_6, q_5, q_4), \\
P_C(q_1, \ldots, q_6) &= (-q_2, -q_1, -q_3, q_5, q_4, q_6), \\
P_D(q_1, \ldots, q_6) &= (q_2, q_3, q_1, q_5, q_6, q_4), \\
P_F(q_1, \ldots, q_6) &= (q_3, q_1, q_2, q_6, q_4, q_5).
\end{aligned} \tag{27.3-33}$$

In this example, A, B, C are rotations through π about axes through a, b, c, respectively; D and F are rotations through $\pm 120°$ about the symmetry axis.

Using (27.3-33) it may be readily verified that $P_E, \ldots, P_F$ acting on U in (27.3-22) leaves U unaltered. Moreover the mass weighted coordinates $q_1, \ldots, q_6$ form a basis for a representation of D_3. This follows from (27.3-33). As an example we note that

$$P_A(q_1, \ldots, q_6) = (-q_1, -q_3, -q_2, q_4, q_6, q_5) \tag{27.3-34}$$

may be written as

$$P_A(q_1, \ldots, q_6) = (q_1, \ldots, q_6) \begin{pmatrix} -1 & 0 & 0 & 0 & 0 & 0 \\ 0 & 0 & -1 & 0 & 0 & 0 \\ 0 & -1 & 0 & 0 & 0 & 0 \\ 0 & 0 & 0 & 1 & 0 & 0 \\ 0 & 0 & 0 & 0 & 0 & 1 \\ 0 & 0 & 0 & 0 & 1 & 0 \end{pmatrix}. \tag{27.3-35}$$

The character of the representation matrix is zero. From insepection of (27.3-33) we can ascertain which qs remain in the same position and thereby obtain the characters. They are

$$\chi(E) = 6, \qquad \chi(A) = 0, \qquad \chi(B) = 0, \qquad \chi(C) = 0, \qquad \chi(D) = 0, \qquad \chi(F) = 0.$$

If we call this representation $\Gamma(q)$, then it is possible, with the aid of a character table of D_3 (Table 5.2), to reduce $\Gamma(q)$:

$$\Gamma(q) = A_1 + A_2 + 2E. \tag{27.3-36}$$

From (27.3-36) we learn that the motion of the three mass points can be described in terms of four modes. Two of these are nondegenerate (one-dimensional) and two are twofold degenerate (two-dimensional). However, these are not all vibratory modes because they contain translational and rotational motions as well. The translation must lie in the xy plane (the plane of the molecule, by hypothesis) and from the character table we learn that (x, y) transform according to E. The rotation of the molecule as a whole is about the z axis (remember that we are not allowing the nuclei to move out of the plane of the molecule) and again, from the character table, we find that R_z—a rotation about the z axis—belongs to the A_2 representation. We therefore subtract E and A_2 from $\Gamma(q)$, i.e., we eliminate them from consideration, and we are left with

$$\Gamma_{\text{vib}} = A_1 + E. \tag{27.3-37}$$

In other words there is a nondegenerate and a twofold degenerate mode for the vibrational motion.

To obtain the normal coordinates we shall project out of the basis set $(q_1, \ldots, q_0)$ those linear combinations which belong to the irreducible representations A_1, A_2, and E. The projection operator has been defined by

$$\rho^{(j)} = (l_j/h) \sum_R \chi^{(j)}(R)^* P_R \tag{27.3-38}$$

where $\chi^{(j)}(R)$ is the character of the irreducible representation matrix $\Gamma^{(j)}(R)$, l_j the dimension of $\Gamma^{(j)}(R)$, and h the order of the group. For A_1, using (27.3-33),

$$\rho^{(1)} q_1 = \tfrac{1}{6} \sum_R P_R q_1 = \tfrac{1}{6}[q_1 - q_1 - q_3 - q_2 + q_2 + q_3] = 0. \tag{27.3-39}$$

Proceeding in this way it is found that

$$\begin{aligned} \rho^{(1)} q_2 &= \rho^{(1)} q_3 = 0, \\ \rho^{(1)} q_4 = \rho^{(1)} q_5 &= \rho^{(1)} q_6 = \tfrac{1}{3}(q_4 + q_5 + q_6). \end{aligned} \tag{27.3-40}$$

Therefore the mode Q_1 belonging to A_1, after normalization, is

$$(A_1){:}\quad Q_1 = (q_4 + q_5 + q_6)/\sqrt{3}. \tag{27.3-41}$$

In the same fashion, the normal coordinate associated with A_2 is found by means of the projection operator to be

$$(A_2){:}\quad Q_2 = (q_1 + q_2 + q_3)/\sqrt{3}. \tag{27.3-42}$$

For the E representation, the projection operator generates six functions which, after normalization, are

$$\begin{aligned} f_1 &= (2q_1 - q_2 - q_3)/\sqrt{6}, & f_2 &= (2q_2 - q_1 - q_3)/\sqrt{6}, \\ f_3 &= (2q_3 - q_1 - q_2)/\sqrt{6}, & f_4 &= (2q_4 - q_5 - q_6)/\sqrt{6}, \\ f_5 &= (2q_5 - q_4 - q_6)/\sqrt{6}, & f_6 &= (2q_6 - q_4 - q_5)/\sqrt{6}. \end{aligned} \tag{27.3-43}$$

These equations are not independent because

$$f_1 + f_2 + f_3 = 0, \qquad f_4 + f_5 + f_6 = 0. \tag{27.3-44}$$

Hence the set (27.3-43) contains four independent functions from which it is possible to construct two pairs of orthogonal functions. There is no unique way of doing this. Here is one choice:

$$\left.\begin{aligned} Q_3 &= [\sqrt{3}(q_2 - q_3) + 2q_4 - q_5 - q_6]/\sqrt{12} \\ Q_4 &= [2q_1 - q_2 - q_3 - \sqrt{3}(q_5 - q_6)]/\sqrt{12} \end{aligned}\right\} E, \tag{27.3-45}$$

$$\left.\begin{aligned} Q_5 &= [-\sqrt{3}(q_2 - q_3) + 2q_4 - q_5 - q_6]/\sqrt{12} \\ Q_6 &= [2q_1 - q_2 - q_3 + \sqrt{3}(q_5 - q_6)]/\sqrt{12} \end{aligned}\right\} E. \tag{27.3-46}$$

Having found the six normal coordinates, the matrix A in (27.3-10) which connects the qs with the Qs is essentially determined. In turn this implies that we have the necessary information with which to diagonalize V and thereby obtain the eigenvalues given in (27.3-24). The normal coordinates are shown in Fig. 27.13.

A_1

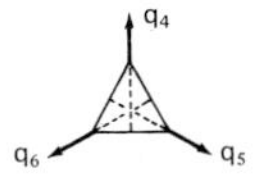

$Q_1 = (q_4 + q_5 + q_6)/\sqrt{3}$, $\omega_1{}^2 = 3k/m$

A_2

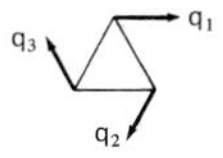

$Q_2 = (q_1 + q_2 + q_3)/\sqrt{3}$ $\omega_2{}^2 = 0$

E_1

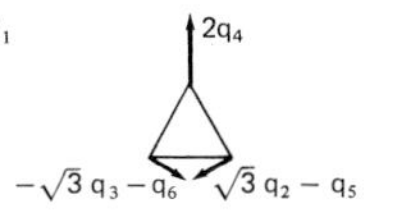

$Q_3 = [\sqrt{3}(q_2 - q_3) + 2q_4 - q_5 - q_6]/\sqrt{12}$, $\omega_{31}^2 = 3k/2m$

E_1

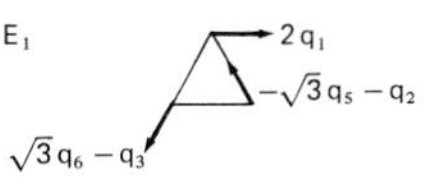

$Q_4 = [2q_1 - q_2 - q_3 - \sqrt{3}(q_5 - q_6)]/\sqrt{12}$, $\omega_{31}^2 \mp 3k/2m$

E_2

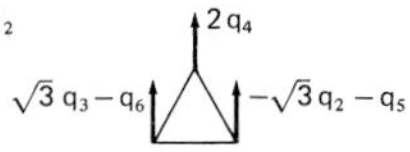

$Q_5 = [-\sqrt{3}(q_2 - q_3) + 2q_4 - q_5 - q_6]/\sqrt{12}$, $\omega_{32}^2 = 0$

E_2

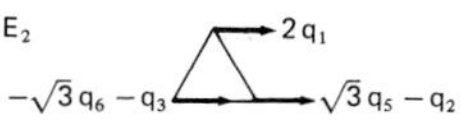

$Q_6 = [2q_1 - q_2 - q_3 - \sqrt{3}(q_5 - q_6)]/\sqrt{12}$, $\omega_{32}^2 = 0$

FIG. 27.13 The normal modes of an A_3 molecule (27.3-41), (27.3-42), (27.3-45), and (27.3-46).

In summary, in order to find the number of vibrational modes and their symmetry type we construct the reducible representation generated by the Cartesian displacements. When this representation is reduced, the modes belonging to translations and rotations may be identified with the help of a character table. The remaining modes are true vibrational modes. The process of finding the characters of the reducible representation is simplified by recognizing that none of the displacement vectors which are shifted to different atoms by a symmetry operation makes any contribution to the character of the matrix corresponding to this operation.

The H_2O molecule provides another simple example of these ideas. The symmetry group is C_{2v} and the representation of the displacement vectors Γ is nine-dimensional. If the molecule lies in the xz plane, the characters are

$$\chi(E) = 9, \qquad \chi(C_2) = -1, \qquad \chi(\sigma^{xz}) = 3, \qquad \chi(\sigma^{yz}) = 1. \tag{27.3-47}$$

The reduction of Γ gives

$$\Gamma = 3A_1 + A_2 + 3B_1 + 2B_2. \tag{27.3-48}$$

The translational modes belong to A_1, B_1, and B_2, and the rotational modes to A_2, B_1, and B_2. Therefore the vibrational modes are

$$\Gamma_v = 2A_1 + B_1 \tag{27.3-49}$$

indicating that there are three nondegenerate vibratory modes, as shown in Fig. 27.14.

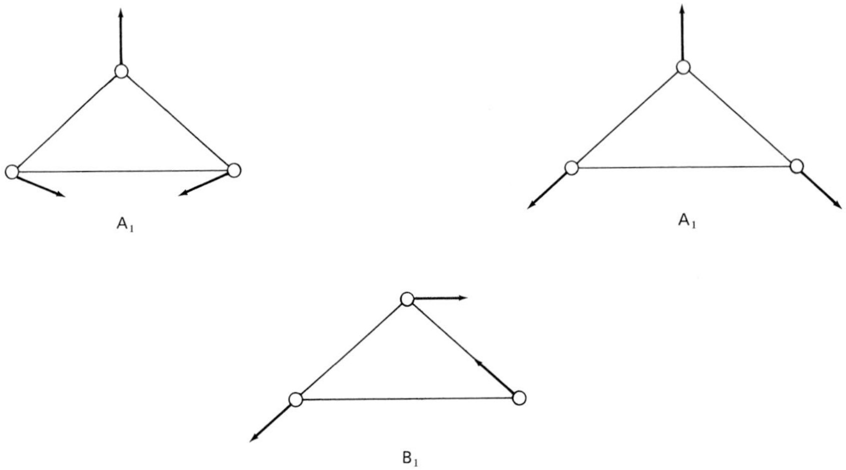

FIG. 27.14 The normal modes of H_2O.

Finally it should be mentioned that the vibration of a molecule may also be described in terms of internal coordinates, i.e., interatomic distances and bond angles. Physical interpretations are often simpler in terms of such coordinates.

27.4 Transitions in Polyatomic Molecules

Selection rules for transitions in polyatomic molecules are derived from the general matrix element theorem of Section 3.7. We shall illustrate the procedure for electronic transitions in benzene. It was previously shown (Section 26.6) that the excitation of an electron from the topmost filled molecular orbital to the lowest vacant orbital resulted in molecular states of symmetry B_{1u}, B_{2u}, and E_{1u}. For dipole transitions from the ground state (A_{1g}) the pertinent matrix elements are $\langle B_{1u}|\mathbf{r}|A_{1g}\rangle$, $\langle B_{2u}|\mathbf{r}|A_{1g}\rangle$, and $\langle E_{1u}|\mathbf{r}|A_{1g}\rangle$ and from the character table for D_{6h} (Table 5.2) which is the symmetry group of benzene, it is ascertained that z belongs to A_{2u} and (x, y) belong to E_{1u}. Forming all the direct products of A_{2u} and E_{1u} with B_{1u}, B_{2u}, and E_{1u} and decomposing them into the irreducible representations of D_{6h} we have

$$A_{2u} \times B_{1u} = B_{2g}, \qquad A_{2u} \times B_{2u} = B_{1g}, \qquad A_{2u} \times E_{1u} = E_{1g},$$
$$E_{1u} \times B_{1u} = E_{2g}, \qquad E_{1u} \times B_{2u} = E_{2g}, \qquad E_{1u} \times E_{1u} = A_{1g} + A_{2g} + E_{2g}.$$

Since A_{1g} is contained in the product $E_{1u} \times E_{1u}$, an electric dipole transition is possible between the ground state and the excited state with E_{1u} symmetry. Also, only the x and y components of $\mathbf{r}$ which, as previously seen, transform according to E_{1u}, give rise to nonvanishing matrix elements. This means that the electric vector of the radiation must lie in the xy plane (perpendicular to the principal axis of the molecule). Transitions between the ground state and B_{1u} or B_{2u} are forbidden (for all polarizations).

To give another example, in molecules with O_h symmetry, $\mathbf{r}$ belongs to T_{1u} and the allowed electric dipole transitions turn out to be

$$\begin{aligned} &A_{1g} \to T_{1u}, \\ &A_{2g} \to T_{2u}, \\ &E_g \to T_{1u}, \quad T_{2u}, \\ &T_{1g} \to A_{1u}, \quad E_u, \quad T_{1u}, \quad T_{2u}, \\ &T_{2g} \to A_{2u}, \quad E_u, \quad T_{1u}, \quad T_{2u}. \end{aligned}$$

Turning now to vibrational transitions it has been shown that the vibrational motion of a complex molecule can be decomposed into a superposition of normal modes. The energy of each mode, whose vibration is that of a simple harmonic oscillator, is

$$E_i = (v_i + \tfrac{1}{2})\hbar\omega_i \tag{27.4-1}$$

where v_i is the vibrational quantum number associated with the mode i. The total vibrational energy is

$$E = \sum_i E_i = \sum_i \hbar\omega_i(v_i + \tfrac{1}{2}) \tag{27.4-2}$$

and the vibrational wave function of the molecule is a product of $3N - 6$ functions

$$\Psi = \prod_i N_{v_i} H_{v_i}(\alpha_i Q_i) \exp(-\alpha_i^2 Q_i^2/2) \tag{27.4-3}$$

where H_{v_i} are Hermite polynomials, $\alpha_i^2 = \omega_i/\hbar$ (αQ is dimensionless), and N_{vi} are normalizing factors $(= \sqrt{\alpha_i/\sqrt{\pi}2^{v_i}(v_i)!}$

We have seen previously that Q_i^2 is invariant under all operations of the symmetry group of the molecule. Since $H_0(x)$ is constant, the wave functions for normal vibrations in the ground state ($v_i = 0$) belong to the totally symmetric (unit) representation of the point group of the molecule. For excited vibrational states, the wave functions have the symmetry of the corresponding Hermite polynomial.

Consider a molecule with k normal modes of vibration. At any time each of these modes will be in a certain quantum state. The total molecular vibrational state, according to (27.4-3), is a product

$$\Psi = \psi_1(v_1)\psi_2(v_2)\cdots\psi_k(v_k) \tag{27.4-4}$$

where $\psi_i(v_i)$ is the wave function of the ith mode in the v_ith state. When each of the v_i is zero, the molecule is in its vibrational ground state. If it absorbs radiation so that the ith normal mode is excited to the state with $v_i = 1$ while the remaining $k - 1$ normal modes remain in their lowest ($v = 0$) states, the molecule is said to have undergone a *fundamental* transition in the ith normal mode. The k different transitions of this kind are called fundamentals. Symbolically, these may be represented by

$$\prod_i \psi_j(0) \rightarrow \psi_i(1) \prod_{j \neq i} \psi_j(0) \tag{27.4-5}$$

or, in abbreviated form,

$$\Psi^{(0)} \rightarrow \Psi^{(i)}. \tag{27.4-6}$$

To obtain the selection rules we employ the matrix element theorem once more. For infrared absorption the operator is $\mathbf{r}$. Since the ground state wave function $\Psi^{(0)}$ belongs to the totally symmetric representation, the requirement for a nonvanishing transition matrix element will be that the product representation to which $\mathbf{r}\Psi^{(i)}$ belongs must contain the totally symmetric representation. Therefore the coordinate (x, y, or z) and $\Psi^{(i)}$ must belong to the same representation. In the first excited vibrational state the wave function is proportional to $H_1(\alpha Q) = \alpha Q$. It therefore follows that a fundamental will be infrared-active (i.e., will absorb infrared radiation) if the excited normal mode belongs to the same irreducible representation of the molecular symmetry group as any one or several of the Cartesian coordinates.

As an illustration of the foregoing we consider the H_2O molecule (Fig. 27.14) whose vibrational modes are given by (27.3-49). Since z belongs to A_1 and x belongs to B_1 infrared absorption is possible with polarization vectors along z and x. However, y belongs to B_2; but B_2 is not one of the vibratory modes. Hence radiation with y polarization (perpendicular to the plane of the molecule) will not be absorbed.

Interaction between the vibrational and electronic motions may result in departures from the selection rules described above. To see how this comes about, consider the electronic Hamiltonian $\mathscr{H}_e$, defined as the Hamiltonian (25.1-1) without the nuclear kinetic energy term, i.e., with the nuclei clamped in fixed positions. Because $\mathscr{H}_e$ depends on the nuclear coordinates through $V(Q_\alpha, q_i)$ in (25.1-2), we may expand $\mathscr{H}_e$ about the equilibrium configuration which we label $Q_\alpha = 0$. To first order, then

$$\mathscr{H}_e = \mathscr{H}_0 + \left(\frac{\partial \mathscr{H}_e}{\partial Q_\alpha}\right)_0 Q_\alpha \tag{27.4-7}$$

where $\mathscr{H}_0$ and $(\partial \mathscr{H}_e/\partial Q_\alpha)_0$ refer to $Q_\alpha = 0$. Writing

$$\mathscr{H}' = \left(\frac{\partial \mathscr{H}_e}{\partial Q_\alpha}\right)_0 Q_\alpha \tag{27.4-8}$$

we shall treat $\mathscr{H}'$ as a perturbation.

If $\psi_k^{(0)}$ is an eigenfunction of $\mathscr{H}_0$ with eigenvalue $E_k^{(0)}$ and $\psi_k^{(1)}$ is the same eigenfunction corrected to lowest order in $\mathscr{H}'$, we have

$$\psi_k^{(1)} = \psi_k^{(0)} + \sum_\alpha \sum_{j\neq k} Q_\alpha V_{kj} \psi_j^{(0)} \tag{27.4-9}$$

where

$$V_{kj} = \frac{\left\langle \psi_k^{(0)} \middle| \left(\frac{\partial \mathscr{H}_e}{\partial Q_\alpha}\right)_0 \middle| \psi_j^{(0)} \right\rangle}{E_j^{(0)} - E_k^{(0)}}. \tag{27.4-10}$$

To develop the symmetry argument it is recognized that $\mathscr{H}'$ must be invariant under the operations of the symmetry group of the molecule, that is, $\mathscr{H}'$ transforms according to the unit representation. Consequently Q_α and $(\partial \mathscr{H}_e/\partial Q_\alpha)_0$ must belong to the same irreducible representation.

Let us now return to the benzene molecule where it was found that the transition $A_{1g} \rightarrow E_{1u}$ was dipole allowed but $A_{1g} \rightarrow B_{2u}$ was forbidden, and let $\varphi_k^{(0)}$ and $\psi_j^{(0)}$ be zero order eigenfunctions belonging to B_{2u} and E_{1u} respectively. If there is a vibrational mode Q_α belonging to E_{2g}, V_{kj} will belong to the product representation $B_{2u} \times E_{2g} \times E_{1u}$ which is readily shown to contain the unit representation A_{1g}. Hence V_{kj} may have a nonvanishing value and we may concentrate on a single nonvanishing term in the sum contained in (27.4-9). Now consider electric dipole matrix elements of $\psi_k^{(1)}$ with the ground state ψ_0:

$$\begin{aligned}\langle\psi_k^{(1)}|\mathbf{r}|\psi_0\rangle &= \langle\psi_k^{(0)}|\mathbf{r}|\psi_0\rangle + Q_\alpha V_{kj}\langle\psi_j^{(0)}|\mathbf{r}|\psi_0\rangle \\ &= \langle B_{2u}|\mathbf{r}|A_{1g}\rangle + Q_\alpha V_{kj}\langle E_{1u}|\mathbf{r}|A_{1g}\rangle \qquad (27.4\text{-}11)\end{aligned}$$

The first matrix element on the right vanishes, as we have already seen, but the second one does not.

We see, then, that although the transition $\psi_0 \to \psi_k^{(0)}$ does not occur because $A_{1g} \to B_{2u}$ is dipole forbidden, the transition $\psi_0 \to \psi_k^{(1)}$ is possible because $\psi_k^{(1)}$ contains an admixture of an E_{1u} electronic state and $A_{1g} \to E_{1u}$ is dipole allowed. The admixture of E_{1u} into $\psi_k^{(1)}$ comes about through the intervention of the vibrational mode E_{2g}. To realize this situation it is sufficient for the electronic transition to be accompanied by a fundamental transition in an E_{2g} mode. Benzene has four inplane E_{2g} modes and a band near the ultraviolet has been interpreted by this mechanism.

Another mechanism whereby electromagnetic radiation interacts with a molecule is Raman scattering whose general formulation has been given in Section 24.2. We shall confine our attention to Raman scattering from molecular vibrational modes. Referring to Fig. 27.15, the two lowest vibrational levels of the ground electronic state are labeled $|\mathrm{i}\rangle$ and $|\mathrm{f}\rangle$ and correspond to $v = 0$ and $v = 1$ respectively; the states $|l\rangle$ are excited states of the molecule and the dashed horizontal lines serve merely as references for the photon energies but do not correspond to actual states of the molecule. An incident light photon provides the perturbation which mixes excited states such as $|l\rangle$ with the states $|\mathrm{i}\rangle$ and $|\mathrm{f}\rangle$. This is accomplished by means of nonvanishing matrix elements of the type $\langle\mathrm{f}|\mathbf{r}|l\rangle$ and $\langle l|\mathbf{r}|\mathrm{i}\rangle$ (assuming dipole transitions). The overall result is that $|\mathrm{i}\rangle$ and $|\mathrm{f}\rangle$ are coupled to the radiation field and it becomes possible for an amount of energy equal to a vibrational quantum to be subtracted or added to the light photon. The former case corresponds to the Stokes line and the latter to the anti-Stokes line. At ordinary temperatures most of the molecules are in the lowest state with $v = 0$ and only a small fraction of the molecules are in the state with $v = 1$. Therefore the intensity of the Stokes line corresponding to the transition $0 \to 1$ is much greater than that of the anti-Stokes line which corresponds to the transition $1 \to 0$.

It is important to recognize the intrinsic distinctions between *fluorescence* and Raman scattering. The former is a two-step process in which a photon is absorbed and after a measurable time delay, typically of the order of 10^{-8} sec, a photon is reemitted with an energy which is never larger (and

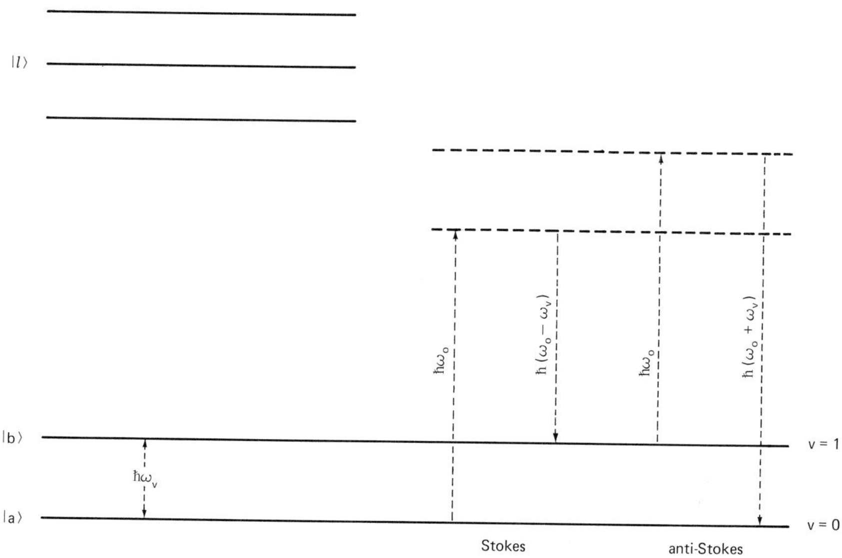

FIG. 27.15 Schematic diagram for vibrational Raman spectra. $|l\rangle$ are intermediate states.

usually smaller) than that of the absorbed photon. Fluorescence can only occur within an absorption band of the molecule, i.e., the energies of both the absorbed and emitted photons must correspond to energy differences between states of the molecule. Each of the two steps in fluorescence is a first order radiative process and is governed by its own matrix element. Raman scattering, on the other hand, is a single step, second order, radiative process. The energies of the incident and scattered photons bear no relation to energy differences in the molecule so that Raman scattering can take place at any energy of incident light and is not confined to any particular band. However the energy *differences* between incident and scattered photons are related to vibrational energies (or, more generally, to the energies of internal degrees of freedom). In this connection it may be remarked that although homonuclear diatomic molecules do not have pure vibration-rotation spectra, such states are amenable to Raman spectroscopy.

Selection rules for Raman scattering (in dipole approximation) may also be obtained from the general expression (24.2-13). Omitting the stimulated portion, the cross section may be written

$$d\sigma(d\Omega = r_0{}^2\omega\omega'^3 m^2\left|\mathbf{e}'\cdot\mathbf{R}\cdot\mathbf{e}\right|^2 \tag{27.4-12}$$

where $r_0 = e^2/mc^2$; m is the mass of the electron; ω, ω' are the frequencies of the incident and scattered photons and

$$\mathbf{R} = \sum_l \frac{\langle \mathrm{f}|\mathbf{r}|l\rangle\langle l|\mathbf{r}|\mathrm{i}\rangle}{E_\mathrm{i} - E_l + h\omega} + \frac{\langle l|\mathbf{r}|\mathrm{i}\rangle\langle \mathrm{f}|\mathbf{r}|l\rangle}{E_\mathrm{i} - E_l - h\omega} \tag{27.4-13}$$

Let us assume that the state $|i\rangle$ belongs to the totally symmetric representation of the symmetry group of the molecule. In that case $\langle l|\mathbf{r}|i\rangle$ will vanish unless $\Gamma(l) = \Gamma(\mathbf{r})$ where $\Gamma(l)$ and $\Gamma(\mathbf{r})$ are the irreducible representations to which the state $|l\rangle$ and the components of $\mathbf{r}$ belong. For $\langle f|\mathbf{r}|l\rangle$ not to vanish it will therefore be necessary for $\Gamma(f)$, the representation to which $|f\rangle$ belongs, to be contained in the product representation $\Gamma(\mathbf{r}) \times \Gamma(\mathbf{r})$. Since the two sets of basis functions, namely, the two vectors $\mathbf{r}$ are identical, basis functions for the antisymmetric product representation (Section 3.6) vanish identically. We conclude that $|f\rangle$ belongs to a representation contained in the reduction of the *symmetric* product representation $(\Gamma(\mathbf{r}) \times \Gamma(\mathbf{r}))^+$. Since the basis functions of $\Gamma(\mathbf{r}) \times \Gamma(\mathbf{r})$ consist of the products $x^2, y^2, z^2, xy, xz, yz$, linear combinations of these products will form bases for the irreducible representations contained in $(\Gamma(\mathbf{r}) \times \Gamma(\mathbf{r}))^+$. Such linear combinations are listed in the character tables.

Confining our attention to fundamental transitions it can now be stated that a transition will be Raman active if the excited normal mode belongs to the same representation as some linear combination of the bilinear products of the coordinates. We note further that such bilinear products have even parity. Thus, the different selection rules governing Raman scattering and absorption make the two techniques complementary to each other.

In H_2O we find from the character table of the group C_{2v} that x^2, y^2, z^2 transform according to A_1 and xz according to B_1. Previously it was shown in (27.3-49) that the vibrational modes in H_2O were $2A_1$ and B_1 and that these modes are infrared active. We now conclude that these modes are also Raman active, i.e., fundamental transitions in water are permitted either by direct absorption or by Raman scattering. Another example is provided by SF_6 which belongs to the group O_h and has $3N - 6 = 15$ vibrational modes; they are

$$A_{1g}, \quad E_g, \quad 2T_{1u}, \quad T_{2g}, \quad T_{2u}.$$

From the character table for O_h we find

$$\begin{aligned} &\text{Infrared-active modes:} && 2T_{1u}, \\ &\text{Raman-active modes:} && A_{1g}, \quad E_g, \quad T_{2g}, \\ &\text{Inactive modes:} && T_{2u}. \end{aligned}$$

This illustrates a general principle: In a centrosymmetric molecule in which parity is a good quantum number those transitions that are Raman active are forbidden in infrared absorption and those that are infrared active are Raman forbidden.

Let us now consider some of the processes whereby an excited molecule relinquishes its excitation energy and relaxes to the ground state. The various pathways may be followed with the aid of Fig. 27.16. We suppose

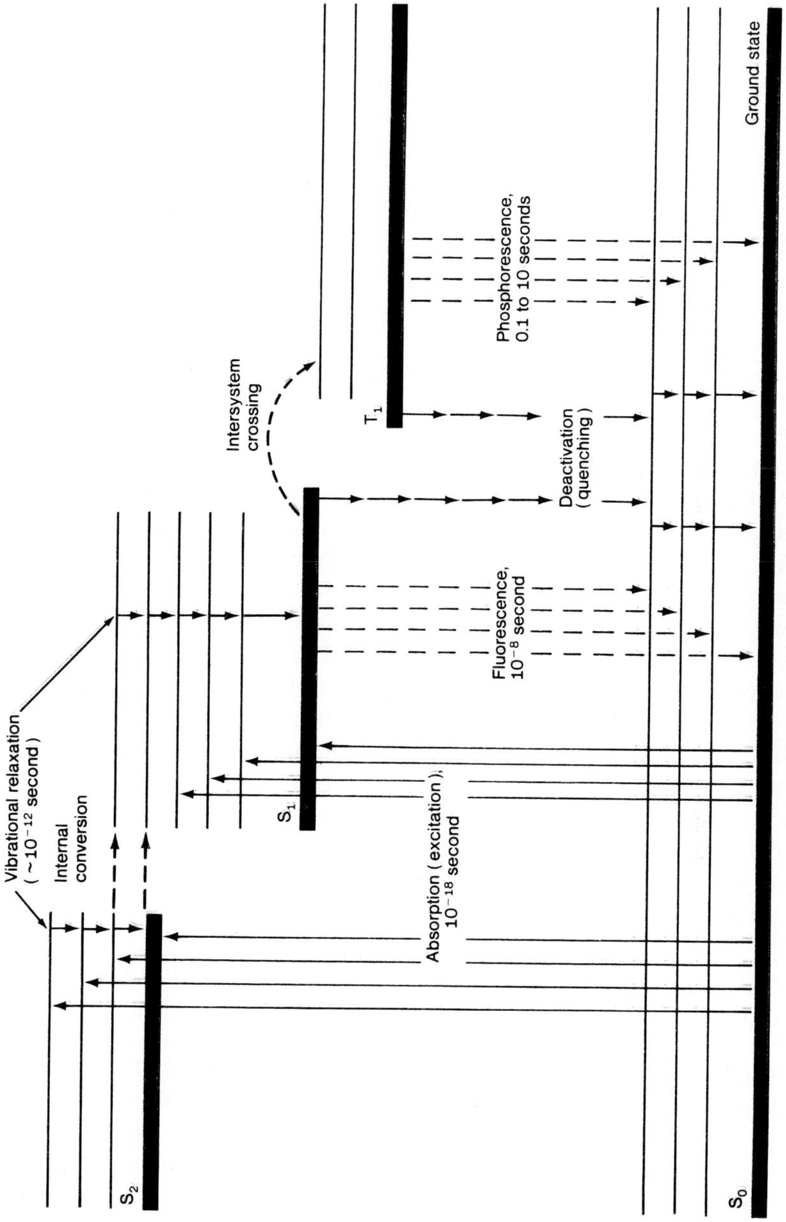

FIG. 27.16 Relaxation pathways for molecules.

that the molecule has been elevated (by absorption of a photon or by other means) to a high vibrational state belonging to an excited electronic singlet state such as S_2. In a time of the order of 10^{-12} seconds the molecule will drop to the lowest vibrational state by nonradiative processes that are mediated by molecular collisions. Quite often there is considerable overlap between the vibrational states of S_2 with the vibrational states belonging to a lower singlet electronic state such as S_1. This occurs when the potential energy surfaces of S_1 and S_2 intersect. In that case the molecule can make a nonradiative transition from S_2 to S_1. This is known as *internal conversion* which is defined as a nonradiative transition between electronic states of the same multiplicity. The rate constant for internal conversion is typically 10^{12} sec^{-1} which is several orders of magnitude greater than the rate constant for photon emission from S_2 to a lower singlet state. The molecule, having crossed over from S_2 to S_1, undergoes further vibrational relaxation and may continue by internal conversion to the ground state S_0 which is also assumed to be a singlet.

Suppose now that S_1 is the first excited state and is well separated from S_0 and that internal conversions, together with vibrational relaxations, have brought the molecule to the lowest vibrational state of S_1. Two pathways are now open. The first is a singlet–singlet transition from S_1 to S_0 with the emission of photons. This is fluorescence. The second pathway is one that can occur if there is a triplet state T_1 whose potential energy intersects that of S_1. In that case a nonradiative transition from S_1 to T_1 is possible with a rate constant comparable to that of fluorescence. Such a transition is known as *intersystem crossing*. Again vibrational relaxation brings the molecule to the lowest vibrational state of T_1 from which it may return to the ground state (S) with the emission of a photon. Photon emission in a triplet–singlet transition is known as *phosphorescence*.

Phosphorescence appears to violate the selection rule $\Delta S = 0$. The reason for the occurrence of phosphorescence is that the wave function of T_1 contains some admixture of singlet states, mainly from S_1 so that, approximately

$$\psi_{T_1} = \psi_{T_1}^0 + a\psi_{S_1}$$

where $\psi_{T_1}^0$ is the unperturbed triplet state and a is a mixing coefficient. The perturbation in question is spin-orbit coupling which has been shown to be capable of coupling singlet and triplet states. Thus, it is the singlet component of the triplet state T_1 which is responsible for the phosphorescence. Since the singlet component is usually small, phosphorescence life-times are usually long—of the order of 0.1 to 10 seconds.

Nonradiative processes, of necessity, depend on mechanisms whereby energy may be transferred from one molecule to another. For example, in a gas or liquid, the time an excited molecule retains its excitation energy depends on the time between collisions. Thus, for relatively small molecules at one atmosphere, collisions occur on the average of every 10^{-10} seconds. When the pressure is reduced, the time between collisions increases and at very low pressures, collisions with the walls become predominant. In a liquid, collisions are more frequent–about 10^{13} sec^{-1}. Most collisions will result in a transfer of rotational energy; the probability for a transfer of vibrational energy is smaller and the transfer of electronic energy occurs much less frequently.

CHAPTER 28

LIGAND FIELDS

28.1 Basic Ideas

There are numerous instances in which a particular atom (or ion) is surrounded by a symmetric distribution of neighboring atoms (or ions). One may cite, for example, the pink ruby crystal which contains small concentrations of the paramagnetic Cr^{3+} ion with each Cr^{3+} surrounded by six O^{2-} ions in an octahedral arrangement. Similar situations occur in molecular complexes such as $(FeF_6)^{3-}$ in which each iron atom is bound to six fluorides. In chemical terminology, ions, atoms, or molecules surrounding a central atom are known as *ligands*. In such circumstances the ligands exert a profound influence on the states of the central (or reference) atom and many new properties emerge which were not present in the free (unliganded) atom.

There are two basic models for the treatment of such systems. One is known as the *crystal field* (or ionic or tight binding) model in which the central atom is assumed to be surrounded by a distribution of charges which produce an electrostatic field at the position of the central atom. In this purely electrostatic model there is no sharing of electronic charge between the central atom and its ligands. The Hamiltonian then consists of $\mathscr{H}_F$, the free atom Hamiltonian, and an additional term $V(\mathbf{r})$ which is the perturbation produced by the surrounding charge distribution. Thus

$$\mathscr{H} = \mathscr{H}_F + V(\mathbf{r}) \tag{28.1-1}$$

where

$$\mathscr{H}_F = \sum_i \left(\frac{p_i^2}{2m} - \frac{Ze^2}{r_i}\right) + \sum_{i<j} \frac{e^2}{r_{ij}} + \sum_i \xi(r_i)\mathbf{l}_i \cdot \mathbf{s}_i. \tag{28.1-2}$$

The objective of crystal field theory is to determine the effect of the perturbation $V(\mathbf{r})$ on the eigenvalues and eigenfunctions of the free atom. With such information one may then ascertain in what manner various physical properties (e.g., optical, magnetic) are altered by the presence of the crystal field.

The second basic model treats the central atom and its ligands as if they were a single molecule. Electrons are permitted to overlap and their probability densities are not confined to specific atoms. In most cases this model is treated by the methods of molecular-orbital theory although valence bond calculations have also been employed.

Regardless of whether one uses the crystal field or the molecular-orbital method or some combination of the two, symmetry considerations are of paramount importance. It is the symmetry of the site in which the reference atom is located that governs the degeneracies of the states and the selection rules pertaining to various matrix elements. We shall now discuss the crystal field model which is usually easier to handle than the molecular model; the latter will be described in Section 28.5.

In practically all cases of interest, the presence of $V(\mathbf{r})$ in (28.1-1) lowers the site symmetry, that is, $V(\mathbf{r})$ is invariant under fewer coordinate transformations than $\mathscr{H}_F$. The effect of the lower symmetry environment produced by $V(\mathbf{r})$ on degenerate states is that the degeneracies are partially or completely lifted. Since $V(\mathbf{r})$ will be treated as a perturbation, it is important to assess its magnitude relative to other terms contained in $\mathscr{H}_F$ (28.1-2). Several possibilities arise:

$$\text{(a)} \quad V(\mathbf{r}) > \sum_{i<j} e^2/r_{ij} > \sum_i \xi(r_i)\mathbf{l}_i \cdot \mathbf{s}_i; \tag{28.1-3}$$

$$\text{(b)} \quad \sum_{i<j} e^2/r_{ij} > V(\mathbf{r}) > \sum_i \xi(r_i)\mathbf{l}_i \cdot \mathbf{s}_i; \tag{28.1-4}$$

$$\text{(c)} \quad \sum_{i<j} e^2/r_{ij} > \sum_i \xi(r_i)\mathbf{l}_i \cdot \mathbf{s}_i > V(\mathbf{r}). \tag{28.1-5}$$

Because crystal field calculations are carried only to a low order in perturbation theory, the strongest perturbation is applied first, followed by the next strongest, etc. In case (a) the effect of $V(\mathbf{r})$ is therefore calculated before or perhaps simultaneously with the imposition of the Coulomb interaction, followed by the spin-orbit interaction. In (b) the sequence is Coulomb interaction, crystal field, and spin-orbit interaction while in (c) crystal field effects are calculated last. Case (a), in which the perturbing action of $V(\mathbf{r})$ is sufficiently strong to produce splittings of the order of $10^4\,\text{cm}^{-1}$, is found most often in complexes containing paramagnetic ions with incomplete 4d and 5d shells. In elements belonging to the 3d group (first transition series) the ocurrence of case (b), with typical splittings of 10^2 to $10^3\,\text{cm}^{-1}$, is more common. In case (c) the splittings are $0.1-1\,\text{cm}^{-1}$ which are typical of the rare earth elements with incomplete 4f shells which are shielded from the influence of the crystal field by outer complete shells ($5s^2$ and $5p^6$). There are also some special situations in which case (c) is applicable as in some actinide compounds and in ferric compounds whose ground term is an S state.

Let us now illustrate how one calculates a crystal field potential. To introduce the formalism we consider a one-electron atom in which the single electron is in a d state. The nucleus is placed at the origin of a coordinate system (Fig. 28.1) and the electron is at the point (x, y, z). A charge q is assumed to be located at each of the positions $(\pm a, 0, 0), (0, \pm a, 0)$, and $(0, 0, \pm a)$. If $\mathbf{r}$ is the position vector of the electron and $\mathbf{a}_i$ the position vector of the ith charge, the potential energy of the electron in the field of the six charges is

$$V(\mathbf{r}) = -e\varphi(\mathbf{r}) \tag{28.1-6}$$

where

$$\varphi(\mathbf{r}) = q \sum_{i=1}^{6} 1/|\mathbf{r} - \mathbf{a}_i|. \tag{28.1-7}$$

The contribution to (28.1-7) from the charge at $(a, 0, 0)$ is

$$\frac{1}{\sqrt{(x-a)^2 + y^2 + z^2}} = \frac{1}{\sqrt{r^2 - 2ax + a^2}} = \frac{1}{a} \frac{1}{\sqrt{1 + [(r^2/a^2) - (2x/a)]}}. \tag{28.1-8}$$

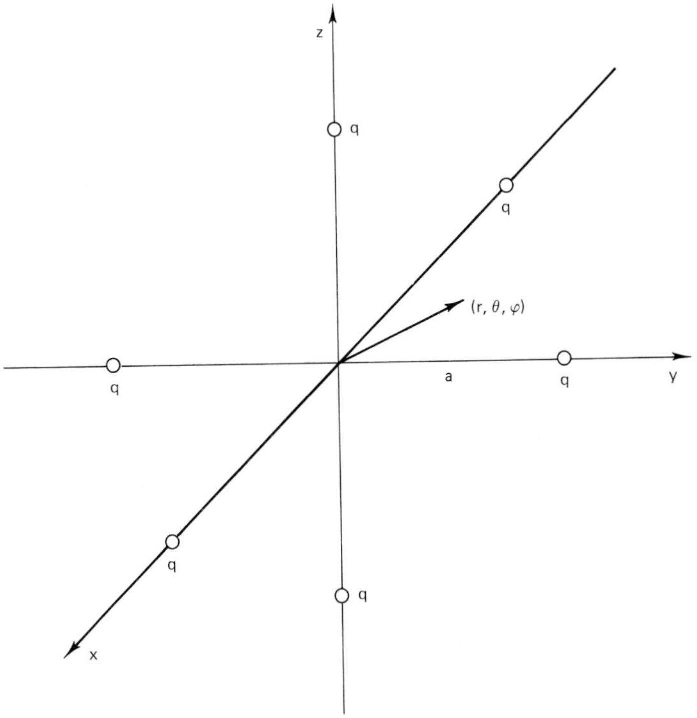

FIG. 28.1 Coordinate system for a cubic field.

Assuming $r \ll a$, the first nonvanishing term in the binomial expansion of $\varphi(\mathbf{r})$, beyond the constant term, is in fourth order; this gives

$$\varphi(\mathbf{r}) = q\left[\frac{6}{a} + \frac{35}{4a^5}\left(x^4 + y^4 + z^4 - \frac{3}{5}r^4\right)\right]. \tag{28.1-9}$$

A more general approach, which does not specifically depend on the point-charge model and which yields an expression more suited to the evaluation of matrix elements, begins with the Laplace equation $\nabla^2\varphi = 0$. Solutions which include $r = 0$ have the form

$$\varphi(\mathbf{r}) = \sum_{L=0}^{\infty}\sum_{M=-L}^{L} A_{LM} r^L Y_{LM}(\theta, \varphi).$$

Since the crystal field is regarded as a perturbation on the free atom, it is necessary to evaluate matrix elements of $\varphi(\mathbf{r})$ with respect to the d orbitals of the free atom. This may be accomplished by means of theorem (1.2-29):

$$\langle Y_{l'm'}|Y_{LM}|Y_{lm}\rangle = (-1)^{m'}\sqrt{\frac{(2l'+1)(2L+1)(2l+1)}{4\pi}} \times \begin{pmatrix} l' & L & l \\ -m' & M & m \end{pmatrix}\begin{pmatrix} l' & L & l \\ 0 & 0 & 0 \end{pmatrix}. \tag{28.1-10}$$

For d orbitals $l = l' = 2$. The right-hand side of (28.1-10) vanishes unless l', l, and L satisfy the triangle conditions and $l' + L + l$ is even. This limits the values of L to $0, 2, 4$.

The same result may be obtained in another way. The integral in

$$\langle Y_{2m'}|Y_{LM}|Y_{2m}\rangle$$

belongs to the product representation $D^{(2)} \times D^{(L)} \times D^{(2)}$. Unless this product representation contains the totally symmetric representation $D^{(0)}$, the matrix element will vanish. Since

$$D^{(2)} \times D^{(2)} = D^{(0)} + D^{(1)} + D^{(2)} + D^{(3)} + D^{(4)}, \tag{28.1-11}$$

the only possible values of L are $0, 1, 2, 3, 4$. Furthermore the integral will vanish unless the integrand has even parity. Hence L is limited to $0, 2, 4$.

Y_{LM} with $L = 0$ is a constant. It will have no effect on the free atom states other than to shift them, and since the reference energy may be chosen arbitrarily, this term may be eliminated. Thus $L = 2, 4$. At this stage

$$\varphi(\mathbf{r}) = \sum_{M=-2}^{2} A_{2M} Y_{2M} r^2 + \sum_{M=-4}^{4} A_{4M} Y_{4M} r^4. \tag{28.1-12}$$

To proceed further we note that the arrangement of charges shown in Fig. 28.1 has the symmetry of the group O. Specifically this means that $\varphi(\mathbf{r})$

is an invariant under all coordinate transformations belonging to the group O. This is equivalent to saying that $\varphi(\mathbf{r})$ transforms according to the A_1 irreducible representation of O. Now Y_{2M} belongs to $D^{(2)}$ of the three-dimensional rotation group $O^+(3)$. With respect to the latter $D^{(2)}$ is irreducible. But with respect to O, which is a subgroup of $O^+(3)$, $D^{(2)}$ is reducible. From Table 5.7,

$$D^{(2)} = E + T_2. \tag{28.1-13}$$

Since $D^{(2)}$ does not contain A_1, it is concluded that Y_{2M} cannot be part of $\varphi(\mathbf{r})$. This leaves Y_{4M} which transforms according to $D^{(4)}$; the reduction with respect to O gives

$$D^{(4)} = A_1 + E + T_1 + T_2. \tag{28.1-14}$$

Since A_1 appears in the decomposition of $D^{(4)}$, $\varphi(\mathbf{r})$ may be expressed in terms of Y_{4M}. We now have

$$\varphi(\mathbf{r}) = \sum_{M=-4}^{4} A_{4M} Y_{4M} r^4. \tag{28.1-15}$$

Further simplification of $\varphi(\mathbf{r})$ results from the fact that $\varphi(\mathbf{r})$ is proportional to those linear combinations of Y_{4M} which constitute a basis set for A_1. From Table 5.8 we find

$$\varphi(\mathbf{r}) = Cr^4[Y_{40} + \sqrt{5/14}(Y_{44} + Y_{4\,-4})] \tag{28.1-16}$$

$$= D(x^4 + y^4 + z^4 - \tfrac{3}{5}r^4) \tag{28.1-17}$$

with

$$D = (15/4\sqrt{\pi})C. \tag{28.1-18}$$

For a six-coordinated point-charge model, the value of D may be obtained by comparing (28.1-17) with (28.1-9) bearing in mind that in deriving (28.1-17) the constant term has been set equal to zero. We then have, for d *electrons*,

$$D = 35q/4a^5. \tag{28.1-19}$$

Upon setting $q = -Ze$ and writing $V_c(\mathbf{r})$ for $V(\mathbf{r})$ to emphasize the cubic symmetry

$$V_c(\mathbf{r}) = \frac{7\sqrt{\pi}}{3}\frac{Ze^2}{a^5} r^4 \left[Y_{40} + \sqrt{\frac{5}{14}}(Y_{44} + Y_{4\,-4}) \right] \tag{28.1-20}$$

$$= \frac{35Ze^2}{4a^5}\left(x^4 + y^4 + z^4 - \frac{3}{5}r^4\right). \tag{28.1-21}$$

One might suppose that there is a discrepancy between (28.1-20) and (28.1-21) since the former is exact while the latter was obtained from a binomial expansion carried, arbitrarily, to fourth order. However, the group theoretical development indicates that there are no nonvanishing matrix elements between d orbitals for terms in $V_c(\mathbf{r})$ higher than the fourth order. It should be kept in mind that the form of the crystal field potential depends on the orientation of the coordinate axes. In (28.1-20) or (28.1-21) the z axis coincided with a fourfold symmetry axis. Were this not the case, the form of the potential would be altered. For example, if the z axis lies along a body diagonal of the cube

$$V_c = A_c \frac{1}{\sqrt{27}} [\sqrt{10}(Y_{43} - Y_{4\,-3}) + \sqrt{7} Y_{40}] \tag{28.1-22}$$

where A_c is a constant.

Some other crystal fields which may be derived by application of the same methods are:

Tetragonal (z axis along the four-fold symmetry axis)

$$V_{\mathrm{TR}} = A(x^2 + y^2 - 2z^2) + B(x^4 + y^4) + Cz^4. \tag{28.1-23}$$

Tetrahedral (coordinate axes parallel to the axes of the enveloping cube)

$$V_{\mathrm{TH}} = A(x^4 + y^4 + z^4) + B(x^2y^2 + x^2z^2 + y^2z^2). \tag{28.1-24}$$

The potentials that have been discussed so far are appropriate for d electron systems. When f electrons are present, as in the rare earths, the potentials contain contributions from Y_{6m}.

In the application of crystal field theory the neighbors surrounding the reference atom may not be arranged in a perfectly symmetrical fashion. One may then use a dominant symmetry and subsequently add lower symmetry terms to take into account the departure from the dominant symmetry. For example, if a system has only a small tetragonal distortion, one may calculate the effect due to cubic symmetry and, at a later stage, apply corrections due to the tetragonal distortion, regarded as perturbations on the cubic terms.

28.2 Single d Electron in an Octahedral and Tetragonal Field

When a d electron is subjected to a field of type (28.1-20), the first-order corrections to the energy will be obtained by diagonalizing the matrix of $V_c(\mathbf{r})$ within the fivefold degenerate set of spatial d orbitals. The matrix is shown in Table 28.1. Since the eigenvalues of the 2×2 matrix are $\frac{3}{5}$ and $-\frac{2}{5}$, it is seen that the fivefold degenerate d level has been split into two levels with degeneracies of 2 and 3 and with energies (Fig. 28.2):

$$E_1 = E_2 = \tfrac{3}{5}\Delta, \qquad E_3 = E_4 = E_5 = -\tfrac{2}{5}\Delta. \tag{28.2-1}$$

TABLE 28.1

Matrix Elements of the Cubic Crystal Field Potential V_c in Units of Δ[a]

m' \ m	0	1	−1	2	−2
0	$\frac{3}{5}$				
1		$-\frac{2}{5}$			
−1			$-\frac{2}{5}$		
2				$\frac{1}{10}$	$\frac{1}{2}$
−2				$\frac{1}{2}$	$\frac{1}{10}$

[a] For a point-charge model

$$\Delta = \frac{5}{3}\frac{Ze^2}{a^5}\langle r^4\rangle.$$

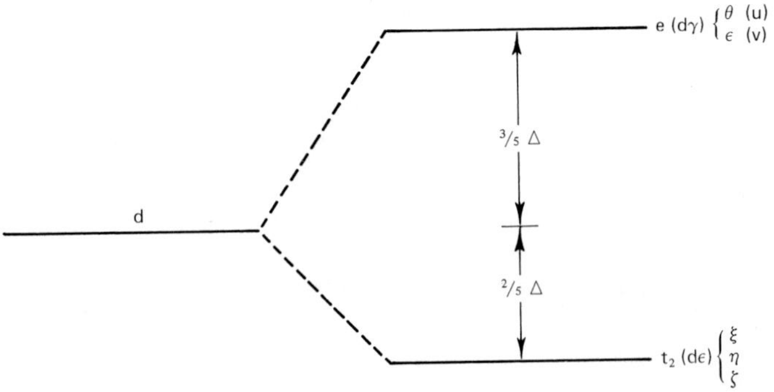

FIG. 28.2 Splitting of a d orbital in a cubic field. θ, ε, ξ, η, ζ are proportional to d_{z^2}, $d_{x^2-y^2}$, d_{yz}, d_{zx}, and d_{xy}, respectively (see also Table 5.8).

Δ is the separation between the two levels; on the point-charge model

$$\Delta = \frac{5}{3}\frac{Ze^2}{a^5}\langle r^4\rangle \tag{28.2-2}$$

where $\langle r^4\rangle$ is the expectation value of r^4 with respect to the radial part of the d orbital. However, the point-charge model should not be taken too literally so that it is the more common practice to obtain Δ experimentally.

Once the eigenvalues are known, it is a straightforward calculation to obtain the eigenfunctions. However, it is instructive to observe that the two levels can be labeled according to the irreducible representations of O. Since d orbitals are basis functions for the $D^{(2)}$ representation of $O^{+(3)}$ and the reduction of $D^{(2)}$ into irreducible representations of O is given by (28.1-13), the two- and threefold degenerate levels can be labeled e and t_2, respectively.† The wave functions associated with the two levels are just the basis functions for e and t_2 constructed as linear combinations of the unperturbed d orbitals. Such combinations are listed in Table 5.8 and shown in Fig. 28.2.

So far no reference has been made to spin properties. There are no spin operators in the crystal field potential. Consequently when spin is included, the fivefold degenerate d state becomes the tenfold degenerate ^2D. Similarly the twofold degenerate e state becomes the fourfold degenerate ^2E and the threefold t_2 state becomes a sixfold $^2T_2$.

When spin-orbit coupling is taken into account we saw in the last section that it is necessary to distinguish between a crystal field that is weak relative to the spin-orbit interaction and one that is strong. For the weak case, the spin-orbit interaction is applied first and has the effect of splitting the ^2D term into $^2D_{3/2}$ and $^2D_{5/2}$, the first having a fourfold and the second a sixfold degeneracy. Upon application of the crystal field it is expected that further splitting may occur. From Table 5.7 it is found that

$$D^{(3/2)} = \Gamma^{(8)}, \tag{28.2-3}$$

$$D^{(5/2)} = \Gamma^{(7)} + \Gamma^{(8)}. \tag{28.2-4}$$

Thus $^2D_{3/2}$ remains unsplit by the crystal field but the designation of the state is changed to conform to the irreducible representations of O. The $^2D_{5/2}$ splits into a twofold and fourfold degenerate state (Fig. 28.3). We note that the order of the levels (on an energy scale) is not given by group theory alone but must be calculated separately.

When the crystal field is strong relative to the spin-orbit interaction it is necessary to reverse the order of the calculations in the preceding paragraph; that is, the crystal field is applied first followed by the spin-orbit interaction. This means that the spin and space parts of the wave function are to be coupled within the framework of the group O. The spatial parts of the wave function, we saw, belong to $E(\Gamma^{(3)})$ and $T_2(\Gamma^{(5)})$; for the spin part Table 5.7 gives

$$D^{(1/2)} = \Gamma^{(6)}. \tag{28.2-5}$$

To couple the space and spin parts of the wave function within the group O, one forms the product representations $\Gamma^{(3)} \times \Gamma^{(6)}$ and $\Gamma^{(5)} \times \Gamma^{(6)}$ which, on

† We use lower case letters to label one-electron states.

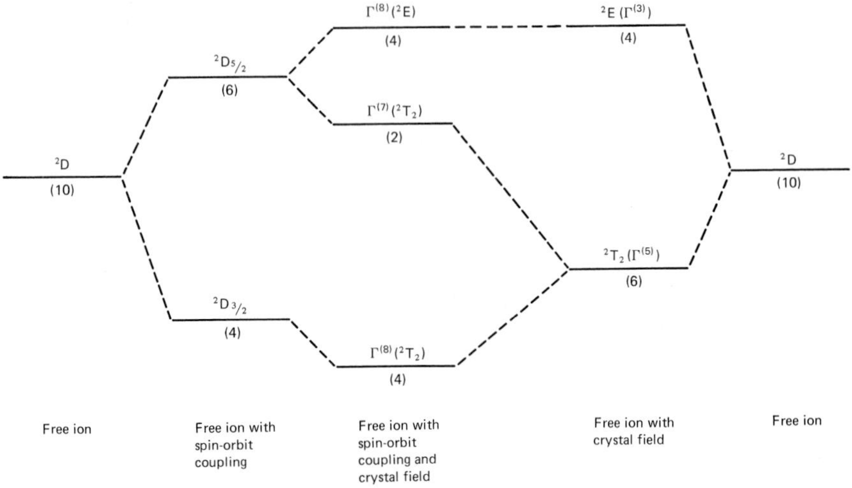

FIG. 28.3 Splitting of a d orbital in a cubic field with spin-orbit coupling.

reduction, give

$$\Gamma^{(3)} \times \Gamma^{(6)} = \Gamma^{(8)}, \qquad \Gamma^{(5)} \times \Gamma^{(6)} = \Gamma^{(7)} + \Gamma^{(8)}. \tag{28.2-6}$$

We see, then, that insofar as the symmetries of the states that are produced by the combined action of the crystal field and the spin-orbit interaction are concerned, it makes no difference in which order the two perturbations are applied. However, the first-order energies are sensitive to the magnitudes of the interactions and hence to the order in which they are applied.

The wave functions in the coupled representation may be written with the help of the tables of coupling coefficients (Table 5.9). For $\Gamma^{(8)}$ arising from $\Gamma^{(3)} \times \Gamma^{(6)}$ we obtain

$$\Gamma^{(8)}(^2\mathrm{E})\begin{cases} \psi^8_{-3/2} = v\alpha, \\ \psi^8_{-1/2} = u\beta, \\ \psi^8_{1/2} = -u\alpha, \\ \psi^8_{3/2} = -v\beta. \end{cases} \tag{28.2-7}$$

The basis functions for $\Gamma^{(7)}$ from the decomposition $\Gamma^{(5)} \times \Gamma^{(6)}$ are

$$\Gamma^{(7)}(^2\mathrm{T}_2)\begin{cases} \psi^7_{-1/2} = -\dfrac{i}{\sqrt{3}}\,\xi\alpha - \dfrac{1}{\sqrt{3}}\,\eta\alpha + \dfrac{i}{\sqrt{3}}\,\zeta\beta, \\[2ex] \psi^7_{1/2} = -\dfrac{i}{\sqrt{3}}\,\xi\beta + \dfrac{1}{\sqrt{3}}\,\eta\beta - \dfrac{i}{\sqrt{3}}\,\zeta\alpha, \end{cases} \tag{28.2-8}$$

and for $\Gamma^{(8)}$ from $\Gamma^{(5)} \times \Gamma^{(6)}$ we have

$$\Gamma^{(8)}(^2\mathrm{T}_2)\begin{cases}\psi^8_{-3/2} = -\dfrac{i}{\sqrt{6}}\,\xi\beta + \dfrac{1}{\sqrt{6}}\,\eta\beta + i\sqrt{\dfrac{2}{3}}\,\zeta\alpha, \\[2ex] \psi^8_{-1/2} = \dfrac{i}{\sqrt{2}}\,\xi\alpha - \dfrac{1}{\sqrt{2}}\,\eta\alpha, \\[2ex] \psi^8_{1/2} = -\dfrac{i}{\sqrt{2}}\,\xi\beta - \dfrac{1}{\sqrt{2}}\,\eta\beta, \\[2ex] \psi^8_{3/2} = \dfrac{i}{\sqrt{6}}\,\xi\alpha + \dfrac{1}{\sqrt{6}}\,\eta\alpha + i\sqrt{\dfrac{2}{3}}\,\zeta\beta.\end{cases} \qquad (28.2\text{-}9)$$

The matrix of $\mathbf{L}\cdot\mathbf{S}$ may be computed in the (e, t_2) basis set as shown in Table 28.2 and subsequently converted into the basis set consisting of the functions (28.2-7), (28.2-8), and (28.2-9) (Table 28.3). We are now in position to compute the various splittings. In the absence of the crystal field, the matrix elements of the spin-orbit interaction are

$$\begin{aligned} E_1 &= \langle {}^2\mathrm{D}_{3/2}|\xi(r)\mathbf{L}\cdot\mathbf{S}|{}^2\mathrm{D}_{3/2}\rangle \\ &= \langle {}^2\mathrm{D}_{3/2}|\tfrac{1}{2}\xi(r)(J^2 - L^2 - S^2)|{}^2\mathrm{D}_{3/2}\rangle \\ &= \tfrac{1}{2}\langle\xi(r)\rangle\left[\tfrac{3}{2}(\tfrac{3}{2}+1) - 2(2+1) - \tfrac{1}{2}(\tfrac{1}{2}+1)\right] = -\tfrac{3}{2}\langle\xi(r)\rangle \end{aligned} \qquad (28.2\text{-}10)$$

in which $\langle\xi(r)\rangle$ is the expectation value of $\xi(r)$ as computed from the radial part of the d function. Similarly

$$E_2 = \langle {}^2\mathrm{D}_{5/2}|\xi(r)\mathbf{L}\cdot\mathbf{S}|{}^2\mathrm{D}_{5/2}\rangle = \langle\xi(r)\rangle. \qquad (28.2\text{-}11)$$

When a crystal field is present, all the necessary information is contained in Table 28.3. If, at first, the off-diagonal terms are ignored—since their contribution is of higher order—it is seen that

$$\begin{aligned} \langle\Gamma^{(8)}(^2\mathrm{E})|\mathbf{L}\cdot\mathbf{S}|\Gamma^{(8)}(^2\mathrm{E})\rangle &= 0, \\ \langle\Gamma^{(7)}(^2\mathrm{T}_2)|\mathbf{L}\cdot\mathbf{S}|\Gamma^{(7)}(^2\mathrm{T}_2)\rangle &= 1, \\ \langle\Gamma^{(8)}(^2\mathrm{T}_2)|\mathbf{L}\cdot\mathbf{S}|\Gamma^{(8)}(^2\mathrm{T}_2)\rangle &= -\tfrac{1}{2}. \end{aligned} \qquad (28.2\text{-}12)$$

The spin-orbit energies are obtained by multiplying these matrix elements by $\langle\xi(r)\rangle$. We now have, for the energies, measured with respect to the unsplit d level,

$$\begin{aligned} &\Gamma^{(8)}(^2\mathrm{E}): && \tfrac{3}{5}\Delta, \\ &\Gamma^{(7)}(^2\mathrm{T}_2): && -\tfrac{2}{5}\Delta + \langle\xi(r)\rangle, \\ &\Gamma^{(8)}(^2\mathrm{T}_2): && -\tfrac{2}{5}\Delta - \tfrac{1}{2}\langle\xi(r)\rangle. \end{aligned} \qquad (28.2\text{-}13)$$

TABLE 28.2

Matrix Elements of $\mathbf{L}\cdot\mathbf{S}$ *in a Cubic Field Using an Uncoupled Representation*

$\mathbf{L}\cdot\mathbf{S}$	$\zeta\alpha$	$\zeta\beta$	$\xi\alpha$	$\xi\beta$	$\eta\alpha$	$\eta\beta$	$u\alpha$	$u\beta$	$v\alpha$	$v\beta$
$\zeta\alpha$				$\frac{1}{2}$		$-\frac{i}{2}$			i	
$\zeta\beta$			$-\frac{1}{2}$		$-\frac{i}{2}$					$-i$
$\xi\alpha$		$-\frac{1}{2}$			$\frac{i}{2}$			$-\frac{\sqrt{3}}{2}i$		$-\frac{i}{2}$
$\xi\beta$	$\frac{1}{2}$					$-\frac{i}{2}$	$-\frac{\sqrt{3}}{2}i$		$-\frac{i}{2}$	
$\eta\alpha$		$\frac{i}{2}$	$-\frac{i}{2}$					$\frac{\sqrt{3}}{2}$		$-\frac{1}{2}$
$\eta\beta$	$\frac{i}{2}$			$\frac{i}{2}$			$-\frac{\sqrt{3}}{2}$		$\frac{1}{2}$	
$u\alpha$				$\frac{\sqrt{3}}{2}i$		$-\frac{\sqrt{3}}{2}$				
$u\beta$			$\frac{\sqrt{3}}{2}i$		$\frac{\sqrt{3}}{2}$					
$v\alpha$	$-i$			$\frac{i}{2}$		$\frac{1}{2}$				
$v\beta$		i	$\frac{i}{2}$		$-\frac{1}{2}$					

The presence of the off-diagonal elements is an indication of mixing between the two $\Gamma^{(8)}$ representations. To calculate the energy we may choose one of the four degenerate states in $\Gamma^{(8)}(^2\mathrm{E})$ and the corresponding one in $\Gamma^{(8)}(^2\mathrm{T}_2)$. The resulting matrix is shown in Table 28.4 and the resulting eigenvalues are

$$E = \begin{cases} \frac{3}{5}\Delta + \sqrt{\frac{3}{2}}\langle\xi(r)\rangle \cot\theta, \\ -\frac{2}{5}\Delta - \frac{1}{2}\langle\xi(r)\rangle - \sqrt{\frac{3}{2}}\langle\xi(r)\rangle \cot\theta, \end{cases} \tag{28.2-14}$$

where

$$\tan 2\theta = -\frac{\sqrt{6}\langle\xi(r)\rangle}{\Delta + \frac{1}{2}\langle\xi(r)\rangle}.$$

TABLE 28.3

Matrix Elements of **L · S** *in a Cubic Field Using a Coupled Representation*

		$\Gamma^{(8)}(^2E)$				$\Gamma^{(7)}(^2T_2)$		$\Gamma^{(8)}(^2T_2)$			
L · S		$\psi^8_{-3/2}$	$\psi^8_{-1/2}$	$\psi^8_{1/2}$	$\psi^8_{3/2}$	$\psi^7_{-1/2}$	$\psi^7_{1/2}$	$\psi^8_{-3/2}$	$\psi^8_{-1/2}$	$\psi^8_{1/2}$	$\psi^8_{3/2}$
$\Gamma^{(8)}(^2E)$	$\psi^8_{-3/2}$							$\sqrt{\frac{3}{2}}$			
	$\psi^8_{-1/2}$								$-\sqrt{\frac{3}{2}}$		
	$\psi^8_{1/2}$									$-\sqrt{\frac{3}{2}}$	
	$\psi^8_{3/2}$										$\sqrt{\frac{3}{2}}$
$\Gamma^{(7)}(^2T_2)$	$\psi^7_{-1/2}$					1					
	$\psi^7_{1/2}$						1				
$\Gamma^{(8)}(^2T_2)$	$\psi^8_{-3/2}$	$\sqrt{\frac{3}{2}}$						$-\frac{1}{2}$			
	$\psi^8_{-1/2}$		$-\sqrt{\frac{3}{2}}$						$-\frac{1}{2}$		
	$\psi^8_{1/2}$			$-\sqrt{\frac{3}{2}}$						$-\frac{1}{2}$	
	$\psi^8_{3/2}$				$\sqrt{\frac{3}{2}}$						$-\frac{1}{2}$

TABLE 28.4

Matrix Elements of **L · S** *Connecting* $\Gamma^{(8)}(^2E)$ *and* $\Gamma^{(8)}(^2T_2)$

L · S	$\psi^8(^2E)$	$\psi^8(^2T_2)$
$\psi^8(^2E)$	$\frac{3}{5}\Delta$	$\pm\sqrt{\frac{3}{2}}\langle\xi(r)\rangle$
$\psi^8(^2T_2)$	$\pm\sqrt{\frac{3}{2}}\langle\xi(r)\rangle$	$-\frac{2}{5}\Delta - \frac{1}{2}\langle\xi(r)\rangle$

Including the first order correction, the wave functions belonging to $\Gamma^{(8)}(^2T_2)$ are of the form

$$\psi^8(^2T_2)' = \psi^8(^2T_2) - \sqrt{\tfrac{3}{2}}\,\frac{\langle\xi\rangle}{\Delta}\,\psi^8(^2E).$$

The conclusions that may be drawn thus far on the effect of crystal fields may be stated as follows: When an atom or ion is placed in a crystal or surrounded by a symmetrical distribution of ligands, the symmetry group of the atom is then no longer the three-dimensional rotation group because all directions in space are no longer equivalent. Instead the symmetry is that of a point group which contains a finite number of elements and is a subgroup of the three-dimensional rotation group. An irreducible representation of the latter becomes reducible with respect to the lower symmetry point group. The effect on the states of the free atom is to remove degeneracies, either partially or completely.

Let us now reduce the symmetry of the environment still further. To be specific, we imagine the atom with its single d electron subjected to a crystal field produced by six equal charges placed at $(\pm a, 0, 0)$, $(0, \pm a, 0)$, and $(0, 0, \pm b)$. If $|a - b| \ll a$, the crystal field potential will consist of V_c (28.1-21) which is invariant under the operations of the group O and V_{TR} (28.1-23) which is invariant with respect to D_4. Since the tetragonal distortion is assumed to be small, $V_{TR} \ll V_c$. We may then use the previous results for the effects of V_c followed by subsequent corrections due to V_{TR}.

D_4 is a subgroup of O; hence irreducible representations of O are (generally) reducible representations of D_4 (Table 5.6). Thus

$$\begin{aligned} \Gamma^{(3)}(O) &= \Gamma^{(1)}(D_4) + \Gamma^{(3)}(D_4), \\ \Gamma^{(5)}(O) &= \Gamma^{(4)}(D_4) + \Gamma^{(5)}(D_4). \end{aligned} \tag{28.2-15}$$

These splittings are shown in Fig. 28.4. To account for spin-orbit interaction we note that $D^{(1/2)}$ which corresponds to $S = \frac{1}{2}$, goes over into $\Gamma^{(6)}$ in O and then to $\Gamma^{(6)}$ in D_4. Hence we simply need the decomposition of the

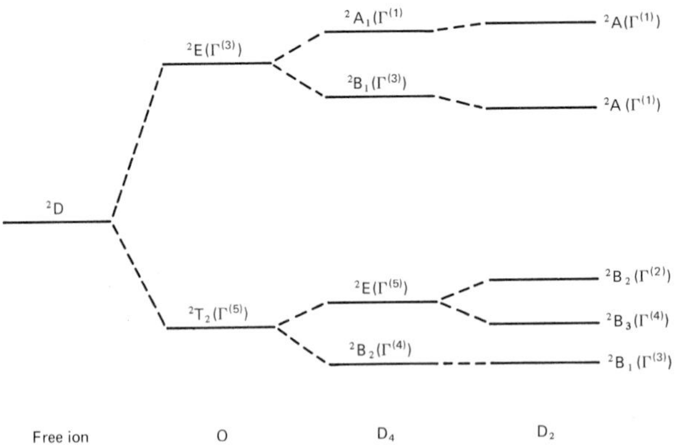

FIG. 28.4 Effect of reduction in symmetry $O \to D_4 \to D_2$ for a single d electron.

product representations of $\Gamma^{(6)}$ with $\Gamma^{(1)}$, $\Gamma^{(3)}$, $\Gamma^{(4)}$, and $\Gamma^{(5)}$ in D_4:

$$\begin{aligned} \Gamma^{(6)} \times \Gamma^{(1)} &= \Gamma^{(6)}, & \Gamma^{(6)} \times \Gamma^{(3)} &= \Gamma^{(7)}, \\ \Gamma^{(6)} \times \Gamma^{(4)} &= \Gamma^{(7)}, & \Gamma^{(6)} \times \Gamma^{(5)} &= \Gamma^{(6)} + \Gamma^{(7)}. \end{aligned} \tag{28.2-16}$$

If the spin-orbit interaction is applied before the crystal field, then $^2\mathrm{D}$ would initially be split into $^2\mathrm{D}_{3/2}$ and $^2\mathrm{D}_{5/2}$. Upon application of the crystal field first with symmetry O and then with symmetry D_4, we have

$$\begin{aligned} \mathrm{D}^{(3/2)} &= \Gamma^{(8)}(O) = \Gamma^{(6)}(D_4) + \Gamma^{(7)}(D_4), \\ \mathrm{D}^{(5/2)} &= \Gamma^{(7)}(O) \quad + \quad \Gamma^{(8)}(O) \\ & \qquad \| \qquad\qquad\qquad \| \\ & \quad \Gamma^{(7)}(D_4) \quad \Gamma^{(6)}(D_4) + \Gamma^{(7)}(D_4). \end{aligned} \tag{28.2-17}$$

The complete genealogy of the states is shown in Fig. 28.5. As in the previous case the crystal field may be dominant over the spin-orbit coupling, in which case it is applied first. This too is shown in Fig. 28.5.

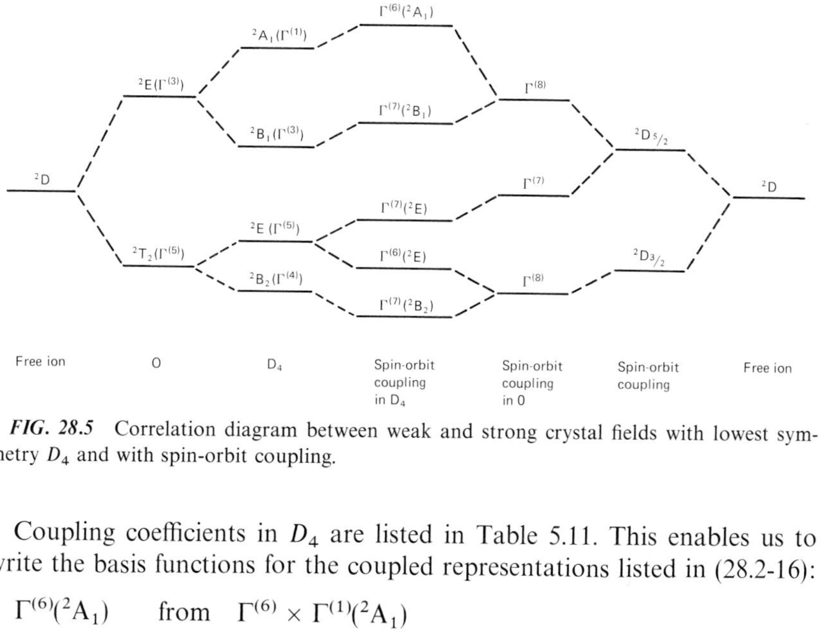

FIG. 28.5 Correlation diagram between weak and strong crystal fields with lowest symmetry D_4 and with spin-orbit coupling.

Coupling coefficients in D_4 are listed in Table 5.11. This enables us to write the basis functions for the coupled representations listed in (28.2-16):

$\Gamma^{(6)}(^2\mathrm{A}_1)$ from $\Gamma^{(6)} \times \Gamma^{(1)}(^2\mathrm{A}_1)$

$$\psi^6_{-1/2} = u\beta, \qquad \psi^6_{1/2} = u\alpha; \tag{28.2-18}$$

$\Gamma^{(7)}(^2\mathrm{B}_1)$ from $\Gamma^{(6)} \times \Gamma^{(3)}(^2\mathrm{B}_1)$

$$\psi^7_{-1/2} = v\beta, \qquad \psi^7_{1/2} = v\alpha; \tag{28.2-19}$$

$\Gamma^{(7)}(^2\mathrm{E})$ from $\Gamma^{(6)} \times \Gamma^{(5)}(^2\mathrm{E})$

$$\psi^7_{-1/2} = \frac{i}{\sqrt{2}}\,\xi\alpha + \frac{1}{\sqrt{2}}\,\eta\alpha, \qquad \psi^7_{1/2} = \frac{i}{\sqrt{2}}\,\xi\beta - \frac{1}{\sqrt{2}}\,\eta\beta; \qquad (28.2\text{-}20)$$

$\Gamma^{(6)}(^2\mathrm{E})$ from $\Gamma^{(6)} \times \Gamma^{(5)}(^2\mathrm{E})$

$$\psi^6_{-1/2} = \frac{i}{\sqrt{2}}\,\xi\alpha - \frac{1}{\sqrt{2}}\,\eta\alpha, \qquad \psi^6_{1/2} = \frac{i}{\sqrt{2}}\,\xi\beta + \frac{1}{\sqrt{2}}\,\eta\beta; \qquad (28.2\text{-}21)$$

$\Gamma^{(7)}(^2\mathrm{B}_2)$ from $\Gamma^{(6)} \times \Gamma^{(4)}(^2\mathrm{B}_2)$

$$\psi^7_{-1/2} = i\zeta\beta, \qquad \psi^7_{1/2} = -i\zeta\alpha. \qquad (28.2\text{-}22)$$

Let us now assume that the tetragonal distortion, without spin-orbit interaction puts the $\Gamma^{(4)}(^2\mathrm{B}_2)$ state at the lowest energy (Fig. 28.5). The expectation value of L_z in this state is given by

$$\langle\zeta|L_z|\zeta\rangle = \left\langle -\frac{i}{\sqrt{2}}(Y_{22} - Y_{2\,-2})|L_z| - \frac{i}{\sqrt{2}}(Y_{22} - Y_{2\,-2})\right\rangle = 0.$$

Similarly,

$$\langle\zeta|L_x|\zeta\rangle = \langle\zeta|L_y|\zeta\rangle = 0 \qquad \text{or} \qquad \langle\zeta|\mathbf{L}|\zeta\rangle = 0. \qquad (28.2\text{-}23)$$

This feature of a crystal field is known as *quenching*. More generally, the quenching phenomenon is a consequence of the theorem that an orbitally nondegenerate eigenfunction ψ must be real, provided the Hamiltonian is real. To show this, assume that $\psi = \psi_1 + i\psi_2$. Then if $\mathscr{H}\psi = E\psi$, we have $\mathscr{H}(\psi_1 + i\psi_2) = E(\psi_1 + i\psi_2)$ or $\mathscr{H}\psi_1 = E\psi_1$ and $\mathscr{H}\psi_2 = E\psi_2$. The level is therefore degenerate in contradiction to the original hypothesis of non-degeneracy. Hence ψ must be real. Since the components of $\mathbf{L}$ are imaginary, the matrix element $\langle\psi|\mathbf{L}|\psi\rangle$ would be imaginary in violation of the theorem that the expectation value of a Hermitian operator must be real. We therefore conclude that if ψ is an orbitally nondegenerate state,

$$\langle\psi|\mathbf{L}|\psi\rangle = 0 \qquad (28.2\text{-}24)$$

and such a state behaves as if it only had spin angular momentum and a magnetic moment associated entirely with the spin.

Now let us apply spin-orbit coupling in addition to the tetragonal field as shown in Fig. 28.5. With the help of Table 28.2 the nonzero matrix elements which affect the states of $\Gamma^{(7)}(^2\mathrm{B}_2)$ are

$$\begin{aligned}
\langle\psi^7_{1/2}(^2\mathrm{B}_2)|\mathbf{L}\cdot\mathbf{S}|\psi^7_{1/2}(^2\mathrm{E})\rangle &= i/\sqrt{2},\\
\langle\psi^7_{-1/2}(^2\mathrm{B}_2)|\mathbf{L}\cdot\mathbf{S}|\psi^7_{-1/2}(^3\mathrm{E})\rangle &= -i/\sqrt{2}, \qquad (28.2\text{-}25)\\
\langle\psi^7_{1/2}(^2\mathrm{B}_2)|\mathbf{L}\cdot\mathbf{S}|\psi^7_{1/2}(^2\mathrm{B}_1)\rangle &= \langle\psi^7_{-1/2}(^2\mathrm{B}_2)|\mathbf{L}\cdot\mathbf{S}|\psi^7_{-1/2}(^2\mathrm{B}_1)\rangle = -1.
\end{aligned}$$

From first-order perturbation theory,

$$\psi_i' = \psi_i - \sum_k \frac{\langle \psi_k | \xi(r) \mathbf{L} \cdot \mathbf{S} | \psi_i \rangle}{E_k - E_i} \psi_k, \tag{28.2-26}$$

where ψ_i' is the corrected wave function. The correction to the ground state $\psi^7_{\pm 1/2}$ due to spin-orbit interaction with higher states, to first order, is then given by

$$\begin{aligned}(\psi^7_{-1/2})' &= \psi^7_{-1/2}(^2\mathrm{B}_2) - \frac{i}{\sqrt{2}} \frac{\langle \xi(r) \rangle}{\delta} \psi^7_{-1/2}(^2\mathrm{E}) + \frac{\langle \xi(r) \rangle}{\varepsilon} \psi^7_{-1/2}(^2\mathrm{B}_1), \\ (\psi^7_{1/2})' &= \psi^7_{1/2}(^2\mathrm{B}_2) + \frac{i}{\sqrt{2}} \frac{\langle \xi(r) \rangle}{\delta} \psi^7_{1/2}(^2\mathrm{E}) + \frac{\langle \xi(r) \rangle}{\varepsilon} \psi^7_{1/2}(^2\mathrm{B}_1),\end{aligned} \tag{28.2-27}$$

in which

$$\begin{aligned}\delta &= E(\Gamma^{(5)}) - E(\Gamma^{(4)}) = E(^2\mathrm{E}) - E(^2\mathrm{B}_2), \\ \varepsilon &= E(\Gamma^{(3)}) - E(\Gamma^{(4)}) = E(^2\mathrm{B}_1) - E(^2\mathrm{B}_2),\end{aligned} \tag{28.2-28}$$

or using (28.2-19), (28.2-20), and (28.2-22),

$$\begin{aligned}(\psi^7_{-1/2})' &= i\zeta\beta - \frac{i}{\sqrt{2}} \frac{\langle \xi(r) \rangle}{\delta} \left[\frac{i}{\sqrt{2}} \xi\alpha + \frac{1}{\sqrt{2}} \eta\alpha \right] + \frac{\langle \xi(r) \rangle}{\varepsilon} v\beta, \\ (\psi^7_{1/2})' &= -i\zeta\alpha + \frac{i}{\sqrt{2}} \frac{\langle \xi(r) \rangle}{\delta} \left[\frac{i}{\sqrt{2}} \xi\beta - \frac{1}{\sqrt{2}} \eta\beta \right] + \frac{\langle \xi(r) \rangle}{\varepsilon} v\alpha.\end{aligned} \tag{28.2-29}$$

It is seen that $(\psi^7_{-1/2})'$ is primarily a state with β spin but has some admixture of α spin. Similarly, $(\psi^7_{1/2})'$ is mainly a state with α spin but which also contains some β spin. Thus, in first order, $(\psi^7_{-1/2})'$ and $(\psi^7_{1/2})'$ are no longer pure spin states.

Having calculated the first-order corrections to the ground state, we shall now recalculate the expectation value of L_z. This is straightforward since ξ, η, ζ are expressible in terms of the spherical harmonics. Thus

$$\begin{aligned}(\psi^7_{-1/2})' &= \frac{1}{\sqrt{2}} \beta \left[\left(1 + \frac{\langle \xi(r) \rangle}{\varepsilon}\right) Y_{22} - \left(1 - \frac{\langle \xi(r) \rangle}{\varepsilon}\right) Y_{2\,-2} \right] - \frac{1}{\sqrt{2}} \frac{\langle \xi(r) \rangle}{\delta} Y_{21}\alpha, \\ (\psi^7_{1/2})' &= \frac{1}{\sqrt{2}} \alpha \left[-\left(1 - \frac{\langle \xi(r) \rangle}{\varepsilon}\right) Y_{22} + \left(1 + \frac{\langle \xi(r) \rangle}{\varepsilon}\right) Y_{2\,-2} \right] - \frac{1}{\sqrt{2}} \frac{\langle \xi(r) \rangle}{\delta} Y_{2\,-1}\beta,\end{aligned} \tag{28.2-30}$$

and

$$\langle (\psi^7_{\pm 1/2})' | L_z | (\psi^7_{\pm 1/2})' \rangle = \mp \left(4 \frac{\langle \xi(r) \rangle}{\varepsilon} + \frac{1}{2} \frac{\langle \xi(r) \rangle^2}{\delta^2} \right). \tag{28.2-31}$$

We see then that the effect of the low symmetry field is to quench orbital angular momentum while the spin-orbit interaction restores it. The two effects work in opposite directions and the net result depends on their relative magnitudes.

28.3 Multielectron Configurations

Crystal field calculations for systems with more than one electron may be illustrated by considering the d^2 configuration. Assuming, at first, that case (b) given by (28.1-4) is applicable, the Coulomb repulsion is the strongest perturbation and has the effect of splitting the degeneracy of the configuration into a set of terms which, in order of increasing energy, are ^3F, ^1D, ^3P, ^1G, and ^1S (Fig. 28.6). Subsequent application of a crystal field will produce further splitting of each term. If the symmetry of the field is that of the group O, the details of the splittings (not including the energies) may be obtained directly from Table 5.7 which relates irreducible representations of $O^+(3)$ to irreducible representations of the subgroup O. Thus, as shown in Fig. 28.6,

$$\begin{aligned} &{}^3F = {}^3A_2 + {}^3T_1 + {}^3T_2, \qquad {}^1D = {}^1E + {}^1T_1, \qquad {}^3P = {}^3T_1, \\ &{}^1G = {}^1E + {}^1T_1 + {}^1T_2 + {}^1A_1, \qquad {}^1S = {}^1A_1 \end{aligned} \tag{28.3-1}$$

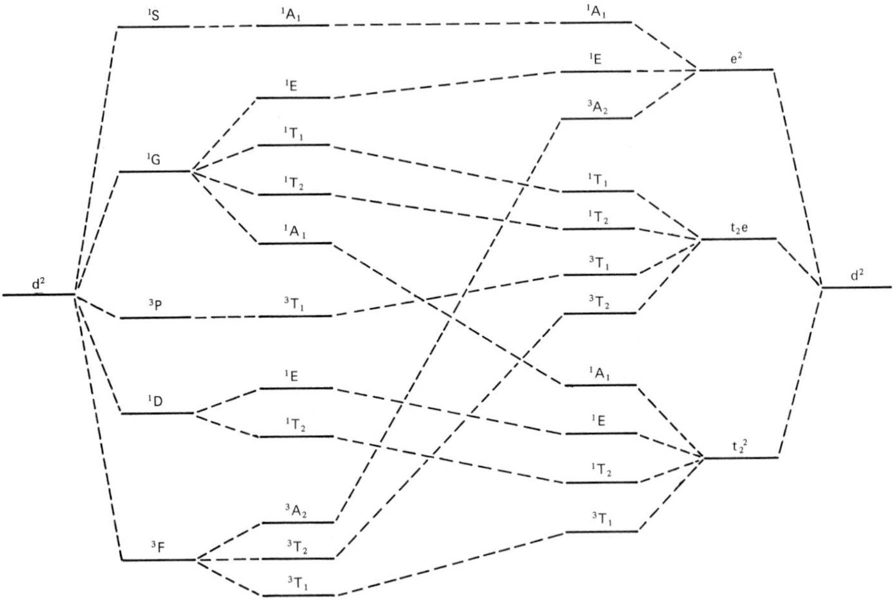

FIG. 28.6 The d^2 configuration in a cubic field with spin-orbit coupling.

To obtain the wave functions one may refer to Table 5.8 which lists the linear combinations of the spherical harmonics that are basis functions for the irreducible representations of O. For example, it is shown that the linear combination $(i/\sqrt{2})(-Y_{32} + Y_{3\,-2})$ is a basis function for $\Gamma^{(2)}(A_2)$. Since F states have the same transformation properties as the Y_{3m}, the linear combination of F states which is a basis function for A_2 is

$$i/\sqrt{2}[-|F, M_L = 2\rangle + |F, M_L = -2\rangle].$$

The crystal field mixes M_L values but has no effect on S and M_S. We may therefore write

$$|^3A_2\rangle = \frac{i}{\sqrt{2}}[-|^3F, M_L = 2\rangle + |^3F, M_L = -2\rangle]$$

in which $|^3A_2\rangle$ represents a state whose wave function belongs to the irreducible representation A_2 of the group O and whose multiplicity $(2S + 1)$ is 3. If the notation is simplified by writing, for example,

$$|^3F, M_L = 2\rangle \equiv |2\rangle,$$

then the linear combinations of the ^3F states that belong to $^3A_2$, $^3T_1$, and $^3T_2$ may be listed as:

$$\begin{aligned}
|^3A_2\rangle &= (i/\sqrt{2})[-|2\rangle + |-2\rangle],\\
|^3T_1x\rangle &= \tfrac{1}{4}[\sqrt{5}(-|3\rangle + |-3\rangle) + \sqrt{3}(|1\rangle - |-1\rangle)],\\
|^3T_1y\rangle &= -\tfrac{1}{4}i[\sqrt{5}(|3\rangle + |-3\rangle) + \sqrt{3}(|1\rangle + |-1\rangle)],\\
|^3T_1z\rangle &= |0\rangle,\\
|^3T_2\xi\rangle &= -\tfrac{1}{4}[\sqrt{5}(-|1\rangle + |-1\rangle) + \sqrt{3}(-|3\rangle + |-3\rangle)],\\
|^3T_2\eta\rangle &= \tfrac{1}{4}i[\sqrt{5}(|1\rangle + |-1\rangle) - \sqrt{3}(|3\rangle + |-3\rangle)],\\
|^3T_2\rangle &= (1/\sqrt{2})(|2\rangle + |-2\rangle).
\end{aligned} \qquad (28.3\text{-}2)$$

Since d orbitals are even under inversion, all states arising from a d^n configuration are even so that a more complete designation of the states in (28.3-2) would be $^3A_{2g}$, $^3T_{1g}$ and $^3T_{2g}$.

To calculate crystal field splittings one may proceed in pedestrian fashion by converting (28.3-2) into expressions which depend on d orbitals (and spin functions). One may then calculate the matrix elements of the crystal field potential with the assistance of Table 28.1. A more efficient method is based on the unit tensor operator as defined by (21.2-12) together with the tabulations of Nielson and Koster (1963). The resulting crystal field energies are

$$E(^3A_2) = \tfrac{6}{5}\Delta, \qquad E(^3T_1) = -\tfrac{3}{5}\Delta, \qquad E(^3T_2) = \tfrac{1}{5}\Delta \qquad (28.3\text{-}3)$$

where Δ is the difference in energy between the e and t_2 orbitals. It is observed that the sum of the energies, weighted by the orbital degeneracies, is zero; hence the crystal field has not shifted the center of gravity of the ^3F state.

Other configurations may be analyzed in similar fashion but it is not necessary to treat every configuration separately. Thus $d^6(^5D)$ which differs from $d^1(^2D)$ by the addition of the closed shell $d^5(^5S)$ has the same splitting as $d^1(^2D)$, i.e.,

$$d^6(^5D) = {}^5T_2 + {}^5E \tag{28.3-4}$$

with $^5T_2$ lower in energy than ^5E. But $d^9(^2D)$, which is equivalent to a single hole in a filled d shell $[d^{10}(^1S)]$ and whose splitting is also governed by (28.3-4), has ^5E as the state of lower energy because the hole behaves as if it were a positively charged particle. The same is true of $d^4(^5D)$ which may be regarded either as a single hole in $d^5(^6S)$ or as six holes in $d^{10}(^1S)$. In either case (28.3-4) is applicable with ^5E lower then $^5T_2$. By the same arguments $d^2(^3F)$ and $d^7(^4F)$ have the same splitting with the same ordering of the levels, namely, the order given by (28.3-3). But $d^3(^4F)$ and $d^8(^3F)$ have the same levels in reverse order (A_2, T_2, T_1).

Suppose now that the crystal field is sufficiently strong to overcome the electrostatic interaction as in case (a) (28.1-3). In that event the crystal field perturbation must be applied first and, as shown in the previous section, a d orbital will be split into a doubly degenerate e and a triply degenerate t_2 orbital (Fig. 28.2). When two electrons are present, the possible crystal field configurations are $t_2{}^2$ (both electrons in t_2), t_2e (one electron in t_2 and one in e), and e^2 (both electrons in e).

The problem now is to compute the terms that may arise from such configurations. These will be determined, in the first place, by the decomposition of the product representations $t_2 \times t_2$, $t_2 \times e$ and $e \times e$ into the irreducible representations of O. From Table 5.5 one finds

$$\begin{aligned} t_2 \times t_2 &= A_1 + E + T_1 + T_2, \\ t_2 \times e &= T_1 + T_2, \\ e \times e &= A_1 + A_2 + E. \end{aligned} \tag{28.3-5}$$

The detailed form of the wave functions are obtained from a table of coupling coefficients such as Table 5.9. One finds, for example, that the basis functions for A_1 which originates from $t_2 \times t_2$ is

$$t_2{}^2A_1 = (1/\sqrt{3})[\xi(1)\xi(2) + \eta(1)\eta(2) + \zeta(1)\zeta(2)].$$

This function is symmetric with respect to an interchange of electrons 1 and 2. Since the total wave function, including the spin, must be antisymmetric, it is necessary to multiply (28.3-5) by the antisymmetric spin function $(1/\sqrt{2})[\alpha(1)\beta(2) - \beta(1)\alpha(2)]$ which is associated with a singlet state. Thus, the complete wave function which is a basis function for A_1 and is antisymmetric with respect to an interchange of the two electrons is

$$|t_2{}^2\,{}^1A_1\rangle = (1/\sqrt{3})[|\xi^+\xi^-\rangle + |\eta^+\eta^-\rangle + |\zeta^+\zeta^-\rangle] \tag{28.3-6}$$

where, for example,

$$|\xi^+\xi^-\rangle = \frac{1}{\sqrt{2}}\begin{vmatrix}\xi(1)\alpha(1) & \xi(1)\beta(1)\\ \xi(2)\alpha(2) & \xi(2)\beta(2)\end{vmatrix}. \tag{28.3-7}$$

In this fashion we find that the basis functions for E and T_2 must also be singlets. For T_1 we find

$$t_2{}^2T_1x = (1/\sqrt{2})[\eta(1)\zeta(2) - \zeta(1)\eta(2)].$$

This function is antisymmetric; it must therefore be multiplied by a symmetric spin function which means that T_1 is a triplet state. Thus, for $M_S = 1$,

$$\begin{aligned}|t_2{}^2\,{}^3T_1 1x\rangle &= (1/\sqrt{2})[\eta(1)\zeta(2) - \zeta(1)\eta(2)]\alpha(1)\alpha(2)\\ &= |\eta^+\zeta^+\rangle.\end{aligned} \tag{28.3-8}$$

This illustrates the general method. In the t_2e configuration the two electrons are in different spatial orbitals; hence both singlet and triplet states are possible. All the terms are listed in Table 28.5.

TABLE 28.5

Wave Functions for the d^2 *Configuration in a Cubic Field*

$t_2 \times t_2 = {}^1A_1 + {}^1E + {}^1T_2 + {}^3T_1$

$$|t_2{}^2\,{}^1A_1\rangle = \frac{1}{\sqrt{3}}[|\xi^+\xi^-\rangle + |\eta^+\eta^-\rangle + |\zeta^+\zeta^-\rangle]$$

$$|t_2{}^2\,{}^1E\theta\rangle = -\frac{1}{\sqrt{6}}[|\xi^+\xi^-\rangle + |\eta^+\eta^-\rangle - 2|\zeta^+\zeta^-\rangle]$$

$$|t_2{}^2\,{}^1E\varepsilon\rangle = \frac{1}{\sqrt{2}}[|\xi^+\xi^-\rangle - |\eta^+\eta^-\rangle]$$

$$|t_2{}^2\,{}^1T_2\xi\rangle = \frac{1}{\sqrt{2}}[|\eta^+\zeta^-\rangle + |\zeta^+\eta^-\rangle]$$

$$|t_2{}^2\,{}^1T_2\eta\rangle = \frac{1}{\sqrt{2}}[|\xi^+\zeta^-\rangle + |\zeta^+\xi^-\rangle]$$

$$|t_2{}^2\,{}^1T_2\zeta\rangle = \frac{1}{\sqrt{2}}[|\xi^+\eta^-\rangle + |\eta^+\xi^-\rangle]$$

$$|t_2{}^2\,{}^3T_1 1x\rangle = |\eta^+\zeta^+\rangle$$

$$|t_2{}^2\,{}^3T_1 1y\rangle = |\zeta^+\xi^+\rangle$$

$$|t_2{}^2\,{}^3T_1 1z\rangle = |\xi^+\eta^+\rangle$$

TABLE 28.5 (*continued*)

$e \times e = {}^1A_1 + {}^3A_2 + {}^1E$

$$|e^2\ {}^1A_1\rangle = \frac{1}{\sqrt{2}}[|\theta^+\theta^-\rangle + |\varepsilon^+\varepsilon^-\rangle]$$

$$|e^2\ {}^3A_2|\rangle = |\theta^+\varepsilon^+\rangle$$

$$|e^2\ {}^1E\theta\rangle = \frac{1}{\sqrt{2}}[|\varepsilon^+\varepsilon^-\rangle - |\theta^+\theta^-\rangle]$$

$$|e^2\ {}^1E\varepsilon\rangle = \frac{1}{\sqrt{2}}[|\theta^+\varepsilon^-\rangle + |\varepsilon^+\theta^-\rangle]$$

$t_2 \times e = {}^1T_1 + {}^3T_1 + {}^1T_2 + {}^3T_2$

$$|t_2e\ {}^1T_1x\rangle = -\frac{1}{\sqrt{8}}[\sqrt{3}|\theta^+\xi^-\rangle - \sqrt{3}|\theta^-\xi^+\rangle + |\varepsilon^+\xi^-\rangle - |\varepsilon^-\xi^+\rangle]$$

$$|t_2e\ {}^1T_1y\rangle = \frac{1}{\sqrt{8}}[\sqrt{3}|\theta^+\eta^-\rangle - \sqrt{3}|\theta^-\eta^+\rangle - |\varepsilon^+\eta^-\rangle + |\varepsilon^-\eta^+\rangle]$$

$$|t_2e\ {}^1T_1z\rangle = \frac{1}{\sqrt{2}}[|\varepsilon^+\zeta^-\rangle - |\varepsilon^-\zeta^+\rangle]$$

$$|t_2e\ {}^3T_11x\rangle = -\frac{1}{2}[\sqrt{3}|\theta^+\xi^+\rangle + |\varepsilon^+\xi^+\rangle]$$

$$|t_2e\ {}^3T_11y\rangle = \frac{1}{2}[\sqrt{3}|\theta^+\eta^+\rangle - |\varepsilon^+\eta^+\rangle]$$

$$|t_2e\ {}^3T_11z\rangle = |\varepsilon^+\zeta^+\rangle$$

$$|t_2e\ {}^1T_2\xi\rangle = \frac{1}{\sqrt{8}}[\sqrt{3}|\varepsilon^+\xi^-\rangle - \sqrt{3}|\varepsilon^-\xi^+\rangle - |\theta^+\xi^-\rangle + |\theta^-\xi^+\rangle]$$

$$|t_2e\ {}^1T_2\eta\rangle = -\frac{1}{\sqrt{8}}[\sqrt{3}|\varepsilon^+\eta^-\rangle - \sqrt{3}|\varepsilon^-\eta^+\rangle + |\theta^+\eta^-\rangle - |\theta^-\eta^+\rangle]$$

$$|t_2e\ {}^1T_2\zeta\rangle = \frac{1}{\sqrt{2}}[|\theta^+\zeta^-\rangle - |\theta^-\zeta^+\rangle]$$

$$|t_2e\ {}^3T_21\xi\rangle = \frac{1}{2}[\sqrt{3}|\varepsilon^+\xi^+\rangle - |\theta^+\xi^+\rangle]$$

$$|t_2e\ {}^3T_21\eta\rangle = -\frac{1}{2}[\sqrt{3}|\varepsilon^+\eta^+\rangle - |\theta^+\eta^+\rangle]$$

$$|t_2e\ {}^3T_21\zeta\rangle = |\theta^+\zeta^+\rangle$$

There is another method for determining whether a given term is a singlet or a triplet. If $\chi_{+}^{(\mu \times \mu)}(R)$ is the character of the symmetric product representation and $\chi_{-}^{(\mu \times \mu)}(R)$ the character of the antisymmetric product representation, we saw in Section 3.6 that

$$\begin{aligned} \chi_{+}^{(\mu \times \mu)}(R) &= \tfrac{1}{2}\{[\chi^{(\mu)}(R)]^2 + \chi^{(\mu)}(R^2)\}, \\ \chi_{-}^{(\mu \times \mu)}(R) &= \tfrac{1}{2}\{[\chi^{(\mu)}(R)]^2 - \chi^{(\mu)}(R^2)\}. \end{aligned} \tag{28.3-9}$$

The symmetric product has basis functions which are symmetric under an interchange of particles and an antisymmetric product has antisymmetric basis functions. Using (28.3-9) the calculation of $\chi_{-}^{(\mu \times \mu)}(R)$ for $t_2 \times t_2$ and for $e \times e$ identifies each set of characters as belonging to T_1 and A_2, respectively. Therefore $t_2{}^2\ T_1$ and $e^2\ A_2$ are both antisymmetric; hence they are triplet states. All the rest are singlets.

Given the detailed form of the terms arising from the d^2 configuration it is a straightforward but somewhat lengthy calculation to obtain the electrostatic matrix elements. In terms of Racah parameters, a few examples are (Griffith, 1961):

$$\begin{aligned} \langle t_2{}^2\ {}^1A_1 | e^2/r_{12} | t_2{}^2\ {}^1A_1 \rangle &= A + 10B + 5C, \\ \langle t_2{}^2\ {}^1A_1 | e^2/r_{12} | e^2\ {}^1A_1 \rangle &= (2B + C)/\sqrt{6} \\ \langle e^2\ {}^1A_1 | e^2/r_{12} | e^2\ {}^1A_1 \rangle &= A + 8B + 4C \\ \langle e^2\ {}^3A_2 | e^2/r_{12} | e^2\ {}^3A_2 \rangle &= A - 8B. \end{aligned}$$

For the general case of d^n electrons the n electrons will fill the t_2 and e orbitals in accordance with the Pauli principle. Within this restriction the distribution of electrons will be governed by three factors: (a) t_2 orbitals lie lower in energy than e orbitals by an amount Δ; this favors occupation of the t_2 orbitals, (b) electrons in the same spatial orbital tend to have higher electrostatic repulsions than electrons in separate orbitals, and (c) exchange energy favors states with high spin but these, according to the Pauli principle, arise from states in which the electrons are distributed in separate orbitals. The energy associated with the last two factors, taken together, is often called the *pairing energy*. Hence the relative magnitude of Δ and the pairing energy will determine the distribution of electrons among the t_2 and e orbitals. When Δ is much smaller than the pairing energy, the electrons tend to distribute themselves so as to achieve maximum spin. Thus, for d^5, the configuration $t_2{}^3\ e^2$ with $S = 5/2$ has the lowest energy. Conversely, when Δ is much larger than the pairing energy, the electrons tend to fill the t_2 orbitals; for d^5 the configuration would be $t_2{}^5$ with $S = \frac{1}{2}$.

Another important consideration in multielectron systems stems from Kramers' theorem which is based on the invariance of real Hamiltonians under time reversal (Section 8.3). In its application to crystal fields, Kramers' theorem distinguishes between systems with an odd number of electrons and those with an even number. In the former case there will always be at least a twofold degeneracy irrespective of the symmetry of the crystal field; that is, there is no distribution of ligands that can remove all the degeneracies. On the other hand, the theorem does not restrict the degeneracy of a system with an even number of electrons and in crystal fields of sufficiently low symmetry the degeneracy may be completely removed. As an example, Fe^{3+} has a configuration of $3d^5$ outside of closed shells. The ground state is ^6S. If Fe^{3+} is immersed in a crystal field, it is impossible to lift the degeneracies beyond three Kramers doublets. Application of an external magnetic field will further split the twofold degeneracies. On the other hand Fe^{2+} has a $3d^6$ configuration. In this case there is an even number of electrons; Kramers' theorem does not apply and there is no restriction concerning the lifting of degeneracies.

28.4 Magnetic Fields and the Spin Hamiltonian

Owing to the close connection between magnetic moments and angular momenta, crystal fields which alter the latter are expected to exert an important influence on magnetic properties. This may again be demonstrated most simply by a single d electron in an octahedral field (symmetry O) with axial distortion (symmetry D_4). We shall assume, at first, that there is no spin-orbit coupling and that the electron is in the orbitally nondegenerate ground state $^2B_2$. In the coupled representation this becomes $\Gamma^{(7)}$ with basis functions

$$\psi^7_{-1/2} = i\zeta\beta, \qquad \psi^7_{-1/2} = -i\zeta\alpha \tag{28.4-1}$$

as in (28.2-22).

In a magnetic field the interaction Hamiltonian is

$$\mathscr{H}_m = \beta\mathbf{B}\cdot(\mathbf{L} + g_e\mathbf{S}) \tag{28.4-2}$$

in which β is the Bohr magneton and $g_e = 2.0023$. But, according to (28.2-23), there are no matrix elements of $\mathbf{L}$ in the basis set (28.4-1), that is, the orbital angular momentum is completely quenched. This leaves matrix elements of $\beta g_e\mathbf{B}\cdot\mathbf{S}$. Assuming the z axis is parallel to $\mathbf{B}$, the nonvanishing matrix elements of $\mathscr{H}_m$ are

$$\begin{aligned} g_e\beta B_z\langle\psi^7_{-1/2}|S_z|\psi^7_{-1/2}\rangle &= -\tfrac{1}{2}g_e\beta B_z, \\ g_e\beta B_z\langle\psi^7_{1/2}|S_z|\psi^7_{1/2}\rangle &= \tfrac{1}{2}g_e\beta B_z. \end{aligned} \tag{28.4-3}$$

Hence the effect of a magnetic field in the z direction is to lift the twofold spin degeneracy (Fig. 28.7) with an energy separation

$$\Delta E = g_e\beta B_z. \tag{28.4-4}$$

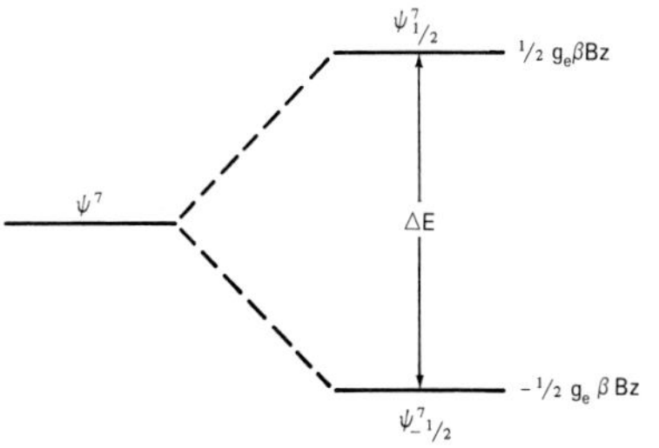

FIG. 28.7 Effect of a magnetic field on $\Gamma^{(7)}(^2B_2)$, the lowest state for a single d electron in D_4 with spin-orbit coupling (see Fig. 28.5).

If g_z is defined by the relation

$$\Delta E = g_z \beta B_z, \tag{28.4-5}$$

it is seen that in the present example

$$g_z = g_e. \tag{28.4-6}$$

A similar calculation may be performed with the magnetic field in the x direction. In that case it is convenient to write

$$S_x = -(1/\sqrt{2})(S_{+1} - S_{-1})$$

from which it is evident that there will be no diagonal elements in the basis set (28.4-1). The off-diagonal elements are

$$\beta g_e B_x \langle \psi^7_{-1/2} | S_x | \psi^7_{1/2} \rangle = \beta g_e B_x \langle \psi^7_{1/2} | S_x | \psi^7_{-1/2} \rangle = -\tfrac{1}{2} g_e \beta B_x \tag{28.4-7}$$

and the eigenvalues of the Hamiltonian matrix are $\pm\frac{1}{2}g_e\beta B_x$. In this case, too, the spin degeneracy has been removed and g_x, defined by a relation analogous to (28.4-5), is also equal to g_e. For a magnetic field in the y direction the results are the same. Thus

$$g_x = g_y = g_z = g_e, \tag{28.4-8}$$

and the system is indistinguishable from a pure $S = \frac{1}{2}$ system despite the fact that the state originates from a ^{2}D. When $g_x = g_y = g_z$, the system is said to have an isotropic g value.

We shall now perform the same calculations but in the basis set (28.2-29) which contains first-order corrections for spin-orbit interaction. With a magnetic field in the z direction the nonvanishing matrix elements, carried to first order, are

$$\begin{aligned} \langle (\psi^7_{-1/2})' | L_z + g_e S_z | (\psi^7_{-1/2})' \rangle &= 4\frac{\langle \xi(r) \rangle}{\varepsilon} - \frac{1}{2} g_e, \\ \langle (\psi^7_{1/2})' | L_z + g_e S_z | (\psi^7_{1/2})' \rangle &= -4\frac{\langle \xi(r) \rangle}{\varepsilon} + \frac{1}{2} g_e, \end{aligned} \tag{28.4-9}$$

from which we obtain

$$g_z = g_e - 8\frac{\langle\xi(r)\rangle}{\varepsilon}. \tag{28.4-10}$$

For a field in the x direction the diagonal matrix elements vanish and

$$\begin{aligned}\langle(\psi^7_{-1/2})'|L_x + g_e S_x|(\psi^7_{1/2})'\rangle &= \langle(\psi^7_{-1/2})'|L_x + g_e S_x|(\psi^7_{1/2})'\rangle \\ &= \frac{\langle\xi(r)\rangle}{\delta} - \frac{1}{2}g_e;\end{aligned}$$

the energy eigenvalues are

$$E = \pm\beta B_x\left(\frac{\langle\xi(r)\rangle}{\delta} - \frac{1}{2}g_e\right) \tag{28.4-11}$$

and

$$g_x = g_e - 2\frac{\langle\xi(r)\rangle}{\delta}. \tag{28.4-12}$$

A similar calculation for B_y shows that

$$g_y = g_x. \tag{28.4-13}$$

To summarize: When spin-orbit coupling is taken into account, the g values are no longer isotropic. The combination of crystal field and spin-orbit coupling leads to

$$\begin{aligned} g_{||} &= g_z = g_e - 8\frac{\langle\xi(r)\rangle}{\varepsilon} \\ g_{\perp} &= g_x = g_y = g_e - 2\frac{\langle\xi(r)\rangle}{\delta}. \end{aligned} \tag{28.4-14}$$

which indicates that the Zeeman splitting in crystals depends on the orientation of the crystal in the magnetic field.

Another example of the manner in which crystal field effects can drastically alter magnetic properties is provided by a single d electron in an octahedral field. In the absence of spin-orbit coupling, the $\psi^8(^2T_2)$ state, whose components are given by (28.2-9), is fourfold degenerate. Using Table 5.8 it is found that $\psi^8_{1/2}$, for example, may be written

$$\psi^8_{1/2} = -\frac{1}{\sqrt{2}}(\eta + i\xi)\beta = Y_{21}\beta.$$

If the magnetic field is in the z direction

$$\langle Y_{21}\beta|L_z + 2s_z|Y_{21}\beta\rangle = 0.$$

The same is true for fields in other directions and for the other components of $\psi^8(^2T_2)$. Therefore this state, despite its fourfold degeneracy, is not split by a magnetic field, i.e., $g = 0$.

The interplay between magnetic fields and spin-orbit interaction may be formulated more generally. Consider the Hamiltonian

$$\mathscr{H} = \lambda \mathbf{L} \cdot \mathbf{S} + \beta \mathbf{B} \cdot (\mathbf{L} + g_e \mathbf{S}) \tag{28.4-15}$$

and an orbitally nondegenerate state ψ_0. If $\mathscr{H}$ is regarded as a perturbation, the correction to the energy eigenvalue of ψ_0 is, to first order,

$$\begin{aligned} E^{(1)} &= \langle\psi_0|\mathscr{H}|\psi_0\rangle \\ &= \lambda[\langle\psi_0|L_x|\psi_0\rangle S_x + \langle\psi_0|L_y|\psi_0\rangle S_y + \langle\psi_0|L_z|\psi_0\rangle S_z] \\ &\quad + \beta[B_x\langle\psi_0|L_x|\psi_0\rangle + B_y\langle\psi_0|L_y|\psi_0\rangle + B_z\langle\psi_0|L_z|\psi_0\rangle] + \beta g_e \mathbf{B} \cdot \mathbf{S}. \end{aligned} \tag{28.4-16}$$

But, since ψ_0 is nondegenerate, $\langle\psi_0|\mathbf{L}|\psi_0\rangle = 0$ so that

$$E^{(1)} = \beta g_e \mathbf{B} \cdot \mathbf{S}. \tag{28.4-17}$$

It should be remarked that $E^{(1)}$ is not actually an energy but an operator which depends on the spin $\mathbf{S}$. This stems from the fact that ψ_0 is a spatial function so that matrix elements of $\mathbf{S}$ remain to be evaluated. We now go to second order:

$$E^{(2)} = \sum_{n \neq 0} \frac{\langle\psi_0|\mathscr{H}|\psi_n\rangle\langle\psi_n|\mathscr{H}|\psi_0\rangle}{E_0 - E_n} \tag{28.4-18}$$

in which

$$\begin{aligned} \langle\psi_0|\mathscr{H}|\psi_n\rangle\langle\psi_n|\mathscr{H}|\psi_0\rangle &= |\langle\psi_0|\mathscr{H}|\psi_n\rangle|^2 \\ &= |\langle\psi_0|\lambda \mathbf{L} \cdot \mathbf{S} + \beta \mathbf{B} \cdot (\mathbf{L} + g_e \mathbf{S})|\psi_n\rangle|^2. \end{aligned} \tag{28.4-19}$$

This expression may be simplified by noting that

$$\langle\psi_0|\mathbf{S}|\psi_n\rangle = \mathbf{S}\langle\psi_0|\psi_n\rangle = 0 \qquad (n \neq 0)$$

because of the orthogonality of the eigenfunctions. Thus

$$\begin{aligned} |\langle\psi_0|\mathscr{H}|\psi_n\rangle|^2 = |&\lambda[\langle\psi_0|L_x|\psi_n\rangle S_x + \langle\psi_0|L_y|\psi_n\rangle S_y + \langle\psi_0|L_z|\psi_n\rangle S_z] \\ &+ \beta[\langle\psi_0|L_x|\psi_n\rangle B_x + \langle\psi_0|L_y|\psi_n\rangle B_y + \langle\psi_0|L_z|\psi_n\rangle B_z]|^2. \end{aligned}$$

Defining

$$\Lambda_{ij} = \sum_{n \neq 0} \frac{\langle\psi_0|L_i|\psi_n\rangle\langle\psi_n|L_j|\psi_0\rangle}{E_n - E_0} \tag{28.4-20}$$

and ignoring terms quadratic in the magnetic field,

$$E^{(2)} = -\sum_{ij} [\lambda^2 \Lambda_{ij} S_i S_j + 2\beta\lambda\Lambda_{ij} S_i B_j] \tag{28.4-21}$$

and

$$E^{(1)} + E^{(2)} = \sum_{ij} \beta(g_e \delta_{ij} - 2\lambda\Lambda_{ij})S_i B_j - \lambda^2 \Lambda_{ij} S_i S_j. \qquad (28.4\text{-}22)$$

We now define two tensors g and D with components

$$g_{ij} = g_e \delta_{ij} - 2\lambda\Lambda_{ij}, \qquad (28.4\text{-}23)$$

$$D_{ij} = -\lambda^2 \Lambda_{ij}, \qquad (28.4\text{-}24)$$

which may be substituted in (28.4-22). The resulting expression

$$\mathscr{H}_s = \beta \mathbf{B} \cdot \mathbf{g} \cdot \mathbf{S} + \mathbf{S} \cdot \mathbf{D} \cdot \mathbf{S} \qquad (28.4\text{-}25)$$

is known as the *spin Hamiltonian.*

What has transpired here is that the original Hamiltonian (28.4-15) has been transformed to form (28.4-25) in which the only operators are spin operators. The other parameters depend on the external magnetic field and the symmetry of the crystal field. The eigenvalues of $\mathscr{H}_s$ are corrections to the energy of the original orbitally nondegenerate ground state in the crystal field due to the combined effect of an external magnetic field and the spin-orbit interaction. The corrections are valid to second order in perturbation theory. It might appear that the requirement of an orbitally nondegenerate ground state for the molecule or complex is a highly restrictive condition. However, if the ground state is orbitally degenerate, it is necessary to take into account the *Jahn–Teller* theorem which states that a symmetric, non-linear, molecule in a degenerate electronic state (other than Kramers degeneracy) distorts in such a way as to reduce the symmetry and therefore remove the degeneracy. Thus orbitally nondegenerate ground states are the rule rather than the exception.

The tensors g and D are symmetric by virtue of the definition of Λ_{ij} (28.4-20). One may then find a principal axis system in which the spin Hamiltonian assumes the form

$$\mathscr{H}_s = \beta[g_x B_x S_x + g_y B_y S_y + g_z B_z S_z] + D_x S_x{}^2 + D_y S_y{}^2 + D_z S_z{}^2 \quad (28.4\text{-}26)$$

with

$$g_i = g_e - 2\lambda \sum_{n \neq 0} \frac{\langle \psi_0 | L_i | \psi_n \rangle \langle \psi_n | L_i | \psi_0 \rangle}{E_n - E_0}, \qquad (28.4\text{-}27)$$

$$D_i = \lambda^2 \sum_{n \neq 0} \frac{\langle \psi_0 | L_i | \psi_n \rangle \langle \psi_n | L_i | \psi_0 \rangle}{E_n - E_0}. \qquad (28.4\text{-}28)$$

An alternative form for the spin Hamiltonian (28.4-26) is obtained by writing

$$D_x S_x{}^2 + D_y S_y{}^2 = \tfrac{1}{2}(D_x + D_y)(S_x{}^2 + S_y{}^2) + \tfrac{1}{2}(D_x - D_y)(S_x{}^2 - S_y{}^2)$$

SERVICE & INSPECTION CALL REQUEST

930 S 2/74 PRINTED IN U.S.A.

[redacted] Litton **MONROE** The Calculator Company [redacted]

Date 19

Firm Name

Street Address

DEPARTMENT	FLOOR & ROOM NO.
PERSON CALLING	PHONE NO.

MODEL **SERIAL NO.**

TROUBLE - REMARKS	TIME CALL TAKEN AM PM	TAKEN BY

$$\left[\frac{\partial}{\partial x}, x^n\right] = ?$$

REFERRED TO	TIME	ATTENDED TO BY	TIME

No. 222950 Signed By

Uarco Business Forms

$$\left(\frac{\partial}{\partial x} + x\right)\left(\frac{\partial}{\partial x} - x\right)$$

$$\frac{\partial^2}{\partial x^2} + x\frac{\partial}{\partial x} - \frac{\partial}{\partial x}x - x^2$$

$$= \frac{\partial^2}{\partial x^2} - x^2 \cancel{- x\frac{\partial}{\partial x}} - 1 \cancel{+ x\frac{\partial}{\partial x}}$$

$$= \frac{\partial^2}{\partial x^2} - x^2 - 1$$

$$\left(\frac{\partial}{\partial x} - x\right)\left(\frac{\partial}{\partial x} + x\right) = \frac{\partial^2}{\partial x^2} + \frac{\partial}{\partial x}x - x\frac{\partial}{\partial x} - x^2$$

$$= \frac{\partial^2}{\partial x^2} + 1 \cancel{+ x\frac{\partial}{\partial x}} \cancel{- x\frac{\partial}{\partial x}} - x^2$$

$$\frac{\partial}{\partial x}x = 1 + x\frac{\partial}{\partial x}$$

=

$$1 - 36 \times 44 \quad \frac{\partial^2}{\partial x^2} =$$

1.65	36.14
	− 2.65
59.63	95.77

and setting

$$\begin{aligned} S^2 - S_z^{\,2} &= S_x^{\,2} + S_y^{\,2} \\ D &= D_z - \tfrac{1}{2}(D_x + D_y) \\ E &= \tfrac{1}{2}(D_x - D_y). \end{aligned} \tag{28.4-29}$$

Then

$$D_x S_x^{\,2} + D_y S_y^{\,2} + D_z S_z^{\,2} + (\tfrac{2}{3}D - D_z)S^2 = D(S_z^{\,2} - \tfrac{1}{3}S^2) + E(S_x^{\,2} - S_y^{\,2}).$$

The quantity $(\frac{2}{3}D - D_z)S^2$ whose eigenvalue is $(\frac{2}{3}D - D_z)S(S+1)$ is a constant and merely adjusts the reference energy for the entire level system so that the unsplit level has zero energy. Hence, in place of (28.4-26) we have

$$\mathscr{H}_s = \beta[g_x B_x S_x + g_y B_y S_y + g_z B_z S_z] + D(S_z - \tfrac{1}{3}S^2) + E(S_x^{\,2} - S_y^{\,2}). \tag{28.4-30}$$

If the crystal has axial symmetry,

$$\begin{aligned} g_x = g_y \equiv g_\perp, \qquad g_z \equiv g_{\|}, \qquad D_x = D_y \equiv D_\perp, \\ D_z \equiv D_{\|}, \qquad D = D_{\|} - D_\perp, \qquad E = 0. \end{aligned}$$

In that case (28.4-30)

$$\mathscr{H}_S = \beta g_{\|} B_z S_z + g_\perp(B_x S_x + B_y S_y) + D[S_z^{\,2} - \tfrac{1}{3}S(S+1)]. \tag{28.4-31}$$

Small splittings have been observed in ions such as Fe^{3+} and Mn^{2+} which have a $3d^5\ {}^6S$ ground state. The spin Hamiltonian previously discussed cannot account for such splitting because there are no first order effects on an S state from the crystal field or from spin-orbit interaction. One must then carry the computation to higher orders in perturbation theory which then results in the inclusion of quartic terms.

More general forms of the spin Hamiltonian include other interactions, e.g., spin-spin and hyperfine interactions (Griffith, 1961; Abragam and Bleaney, 1970).

28.5 Molecular Orbitals

Previously we considered an octahedral arrangement of charges about a central atom. The charges produced a potential in the vicinity of the atom which perturbed the motion of the electrons. We shall now view the situation from the standpoint of molecular orbital theory. Again we assume an octahedral arrangement with the ligands located at $x = \pm a$, $y = \pm a$, $z = \pm a$ and the central atom at the origin (Fig. 28.8). The molecular orbitals have the form

$$\psi = \alpha\psi(\Gamma) + \beta \sum_i a_i \psi_i \tag{28.5-1}$$

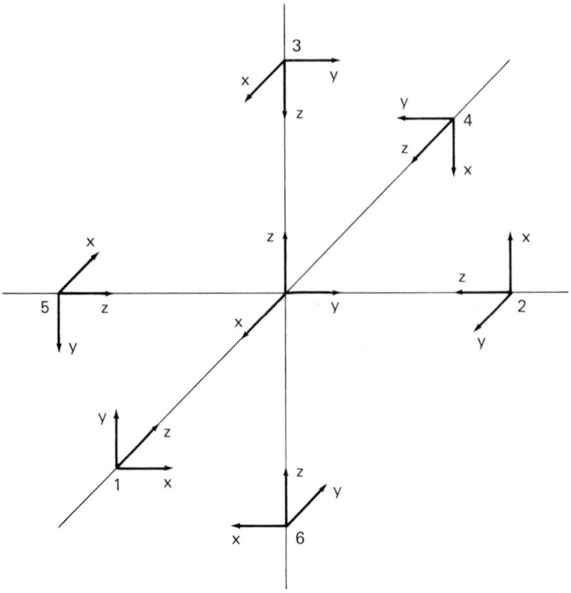

FIG. 28.8 Coordinate system for molecular orbital calculations of an octahedral complex AB_6.

where $\psi(\Gamma)$ is a wave function of the central atom transforming in the molecular point group as the irreducible representation Γ and $\sum_i a_i\psi_i$ is a linear combination of wave functions centered on the ligands and transforming as the same irreducible representation Γ.

In the general case the central atom and the ligands will form both σ and π bonds. The transformation properties of these bonds lead to the characters shown in Table 28.6 and with this information one obtains the reduction of Γ_σ and Γ_π:

$$\begin{aligned} \Gamma_\sigma &= \mathrm{A_{1g}} + \mathrm{E_g} + \mathrm{T_{1u}}, \\ \Gamma_\pi &= \mathrm{T_{1g}} + \mathrm{T_{1u}} + \mathrm{T_{2g}} + \mathrm{T_{2u}}. \end{aligned} \qquad (28.5\text{-}2)$$

The corresponding basis functions are shown in Table 28.7 in which z_i stands for s, p_z, d_{z^2}, ... and x_i, y_i stand for orbitals which transform like p_x and p_y,

TABLE 28.6

Characters of σ and π Orbitals in O_h

O_h	E	$8C_3$	$3C_2$	$6C_4$	$6C_2'$	i	$8iC_3$	$3iC_2$	$6iC_4$	$6iC_2'$
σ	6	0	2	2	0	0	0	4	0	2
π	12	0	−4	0	0	0	0	0	0	0

TABLE 28.7

Symmetry Adapted Orbitals for the Complex AB_6 in an Octahedral Configuration

Symmetry	Central atom orbitals	Ligand orbitals	
		σ	π
a_{1g}	s	$\frac{1}{\sqrt{6}}(z_1 + z_2 + z_3 + z_4 + z_5 + z_6)$	
e_g	$d_{x^2-y^2}$	$\frac{1}{2}(z_1 - z_2 + z_4 - z_5)$	
	d_{z^2}	$\frac{1}{3\sqrt{2}}(2z_3 + 2z_6 - z_1 - z_2 - z_4 - z_5)$	
t_{1g}			$\frac{1}{2}(x_3 - y_1 - x_4 + y_6)$
			$\frac{1}{2}(x_2 - y_3 - x_6 + y_5)$
			$\frac{1}{2}(x_1 - y_2 - x_5 + y_4)$
t_{1u}	p_x	$\frac{1}{\sqrt{2}}(z_1 - z_4)$	$\frac{1}{2}(x_3 + y_2 - x_5 - y_6)$
	p_y	$\frac{1}{\sqrt{2}}(z_2 - z_5)$	$\frac{1}{2}(x_1 + y_3 - x_6 - y_4)$
	p_z	$\frac{1}{\sqrt{2}}(z_3 - z_6)$	$\frac{1}{2}(x_2 + y_1 - x_4 - y_5)$
t_{2g}	d_{zx}		$\frac{1}{2}(x_3 + y_1 + x_4 + y_6)$
	d_{yz}		$\frac{1}{2}(x_2 + y_3 + x_6 + y_5)$
	d_{xy}		$\frac{1}{2}(x_1 + y_2 + x_5 + y_4)$
t_{2u}			$\frac{1}{2}(x_3 - y_2 + x_5 - y_6)$
			$\frac{1}{2}(x_1 - y_3 + x_6 - y_4)$
			$\frac{1}{2}(x_2 - y_1 + x_4 - y_5)$

respectively. The central atom orbitals that participate in σ bonds are $(s, d_{x^2-y^2}, d_{z^2}, p_x, p_y, p_z)$ which may be abbreviated to (sd^2p^3). We note that there are no central atom orbitals (within the set s, p, d) that transform according to T_{1g} or T_{1u}.

We may now construct molecular orbitals of the type (28.5-1); thus, for example,

$$\psi_{x^2-y^2} = \alpha d_{x^2-y^2} + \beta \tfrac{1}{2}(z_1 - z_2 + z_4 - z_5) \tag{28.5-3}$$

is a molecular orbital having e_g symmetry. α and β are mixing coefficients obtained from a variational calculation. If we neglect the overlap between central atom and ligand orbitals, $\alpha^2 + \beta^2 = 1$. For $\alpha \approx 1$ we have the crystal field case whereas $\alpha = \beta = \sqrt{\frac{1}{2}}$ corresponds to the covalent case.

π orbitals produce much weaker bonding than the σ orbitals. In many cases it is possible to neglect the π bonds and consider only the σ bonds. If this is done, the π orbitals are treated as nonbonding. A typical situation for an octahedral complex is shown in Fig. 28.9. The crystal field splitting is here interpreted as the difference in energy between the antibonding e_g and t_{2g} level.

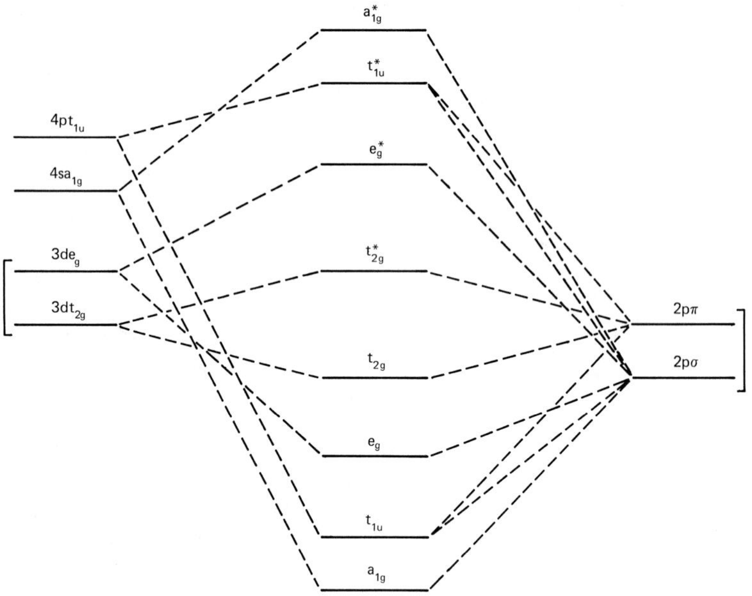

FIG. 28.9 Molecular orbital energy level diagram for octahedral complexes.

APPENDIX 1

DIRAC NOTATION

If f and g depend on a discrete index, $\langle f|g\rangle$ is defined by

$$\langle f|g\rangle = f_1{}^*g_1 + f_2{}^*g_2 + \cdots . \tag{A1-1}$$

We may then interpret the *ket* $|g\rangle$ as a column matrix (vector)

$$g = \begin{pmatrix} g_1 \\ g_2 \\ \vdots \end{pmatrix} \tag{A1-2}$$

with components $g_1, g_2, \ldots$ and the *bra* $\langle f|$ as a row matrix (dual vector)

$$\langle f| = (f_1{}^*, f_2{}^*, \ldots) \tag{A1-3}$$

with components $f_1{}^*, f_2{}^*, \ldots$.

If f and g depend continuously on a set of variables, say $f = f(\mathbf{r})$ and $g = g(\mathbf{r})$,

$$\langle f|g\rangle = \int f^*(\mathbf{r})g(\mathbf{r})\,d\mathbf{r}. \tag{A1-4}$$

In this case the ket $|g\rangle$ is regarded as a vector in a function space and the bra $\langle f|$ a vector in the dual space. For both the discrete and continuous case a bra $\langle f|$ is related to a ket $|f\rangle$ by

$$\langle f| = (|f\rangle)^*. \tag{A1-5}$$

The scalar product is a pure number that satisfies

$$\langle f|g\rangle = \langle g|f\rangle^*, \tag{A1-6}$$

$$\langle f|f\rangle \geq 0. \tag{A1-7}$$

If $\langle f|g\rangle = 0$, f and g are said to be orthogonal. Since the Hermitian conjugate of a pure number is identical with its complex conjugate,

$$\langle f|g\rangle^\dagger = \langle f|g\rangle^*. \tag{A1-8}$$

APPENDIX 2

OPERATORS

Let A be a linear operator acting on the ket $|f\rangle$. This may be written in either of two ways:

$$A|f\rangle \qquad \text{or} \qquad |Af\rangle \tag{A2-1}$$

The operator A can be quite general, e.g., the Laplacian operator, an angular momentum operator, another function which multiplies $|f\rangle$, a matrix, etc. From (A1-5),

$$(A|f\rangle)^+ = (|f\rangle)^+ A^+ = \langle f|A^+ \tag{A2-2}$$

and

$$(A|f\rangle)^+ = (|Af\rangle)^+ = \langle Af|. \tag{A2-3}$$

Therefore

$$\langle Af| = \langle f|A^+. \tag{A2-4}$$

Equation (A2-4) is often called the "turn-over" rule.

When f and g depend continuously on a set of variables, $\langle f|A|g\rangle$ is a matrix element of the operator A and is given by

$$\langle f|A|g\rangle \equiv \langle f|Ag\rangle = \int f^*(\mathbf{r})Ag(\mathbf{r})\,d\mathbf{r} \tag{A2-5}$$

with the understanding that the operator A acts on the function that stands to the right; $\langle f|A|f\rangle$ is known as a diagonal matrix element or the expectation value of the operator A in the state f. For a product of two operators AB

$$\langle f|AB|g\rangle = \sum_k \langle f|A|k\rangle\langle k|B|g\rangle. \tag{A2-6}$$

Matrix elements are pure numbers. Using (A2-5) and the "turn-over" rule (A2-4)

$$\langle f|A|g\rangle^* \equiv \langle f|A|g\rangle^\dagger = \langle f|Ag\rangle^\dagger = \langle Ag|f\rangle$$
$$= \langle g|A^\dagger f\rangle \equiv \langle g|A^\dagger|f\rangle. \qquad \text{(A2-7)}$$

When A is given in matrix form, $A^\dagger$ is directly obtained from the definition $(A^\dagger)_{ij} = A^*_{ji}$. If A is an operator which depends continuously on a set of variables, $A^\dagger$ is obtained by means of (A2-7).

If A is Hermitian, $A = A^\dagger$ and

$$\langle f|A|g\rangle = \langle A^\dagger f|g\rangle = \langle Af|g\rangle. \qquad \text{(A2-8)}$$

This means that a Hermitian operator can operate on functions standing either to the right or to the left. Expressed as an integral (A2-8) states that

$$\int f^*(\mathbf{r})[Ag(\mathbf{r})]\,d\mathbf{r} = \int [Af(\mathbf{r})]^* g(\mathbf{r})\,d\mathbf{r}. \qquad \text{(A2-9)}$$

APPENDIX 3

EIGENVALUES AND EIGENFUNCTIONS

The general eigenvalue equation associated with an operator A has the form

$$A|f\rangle = \lambda|f\rangle. \tag{A3-1}$$

The values of λ and the corresponding kets $|f\rangle$ which satisfy (A3-1) are known, respectively, as the eigenvalues and eigenkets of the operator A. Depending on the context, the eigenkets are also known as eigenvectors, eigenfunctions, eigenstates, or state vectors.

Let S be a linear operator and let

$$|g\rangle = S|f\rangle \qquad \text{or} \qquad |f\rangle = S^{-1}|g\rangle. \tag{A3-2}$$

Substitution of (A3-2) into (A3-1) leads to

$$SAS^{-1}|g\rangle = \lambda|g\rangle. \tag{A3-3}$$

Letting

$$A' = SAS^{-1}, \tag{A3-4}$$

we say that A' is the *transform* of A by the operator S. Alternatively, one refers to the transformation (A3-4) as a *similarity* or *equivalence* transformation. The important characteristics of such a transformation are:

$$\begin{aligned}&A' \text{ and } A \text{ have the same eigenvalues,}\\ &\operatorname{Tr} A' = \operatorname{Tr} A,\\ &\det A' = \det A.\end{aligned} \tag{A3-5}$$

If A contains both diagonal and off-diagonal matrix elements (in some basis) and A', given by (A3-4), contains only diagonal elements, we say that A has been *diagonalized* by S. A *unitary* transformation of an operator A is one in which A is transformed according to (A3-4) and S is a unitary operator. A unitary transformation of A may or may not diagonalize A.

If A is Hermitian:

(a) A can be diagonalized by some unitary operator S,
(b) the eigenvalues of A are real,
(c) the eigenfunctions of A belonging to different eigenvalues are orthogonal.

Hermitian operators that represent physical quantities are called *observables*. It is assumed that observables possess a complete, orthonormal set of eigenfunctions. This makes it possible to expand an arbitrary function in terms of the eigenfunctions of an observable. Let these eigenfunctions be represented by $|i\rangle$. Then

$$|\psi\rangle = \sum_i a_i |i\rangle \tag{A3-6}$$

where $|\psi\rangle$ is an arbitrary function and a_i are coefficients. If $|j\rangle$ is a particular eigenfunction,

$$\langle j|\psi\rangle = \sum_i a_i \langle j|i\rangle = a_j,$$

where we have used the orthonormality property of the eigenfunctions. Thus

$$|\psi\rangle = \sum_i |i\rangle\langle i|\psi\rangle$$

or

$$\sum_i |i\rangle\langle i| = 1. \tag{A3-7}$$

Equation (A3-7) is the completeness or closure condition. Note that $|i\rangle\langle i|$ is an operator and has a totally different meaning from $\langle i|i\rangle$.

If A is unitary, it can be diagonalized by a similarity transformation with some other unitary matrix.

APPENDIX 4

RELATIONSHIPS AMONG UNIT VECTORS

$$\hat{\mathbf{i}} \equiv \hat{\mathbf{e}}_x = \begin{pmatrix} 1 \\ 0 \\ 0 \end{pmatrix} = -\frac{1}{\sqrt{2}}(\hat{\mathbf{e}}_{+1} - \hat{\mathbf{e}}_{-1}),$$

$$\hat{\mathbf{j}} \equiv \hat{\mathbf{e}}_y = \begin{pmatrix} 0 \\ 1 \\ 0 \end{pmatrix} = \frac{i}{\sqrt{2}}(\hat{\mathbf{e}}_{+1} + \hat{\mathbf{e}}_{-1}), \qquad \text{(A4-1)}$$

$$\hat{\mathbf{k}} \equiv \hat{\mathbf{e}}_z = \begin{pmatrix} 0 \\ 0 \\ 1 \end{pmatrix} = \hat{\mathbf{e}}_0,$$

$$\hat{\mathbf{e}}_{+1} = -\frac{1}{\sqrt{2}}\begin{pmatrix} 1 \\ i \\ 0 \end{pmatrix} = -\frac{1}{\sqrt{2}}(\hat{\mathbf{e}}_x + i\hat{\mathbf{e}}_y),$$

$$\hat{\mathbf{e}}_0 = \begin{pmatrix} 0 \\ 0 \\ 1 \end{pmatrix} = \hat{\mathbf{e}}_z, \qquad \text{(A4-2)}$$

$$\hat{\mathbf{e}}_{-1} = \frac{1}{\sqrt{2}}\begin{pmatrix} 1 \\ -i \\ 0 \end{pmatrix} = \frac{1}{\sqrt{2}}(\hat{\mathbf{e}}_x - i\hat{\mathbf{e}}_y).$$

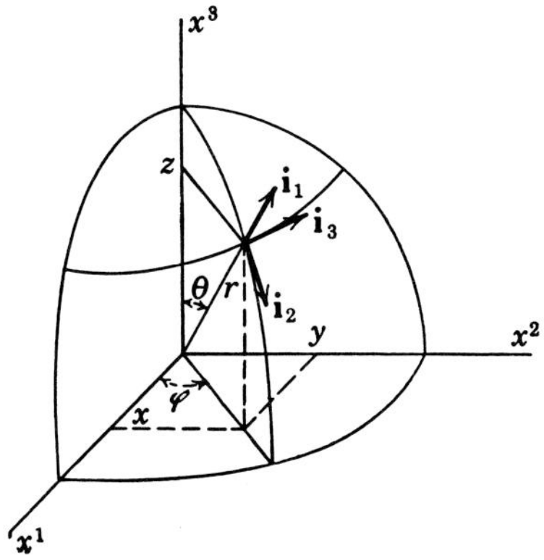

FIG. A4.1 Unit vectors in spherical coordinates (Stratton, 1941).

Referring to Fig. A4.1 for the definition of $\mathbf{i}_1$, $\mathbf{i}_2$, and $\mathbf{i}_3$,

$$\begin{aligned}\hat{\mathbf{i}}_1 &= \hat{\mathbf{e}}_x \sin\theta\cos\varphi + \hat{\mathbf{e}}_y \sin\theta\sin\varphi + \hat{\mathbf{e}}_z \cos\theta,\\ \hat{\mathbf{i}}_2 &= \hat{\mathbf{e}}_x \cos\theta\cos\varphi + \hat{\mathbf{e}}_y \cos\theta\sin\varphi - \hat{\mathbf{e}}_z \sin\theta,\\ \hat{\mathbf{i}}_3 &= -\hat{\mathbf{e}}_x \sin\varphi + \hat{\mathbf{e}}_y \cos\varphi,\end{aligned} \tag{A4-3}$$

$$\begin{aligned}\hat{\mathbf{e}}_x &= \hat{\mathbf{i}}_1 \sin\theta\cos\varphi + \hat{\mathbf{i}}_2 \cos\theta\cos\varphi - \hat{\mathbf{i}}_3 \sin\varphi,\\ \hat{\mathbf{e}}_y &= \hat{\mathbf{i}}_1 \sin\theta\sin\varphi + \hat{\mathbf{i}}_2 \cos\theta\sin\varphi + \hat{\mathbf{i}}_3 \cos\varphi,\\ \hat{\mathbf{e}}_z &= \hat{\mathbf{i}}_1 \cos\theta - \hat{\mathbf{i}}_2 \sin\theta,\end{aligned} \tag{A4-4}$$

$$\begin{aligned}\hat{\mathbf{e}}_{+1} &= -\frac{e^{i\varphi}}{\sqrt{2}}[\sin\theta\,\hat{\mathbf{i}}_1 + \cos\theta\,\hat{\mathbf{i}}_2 + i\,\hat{\mathbf{i}}_3],\\ \hat{\mathbf{e}}_0 &= \cos\theta\,\hat{\mathbf{i}}_1 - \sin\theta\,\hat{\mathbf{i}}_2,\\ \hat{\mathbf{e}}_{-1} &= \frac{e^{-i\varphi}}{\sqrt{2}}[\sin\theta\,\hat{\mathbf{i}}_1 + \cos\theta\,\hat{\mathbf{i}}_2 - i\,\hat{\mathbf{i}}_3].\end{aligned} \tag{A4-5}$$

The unit vectors satisfy orthogonality conditions:

$$\hat{\mathbf{e}}_i \cdot \hat{\mathbf{e}}_j = \delta_{ij} \qquad (i, j = x, y, z), \tag{A4-6}$$

$$= \hat{\mathbf{e}}_p{}^* \cdot \hat{\mathbf{e}}_q = (-1)^p \hat{\mathbf{e}}_{-p} \cdot \hat{\mathbf{e}}_q = \delta_{pq} \qquad (p, q = 1, 0, -1). \tag{A4-7}$$

An arbitrary vector may be expressed in terms of the unit vectors. Thus

$$\begin{aligned}\mathbf{A} &= A_x\hat{\mathbf{e}}_x + A_y\hat{\mathbf{e}}_y + A_z\hat{\mathbf{e}}_z\\ &= \sum_q (-1)^q A_q \hat{\mathbf{e}}_{-q} \qquad (q = 1, 0, -1),\end{aligned} \tag{A4-8}$$

where

$$\begin{aligned} A_{+1} &= -\frac{1}{\sqrt{2}}(A_x + iA_y), \\ A_0 &= A_z, \\ A_{-1} &= \frac{1}{\sqrt{2}}(A_x - iA_y). \end{aligned} \tag{A4-9}$$

For two vectors **A** and **B**

$$\begin{aligned} \mathbf{A} \cdot \mathbf{B} &= -A_{+1}B_{-1} + A_0B_0 - A_{-1}B_{+1} \\ &= \sum_q (-1)^q A_q B_{-q} \\ &= \sum_q A_q{}^* B_q \qquad (q = 1, 0, -1). \end{aligned} \tag{A4-10}$$

APPENDIX 5

BESSEL FUNCTIONS

The Bessel equation is

$$\frac{d^2y}{dx^2} + \frac{1}{x}\frac{dy}{dx} + \left(1 - \frac{n^2}{x^2}\right)y = 0. \qquad \text{(A5-1)}$$

In terms of $x = kr$, (A5-1) transforms into

$$\frac{d^2y}{dr^2} + \frac{1}{r}\frac{dy}{dr} + \left(k^2 - \frac{n^2}{r^2}\right)y = 0. \qquad \text{(A5-2)}$$

The solutions to the Bessel equation are

$$y = \begin{cases} AJ_n(x) + BN_n(x) & (n \text{ integral}), \\ AJ_n(x) + BJ_{-n}(x) & (n \text{ nonintegral}), \end{cases} \qquad \text{(A5-3)}$$

in which J_n and N_n are Bessel functions of the first and second kind, respectively.

Hankel functions:

$$\begin{aligned} H_n^{(1)}(x) &= J_n(x) + iN_n(x), \\ H_n^{(2)}(x) &= J_n(x) - iN_n(x). \end{aligned} \qquad \text{(A5-4)}$$

Spherical Bessel Functions:

$$\begin{aligned} j_l(x) &= \sqrt{\pi/2x}J_{l+1/2}(x), \\ n_l(x) &= \sqrt{\pi/2x}N_{l+1/2}(x), \end{aligned} \qquad \text{(A5-5)}$$

with l integral.

Spherical Hankel Functions:

$$h_l^{(1)}(x) = \sqrt{\pi/2x}H_{l+1/2}^{(1)}(x) = \sqrt{\pi/2x}[J_{l+1/2}(x) + iN_{l+1/2}(x)]$$
$$= j_l(x) + in_l(x), \tag{A5-6}$$

$$h_l^{(2)}(x) = \sqrt{\pi/2x}H_{l+1/2}^{(2)}(x) = \sqrt{\pi/2x}[J_{l+1/2}(x) - iN_{l+1/2}(x)]$$
$$= j_l(x) - in_l(x). \tag{A5-7}$$

Some examples of spherical functions:

$$j_0(x) = \frac{\sin x}{x}, \qquad h_0^{(1)}(x) = \frac{e^{ix}}{ix},$$
$$n_0(x) = -\frac{\cos x}{x}, \qquad h_0^{(2)}(x) = -\frac{e^{-ix}}{ix}, \tag{A5-8}$$

$$j_1(x) = \frac{\sin x}{x^2} - \frac{\cos x}{x}, \qquad h_1^{(1)}(x) = \frac{e^{ix}}{x}\left(1 + \frac{i}{x}\right),$$
$$n_1(x) = -\frac{\cos x}{x^2} - \frac{\sin x}{x}, \qquad h_1^{(2)}(x) = \frac{e^{-ix}}{x}\left(1 - \frac{i}{x}\right), \tag{A5-9}$$

$$j_2(x) = \left(\frac{3}{x^3} - \frac{1}{x}\right)\sin x - \frac{3\cos x}{x^2},$$
$$n_2(x) = -\left(\frac{3}{x^3} - \frac{1}{x}\right)\cos x - \frac{3\sin x}{x^2}, \tag{A5-10}$$

$$h_2^{(1)}(x) = \frac{ie^{ix}}{x}\left(1 + \frac{3i}{x} - \frac{3}{x^2}\right),$$
$$h_2^{(2)}(x) = \frac{-ie^{-ix}}{x}\left(1 - \frac{3i}{x} - \frac{3}{x^2}\right). \tag{A5-11}$$

Recursion Relations: Let $f_l(x)$ represent any of the functions $j_l(x), n_l(x)$, $h_l^{(1)}(x), h_l^{(2)}(x)$:

$$f_{l-1}(x) + f_{l+1}(x) = \frac{(2l+1)}{x} f_l(x), \tag{A5-12}$$

$$lf_{l-1}(x) - (l+1)f_{l+1}(x) = (2l+1)\frac{d}{dx} f_l(x), \tag{A5-13}$$

$$\frac{l+1}{x} f_l(x) + \frac{d}{dx} f_l(x) = f_{l-1}(x), \tag{A5-14}$$

$$\frac{l}{x} f_l(x) - \frac{d}{dx} f_l(x) = f_{l+1}(x). \tag{A5-15}$$

Differential Equation: $f_l(x)$ satisfy

$$\left[\frac{d^2}{dx^2} + \frac{2}{x}\frac{d}{dx} + \left(1 - \frac{l(l+1)}{x^2}\right)\right] f_l(x) = 0 \qquad \text{(A5-16)}$$

where l is an integer.
Asymptotic forms:

$$\begin{aligned} j_l(x) &\to x^l/(2l+1)!! \\ n_l(x) &\to (2l-1)!!/x^{l+1} \end{aligned} \qquad (x \ll l) \qquad \text{(A5-17)}$$

where $(2l+1)!! = (2l+1)(2l-1)(2l-3)\ldots$.

$$\begin{aligned} j_l(x) &\to \frac{1}{x}\sin\left(x - \frac{l\pi}{2}\right) \\ n_l(x) &\to \frac{1}{x}\cos\left(x - \frac{l\pi}{2}\right) \end{aligned} \qquad (x \gg l) \qquad \text{(A5-18)}$$

$$\begin{aligned} h_l^{(1)}(x) &\to (-i)^{l+1}(e^{ix}/x) \\ h_l^{(2)}(x) &\to (i)^{l+1}(e^{-ix}/x). \end{aligned}$$

APPENDIX 6

LAGUERRE POLYNOMIALS

$$L_{n+1}^{2l+1}(x) = -[(n+l)!]^2 \sum_{k=0}^{n-l-1} \frac{(-x)^k}{k!(n-l-k-1)!(2l+k+1)!}$$

$$= B_0 + B_1 x + B_2 x^2 + \cdots + B_{n-l-1} x^{n-l-1} \qquad \text{(A6-1)}$$

where

$$B_k = -B_{k-1} \frac{n-l-k}{k(2l+k+1)}, \qquad \text{(A6-2)}$$

$$B_{n-l-1} = (-1)^{n-l} \frac{(n+l)!}{(n-l-1)!}. \qquad \text{(A6-3)}$$

Alternative forms for the Laguerre polynomials are

$$L_{n+l}^{2l+1}(x) = (-1)^{2l+1} \frac{(n+l)!}{(n-l-1)!} e^x x^{-2l-1} \frac{d^{n-l-1}}{dx^{n-l-1}} e^{-x} x^{n+l} \qquad \text{(A6-4)}$$

$$L_{n+l}^{2l+1}(x) = (-1)^{2l+1} \frac{[(n+l)!]^2}{(2l+1)!(n-l-1)!} F[l+1-n, 2l+2, x], \qquad \text{(A6-5)}$$

in which the confluent hypergeometric series $F(\alpha, \beta, x)$ is defined by

$$F(\alpha, \beta, x) = 1 + \frac{\alpha}{\beta \cdot 1!} x + \frac{\alpha(\alpha+1)}{\beta(\beta+1) \cdot 2!} x^2 + \frac{\alpha(\alpha+1)(\alpha+2)}{\beta(\beta+1)(\beta+2) \cdot 3!} x^3 + \cdots . \qquad \text{(A6-6)}$$

APPENDIX 7

HERMITE POLYNOMIALS

$$H_n(x) = (-1)^n e^{x^2} \frac{d^n}{dx^n} e^{-x^2}, \tag{A7-1}$$

$$\frac{d^2 H_n}{dx^2} - 2x \frac{dH_n}{dx} - 2nH_n = 0, \tag{A7-2}$$

$$H_n(x) = (-1)^n e^{x^2} \left(\frac{d^n}{dx^n} e^{-x^2} \right), \tag{A7-3}$$

$$\frac{dH_n}{dx} = 2nH_{n-1}, \tag{A7-4}$$

$$\left(2x - \frac{d}{dx} \right) H_n = H_{n-1}, \tag{A7-5}$$

$$2xH_n = H_{n+1} + 2xH_{n-1}, \tag{A7-6}$$

$$\int_{-\infty}^{\infty} H_n(x) H_p(x) e^{-x^2} \, dx = 0 \qquad (n \neq p) \tag{A7-7}$$

$$\int [H_n(x)]^2 e^{-x^2} \, dx = 2^n n! \sqrt{\pi}. \tag{A7-8}$$

The first six Hermite polynomials are

$$H_0(x) = 1, \qquad H_1(x) = 2x, \qquad H_2(x) = -2 + 4x^2, \qquad H_3(x) = -12x + 8x^3,$$
$$H_4(x) = 12 - 48x^2 + 16x^4, \qquad H_5(x) = 120x - 160x^3 + 32x^5. \tag{A7-9}$$

APPENDIX 8

DIRAC δ-FUNCTIONS

One-Dimensional δ functions. Definition:

$$\int F(x)\,\delta(x - x_0)\,dx = F(x_0) \tag{A8-1}$$

provided the integration includes $x = x_0$.
Properties:

$$\delta(x' - x) = \delta(x - x'), \tag{A8-2}$$

$$x\,\delta x = 0, \tag{A8-3}$$

$$\delta(\alpha x) = \frac{1}{|\alpha|}\,\delta(x), \tag{A8-4}$$

$$f(x)\,\delta x = f(0)\,\delta(x), \tag{A8-5}$$

$$\delta[(x - x_1)(x - x_2)] = \frac{\delta(x - x_1) + \delta(x - x_2)}{|x_1 - x_2|}, \tag{A8-6}$$

$$\delta(x^2 - a^2) = \frac{1}{2a}\,[\delta(x - a) + \delta(x + a)] \qquad (a > 0), \tag{A8-7}$$

$$\delta(x - x_1)\,\delta(x - x_2) = \delta(x - x_1)\,\delta(x_1 - x_2) = \delta(x - x_2)\,\delta(x_1 - x_2), \tag{A8-8}$$

$$\int_{-\infty}^{\infty} \delta(x)\,dx = 1, \tag{A8-9}$$

$$\delta'(-x) = -\delta'(x), \tag{A8-10}$$

$$x\,\delta'(x) = -\delta(x), \tag{A8-11}$$

$$x^2 \, \delta'(x) = 0, \tag{A8-12}$$

$$\int \delta^{(n)}(x) f(x) \, dx = (-1)^n f^{(n)}(0). \tag{A8-13}$$

If $u_n(x)$ form a complete set,

$$\begin{aligned} f(x) &= \sum_n c_n u_n(x) \\ &= \sum_n u_n(x) \int u_n^*(x') f(x') \, dx' \\ &= \int \left[\sum_n u_n^*(x') u_n(x) \right] f(x') \, dx' \\ &= \int \delta(x' - x) f(x') \, dx'. \end{aligned}$$

We then have the *closure* property

$$\sum_n u_n^*(x') u_n(x) = \delta(x' - x). \tag{A8-14}$$

Representations:

$$\delta(x) = \frac{1}{2\pi} \int_{-\infty}^{\infty} e^{ikx} \, dk, \left(= \frac{1}{2\pi} \lim_{\alpha \to 0} \int_{-\infty}^{\infty} e^{ikx} e^{-\alpha |x|} \, dx \right) \tag{A8-15}$$

$$= \frac{1}{2\pi} \int_{-\infty}^{\infty} \cos(kx) \, dk \tag{A8-16}$$

$$= \lim_{\alpha \to \infty} \frac{\sin(\alpha x)}{\pi x} \tag{A8-17}$$

$$= \lim_{\alpha \to \infty} \frac{1 - \cos(\alpha x)}{\pi \alpha x^2} = \lim_{\alpha \to \infty} \frac{\sin^2 \alpha (x/2)}{2\pi \alpha (x/2)^2} \tag{A8-18}$$

$$= \lim_{\alpha \to 0} \frac{\alpha}{\pi(\alpha^2 + x^2)} \tag{A8-19}$$

$$= \lim_{\alpha \to 0} \frac{e^{-x^2/4k\alpha}}{\sqrt{4\pi k \alpha}} \qquad (k > 0) \tag{A8-20}$$

$$= \lim_{\alpha \to \infty} \frac{\alpha e^{-\alpha^2 x^2}}{\sqrt{\pi}} \tag{A8-21}$$

$$= \lim_{\alpha \to 0} \frac{1}{\pi} \int_0^{\infty} e^{-\alpha t} \cos kx \, dk. \tag{A8-22}$$

Three-dimensional δ functions:

$$\delta(r) = \delta(x)\,\delta(y)\,\delta(z) \tag{A8-23}$$

$$= \begin{cases} \dfrac{1}{2\pi}\dfrac{\delta(r)}{r^2} & \text{(A8-24)} \\[2ex] -\dfrac{1}{2\pi}\dfrac{\delta'(r)}{r} & \text{(A8-25)} \\[2ex] \dfrac{1}{(2\pi)^3}\displaystyle\int_{-\infty}^{\infty} e^{i\mathbf{k}\cdot\mathbf{r}}\,d^3k, & \text{(A8-26)} \end{cases}$$

$$\delta(\mathbf{r}_1 - \mathbf{r}_2) = \frac{1}{{r_1}^2}\,\delta(r_1 - r_2)\,\delta(\cos\theta_1 - \cos\theta_2)\,\delta(\varphi_1 - \varphi_2), \tag{A8-27}$$

$$\delta(\mathbf{k} - \mathbf{k}') = \frac{1}{(2\pi)^3}\int e^{i(\mathbf{k}-\mathbf{k}')\cdot\mathbf{r}}\,d\mathbf{r}, \tag{A8-28}$$

$$\int f(\mathbf{r}')\,\delta(\mathbf{r} - \mathbf{r}')\,d\mathbf{r}' = f(\mathbf{r}) \qquad \text{(closure)}, \tag{A8-29}$$

$$\nabla^2(1/r) = -4\pi\,\delta(\mathbf{r}). \tag{A8-30}$$

REFERENCES

Part I

Abragam, A., and Bleaney, B. (1970). "Electron Paramagnetic Resonance of Transition Ions." Oxford Univ. Press (Clarendon), London and New York.

Abramowitz, M., and Stegun, I. A. (1965). "Handbook of Mathematical Functions." US Dept. Commerce, Nat. Bur. Std., Appl. Math. Ser. 55.

Akhiezer, A. I., and Berestetskii, V. B. (1965). "Quantum Electrodynamics." Wiley (Interscience), New York.

Atkins, P. W., Child, M. S., and Phillips, C. S. G. (1970). "Tables for Group Theory." Oxford Univ. Press, London and New York.

Bethe, H. (1929). *Ann. Physik* **3**, 135.

Chestnut, D. B. (1974). "Finite Groups and Quantum Theory." Wiley, New York.

Cotton, F. A. (1971). "Chemical Applications of Group Theory." Wiley (Interscience), New York.

de Shalit, A., and Feshbach, H. (1974). "Theoretical Nuclear Physics, Vol. 1, Nuclear structure." Wiley, New York.

de Shalit, A., and Talmi, I. (1963). "Nuclear Shell Theory." Academic Press, New York.

Edmonds, A. R. (1960). "Angular Momenta and Quantum Mechanics." Princeton Univ. Press, Princeton, New Jersey.

Fano, U., and Racah, G. (1959). "Irreducible Tensorial Sets." Academic Press, New York.

Ferraro, J. P., and Ziomek, J. S. (1969). "Introductory Group Theory." Plenum, New York.

Fonda, L., and Ghirardi, G. C. (1970). "Symmetry Principles in Quantum Physics." Dekker, New York.

Gelfand, I. M., Minlos, R. A., and Shapiro, Z. Ya. (1963). "Representation of the Rotation and Lorentz Groups and Their Applications." Pergamon, Oxford.

Gilmore, R. (1974). "Lie Groups, Lie Algebras and Some of Their Applications." Wiley, New York.

Griffith, J. S. (1961). "Theory of Transition Metal Ions." Cambridge Univ. Press, London and New York.

Griffith, J. S. (1962). "The Irreducible Tensor Method for Molecular Symmetry Groups." Prentice-Hall, Englewood Cliffs, New Jersey.

Hall, G. G. (1967). "Applied Group Theory." Longmans, Green, New York.
Hamermesh, M. (1962). "Group Theory." Addison-Wesley, Reading, Massachusetts.
Harnung, S. E., and Schäffer, C. E. (1972). *Structure and Bonding* **12**, 201, 257.
Heine, V. (1960). "Group Theory in Quantum Mechanics." Pergamon, Oxford.
Herzberg, G. (1966). "Electronic Spectra and Electronic Structure of Polyatomic Molecules." Van Nostrand, New York.
Hochstrasser, R. M. (1966). "Molecular Aspects of Symmetry." Benjamin, New York.
Jackson, J. D. (1962). "Classical Electrodynamics." Wiley, New York.
Judd, B. R. (1963). "Operator Techniques in Atomic Spectroscopy." McGraw-Hill, New York.
Kaplan, I. G. (1975). "Symmetry of Many-Electron Systems." Academic Press, New York.
Koster G. F., Dimmock, J. O., Wheeler, R. G., and Statz, H. (1963). "Properties of the Thirty-Two Point Groups." MIT Press, Cambridge, Massachusetts.
Landau, L. D., and Lifshitz, E. M. (1965). "Quantum Mechanics, Non-Relativistic Theory." Pergamon, Oxford.
Lax, M. (1974). "Symmetry Principles in Solid State and Molecular Physics." Wiley, New York.
Levine, I. N. (1970). "Quantum Chemistry." Allyn & Bacon, Rockleigh, New Jersey.
Löwdin, P. O. (1964). *Rev. Modern Phys.* **36**, 466.
McWeeny, R. (1963). "Symmetry." Pergamon, Oxford.
Mizushima, M. (1970). "Quantum Mechanics of Atomic Spectra and Atomic Structure." Benjamin, New York.
Racah, G. (1942a). *Phys. Rev.* **61**, 186.
Racah, G. (1942b). *Phys. Rev.* **62**, 438.
Racah, G. (1943). *Phys. Rev.* **63**, 367.
Racah, G. (1949). *Phys. Rev.* **76**, 1352.
Rose, M. E. (1957). "Elementary Theory of Angular Momentum." Wiley, New York.
Rotenberg, M., Bivins, R., Metropolis, N., and Wooten, Jr., K. (1959). "The $3j$ and $6j$ Symbols." MIT Press, Cambridge, Massachusetts.
Shore, B. W., and Menzel, D. H. (1968). "Principles of Atomic Spectra." Wiley, New York.
Sobelman, I. I. (1972). "Theory of Atomic Spectra." Pergamon, Oxford.
Tinkham, M. (1964). "Group Theory and Quantum Mechanics." McGraw-Hill, New York.
Watanabe, H. (1966). "Operator Methods in Ligand Field Theory." Prentice-Hall, Englewood Cliffs, New Jersey.
Weyl, H. (1931). "The Theory of Groups and Quantum Mechanics." Dover, New York.
Wigner, E. P. (1959). "Group Theory." Academic Press, New York.
Wybourne, B. G. (1974). "Classical Groups for Physicists." Wiley, New York.

Part II

Abragam, A., and Bleaney, B. (1970). "Electron Paramagnetic Resonance of Transition Ions." Oxford Univ. Press (Clarendon), London and New York.
Abrikosov, A. A., Gorkov, L. P., and Dzyaloshinski, I. E. (1963). "Methods of Quantum Field Theory in Statistical Mechanics." Prentice-Hall, Englewood Cliffs, New Jersey.
Anderson, P. W. (1963). "Concepts in Solids." Benjamin, New York.
Avery, J. (1972). "The Quantum Theory of Atoms, Molecules and Photons." McGraw-Hill, New York.
Baldwin, B. (1973). *Amer. J. Phys.* **41**, 678.
Bjorken, J. D., and Drell, S. D. (1964). "Relativistic Quantum Mechanics." McGraw-Hill, New York.
Fano, U. (1957). *Rev. Modern Phys.* **29**, 74.

Hameka, H. F. (1965). "Advanced Quantum Chemistry." Addison-Wesley, Reading, Massachusetts.
Harris, E. G. (1972). "A Pedestrian Approach to Quantum Field Theory." Wiley, New York.
Heine, V. (1960). "Group Theory in Quantum Mechanics." Pergamon, Oxford.
Landau, L. D., and Lifshitz, E. M. (1958). "Statistical Physics." Pergamon, Oxford.
Landau, L. D., and Lifshitz, E. M. (1965). "Quantum Mechanics, Non-Relativistic Theory." Pergamon, Oxford.
Levine, I. N. (1970). "Quantum Chemistry." Allyn & Bacon, Rockleigh, New Jersey.
Louisell, W. H. (1973). "Quantum Statistical Properties of Radiation." Wiley, New York.
Löwdin, P. O. (1955). *Phys. Rev.* **97**, 1474.
McWeeny, R., and Sutcliffe, B. T. (1969). "Methods of Molecular Quantum Mechanics." Academic Press, New York.
Merzbacher, E. (1970). "Quantum Mechanics." Wiley, New York.
Messiah, A. (1962). "Quantum Mechanics." Wiley, New York.
Mizushima, M. (1970). "Quantum Mechanics of Atomic Spectra and Atomic Structure." Benjamin, New York.
Moss, R. E. (1973). "Advanced Molecular Quantum Mechanics." Chapman & Hall, London.
Rodberg, L. S., and Thaler, R. M. (1967). "Quantum Theory of Scattering." Academic Press, New York.
Slater, J. C. (1960). "Quantum Theory of Atomic Structure," McGraw-Hill, New York.
Slichter, C. P. (1963). "Principles of Magnetic Resonance." Harper, New York.
Sugano, S., Tanabe, Y., and Kaminura, H. (1970). "Multiplets of Transition-Metal Ions in Crystals." Academic Press, New York.
Ter Haar, D. (1961). *Rep. Progress Phys.* **24**, 304.
Wigner, E. P. (1959). "Group Theory," Academic Press, New York.
Ziman, J. M. (1969). "Elements of Advanced Quantum Theory." Cambridge Univ. Press, London and New York.

Part III

Abragam, A. (1961). "The Principles of Nuclear Magnetism." Oxford Univ. Press (Clarendon), London and New York.
Abramowitz, M., and Stegun, I. A. (1965). "Handbook of Mathematical Functions." Dover, New York.
Avery, J. (1972). "The Quantum Theory of Atoms, Molecules and Photons," McGraw-Hill, New York.
Bethe, H. A., and Salpeter, E. E. (1957). "Quantum Mechanics of the One- and Two-Electron Atoms." Academic Press, New York.
Bethe, H. A., and Jackiw, R. W. (1968). "Intermediate Quantum Mechanics." Benjamin, New York.
Bjorken, J. D., and Drell, S. D. (1964). "Relativistic Quantum Mechanics." McGraw-Hill, New York.
Blinder, S. M. (1960). *J. Mol. Spec.* **5**, 17.
Condon, E. U., and Shortley, G. H. (1967). "Theory of Atomic Spectra." Cambridge Univ. Press, London and New York.
Fano, U., and Fano, L. (1972). "Physics of Atoms and Molecules." Univ. of Chicago Press, Chicago, Illinois.
Fermi, E. (1930). *Z. Phys.* **60**, 320.
Feynman, R. P. (1961). "Quantum Electrodynamics." Benjamin, New York.
Foldy, L. L., and Wouthuysen, S. A. (1958). *Phys. Rev.* **78**, 29.

Kuhn, H. G. (1962). "Atomic Spectra." Academic Press, New York.
Levine, I. N. (1970). "Quantum Chemistry." Allyn & Bacon, Rockleigh, New Jersey.
Mizushima, M. (1970). "Quantum Mechanics of Atomic Spectra and Atomic Structure." Benjamin, New York.
Sakurai, J. J. (1967). "Advanced Quantum Mechanics." Addison-Wesley, Reading, Massachusetts.
Shore, B. W., and Menzel, D. H. (1968). "Principles of Atomic Spectra," Wiley, New York.
Slater, J. C. (1960). "Quantum Theory of Atomic Structure." McGraw-Hill, New York.
Slichter, C. P. (1963). "Principles of Magnetic Resonance." Harper, New York.
Sobelman, I. I. (1972). "Theory of Atomic Spectra," Pergamon, New York.
Townes, C. H. and Schawlow, A. L. (1975). "Microwave Spectroscopy," Dover, New York.
Zeiger, H. J., and Pratt, G. W. (1973). "Magnetic Interactions in Solids." Oxford Univ. Press (Clarendon), London and New York.

Part IV

Anderson, P. W. (1963). "Concepts in Solids." Benjamin, New York.
Avery, J. (1972). "The Quantum Theory of Atoms, Molecules and Photons," McGraw-Hill, New York.
Bearden, J. A. (1967). "X-ray Wavelengths and X-ray Atomic Energy Levels," US Dept. Commerce, Nat. Bur. Stand., NSRDS-NBS 14.
Bethe, H. A. (1964). "Intermediate Quantum Mechanics." Benjamin, New York.
Bethe, H. A. and Jackiw, R. W. (1968). "Intermediate Quantum Mechanics." Benjamin, New York.
Bethe, H. A., and Salpeter, E. E. (1957). "Quantum Mechanics of the One- and Two-Electron Atoms," Academic Press, New York.
Clementi, E. (1965). *IBM J. Res. Develop.* **9**, 2.
Clementi, E., and Roetti, C. (1974). Atomic Data and Nuclear Data Tables, **14**, 177.
Condon, E. U., and Shortley, G. H. (1967). "Theory of Atomic Spectra," Cambridge Univ. Press, London and New York.
de Shalit, A., and Feshbach, H. (1974). "Theoretical Nuclear Physics, Vol. 1, Nuclear Structure," Wiley, New York.
de Shalit, A., and Talmi, I. (1963). "Nuclear Shell Theory," Academic Press, New York.
Fano, U., and Fano, L. (1972). "Physics of Atoms and Molecules." Univ. of Chicago Press, Chicago, Illinois.
Fock, V. A. (1935). *Z. Physik*. **98**, 145.
Froese-Fisher, C. (1972). *Atomic Data* **4**, 301.
Hamermesh, M. (1962). "Group Theory." Addison-Wesley, Reading, Massachusetts.
Hartree, D. R. (1957). "The Calculation of Atomic Structures." Wiley, New York.
Heine, V. (1960). "Group Theory in Quantum Mechanics." Pergamon, Oxford.
Herman, F. and Skillman, S. (1963). "Atomic Structure Calculations." Prentice-Hall, Englewood Cliffs, New Jersey.
Judd, B. R. (1963). "Operator Techniques in Atomic Spectroscopy," McGraw-Hill, New York.
Kaplan, I. G. (1975). "Symmetry of Many-Electron Systems," Academic Press, New York.
Kuhn, H. G. (1962). "Atomic Spectra," Academic Press, New York.
Levine, I. N. (1970). "Quantum Chemistry," Allyn & Bacon, Boston, Massachusetts.
Löwdin, P. O. (1955). *Phys. Rev.* **97**, 1474.
Mann, J. B. (1967). "Atomic Structure Calculation," Los Alamos Scientific Laboratory, LA 3690.
Mizushima, M. (1970). "Quantum Mechanics of Atomic Spectra and Atomic Structure," Benjamin, New York.

Moore, C. E. (1971). "Atomic Energy Levels," US Dept. Commerce, Nat. Bur. Std. NSRDS-NBS 35.
Nielsen, C. W., and Koster, G. F. (1963). "Spectroscopic Coefficients for p^n, d^n and f^n Configurations," M.I.T. Press, Cambridge, Massachusetts.
Racah, G. (1942). *Phys. Rev.* **61**, 186,
Racah, G. (1942). *Phys. Rev.* **62**, 438,
Racah, G. (1943). *Phys. Rev.* **63**, 367.
Racah, G. (1949). *Phys. Rev.* **76**, 1352.
Schaefer, III, H. F. (1972). "Electronic Structure of Atoms and Molecules," Addison-Wesley, Reading, Massachusetts.
Shore, B. W., and Menzel, D. H. (1968). "Principles of Atomic Spectra," Wiley, New York.
Slater, J. C. (1960). "Quantum Theory of Atomic Structure," McGraw-Hill, New York.
Sobelman, I. I. (1972). "Theory of Atomic Spectra," Pergamon Oxford.
Stevenson, R. (1965). "Multiplet Structure of Atoms and Molecules," Saunders, Philadelphia, Pennsylvania.
Tinkham, M. (1964). "Group Theory and Quantum Mechanics," McGraw-Hill, New York.
Townes, C. H. and Schawlow, A. L. (1975). "Microwave Spectroscopy," Dover, New York.
Zeiger, H. J. and Pratt, G. W. (1973). "Magnetic Interactions in Solids," Oxford Univ. Press (Clarendon), London and New York.

Part V

Akhiezer, A. I., and Berestetskii, V. B. (1965). "Quantum Electro-dynamics," Interscience, New York.
Avery, J. (1972). "The Quantum Theory of Atoms, Molecules and Photons," McGraw-Hill, New York.
Bethe, H. A., and Salpeter, E. E. (1957). "Quantum Mechanics of the One- and Two-Electron Atoms," Academic Press, New York.
Bethe, H. A. and Jackiw, R. W. (1968). "Intermediate Quantum Mechanics," Benjamin, New York.
Bloembergen, N. (1965). "Nonlinear Optics," Benjamin, New York.
Butcher, P. N. (1965). "Nonlinear Optical Phenomena," Bulletin 200, Engineering Experiment Station, Ohio State Univ., Columbus, Ohio.
de Shalit, A., and Feshbach, H. (1974). "Theoretical Nuclear Physics, Vol. 1, Nuclear Structure," Wiley, New York.
Dirac, P. A. M. (1958). "Principles of Quantum Mechanics," Oxford Univ. Press, London and New York.
Ducuing, J. (1969). In "Quantum Optics." (R. J. Glauber, ed.) Academic Press, New York.
Fain, V. M., and Khanin, Ya. I. (1969). "Quantum Electronics," MIT Press, Cambridge, Massachusetts.
Feynman, R. P. (1961). "Quantum Electrodynamics," Benjamin, New York.
Grandy, W. T. (1970). "Introduction to Electrodynamics and Radiation," Academic Press, New York.
Hameka, H. F. (1965). "Advanced Quantum Chemistry," Addison-Wesley, Reading, Massachusetts.
Harris, E. G. (1972). "A Pedestrian Approach to Quantum Field Theory," Wiley (Interscience), New York.
Heitler, W. (1954). "Quantum Theory of Radiation," Oxford Univ. Press (Clarendon) London and New York.
Kuhn, H. G. (1962). "Atomic Spectra," Academic Press, New York.
Levine, I. N. (1970). "Quantum Chemistry," Allyn Bacon, Rockleigh, New Jersey.

Loudon, R. (1973). "The Quantum Theory of Light," Oxford Univ. Press (Clarendon), London and New York.
Louisell, W. H. (1973). "Quantum Statistical Properties of Radiation," Wiley, New York.
Mizushima, M. (1970). "Quantum Mechanics of Atomic Spectra and Atomic Structure," Benjamin, New York.
Moore, C. E. (1971). "Atomic Energy Levels," Vol. I, II, III. US Dept. Commerce, Nat. Bur. Std., NSRDS-NBS 35.
Sargent III, M., Scully, M. O., and Lamb, Jr., W. E. (1974). "Laser Physics," Addison-Wesley, Reading, Massachusetts.
Shen, Y. R. (1969). In "Quantum Optics." (R. J. Glauber, ed.), Academic Press, New York.
Shore, B. W. and Menzel, D. H. (1968). "Principles of Atomic Spectra," Wiley, New York.
Townes, C. H., and Schawlow, A. L. (1975). "Microwave Spectroscopy," Dover, New York.
Wallace, R. (1966). *Molecular Phys.* **11**, 457.
Wiese, W. L., Smith, M. W., and Glennon, B. M. (1966). "Atomic Transition Probabilities, Vol. I Hydrogen Through Neon," US Dept. Commerce, Nat. Bur. Std., NSRDS-NBS 4.
Wiese, W. L., Smith, M. W., and Miles, B. M. (1969). "Atomic Transition Probabilities, Vol II Sodium Through Calcium," US Dept. Commerce, Nat. Bur. Std., NSRDS-NBS 4.

Part VI

Abragam, A., and Bleaney, B. (1970). "Electron Paramagnetic Resonance of Transition Ions," Oxford Univ. Press (Clarendon), London and New York.
Avery, J. (1972). "The Quantum Theory of Atoms, Molecules and Photons," McGraw-Hill, New York.
Ballhausen, C. J. (1962). "Ligand Field Theory," McGraw-Hill, New York.
Ballhausen, C. J., and Gray, H. B. (1965). "Molecular Orbital Theory," Benjamin, New York.
Bersohn, M. and Baird, J. C. (1966). "Introduction to Electron Paramagnetic Resonance," Benjamin, New York.
Bethe, H. A. (1929). *Ann. Physik* **3**, 135.
Bleaney, B., and Stevens, K. W. H. (1953). *Rep. Prog. Phys.* **16**, 108.
Born, M., and Oppenheimer, J. R. (1927). *Ann. Physik* **84**, 457.
Carrington, A., and McLachlan, A. D. (1967). "Introduction to Magnetic Resonance," Harper, New York.
Condon, E. U. (1947). *Am. J. Phys.* **15**, 365.
Cotton, F. A. (1971). "Chemical Applications of Group Theory," Wiley (Interscience), New York.
Daudel, R., Lefebvre, R., and Moser, C. (1959). "Quantum Chemistry," Wiley (Interscience), New York.
Davies, D. W. (1967). "The Theory of the Electric and Magnetic Properties of Molecules," Wiley, New York.
Dunford, H. B. (1968). "Elements of Diatomic Molecular Spectra," Addison-Wesley, Reading, Massachusetts.
Fano, U., and Fano, L. (1972). "Physics of Atoms and Molecules," Univ. Chicago Press, Chicago, Illinois.
Feynman, R. P. (1939). *Phys. Rev.* **56**, 340.
Goodisman, J. (1973). "Diatomic Interaction Potential Theory," Academic Press, New York.
Griffith, J. S. (1961). "Theory of Transition Metal Ions," Cambridge Univ. Press, London and New York.
Griffith, J. S. (1962). "The Irreducible Tensor Method for Molecular Symmetry Groups, Prentice-Hall, Englewood Cliffs, New Jersey.

Hameka, H. F. (1965). "Advanced Quantum Chemistry," Addison-Wesley, Reading, Massachusetts.
Hellmann, H. (1937). "Einfühurng in die Quantenchemie," Franz Deuticke, Leipzig and Vienna.
Heine, V. (1960). "Group Theory in Quantum Mechanics," Pergamon, Oxford.
Herzberg, G. (1950). "Spectra of Diatomic Molecules," Van Nostrand-Reinhold, Princeston, New Jersey.
Herzberg, G. (1966). "Electronic Spectra and Electronic Structure of Polyatomic Molecules," Van Nostrand-Reinhold, Princeton, New Jersey.
Hochstrasser, R. M. (1966). "Molecular Aspects of Symmetry," Benjamin, New York.
Holden, A. (1971). "Bonds Between Atoms," Oxford Univ. Press, London and New York.
Jug, K. (1969). *Theoret. Chim. Acta* **14**, 91.
Kaplan, I. G. (1975). "Symmetry of Many-Electron Systems," Academic Press, New York.
Karplus, M., and Porter, R. N. (1970). "Atoms and Molecules," Benjamin, New York.
Kauzmann, W. (1957). "Quantum Chemistry," Academic Press, New York.
Kovacs, I. (1969). "Rotational Structure in the Spectra of Diatomic Molecules," Amer. Elsevier, New York.
Lax, M. (1974). "Symmetry Principles in Solid State and Molecular Physics," Wiley, New York.
Levine, I. N. (1970). "Quantum Chemistry," Allyn & Bacon, Rockleigh, New Jersey.
Levine, I. N. (1975). "Molecular Spectroscopy," Wiley & (Interscience), New York.
McLean, A. D., and Yoshimine, M. (1967). "Tables of Linear Molecule Wave Functions," Int. Bus. Mach. Corp., San Jose, California.
McWeeny, R. and Sutcliffe, B. T. (1969). "Methods of Molecular Quantum Mechanics," Academic Press, New York.
Mizushima, M. (1975). "The Theory of Rotating Diatomic Molecules," Wiley, New York.
Moss, R. E. (1973). "Advanced Molecular Quantum Mechanics," Chapman & Hall, London.
Mulliken, R. S. (1955). *J. Chem. Phys.* **23**, 1833, 1841.
Parr, R. G. (1963). "Quantum Theory of Molecular Electronic Structure," Benjamin, New York.
Pimentel, G. C., and Spratley, R. D. (1969). "Chemical Binding Clarified Through Quantum Mechanics," Holden-Day, San Francisco, California.
Pople, J. A., and Beveridge, D. L. (1970). "Approximate Molecular Orbital Theory," McGraw-Hill, New York.
Richards, W. G., and Horsley, J. A. (1970). "Ab Initio Molecular Orbital Calculations for Chemists," Oxford Univ. Press (Clarendon), London and New York.
Roberts, J. D. (1962). "Molecular Orbital Calculations," Benjamin, New York.
Roothaan, C. C. J. (1951). *Rev. Mod. Phys.* **23**, 69.
Salem, L. (1966). "The Molecular Orbital Theory of Conjugated Systems," Benjamin, New York.
Schutte, C. J. H. (1971). *Structure and Bonding* **9**, 213.
Slater, J. C. (1963). "Quantum Theory of Molecules and Solids, Vol. I. Electronic Structure of Molecules," McGraw-Hill, New York.
Slater, J. C. (1974). "Quantum Theory of of Molecules and Solids, Vol. IV. The Self-Consistent Field for Molecules and Solids," McGraw-Hill New York.
Streitwieser, A. (1961). "Molecular Orbital Theory," Wiley, New York.
Stratton, J. A. (1941). "Electromagnetic Theory," McGraw-Hill, New York.
Sugano, S., Tanabe, Y., and Kamimura, H. (1970). "Multiplets of Transition-Metal Ions in Crystals," Academic Press, New York.
Tinkham, M. (1964). "Group Theory and Quantum Mechanics," McGraw-Hill, New York.
Townes, C. H., and Schawlow, A. L. (1975). "Microwave Spectroscopy," Dover, New York.
Watanabe, H. (1966). "Operator Methods in Ligand Field Theory," Prentice-Hall, Englewood Cliffs, New Jersey.

Wilson, Jr., E. B., Decius, J. C., and Cross, P. C. (1955). "Molecular Vibrations," McGraw-Hill, New York.

Wollrab, J. E. (1967). "Rotation Spectra and Molecular Structure," Academic Press, New York.

Woodward, R. B., and Hoffman R. (1970). "The Conservation of Orbital Symmetry," Academic Press, New York.

Zeiger, H. J., and Pratt, G. W. (1973). "Magnetic Interactions in Solids," Oxford Univ. Press, London and New York.

INDEX

C

D

H

I

J

K

L

M

N

O

P

T

U

V

W

X

Y

Z

A
B
C 8
D 9
E 0
F 1
G 2
H 3
I 4
J 5